21st International Conference of the International Society for Terrain Vehicle Systems and 12th Asia-Pacific Regional Conference of the International Society for Terrain-Vehicle Systems (ISTVS 2024)

Yokohama, Japan
28-31 October 2024

ISBN: 979-8-3313-1394-4

Printed from e-media with permission by:

Curran Associates, Inc.
57 Morehouse Lane
Red Hook, NY 12571

Printed with permission by Curran Associates, Inc. (2025)

International Society for Terrain-Vehicle Systems
c/o Cold Regions Research and Engineering Laboratory
72 Lyme Road Hanover, New Hampshire 03755-1290
USA

http://www.istvs.org

Additional copies of this publication are available from:

Curran Associates, Inc.
57 Morehouse Lane
Red Hook, NY 12571 USA
Phone: 845-758-0400
Fax: 845-758-2634
Email: curran@proceedings.com
Web: www.proceedings.com

Proceedings of the 21st International and 12th Asia-Pacific Regional Conference of the ISTVS

ISTVS2024
October 28-31, 2024
Yokohama, Japan
https://2024.istvs.org

Proceedings of the 21st International and 12th Asia-Pacific Regional Conference of the ISTVS

Editors

Junya Yamakawa, Editor-in-Chief
National Defense Academy, Japan

Genya Ishigami
Keio University, Japan

Shingo Ozaki
Yokohama National University, Japan

Ryosuke Eto
National Defense Academy, Japan

Massimo Martelli
National Research Council, Italy

Contents

Conference overview

The 21st International and 12th Asia-Pacific Regional Conference of the ISTVS is an opportunity to present and disseminate the latest scientific and technological achievements in the field of vehicle terrain systems.

Both representatives of science in the field of mechanical engineering, vehicles and machines, mechatronics and automation, soil science, agriculture and environmental engineering, as well as professionals from industry, business environment and government agencies for whom the issues discussed at the conference are an important merit are invited to participate. The organizers also count on the participation of students from under- and postgraduate courses.

The intention of the organizers is to create a friendly atmosphere and comfortable conditions for holding plenary sessions. The conference includes the following thematic tracks:

- soil mechanical characterization;
- off-road mobility modeling;
- soil compaction;
- driving systems of off-road vehicles and machines;
- innovative concepts of tires, wheels, and tracks;
- propulsion systems and engines;
- vetronics;
- autonomous and robotic systems;
- metrology in terramechanics.

Conference Chair

Junya Yamakawa | National Defense Academy, Japan

Conference co-chair

Taizo Kobayashi | Ritsumeikan University, Japan

Organizing Committee

Shingo Ozaki | Yokohama National University, Japan
Genya Ishigami | Keio University, Japan
Ryosuke Eto | National Defense Academy, Japan
Kojiro Iizuka | Shibaura Institute of Technology, Japan
Shinichiro Miyai | Komatsu Ltd., Japan
Takeshi Hashimoto | Public Works Research Institute, Japan
Massimo Martelli | ISTVS General Secretary, Italy
Jenna Dixon | ISTVS Digital Operations, USA

Scientific Committee

Schalk Els | University of Pretoria, South Africa
Vilas Salokhe | Kaziranga University, India
Corina Sandu | Virginia Tech, USA
Dror Rubinstein | Ariel University, Israel
Jarosław Pytka | Lublin University of Technology, Poland
József Kövecses | McGill University, Canada
Peter Kiss | Hungarian University of Agriculture and Life Sciences, Hungary
Massimo Martelli | National Research Council, Italy
Vladimir Vantsevich | Worcester Polytechnic Institute, USA
Lutz Richter | SoftServe, Inc., Germany

Conference venue

APA HOTEL & RESORT YOKOHAMA BAY TOWER
5-25-3 Kaigan-dori, Naka-ku, Yokohama,
Kanagawa 231-0002, Japan

Sponsors

NISHIO RENT ALL CO., LTD. | Pioneers in the construction machinery rental business. Since we began renting road work machines in 1965, we have been one of the leading companies in this industry. Expanding our network services, we have over 200 domestic branches, with additional branches and subsidiaries in Malaysia, Thailand, Singapore, Vietnam, China and Australia.
https://nishio-grp.co.jp/english/

SAKAI
MASTERS OF COMPACTION

SAKAI HEAVY INDUSTRIES, LTD. | Manufacture and sale of construction equipment and industrial machinery, and sale of out sourced industrial machinery.
https://www.sakainet.co.jp/en/

KOMATSU LTD. | Manufacture and sale of construction and mining equipment, utility equipment (compact machines), forest machines, industrial machinery and others.
https://www.komatsu.jp/en

St. Christopher Lecture

ROVING THE MOON - STEEPER, FASTER AND SMARTER

Professor Kazuya Yoshida
Department of Aerospace Engineering, Tohoku University, Sendai, Japan

In 2011, Professor Kazuya Yoshida established the International Center for Extreme Robotics Research at Tohoku University as the center's director. He also built an exploration robot for the Fukushima Daiichi Nuclear Power Plant accident that occurred in the same year. He specializes in space robotics, including the development of SPRITE-SAT, and was the technical director of the Google Lunar XPRIZE team "HAKUTO". For Hayabusa2, he developed Minerva II2, a small robot to be dropped into Ryugyu.

He has been working on lunar and planetary robotics research since 1997. From the beginning, he demonstrated the terramechanics, specifically the wheel-soil traction mechanics, of wheeled rovers on soft and deformable terrain. He conducted a series of experiments and simulations, using Particle Image Velocimetry (PIV) to observe soil particle flow around the wheels and Discrete Element Method (DEM) to simulate soil particle interactions. These methods provided insights into how rovers could enhance their mobility, particularly on steep and challenging lunar slopes. By investigating soil compaction, shear forces, and slip ratios under various loads and conditions, Yoshida's work has advanced traction models essential for navigating rugged extraterrestrial surfaces.

More recent research focuses on two key aspects of lunar rover design: faster locomotion and smarter navigation. By increasing rover speed, Yoshida observed a more dynamic soil flow, which demands refined terramechanics models to maintain traveling performance on loose terrain. Additionally, he and his team have developed a vision-based AI system for smarter navigation in unstructured environments. This AI system interprets complex visual information to identify obstacles, optimize path selection, and enhance rover safety on unpredictable lunar surfaces. Together, these innovations in high-speed terramechanics and AI-powered navigation will shape the future of lunar exploration, equipping rovers to operate reliably and autonomously across extreme terrains.

Keynote Speech

PRESSURIZED ROVER AS THE JAPANESE CONTRIBUTION TO ARTEMIS PROGRAM

Naoki Sato
Senior Advisor, Pressurized Rover Engineering Center, Human Spaceflight Technology Directorate, Japan Aerospace Exploration Agency, Tsukuba, Japan

Naoki Sato graduated Aeronautics Engineering Department, Kyusyu University in 1986, and gained master degree of applied engineering of Kyusyu University in 1988. At the same year, he entered in National Space Development Agency of Japan (predecessor of JAXA). Since 1990, he had been involved in the International Space Station program for about 16 years. After that, he started to work for the international space exploration program formulation. He is currently working as a senior advisor of the Pressurized Rover Engineering Center at JAXA.

Abstract

Recognition of great success of International Space Station (ISS) program from the view of technology development and international relationship maturation, space agencies around the world started discussions for promoting this success toward the human space exploration beyond Low Earth Orbit (LEO) in early 2000's. After long discussion and coordination among agencies, it was agreed to go to the moon first and demonstrate many technologies and operation, then go to the Mars. Especially, United States has begun Moon to Mars program, so called, Artemis program. Japanese government decided to participate in the Artemis Program in 2021, and then concluded an agreement with NASA to provide a Pressurized Rover (PR), and to get a right to send two Japanese astronauts onto the moon in 2024. The PR is like a camping car providing living environments with shirts sleeve and traversing capability on lunar surface for more than 10 years. Because the lunar environment is much harsh than LEO such as gravity, strong radiation, long night, dust, etc., the development of the PR is a huge challenge.

Conference program

Monday, October 28, 2024

14.00 – 17.00 | **ISTVS Board of Directors meeting** Room B

17.00 – 20.00 | **Registration** Foyer

18.00 – 20.00 | **Ice breaker party** Room A

Tuesday, October 29, 2024

08.30 – | **Registration** Foyer

09.00 – 09:30 | **Opening ceremony** Room CD

Chair: Prof. Junya Yamakawa

Conference chair welcome — Prof. Junya Yamakawa

President's address — Prof. Corina Sandu

Award ceremony:

Bekker-Reece-Radforth award

St. Christopher lecture

09.30 – 10.30 | **St. Christopher lecture** Room CD

Chair: Prof. Junya Yamakawa

ROVING THE MOON - STEEPER, FASTER AND SMARTER
Professor Kazuya Yoshida, Tohoku University

10.30 – 11.00 | **Conference photo & Break** Room CD

11.00 – 12.00 | **Keynote** Room CD

Chair: Prof. Taizo Kobayashi

PRESSURIZED ROVER AS THE JAPANESE CONTRIBUTION TO ARTEMIS PROGRAM
Naoki Sato, Japan Aerospace Exploration Agency

12.00 – 13.30 | **Lunch break** REVO BREWING

13.30 – 15.00 | **Parallel sessions**

Room C
Session 1A:
Extraterrestrial Terramechanics 1
Chair: Prof. Shingo Ozaki

8506 | INTRODUCTION OF HOURGLASS MISSION TO INVESTIGATE CHARACTERISTICS OF GRANULAR MATERIALS IN LOW GRAVITY ENVIRONMENT

Masatsugu Otsuki, Shingo Ozaki, Genya Ishigami, Takao Maeda, Masataku Sutoh and Taizo Kobayashi

2800 | GRANULAR FLOW IN REDUCED GRAVITY: ANALYSIS AND INSIGHTS FROM CENTRIFUGE EXPERIMENTS AT THE INTERNATIONAL SPACE STATION

Genya Ishigami, Shingo Ozaki, Masatsugu Otsuki, Hideaki Miyamoto, Koji Wada, Masataku Sutoh, Takao Maeda and Taizo Kobayashi

9295 | GRANULAR SCALING LAWS FOR ACCURATE PREDICTION OF WHEEL MOBILITY ON SLOPES IN LOW-GRAVITY ENVIRONMENTS

Takuya Omura and Genya Ishigami

Room D
Session 1B:
Machine-terrain Interaction and Vehicle Engines
Chair: Prof. Peter Kiss

1204 | MEASUREMENT AND VISUALIZATION OF SOIL CUTTING AND THROWING BEHAVIOR BY A ROTARY TILLAGE BLADE

Toyohiro Katou, Yuta Hirayama, Yasumaru Hirai, Muneshi Mitsuoka, Koki Nakatani, Kenji Kaida, Mutsumi Hidekawa and Takashi Okayasu

1241 | STUDY ON CLASSIFICATION OF EXCAVATED SOIL USING INTERNAL SENSOR DATA OF HYDRAULIC EXCAVATOR

Naoki Morisawa, Masaya Imanishi, Masaki Yanagishita, Teiichirou Chiba, Hiroshi Yamamoto, Takeshi Hashimoto and Daisuke Endo

2369 | EXPERIMENTAL VERIFICATION OF PARTICLE BEHAVIOR DURING CRUSHING AND MIXING OF DETERIORATED ASPHALT PAVEMENT LAYERS BY STABILIZER

Takatomo Fujii, Takashi Kurosu, Osamu Oikawa, Takashi Okayasu, Toyohiro Katou, Yuta Hirayama and Yukinori Tsukimoto

3773 | EVALUATION OF THE MULTI-PASS EFFECT OF AN EXPLORATION ROVER BY SINGLE WHEEL TESTING ASSUMING LUNAR GRAVITY AND SOIL
Tomomi Tanaka, Keisuke Takehana, James Hurrell, Kentaro Uno and Kazuya Yoshida

8590 | MODELING OF TERRAIN DEFORMATION BY A GROUSER WHEEL FOR LUNAR ROVER SIMULATOR
Junnosuke Kamohara, Vinicius Ares, James Hurrell, Keisuke Takehana, Antoine Richard, Shreya Santra, Kentaro Uno, Eric Rohmer and Kazuya Yoshida

0412 | EVALUATION OF GROUSER WHEEL TRACTION PERFORMANCE AT HIGH-SPEED BY SINGLE-WHEEL TEST
Keisuke Takehana, Kenta Sawa, Kentaro Uno, Shreya Santra and Kazuya Yoshida

1902 | THE COMBINATION OF EXHAUST GAS RECIRCULATION AND WATER INJECTION IN A MODERN DIESEL ENGINE
Dániel Szőllősi and Peter Prof. Kiss

15.00 – 15:30 | **Poster presentation** Foyer

0178 | MODEL TO ESTIMATE THE FORCE EXERTED ON THE SPHERE ROLLING ON THE SANDY ROAD
Takeshi Fukumoto, Ken Yamamoto, Makoto Katsura and Hiroaki Katsuragi

1649 | DYNAMIC MODE DECOMPOSITION FOR PIV-BASED SAND FLOW FIELD BENEATH TRAVELING WHEEL
Ohta Kuramoto, Jorge Ruben Casir Ricano and Kenji Nagaoka

4046 | STUDY ON MOBILITY AND STRATEGY OF MARS ROVER WITH FAULTY-DRIVEN WHEEL
Zhicheng Jia, Jingfu Jin, Xinju Dong, Yingchun Qi, Meng Zou and Lianbin He

4142 | ROOT SYSTEM ANALYSIS AFTER A VIBRATING ROLLER COMPACTION IN DRY DIRECT SEEDING OF RICE FIELD
Koichiro Fukami, Naoki Matsuo, Keiko Nakano and Kimiyasu Takahashi

4375 | DEVELOPMENT OF A MULTI-LEGGED ROBOT CAPABLE OF GROUND STIFFNESS DETECTION
Tomoya Nishiyama, Tatsuki Honjo, Yugo Hosizawa, Ryota Hayashi, Koji Yoshida and Tetsuya Kinugasa

6086 | TRAVELING ANALYSIS OF WHEEL FOR LUNAR EXPLORATION ROVER BASED ON EXTENDED TERRAMECHANICS MODEL: EXAMINATION OF SIMILARITY LAW OF GRAVITY
Mai Shimizu and Shingo Ozaki

6422 | SINKAGE STUDY IN GRANULAR MATERIAL FOR SPACE EXPLORATION LEGGED ROBOT GRIPPER
Arthur Candalot, James Hurrell, Manel Malik Hashim, Brigid Hickey, Mickael Laine and Kazuya Yoshida

9028 | UNCERTAINTY QUANTIFICATION FOR WHEELED LOCOMOTION MACHINE LEARNING PREDICTIONS ON SOFT SOIL
Vladyslav Fediukov, Felix Dietrich, Fabian Buse and Jana Huhne

15.30 – 17.00 | **Parallel sessions**

Room C
Session 2A:
Extraterrestrial Terramechanics 2
Chair: Prof. Taizo Kobayashi

1499 | DEVELOPMENT OF EVALUATION TECHNOLOGY FOR TIRE TRAVELING PERFORMANCE ON SOFT SOIL OF LUNAR SURFACE
Kei Tsuchiya, Kon Seiji, Masahiro Katayama, Tomoya Arai and Shingo Ozaki

4695 | AN IMAGE ANALYSIS METHOD FOR OBTAINING WHEEL PERFORMANCE OF ROVER USING WHEEL TREAD TRACES REMAINED ON LUNAR SIMULANT SOIL
Yujin Lim and Viet D. Le

4213 | DEM-BASED ANALYSIS AND OPTIMIZATION OF AN EXCAVATION BUCKET DRUM FOR IN-SITU RESOURCE UTILIZATION
Tomoyasu Nakano, Takuya Omura and Genya Ishigami

Room D
Session 2B:
Driving Systems
Chair: Dr. Paweł Tomiło

4998 | DYNAMICAL MODELING OF THE POWER HOP PHENOMENON IN AN AGRICULTURAL TRACTOR WITH FRONT AXLE SUSPENSION
Masahisa Watanabe, Keisuke Kazama and Kenshi Sakai

5334 | ENHANCED OPEN-LOOP CONTROL OF AUTOMATIC GEAR SHIFTING IN HYDROMECHANICAL CVT FOR AGRICULTURAL TRACTORS
Massimo Martelli, Pietro Marani, Silvia Gessi and Damiano Chiarabelli

7829 | DEVELOPMENT OF FOOT IN BALLOON BIPED ROBOT USING BUOYANCY FORCE FOR TRAVELING SOFT GROUND
Noriaki Mizukami, Masanori Aragane and Mami Nishida

0815 | USING SIMULATION TO CHARACTERIZE RASSOR EXCAVATION IN LUNAR ENVIRONMENTS

Luning Bakke, Zhenhao Zhou, Alexandra Kissel, Ruochun Zhang and Dan Negrut

0356 | HAMMERING ENERGY REQUIREMENTS FOR SURVEYING LUNAR SURFACE WITH DYNAMIC CONE PENETROMETERS AND SEISMIC METHODS

Karlis Slumba, Brendan Scott, Mark Jaksa and Samuel Ximenes

3304 | DEVELOPMENT OF AN UNMANNED EXPLORATION ROBOT FOR LUNAR SURFACE GEOTECHNICAL INVESTIGATION

Taizo Kobayashi, Atsushi Kakogawa, Shinichi Ito, Masafumi Nakagawa, Takeshi Tsuji, Shingo Ozaki, Satoshi Matsumura, Akihiko Kondo, Masanori Takigawa, Keitaro Kitamura, Takahito Hiramatsu, Hisatoshi Sano, Takeshi Yoneoka, Junichiro Odaka, Kenji Hosobori, Taichi Ikenaga and Ryu Taniguchi

8812 | ENERGY METHOD TO COMPARE PERFORMANCE OF NEW TYPES OF SUGAR CANE TRANSPORT EQUIPMENT

Jorge Lopera, Fernando Casanova, Adolfo Leon Gomez Perlaza and Carlos Daniel Muñoz Delgado

9747 | APPLICATION OF A ROCKIE BOGIE SUSPENSION FOR A NEW TYPE OF SUGAR CANE TRANSPORT SYSTEM

Jorge Lopera

17.00 – 18.00 | **Break**

18.00 – 20.00 | **Banquet** Room A

Wednesday, October 30, 2024

08.30 – | **Registration** Foyer

09.00 – 10.15 | **Parallel sessions**

Room A
Session 3A:
Extraterrestrial Terramechanics 3
Chair: Prof. Kojiro Iizuka

2411 | PROPOSING TURNING MOTION OF THE SMALL AND LIGHTWEIGHT PUSH-ROLLING ROVER WITH MIMIMAL CONFIGURATION

Daisuke Fujiwara and Kojiro Iizuka

0906 | EXPERIMENTAL INVESTIGATION OF INCREASED BEARING CAPACITY WHEN IMPARTING VIBRATION TO LOOSE GROUND IN LOW ATMOSPHERIC PRESSURE

Tomohiro Watanabe and Kojiro Iizuka

5720 | SUPPRESS SLIP WHILE CROSSING LOOSE SLOPES USING REVERSE ROTATION BEHAVIOR OF ROVERS WITH FUNCTION OF INDEPENDENT CONTRACTION/EXPANSION MECHANISM

Tsukasa Mochizuki, Daisuke Fujiwara, Kojiro Iizuka, Takaaki Sakata and Tatsuhiko Suga

5295 | MODEL-BASED ONLINE OPTIMAL CONTROL FOR VEHICLES IN REDUCED GRAVITY

Chaitanya Sonalkar, Adwait Verulkar and Corina Sandu

8774 | PROPOSAL OF HYBRID LOCOMOTION LUNAR ROVER WITH CRAWLING MECHANISM

Paweł Tomiło

Room CD
Session 3B:
Autonomous Robotic Systems
Chair: Massimo Martelli

5620 | STEP-CLIMBING MOTION ACQUISITION OF TRACKED ROBOT WITH FLIPPERS WITHOUT USING ENVIRONMENT INFORMATION BY REINFORCEMENT LEARNING

Ryosuke Eto, Hayatake Sato and Junya Yamakawa

4966 | REAL-TIME IMPLEMENTATION OF NON-LINEAR CONTROLLERS AND PREDICTORS FOR OFF-ROAD VEHICLE DYNAMICS ON EMBEDDED SYSTEMS

Andries J. Peenze, A. Glenn Guthrie and P. Schalk Els

8084 | LOG DETECTION FOR AUTONOMOUS FORWARDING USING AUTO-ANNOTATED DATA FROM A REAL-TIME VIRTUAL ENVIRONMENT

Mattias Lehto, Håkan Lideskog and Magnus Karlberg

4715 | UAV-BASED THREE-DIMENSIONAL ROUGH TERRAIN MODELLING

Herman Hamersma, Christian van Aswegen, Glenn Guthrie and Schalk Els

2812 | AN INTERACTION-AWARE TWO-LEVEL ROBOTIC PLANNING AND CONTROL SYSTEM FOR VEGETATION OVERRIDE

Charles Noren, Burhanuddin Shirose, Bhaskar Vundurthy, Sebastian Scherer and Matthew Travers

10.15 – 10.45 | **Break**

10.45 – 12.00 | **Parallel sessions**

Room A
Session 4A:
DEM-FEM Simulation
Chair: Dr. Shinichiro Miyai

0677 | ADAPTIVE PARTICLE REFINEMENT IN TERRAMECHANICAL DEM SIMULATION
Markus Pogulis and Martin Servin

7603 | EVALUATION OF OFF-ROAD UNINHABITED GROUND VEHICLE MOBILITY USING DISCRETE ELEMENT METHOD AND SCALABILITY INVESTIGATION
Dr. Ayush Nuwal, Ajay Kumar and John Economou

7116 | DEM-SPH ANALYSIS FOR INTERACTION MECHANICS OF TRACKED VEHICLE ON WET SAND
Hiroki Yanagawa and Genya Ishigami

6412 | TRAVERSING ABILITIES SIMULATION OF A BIOMIMETIC ROBOT ON GRANULAR SOIL TERRAIN
Zhenwen Zhou, Xiang Lei, Dehai Guan, Yan Zhang, Gabriel Lodewijks and Guangming Chen

7555 | PRACTICAL APPLICATIONS OF HYBRID TERRAMECHANICS MODEL USING MACHINE LEARNING
Eric Karpman, Jozsef Kovecses and Marek Teichmann

Room CD
Session 4B:
Wheel/Tire-terrain Interaction 1
Chair: Prof. Schalk Els

3303 | SOIL INSTRUMENTATION FOR MEASURING NORMAL STRESS DISTRIBUTION UNDER OFF-ROAD TIRE
Alexandru Vilsan, Nikhil Ravichandran, Chaitanya Sonalkar and Corina Sandu

1882 | SPATIO-TEMPORAL ANALYSIS OF SAND-DENSITY DISTRIBUTION BENEATH TRAVELING WHEEL BASED ON PARTICLE IMAGE VELOCIMETRY
Kenji Nagaoka and Seigo Sakai

1928 | A NOVEL SOIL STRESS ESTIMATION METHOD OF WHEEL-SOIL INTERACTION USING PHOTOELASTICITY
Kenji Nagaoka and Yuto Yoshida

7322 | LATERAL TYRE CHARACTERIZATION: ROLLING TYRE VS STATIC TYRE TESTING
Carl Becker and Schalk Els

1710 | A NOVEL MEASUREMENT METHOD TO AID DEVELOPMENT OF SOFT SOIL TYRE-TERRAIN INTERACTION MODELS
Thomas Bernd Kabutz and Schalk Els

12.00 – 13.00 | **Lunch break** — Room A and Room CD
(Bento box will be reserved)

13.00 – 13.30 | **Poster presentation** — Foyer
Same as the poster presentation on October 29

13.30 – 15.00 | **Parallel sessions**

Room A
Session 5A:
Mobility & Trafficability Characterization 1
Chair: Dr. Lutz Richter

3769 | MOBILITY DURING THE TRANSITION SEASONS IN THE ARCTIC
Michael Parker, Clifford Witte and Susan Frankenstein

4476 | SPECTRAL DETERMINATION OF SOIL MOISTURE CONTENT BASED ON THE DRY COLOUR OF THE SOIL
György Pillinger, Ahmed Elawad Eltayeb Ahmed and Peter Kiss

4859 | TRAFFICABILITY CONDITIONS FOR MILITARY WHEELED TRUCKS ON CULTIVATED FIELDS
Kersti Vennik, Priit Põdersalu, Tamur Kusnets and Tõnu Tõnutare

6476 | IMPROVED TRAFFICABILITY OVER SOFT SOILS USING GROUND MATTING
John Rushing, Brad Hansen and Michael Parker

7453 | THE IMPACT OF CHANGES IN THE RIVER REGIME ON THE MOBILITY OF OFF-ROAD VEHICLES
Marian Rybansky, Filip Dohnal, Martin Hubacek, Jaromir Capek and Vladimir Kovarik

Room CD
Session 5B:
Wheel/Tire-terrain Interaction 2
Chair: Prof. Corina Sandu

4093 | THE ROLE OF TIRE-SOIL INTERFACE CHARACTERISTICS ON PERFORMANCE PARAMETERS THROUGH EXPERIMENTAL AND NUMERICAL INVESTIGATION
Jasleen Kaur Bheora, Varsha S Swamy, Mason Destiny, Alba Yerro, Corina Sandu, Katherine Sebeck and David Gorsich

3181 | ASSESSING SENSITIVITIES OF OFF-ROAD PNEUMATIC TIRE ON CLAY: A FINITE ELEMENT INVESTIGATION ON TIRE OPERATIONAL AND DESIGN PARAMETERS
Destiny Mason, Varsha S Swamy, Corina Sandu, Alba Yerro, David Gorsich and Katie Sebeck

2919 | ACCURATE ROVER MOBILITY ANALYSIS USING HILS-DRFT WITH REAL-TIME PARAMETER TUNING APPROACH
Takahiro Fuke, Sora Ishikawa and Genya Ishigami

2520 | EFFICIENT TIRE-TERRAIN INTERACTION MODELLING: EFFECT OF FLEXIBILITY ON TRACTION
Mahdi Maleki and Jozsef Kovecses

2651 | RAPID ASSESSMENT TOOLS FOR ESTIMATING TRAFFICABILITY ON LOW-VOLUME ROADS
John Rushing, Lulu Edwards, Haley Bell and Margarita Ordaz

6865 | EXTENDED TERRAMECHANICS MODEL CONSIDERING GROUND SURFACE DEFORMATION AND ITS APPLICATION TO WHEEL TRAVELING ANALYSIS
Shingo Ozaki, Tomoya Arai and Mai Shimizu

3071 | STUDY ON APPLICABILITY OF EXTENDED TERRAMECHANICS MODEL TO VARIOUS TRAVELING MODES OF WHEELS
Tomoya Arai, Kei Tsuchiya, Seiji Kon, Masahiro Katayama and Shingo Ozaki

15.00 – 15.30 | **Break**

15.30 – 16.45 | **Parallel sessions**

Room A
Session 6A:
Mobility & Trafficability Characterization 2
Chair: Dr. Genya Ishigami

0376 | IMPROVEMENT OF FORMULATIONS OF CLEGG IMPACT HAMMER AND RAMMSONDE PENETROMETER FOR USE IN COMPACTED SNOW
Mohit Shenvi, Corina Sandu and Costin Untaroiu

4927 | TIP ANGLE DEPENDENCE FOR RESISTIVE FORCE INTO DRY GRANULAR MATERIALS AT SHALLOW CONE PENETRATION
Naoki Iikawa and Hiroaki Katsuragi

6325 | PROPOSAL OF A NEW MANUAL FOR TELESCOPIC PENETROMETER
Klara Cibulova and Martin Priesner

9663 | YEAR-ROUND MEASUREMENTS OF THE SOIL CONE INDEX ON GRASS AIRFIELDS FOR GROUND PERFORMANCE OF AIRPLANE
Jarosław Pytka, Anna Zalewska Tytłak, Paweł Tomiło, Michal Kuszneruk and Dawid Tatarynow

Room CD
Session 6B:
Modeling of Off-road Vehicles
Chair: Prof. Dror Rubinstein

8330 | SEMI-EMPIRICAL TERRAMECHANICS MODEL FOR VARIABLE TERRAIN HEIGHT IN 3D
Eric Karpman, Wing Hang Ho, Jozsef Kovecses and Marek Teichmann

9688 | EVALUATION OF THE DYNAMIC SINKAGE AND ITS EFFECT ON THE COMPACTION FORCE PREDICTION OF OFF-ROAD VEHICLES
Yang Jiao, Jozsef Kovecses and Marek Teichmann

3649 | MODELING THE RESISTIVE FORCES ON VEHICLES IN DEEP SNOW
Micah M. Borrero and Orian Welling

8861 | AN EFFICIENT AND HIGH-FIDELITY TRACK MODEL FOR DYNAMIC SIMULATION OF OFF-ROAD TRACKED VEHICLES
Oz Ben-Yosef and Dror Rubinstein

1525 | PHOTOGRAMMETRY BASED MOBILITY MAPPING FOR SMALL SCALE MODEL VEHICLE
Dávid Körmöczi and Peter Kiss

16.45 – 17.00 | **Break**

17.00 – 17.30 | **Closing ceremony** Room CD

Chair: Prof. Junya Yamakawa

Best paper awards — Prof. Kojiro Iizuka

Introduction of new ISTVS officers — Massimo Martelli

New President address — Dr. Lutz Richter

Next ISTVS conference presentation — Michael Parker

Closing Remarks

Thursday, October 31, 2024

07.45	**Meeting**		B1F entrance lobby
08.00	**Departure**		
10.00 – 12.00	**Technical visit 1**		
	Group A Construction Technology Pavillion	Group B Komatsu IoT Center	
12.30 – 13.30	**Lunch**		ASIAN BUFFET FOOD SAGAR
14.00 – 16.00	**Technical visit 2**		
	Group A Komatsu IoT Center	Group B Construction Technology Pavillion	
18.00	**Arrival at APA hotel**		

List of submissions

Full papers

Abstract-only

ISTVS

Conference papers

0356 | Dynamic cone penetration in lunar regolith simulant

Kārlis Šļumba [a,*], Brendan T. Scott [a], Mark B. Jaksa [a], Samuel W. Ximenes [b]

[a] *Lunar Construction Group, Andy Thomas Centre for Space Resources, University of Adelaide, SA 5005, Australia*
[b] *Exploration Architecture Corp., San Antonio, TX 78205, United States*

* *Corresponding author: karlis.slumba@adelaide.edu.au*

ABSTRACT

Ground investigation and survey is necessary before anything can be built on the Moon. Exploration Architecture Corporation (XArc) proposed that geotechnical surveying on the Moon could be performed with a SurveyorBot; a robot equipped with cone penetrometers and seismic instruments. This research consists of development and testing of a mini dynamic cone penetrometer (Mini-DCP) with variable impact energy. The Mini-DCP will act both as a penetrometer and a seismic source, used in tandem with seismic instruments.

The purpose of this study is to investigate an appropriate range of hammering energies for a Mini-DCP to penetrate through regolith simulant. Hammering energies that are too large penetrate through the soil too fast, not providing enough data about the layering. However, hammering energies that are too low risk test stagnation and producing impact blows that may be too weak to be detected by a seismic source. This study uses bespoke Mini-DCP equipment to quantify the effects of both hammering energy and momentum. Mini-DCP test results are compared to mini cone penetration tests (Mini-CPT). Experiments are performed using engineering grade lunar highlands simulant (LHS-1E) in the regolith pit at the Extraterrestrial Environmental Simulation (EXTERRES) laboratory at the Andy Thomas Centre for Space Resources at the University of Adelaide, Australia.

Keywords
Lunar regolith
Geotechnical survey
Cone penetrometers
Soil characterization

1. Introduction

Within this decade humans will return to the Moon with National Aeronautics and Space Administration's (NASA's) Artemis program (NASA, 2020). To achieve a permanent human presence on the Moon, it is necessary to build permanent infrastructure. Before building any infrastructure projects on Earth, ground investigation and survey of the site must be performed (Rix et al., 2019), and the same will be true on the Moon. While civil engineering and construction use geotechnical data derived from well-established instruments and practices, lunar construction practices remain at low Technology Readiness Levels (TRL).

Exploration Architecture Corporation (XArc) subsidiary, Astroport Space Technologies is developing construction technologies to define a Landing/Launch Pad (LLP) construction concept of operations and a system architecture of surface assets needed for LLP site preparation and pad construction (https://explorationarchitecture.com/astroport/). Before these construction equipment assets are delivered to the lunar surface, the site needs to be surveyed for planning excavation and fill operations.

XArc is developing a lunar pathfinder called SurveyorBot under a NASA Small Business Technology Transfer (STTR) contract. The SurveyorBot will be equipped with cone penetrometers and seismic sensors to measure soil penetration resistance among other subsurface characteristics that are critical items in site leveling (Ximenes, n.d.). The idea is ultimately to deploy a swarm of semi-autonomous SurveyorBot rovers that would survey the lunar surface and immediate subsurface.

The lunar surface is covered with an unconsolidated particulate surface layer called regolith. Regolith is made by physical destruction of rock by meteorite impacts, radiation and thermal cycling called space weathering (Pieters and Noble, 2016). As opposed to the weathering processes on Earth, space weathering produces sharp particles with smaller particle sizes. A large fraction of the regolith is toxic, and electrostatically active dust adheres to all surfaces and degrades equipment (Pohlen et al., 2022). Larger meteorite impacts and landslides reveal fresh unweathered surfaces, and this process over millions of years creates a progressively deeper regolith layer. Regolith thickness can be about 10-15 m deep in lunar highlands regions, and about 4-5 m deep in mare regions (Plescia et al., 2023). Depth correlates with the age of the surface and it can differ significantly depending upon the landing site (Carrier et al., 1991). Highlands are the oldest regions of the Moon and are made from the original lunar crust. Mare regions are significantly younger and were made by volcanic eruptions of the basaltic magma, hence the regolith layer is not as thick (Heiken et al., 1991). Regolith mineralogical composition slightly differs between mare and highlands regions with mare regions being mostly basaltic (pyroxene), whereas highlands regions are mostly feldspar (Vaniman et al., 1991). Visually, highlands have higher albedo and much rougher terrain as compared to mare regions. NASA's Artemis program intends to explore the regions of the lunar south pole which are primarily made of highlands type material (Lemelin et al., 2022). Lunar highlands type material is therefore most important for this research project.

While Apollo missions returned 382 kg of lunar samples to Earth (Heiken et al., 1991), they are not accessible in large amounts. For most research projects that require testing in a lunar environment, some sort of regolith simulant is necessary. Engineering projects like this can use regolith simulants that are made for engineering purposes, and these simulants simulate

https://doi.org/10.56884/NB365QQR

geotechnical properties well, but the mineralogy is only approximate (Long-Fox et al., 2023).

Regolith is not homogeneous; it has many local layers from smaller impacts and more regional layers from large impacts. Impact debris thickness can vary substantially. From penetration and sampling experiments during Apollo, it can be seen that this heterogeneity can influence experiments. In some places, the regolith provided very little resistance to penetration (especially around crater edges), but in some places astronauts had to give up the attempts to penetrate it (especially in fresh impact ejecta) (Carrier et al., 1991). For example, penetration resistance at 40 cm depth varied between 0.3 to 1.6 MPa at Apollo 16, Station 4 and to an even higher value at Station 10.

Regolith properties might differ even further in polar regions (Sargeant et al., 2020), but the authors are not aware of a better regolith model than from the Lunar Sourcebook (Carrier et al., 1991). It has been shown that when water evaporates from regolith simulant, it can produce very loose and porous material (Šljumba et al., 2024), called a 'fairy-castle structure' (Hapke and van Horn, 1963). This is why the authors speculate that at the lunar polar region, regolith could be less compacted to deeper depths as compared to equatorial regions. This may be applicable to extinct permanently shadowed regions (PSRs) or near PSRs in the polar region or the region where the Artemis missions are aiming for.

The experiments detailed herein are performed at the Extraterrestrial Environmental Simulation (EXTERRES) laboratory at the Andy Thomas Centre for Space Resources (ATCSR) at the University of Adelaide, Australia. The EXTERRES laboratory is equipped with modern equipment (e.g. sand pit, regolith pit, vacuum chambers) to simulate the surface environment on the Moon ("Andy Thomas Centre for Space Resources | University of Adelaide," n.d.). The regolith pit is 3.1 by 3.2 m in plan dimensions and filled with 8 metric tons of engineering grade lunar highlands simulant (LHS-1E). The simulant was made by Space Resource Technologies, previously known as Exolith Lab. ("Space Resource Technologies.," n.d). The regolith simulant is on average 0.55 m deep, but the pit has some volume exclusions, so the average density can't be calculated in a straightforward manner.

2. Cone penetration tests

There are many geotechnical engineering methods for soil penetration. Some of the cone penetration methods are used in this research. The cone penetration test (CPT) consists of a cone at the end of a rod that is pushed into the soil at constant rate of 20 mm/s. By measuring the resistance of the penetration, it is possible to characterize layering and geotechnical properties of the soil. The CPT was first developed in 1932 by P. Barensten in the Netherlands (Lunne et al., 1997). The cone shape itself has not seen significant changes and the main purpose of the test remains the same. The CPT nowadays, however, can provide much more data with the additional equipment added to the test, for example, pressure, seismic, and thermal sensors (Lunne et al., 1997; White, 2022).

2.1. Dynamic cone penetration test

When comparing a dynamic cone penetrometer test (DCPT), the major difference from the CPT is that the rod is hammered into the soil instead of statically advancing it. This test is most applicable to sandy soils but is also used in silts and clays. As lunar regolith is best described as silty sand (Carrier et al., 1991), it can be expected that the DCPT will be applicable to lunar regolith. The DCPT was developed by George F. Sowers in 1959 as a simple and quick method to characterize soil compaction. The DCPT overcomes one of the most significant drawbacks of the CPT, which is the problem when encountering a stiff soil layer. For a CPT, one of the ways to penetrate through a harder layer is to increase the reaction load, but for a DCPT, this can be achieved by increasing the number of hammer drops (Sowers and Hedges, 1966).

2.2. Cone penetration on the Moon

While NASA's Apollo missions weren't the first landers on the Moon, they landed the first astronauts on 6 successful Moon landings between 1969 and 1972. The Apollo missions provided the most valuable information about the Moon to date. A detailed description of the Apollo missions and their findings is provided in the Lunar Sourcebook (Carrier et al., 1991) and references therein. Apollo 11 and 12 did not include any penetrometers, so the geotechnical properties were inferred using other indirect approaches, for example, the flagpole, core samples, digging, and photographs of the astronauts' footprints. These measurements were more qualitative in nature, but scientists were able to draw a general picture of the lunar regolith's properties. Apollo 14 incorporated a simplified cone penetrometer, and this provided better geotechnical data as compared to previous missions, but it was still very approximate. Apollo 15 and 16 had a self-recording cone penetrometer, which provided the most valuable penetration data to date.

During the Apollo missions on the Moon, astronauts encountered significant issues when trying to penetrate the lunar regolith; the biggest issue was the reduced gravity and the heterogeneous nature of the regolith layering (Carrier et al., 1991). These issues are even harder to overcome for a robotic mission, especially for a small rover. Increased penetration capability is especially important for applications on other planetary bodies like Moon and Mars where the gravity is approximately 6 and 3 times lower than on Earth respectively.

Lessons learned from the Apollo missions are the most important and most directly applicable to this research project. While Apollo missions didn't have dynamic cone penetrometers (DCPs), they still had cone penetrometers that provided useful geotechnical data and information about how the instruments operated. The Apollo findings show that the DCP would be more applicable for a geotechnical survey of the Moon, especially for a lightweight robot that can only provide a light reaction to push a penetrometer into the regolith.

The lunar regolith properties can change significantly both spatially and with depth (Plescia et al., 2023). To advance the DCP within a reasonable number of blows, an appropriate hammering energy must be quantified.

2.3. Mini dynamic cone penetrometer

Typical DCPs that are used in civil engineering, field geology and military applications are robust, large and heavy. For laboratory experiments in loose regolith simulants, the self-mass of the regular DCP is sufficient to penetrate the uncompacted soil. To tackle this issue, a mini dynamic cone penetrometer (Mini-DCP) was developed (Fig. 1). The instrument is lightweight and incorporates other parameters that can be varied. The mass of the Mini-DCP is 3 kg, excluding the mass of the hammer. This is not a flight model or a prototype, but it is made explicitly for laboratory experimentation.

Fig. 1. Mini dynamic cone penetrometer (Mini-DCP) demonstration in the sand pit, EXTERRES lab, University of Adelaide.

The Mini-DCP has an option to change the hammer mass by adding mass in increments of 500 g. The hammer can be dropped from various heights of up to a maximum of 50 cm.

The hammer impact energy comes from its potential energy E_P in Eq. 1.

$$E_P = m \cdot g \cdot h \tag{1}$$

Where m is hammer mass, g is acceleration from gravity, and h is the release height. The potential energy is transformed into kinetic energy E_K in Eq. 2.

$$E_K = \frac{1}{2} \cdot m \cdot v^2 \tag{2}$$

Where v is the impact velocity. At the time of hammer release, $E_{P,0}$ is maximum, but $E_{K,0}$ is zero, and it is opposite at the impact. As total energy of the system is constant, we can derive Eq. 3.

$$E_{P,0} + E_{K,0} = E_{P,1} + E_{K,1} \tag{3}$$

Another aspect is that it is possible to have the same energy impact with different hammer masses released at different heights. For example, hammer with mass of 0.5 kg released from 0.2 m height would have an impact energy of 0.98 J. The same energy would be produced with a 1 kg hammer that is released from a 0.1 m height.

Taking 50 mm as the minimum hammer release height, the impact energy can be varied from 0.25 to 7.5 J and momentum from 0.354 to 3.354 kg·m/s. The hammer could be dropped from a smaller height, but then the produced momentum might be too small to overcome the inertia of the Mini-DCP and to penetrate the soil. The proposed design can cover around one order of magnitude in energy and momentum levels.

The Mini-DCP cone is changeable, but for the experiments described here, the same cone from the regular DCP was used. The apex angle is 30° and the diameter of the cone is 20 mm.

3. Penetration test methodology

Filling of the regolith pit was not achieved in a controlled manner hence the density distribution is quite variable. Various individuals have been walking over the regolith simulant in recent times. The regolith simulant has been compacted by its own weight and from foot traffic, with the latter extending to a shallow depth. It was known that the simulant is not homogeneous with depth. It was hypothesized that the simulant shouldn't have much variation when comparing locations that are close to each other. Previous experiments showed that the simulant is not compacted near the edges of the regolith pit, as no one has been walking there. On the other hand, near the entrance in the pit the simulant was much more compact. The site in the central part of the pit was selected, where the conditions were expected to be similar.

A 1.2 by 1.2 m sub-section of the regolith pit was selected in the middle of the pit for experiments. As shown in Figs. 2 and 3 the sub-section was divided into 49 locations at 20 cm spacings. Based on a hypothesis that 40 by 40 cm regions should have similar conditions, locations were selected so that Mini-DCPs and Mini-CPTs would be undertaken in each region. Within each region, 9 measurements were planned, one Mini-CPT in the centre, and Mini-DCP and nuclear density gauge measurements around. As is explained below, not all nuclear density gauge measurements were able to be performed, so there were between 6 and 8 Mini-DCP measurements in each region.

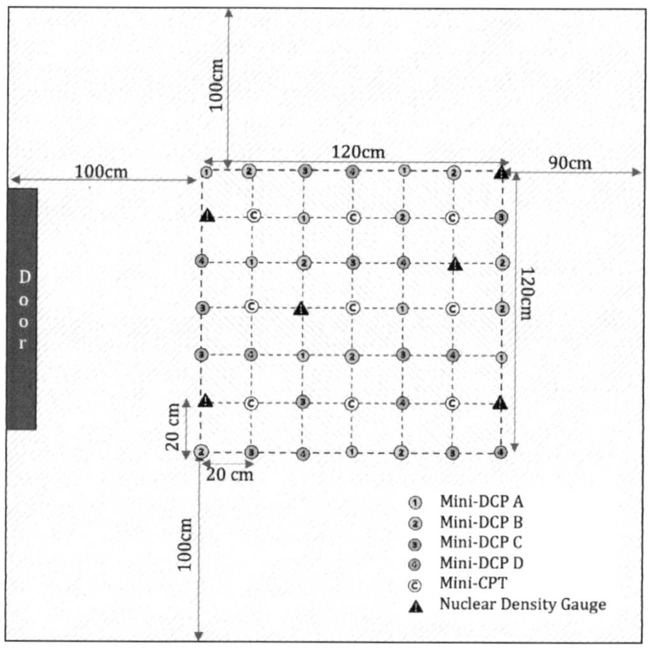

Fig. 2. Mini-DCP, Mini-CPT, and nuclear density gauge testing plan in the Regolith pit at EXTERRES lab.

Four different setups were used for the Mini-DCP, as shown in Table 1. They were scattered over the site so that each region would have at least one measurement from each setup.

Table 1. Mini-DCP setup for experiments.

Setup	Hammer mass, g	Drop height, cm	Energy, J	Momentum, kg·m/s
Mini-DCP A	477	21	0.982	0.968
Mini-DCP B	477	5	0.234	0.472
Mini-DCP C	998	10	0.978	1.397
Mini-DCP D	1497	7	1.027	1.753

The Humboldt nuclear density gauge HS-5001SD (Humboldt Nuclear Density Gauges., n.d.), normally being a very robust instrument, malfunctioned because regolith simulant found its way into the instrument and jammed the source rod mechanism. As a result, it was not possible to perform the density tests as originally intended. This highlights the important and often dismissed dust problem that exists in the lunar environment.

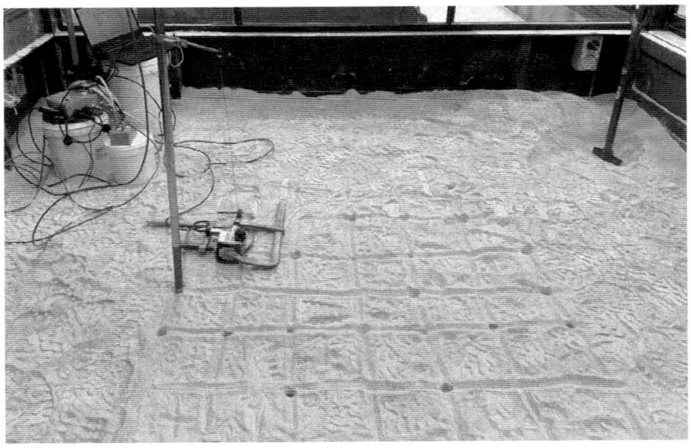

Fig. 3. Cone penetration resistance measurements in the Regolith pit at EXTERRES lab.

Mini-DCP measurements were performed using the standard DCP testing technique (Standards Australia, 1997). The number of hammer strikes required to penetrate 50 mm depth were registered. The Mini-CPT measurements were performed using a Geomil Digital CPT cone ("Geomil," n.d.). The instrument was pushed into the ground by hand, at a speed of approximately 20 mm/s.

4. Results

Mini-CPT measured penetration resistance of the regolith simulant at 9 locations in the regolith pit. Each location, together with the Mini-DCP measurements around it, accounts for a single region. Figure 4 shows a map of Pearson correlation coefficients between the penetration resistance measured by the Mini-CPT and the number of hammer strikes per 50 mm penetration. In this figure, Region 2 is identified by a red rectangle. The single blue dot in the centre is the Mini-CPT measurement for which the correlation was calculated, whereas the red dots are all Mini-DCP measurements. The highest correlations are between the measurements that are closest, meaning that the regolith structure is similar at this scale.

The Pearson correlation coefficient (r) was calculated using Eq. 4 to test the hypothesis that regions at this scale are similar, and all measurements in this region can be considered as if they sampled the same soil. Values of x and y correspond to Mini-CPT and Mini-DCP measurements at each depth, while $\bar{x}$ and $\bar{y}$ correspond to their average values.

$$r = \frac{\Sigma\big((x-\bar{x})\cdot(y-\bar{y})\big)}{\sqrt{\Sigma(x-\bar{x})^2\cdot\Sigma(y-\bar{y})^2}} \tag{4}$$

The Pearson correlation was considered high where it was higher than 0.6, average where it was between 0.4 and 0.6, and low where it was between 0.2 and 0.4. Only measurements with correlations above 0.4 were considered in the calculations.

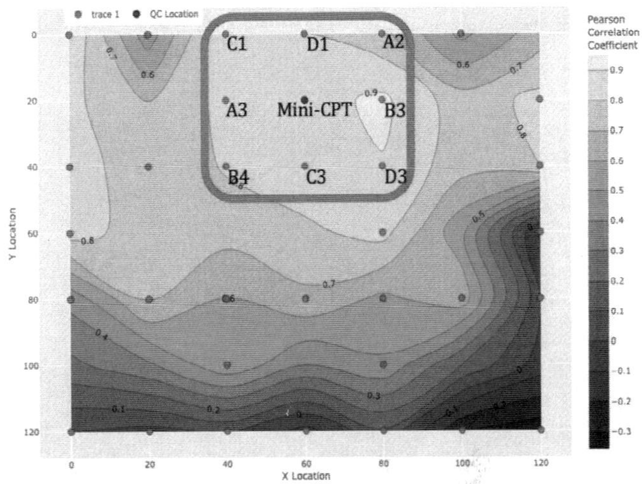

Fig. 4. Map of Pearson correlation coefficients for Mini-DCP and Mini-CPT at region 2.

A function was calculated between measurements with average and high Pearson coefficients using Eq. 5. Values of x and y correspond to Mini-CPT and Mini-DCP measurements at each depth, m is the slope and b the y-intercept.

$$y = m\cdot x + b \tag{5}$$

In this case b is set to 0, so that m becomes a multiplier or a constant by which Mini-CPT measurements should be multiplied to transform the data into units of Mini-DCP. Eq. 5 can be divided by m to arrive at Eq. 6. From this equation, m is a value by which Mini-DCP measurements should be divided to transform into units of Mini-CPT, in this case megapascal (MPa). The multiplier is in units of MPa^{-1}.

$$x = \frac{y}{m} \tag{6}$$

The multiplier m was calculated by fitting a line using least squares regression for all measurements giving 16 eligible multiplier values to transform between Mini-DCP A and Mini-CPT and 13 eligible multiplier values for each of the other Mini-DCPs. This is greater than the number of measurements because most of the Mini-DCP measurements fall into several regions. For example, Mini-DCP B4 was compared with 4 Mini-CPT measurements that are within the range.

The average multiplier values $\overline{m}$ and standard deviation σ were calculated for each setup and are shown in Table 2. One measurement of Mini-DCP B had a multiplier value that deviated from average by 3σ, this measurement was excluded from the calculations.

When all average multipliers are calculated, then all data from the Mini-DCP can be divided by the corresponding average multiplier and data in each region can be plotted on the same graph (Fig. 5). Error bars are not included on the graph because they obscure the plot. Within the standard deviation range, all measurements yield the same result.

Table 2. Multiplier values for different Mini-DCP setups.

Setup	Measurements	$\bar{m}$, MPa^{-1}	σ, MPa^{-1}	σ, %
Mini-DCP A	16	37.8	3.5	9.2
Mini-DCP B	12	114.4	12.5	10.9
Mini-DCP C	13	24.7	4.6	18.6
Mini-DCP D	13	21.0	3.9	18.7

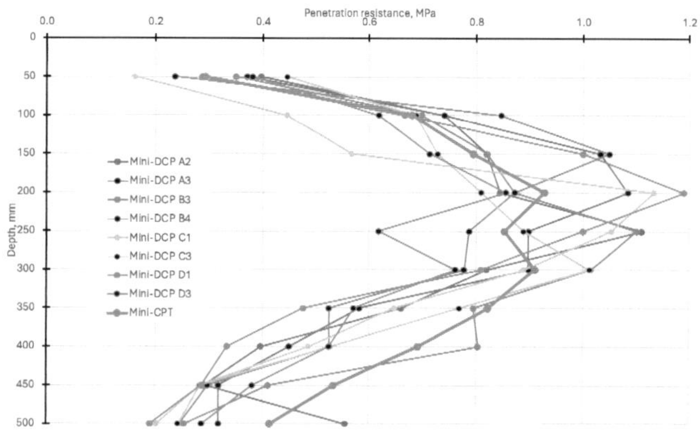

Fig. 5. Penetration resistance measured by Mini-CPT and calculated from Mini-DCP measurements.

5. Discussion

5.1. Sowers and Hedges criterion

During the preliminary experiments it was shown that the regular DCP is too heavy for this application. Sowers and Hedges (1966) explain that for the best correlation hammer drop count, there should be no fewer than 4 and no more than 30 per 4.5 cm penetration (Sowers and Hedges, 1966). With a hammer drop count that is smaller, the penetration does not provide enough data. On the other hand, where the count is too large, there is a risk of stagnation and damage to the equipment. This criterion allows to create a custom DCP that lies within this range. The previously calculated multiplier (Table 2) can be regarded as the number of hammer strikes needed to penetrate 50 mm depth at 1 MPa penetration resistance.

Penetration resistances measured during Apollo 14, 15, and 16 varies from virtually 0 to 1 MPa at 10 cm depth, from 0.3 to 1.9 MPa at 40 cm depth, and reaches more than 3 MPa at 60 cm depth. The heterogeneous regolith and variability of penetration resistance makes it difficult to determine a hammering energy that would lie within the Sowers and Hedges criterion over the entire range. Comparing the multiplier results to the Sowers and Hedges criterion, it is obvious that Mini-DCP B does not provide enough penetration at 1 MPa penetration resistance, but it would fall in the range if penetration resistance would be between 0.05 and 0.25 MPa. On the other hand, Mini-DCP D satisfies the criterion while the penetration resistance of the regolith is between 0.2 and 1.5 MPa.

5.2. Penetration in low relative density

Mini-DCP A and B setups were also tested in regolith simulant with lower relative density, where the penetration resistance is between 0.05 and 0.15 MPa. It was found that at penetration resistance below 0.12 MPa Mini-DCP sinks under its own weight. The impact of self-weight is important at low relative densities.

In this study, a multiplier was used because that is the same approach that was used by Hajduk et al. (2012). They found that sufficient correlation between DCP and CPT existed only in sand below the water table. However, they couldn't find a correlation in silt, clay or dry sand. They pointed out that one of the reasons for the large scatter in their data was the inability of DCP to detect thin layers, of which there were plenty in their soil. With the Mini-DCP, this problem is partly overcome by incorporating smaller energies.

There might be some constant values that are especially important where relative density is low, in other words, constant b in Eq. 5 might not be zero. More testing in a range of relative densities is necessary to evaluate this.

5.3. Effects of dynamic force

Loose, cohesionless soil will have smaller dynamic penetration resistance and compacted, cohesionless soil will have a larger dynamic penetration resistance as compared to a static resistance (Sowers and Hedges, 1966). This might mean that the conversion between the Mini-DCP and Mini-CPT would be linear only in a relatively narrow relative density range. More research is necessary in both lower and higher relative densities to test if this applies to regolith simulant that is not cohesionless.

Mini-DCP A, C and D all were set up to have approximately 1 J of hammer energy per strike, but different momentums, as indicated in Table 1. When comparing to the multiplier, as shown in Table 2, the multiplier decreases as the momentum increases. Testing over a larger dynamic range is necessary to find the relationship between multiplier, energy and momentum.

5.4. Boundary effects

A commonly adopted rule-of-thumb for boundary effects in penetration experiments is to keep the distance between measurements to 10 times the diameter of the penetrometer, but this can be somewhat less for loose soil. The Mini-DCP cone is 20 mm, and the Mini-CPT is 25 mm in diameter. Based on the rule-of-thumb, 200 to 250 mm distance should be necessary between measurements. The boundary effects are dependent on the internal angle of friction (ϕ) by Eq. 7 (Puech and Foray, 2002) where L is the dimension of the boundary effect, B is diameter of the cone and ϕ is the internal angle of friction.

$$L = B \cdot e^{\left(\frac{\pi}{2}\cdot\tan(\phi)\right)} \cdot \tan\left(\frac{\pi}{4} + \frac{\phi}{2}\right) \tag{7}$$

The internal angle of friction has been measured for LHS-1E to be 46° at 40% relative density and 51° at 70% relative density (Agarwal et al., 2023). Density (ρ) was measured with the nuclear density gauge to be between 1675 and 1800 kg/m^3 from the surface to 30 cm depth. Relative density (RD) was calculated using Eq. 8 where minimum (ρ_{min}) and maximum (ρ_{max}) densities of the LHS-1E are 1450 and 2000 kg/m^3 respectively (Agarwal et al., 2023).

$$RD = \frac{\rho_{max}}{\rho} \cdot \frac{\rho - \rho_{min}}{\rho_{max} - \rho_{min}} \cdot 100\% \tag{8}$$

These measurements reveal that relative density in the regolith pit was between 50 and 70%. Since the nuclear density gauge malfunctioned, additional tests were not carried out, but it is expected that relative densities as low as 40% should exist in the regolith pit. The boundary effect can be found by dividing L by B from Eq. 7. Substituting the numbers yields 12.6 for 46° and 19.6

for 51° internal angle of friction. This implies that the penetration measurements adopted in this study were too close to each other. On the other hand, the penetration holes stayed intact even when adjacent tests were performed 20 cm away; this would suggest that the influence of nearby tests was not strong enough to noticeably disturb the surface. The relative density was measured after the penetration tests were performed. This finding should be explored further and considered when planning future experiments.

5.5. Lunar analogue environment

It has been shown that the Mini-DCP can be used as a geotechnical investigation tool when operated by a human in a laboratory environment, in room temperature, and under Earth's gravity and atmosphere. However, the lunar environment provides many challenges. Laboratory experiments in a lunar analogue environment are necessary to test if this technique would work on the Moon. The regolith compaction chamber at the EXTERRES lab will allow one to simulate lunar surface properties by having a controlled density and layering of the regolith simulant. At a later stage, the Mini-DCP will be upgraded further to automatize the experiment and to make the instrument vacuum compatible. The EXTERRES lab provides a capability to use a large regolith thermal vacuum chamber, further experiments will be performed there. Simulating lunar gravity is beyond the scope of this research project, however, this experiment could be upgraded further to run it on a parabolic flight, under simulated lunar gravity.

6. Conclusions and future work

Penetration in lunar regolith simulant was measured using a Mini-DCP and a Mini-CPT. Results from both instruments were successfully correlated. A multiplier for each Mini-DCP setup was found that allows one to convert the results from one instrument to another. This research indicates that the dynamic cone penetrometer could be used to measure penetration resistance on the Moon, although more experiments are necessary to improve the reliability of this statement.

It was shown that penetration resistance of regolith simulant can be measured by a Mini-DCP. It is also possible to correlate the Mini-DCP data to other geotechnical parameters like relative density, cohesion, internal angle of friction, and compressibility.

Future work will involve equipment upgrades, for example, adding changeable cones at variety of sizes and reducing the Mini-DCP size and mass. Environmental upgrades will include controlling regolith simulant relative density from loose to dense, as well as considering layering effects. Another set of experiments will be undertaken in the regolith pit, where the Mini-DCP will be used as a seismic source.

Boundary effects depend on angle of internal friction, which is relatively high for lunar regolith simulants. It was found that spacing for penetration tests should be larger than what is typically used as a rule-of-thumb.

During this study, regolith simulant particles found their way into the instrument and jammed the source rod mechanism of the nuclear density gauge causing the equipment to malfunction. Harsh lunar environment and dangers to the equipment shouldn't be ignored.

7. Acknowledgements

This work is a part of a PhD project that is sponsored by the Andy Thomas Centre for Space Resources (ATCSR) and a University of Adelaide Research Scholarship.

Material discussed in this document is based upon work supported by NASA under contract award Number 80NSSC23PB427. Any opinions, findings, and conclusions or recommendations expressed in this material are those of the authors and do not necessarily reflect the views of the NASA.

8. Declaration of competing interest

The authors declare that they have no known competing financial interests or personal relationships that could have appeared to influence the work reported in this paper.

9. References

Agarwal, A.K., Kuo, Y.L., Jaksa, M., Scott, B., 2023. Density, strength and compressibility characteristics of lunar regolith simulant, in: Proceedings of the 14th Australia and New Zealand Conference on Geomechanics. Presented at the 14th Australia and New Zealand Conference on Geomechanics (ANZ2023), Cairns, Australia.

Andy Thomas Centre for Space Resources | University of Adelaide. n.d. EXTERRES Laboratory. https://set.adelaide.edu.au/atcsr/space-research/exterres-laboratory (accessed 24.05.24).

Carrier, W.D., III, Olhoeft, G.R., Mendell, W., 1991. Physical Properties of the Lunar Surface, in: Lunar Sourcebook, A User's Guide to the Moon. Cambridge University Press, pp. 475–594.

Geomil. n.d. Digital CPT. https://www.geomil.com/products/d-cone (accessed 24.05.24).

Hajduk, E.L., Shiver, B.T., Meng, J., 2012. A Comparison of the Sowers Dynamic Cone Penetrometer Test with Cone Penetration and Flat Blade Dilatometer Testing, in Problematic Soils and Rocks and In Situ Characterization. ASCE, pp. 1-10. https://doi.org/10.1061/40906(225)13

Hapke, B., van Horn, H., 1963. Photometric studies of complex surfaces, with applications to the Moon. Journal of Geophysical Research (1896-1977) 68, 4545–4570. https://doi.org/10.1029/JZ068i015p04545

Heiken, G., Vaniman, D., French, B.M., 1991. Lunar sourcebook: A user's guide to the Moon. Cambridge University Press.

Humboldt Nuclear Density Gauges. https://www.humboldtscientific.com/ (accessed 27.05.24).

Lemelin, M., Lucey, P.G., Camon, A., 2022. Compositional Maps of the Lunar Polar Regions Derived from the Kaguya Spectral Profiler and the Lunar Orbiter Laser Altimeter Data. Planet. Sci. J. 3, 63. https://doi.org/10.3847/PSJ/ac532c

Long-Fox, J.M., Landsman, Z.A., Easter, P.B., Millwater, C.A., Britt, D.T., 2023. Geomechanical properties of lunar regolith simulants LHS-1 and LMS-1. Advances in Space Research. https://doi.org/10.1016/j.asr.2023.02.034

Lunne, T., Robertson, P.K., Powell, J.J.M., 1997. Cone Penetration Testing In Geotechnical Practice. E&FN SPON.

NASA, 2020. NASA's Lunar Exploration Program Overview. https://www.nasa.gov/specials/artemis/

Pieters, C.M., Noble, S.K., 2016. Space weathering on airless bodies. Journal of Geophysical Research: Planets 121, 1865–1884. https://doi.org/10.1002/2016JE005128

Plescia, J.B., Cahill, J., Greenhagen, B., Hayne, P., Mahanti, P., Robinson, M.S., Spudis, P.D., Siegler, M., Stickle, A., Williams, J.P., Zanetti, M., Zellner, N., 2023. Lunar Surface Processes. Reviews in Mineralogy and Geochemistry 89, 651–690. https://doi.org/10.2138/rmg.2023.89.15

Pohlen, M., Carroll, D., Prisk, G.K., Sawyer, A.J., 2022. Overview of lunar dust toxicity risk. npj Microgravity 8, 1–9. https://doi.org/10.1038/s41526-022-00244-1

Puech, A., Foray, P., 2002. Refined Model for Interpreting Shallow Penetration CPTs in Sands. Presented at the Offshore Technology Conference, Houston, Texas. https://doi.org/10.4043/14275-MS

Rix, G.J., Wainaina, N., Ebrahimi, A., Bachus, R.C., Limas, M., 2019. Manual on Subsurface Investigations. National Academy of Sciences.

Sargeant, H.M., Bickel, V.T., Honniball, C.I., Martinez, S.N., Rogaski, A., Bell, S.K., Czaplinski, E.C., Farrant, B.E., Harrington, E.M., Tolometti, G.D., Kring, D.A., 2020. Using Boulder Tracks as a Tool to Understand the Bearing Capacity of Permanently Shadowed Regions of the Moon. Journal of Geophysical Research: Planets 125, e2019JE006157. https://doi.org/10.1029/2019JE006157

Šļumba, K., Sargeant, H.M., Britt, D.T., 2024. Development of icy regolith simulant for lunar permanently shadowed regions. Advances in Space Research 73, 3222–3234. https://doi.org/10.1016/j.asr.2024.01.014

Sowers, G.F., Hedges, C.S., 1966. Dynamic Cone for Shallow In-Situ Penetration Testing. https://doi.org/10.1520/STP44629S

Space Resource Technologies. n.d. (LHS-1E) Engineering Grade Lunar Highlands Simulant. https://spaceresourcetech.com/products/lhs-1e-simplified-lunar-highlands-simulant (accessed 24.05.2024).

Standards Australia, 1997. AS 1289.6.3.2-1997. https://www.standards.org.au/standards-catalogue/standard-details?designation=as-1289-6-3-2-1997

Vaniman, D., Dietrich, J., Taylor, G.J., Heiken, G., 1991. Exploration, Samples, and Recent Concepts of the Moon, in: Lunar Sourcebook. Cambridge University Press, pp. 5–26.

White, D.J., 2022. CPT equipment: Recent advances and future perspectives, in: Cone Penetration Testing 2022. CRC Press.

Ximenes, S. n.d. NASA SBIR STTR Contract T7.04-1837 - SurveyorBot: A Proof of Value Precursor Mission for Launch / Landing Pad Geotechnical Assessment https://sbir.nasa.gov/SBIR/abstracts/23/sttr/phase1/STTR-23-1-T7.04-1837.html (accessed 10.23.23).

0412 | Evaluation of Grouser Wheel Traction Performance at High-Speed by Single-Wheel Test

Keisuke Takehana [a,*], Kenta Sawa [a], Kentaro Uno [a], Shreya Santra [a], Kazuya Yoshida [a]

[a] *Department of Aerospace Engineering, Tohoku University, Sendai, Miyagi, Japan*

* *Corresponding author: keisuke.takehana.s1@dc.tohoku.ac.jp*

ABSTRACT

This study focuses on the locomotion characteristics of a high-speed lunar exploration rover. On the loose soil such as the lunar regolith, wheeled mobility systems often encounter wheel sinkage and slippage, leading to trafficability challenges. While conventional exploration rovers are designed to improve mobility using grouser wheels, their exploration speed remains relatively slow (~10^-2 m/s). High-speed traversal is essential for expanding exploration areas in future missions.

In this paper, we conduct single-wheel tests under high-speed conditions and provide detailed results from force measurements. Three types of wheels are prepared: no grouser, low grouser, and high grouser, and their traction performance is compared. The testbed allows speeds of 1 m/s, which is about a hundred times faster than conventional exploration speeds.

We calculated traction coefficients (the ratio of drawbar pull to vertical load) and traction efficiency (the ratio of input energy to output energy) for variable slip ratios, and these are evaluated as traction performance. Our results revealed a trend where traction performance decreases as driving velocity increases. This behavior of performance at various speeds differs from that at low speeds. Each performance exhibited substantial dependence on not only the slip ratio but also the magnitude of the wheel circumference velocity. Additionally, experiments varying grouser height under identical driving conditions demonstrated improved traction performance with higher grouser wheels. Through this experimental evaluation, we confirmed the effectiveness of grouser wheels even at high speeds.

These terramechanical findings contribute to the wheel design and locomotion control of next-generation exploration rovers.

Keywords
High-speed mobility
Single-wheel test
Lunar exploration rover
Grouser wheel

1. Introduction

Space exploration significantly contributes to uncovering the origins and evolution of the solar system, expanding scientific knowledge, and obtaining crucial geological samples for the construction of future manned bases (Colaprete, 2010; Li, 2018). Particularly, environments like those of the Moon and Mars are subject to extreme temperatures and intense radiation, making them unsuitable for direct human activity. To address this, unmanned exploration utilizing robotic technology has proven effective, with wheeled rovers being the primary means employed. These rovers, due to their high mobility, have conducted numerous surface explorations of celestial bodies.

The surfaces of the Moon and Mars are covered with a very fine dust known as regolith, presenting a substantial challenge to rovers in transit. This soft terrain is prone to destruction, frequently causing slippage and subsidence. Moreover, the lower gravity compared to Earth complicates driving control even further.

The field of study that investigates the phenomena occurring between the soft terrain and the rover wheels is known as terramechanics. Terramechanics specializes in the mechanical interactions between terrain and machinery, contributing to the development of technologies that enable movement in various environments, including lunar and planetary exploration rovers, construction machinery, and agricultural machinery (Bekker, 1956, 1960, 1969; Wong, 1967, 2008). Rovers traversing planets like the Moon or Mars require advanced driving control technologies based on terramechanics to efficiently collect data and explore resources.

Terramechanics in lunar and planetary exploration plays a crucial role in the efficient operation of rovers under harsh conditions. Particularly, the slip phenomena and traction performance of rover wheels have been the focus of numerous studies. Research has included investigating the wheel performance using single-wheel testbed (Higa, 2017; Tsubaki 2021; Girija, 2023), using actual rover prototypes combined with various sensors to estimate slip conditions and subsidence in real-time (Higa, 2016; Gonzalez, 2018), and studies that feed these estimations back into the driving control. (Ishigami, 2009; Ding, 2010).

While high traversability is required from rovers, mission efficiency necessitates the completion of objectives in a shorter duration (Potts, 2015; Smith, 2020; Flores, 2021). Traditional rover speeds are very low, commonly constrained by safety concerns in harsh environments, the need for stable environmental recognition, limited computational capacity of onboard equipment, and thermal energy consumption. This is especially problematic for missions requiring broader exploration areas, such as polar lunar explorations.

To reduce travel time to lunar poles and expand exploration areas, development has progressed towards rovers capable of moving at speeds tens to hundreds of times faster than traditional models (Rodriguez, 2017,2019; Hamabe, 2021). In research conducted at the German Aerospace Center (DLR), a single-wheel test device known as Terramechanics Robotics Locomotion Lab "TROLL" was developed, which clarified differences in driving performance characteristics up to speeds of 1 m/s (3.6 km/h) (Buse, 2019; Rodriguez, 2023). Although this high-speed test device faced challenges in extending the data acquisition range due to its size, Nakagoshi developed a new type of single-wheel test apparatus capable of infinite running on a circular track, enabling more practical data acquisition (Nakagoshi, 2020).

https://doi.org/10.56884/94WT27WY

Fig. 1. Overview of High-speed single-wheel testbed.

This study's objective is to evaluate wheel performance with a focus on high-speed traversal expected in future lunar exploration rovers. One reason for the scarcity of research in high-speed traversal is the limitation of experimental facilities. Typically, single-wheel test apparatuses have a traversal section limited to 1-2 meters, which is insufficient for achieving high speeds. Our research group has developed a new high-speed single-wheel test apparatus capable of measuring speeds exceeding 1 m/s, thus enabling experiments in velocity ranges that have not been extensively explored in previous studies.

This paper provides a detailed discussion of the results and analysis obtained from the high-speed single-wheel test apparatus. Given the high performance required during high-speed traversal, this study not only tests slick wheels but also wheels equipped with grousers. A meticulous examination is conducted with wheels of varying designs under multiple speed and slip conditions to thoroughly evaluate wheel traction. The study measures differences in traction performance across different traversal conditions and wheel types.

This research is expected to provide new foundational data for rover wheels that will enhance the efficiency of future lunar exploration activities.

2. High-speed single-wheel Testbed

This section describes the single-wheel testbed used in this study, which was designed to analyze the performance of specific wheel designs in detail.

2.1. Overview of the testbed

Figure 1 and Table 1 show the overview and specifications of the high-speed single-wheel testbed developed by our research group. The entire setup measures 6.85 m in length, 2.0 m in width, and 2.3 m in height, featuring a sandpit as the testing field with dimensions of 6.85 m by 0.5 m by 0.4 m. A single wheel is mounted to perform translational movement along the longitudinal axis above this sandpit.

The movement direction is controlled by sending current from a controller to a motor, enabling a linear actuator to reach predetermined speeds, thus allowing the wheel to traverse forcibly along a guide rail. Similarly, the wheel's rotation is controlled by currents sent to a motor to achieve specified rotational speeds. The vertical movement component is also controlled through currents sent from the controller to the motor, activating a linear actuator that adjusts the vertical load applied to the wheel. An encoder measures the amount of wheel rotation, translational movement, and sinking depth. Additionally, sensors mounted at the wheel's center measure forces and torques.

Thanks to this large apparatus, the maximum traversal distance on sand is approximately 3.5 meters, and it is capable of conducting single-wheel tests at speeds '1.0 m/s'.

The device includes a component called the "Moving Unit," which controls the translational speed along the guide rail and adjusts the vertical load and rotational speed at the wheel attachment point. This system allows for experiments under any specified speed and slip conditions, enabling detailed evaluation of critical wheel performance parameters such as slip-traction characteristics.

Table 1

Specifications of High-speed single-wheel testbed.

Property	Value
Sandbox size L – W - H (m)	6.85 - 0.5 – 0.4
Sand height (m)	0.3
Test distance (m)	3.5
Max rotational speed (m/s)	1.0
Max translational speed (m/s)	1.0
Sand type	Tohoku Silica No.5

2.2. Property of the sand

The sand used in the single-wheel testing apparatus employed in this study is "Tohoku Silica Sand No. 5 (Tohoku-Keisa, 2024)" shown in Fig. 2. This type of sand is rich in silicon dioxide (SiO2), exhibiting excellent wear resistance and acid resistance, and is whitish natural silica sand. The sand is derived from a mixture of frog-eye clay and silica sand, from which quartz sand is separated through washing, followed by drying and sieving. This process ensures that the sand contains very few impurities and maintains very stable granular size.

The grain size distribution ranges from 0.106 to 0.6 mm. The particles are round and highly fluid, facilitating consistent conditions during testing. Being hard quartz sand, the particles have no cracks and exhibit low crushability, which is crucial for obtaining consistent results in wheel friction and wear tests.

Table 2 provides a detailed description of the grain size distribution and chemical composition of Tohoku Silica Sand No. 5. This information is essential to enhance the reproducibility of the research and improve the accuracy of the results.

Fig. 2. Tohoku Silica Sand No. 5.

Table 2

Sand mechanical properties of Tohoku Silica Sand No. 5.

Grain diameter (mm)	0.85	0.6	0.425	0.3	0.212	0.15	0.106	0.075
Percentage (%)	0	27.3	51.9	15.2	4.2	0.3	0.1	0
Chemical component	SiO^2	Al^2O^3	Fe^2O^3	CaO	Na^2O	K^2O	TiO^2	Ig.loss
Percentage (%)	95.11	2.75	0.09	0.56	0.71	0.55	0.03	0.20

3. Grouser wheel

This study focuses on the design of wheels that play a crucial role in rover traversal on lunar and Martian surfaces. We particularly elaborate on the design of grouser wheels, optimized for interaction with the soil to enhance performance.

3.1. Wheel design

Grouser wheels feature plate-like protrusions extending radially from the cylindrical part of the wheel, where the shape and arrangement significantly impact performance. Grousers are designed to improve ground contact and prevent slippage, particularly on soft terrains. When small rovers traverse such terrains with smooth, non-groused wheels, the wheels, due to their relatively small diameter, tend to sink into the soil, resulting in increased slippage.

To address this issue, grouser wheels are equipped with multiple grousers, and the influence of the number and length of these grousers on traction has been studied (Ding, 2011; Sutoh, 2012; Nagaoka, 2020). It has been observed that more grousers, as well as longer ones, demonstrate superior slip-traction characteristics. Moreover, incorporating angled or V-shaped grousers has also been shown to enhance traction performance on sandy terrain. This arrangement increases the contact area with the sand, allowing more sand to be displaced and consequently increasing traction.

This section describes a detailed analysis of the design parameters of grouser wheels and their specific impacts on traversal performance, proposing optimal shapes and arrangements for the grousers.

3.2. Grouser spacing equation

In this study, the wheel design is based on the "Grouser Spacing Equation" proposed by Skonieczny et al (Skonieczny, 2012) represented as equation 1 (Each parameter is referred to in section 7, Nomenclature).

$$\emptyset < \frac{1}{1-i}\left(\sqrt{(1+h)^2 - (1-z)^2} - \sqrt{1 - (1-z)^2}\right) \quad (1)$$

This equation integrates geometric parameters of the wheel (grouser height: h, and grousers' spacing: Φ) with terramechanics parameters (slip ratio: i and sinkage: z) to predict the maximum permissible grouser angle.

Although slip ratio and sinkage values may vary during rover operations, the procedure to calculate the number of grousers using these parameters is as follows. First, set the slip ratio at 20% (nominal slip ratio) and treat sinkage according to Lyasko's reports (Lyasko, 2010). Based on this, solve the following equation to determine the minimum necessary number of grousers: n:

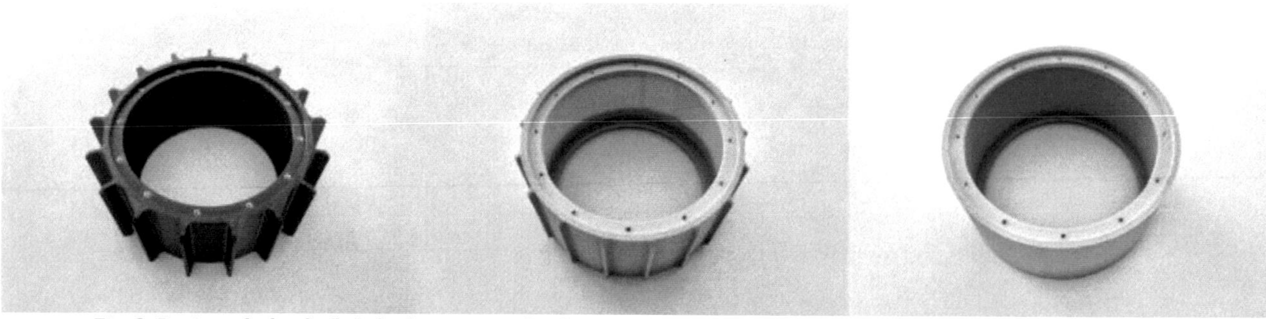

Fig. 3. Design of wheels (left; high-grouser wheel, center; low-grouser wheel, right; non-grouser (slick) wheel).

$$n > \frac{4\pi\,(1-i)}{\sqrt{4(1-h)^2-(2-i)^2}-\sqrt{4-(2-i)^2}} \tag{2}$$

According to this equation, the required number of grousers decreases as slip increases but increases as sinkage increases. This calculation enables the design of an optimal grouser layout.

Based on the equations, a total of three types of wheels are designed for this study: a high-grouser wheel that does not cause slippage ("high-grouser wheel"), a wheel with lower performance and increased potential for slippage ("low-grouser wheel"), and a slick wheel without grousers.

Figure 3 shows the three wheels created. The detailed designs are also described next.

3.3. High-grouser wheel

The high-grouser wheel is designed with an optimal number of grousers based on the nominal slip ratio and calculated sinkage. The nominal slip ratio is set at 20%. The wheel diameter is set at 180mm, and the grouser height at 15mm. This height is chosen considering the physical limits and strength of wheels manufactured using PLA (Polylactic Acid) resin with 3D printing. Since its strength is inferior to that of metal wheels, excessively high grousers increase the risk of damage.

The number of grousers is calculated based on the grouser spacing equation and optimized to achieve ideal traction and slip prevention, requiring 17 grousers. This number is optimized based on the equation (2).

3.4. Low-grouser wheel

The low-grouser wheel is designed to intentionally experience higher slip ratio. The nominal slip ratio is set at 80%, and the design process aims to determine the most suitable number of grousers and height under these conditions.

By maintaining a constant number of 17 grousers and setting the grouser height at 5.4mm, the design achieves the optimum number of 17 grousers for the low-grouser wheel as well. This design is used to simulate conditions where slip and sinkage occur most effectively.

3.5. Non-grouser (slick) wheel

The slick wheel is not fitted with a grouser. It has a wheel diameter of 180 mm at the grouser-free portion such as other wheels. This wheel is expected to be the most likely to cause slippage.

The weight of all wheels is 360 g, adjusted for the thickness of the cylindrical part.

4. Test procedure

4.1. Evaluation parameter

In this study, the performance of wheel traversal is evaluated using two primary metrics: traction coefficient and traction efficiency. These measures are calculated based on the drawbar pull, vertical load, and torque, which are recorded during single-wheel traversal experiments using a force sensor attached to the wheel drive assembly.

Drawbar pull is generated from the stress between the wheel and the terrain. As the vertical load increases, so does the drawbar pull. To compare performance under varying loads, drawbar pull (F_x) is normalized by the vertical load (F_z), allowing for a dimensionless comparison. This ratio, given by F_x divided by F_z, is defined as the traction coefficient, and is calculated as follows:

$$\text{Traction coefficient} = \frac{F_x}{F_z} \tag{3}$$

Traction efficiency measures the proportion of input energy that is converted into effective driving energy. By using drawbar pull (F_x), driving torque (T), angular speed (ω), and translational speed (v_x), it is defined as:

$$\text{Traction efficiency} = \frac{F_x\, v_x}{T\,\omega} \tag{4}$$

Both the traction coefficient and traction efficiency are critical indicators of wheel performance. A higher value in these metrics signifies a better performance, indicating that the wheel can generate more traction and efficiently convert energy into movement.

These indicators are measured in detail for each slip ratio. A common formulation in the field of terramechanics for slip ratio is expressed as the ratio of rotational speed to translational speed as follows:

$$\text{slip ratio} = \frac{r\omega - v_x}{r\omega} \tag{5}$$

4.2. Experimental method

To conduct the experiment efficiently and ensure reliable results, the following procedural steps are followed:

Prepare the Sand Surface: Begin by tilling the sandy terrain using a shovel, and then level it with a flat board or a screed. After leveling, use a level to confirm that the surface of the sand is uniform in height.

Verify Sand Cohesion: Confirm that the vertical shear torque of the sand remains consistent by conducting a vane shear test. This step ensures that the sand properties do not vary significantly throughout the experiment, which could affect the results.

Table 3

Experimental condition.

Parameter	Value
Vertical load (N)	50
Speed (m/s)	0.2. 0.5, 1.0
Slip ratio (-)	0, 0.05, 0.1, 0.15, 0.2, 0.25, 0.3, 0.35, 0.4, 0.45, 0.5

Control Wheel Movement: Adjust the rotational and translational speeds of the wheel to maintain a constant slip ratio, ensuring that the wheel moves straight ahead without deviating.

Measure Performance: During the wheel's traversal, use encoders to measure the depth of wheel sinkage and the various speeds. Simultaneously, measure the forces and torque applied to the wheel using a force/torque (F/T) sensor.

Repeat the Experiment: Return to step 1 to repeat the experiment multiple times.

4.3. Experimental condition

This study focuses on the high-speed four-wheel rover, EX1 (Rodriguez and Uno, 2024), developed by our research group. The total weight of the rover is 24 kg. Assuming that the weight is distributed equally among the wheels, each wheel bears approximately 50 N. The wheel rotation speeds were set at three different levels: 0.2 m/s for low speed, 0.5 m/s for medium speed, and 1.0 m/s for high speed. Additionally, the translational speeds were adjusted to match the rotational speeds to manipulate the slip ratio systematically. The slip ratio was incrementally increased from 0 to 0.5, in steps of 0.05, facilitating a comprehensive series of traction tests across multiple slip environments. Table 3 shows all experimental conditions.

5. Test result and discussion

This section discusses the results obtained from the single-wheel tests, specifically evaluating the traction performance across different slip ratios.

5.1. Overview of data measurement and evaluation

The experimental runs were conducted at the maximum speed of 1.0 m/s using wheels without grousers. This section will outline the method used to evaluate these results. Only the data from the steady-state portion of each test run was extracted for analysis. The mean and standard deviation of these data points were calculated and plotted with error bars to represent the outcome of each experiment, as shown in Fig. 4.

As indicated by the experimental conditions, both rotational and translational speeds were independently controlled to set specific slip ratios, under which drawbar pull, vertical load, and torque were measured. These values were then used to calculate the traction coefficient and efficiency based on predefined formulas.

Figure 4 display the traction coefficients and efficiencies for each slip ratio, with fitted curves showing the relationship. This approach allows for the calculation of each performance index as a function of the slip ratio, facilitating a comprehensive analysis of the traction characteristics under varying conditions.

Figure 4 on the left displays the results for the traction coefficient as a function of slip ratio. When the slip ratio is between 0 and approximately 0.25, the traction coefficient does not exceed zero. A traction coefficient of zero or less indicates that the resistance from the ground exceeds the drawbar pull needed for forward movement, implying that under these conditions, the wheel cannot move forward. In this experiment, translational speed is forcibly applied in conjunction with the advancement of the guide rail, indicating that although the wheel moves, it would not be capable of self-propelled under these conditions. As the slip ratio increases, the traction coefficient gradually rises, surpassing zero near a slip ratio of 0.25 and maintaining positive values thereafter. However, this increase is minimal; even at a slip ratio of 0.5, the traction coefficient remains very low at approximately 0.02. Considering that in low-speed experiments (with grouser-equipped wheels) conducted by our research group, the traction coefficient typically ranged between 0.3 and 0.4, it is evident that wheels without grousers perform extremely poorly at high speeds, indicating a high risk of becoming stuck and unable to move forward effectively.

Figure 4 on the right presents the results for traction efficiency as a function of slip ratio. As efficiency is defined as the ratio of output energy to input energy, it is expected to range from 0 to 1. In the regions where the traction coefficient is negative, as shown in Fig. 4 on the left, the traction efficiency is also negative. This numerical outcome is valid since the drawbar pull is negative; however, it is thought to occur because the wheel is forcibly translated along the guide rail. Even under conditions where self-propelled might be feasible, at slip ratios above 0.25, the efficiency remains very low, around 0.03. This indicates that, given the speed and wheel conditions, very little drawbar pull is generated relative to the torque required for propulsion.

From these results, it can be concluded that wheels without grousers perform exceedingly poorly at high speeds and are not advisable for use in situations where getting stuck is a concern, as they cannot effectively generate sufficient traction.

5.2. Non-grouser (slick) wheel results

In addition to the data shown in the previous section, results from two other speed ranges have been integrated, presenting the outcomes for wheels without grousers at three different speeds as results for slick wheels in Fig. 5.

The left Fig. 5 shows the slip-traction coefficient characteristics. Each of the three speeds is color-coded: the fastest speed of 1.0 m/s in red, the next speed of 0.5 m/s in green, and the slowest speed of 0.2 m/s in blue. Figure 5 shows a clear trend where lower speeds result in higher traction coefficients, and higher speeds lead to lower traction coefficients, thus confirming the presence of speed dependency. The right Fig. 5 follows a similar pattern, where the notably low traction efficiency at 1.0 m/s increases at slower speeds, particularly at 0.2 m/s.

These results elucidate that reducing speed can effectively enhance performance in cases where high-speed travel significantly degrades performance.

Comparing the lowest (blue) and highest (red) speeds in Fig.5, while a slip ratio of 0.25 was required to exceed a traction coefficient of zero at 1.0 m/s, at 0.2 m/s, the traction coefficient surpasses zero at a slip ratio of approximately 0.1. This indicates a significant increase in the area where self-propelled is possible. Furthermore, subsequent rises show that the traction coefficient generally exceeds 0.1, suggesting that the minimum

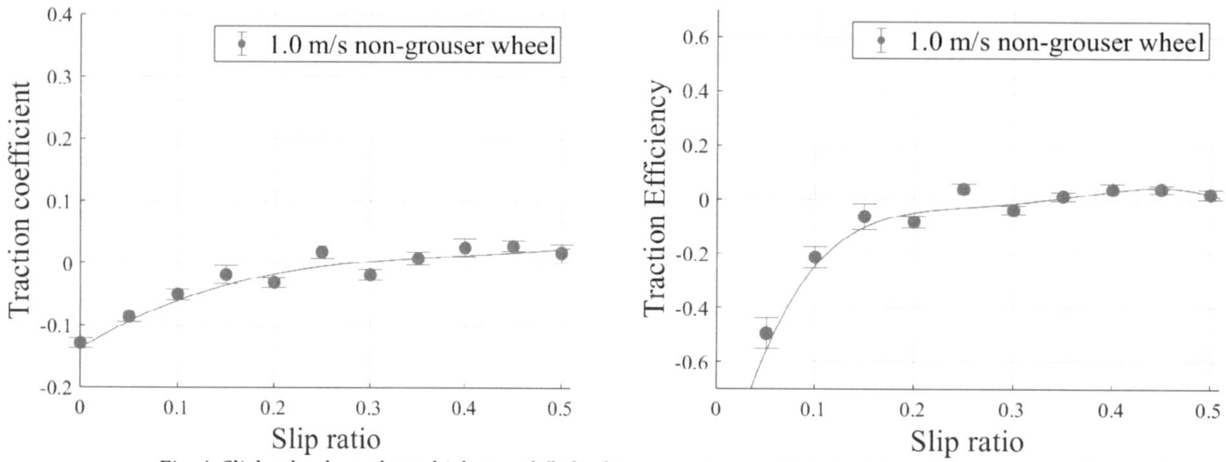

Fig. 4. Slick wheel results at high speed (left; slip vs traction coefficient, right; slip vs traction efficiency).

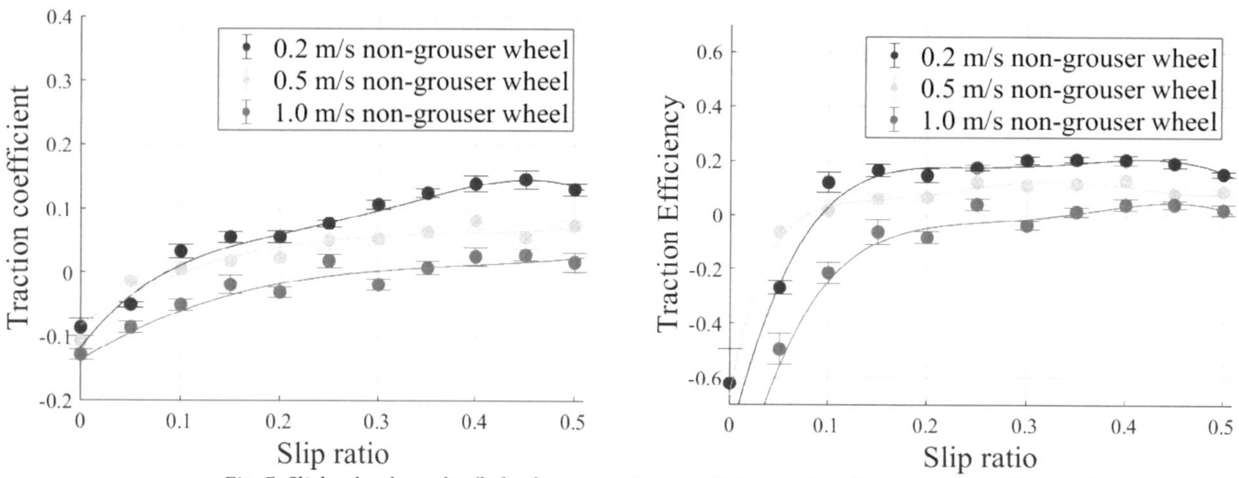

Fig. 5. Slick wheel results (left; slip vs traction coefficient, right; slip vs traction efficiency).

drawbar pull required for self-propelled is being produced.

In the right Fig. 5, the traction efficiency that was almost negligible at 1.0 m/s shows a marked improvement, reaching approximately 0.1 during travel at 0.5 m/s, and doubling to around 0.2 at 0.2 m/s. This indicates that for both performance indicators, traveling at lower speeds is beneficial for enhancing performance and effective for ensuring self-propelled and avoiding getting stuck.

Regarding this speed dependency, previous studies by our research group asserted tahat rover traversal exhibited no dependency on speed (Takehana, 2023). However, those studies only compared extremely low speeds ranging from 0.02 m/s to 0.04 m/s, suggesting that the absence of observed effects was due to the limited speed range analyzed. In contrast, the current study highlights the need to consider speed dependency in the operational range of high-speed rovers expected in future missions. This insight is crucial for optimizing performance and designing strategies to mitigate potential mobility issues in more demanding terrain and speed conditions.

5.3. Low-grouser wheel results

Figure 6 shows the slip-traction coefficient characteristics and slip-traction efficiency characteristics of the low-grouser wheels. From the left Fig. 6, it is evident that lower speeds result in higher traction coefficients, while higher speeds correspond to lower traction coefficients, confirming the presence of speed dependency similar to that observed with slick wheels. The right Fig. 6 also demonstrates a similar trend, where lower speeds enhance traction efficiency, and higher speeds reduce it, further confirming speed dependency for low-grouser wheels as well.

Particularly notable in the slip-traction efficiency characteristics of the low-grouser wheels is the rate of change. Until a slip ratio of 0.15, the traction efficiency increases sharply, and beyond this slip ratio, the increase tapers off, converging towards a constant value. This rate of change also mirrors the trend observed with slick wheels, indicating a consistent speed-dependent pattern across different wheel types. This consistency underscores the critical influence of speed on the traction properties of wheels, suggesting that adjustments in speed could be strategically utilized to optimize traction efficiency in practical rover applications.

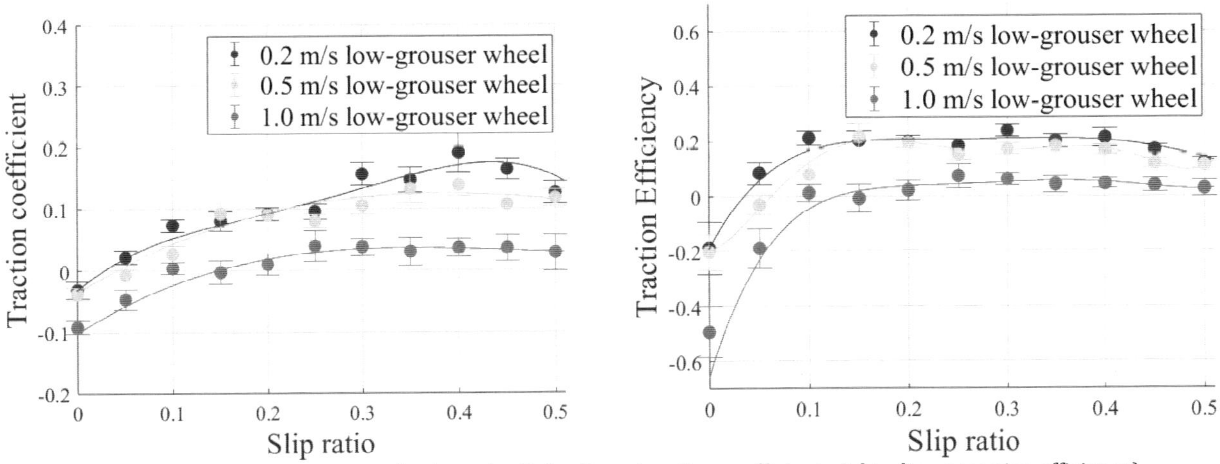

Fig. 6. Low-grouser wheel results (left; slip vs traction coefficient, right; slip vs traction efficiency).

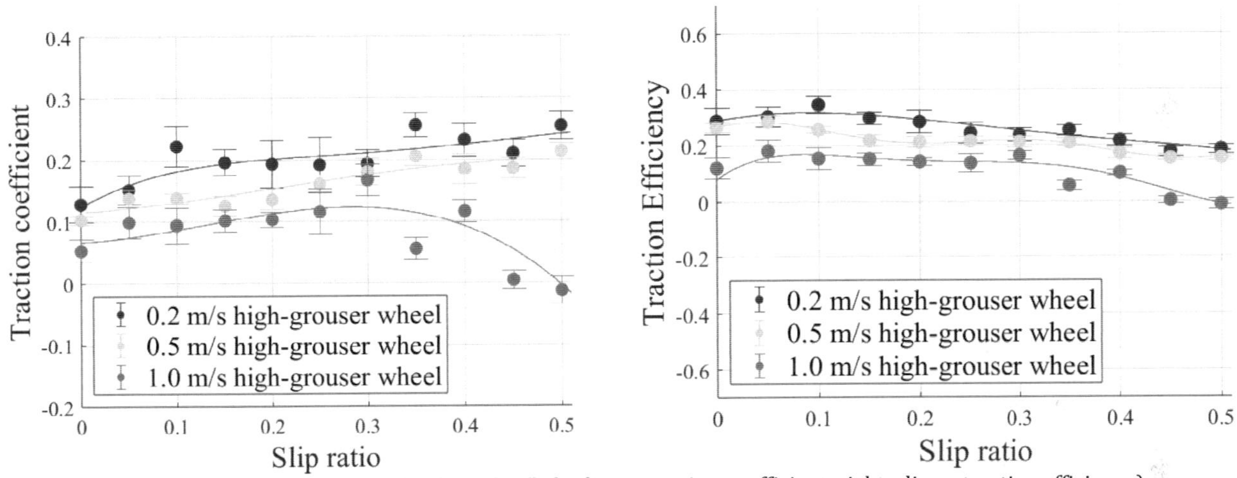

Fig. 7. High-grouser wheel results (left; slip vs traction coefficient, right; slip vs traction efficiency).

5.4. High-grouser wheel results

Figure 7 presents the slip-traction coefficient characteristics and slip-traction efficiency characteristics for the high-grouser wheel. As observed with the slick and low-grouser wheels, the speed dependency is confirmed for the high-grouser wheel as well, where lower speeds result in higher traction coefficients, and higher speeds lead to lower coefficients.

In the left Fig. 7, assessing traction coefficients, it is noted that at lower speeds, similar to other wheel types, the traction coefficient increases monotonically with rising slip ratios. However, the rate of increase for the high-grouser wheel is less pronounced compared to other wheels. At a speed of 1.0 m/s, after a slight increase in the traction coefficient, there is a sharp decline starting from a slip ratio of 0.3 onwards.

Regarding traction efficiency, the rate of change shows differences compared to other wheels. For the slick and low-grouser wheels, traction efficiency quickly increases up to a certain slip ratio, after which the increase slows down and converges towards a stable value. In contrast, for the high-grouser wheel, at each speed, traction efficiency peaks between a slip ratio of approximately 0.05 to 0.10, and then slightly decreases. Notably, the rate of decrease at 1.0 m/s is more significant than at the other two speeds. These observations underline the nuanced performance differences among the wheel types, emphasizing the impact of both speed and slip ratio on traction dynamics. The high-grouser wheel, particularly at higher speeds, demonstrates unique characteristics that may require careful consideration in practical applications to optimize rover mobility and efficiency.

5.5. Comparison between all wheels

In this section, we discuss the comparative results of the three types of wheels used in the study—non-grouser, low-grouser, and high-grouser wheels. Figure 8 presents the slip-traction coefficient characteristics and slip-traction efficiency characteristics at a high-speed condition of 1.0 m/s for all three wheels, plotted on the same graph. It is important to note that the color legend in this graph differs from the one based on speed in the previous section; here, the high-grouser wheel is depicted in blue, the low-grouser wheel in green, and the slick wheel in red.

Although experiments were also conducted at two other speeds (0.5 m/s and 0.2 m/s), these results are not shown in the graph as they were consistent with the findings at 1.0 m/s. This information suggests that it is possible to perform a similar

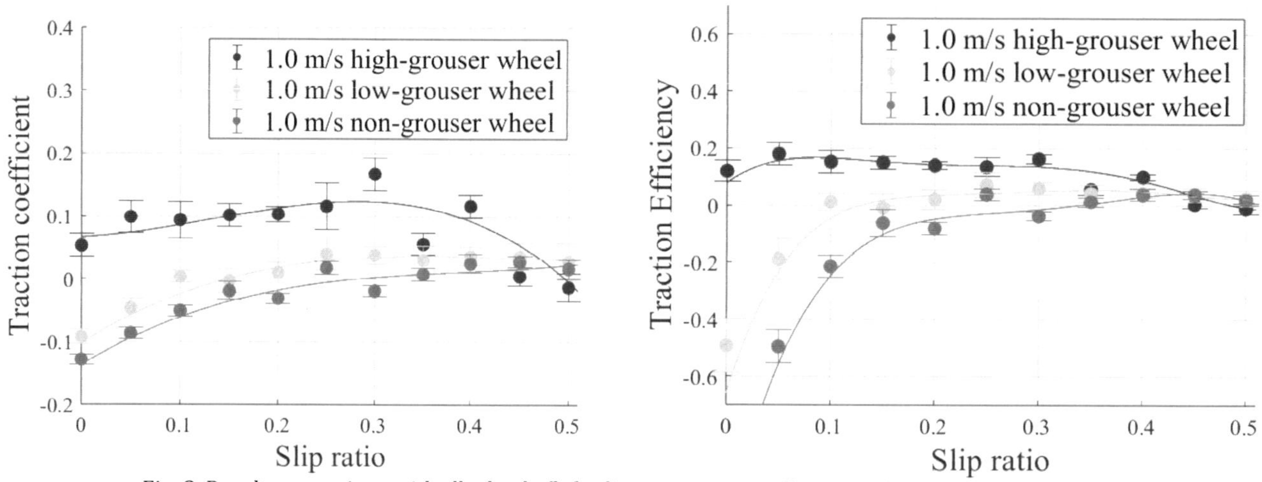

Fig. 8. Result comparison with all wheels (left; slip vs traction coefficient, right; slip vs traction efficiency).

evaluation by comparing Fig. 8 presented. By analyzing the performance of each wheel type in detail, we can gain insights into their relative efficiencies and characteristics under various operational conditions.

In the left Fig. 8, examining the slip-traction coefficient characteristics of each wheel, the traction coefficients typically follow the order: non-grouser wheel < low-grouser wheel < high-grouser wheel across most slip ratios. This pattern aligns with the wheel design methods described in Section 3, where it was noted that wheels with higher grouser heights can generate more traction by engaging more soil during traversal. This confirms that a greater grouser height enhances wheel performance by increasing the drawbar pull.

However, an exception is observed at a speed of 1 m/s and slip ratios of 0.45 and 0.50, where the high-grouser wheel demonstrates lower traction coefficients than the low-grouser wheel at the same slip ratios. This could be attributed to the high-speed and high slip conditions, which causes the high grouser wheel to dig into the soil rather than effectively pull it through. As the grouser penetrates and digs into the sand, more torque, or input energy, is required. As a result, this condition causes a reduction in traction efficiency. This phenomenon suggests that, in general, grouser height increases traction, but at higher slip ratios and speeds, it can have the counterproductive effect of destroying soil structure, increasing input energy, and decreasing overall traction efficiency.

In the right Fig. 8, examining the slip-traction efficiency characteristics of each wheel reveals a similar pattern to the slip-traction coefficient characteristics, where the traction efficiency values also follow the order: non-grouser wheel < low-grouser wheel < high-grouser wheel across various slip ratios. Notably, while the slick wheel demonstrated negligible traction efficiency, the high-grouser wheel consistently achieved a significant amount of traction efficiency, highlighting the impact of grousers.

Focusing on the increase in traction efficiency with rising slip, as discussed in the previous section, it was anticipated that the slick and low-grouser wheels, designed to perform poorly, would show an increase in traction efficiency with increasing slip, plateauing at a certain slip ratio. In contrast, the high-grouser wheel, while achieving higher traction efficiency than the other wheels, shows an increase in efficiency with slip but only up to a slip ratio of about 0.1, after which the efficiency begins to decrease.

As shown in Fig. 8, at high slip states, the differences in traction efficiency among the wheels are not pronounced. This suggests that in high-speed and highly slippery conditions, there is a limit to the traction efficiency that can be achieved, regardless of whether grousers are employed. This limitation indicates that under adverse conditions, even the mechanical advantages provided by grousers have their bounds, highlighting the need for careful consideration of wheel design relative to expected operational environments.

From the analysis, it is evident that there is a speed dependency in traction performance relative to the slip ratio for wheels operating on soft terrain at high speeds. This may be due to sand dynamics, which is not considered in the conventional model of terramechanics. We consider that the sand movement is greater in the high velocity conditions and that this action affected the traction performance of the wheels. In that sense, it is considered that the traction force did not differ for several different speed cases with less influence of sand movement.

Particularly, wheels like the slick wheel, which are not specifically designed with innovative features, demonstrate significantly reduced traction capabilities during high-speed operation. In such cases, there are two potential directions to enhance drawbar pull.

Considering speed dependency, reducing speed is one straightforward method that can significantly enhance traction performance. As shown in Fig. 8, both traction coefficient and traction efficiency can be improved multiple times by lowering the speed. Leveraging this speed dependency allows for the potential construction of driving control strategies for rover wheels that aim to achieve higher traction efficiency more quickly.

Another method to increase traction performance is through wheel design, specifically by incorporating grousers. Research on enhancing wheel performance with grousers has been extensively conducted by many institutions, and their effectiveness in high-speed operations has been clearly demonstrated in this study. Generally, wheels with longer grousers, such as the high-grouser wheel, are most suitable for increasing traction efficiency, showing superior performance compared to wheels with shorter grousers. The effectiveness of grouser-equipped wheels over slick wheels is apparent, indicating a clear improvement in traction performance.

However, focusing on the rate of change in traction efficiency from Fig. 8, it is evident that the effectiveness of grousers does not always apply in high-slip conditions. Under adverse operational conditions, there are limitations to what grousers can achieve, highlighting that even with optimal grouser design, there are boundaries to their effectiveness in extremely challenging environments. This suggests that while grousers significantly enhance traction under normal conditions, their performance can degrade in highly adverse conditions, thus necessitating a balanced approach in their application depending on the expected terrain and speed conditions.

6. Conclusions and future work

In this study, we evaluated wheel performance during high-speed travel on sandy terrain using a developed high-speed single-wheel testbed. This testbed allows simultaneous control of both the translational and rotational speeds of the wheel and adjustment of the vertical load, enabling detailed investigations of slip and traction characteristics across various wheel designs.

The experimental results demonstrated that traction performance is highly dependent on speed, with significant performance decreases at higher speeds. This finding underscores the necessity of optimizing wheel design for high-speed lunar rover applications. Notably, wheels equipped with grousers performed better than slick wheels, but even grousers reached performance limits at high speeds and high slip conditions.

Future research will focus not only on the height of the grousers but also on their number, angle, and shape to further optimize wheel design. Additionally, we plan to incorporate simulation approaches along with experimental methods to more accurately assess traction characteristics during high-speed travel. Furthermore, determining the equation relating the slip traction coefficient and slip traction efficiency, represented here as a graph, would be beneficial for future rover control.

This study provides foundational data for wheel design that can significantly enhance the efficiency and effectiveness of future planetary exploration missions.

7. Nomenclature

L	Length	[m]
W	Width	[m]
H	Height	[m]
r	Wheel radius	[m]
ω	Angular speed	[rad/s]
v_x	Translational speed	[m/s]
F_x	Drawbar pull	[N]
F_z	Vertical load	[N]
T	Driving torque	[Nm]
h	Grouser height	[m]
i	Wheel Slip	[-]
z	Sinkage	[m]
Φ	Grouser spacing	[rad]
n	Grouser number	[-]

8. Acknowledgements

The experimental testbed was developed through a joint research project with TOYOTA MOTOR CORPORATION. I would like to express my deep gratitude to Mr. Tsutomu Hamabe, Mr. Fumito Kurata, and Mr. Kenji Niwa for the extensive discussions that greatly contributed to this work.

The authors also would like to acknowledge Arthur Candalot for kindly helping to improve the hardware part of the high-speed single-wheel testbed.

9. Declaration of competing interest

The authors declare that they have no known competing financial interests or personal relationships that could have appeared to influence the work reported in this paper.

10. References

Bekker, M.G., 1956. Theory of land locomotion. University of Michigan Press, Ann Arbor.

Bekker, M.G., 1960. Off-the-road locomotion. University of Michigan Press, Ann Arbor.

Bekker, M.G., 1969. Introduction to terrain-vehicle systems. University of Michigan Press, Ann Arbor.

Buse, F., 2019. Fully automated single wheel testing with the DLR Terramechanics Robotics Locomotion Lab (TROLL).

Colaprete, A., Schultz, P., Heldmann, J., Wooden, D., Shirley, M., Ennico, K., Hermalyn, B., Marshall, W., Ricco, A., Elphic, R.C., et al., 2010. Detection of water in the LCROSS ejecta plume. Science 330 (6003), 463–468.

Ding, L., Gao, H., Deng, Z., Liu, L., 2010. Slip-ratio-coordinated control of planetary exploration robots traversing over deformable rough terrain. Proceedings of the 2010 IEEE/RSJ International Conference on Intelligent Robots and Systems, 4958–4963.

Ding, L., Gao, H., Deng, Z., Nagatani, K., Yoshida, K., 2011. Experimental study and analysis on driving wheels' performance for planetary exploration rovers moving in deformable soil. J. Terramech. 48 (1), 27–45.

Flores, G., Harris, D., McCauley, R., Canerday, S., Ingram, L., Herrmann, N., 2021. Deep space habitation: Establishing a sustainable human presence on the Moon and beyond. Proceedings of the 2021 IEEE Aerospace Conference, 1–7.

Girija, A.P., Agrawal, R., Lu, Y., Arora, A., de Jong, M., Saikia, S.J., Longuski, J.M., 2023. A single wheel test rig for ocean world rovers. J. Terramech. 109, 101–119.

Gonzalez, R., Iagnemma, K., 2018. Slippage estimation and compensation for planetary exploration rovers: State of the art and future challenges. J. Field Robot. 35 (4), 564–577.

Hamabe, T., Niwa, K., Yoshida, K., Tanaka, K., 2021. Law of dynamical similarity on traction performance for lunar vehicles. 1st Report: Overview of this study. Proceedings of the 65th Space Sciences and Technology Conference.

Higa, S., Nagaoka, K., Yoshida, K., 2016. Online estimation of wheel sinkage and slippage using a ToF camera on loose soil. Proceedings of the ISTVS 8th Americas Regional Conference, Troy, MI, USA.

Higa, S., Nagaoka, K., Yoshida, K., 2017. Reaction force/torque sensing wheel system for in-situ monitoring on loose soil. Proceedings of the 19th International and 14th European-African Regional Conference of the International Society for Terrain-Vehicle Systems (ISTVS).

Ishigami, G., Nagatani, K., Yoshida, K., 2009. Slope traversal controls for planetary exploration rover on sandy terrain. J. Field Robot. 26 (3), 264–286.

Li, S., Lucey, P.G., Milliken, R.E., Hayne, P.O., Fisher, E., Williams, J.P., Hurley, D.M., Elphic, R.C., 2018. Direct evidence of surface exposed water ice in the lunar polar regions. Proc. Natl. Acad. Sci. 115 (36), 8907–8912.

Lyasko, M., 2010. Slip sinkage effect in soil-vehicle mechanics. J. Terramech. 47 (1), 21–31.

Nagaoka, K., Sawada, K., Yoshida, K., 2020. Shape effects of wheel grousers on traction performance on sandy terrain. J. Terramech. 90, 23–30.

Nakagoshi, K., Rodriguez-Martinez, Yoshida, K., 2020. A new single-wheel testbed for fast-moving planetary robots. Proceedings of the 2020 Aerospace Europe Conference, Association Aéronautique et Astronautique de France.

Potts, N.J., Gullikson, A.L., Curran, N.M., Dhaliwal, J.K., Leader, M.K., Rege, R.N., Klaus, K.K., Kring, D.A., 2015. Robotic traverse and sample return

strategies for a lunar farside mission to the Schrödinger Basin. Adv. Space Res. 55 (4), 1241–1254.
Rodriguez-Martinez, D., Van Winnendael, M., Yoshida, K., 2019. High-speed mobility on planetary surfaces: A technical review. J. Field Robot. 36 (8), 1436–1455.
Rodriguez-Martinez, D., Yoshida, K., 2017. High-speed lunar rovers. ROOM: The Space Journal, 1 (12), 53–56.
Rodriguez-Martinez, D., Buse, F., Van Winnendael, M., Yoshida, K., 2023. The effects of increasing velocity on the tractive performance of planetary rovers.
Rodriguez-Martinez, D., Uno, K., Sawa, K., Uda, M., Kudo, G., Diaz, G., Umemura, A., Santra, S., Yoshida, K., 2024. Enabling faster locomotion of planetary rovers with a mechanically-hybrid suspension. IEEE Robotics and Automation Letters (RA-L), 9 (1), 619-626.
Skonieczny, K., Moreland, S.J., Wettergreen, D.S., 2012. A grouser spacing equation for determining appropriate geometry of planetary rover wheels. Proceedings of the 2012 IEEE/RSJ International Conference on Intelligent Robots and Systems, 5065–5070.
Smith, M., Craig, D., Herrmann, N., Mahoney, E., Krezel, J., McIntyre, N., Goodliff, K., 2020. The Artemis program: An overview of NASA's activities to return humans to the Moon. Proceedings of the 2020 IEEE Aerospace Conference, 1–10.
Sutoh, M., Nagaoka, K., Nagatani, K., Yoshida, K., 2012. Evaluation of influence of surface shape of locomotion mechanism on traveling performance of planetary rovers. Proceedings of the 2012 IEEE International Conference on Robotics and Automation, 3419–3424.
Takehana, K., Kizaki, S., Uno, K., Tanaka, T., Suda, G., Yoshida, K., 2023. Evaluation and comparison of driving performance of a lunar exploration rover wheel in different soils. Proceedings of the 16th European-African Regional Conference of the International Society for Terrain-Vehicle Systems (ISTVS), 151-158.
"Tohoku-Keisa", available at https://www.ktsangyo.co.jp/data1.htm (accessed on 31/5/2024).
Tsubaki, H., Ishigami, G., 2021. Experimental study on wheel-soil interaction mechanics using in-wheel sensor and particle image velocimetry Part I: Analysis and modeling of normal stress of lightweight wheeled vehicles. J. Terramech. 93, 23–39.
Wong, J.Y., 2008. Theory of ground vehicles. John Wiley & Sons Inc, Hoboken, New Jersey.
Wong, J.Y., Reece, A., 1967. Prediction of rigid wheel performance based on the analysis of soil-wheel stresses part i. performance of driven rigid wheels. J. Terramech. 4, 81–98.

0906 | Experimental Investigation of Increased Bearing Capacity When Imparting Vibration to Loose Ground in Low Atmospheric Pressure

Tomohiro Watanabe [a,*], Kojiro Iizuka [b]

[a] *Department of Engineering, Mechanical Engineering Program, Niigata University, Niigata City, Niigata Prefecture, Japan*

[b] *Department of Science and Engineering, Machinery and Control Systems, Shibaura Institute of Technology, Saitama City, Saitama Prefecture, Japan*

** Corresponding author: watanabe@eng.niigata-u.ac.jp*

ABSTRACT

Recently, legged robots have attracted considerable attention as highly mobile rovers for planetary exploration. However, the surfaces of celestial bodies such as Mars and the Moon are primarily loose, causing slippage due to surface deformation from the rover's leg movements. To mitigate this, a walking method aimed at preventing slippage was proposed. In this walking method, the bearing capacity is increased by imparting vibration from the rover's leg to the loose soil on the ground. This walking method improved the mobility performance of the legged rover. It is crucial to investigate the increase in bearing capacity by imparting vibration in a realistic space environment to validate the effectiveness of the proposed method. This study examines the change in bearing capacity due to vibration under low atmospheric pressure, a condition found on Mars and the Moon. This condition affects the characteristics of ground sand due to decreased air resistance. Our findings provide valuable insights into using vibration in planetary exploration to enhance the performance of legged rovers. The experimental results indicate that the bearing capacity under low atmospheric pressure is nearly identical to that under standard atmospheric pressure, suggesting that atmospheric pressure conditions have minimal impact on the bearing capacity when vibration is applied.

Keywords
Loose soil
Bearing capacity
Vibration
Terramechanics
Low atmospheric pressure

1. Introduction

Many missions are undertaken to explore the planets for human development. Perseverance, the rover developed by NASA, reached Mars in 2021 and investigated the ground environment and sources of life (NASA, 2024b). ispace, Japanese private space development company, is developing and launching the luna lander (ispace, 2024). As part of the feature missions, NASA has outlined ambitious plans for missions on the lunar surface to prepare for long-term extraterrestrial activities and expand the horizons of space exploration (NASA, 2024a). In addition, the Smart Lander for Investigating Moon (SLIM) was launched in September 2023. The robots were carried internally and delivered to the surface of the Moon (JAXA, 2024). In these backgrounds, the demand for exploration robots operating on Mars and lunar surfaces could be expected to increase. As stated above, the rovers, the robots that move on extraterrestrial bodies, are used in the space mission. Recently, robots with leg mechanisms have been investigated as exploration rovers with outstanding running performance (Komizunai et al., 2010; Yeomans et al., 2013). They can go across rugged terrains, facilitated by each leg's ability to select points of ground contact precisely. However, the lunar surface is characterized by its rough and loose soil, known as regolith. This type of ground is particularly susceptible to deformation under external forces. Owing to the inherent characteristics of loose ground, legged rovers mostly slip because the motion of their legs deforms the surface of loose ground. To solve this problem, a walking method was proposed for legged rovers that reduces the distance they slip on the loose ground using vibration (Watanabe et al., 2020a). The effectiveness of the proposed walking method was evaluated by conducting an experiment in which a legged testbed walked on loose ground.

In the proposed walking method, the bearing capacity is increased by vibrating the loose soil on the ground. The bearing capacity is the force that the leg receives from the ground. Vibration increases bearing capacity because it increases the ground's shear strength and the sinkage of the rover's leg. In a previous study, the relationship between the bearing capacity and vibration was investigated to improve the effectiveness of the proposed walking method (Watanabe et al., 2020b; Watanabe et al., 2023b). We did not investigate the influence of low atmospheric pressure, a condition found on Mars and the Moon, for increasing the bearing capacity by vibration. Investigating the ground change by vibration in the real exploration environment is important for suggesting the usefulness of the proposed walking method.

This study examined the change in bearing capacity due to vibration under low atmospheric pressure, a condition found on Mars and the Moon. In this study, the measurement setup was in the case with low interior atmospheric pressure, and the bearing capacity was measured. In the experiment, a rod was lowered to the ground and vibrated. The bearing capacity was measured when the rod was dragged on the ground. We measured and compared the bearing capacity at standard and low atmospheric pressures. Finally, the influence of low atmospheric pressure on increasing the bearing capacity by vibration was confirmed.

https://doi.org/10.56884/I6EQREI6

2. Experimental method for measuring bearing capacity

2.1. Explanation of increasing bearing capacity by vibration

Figure 1 shows the change in bearing capacity caused when the leg is dragged on the ground. The bearing capacity increases immediately after dragging the leg because the movement of the leg compacts the ground. After the bearing capacity reaches a peak value, it decreases because the ground crumbles. The peak value of the bearing capacity is increased by vibration (*Fig. 1*). This factor is considered that the ground is compacted, and the density of the ground is increased by vibration.

2.2. Introduction of a related study that investigated terramechanics in vacuum condition

The relationship between the ground condition and the vacuum environment is essential to clarify for developing the exploration rover. Some study groups investigated the moving mechanism of the exploration rover in the chamber with vacuum interior atmospheric pressure. Sutoh investigated the mobility of the planetary rover's wheel in the vacuum atmospheric pressure condition(Sutoh, 2021). He stored the experimental setup in the chamber with vacuum interior atmospheric pressure. In this experiment, the relationship between slip late and the traction force of the wheel is confirmed in some kinds of sand. Reiss et al. evaluated the hopper's performance in transporting the Moon and Mars soil resources (Reiss et al., 2014). They stored the experimental setup in the chamber with vacuum interior atmospheric pressure. The chamber rode in the plane to create a low-gravity environment by using flite. Sakatani et al. investigated the thermal conductivity of lunar regolith in vacuum atmospheric pressure using regolith simulant (JSC-1A) and the vacuum chamber (Sakatani et al., 2018). As the related studies above described, the experimental setup method is stored in the vacuum chamber. Therefore, we measured the bearing capacity by storing the experimental setup in the vacuum chamber.

2.3. Experimental method

Figure 2 (a) shows the experimental environment, which consists of a vacuum pump, an experimental setup, and a chamber. *Figure 2 (b)* shows the experimental setup that consists of a soil tank, rod, and force sensor. *Figure 3* shows the measuring flow of the bearing capacity. First, the ground was mixed and flattened. Then, the rods were placed on the ground. The sinkage of the rod was 30 mm. Next, the vacuum pump exhausted air which is in the chamber. Atmospheric pressure was decreased to 13 hPa. The bearing capacity was compared with and without vibration. In the case of imparting vibration, the vibration was generated for 30 seconds. Subsequently, the rod was dragged after the vibration ceased. The force sensor measured the supporting force value. The rod was moved at a speed of 1.67 mm/s and dragged for 30 seconds. The soil tank had a length, width, and height of 90, 150, and 70 mm, respectively. The dragging direction of the rod was parallel to the length of the soil tank. The ground material consisted of Silica No. 5. Vibration was generated in the vibration motor, which is mounted in the rod. This experiment measured the bearing capacity under standard and vacuum atmospheres. Five trials were performed for each experimental condition. Table 1 shows the components of the experimental environment. Table 2 presents the experimental conditions.

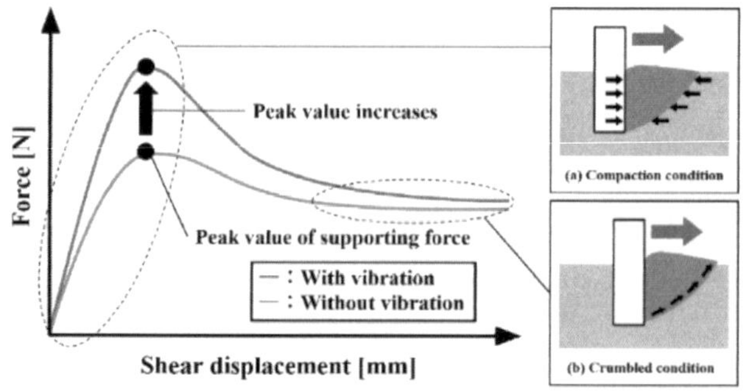

Fig. 1. Bearing capacity vs shear displacement (Higashiyama et al., 2023)

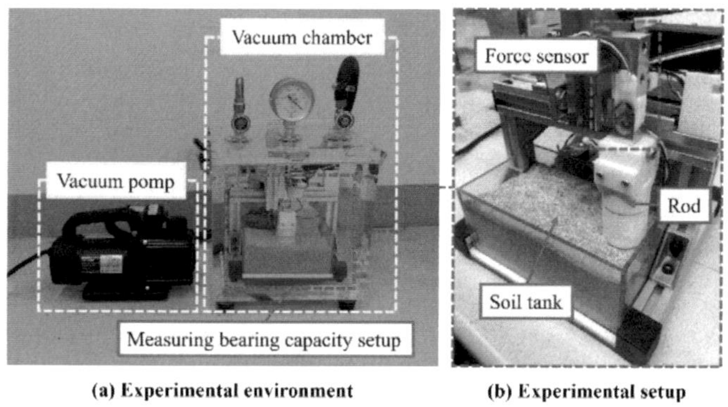

Fig. 2. Experimental setup for measuring bearing capacity

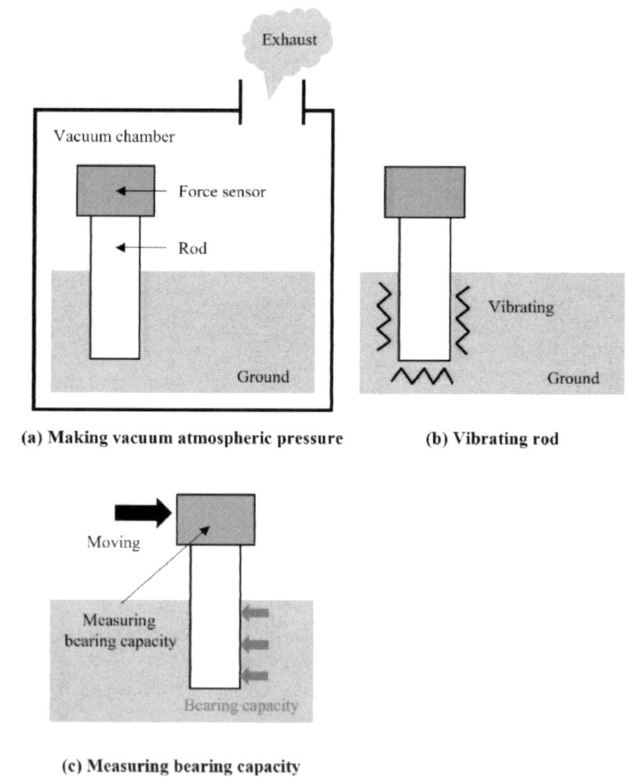

Fig. 3. Experimental flow for measuring bearing capacity (Watanabe et al., 2023a)

Table 1
Element list of experimental setup

Item	Condition (value)
Vacuum chamber	1-5800-01(Azuwan)
Vacuum pump	TA150SW-K
Force sensor	Load cell SC133-2kg
Vibration motor	TP-2528C-24

Table 2
Experimental conditions for measuring bearing capacity

Item	Condition (value)
Number of trials	5
Sinkage of arod	30 mm
Vibration time	30 s
Traction speed	1.67 mm/s
Traction time	30 s
Kind of sand	Silica No. 5
Voltage for vibration motor	36 V
Vibration force	3.66 N
Vibration frequency	253 Hz
Atmospheric pressure	Standard atmosphere (1013hPa) Vacuum atmosphere (13 hPa)

3. Experimental results

Figure 4 (without vibration) and *Figure 5* (with vibration) show the variation in the bearing capacity against the shear displacement. These figures show the average data of five trials. Without vibration (*Fig. 4*), the bearing capacity initially increased, whereas after, it converged. With vibration (*Fig. 5*), the bearing capacity initially increased, after which it decreased and converged. In these figures, the variations in the bearing capacity are similar in standard and vacuum atmosphere pressure conditions.

Figure 6 compares the peak value of bearing capacity in each experimental condition. This figure shows the standard errors calculated from the five experimental trials. The bearing capacity with vibration is larger than one without vibration in all atmosphere pressure conditions. The difference between standard and vacuum atmospheric pressure conditions is obtained to compare the bearing capacity in each atmospheric pressure condition. In the result without vibration, the difference is 0.23 N. In the result with vibration, the difference is 0.31 N. Therefore, the bearing capacities in standard and vacuum atmospheric pressure are almost the same because these different values are small.

Table 3 shows the RMSE value in each experimental condition. The RMSE value without vibration is 0.185 N, and the one with vibration is 0.274 N. This result suggests that the variations in the bearing capacity against the shear displacement in standard and vacuum atmosphere pressure conditions are almost the same because the RMSE values are small.

These experimental results show that the influence of vacuum atmospheric pressure on increasing the bearing capacity by vibration is small. The knowledge obtained in this study suggests that vibration is effective for moving on loose ground.

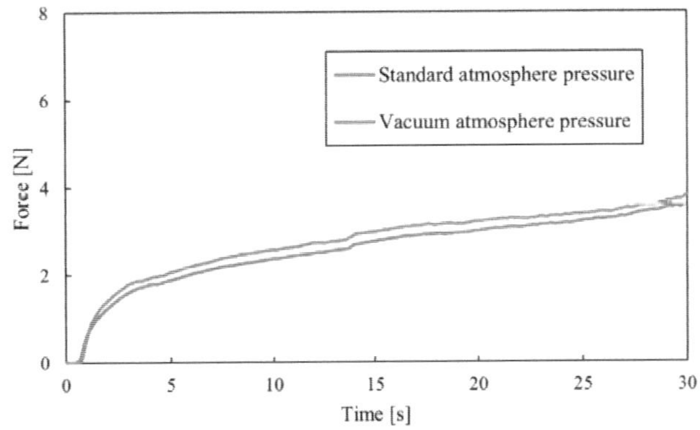

Fig. 4. Bearing capacity vs distance (without vibration)

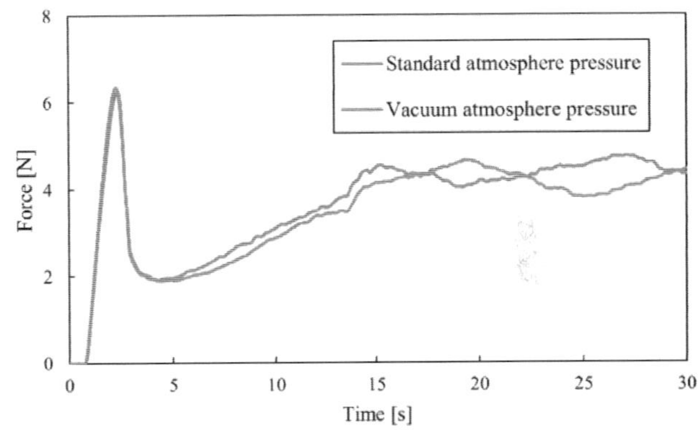

Fig. 5. Bearing capacity vs distance (with vibration)

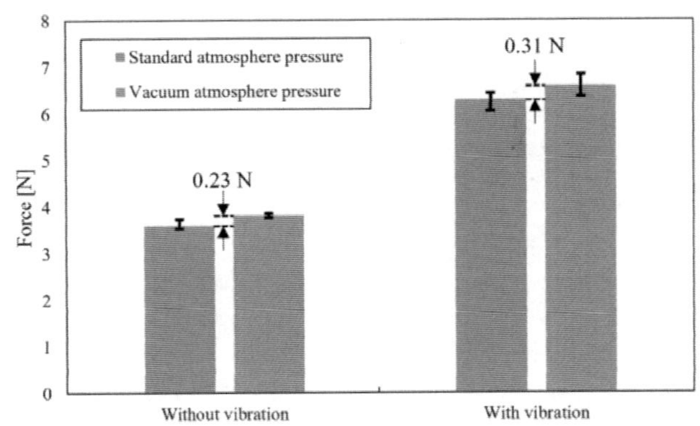

Fig. 6. Comparison of peak value of bearing capacity

Table 3
RMSE value in each experimental condition

Condition	Value
Without vibration	0.185 N
With vibration	0.274 N

4. Conclusion

This study examined the change in bearing capacity due to vibration under low atmospheric pressure, a condition found on Mars and the Moon. In this study, the measurement setup was in the case with low interior atmospheric pressure, and the bearing capacity was measured. In the experiment, a rod was lowered to the ground and vibrated. The bearing capacity was measured when the rod was dragged on the ground. From the experimental result, increasing the bearing capacity by vibration was confirmed in vacuum atmospheric pressure. Moreover, the bearing capacities in standard and vacuum atmospheric pressure are almost the same. Therefore, we consider that the influence of vacuum atmospheric pressure on increasing the bearing capacity by vibration is small. The knowledge obtained in this study suggests that vibration is effective for moving on loose ground.

In future study, the model that predicts the bearing capacity when imparting vibration will be constructed. This model will be constructed to be able to predict the bearing capacity in vacuum atmospheric by considering the influence of vacuum atmospheric pressure which is obtained in this study. Moreover, the investigation of the bearing capacity in vacuum atmospheric pressure will be conducted using more experimental parameters, such as different kinds of soil.

5. Acknowledgements

This study was supported by JSPS KAKENHI (Grant Number JP 21J11944).

6. Declaration of competing interest

The authors declare that they have no competing interests.

7. References

Higashiyama R., Watanabe T., Iizuka K., Watanabe D., 2023. Simulation of distinct element method to analyses of increased supporting force by imparting vibration for improving walking performance of lunar exploration legged rover, in: Nagoya, Japan, JSME annual Conference on Robotics and Mechatronics.

Ispace. (2024). https://ispace-inc.com/jpn/. Accessed 28 April 2024.

JAXA, International space exploration, https://www.exploration.jaxa.jp/e/program/#slim (accessed 28 April 2024).

Komizunai S., Konno A., Abiko S., and Uchiyama M. (2010). Development of a static sinkage model for a biped robot on loose soil, in: Sendai, Japan, 2010 IEEE/SICE International Symposium on System Integration.

NASA, Artemis. https://www.nasa.gov/specials/artemis/. Accessed 28 April 2024.

NASA, MARS 2020 MISSION PERSEVERANCE ROVER. (2024). https://mars.nasa.gov/mars2020/mission/overview/. Accessed 28 April 2024.

Reiss P., Hager P., Hoehn A., Rott M., and Walter U., 2014, Flowability of lunar regolith simulants under reduced gravity and vacuum in hopper-based conveying devices, J. Ter., 55, 61-72.

Sakatani N., Ogawa K., Arakawa M., and Tanaka S., 2018, Thermal conductivity of lunar regolith simulant JSC-1A under vacuum, Icarus, 309, 13-24.

Sutoh M., 2021, Traveling performance analysis of planetary rovers using a repeatable test system in vacuum, J. Ter., 95, 15-24.

Watanabe, T., Iizuka, K., 2020a. Proposal of walking to reduce slipping behavior using compaction effect of loose soil caused by a propagation of vibration for small light lunar planetary exploration rovers with legs, Tra. JMSE, 86.

Watanabe, T., Iizuka, K., 2020b. Study on Connection between Kind of Vibration and Influence of Ground for Leg Typed Rovers, Int. J. Mech. Eng. Rob. Res., 9.

Watanabe T., Higashiyama R., Iizuka K., 2023a. Experimental Investigation of Increased Supporting Force by Vibration on Loose Ground in Low Atmospheric Pressure, in: Nagoya, Japan, JSME annual Conference on Robotics and Mechatronics.

Watanabe, T., Iizuka, K., 2023b. Experimental investigation of relationship between bearing capacity and vibration parameters for planetary exploration legged rovers, Int. ROBOMECH J., 10.

Yeomans B., Saaj C. M., and Winnendael M. V., 2013. Walking planetary rovers – experimental analysis and modelling of leg thrust in loose granular soils, J. Ter., 50, 107-120.

1204 | Measurement and visualization of soil cutting and throwing behavior by a rotary tillage blade

Toyohiro Kato[a], Yuta Hirayama[a], Yasumaru Hirai[b], Muneshi Mitsuoka[b], Kouki Nakatani[c], Kenji Kaida[c], Mutsumi Hidekawa[c], Takashi Okayasu[b]*

[a] *Graduate School of Bioresource and Bioenvironmental Sciences, Kyushu University, Japan*
[b] *Faculty of Kyushu, Fukuoka, Japan*
[c] *KOBASHI INDUSTRIES CO., LTD, Okayama, Japan*

* *Corresponding author: okayasu@bpes.kyushu-u.ac.jp*

ABSTRACT

The condition and quality of soil after tillage operations are strongly influenced by soil condition, machine performance, and operating conditions. For this reason, various studies have been conducted on the changes in soil mechanical properties due to tillage. In this study, a measurement method was developed to evaluate soil cutting and throwing behavior during rotary tillage operations using a motion sensor. The accuracy was verified by comparison with the results obtained from a motion sensor and high-speed camera images. The soil cutting and throwing behavior estimated from the motion sensor data treated by a low-pass filter was consistent with that obtained from the high-speed camera images. Moreover, 3D DEM simulation was also conducted to predict its behavior, and the prediction accuracy of soil cutting and throwing behavior was verified by comparing it with the test results. However, relatively large differences were observed between the results obtained by the 3D DEM simulation and the motion sensor data. These results indicate that the material constants between particles and between particles and objects in the DEM simulation should be investigated carefully.

Keywords
Inertial Measurement Unit (IMU)
Discrete Element Methods (DEM)
Edinburgh Elasto-Plastic Adhesion Model
Tillage blade
Image analysis

1. Introduction

Soil surface in farmland is compacted well due to fluctuations of effective stresses in the soil caused by rainfall and the repetition of traveling loads of farm machinery. An excessively compacted soil is not suitable for crop cultivation. For this reason, tillage is performed before cultivating crops. The purpose of tillage is to dispose of weeds and residues, improve the physical properties of the soil, especially aeration, water retention, and permeability, mix fertilizer and soil, and prevent pests and diseases. Typical tillage and soil preparation machinery include plows, rotary tillers, and compactors. Among them, a rotary tiller is a work machine with multiple tillage blades attached to a rotating drive shaft, and it is used to connect to a tractor's PTO shaft. The working performance is influenced by the following conditions: (1) soil types (sandy soil, cohesive soil), (2) cutting properties, and (3) operating conditions (Sakai, 1978; Lee et al., 2003). Therefore, the design of rotary tiller blades was conducted based on prototyping with a trial and error method in the past. However, this method is disadvantageous in terms of time and cost when designing tillage blades under various conditions.

On the other hand, recent developments in computers and computational mechanics have led to the introduction of mechanical design aided by numerical simulation. Among them, a discrete element method (DEM) proposed by Cundall and Starck (1979). is a computational method that models the mechanical interaction between particles and is highly effective for analyzing and visualizing various mechanical behaviors of granular materials, such as flow, storage, cutting, and mixing. It has already been applied to the analysis of soil tillage behavior by plows and rotary tillers, and then various valuable results have been obtained up to the present (Okayasu et al., 2012; Ucgul et al., 2017; Zhao et al., 2020). Since particles' mechanical properties are strongly dependent on the material constants used, rigorous comparison and verification with actual measurement results are required to evaluate the accuracy of the analysis. However, studies to evaluate the dynamic behavior of particles are not enough yet.

In this study, we developed an evaluation method for tilling behavior by using a motion sensor to accurately measure the soil cutting and throwing behavior by a rotary tillage blade and verified the validity by comparison with the results of 3D DEM simulation.

2. Material and Methods

2.1. Motion Sensor for Soil Cutting and Throwing Test

Figure 1 shows the motion sensor used to track the soil mass behavior. The sensor is a DSP wireless 9-axis motion sensor (SS-MS-HMA200G60, Sports Sensing Co., Ltd.) capable of measuring 3 accelerations (max. 200 G), 3 angular velocities (max. 6000 dps), and 3 geomagnetic fields. The maximum sampling frequency is 1000 Hz. The measured data is transferred by wireless communication. The sensor was covered with silicon lubber mold created by a liquid-type of 3D printer (Saturn2, ELEGOO). The average value of the first 2000 data (for 2 seconds) was used as the offset value since the drift of angular velocity was observed.

The measured data is stored in the sensor's internal memory (128 MB) to avoid data losses due to direct data transmission. The sensor data can be transmitted through USB or wireless network communications after the soil cutting and throwing test.

https://doi.org/10.56884/TNVE38T5

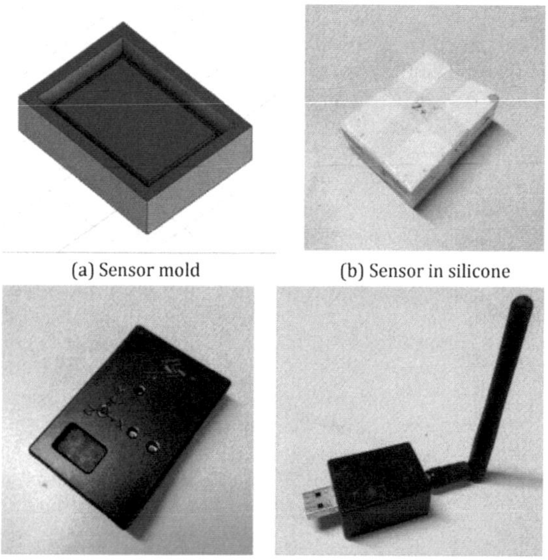
(a) Sensor mold (b) Sensor in silicone

(c) Motion sensor (d) Transmitter and receiver

Fig.1 Motion sensor

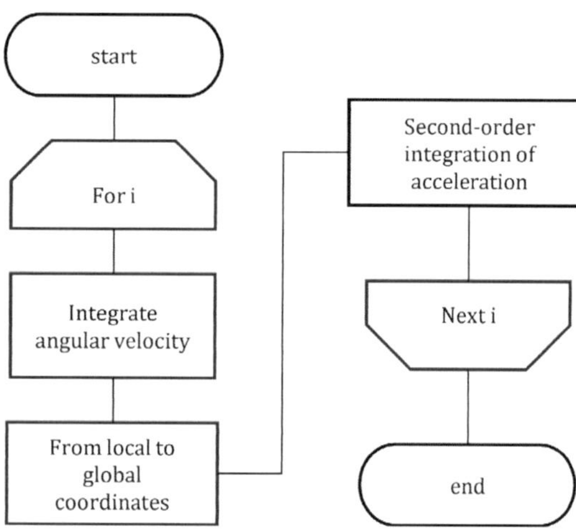

Fig.2 Flowcharts for calculating displacement and angle

2.2. Calculation of displacement and angle

Figure 2 shows a flowchart of the program for calculating displacement and angle. The following calculations were performed using MATLAB (MathWorks Inc.). First, acceleration, angular velocity, and elapsed time were read as the measured data. Noise included in the raw data was removed from the acceleration using a low path filter with a cutoff frequency of 10Hz. The total displacement, defined as the sensor's motion, was given by time integration of the acceleration and the angular velocity.

2.3. Soil Cutting and Throwing Test

Soil cutting and throwing tests were conducted using a single rotary tillage blade to measure the rotation and displacement of a soil clod. Soil cutting and throwing were conducted by rotating the rotary tillage blade by using a 3-phase electrical motor (200 V, 20 HP, Mitsubishi Electric Co., Ltd.). The soil used in each test was

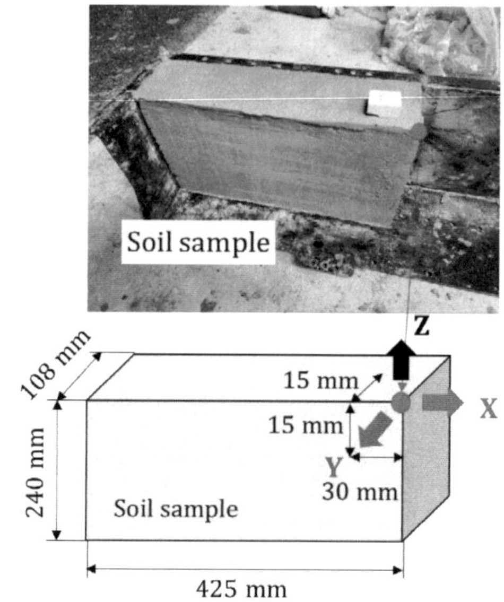

Fig.3 Soil sample used in soil cutting and throwing test

prepared by mixing the Kasaoka clay (Kasanen Kogyo Co., Ltd.) completely dried in an oven for one day with water using a soil mixer (KM-200, Hayashida Tekko Co., Ltd.). The moisture content of the soil was maintained at around 25%/w.b. in all the tests. The soil sample with 425 mm (width), 108 mm (length), and 240 mm (height) was created by filling the soil into the metal-frame box, as shown in Fig. 3.

The motion sensor was placed at the location shown in Fig. 3. To evaluate the accuracy of measured motion data, the high-speed movie (480 fps) of the soil clod cut by the blade was also recorded by the camera (Tough TG-6, OM Digital Solutions Co., Ltd.). Markers made of white paint were created to track the soil clod motion by directly spraying the paint on the side and top face of the soil sample. The position of the markers

The soil clod motion in the video was measured by tracking the markers using MATLAB. Both the origins of each displacement obtained by the sensor and the marker's tracking were set up manually. We compared both results to verify the accuracy of the measurement method using the motion sensor.

2.4. Discrete Element Method

The soil cutting and throwing test simulations were conducted using a commercial three-dimensional discrete element method (3D DEM) called EDEM (Altair Engineering, Inc.). EDEM has many functions and material models to describe the mechanical behavior and handling for granular media. Extension of the material, models, and multi-physical simulations coupled with finite element method, computational fluid dynamics, and multi-body dynamics are also incorporated in the latest version of EDEM. For the above reasons, we adopted it as the analytical software.

A typical computational procedure of DEM simulation is shown in Fig. 4. First, contact with neighboring particles and objects is determined, and if there is contact, the contact forces are calculated. In this stage, all the external forces, such as gravity and body force are also calculated. After the calculation of the forces acting on the particles is finished, acceleration, angular velocity, displacement, and rotation are calculated and updated by applying Newton's second law. The time step is updated when the

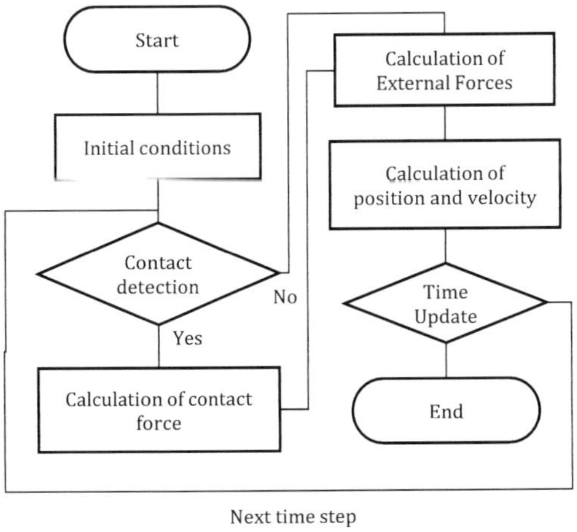

Figure 4 Flowchart of a typical computational procedure of DEM

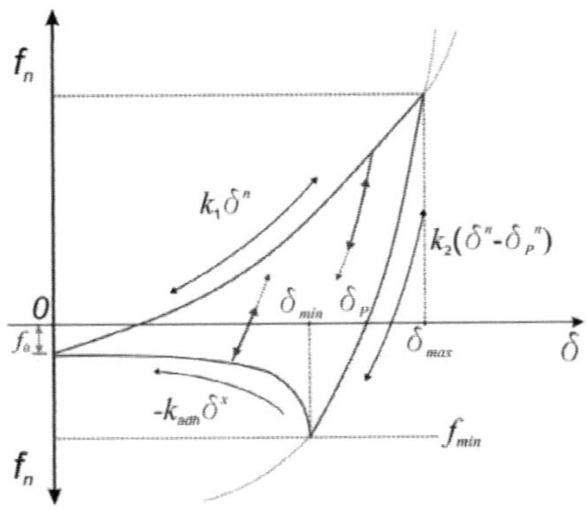

Fig 5 Conceptual diagram of Edinburgh elasto-plastic adhesion model

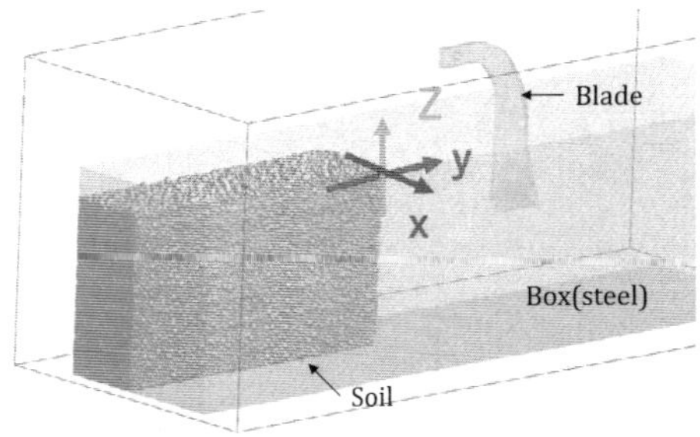

Fig 6 3D DEM simulation model for soil cutting and throwing test

Table 1 Soil particle size and proportion

Type	Real particle size [μm]	DEM particle size [mm]	Mass percentage [%]
Small	3.0	378.4	56
Medium	4.5	347.5	6.5
Large	6.0	178.3	37.5

Table 2 Material Parameter

Material	Poisson ratio	Density [kg/m^3]	Elasticity coefficient [GPa]
Soil	0.3	2600	0.005
Steel	0.3	7800	80
Blade	0.3	7800	80

Table 3 Contact Parameter

Material	Poisson ratio	Coefficient of static friction	Coefficient of rolling friction
Soil	0.6	0.9	0.01
Steel	0.6	0	0
Blade	0.9	0.2	0.01

calculations for all particles are completed. The same calculations are conducted continuously up to a prescribed time.

2.5. Contacts Models between Particles

Two types of material models were used in the soil cutting and throwing test simulations. These models are explained briefly.

Edinburgh Elasto-Plastic Adhesion Model

The Edinburgh Elasto-Plastic Adhesion (EEPA) model (Morrissey, 2013) was employed to describe the mechanical interactions between particles. This model is capable of describing the elasto-plastic mechanical behavior between particles, including irreversible deformation, friction, and cohesion. An overview of the EEPA model is given below. The amount of overlap between particles is shown on the horizontal axis and the forces acting between particles on the vertical axis. First, the repulsive force between particles increases as the amount of overlap increases. Next, when the force is unloaded, cohesive forces are generated between the particles. Finally, by eliminating the overlap between the particles, the initial position is returned.

Hertz-Mindlin with JKR Model

The Hertz-Mindlin with JKR model (Johnson et al., 1971) is employed for particle-geometry contact. This model is a method that adds the concept of adhesion to Hertz's contact theory. In this study, the model is used to reproduce the friction between tillage blades and soil particles, and to reproduce the rotation of soil clods.

2.6. DEM Analysis

Figure 6 shows the cutting and throwing test simulation. Soils classified as heavy clay by the International Union of Soil Sciences (IUSS) were used in the soil cutting and throwing tests. However, particle size used in the DEM simulation has a significant influence on the computational cost, so three particle sizes were selected as representative of the grain size accumulation curve, and these were multiplied by 2000 and used in the analysis. The soil layer for the DEM simulation was generated with 425 mm wide, 108 mm long, and 240 mm high, which are the same size as in the experiment. To ensure even compaction of the constituent particles, the soil layer was divided into eight layers and generated by repeated filling and compaction. The vertical compaction stress for each layer was 125 kPa, and there was one compaction cycle for each layer. A single rotary tillage blade, created by 3D CAD, was placed on top of the soil layer model. The material constants between each soil particle and between each soil particle and object used in the analysis are given in Tables 1-4. These constants were determined so as to fit the soil cutting test result using a single tillage blade. The tillage blade and soil layers were used as objects with surrounding steel frames. Cutting conditions were as follows: tillage depth: 150 mm, tillage pitch: 110 mm, tillage jaw rotation speed: 200 rpm.

EDEM simulations were performed on a Windows PC equipped with an Intel(R) Xeon(R) Gold6230CPU@2.10GHz, 128GB RAM, and an NVIDIA Quadro RTX 5000 GPU. CUDA, a parallel computing

Table 4 Parameter of EEPA model and JKR model

Item	Value
EEPA Contact pull-off force(N)	0
EEPA Surface Energy(J/m2)	700
EEPA Contact Plasticity Ratio	0.9
EEPA Slope Exponent	1.5
EEPA Tensile Exponent	3
EEPA Tangential Stiff Multiplier	0.7
JKR Soil-steel Surface Energy(J/m2)	0
JKR Soil-Blade Surface Energy(J/m2)	100

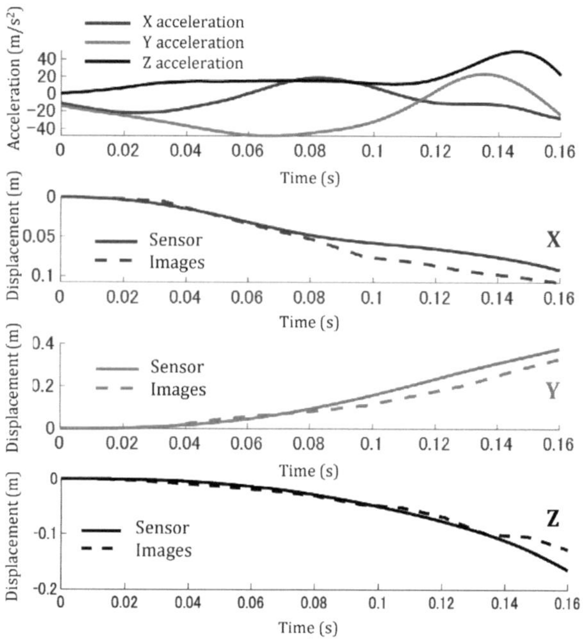

Fig 7. Comparison of soil cutting and throwing behavior by motion sensor and image analysis

programming library and environment for the NVIDIA GPU platform, was used to improve the computation speed. Computed results for visualizing soil cutting and throwing behavior were stored at constant time intervals.

3. Results and Discussions

3.1. Soil Cutting and Throwing Test

A comparison of the soil cutting and throwing behavior obtained by the motion sensor's data and the tracking data given from the test images is shown in Fig. 7. A low-pass filter (10 Hz) was applied for the acceleration data to eliminate the high-frequency acceleration data due to the impact load at the initiation of soil cutting. In this treatment, the soil cutting and throwing behavior estimated from the motion sensor data was consistent well with that obtained from the high-speed camera images. However, for more accurate measurement with motion sensors, the filter to be introduced needs to be improved.

Figure 8 shows the comparison of soil throwing behavior over time. The image at 0.06 seconds indicated that the soil cutting by the rotary tillage blade had just been completed. Immediately after that, the soil clod cut by the blade was found to be moved downward with a large rotation. The soil clod behavior obtained by the motion sensor was generally consistent with the motion shown in the image. It was verified that the tracking method for the soil clod by the motion sensor was effective and had high accuracy

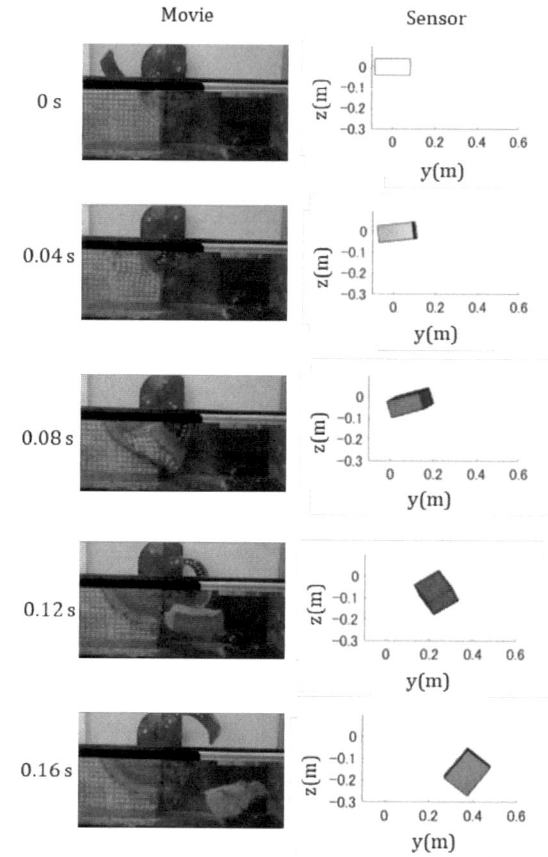

Fig 8. Comparison of soil throwing behavior

3.2. 3D DEM simulation result and its comparison with motion sensor data

Figure 9 shows the comparison of soil cutting and throwing behavior between image analysis and DEM simulation. Soil behavior obtained from DEM simulations did not agree with measured results. This result can be also confirmed from the graph in Fig. 10. The difference between the two behaviors is mainly due to the fact that in the DEM simulation, the soil was significantly thrown in the x-axis direction, resulting in a collision with the wall (after 0.06s). Therefore, the material constants that represent the contact between soil particles and the tillage blade need to be reexamined in DEM simulations.

4. Concluding Remarks

A measurement method using a motion sensor was developed to evaluate the soil cutting and throwing behavior during rotary tillage operations, and its accuracy was evaluated. Moreover, a 3D DEM simulation was conducted to predict its behavior, and the prediction accuracy of soil cutting and throwing behavior was verified by comparing it with the test results. The findings given in this study are as follows:

1. A 10 Hz low-pass filter was introduced to remove high-frequency acceleration data caused by the impact load at the

start of soil cutting. This treatment improved the reliability of the acceleration data.

2. The soil cutting and throwing behavior estimated from the motion sensor data was generally consistent with that obtained from the high-speed camera images.
3. Relatively large differences were observed between the results of the 3D DEM simulation and the results of the motion sensor measurements. These results indicate that the material constants between particles and between particles and objects in the DEM simulation should be investigated carefully.

Soil cutting and throwing tests under different tillage conditions will be required to contribute to the advancement of the mechanical design of rotary tillage blades and the optimization of tillage conditions. Additionally, the prediction accuracy of the DE simulations, we will also try to develop a determination method for material constants of sol models used in the DE simulations.

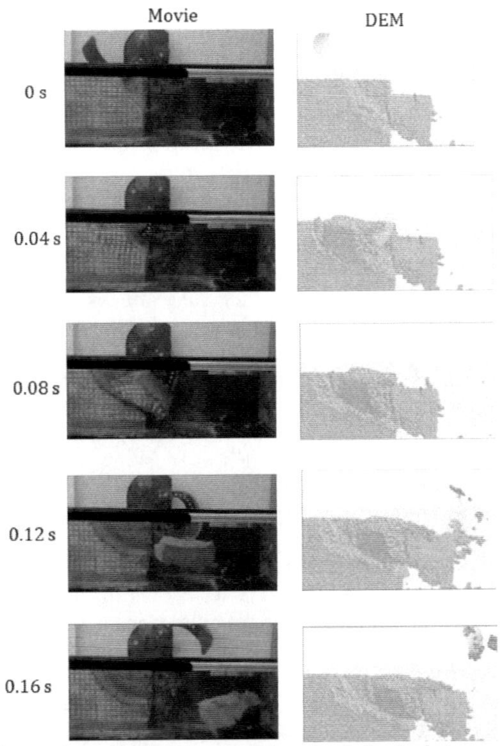

Fig 9. Comparison of soil cutting and throwing behavior between image analysis and DEM simulation

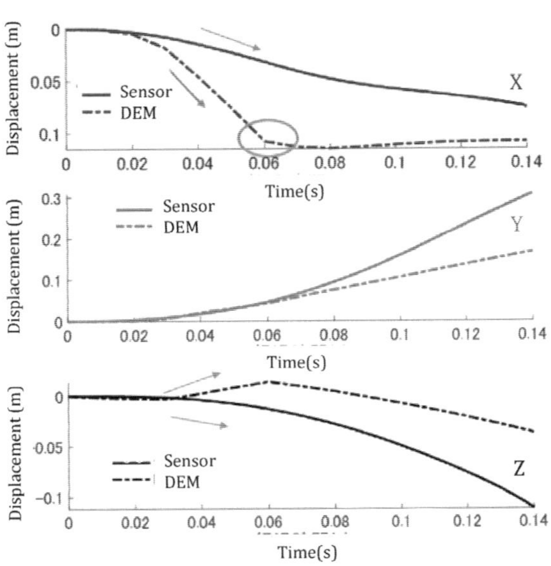

Fig 10. Difference in the displacement of each axis between DEM analysis and actual measurement

5. Acknowledgements

Parts of this research were conducted as corroborative research with KOBASHI INDUSTRIES CO., LTD, and the JSPS, KAKENHI No. JP23K05468, Japan. We want to express our gratitude for their valuable support.

6. References

Cundall, P.A., Starck, O.D.L., 1979. A discrete numerical model for granular assemblies. Geotechnique. 29(1), 47-65.

Johnson, K.L., Kendall, K., Roberts. A.D., 1971. Surface energy and the contact of elastic solids. Proc. the Royal Soc. of London. Series A, 324, pp.301-313

Lee, K.S., Park, S.H., Park, W.Y., Lee., C.S., 2003. Strip tillage characteristics of rotary tiller blades for use in a dryland direct rice seeder. Soil and Tillage Res.., 71, 25-32.

Morrissey, J.P., 2024. Discrete element modelling of iron ore pellets to include the effects of moisture and fines. 2013. https://era.ed.ac.uk/handle/1842/8270 , Accessed June. 9,

Okayasu, T., Morishita, K., Terao, H., Mitsuoka, M., Inoue, E., Fukami, K. 2012. Modeling and prediction of soil cutting behavior by a plow. Valencia, Spain: Int. Conf. of Agr. Engng.

Sakai, J.., 1978. Designing process and theories of rotary blades for better tillage (part-I). Jpn. Agric. Res. Quart., 12 (2), 86-93.

Ucgul, M., Saunders, C., Fielke, J. M., 2017. Discrete element modelling of top soil burial using a full scale moldboard plough under field conditions. Biosys. Engng., 160, 140-153.

Zhao, H., Li, H., Ma, S., He, J., Wang, Q., Lu, C., Zheng, Z,, Zhang, C., 2020. The effect of various edge-curve types of plain-straight blades for strip tillage seeding on torque and soil disturbance using DEM. Soil and Tillage Res., 202, 104674.

1241 | Study on Classification of Excavated Soil Using Internal Sensor Data of Hydraulic Excavator

Naoki Morisawa [a, *], Masaya Imanishi [a], Masaki Yanagishita [a], Teiichirou Chiba [b], Hiroshi Yamamoto [b], Takeshi Hashimoto [c], Daisuke Endo [c]

[a] *Komatsu.Ltd., Test Engineering Center, Development Division, Japan*
[b] *Komatsu.Ltd., Corporate Technology Department, Office of the CTO, Japan*
[c] *Advanced Technology Research Team, National Research and Development Agency Public Works Research Institute, Japan*

** Corresponding author: naoki_morisawa@global.komatsu*

ABSTRACT

In these days, construction sites are facing workforce shortages caused by declining labor and lack of young workers. Therefore, "Improve Productivity" is very important. Automation and autonomy of construction machinery is one of the solutions.

As above, ICT hydraulic excavators and bulldozers are widely used in many sites, but the scope of the application is limited. In particularly, automated excavating in hydraulic excavators still has many difficulties in actual use. The root cause is needed to the complicated operation during excavation affected by the soil conditions. Since the soil conditions change anytime, if the operation is not adjusted to match to them, the efficiency will be decreased significantly. In other words, to achieve automated excavating, the hydraulic excavator must recognize variation of soil conditions first.

While digging, the operator has been sensing the machine's response affected by the soil conditions to consider it and adjusting machine operation simultaneously. In other words, human senses were used to recognized soil conditions to perform efficient operations. In that case, we thought that if we could replace human senses with internal sensors of the machine, the machine could recognize the soil conditions by itself.

In this study, based on this hypothesis, we conducted measuring sensor data during excavating in various soil conditions (e.g. hardness, materials...).

The test results indicated that the soil hardness measured by simplified N-value meter, has a high correlation with the work efficiency and bucket tooth speed calculated from internal sensor data. This implicated that the sensor data from the hydraulic excavator can be used to be determine soil conditions. In addition, there is a possibility that excavator could recognize soil conditions by itself.

Keywords
Soil Classification
ICT
Hydraulic Excavator
Internal Sensor
Automation
Improve Productivity

1. Introduction

Due to the recent decline in the working-population and the shortage of young workers, construction sites are facing a serious labor shortage. And the construction industry has implemented work style reforms aimed at limitations on overtime work in Japan. Even the aspects that were previously compensated for through extensive overtime work will now be subject to stringent regulation (*Ministry of Health, Labor, and Welfare, Japan: About the upper limit regulation of overtime work*). Consequently, there is apprehension regarding a term of construction delays.

Therefore, Improving Productivity at construction sites has become an urgent priority. In Japan, there are nationwide efforts such as "i-Construction 2.0" that aim to enhance productivity through the full utilization of information and Communication Technology (ICT) in construction industry(*Ministry of Land, Infrastructure and Transport, Japan: i-construction 2.0*). Automation and autonomy of construction machinery is one of the methods of improving productivity.

One representative example of this is the use of ICT bulldozers and ICT hydraulic excavators in the construction sites. Not only be eliminating pre-marking and other auxiliary tasks but also pre-automation in construction machinery, particularly in the forming and finishing processes, has enabled to be carried out with the same precision without relying on the operator's skills. This has greatly improved construction efficiency (*Shimano et al., 2014*). However, there are still limitations to the applicable processes, and the excavating process which is most frequently usage of hydraulic excavator has not well progressed yet.

One of the root causes of this is believed to be the complex operational procedures required to accommodate the soil conditions. Since the internal soil conditions cannot be recognize based on the surface condition before excavating, operators must perform appropriate adjustment of operations depend on the soil conditions during excavation. Failure to achieve the flexible operations carried out by operators during excavating would result in a decrease in the efficiency of automated excavation. If we aim to replicate the actions of operators, it is believed that the excavator must recognize the soil conditions itself first during excavation and reflect them in its control.

Traditionally, in the excavating process, operators have improved their operational skills and senses to enhance excavating efficiency (*Sakaida et al., 2008*) (*Koiwai et al., 2018*). In response to the ever-changing soil conditions of the excavating site, operators have been performing the proper excavating operations while perceiving the machine's response to their operations with their human senses and making instant judgments to carry out the proper operations based on experience(*Fujita et al., 2001,a*).

On the other hand, from the perspective of ease of operation for the operator, the hydraulic excavator has continued to evolve through improvements in operability and excavating force. However, in the future, attention is also being paid to the evolution of the excavator itself, independent of the machine operator. It is believed that this could potentially lead to higher efficiency automation than human-operated operations.

If the machine can accurately replicate the human senses and the operator's operational actions based on human process, it could enable the machine to autonomously adapt to various changes in the external environment, such as soil conditions.

To initiate the research from this perspective, we conducted interviews with several experienced operators within the company to gather insights on key aspects to focus on regarding the machine's behavior during excavation and how they perceive soil conditions through their senses. *Table 1* shows frequently

https://doi.org/10.56884/76LVS50I

mentioned items identified through these interviews. It was confirmed that the operators recognize the soil conditions of the excavating site by observing the phenomena that appear on the excavator's body, which are influenced by the properties of the soil. In other words, they are making soil classifications based on their human senses.

We believe that enabling the substitution of the operators' sensory cognition of soil conditions with machine sensor data is the first step towards advancing automated excavation.

On the other hand, soil classification is currently being carried out during the pre-construction geological surveys and when utilizing the excavated soil. Another objective of this research, in addition to advancing automated excavation, is that by enabling soil classification during the excavating process, it becomes possible to have consistent soil management from construction planning to completion. This is expected to contribute not only to the productivity improvement of the excavator itself, but also to the overall productivity improvement of the construction process.

In addition to various sensors being installed on construction machinery to accommodate ICT construction(*Harada., 2016*), recent advancements in ICT technology allow for the replacement of the operator's senses with sensor data. This hypothesis suggests that the soil classification, which was previously carried out by the operator based on their senses, can now be quantitatively performed by the construction machinery.

Studies have been conducted to estimate soil conditions from excavation experiments on a laboratory-scale model (*Satomi et al., 2020*) (*Chen et al., 2014*) (*Yousefi Moghaddam et al., 2012*) and to estimate soil conditions from actual excavator data under limited conditions (*Tani et al., 2017*) (*Satomi et al., 2022*). However, there are few studies that involve excavation experiments conducted by operators using actual hydraulic excavators.

In this study, to validate this hypothesis, verification experiments were conducted using actual excavator and operator.

Table 1
Example of operator's focus points during excavation.

Phenomena Occurred in Machine	Methods of Grasping Phenomena by Operator	
	The Human Sense	Recognition
Movement and speed variations of work equipment	Visual cognition	(In response to the input) Slow ->Hard, Fast ->Soft
Penetration and excavating depth of bucket tooth	Visual cognition	(In response to the input) Sallow ->Hard, Deep ->soft
Volume of excavating soil (Bucket fill factor)	Visual cognition	Less->Hard, Much->Soft
Noise of load fluctuations from the engine and pump	Auditory cognition	Volume and timbre of sound
Vibration of machine	Sensory cognition	Large-> Hard, Few->Soft

2. The verification method for the hypothesis

Considering that the operator gathers a lot of information through visual cognition(*Fujita et al., 2001,b*) as shown in *Table 1*. Therefore, camera image was considered as the first approach to use. However, considering the operating environment of construction machinery, there are concerns about the impact of various disturbances. Therefore, we aimed to achieve the implementation using only internal sensors without relying on external sensors, as it is difficult to ensure robustness. *Table 2* shows the correlation between the operator's focal points during excavation and the internal sensors of the excavator.

We will conduct verification focusing on the hardness of the soil. The hardness of the soil has a significant effect on excavating efficiency. Therefore, operators were naturally prioritized grasping the feature of soil hardness during excavation.

We also considered a measurement instrument that could provide a simple and relatively accurate measurement of soil hardness, assuming that it could evaluate the same indicator across different levels of experimental conditions. As a result, we decided to use a simplified N-value instrument.

Standy soil and Clay soil were selected to digging materials, which are frequently excavated by 20-ton class hydraulic excavator.

To focus on the differences in hardness, we have decided to prepare a lot of experimental conditions with varying simplified N-values.

Furthermore, we have decided to use the same operator and simulate the usual excavating operation while conducting measurements. We will then organize the correlation between the sensor data, which is believed to strongly correlate with the operator's cognition, and the simplified N-values of the soil at that time.

Table 2
Example of relationship between operator's focus points during excavation and internal sensor data of excavator .

Focus Points	Alternative Machine Internal Sensors	Quantitative Value
Movement and speed of work equipment	IMU Potentiometer Cylinder stroke sensor Cylinder speed sensor	Bucket tooth speed Cylinder speed Variation of coordinate of tooth
Penetration of bucket tooth		
Depth of excavation		
Volume of excavated soil	Payload meter	Weight of excavated soil

3. Overview of excavating experiments with actual excavator

3.1. Excavator in use

A 20-ton class ICT hydraulic excavator PC200i-11 manufactured by Komatsu Ltd. was used in this study. The experiment was performed using a standard bucket with a capacity of 0.8m^3 (based on the new JIS standard). *Figure 1* shows external view of the machine.

Fig. 1. External view of the PC200i-11

3.2. Method of preparation of excavating pit and measurement of hardness of the soli

Experiments were conducted at Komatsu's outdoor testing fields using three types of materials: gravelly sand (hereinafter, called "S-G"), sandy soil mixed with fine particles (hereinafter, called "S-F"), and clayey soil (hereinafter, called "Cs"). *Table 3* shows the particle size distribution of these three materials measured by the metal mesh method. An excavating pit was created with a width of approximately 3m in the left-right direction, a length of approximately 5m in the front-back direction, and a depth of approximately 2m. When preparing the hard soil conditions, the soil was backfilled to a thickness of approximately 20-30cm and leveled evenly. Bucket compaction was using for soil compaction with caution in terms of the depth direction also became as homogeneous as possible.

There is a limit to the hardness that can be achieved through bucket compaction, so for more hard conditions, we utilized naturally compacted areas that had not been disturbed for a long period. *Figure 2* shows a schematic diagram of the excavating pit.

Simplified N-Value was measured using the soil penetration instrument made by Nishinihon-Shikenki. We measured simple N-value on the surface of the excavating pit at three locations along the longitudinal direction. These average values were used as an indicator of the hardness for the test condition.

Table 3
Particle size distribution of test materials

Classification	Gravelly sand	Sandy soil mixed with fine particles	Clay
Pebble (%)	15.4	16.5	9.0
Sand (%)	75.4	37.1	21.6
Silt (%)	2.2	20.3	47.2
Clay (%)	7.0	26.1	22.0

Fig. 2. A schematic diagram of the excavating pit

3.3. Start position of work equipment and method of excavation.

With the aim of minimizing data variability, we conducted excavating experiments for each soil condition using the excavating start position where the arm was extended to its maximum and the bucket angle θ (0.105 [rad])was fixed (see Fig. 3). This value was determined based on confirmation of several test operators in advance and there were no significant differences between the operators.

The operation of the excavator was carried out by two test operators (Operator A with 9 years of experience within the company and Operator B with 10 years of experience) belonging to Komatsu's testing department.

The excavation was initiated from the excavating start position and the measurement was taken until the excavator came to a stop after the lifting operation following the completion of excavating, as shown in *Figure 4*. This entire process was considered as one cycle of excavating data.

No specific instructions were given to the operators regarding the excavating method, and they were allowed to excavate freely at their discretion.

Fig. 3. External view of initial position of each work equipment

Fig. 4. External view of completion of excavation

3.4. Data measurement environment of excavator sensors

The sensor data from the excavator was connected to the excavator's controller and measured using a laptop via CAN (Controller-Area-Network). *Figure 5* shows the main excavator sensor items that were measured in this study, along with their respective sampling rates.

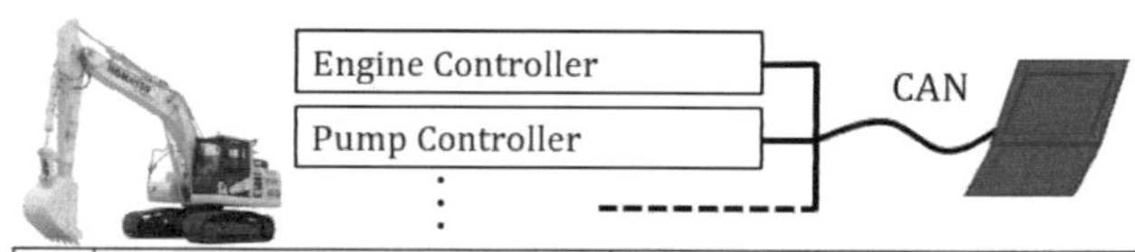

No.	Measurement items	Unit	Sampling rate[Hz]
1	Displacement of cylinder	mm	100
2	Speed of cylinder	mm/s	100
3	Angle of work equipment	rad	100
4	Input lever rate	%	100
5	Payload	kg	10
6	Coordinates of the bucket tooth	-	100

Fig. 5. Measurement method and contents

4. Results of excavating experiment with actual excavator

Based on the focus points of the operator during excavation (see *Table 1*), the experimental results were organized by considering the bucket tooth speed during excavation and the excavated weight as the amount of soil excavated in one cycle.

The relationship between the simplified N-value and the average bucket tooth speed during excavation is shown in *Figure 6*, while the relationship between the simplified N-value and the excavated weight per hour (excavating efficiency) calculated from the excavated weight and excavating time in one cycle is shown in *Figure 7*, categorized by material and operator.

The average bucket tooth speed during excavation was calculated using the position coordinates of the bucket tooth. It was calculated by taking the difference between the position of the bucket tooth at the previous time and the position of it at the current time, and then dividing it by the sampling period.

In addition, the excavated weight is defined as the weight measured by the payload meter on the excavator when it comes to a standstill after excavated. Excavating time defined the time from start of excavation to when the arm lever was returned from excavating side to neutral side. The excavated weight and excavating time were used to calculate the excavated weight per hour, which is equivalent to excavating efficiency.

The average bucket tooth speed (*Fig.6*) and the excavating efficiency calculated from the excavated weight (*Fig.7*) showed a high correlation with the simplified N-value of the excavating soil measured beforehand.

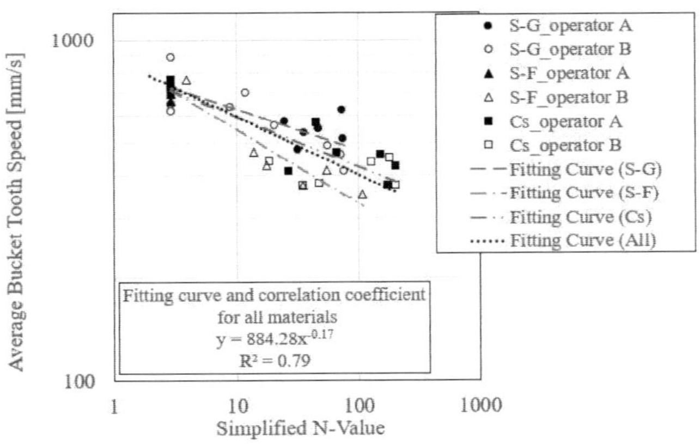

Fig. 6. Relationship between the simplified N-value and average bucket tooth speed during excavation

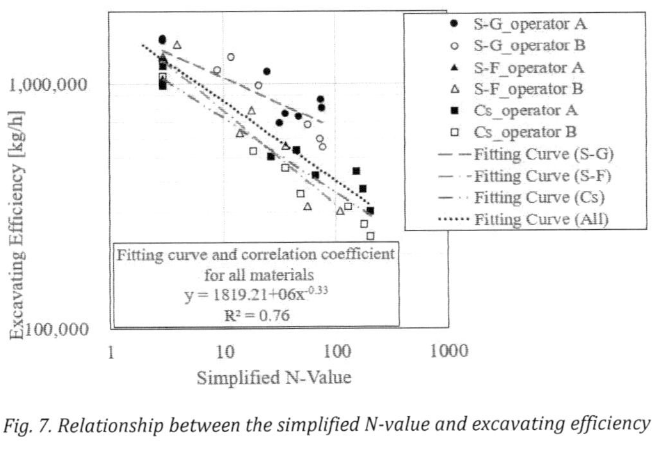

Fig. 7. Relationship between the simplified N-value and excavating efficiency

4.1. Discussion of relationship between simplified N-value and average s bucket tooth speed during excavation

A tendency of decreasing average bucket tooth speed during excavation was observed as simplified N-value increases. This result was consistent with the operator's focal points identified pre-interview.

To further investigate the above trend, we will conduct an analysis focusing on the trajectory of the bucket tooth during excavation. As an example, *Figure 8* shows the difference in the trajectory of the bucket tooth during excavation at simplified N-value 3 and 49 in clayey soil. At a simplified N-value of 3, the maximum excavating depth recorded was 952mm, whereas at a simplified N-value of 49, it was 360mm. This represents a significant difference of approximately 2.6 times.

The trajectory observed for a simplified N-value of 3 exhibits fewer irregularities, enabling smoother progress during excavation. Conversely, the trajectory observed for a simplified N-value of 49 demonstrates more irregularities.

In case of soft soil with a simplified N-value of approximately 3, the excavator's excavating capacity enables easy to penetrate and fill in the bucket. By being cautious about excessive penetration and excessive load in bucket, it is possible to minimize the decrease in the excavating speed even without relying on precise lever control. In other words, the operator's excavating technique is not so required in soft soil conditions.

On the other hand, as the soil becomes harder, the resistance during penetration increases. This causes the excavating depth to become shallower. Also, due to the higher reactive force exerted by the soil, the bucket tooth is more prone to stopping during excavation. Therefore, the same operational approach used for soft soil cannot be applied when dealing with hard soil. The operator adjusted the bucket tooth angle and excavating depth during excavation using more precise lever controls to acquire as much soil as possible in one path. It is speculated that the operator had the intention to maximize the amount of soil, even if sacrificing excavating speed. As a result, the bucket tooth speed during excavation decreased in harder soil.

The excavating speed of gravelly sand was observed to be faster compared to other materials. This was caused by the decrease of cohesion as the proportion of sand particles increase, the excavating resistance to the bucket also decreased.

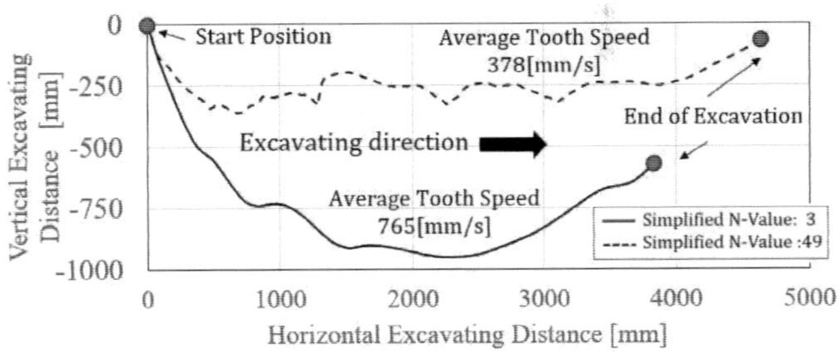

Fig. 8. Difference in trajectories of bucket tooth

4.2. Discussion of relationship between simplified N-value and excavating efficiency

As the simplified N-value increases, the excavating speed decreases, as a results, longer excavating time was needed. *Figure 9* shows one example of external view of backet fill conditions after excavated. Differences in simplified N-values resulted in variations in the bucket fill conditions. As a result, excavating efficiency also decreased with the increase in N-values.

In this study, we compared the excavating volumes per cycle using the excavated weight measured by the payload meter. However, it should be noted that in the case of materials with significantly different densities, this evaluation method may not accurately evaluate efficiency. Therefore, it is a future task to consider evaluation methods, such as measuring and comparing both the excavated weight and the excavated volume simultaneously.

Furthermore, since differences in operator skills can also have a significant impact on the results, we plan to conduct comparisons and verifications considering the skill levels of the operators.

Fig. 9. External view of backet fill conditions after excavated.

5. Excavating experiment excluding the influence of operator skills

In the previous experiments, only the initial excavation posture was fixed, while the operations of excavating were performed at the discretion of each operator. Therefore, it cannot be denied that the variability due to the operator's specific manipulations is reflected in the obtained data.

Therefore, in this chapter, we have considered examining a new experimental method (hereinafter, called simplified operation method), which aims to minimize the variability caused by operator manipulation. We compared the measurement results obtained this method with the results obtained in Chapter 4 in order to validate the results.

Furthermore, we considered that enabling discrimination of soil conditions using a simplified method that is independent of operator manipulation could be one way to enhance the reliability of automated discrimination conducted by excavator itself.

5.1. Discussion and determination of operation method

Figure. 10 shows the time-series data of the input status of each operation levers for two test operators in the previous experiments with a simplified N-value of around 50. Here, the positive range of the arm and bucket indicates the excavating side, while the positive range of the boom indicates the lifting side.

First, regarding the arm lever, each operator maintained a consistent state of excavating position throughout the excavating process. Also, at the start of excavation, the timing moves neutral position to excavating position was almost same. The differences in lever input state due to operator were minimal throughout the excavating process in arm lever.

On the other hand, there were significant differences in lever input state of the boom and bucket among operators. Especially, at the start of excavation, operator B moved the boom lever to down position as attempt to penetrate using downward force of boom. As a bucket lever, the timing for inputting the lever on the excavating side was the almost same at the start of excavation. However, significant differences were observed both of levers thereafter in terms of input amount and timing.

At the start of excavation, the operator performs an action to penetrate the bucket's tooth and based on the excavator's response in terms of the penetration depth and the work equipment's speed, they quickly assess the approximate condition of the soil being excavated. Thereafter, the operator proceeds with excavation by adjusting the lever input accordingly to the estimated soil conditions. Therefore, operations on the way to excavation are often performed based on the operator's own operating skills, which can easily lead to variations in operations.

Then, focusing on the operation of bucket tooth penetration at the start of excavation and limiting the operator's lever input to only rapid inputs for arm excavating side, we could control the variations in operations.

Based on the results from previous chapter (refer to *Fig. 8*), changing the simplified N-value results in variations in the penetration amount at the start of excavation.

When the arm is operated independently, it moves in an arc motion. To minimize resistance and achieve a more favorable penetration, we determined that aligning the bucket tooth with the tangent of this arc motion would be beneficial. So, we defined the excavating start posture as shown in *Fig.11*. The arm is fully extended, and the angle between the line connecting the boom top pin to arm top pin with the bottom surface of the bucket was 0.5π [rad].

The operator conducted measurements by executing only the arm lever move to excavating side quickly (Simplified operating method) from this posture and held the lever until the bucket tooth came to a belief stop after reaching hydraulic relief. The other experimental methods and conditions are the same as in Chapter 3.

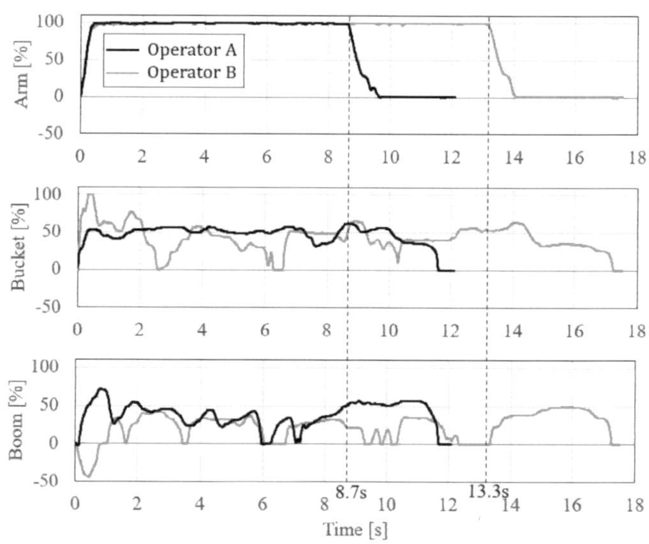

Fig. 10. Comparison of lever inputs during excavation.

Fig. 11. External view of start position of this simplified operation method

5.2. Results of simplified operation method

Figure 12 shows the time series data of arm lever inputs for two operators when performing this simplified operation method in same simplified N-value. As intended with this simplified operation method, there was no variation in terms of input and holding the lever state. The difference in finishing time is due to individual variations in determining the end of operation after the bucket tooth has stopped, so it does not have any impact on the results.

Figure 13 shows the relationship between simplified N value and the maximum bucket tooth speed until it comes to a stop. *Figure 14* shows the relationship between simplified N-value and the excavated volume of the bucket until the bucket tooth stops, categorized by material. Here, in *Figure 6* of Chapter 4, the average bucket tooth speed during excavation was used for organized, while in *Figure 13*, the maximum speed during penetration was used. This choice was made based on the belief that the response of the excavator at the moment of input, as shown in *Figure 16* mentioned later, effectively represents the soil conditions.

Furthermore, in *Figure 7* of Chapter 4, excavating efficiency was used for organized, while in *Figure 14,* excavated volume was used. This is because the payload meter cannot accurately measure the excavated weight unless it is lifted. The excavated volume was calculated based on the penetration depth at the bucket tooth and the bucket shape, and this volume was used as the excavated volume (refer to the schematic diagram of the excavated volume in *Figure 14*).

Here, we also considered converting the time until the bucket tooth stops into excavating time and using it to calculate efficiency. However, we found that when the simplified N-value is low, the excavated volume increases, but the time until the bucket tooth stops also tends to be longer. This tends to cause difficult to distinguish the characteristics affected by the simplified N-value, so we decided to use excavated volume which is easy to understand.

Figure 15 shows external view of bucket penetration conditions in different simplified N-values after the bucket tooth movement stops.

Compared with *Fig.13* and *Fig.14* with *Fig. 6* and *Fig.7* in chapter 4, there are similar correlation between the simplified N-values and the results obtained from operator's discretionary operation and obtained from simplified operation method in this chapter.

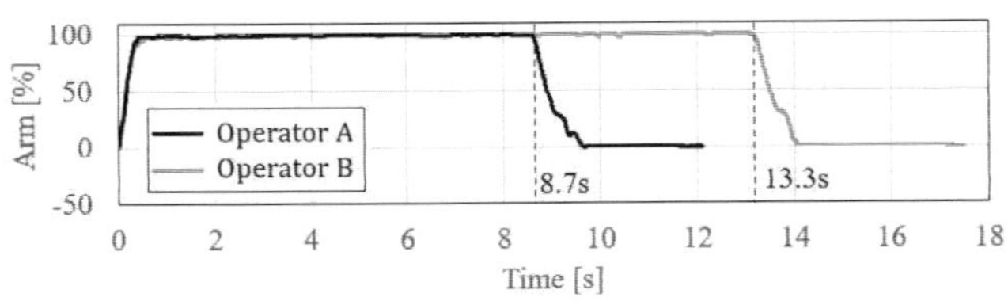

Fig. 12. Time series data of arm lever inputs in simplified operation method

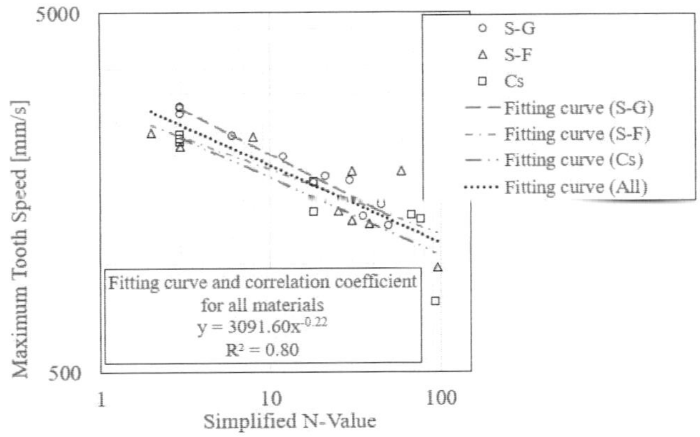

Fig. 13. Relationship between the simplified N-value and maximum speed.

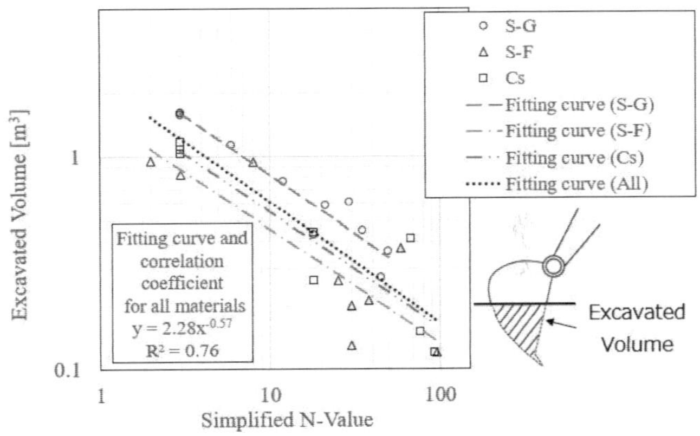

Fig. 14. Relationship between the simplified N-value and excavated volume.

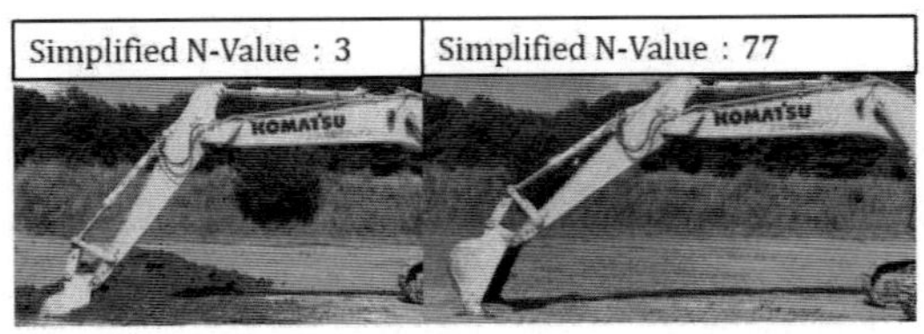

Fig. 15. External view of penetration conditions in different simplified N-value

5.3. Discussion of results in simplified operation method

Figure 16 shows the time-series data of the bucket tooth speed for simplified N-values of 3 and 77 measured in this simplified operation, and vertical displacement of the bucket tooth was shown on the secondary axis. The bucket tooth speed rapidly increased after the start of penetration and reaching a peak, regardless of the value of the simplified N-value.

In case of the simplified N-value is low, the bucket tooth speed gradually decreases after penetration as the soil that enters the bucket creates resistance. Eventually, this load exceeds the excavating capacity of the arm, resulting in a complete stop finally. However, the amount of penetration was deep. On the other hand, in case of simplified N-value was high, the bucket tooth was unable to penetrate excavating soil adequately and came to stop.

The maximum speed and excavated volume obtained from this simplified operation method effectively reflected the differences in penetrability caused by variations in the hardness of the excavating soil.

This suggests that even though the data obtained from manual operations by operator, by focusing on the behavior of penetration at the start of excavation, it may be possible to achieve soil classification in a similar manner. From the perspective of automating the excavating process, it is convenient to be able to perform soil classification at an early stage of the excavating process. This allows the subsequent control of the excavating process to be reflected accordingly.

In the future, with the advance of automation, there is a possibility that effective utilization based on specific applications can improve the accuracy of soil classification. One example could be using this simplified operation method to assess the solid feasibility of entry of hydraulic excavator by investigate the soil hardness in advance.

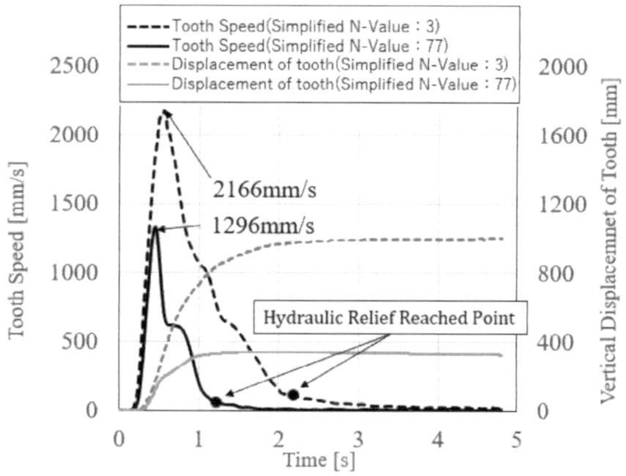

Fig. 16. Time series data of bucket tooth speed and depth.

6. Conclusions

We conducted excavating experiments under varying conditions, including three different materials and soil hardness, with the aim of replacing the operator's sensory-based soil classification during excavating with quantitative data obtained from sensors on the construction machinery.

As a result, a correlation was observed between the simplified N-value, which is one of the indicators of soil hardness, and the average bucket tooth speed during excavation and the excavating efficiency calculated from the excavated weight. These data were obtained from the internal sensor on excavator. This suggests the possibility of soil classification using only the internal sensor data of the construction machinery.

Furthermore, we conducted experiments using simplified operating method to control the variation in operator's specific manipulations. The results obtained using this method showed same trend as data obtained from manual operation.

This suggests that it is possible to determine the soil conditions even in the excavating data that includes variations in operator's actions, and it also implies the possibility of machine-based recognition independent of operator's actions.

In this study, we focused on soil hardness, but there are various other soil parameters related to the operator's cognition during excavation. If replace them with sensor data what we did in this study, it would be possible to classify other soil parameters as well. We will continue to verify and examine further.

7. Declaration of competing interest

The authors have no conflicts of interest directly relevant to the contents of this article.

8. References

8.1. Reference to website

Ministry of Health, Labor, and Welfare, Japan: About the upper limit regulation of overtime work https://www.mhlw.go.jp/stf/seisakunitsuite/bunya/koyou_roudou/roudoukijun/gyosyu/topics/01.html (accessed 10 May 2023, in Japanese)

Ministry of Land, Infrastructure and Transport, Japan: i-construction 2.0, https://www.mlit.go.jp/tec/constplan/content/001738240.pdf (accessed 10 May 2023, in Japanese)

8.2. Reference to a journal publication

Y. Shimano, Y. Kami, K. Shimokaze, "Development of PC210LCi-10PC200i-10 Machine Control Hydraulic Excavator ", KOMATSU TECHNICAL REPORT4, Vol.60 No.167, 2014 (in Japanese)

Y. Sakaida, D. Chugo, H. Yamamoto, H. Asama, "The Analysis of Excavator Operation by Skillful Operator - Extraction of common skills -"SICE Annual Conference 2008, pp. 538-542, 2008

K. Koiwai, H. Imaji, T. Yamamoto, K. Ueda, Y. Yazaki, "A Proposal on Feature Analysis for Digging Operation Excavator by Machine Learning", Transactions of the Institute of Systems, Control and Information Engineers, Vol.31, No.9, pp.328-335,2018(in Japanese)

M. Fujita, M, Kamata, K. Miyata, "A Study on Skill Extraction in Machine Operation(1st Report, The Extraction of Recognition Ability on Visual Information)", Transactions of the Japan society of mechanical engineers. C, Vol.67, No.653, pp.201-208, 2001(in Japanese)

J. Harada, "Development of ICT Bulldozer and ICT Hydraulic Excavator [Translated from Japanese.]", Journal of the Society of Instrument and Control Engineers, Vol.55, No.6, 2016(in Japanese)

T. Satomi, Y. Takahashi and H. Takahashi, "Investigation on Neural Network Model for Estimating Soil Strength by Bucket Excavation", Proc. of the 7th Asian-Pacific Symposium on Structural Reliability and Its Applications, pp.329-334, 2020.

M. Chen, T. Satomi, H. Takanashi, "A study on Estimation of Soil Strength Characteristics by using Resistive Forces Acting on Bucket",Jornal of JCMA, vol.66, No.9, 2014 (in Japanese)

R. Yousefi Moghaddam, A. Kotchon, M.G. Lipsett, "Method and apparatus for on-line estimation of soil parameters during excavation", Journal of Terramechanics 49, pp. 173–181, 2012

H. Tani, S. Hayakawa, R. Ikeura, T. Ooi and T. Nakamura, "N value estimation method for hydraulic excavator in real time by using logistic regression model," 2017 IEEE International Conference on Robotics and Biomimetics (ROBIO), pp. 1352-1357, 2017

T. Satomi, H .Nakao, X-D. Liu, H. Takahashi, "The relationship between the cone index of sandy soil and the force acting on the cylinder of hydraulic excavator during excavation [Translated from Japanese.]", THE 40th Anural Conference of the Robotics Society of Japan, RONBUN No.4A2-01, 2022 (n Japanese)

M. Fujita, M, Kamata, K. Miyata, "A Study on Skill Extraction in Machine Operation(2nd Report, Clarification of Cognition Ability in Machine Operation and a Proposal of Its Application Theory to Assist Operators)", Transactions of the Japan society of mechanical engineers. C, Vol.68, No.666, pp.547-554, 2001(in Japanese)

1525 | Photogrammetry based mobility mapping for small scale model vehicle

Dávid Körmöczi [a,*], Péter Kiss [b]

[a] *Doctoral School of Mechanical Engineering, Hungarian University of Agriculture and Life Sciences, Gödöllő, Hungary*
[b] *Department of Vehicle Technology, Hungarian University of Agriculture and Life Sciences, Gödöllő, Hungary*

** Corresponding author: kormoczi.david@uni-mate.hu*

ABSTRACT

Mobility on terrain is one of the key factors of off-road vehicles. In conventional vehicle technology, an experienced driver can assess the obstacle negotiation capability of their vehicle, hence mobility models for manned vehicles usually have smaller resolution and are less detailed. However, unmanned vehicles can only rely on an algorithm both for global path planning and local vehicle control, thus such an algorithm for mobility mapping needs to be more detailed and have higher resolution than conventional mobility models. In this paper, a method is shown to create a mobility map for a ClearPath A200 small scale test vehicle based on a drone-borne photogrammetry survey of the test area. The vehicle specific mobility model examines the vehicle-terrain interaction from the perspective of obstacle negotiation capability, while also taking into account the soil deformation and the stability of the vehicle. The vehicle-wheel deformations are characterized by a parallel element rheological model. Apart from the perspectives of off-road mobility, the computational capacity requirement of the model is also examined, which is a key factor for real time applications of unmanned vehicles. At the end, a method and test setup is proposed for the validation of the theoretical mobility model via field measurements.

Keywords
Off-road mobility
Soil characterization
Autonomous systems
Propulsion systems
Agricultural tires

1. Introduction

Mobility modeling and mapping is an integral step of navigation and path planning for off-road vehicles. Existing mobility modeling methods, which are created for conventional off-road vehicles, mostly are only useable for global navigation purposes and for planning an estimated route, as they rely on the driver to make local corrections based on their real-time observation of the surrounding area. However, for robotized and autonomous vehicles, this task too has to be fulfilled by the vehicles on-board (or cloud based) computer, and the related algorithm. For this reason, a mobility model intended for autonomous vehicle operations needs to take into account every aspect of the vehicle-terrain interaction which have an effect of both global and local routing, and also either be continuous or have a resolution at least a magnitude smaller than the vehicles characteristic dimensions. Apart from this, to implement a sophisticated route planning algorithm, a mobility map is required that can describe a vehicle terrain interaction with continuous values instead of binary go/no go classification, and it should be able to do so in multiple directions at each examined point, instead of describing the mobility at each point with a single variable regardless of the direction. In this paper, we introduce a mobility modeling method that meets these requirements. The model characterizes the interaction between the vehicle and the terrain based on the parallel element tire-soil deformation model. The mobility of the vehicle on every examined point of the terrain is characterized by a mobility index value. The mobility index has a value of 0 when the vehicle can traverse in the given point and direction without any effort, and a value of 1 or 100 % when in the given point, the capabilities of the vehicle are fully utilized to overcome an obstacle, but it is still able to do so. This means that any point of the terrain with a mobility index of less than one is traversable, while points with a mobility index greater than one are not.

2. Literature review

There are several already existing off-road mobility models, which feature the main aspects of terrain mobility. On Table 1., a (not exhaustive) list of different mobility models can be seen.

One of the most well-known and frequently cited mobility model is the NRMM (NATO Reference Mobility Model). Created for global mobility mapping purposes, this large-scale model takes into account all main aspects of the terrain-vehicle interaction, but its usefulness is quite limited for autonomous vehicles. The resolution of the model is approximately 100 m, which means that every hectare of the examined area is considered to be theoretically homogeneous and is characterized by a set of variable. Besides that, while the model takes various parameters into account, many of them are not clearly defined. For example, the microscopic terrain profile is characterized by a „roughness" value. However, there is no explanation how exactly this value can be derived from a known terrain profile, or what mathematical formula can be used to distinguish the terrain profiles roughness from the macroscopic features.

https://doi.org/10.56884/RM06HALP

Table 1: Review of existing mobility models (Chen, 2022; Laib, 2002; Lessem, 1996)

Model / Author	Resolution	Model inputs	Remark
NRMM	~100 m	Topography; Soil mechanical parameters; Vegetation; Terrain profile roughness	Uses emphirical data
VSE	-	Terrain geometry	Only applicable for discrete, regular shaped obstacles
NG-NRMM	30+ m	Topography; Soil mechanical parameters; Vegetation; Terrain profile roughness	Adaptive resolution
Chen et al.	10 m (30 m)	Topography; Soil mechanical parameters (cone index / load bearing capacity)	

All other listed mobility models have similar limits. They usually have a resolution that is too large, and often only take a limited number of factors into account. For example, the mobility model created by Chen et al. only examines the slope climbing capability of the vehicles from the perspective of energetics, while ignoring the limits of rollover stability.

Another commonly referenced modeling method is the VSE (Vehicle Slope Elevation) method (Laib, 1997). Compared to other mobility models, the VSE does not rely on a fixed resolution of the examined terrain, as its inputs are the geometrical parameters of each obstacle. However, while the method is useful for generally characterizing a vehicles obstacle negotiation capability, or comparing such capabilities of different vehicles, its applicability for actual navigation purposes is limited by the fact that it can only handle a few types of theoretical obstacles. When examining a real terrain profile, every obstacle would have to be typified to be described by a limited number of characteristic values. For example, the slope shown on Fig. (1) would have to be approximated as two horizontal surfaces connected by a flat sloped surface. Such approximations would require complicated mathematical methods (which, in real applications, translates to high computational capacity requirements) and even then would introduce significant errors into the model.

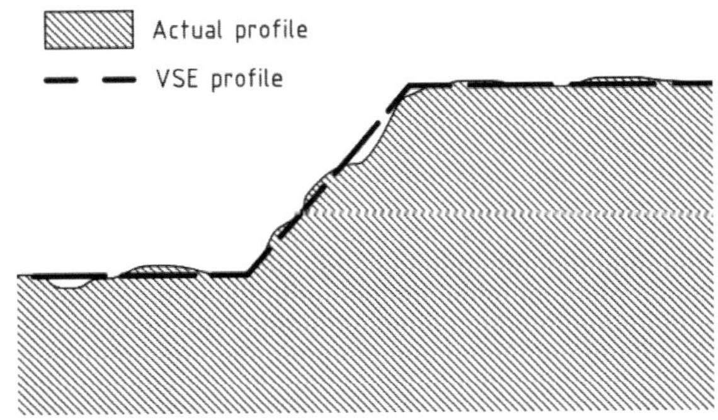

Fig. 1: VSE obstacle classification

3. Mobility model

Any off-road mobility modeling method relies on the accurate characterization of the tire-soil interaction. The requirement of real time applicability means that a modeling method does not only need to meet the requirements of accuracy, but also the limits of the available computational capacity. Furthermore, the tire-terrain deformation model – contrary to most similar models- must not only be applicable on a theoretical flat soil surface, but instead take into account any soil geometry (He, 2019; Kotrocz, 2019). To fulfill these criteria, the mobility modeling method introduced in this paper relies on the parallel element tire-soil deformation model to characterize the interaction between the vehicle and the terrain. The parallel element rheological model (Fig. 2) consists of independently acting elastoplastic elements. The parameter identification of the model is explained in detail in the authors' earlier article about tire-terrain interaction modeling (Körmöczi and Kiss, 2023). The soil mechanical properties which give the input variables of the parameter identification will be supposedly obtained by remote sensing of the terrain. However, while the theoretical possibilities for indirect remote sensing of the soil mechanical properties have been shown by many authors (Ahmed, 2023; de Souza, 2022; Wu, 2019), it was deemed that using conventional soil mechanical measurement methods (cone penetrometer and torsion shear measurement) would be more suitable for the testing of these tests. For the creation of the initial mobility map shown in this article, preliminary field measurements have been conducted using a FieldScout SC900 soil compaction meter, but naturally the mobility map has to be updated based on the actual soil properties measured at the time of each test.

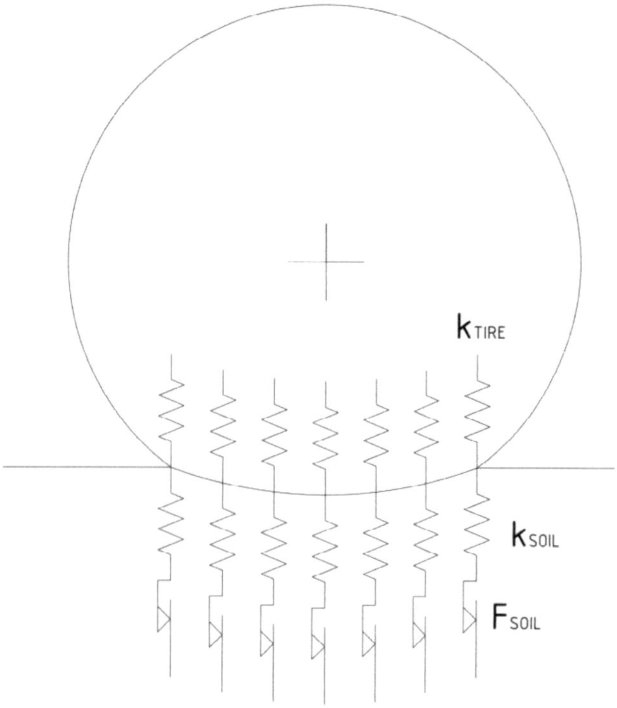

Fig. 2: Parallel element tire model

The physical position of the vehicle is calculated based on the tire-terrain deformation and the suspension deformation of the vehicle. As the tire-soil deformation if affected by the wheel load, and the wheel load is affected by the geometrical position (pitch / roll) of the vehicle due to the change of load between the axles and tires, this could only be calculated using an iterative process. Since it is expected that the effect of deformation on the load distribution between the wheels are negligible compared to the pitch and roll due to the macroscopic topography, the wheel loads are calculated in a static undeformed state (Fig. 3), and then the tire-soil and suspension deformation is modeled based on this constant load value (Fig. 4).

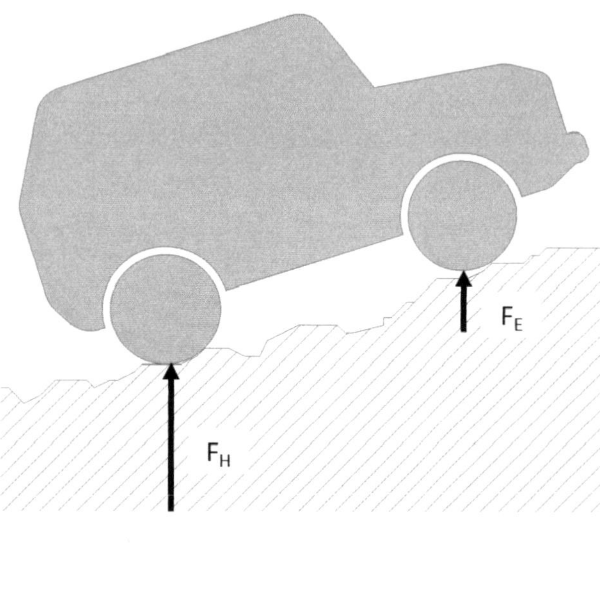

Fig. 3.: Load distribution in the undeformed state

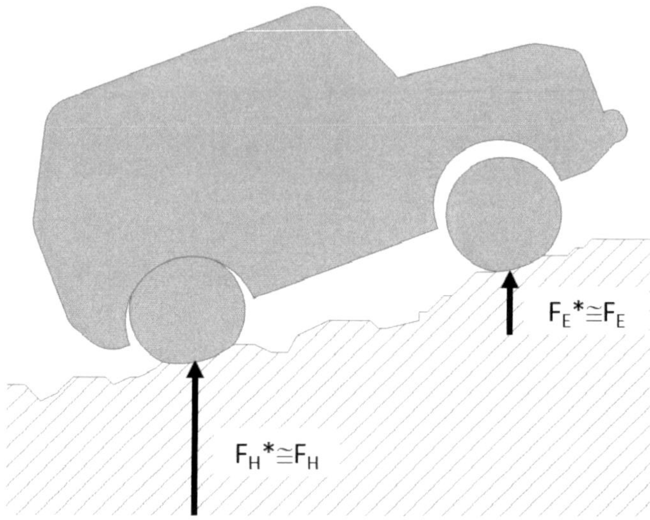

Fig. 4: Load distribution in the deformed state

The undeformed geometrical position of the vehicle is offset by the wheel-tire and suspension deformation values, as explained in the earlier segment. The suspension is considered to have a perfectly linear characteristic.

The obstacle negotiation capability of the vehicle is classified by calculating the vertical distance between the selected features of the vehicle geometry and the points of the terrain profile. Such features typically can be the floor panel of the vehicle, and the front and rear overhangs. The geometrical coordinates of these selected features are transformed from the base values for the vehicle (which are an input for the model) using the deformation calculations described earlier. As it is expected that the selected points of the vehicle are not directly above the discrete points of the terrain model, an interpolation of either the terrain profile or the vehicle features is necessary. For discrete points or axes of the vehicle geometry, such as the front overhang, an interpolation of the terrain profile is used. For larger surfaces, like the floor panel, the surface is sampled using the horizontal resolution of the terrain profile. This can be seen on Fig. 5.

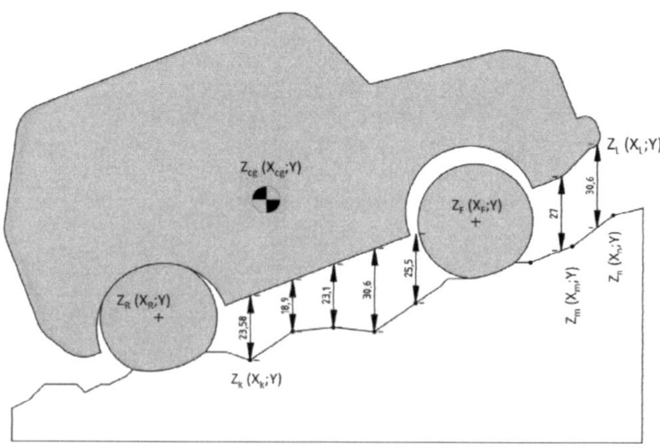

Fig. 5: Obstacle negotiation classification

The clearance at each examined point (point "L" in the example) can be calculated, after interpolation, using the following equation:

$$H_L = Z_L - \left[Z_m + \frac{Z_n - Z_m}{X_n - X_m}(X_L - X_m)\right] \quad (1)$$

As explained earlier, the mobility model needs to have a mobility index, which characterizes the utilization of the vehicle's capabilities. For the obstacle negotiation capability, this means that the mobility index on a flat terrain is 0, and when an obstacle has the exact same dimensions as the vehicles limit (for example ground clearance) the mobility index takes a value of 1. However, on a terrain with a complex geometry, the obstacle height is not easily describable, since it cannot be differentiated from the overall terrain profile. For these reasons, the mobility index is calculated by the method shown on Fig. 5., using the vertical clearance between the vehicles body and the terrain profile instead of the obstacle height value.

Apart from the obstacle negotiation capability, the vehicles geometric position is also examined from the perspective of stability. Here a simplification is used that the rollover limits for pitch and roll angles are only applied for the vehicle body, while neglecting the small effects of the unsprung mass displacement due to the suspension deformation.

The mobility map takes into account, apart from the parameters explained earlier, the maximal wheel force, which is usually limited by the shear force of the soil-tire interaction. This is calculated based on the parallel element rheological model, taking into consideration the approximation that the shear force is proportional to the shearing are, and thus, the number of connection rheological element pairs. The shear strength of the soil, which is used for these calculations, is considered to be a known value at each point of the terrain.

The mobility map has been created based on a photogrammetrical mapping of the off-road test area (Fig 6.) on the Hungarian University of Agriculture, Gödöllő. A 3D model of the terrain (with a 10 cm sampling distance) can be seen on figure 6. As it is explained earlier, the mobility model characterizes a set number of directions at each point (in this case 4), but for visualization, the maximal value for each point has been selected, which can be easily plotted on a grayscale mobility map shown on Fig. 7.

Fig. 6. Test area

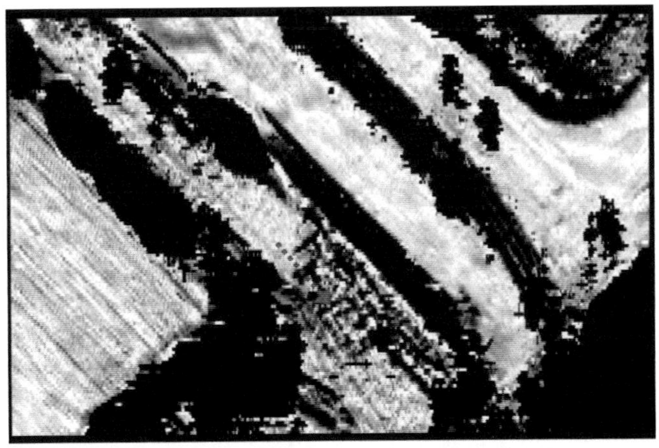

Fig. 7: Grayscale mobility map and direction of the example test

4. Proposed testing methods and expected results

The testing of the mobility model requires a test vehicle fitted with necessary sensors to measure every aspect of the mobility, which is characterized by the validated model. For these tests, a ClearPath A200 UGV will be used, which can be seen on Fig. 8.

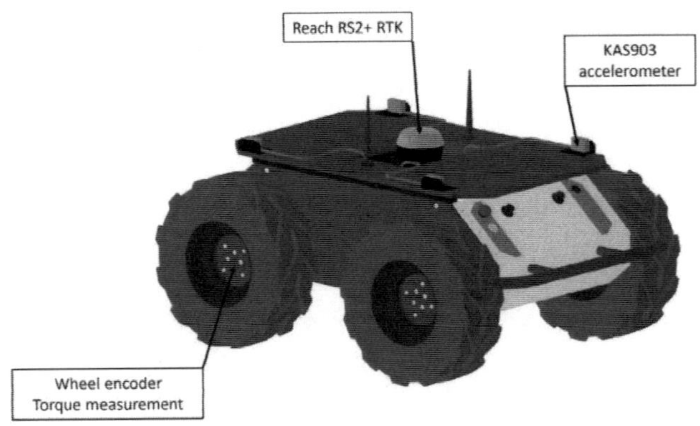

Fig. 8: Test vehicle (Clearpath A200 Datasheet, 2024)

The test vehicle is fitted with RTK receiver to accurately measure the geometric position of the vehicle, as well as inertial measurement for pitch and roll values. The applied wheel force is logged by the vehicles own motor control unit, based on the voltage and current of the motors.

Apart from testing on the whole off-road test are, a small obstacle course matching the scale of the test vehicle has been created by building discrete obstacle, which can be seen on Fig. 9.

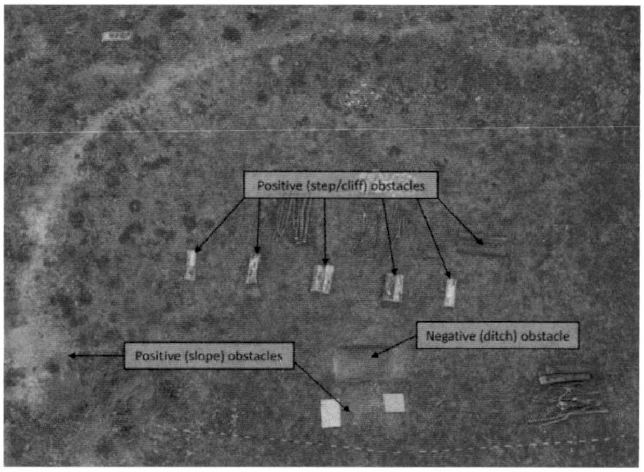

Fig. 9: Obstacle course on the test area and the types of obstacles

Based on the measurements, the real and modeled mobility index values will be comparable. On Fig. 10. the simulated mobility index and the expected results for the section marked on Fig. 7. can be seen. Due to the nature of the mobility index, any value over 1 is not measurable, as it is shown on Fig. 10. However, these values are not necessary for mobility classification, as these are all in the no go category, where it is not important how much of each capacity of the vehicle (engine power, stability, obstacle negotiation capability) would be theoretically required to traverse, as it is higher than the actually available capabilities of the vehicle.

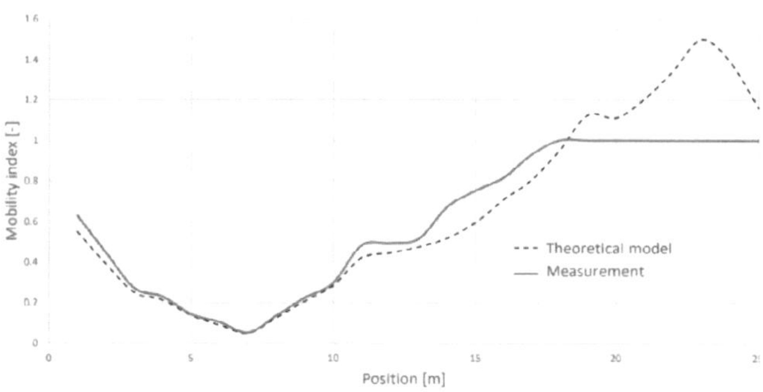

Fig. 10: Expected results of the tests

5. Conclusions

In this paper, a mobility modeling method was introduced, which is created based on the needs for autonomous navigation for unmanned ground vehicles. The modeling method characterizes the interaction between the vehicle and the terrain using a parallel element rheological tire model, and examines the mobility from the perspectives of stability, traction, and obstacle negotiation capability using a coordinate geometry method. A test area was selected for the validation measurements, featuring both undisturbed natural terrain sections and an obstacle course with different types of obstacles that match the scale of the vehicle. The mobility of the selected test vehicle on this area was classified using a mobility index, which was visualized on a grayscale mobility map. A measurement setup was assembled for the validation tests: a ClearPath A200 UGV outfitted with RTK GPS receiver, wheel acceleration sensors, and an inertial measurement unit. At the end, the form of the expected results was shown, including the limitation that on areas with mobility index greater than 1 (which means the are is not traversable) no measurements can be done, which means in this case the validation is limited to classifying the area as not traversable, instead of the exact value of the mobility index.

6. Nomenclature

K_{TIRE}	Elastic tire parameter	[N/m]
F_E	Front wheel load (undeformed)	[N]
F_E^*	Front wheel load (deformed)	[N]
F_H	Rear wheel load (undeformed)	[N]
F_H^*	Rear wheel load (deformed)	[N]
F_{SOIL}	Plastic soil parameter	[N]
H_L	Clearance at point "L"	[m]
k_{SOIL}	Elastic soil parameter	[N/m]
X_k; X_m etc.	Longitudinal coordinate of points „k"; „m" etc.	[m]
Y	Lateral coordiante of the examined section	[m]
Z_k; Z_m etc.	Vertical coordinate of points „k"; „m" etc.	[m]

7. Acknowledgements

This research was supported by the Ministry of Culture and Innovation and the National Research, Development and Innovation Office within the Cooperative Technologies National Laboratory of Hungary (Grant No. 2022-2.1.1-NL-2022-00012)

8. Declaration of competing interest

The authors declare that they have no known competing financial interests or personal relationships that could have appeared to influence the work reported in this paper.

9. References

Ahmed, A. E. E., El Hariri, A., & Kiss, P., 2023. Soil Spectral Behavior Related to Its Load-Bearing Capacity Based on Moisture Content. Applied Sciences, 13(6), 3498.

Chen, H., Niu, R., Yu, B., Zheng, X., Bai, R., & Zhang, S., 2022. A Global Path Planning Method for Unmanned Ground Vehicles in Off-Road Environments Based on Mobility Prediction. Machines, 10(5), 375.

Clearpath A200 UGV Datasheet https://www.clearpathrobotics.com/wp-content/uploads/2013/02/HUSKY_A200_UGV_2013_TEASER_email.pdf (accessed 2024.03.17)

de Souza, I. R. M., Sano, E. E., de Lima, R. P., & da Silva, A. R., 2022. A remote sensing approach to estimate the load bearing capacity of soil. Information Processing in Agriculture.

He, R., Sandu, C., Khan, A. K., Guthrie, A. G., Els, P. S., & Hamersma, H. A., 2019. Review of terramechanics models and their applicability to real-time applications. Journal of Terramechanics, 81, 3-22.

Körmöczi, D., Kiss, P., 2023. Comparison of Modeling Methods for Off-Road Tires. Machines, 11(6), 658.

Kotrocz, K., 2019. Talaj Diszkrét Elemes Modellezése Talaj-Kerék Kölcsönhatásának Elemzéséhez. Ph.D. Thesis. Budapest University of Technology and Economics, Budapest, Hungary

Laib, L. (1997). On the mobility of cross-country vehicles, with emphasis on modelling their dynamic motion. International Journal of Heavy Vehicle Systems, 4(1), 49-71.

Laib, L., 2002. Terepen Mozgó Járművek; Szaktudás Kiadó Ház: Budapest, Hungary

Lessem, A., Mason, G., & Ahlvin, R., 1996. Stochastic vehicle mobility forecasts using the NATO reference mobility model. Journal of Terramechanics, 33(6), 273-280.

McCullough, M., Jayakumar, P., Dasch, J., & Gorsich, D. (2017). The next generation NATO reference mobility model development. Journal of Terramechanics, 73, 49-60.

Wu, K., Rodriguez, G. A., Zajc, M., Jacquemin, E., Clément, M., De Coster, A., & Lambot, S., 2019. A new drone-borne GPR for soil moisture mapping. Remote Sensing of Environment, 235, 111456.

1649 | Dynamic mode decomposition for PIV-based sand flow field beneath traveling wheel

Ohta Kuramoto[a], Jorge Ruben Casir Ricano[a], Kenji Nagaoka[a,*]

[a] *Department of Mechanical and Control Engineering, Kyushu Institute of Technology, Japan*

** Corresponding author: nagaoka.kenji572@mail.kyutech.jp*

ABSTRACT

Grouser wheels have demonstrated effectiveness for robotic mobility in soft, loose terrain. Clarifying the relationship between the grouser wheel and sand flow will improve wheel running performance. So far, particle image velocimetry (PIV) has been widely used to observe sand flow beneath a traveling wheel. Although the PIV-based spatial velocity field has been discussed, few studies have examined its time evolution or spatiotemporal dynamics. This could be attributed to the sand flow vector field being a complex high-dimensional dynamics field and a mixture of spatial and temporal information. Thus, we apply dynamic mode decomposition (DMD) to the sand flow data obtained by the PIV method to reduce its dimension and divide it into major modes. A mode is a characteristic pattern or correlation. The DMD can extract temporal and spatial information separately, allowing the dynamics to be analyzed by the major modes. We first conducted PIV experiments using the traveling wheel on sand at high-slip (slip ratio is approximately 70%) and low-slip (slip ratio is approximately 20%) conditions. Then, the DMD was applied to decompose the PIV-based sand velocity data into major modes. As a result, the obtained modes were mainly divided into two dynamics, one representing the inter-grouser sand flow trapped in the grouser motion and the other representing soil compaction and failure, affecting deeper soil. It was found that the low-slip condition included both modes, whereas the high-slip one did only inter-grouser sand flow. The DMD analysis confirmed that two major modes can explain wheel-soil interaction.

Keywords
Dynamic mode decomposition
PIV
Wheel-soil interaction

1. Introduction

One of the challenges of wheel driving in loose soil has been excessive slippage. Research is being conducted on wheel geometry to reduce wheel slippage. One example is the grouser wheel, which has plate-like protrusions on the wheel surface. Research on grouser wheels has shown that they improve running performance on soft ground (Sutoh et al., 2013). Studies have focused on the earth removal action of the grouser and wheel earth pressure (Irani et al., 2011; Ding et al., 2015). Studies have also been conducted using the Discrete Element Method (DEM) to simulate sand flow when a grouser wheel travels on soft ground and to study ground stresses acting on the grouser and wheel surface (Lichtenheldt et al., 2013; Higa et al., 2016).

Fig. 1. Grouser wheel

Particle Image Velocimetry (PIV) is often used to study grouser wheels running on soft ground (Skonieczny et al., 2012). PIV was also used to investigate the relationship between changes in grouser geometry and sand flow and the direction of the wheel width caused by grouser wheel running. In these studies, the area affected by the grouser on the sand flow field was used as the main evaluation index, and the spatial distribution of sand flow and its temporal variation were not studied. One of the challenges of PIV analysis is integrating spatiotemporal information in the results of a single analysis. In addition, sand flow with a grouser wheel is a high-dimensional phenomenon in space and time, making it difficult to understand the PIV analysis data intuitively.

Clarifying the sand flow caused by the grouser wheel is essential to constructing an effective mechanical model to generate the optimal design of the grouser wheel and control rover travel using the grouser wheel. PIV is used for this purpose, but the analytical results for high-dimensional flows are difficult to understand quantitatively. This study uses dynamic mode decomposition (DMD) of PIV results to find characteristic patterns and correlations in sand flow. In addition, a more detailed analysis is performed by extracting temporal and spatial information separately.

2. PIV against single-wheel driving

2.1. Single-wheel experimental system

The sand flow by a grouser wheel is filmed, and the velocity of sand grains is quantified using PIV, an image processing technique that tracks the sand flow as tracer particles.

Fig. 2 shows the schematic representation of the experimental setup, while Fig. 3 shows the real experimental environment. The translational drive unit consists of a linear guide, plate, and traction mechanism connected to the wheel drive unit via a shaft fixed to the plate. The traction mechanism is connected to a powder brake. A constant controller is used to apply a certain amount of current to the powder brake, and the load torque can be calculated by measuring the current value with a current meter. The ends of the strings are connected to a plate fixed to the linear guide, and a traction load can be applied to the wheel drive unit. This allows the system to reproduce running on a slope and simulate wheel slip. The wheel drive unit combines a motor, gearhead, and an encoder attached to a fixture on the shaft

https://doi.org/10.56884/I7J6SFB3
ISBN 978-1-942112-55-6

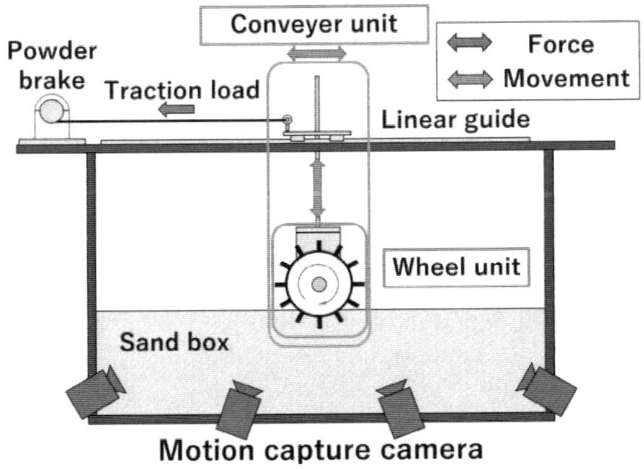

Fig. 2. Outline drawing of experimental device

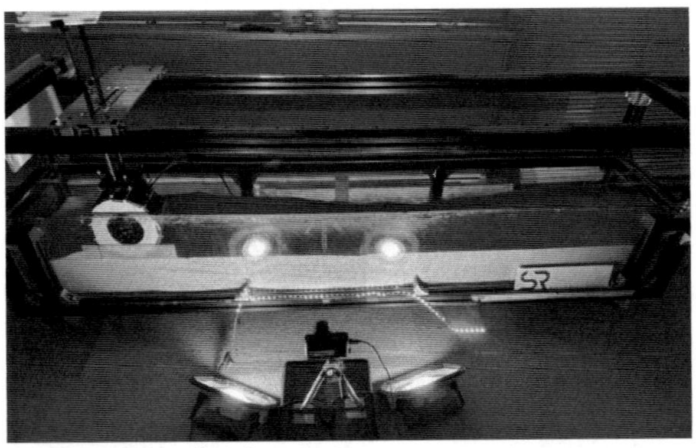

Fig. 3. PIV experimental environment

connected to the translation drive unit. The reduction ratio of this motor is 679 : 1, and the shaft holder connects the motor end and the wheel. A regulated DC power supply powers the drive motor, and the motor's rated voltage, 24 [V], is applied.

Two types of cameras are used: motion-tracking cameras and cameras for external recording. Motion-tracking cameras can measure the three-dimensional position of wheels with high accuracy, and four cameras are used in this experiment. The motion-tracking cameras by OptiTrack and Motive, a dedicated software that can measure the movement of markers in the measurement area and calculate the translational velocity of the wheels. The external recording camera acquires video of the wheel's movement and sand during wheel travel. The sand flow under the wheel cross-section can be observed by running the wheel. The shooting conditions are shown below. The video resolution is 1980 × 1080 [pixel], the aperture is 8.0, the frame rate is 30 [fps], and the shooting range is 400 × 200 (W × H) [mm].

2.2. Grouser wheel running experiment

The wheels were fabricated using a 3D printer. The wheel is designed with a strength that is sufficient in the experiment. First, the parameters of the grouser wheel are described. The number of grousers is defined as $N = 12$, grouser outer radius as $R = 100$ [mm], grouser inner radius as $r = 75$ [mm], and grouser thickness as $b = 5$ [mm].

The grouser wheel described above is attached to the single-wheel running test apparatus and is run on the sandbox. A motion-tracking camera measures the amount of wheel subsidence and translational movement, and an encoder measures the angular

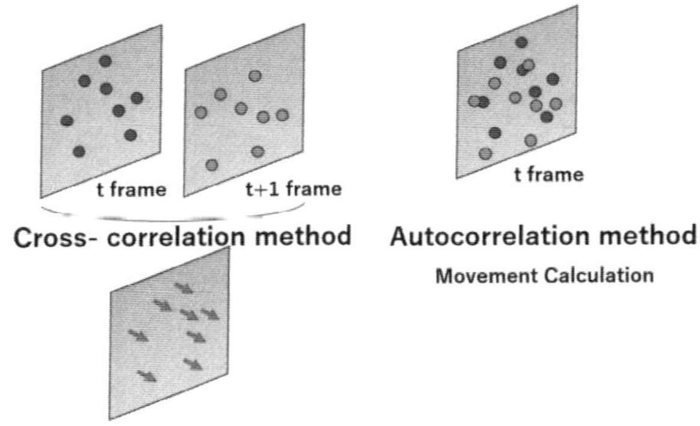

Fig. 4. Cross-correlation and autocorrelation methods

velocity of wheel rotation. A camera films the running of the grouser wheel for PIV analysis. The powder brake was used to apply traction load and run in high-slip and low-slip conditions. A slip rate of about 20% was considered low-slip, and a slip rate of about 70% was considered high-slip. The average of the five runs was analyzed, where one grouser wheel penetrated vertically into the sand surface. The rotation angle of the grouser wheel is θ. The wheel rotation speed is controlled with a target speed of 5 [rpm], assuming the wheel circumferential speed is 5 [cm/s]. The slip ratio is calculated as follows. The slip ratio is defined by Eq. (1), where the body speed v, the effective wheel radius r_e, the grouser radius R from the center to the tip of the grouser, and the angular velocity of wheel rotation ω are assumed.

$$s = 1 - \frac{v}{r_e \omega} \qquad (0 \leq s \leq 1) \tag{1}$$

In this experiment, $r_e = R$.

The experiment followed the following steps: First, the sand surface is leveled to a constant height. Next, the grouser wheel is set to the initial angle and slowly placed on the initial position of the sand surface. Then, the wheel's position and speed are measured, and a video is recorded. Finally, the wheel can run straight ahead at a constant angular wheel rotation velocity. These procedures were performed five times for each condition.

2.3. PIV (Particle Image Velocimetry)

Many researchers, such as Prandtl and Reynolds, have significantly contributed to elucidating the flow field (Prandtl et al., 1934; Reynolds 1883). PIV (Particle Image Velocimetry) is a non-contact measurement method to obtain the velocity of tracer particles in a flow field by image processing. The principle of PIV is to place an indicator tracer particle in the flow, obtain the displacement vector of the particle at small time intervals between frames captured by a camera, and calculate the local vector of the flow field by image processing. This method of visualizing the flow field using particle image measurement, called PIV, is currently attracting attention. In this study, since sand grains are tracer particles, local vectors in the flow field are calculated by following the sand grains. PIV can be broadly classified into two methods: the autocorrelation method and the cross-correlation method. The autocorrelation method calculates the local vectors by taking images of particles at minute intervals in a single image, i.e., by double exposure, setting an inspection region in the image, and calculating a two-dimensional autocorrelation function of the luminance distribution in that region. Fig. 4 shows a conceptual diagram of the two correlation methods. manner, the ease of understanding the detailed structure of complex flows, and the

Fig. 5. PIV analysis result

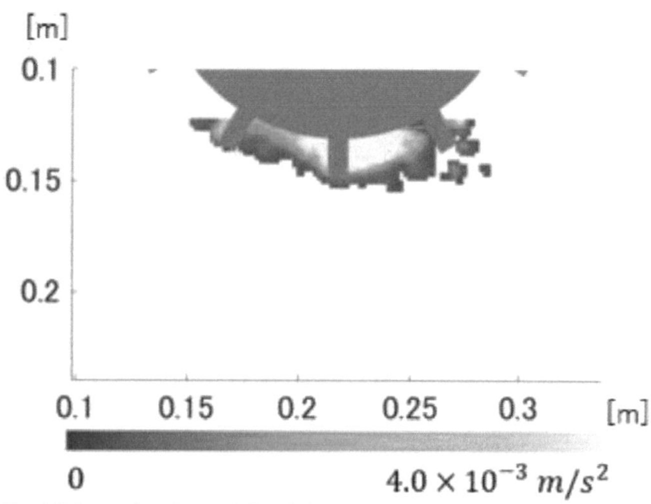

Fig. 6. Velocity data for sand flow field

ability to obtain quantitative values. The advantages of PIV are the ability to visualize the flow at multiple points in a non-contact. As an example of the application of PIV to terramechanics by tracking sand itself as tracer particles (Senatore et al., 2013; Yamakawa et al., 2015), a study that qualitatively analyzed sand displacement by measuring the displacement field of sand under the wheel (Moreland et al., 2014; Saengprachatanarug et al., 2013), a study that conducted a wheel running experiment, analyzed sand flow by interaction with the wheel using PIV (Yamano et al., 2018; Nakamura et al., 2020), and reproduced the running experiment by numerical analysis based on the finite element method, and a study that visualized sand flow under the wheel from images taken from the side of the wheel while the grouser wheel was running.

In this study, we perform PIV measurements using the latter cross-correlation method and apply dynamic mode decomposition (DMD) to the results for a more detailed analysis. This chapter first describes the setup and analysis conditions of the PIV measurement. Figs. 5 and 6 show one frame of the PIV-analyzed video and graphs plotting the absolute values of the sand particle velocities obtained from the PIV results.

3. DMD-based analysis methods

In the modern era, numerical simulations and experimental measurements are more precise than ever before. Still, despite the abundance of such faithful data, modeling is often elusive. One reason for this is that the system consists of a high-dimensional nonlinear dynamical system that exhibits multiscale phenomena in both space and time. Dynamic mode decomposition (DMD) is effective for modeling high-dimensional phenomena because it separates complex flows into simple representations and shows a superposition of these representations. The vibration components with a specific frequency, which are the elements of the superposition, are called modes. Each mode can be further decomposed into DMD modes, representing a temporal feature and Time Dynamics, which is its time evolution. This chapter describes the data and theory handled by DMD, based on reference (Kutz et al., 2016). It also describes data preprocessing for applying DMD.

3.1. Data handled

The data collection process in DMD involves two parameters: the number of spatial points stored per snapshot, n, and the number of snapshots taken, m. The data obtained by the measurement is denoted by $\mathbf{X}$ below.

$$\mathbf{X} = \begin{bmatrix} \vdots & \vdots & & \vdots \\ \mathbf{x}_1 & \mathbf{x}_2 & \cdots & \mathbf{x}_{m-1} \\ \vdots & \vdots & & \vdots \end{bmatrix} \tag{2}$$

$$\mathbf{X}' = \begin{bmatrix} \vdots & \vdots & & \vdots \\ \mathbf{x}_2 & \mathbf{x}_3 & \cdots & \mathbf{x}_{m} \\ \vdots & \vdots & & \vdots \end{bmatrix} \tag{3}$$

$\mathbf{X}'$ is one step forward in time from $\mathbf{X}$ and can be expressed as $\mathbf{X} = \mathbf{A}\mathbf{X}'$ using linear operator $\mathbf{A}$. The DMD guideline is to compute the eigenvalue decomposition of the linear operator $\mathbf{A}$ that is optimal.

$$\mathbf{A} = \mathbf{X}'\mathbf{X}^{\dagger} \tag{4}$$

3.2. Algorithm

When the dimension n of $\mathbf{X}$ is large, it may be difficult to decompose the matrix $\mathbf{A}$ directly into eigenvalues. Therefore, DMD avoids direct eigenvalue decomposition of matrix $\widetilde{\mathbf{A}}$ by considering a rank-reduced representation of the POD-projected matrix $\mathbf{A}$. The DMD procedure is described below.

I. Perform singular value decomposition (SVD) of $\mathbf{X}$

$$\mathbf{X} \approx \mathbf{U\Sigma V}^* \tag{5}$$

$*$ denotes the conjugate transpose, where $\mathbf{U} \in \mathbb{R}^{n\times r}$, $\mathbf{\Sigma} \in \mathbb{R}^{n\times r}$, and $\mathbf{V} \in \mathbb{R}^{n\times r}$. At this stage, the dimensionality of $\mathbf{U}$, $\mathbf{\Sigma}$, and $\mathbf{V}$ is reduced to a low-dimensional approximation of the data. When a low-dimensional structure exists in the data, the singular values of $\mathbf{\Sigma}$ are sharply reduced, leaving only a limited number of dominant modes.

II. The matrix $\mathbf{A}$ in Eq. (4) is obtained using the pseudo-inverse matrix $\mathbf{X}^{\dagger}$ of $\mathbf{X}$ obtained from SVD.

$$\mathbf{A} = \mathbf{X}'\mathbf{V\Sigma}^{-1}\mathbf{U}^* \tag{6}$$

Matrix $\widetilde{\mathbf{A}}$ projected into POD mode Matrix $\mathbf{A}$ is computed.

$$\widetilde{\mathbf{A}} = \mathbf{U}^*\mathbf{A}\mathbf{U} = \mathbf{U}^*\mathbf{X}'\mathbf{V\Sigma}^{-1} \tag{7}$$

III. Compute the eigenvalue decomposition of $\widetilde{\mathbf{A}}$.

$$\widetilde{\mathbf{A}}\mathbf{W} = \mathbf{W\Lambda} \tag{8}$$

where the columns of $\mathbf{W}$ are eigenvectors and $\mathbf{\Lambda}$ is a diagonal matrix containing the corresponding eigenvalues $\boldsymbol{\lambda}_k$.

IV. Finally, $\mathbf{W}$ reconstructs the eigenvalue decomposition of $\mathbf{A}$ from $\mathbf{\Lambda}$. In particular, the eigenvalues of $\mathbf{A}$ are given by $\mathbf{\Lambda}$, and the eigenvectors (DMD modes) of $\mathbf{A}$ are given by the sequence of $\mathbf{\Phi}$.

$$\mathbf{\Phi} = \mathbf{X}'\mathbf{V}\mathbf{\Sigma}^{-1}\mathbf{W} \tag{9}$$

We can construct predictive solutions for all times using low-rank approximations of both eigenvalues and eigenvectors. First, rewriting $\boldsymbol{\omega}_k = \ln \boldsymbol{\lambda}_k$ for convenience, the approximate solution for all times, including the future, is given by

$$\mathbf{x}(\mathrm{t}) \approx \sum_{\mathrm{k}=1}^{\mathrm{r}} \boldsymbol{\phi}_k \exp(\boldsymbol{\omega}_k \mathrm{t}) \boldsymbol{b}_k = \mathbf{\Phi} \exp(\mathbf{\Omega}\mathrm{t})\, \mathbf{b} = \mathbf{\Phi}\mathbf{\Psi} \tag{10}$$

where $\boldsymbol{b}_k$ is the initial amplitude of each mode, $\mathbf{\Phi}$ is a matrix with $\boldsymbol{\phi}_k$ as columns, and $\mathbf{\Omega} = \mathrm{diag}(\boldsymbol{\omega})$ is a diagonal matrix with eigenvalues $\boldsymbol{\omega}_k$ as elements. The eigenvector $\boldsymbol{\phi}_k$ has the same magnitude as $\mathbf{x}$, and b is a vector of coefficients **bk**, where $\mathbf{\Psi} = \exp(\mathbf{\Omega}\mathrm{t})\,\mathbf{b}$ and each row of it is $\boldsymbol{\psi}_k$. As mentioned earlier, Eq. (10) can be interpreted as a least-squares fit, or regression, of the linear dynamical system $\boldsymbol{x}_{k+1} = \mathbf{A}\boldsymbol{x}_k$ on the sampled data. The matrix $\mathbf{A}$ is constructed so that $\|\boldsymbol{x}_{k+1} - \mathbf{A}\boldsymbol{x}_k\|_2$ is minimized at all snapshots. It is necessary to find the initial values of the coefficients bk. $\mathbf{x}_1$ will be the initial snapshot state at the time $\mathbf{t}_1 = 0$. Eq. (10) $\boldsymbol{x}_1 = \mathbf{\Phi}\mathbf{b}$. Since the matrix of eigenvectors $\mathbf{\Phi}$ is generally not a square matrix. The initial condition is:

$$\mathbf{b} = \mathbf{\Phi}^{\dagger}\boldsymbol{x}_1 \tag{11}$$

It can be obtained using the pseudo-inverse matrix $\mathbf{\Phi}^{\dagger}$.

3.3. Interpretation of each value

In DMD, the number of ranks after dimensionality reduction, i.e., the number of modes used for reconstruction, is sometimes determined by the contribution ratio or cumulative contribution ratio. The singular value σ_k $(k = 1, \cdots, m-1)$, which is the diagonal element of $\mathbf{\Sigma}$ obtained by singular value decomposition of the experimental data $\mathbf{X}$, is used to calculate the cumulative contribution ratio C. The cumulative contribution ratio C is calculated using the singular value σ_k $(k = 1, \cdots, m-1)$, which is the diagonal element of $\mathbf{\Sigma}$ obtained when decomposing the experimental data $\mathbf{X}$.

$$C = \frac{\sum_{k=1}^{i} \sigma_k}{\sum_{k=1}^{m-1} \sigma_k} \qquad (i = 1, \cdots, m-1) \tag{12}$$

However, since there is no indicator for the cumulative contribution rate up to a certain percentage, the number of modes should be determined based on a comprehensive consideration of the characteristics of the data and the purpose of the analysis. The relationship between the cumulative contribution ratio and the "error between the original data and the reconstructed data" is shown in the actual analysis results. Explain the characteristics of each mode that can be read from its values. Each column $\boldsymbol{\phi}_k$ of $\mathbf{\Phi}$ is called a DMD mode and represents the spatial characteristics of each mode. Each column $\boldsymbol{\psi}_k$ of $\mathbf{\Psi}$ is called time dynamics and represents the spatial characteristics of each mode, corresponding to the time evolution of the corresponding DMD mode.

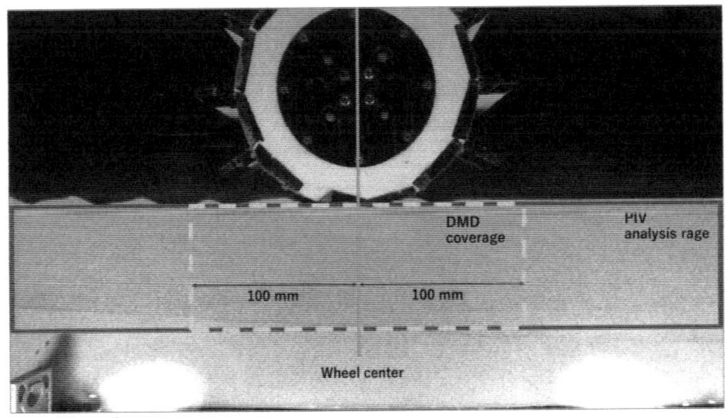

Fig. 7. Analysis range

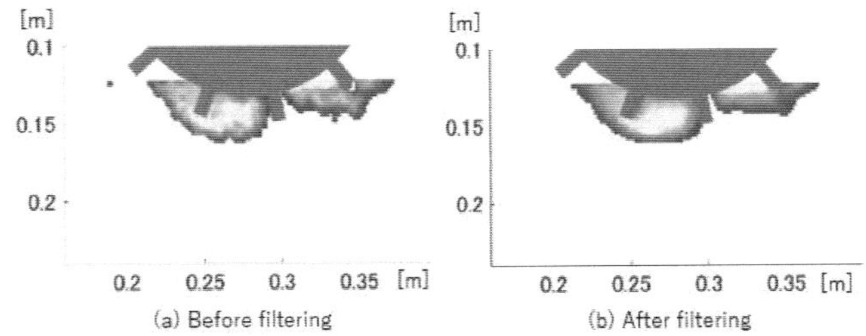

Fig. 8. Filtering experiment data

The more dominant modes are extracted in order. $\boldsymbol{\omega}_k$ can be calculated for the frequency [Hz] of each mode. The absolute value of the imaginary part of f obtained from $f = \omega/2\pi$ represents the frequency of that mode. The growth/decay rate of each mode can be read off from the eigenvalue λ_k.

$|\lambda_k| < 1$: the mode attenuates
$|\lambda_k| = 1$: the mode neither attenuates nor amplifies.
$|\lambda_k| > 1$: the mode amplifies

3.4. Data processing before DMD

When applying DMD, some preprocessing is performed on the data to obtain appropriate analysis results. The necessity of such preprocessing depends on the study's purpose and the data's characteristics. This study uses three preprocessing: limitation of analysis scope, filtering, and subtracting on average.

To have periodicity in the data for each coordinate, data is excerpted on a wheel basis. In DMD, noise in the experimental data is extracted as modes and affects the analysis results. Therefore, it is necessary to remove noise by filtering in advance. Fig. 8 shows an example of filtering on velocity data of a flow field caused by wheel running. Applying DMD by subtracting the mean value from the experimental data allows mode decomposition to focus on the variability, making it easier to consider the analysis results. In addition, since the center of the data is the origin, the influence of noise, such as outliers, can be reduced. In this study, when comparing the experimental data with the data reconstructed by DMD, the mean values of the reconstructed data are added again and compared. The data are compared after adding the mean values to the reconstructed data again.

4. DMD analysis results

4.1. Restructuring with dimensional reduction

The results of applying DMD to the low-slip and high-slip data and reconstructing them with reduced dimensionality are shown. The

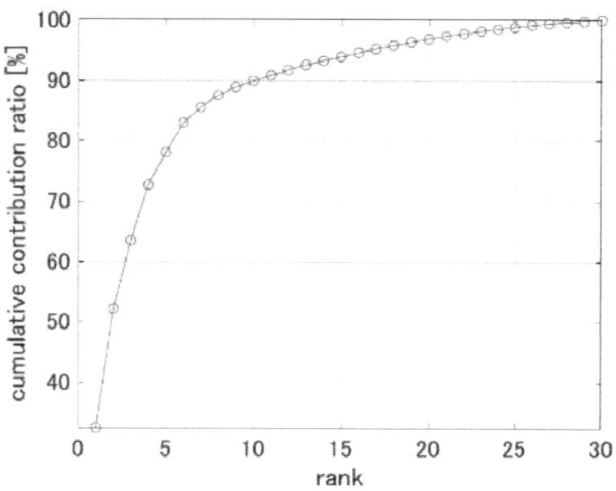

Fig. 9. Cumulative contribution ratio: low slip

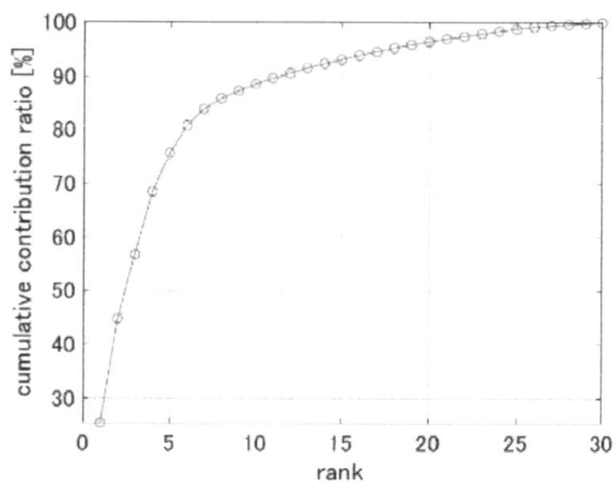

Fig. 11. Cumulative contribution ratio: high slip

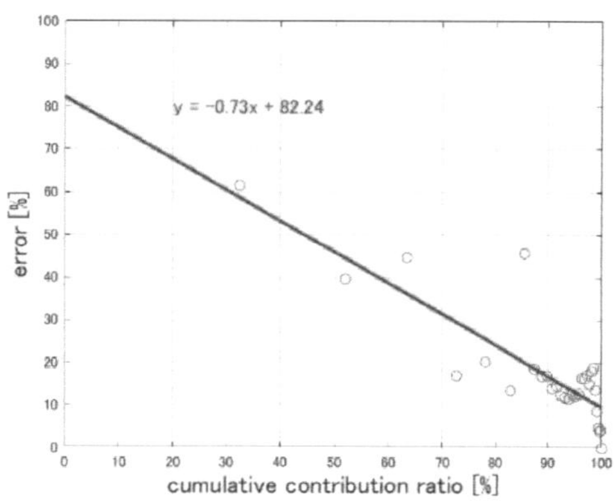

Fig. 10. Error and cumulative contribution ratio: low slip

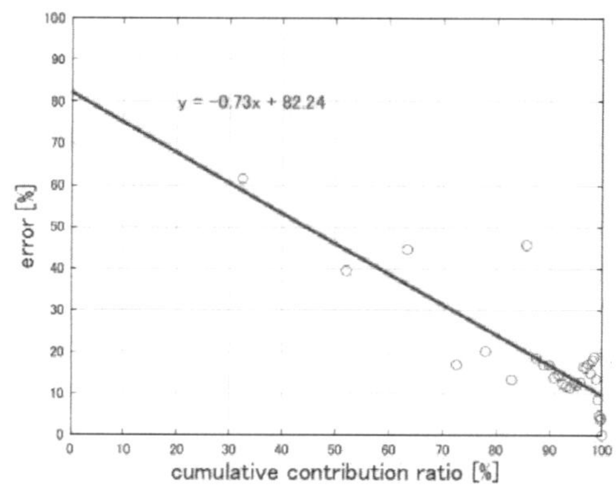

Fig. 12. Error and cumulative contribution ratio: high slip

error between the reconstructed and original data is calculated to determine the number of ranks after the reduction considered sufficient for the reconstruction.

The relationship between the cumulative contribution ratio of the error and the singular value is shown to verify the validity of using the cumulative contribution ratio as a criterion for determining the number of ranks. Then, by focusing on the number of ranks, it is possible to compare the spatiotemporal characteristics of the ranks for each condition.

Fig. 9 shows the cumulative contribution of the singular values obtained by applying DMD to the low-slip running data. The larger the number of ranks after reduction, the larger the cumulative contribution ratio. However, the rate of increase of the cumulative contribution ratio is more significant when the number of post-reduction ranks is smaller. The cumulative contribution ratio is considerably higher at the stage where the number of post-reduction ranks is small. Fig. 10 shows the relationship between the cumulative contribution ratio and the error in the reconstructed data. It can be seen that there is a negative correlation between the error and the cumulative contribution ratio. Fig. 11 shows the cumulative contribution of the singular values obtained by applying DMD to the high-slip running data.

Fig. 12 shows the relationship between the cumulative contribution ratio and the error in the reconstructed data. The same trend was observed in the results for high-slip runs as in the results for low-slip runs. Since there is a correlation between the cumulative contribution rate and the error, it is reasonable to determine the number of ranks after the reduction from the cumulative contribution rate. It can also be determined that a small number of post-reduction ranks is sufficient.

4.2. Spatiotemporal decomposition

Based on the previous section's cumulative contribution and error results, this study focuses on ranks 4, 6, and 8 and decomposes the modes spatiotemporally. When DMD is applied, the resulting modes are conjugate, and in many cases, the reduction can be represented by half the number of modes in effect. Since conjugate modes also exist in the analysis of this study, the number of modes can be reduced to 2, 3, and 4 for ranks of 4, 6, and 8, respectively. The DMD mode and Time Dynamics of each mode, i.e., the spatial distribution and its time evolution, are shown separately to discuss each mode's temporal and spatial characteristics.

This section explains how to interpret the results of DMDmode and Time Dynamics obtained by DMD analysis: the velocity at the time and position where the product of DMD mode and Time Dynamics is the minimum value is 0, and the velocity at the time and position where the product is the maximum value of the velocity in that mode. The velocity at the time and location where the product is the maximum is the maximum value of the velocity in that mode.

The results of spatiotemporal decomposition by applying DMD are shown in Fig. 13 and Fig. 14 for the case with rank 4, Fig. 15 and Fig. 16 for the case with rank 6, and Fig. 17 and Fig. 18 for the case with rank 8. The position of the grouser at the start and end of the analysis is plotted as a red line. The growth/decay rate of each mode and the frequency [Hz] per grouser period (1 second) are shown in Table 1, Table 2, Table 3, Table 4, Table 5, and Table 6. In this analysis, all modes' damping/amplification rate is approximately 1, indicating that the vibration is neither damped

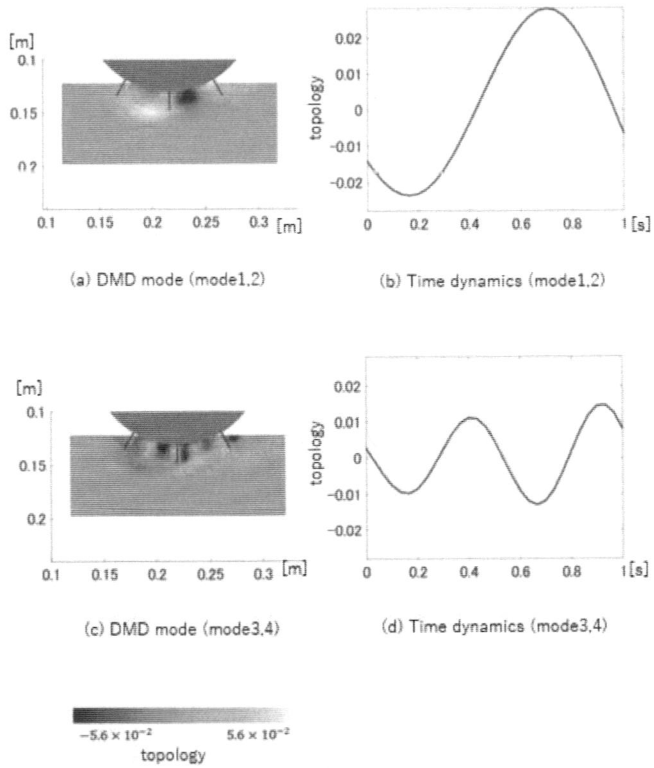

Fig. 13. Spatiotemporal decomposition (rank=4): low slip

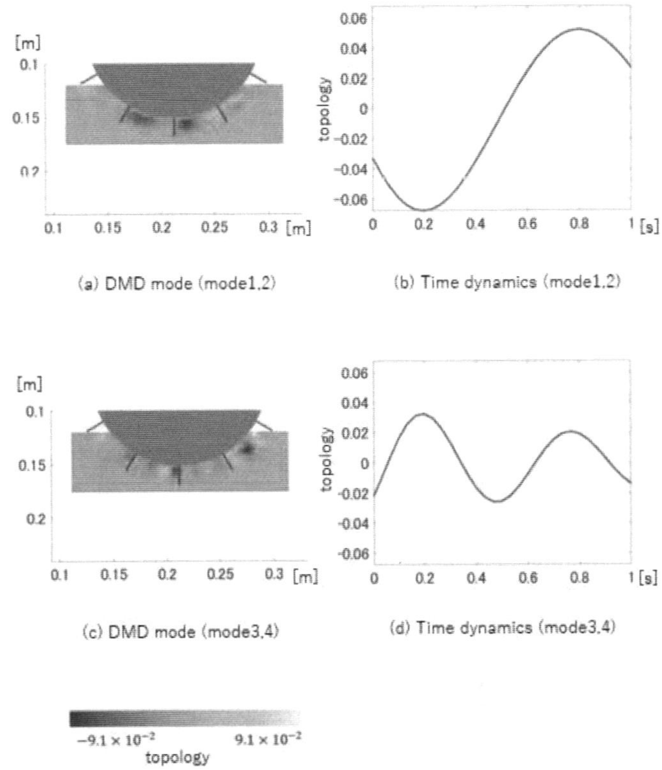

Fig. 14. Spatiotemporal decomposition (rank=4): high slip

Table 1
Mode information (rank=4): low slip

Mode	Growth/Decay Rate	Frequency [Hz]
1,2	1.0	0.9
3,4	1.0	2.0

Table 2
Mode information (rank=4): high slip

Mode	Growth/Decay Rate	Frequency [Hz]
1,2	1.0	0.8
3,4	1.0	1.7

nor amplified.

Fig. 13 shows that in the DMD of modes 1 and 2, the shallow area in front of the wheel has a negative value (blue), and the deep area behind the wheel has a positive value (yellow). The time dynamics in modes 1 and 2 transition from negative to positive values. Therefore, modes 1 and 2 represent alternating increases in sand flow velocity in the shallow part in front of the wheel and the deep part behind the wheel. In the DMD mode of modes 3 and 4, positive and negative values are striped across the grouser pass range. The number of stripes in the range of one grouser pass is 2, corresponding to the frequency of the time dynamics in modes 3 and 4. Therefore, modes 3 and 4 represent that the sand is moving in the same way as the grouser.

Fig. 14 shows the results of DMD at rank number 4 after reduction for high-slip run. The results show that modes 1, 2, 3, and 4 DMD modes had a striped distribution of positive and negative values in the grouser passband. The number of stripes in the range through which one grouser passes corresponds to the frequency of the time dynamics of each mode.

Fig. 15 shows the results of applying DMD at rank number 6 after reduction to low-slip running. The rank number increases by two, and modes 5 and 6 are added. It reads as if the number of ranks is 4, and modes 1 and 2 represent alternating increases in sand flow velocity in the shallow part in front of the wheel and the deep part behind the wheel. Mode 3, 4, and mode 5, 6 represent that the sand is moving in the same way as the grouser.

In this type of mode, the higher the frequency, the more sensitive the mode is to sand movement. In this result, modes 3 and 4 have more minor frequencies and are more dominant. Even though the number of modes increased, the most dominant modes for low-slip driving represented periodic sand flow in front of and behind the wheel.

Fig. 16 shows the results of applying DMD at rank number 6 after reduction to high-slip running. In high-slip driving, modes representing periodic sand flow ahead and behind the wheel do not appear even if the number of modes is increased by one. In high-slip driving, the mode representing periodic sand flow ahead and behind the wheel does not appear even if the number of modes is increased by one. All modes represent sand movement similar to that of the grouser. The most dominant mode is the one with the lowest frequency. However, looking at mode 3, 4 and mode 5, 6, the smaller frequency is not necessarily more dominant.

Fig. 17 shows the results of applying DMD at rank number 8 after reduction to low-slip running. In this result, only the most dominant mode represents periodic sand flow ahead and behind the wheel. The other modes represent sand flow similar to the movement of the grouser. There is no relationship between frequency and the strength of dominance.

Fig. 18 shows the results of applying DMD at rank number 8 after reduction to high-slip running. Even though the number of modes is increased to four, the mode representing periodic sand flow in front of and behind the wheel does not appear.

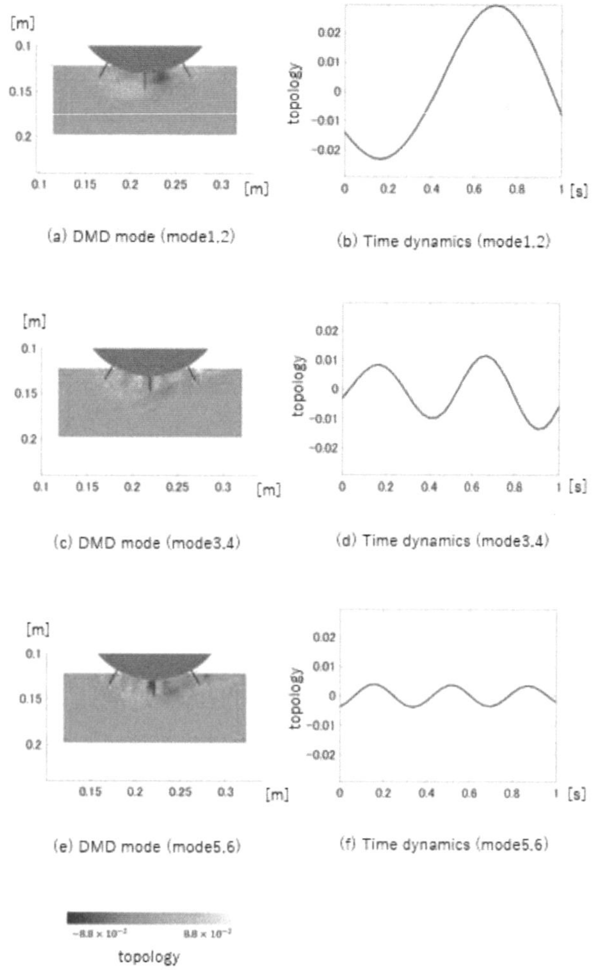

Fig. 15. Spatiotemporal decomposition (rank=6): low slip

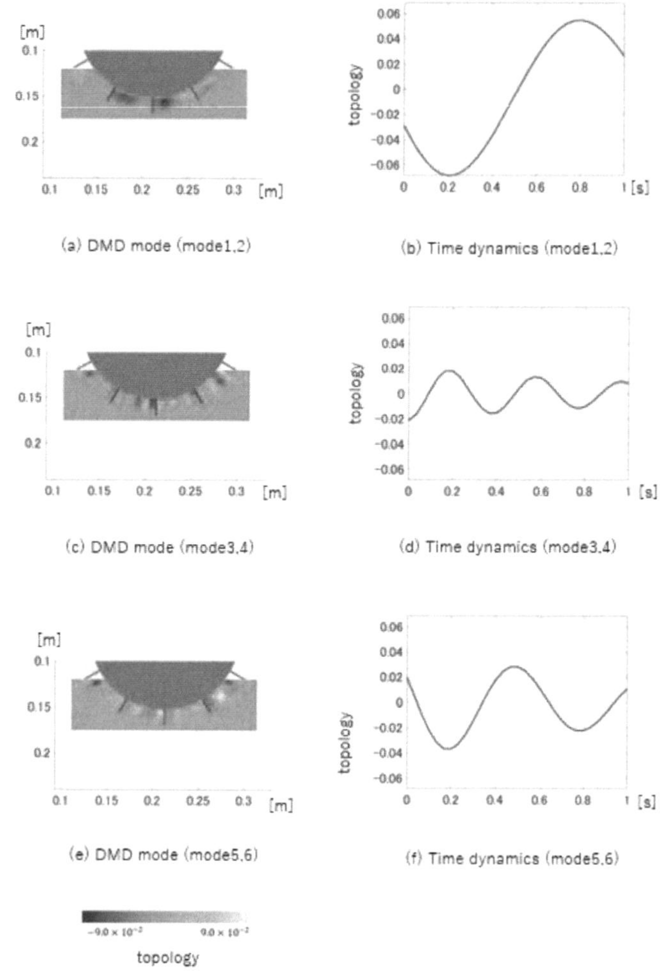

Fig. 16. Spatiotemporal decomposition (rank=6): high slip

Table 3
Mode information (rank=6): low slip

Mode	Growth/Decay Rate	Frequency [Hz]
1,2	1.0	0.9
3,4	1.0	2.0
5,6	1.0	2.8

Table 4
Mode information (rank=6): high slip

Mode	Growth/Decay Rate	Frequency [Hz]
1,2	1.0	0.9
3,4	1.0	2.6
5,6	1.0	1.7

4.3. Summary of results and discussion

Focusing on the DMD mode, the mode can be divided into two types. DMD mode at mode 1 and 2 at low slip are negative values in front of the wheel and positive values behind the wheel. The Time Dynamics of these modes show that positive and negative values are cyclically repeated. Thus, it can be seen that these modes represent alternating sand flows in front of and behind the wheel. Sand compaction occurs alternately in the shallower blue areas of the spatial distribution, and soil breakdown occurs in the deeper yellow areas. The DMD modes for the three subsequent modes at low slip and all modes at high slip are distributed in a stripe pattern over the grouser passband. The number of stripes in the range through which one grouser passes coincides with the frequency of Time Dynamics. Thus, we can see that this mode represents the sand between the grousers moving with the grousers. Two dominant modes existed during low-slip, whereas only one dominant mode existed during high-slip. Since sand collapse occurs during high-slip, the dominant mode was likely taken to represent the movement of sand between grousers. A comparison of the DMD modes during low-slip and high-slip shows that the low-slip mode is distributed at a greater depth.

5. Conclusion

In this study, DMD was applied to the PIV results to provide low-dimensional and spatiotemporal separation of the sand flow caused by the running of the grouser wheels. After separating the spatiotemporal features, each mode could be divided into two types based on the distribution trend of the DMD mode. When focusing only on the dominant mode, the low-slip data contained two types, whereas the high-slip data contained only one type. A comparison of the DMD modes between the low-slip and high-slip modes showed that the low-slip mode was distributed at a deeper

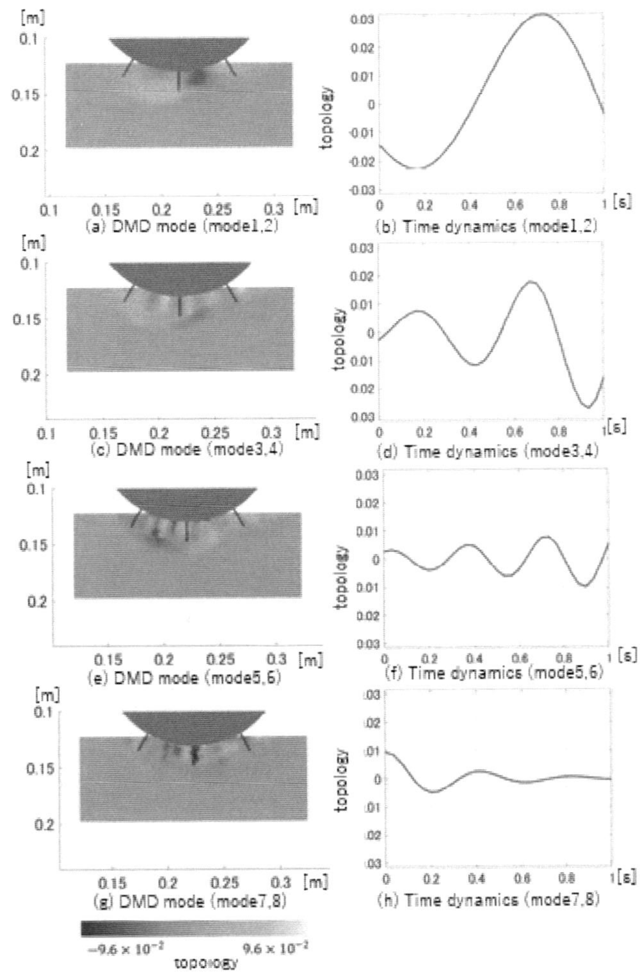

Fig. 17. Spatiotemporal decomposition (rank=8): low slip

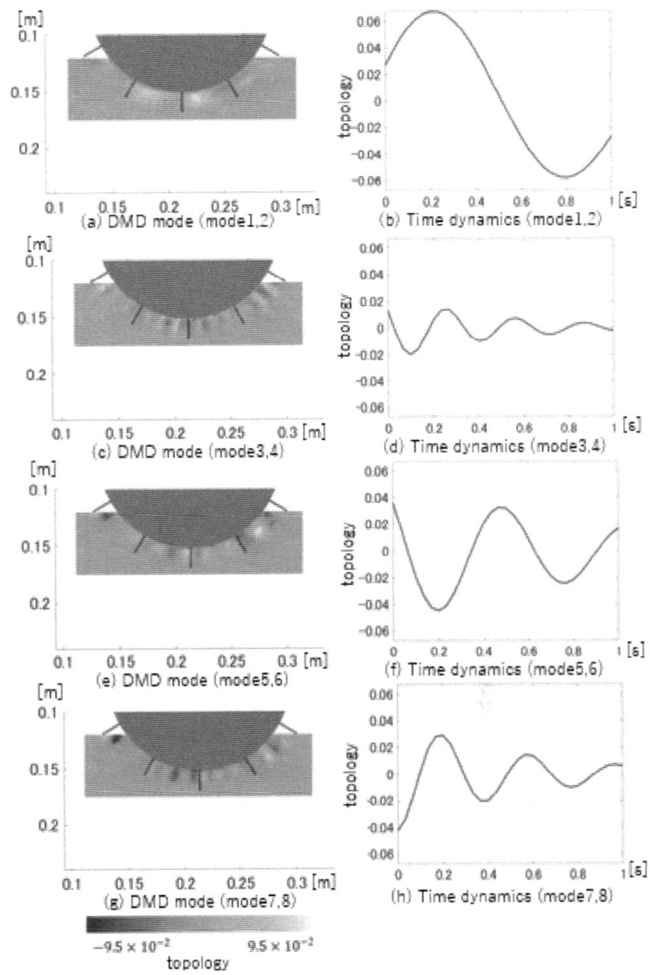

Fig. 18. Spatiotemporal decomposition (rank=8): high slip

Table 5
Mode information (rank=8): low slip

Mode	Growth/Decay Rate	Frequency [Hz]
1,2	1.0	0.9
3,4	1.1	2.0
5,6	1.0	2.9
7,8	0.9	2.5

Table 6
Mode information (rank=8): high slip

Mode	Growth/Decay Rate	Frequency [Hz]
1,2	1.0	0.9
3,4	0.9	3.3
5,6	1.0	1.8
7,8	0.9	2.6

depth. In this study, wheel geometry was not changed and only slip rate was used as a comparison condition. In the future, it is necessary to clarify which part of the wheel structure is responsible for each mode by varying the number of grouser wheels, grouser length, and wheel rotation speed.

6. References

Ding, L., Deng, Z., Gao, H., Tao, J., Iagnemma, K. D., Liu, G., 2015. Interaction mechanics model for rigid driving wheels of planetary rovers moving on sandy terrain with consideration of multiple physical effects. Journal of Field Robotics 32, 827-859.

Higa, S., Sawada, K., Nagaoka, K., Nagatani K., Yoshida, K., 2016. Measurement of stress distributions of a wheel with grousers traveling on loose soil. Proc. 2016 IEEE International Conference on Robotics and Automation (ICRA). 2828-2833.

Irani, R. A., Bauer R. J., Warkentin, A., 2011. A dynamic terramechanic model for small lightweight vehicles with rigid wheels and grousers operating in sandy soil. Journal of Terramechanics 48, 307-318.

Kutz J. Nathan, Brunton Steven L., Brunton Bingni W., Proctor Joshua L., 2016. Dynamic mode decomposition, data-driven modeling of complex systems.

Lichtenheldt, R., Schafer, B., 2013. Locomotion on soft granular soils: a discrete element based approach for simulations in planetary exploration. Proc. 12th Symposium on Advanced Space Technologies in Robotics and Automation. ASTRA. 1-8.

Moreland, S. J., Skonieczny, K., Wettergreen, D. S., 2014. Motion analysis system for robot traction device evaluation and design. Proc. Field and Service Robotics, 449-463.

Nakamura, H., Nagaoka, K., Yoshida, K., 2020. Soil flow analysis beneath a grouser wheel based on a particle image velocimetry method. Journal of Terramechanics 91, 233-241.

Prandtl, L., Tietjens, O., 1934. Fundamentals of hydro and aeromechanics. Dover Publications Inc.

Reynolds, O., 1883. Experimental investigation of the circumstances which determine whether the motion of water shall be direct or sinuous, and of the law of resistance in parallel channels. Philosophical Transactions of the Royal Society of London 174, 935-982.
Saengprachatanarug, K., Okayasu, T., 2013. Modeling of soil displacement and soil strain distribution under a traveling wheel. Journal of Terramechanics 50, 5-16.
Senatore, C., Wulfmeier, M., Vlahinic, I., Andrade, J., Iagnema, K., 2013. Design and implementation of a particle image velocimetry method for analysis of running gear-soil interaction. Journal of Terramechanics 50, 311-326.
Skonieczny, K., Moreland S. J., Wettergreen, D. S., 2012. A grouser spacing equation for determining appropriate geometry of planetary rover wheels. Proc. 2016 IEEE/RSJ International Conference on Intelligent Robots and Systems 10, 5065-5070.
Sutoh, M., Nagaoka, K., Nagatani K., Yoshida, K., 2013. Design of wheels with grousers for planetary rovers traveling over loose soil. Journal of Terramechanics 50, 345-353.
Yamakawa, J., Goto, M., 2015. Experiment and numerical analysis of driving wheel on soft sandy soil. Proc. 13th ISTVS European Regional Conference, 500-508.
Yamano, Y., Nagaoka, K., Yoshida, K., 2018. PIV analysis of soil deformation beneath a grouser wheel. Proc. 10th ISTVS Asia-Pacific Regional Conference, 39.

1710 | A novel measurement method to aid development of soft soil tyre-terrain interaction models

Thomas B. Kabutz [a, *], P. Schalk Els [a]

[a] Department of Mechanical and Aeronautical Engineering, University of Pretoria, Pretoria, South Africa

* Corresponding author: thomas.kabutz@tuks.co.za (T.B. Kabutz), schalk.els@up.ac.za (P.S. Els)

ABSTRACT

Most forces on a ground vehicle go through the tyre contact patch. Because the tyre interfaces with the road surface, it is not possible to directly measure the contact patch. This presents a significant challenge in assessing soil characteristics. Predictions of the tread deformation allow estimates of total soil deformations. This can be used to determine elastic and plastic deformation of soil under moving vehicles, giving insight into soil parameters. Soil deformations and parameters are beneficial for vehicle control as well as evaluating the environmental impact. Previous studies utilized digital image correlation techniques to measure tyre deformation on the inner surface of the tyre carcass.

This paper investigates the feasibility of developing a measurement system that uses deformations on the inner surface of a tyre to predict the deformation on the outside. The proposed method involves offsetting the inner surface along its normal directions by the tread thickness to obtain the outer surface of the tyre.

A 2D proof of concept was developed, showing the ability to predict tread deformation based on inner surface measurements. Reasonable accuracy was achieved, thus confirming the feasibility of predicting tyre tread deformations from inner surface measurements by a geometry offset. The identified errors are deemed acceptable for the given problem, with the errors due to the simplification of the problem proving significantly smaller than measurement-induced errors. A full 3D model was consequently developed to predict tread deformation over the full contact patch region.

The findings pave the way for the development of a real-time system to predict soil volume displacement, providing crucial insights for vehicle control and environmental impacts in offroad scenarios.

Keywords
Tyre deformation
Digital image correlation
Tyre-Terrain interaction

1. Introduction

1.1. Background

The area where all non-aerodynamic forces are applied on a vehicle is the contact patch between the tyres and the terrain. However, since the tyre interfaces with the terrain, it is not possible to measure the contact patch directly. When dealing with deformable terrain, the contact patch region becomes more complex to model since the terrain deforms under the tyre. This presents a challenge in assessing soil characteristics. One way to get around this problem is to measure the deformation on the inside of the tyre and then model what happens on the outside. This could then be used to determine elastic and plastic deformation of the soil under a moving vehicle. Which in turn can give insight into soil parameters. Soil parameters are useful for vehicle control and evaluating the environmental impact of vehicles.

Various studies have suggested solutions to the measurement of the displacement of the tyre contact patch such as ultrasonic sensors (Longoria et al., 2019) or digital image correlation with cameras (Guthrie et al., 2017a). The use of non-contact 3D measurement in the field of Terramechanics has become popular in recent years (Farhadi et al., 2018, Farhadi et al., 2019, Phakdee and Suvanjumrat, 2023, Kenarsari et al., 2017, Pytka et al., 2018, Medeiros Jr et al., 2023). Most studies focus on static measurements with either a handheld laser scanner or a camera and photogrammetry. Botha et al. (2019) used stereo cameras to measure the rut profile in real time. However, so far only Guthrie et al. (2017b) attempted to estimate elastic deformation of the soil for the entire contact patch region.

1.2. T2Cam

The Vehicle Dynamics Group (VDG) at the University of Pretoria has developed a measurement system that can capture the deformation of the contact patch inside of a tyre. The Tyre-Terrain Camera System (called T2Cam) uses two mechanically stabilized cameras inside of a tyre that are pointed at the contact patch (shown in Fig 1). With Digital Image Correlation (DIC) techniques, point clouds representing the deformation of the inner surface of the tyre can be calculated. Guthrie et al. (2017a) developed the initial system and illustrated the feasibility of using digital image correlation (DIC) techniques for measuring tyre deformation.

Feldesi et al. (2020) further refined the system to measure strain in addition to the deformation. The side-by-side camera setup was replaced with two opposing cameras improving the depth resolution by a factor of 3.

Pegram et al. (2021) continued the investigation into the strain measurement, including the effect of slip angle and validating the calculated strain against strain gauges mounted on the tyre for static and dynamic conditions.

Gast (2022) investigated the relationship between the strain and displacement measurements on the inside of the tyre and the external loads that caused them. Ultimately a model was developed to predict external forces from the internal displacements.

Guthrie et al. (2017b) attempted offsetting the inner surface by an even amount to obtain an estimate of the outer surface of the tyre. However, the varying tread thickness with the tread blocks was not accounted for, thus the elastic deformation of the soil resulted in artificially high values. This paper investigates a method for estimating the deformed outer surface of a tyre accounting for the varying tread thickness.

https://doi.org/10.56884/KD1BZC7K

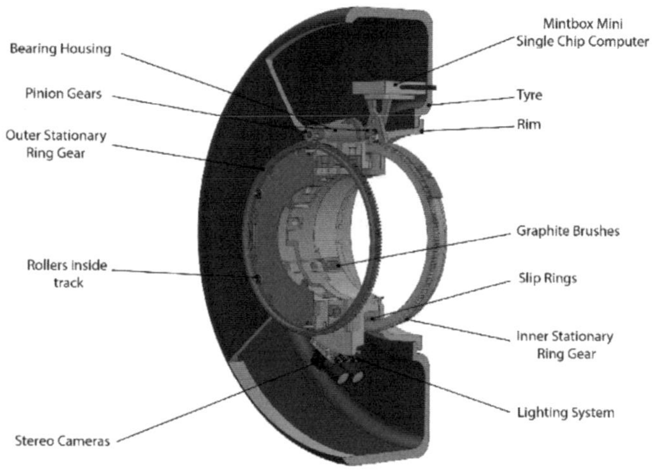

Fig 1. The in-tyre camera stabilisation rig T2Cam (Guthrie, 2016)

1.3. Proposed method

This paper proposes tracking the location of the inner surface and offsetting the inner surface along its normal directions by the correct tread thickness to obtain an estimate of the outer surface of the tyre.

The method assumes that the rubber of the tyre is incompressible and as such the volume should remain constant. Considering the deformable terrain and the fact that the Poisson's ratio of tyre rubber is close to 0.5, this is a reasonable assumption. The implementation of the proposed method went one step further and assumes that the total area of the inner surface remains constant. This is not true since the neutral bending plane of the tyre is closer to the middle of the rubber. However, it is expected that the error introduced by this assumption is small compared to the possible available measurement accuracy. It is further assumed that the tyre condition remains constant (i.e., no wear).

The advantage with the proposed method is that simple displacement maps can be used to offset the inner surface. Compared to using Finite Element Methods (FEM) to calculate the outer surface this is computationally inexpensive and could pave the way for real-time deformation measurements.

The proposed solution consists of the following steps. First a displacement map of the tread thickness needs to be baked. In the context of 3D computer graphics, the term "baking" refers to the process of calculating a displacement or texture map. For this a 3D reference model of the tyre is needed. The inner surface is mapped into UV coordinates, reducing the three dimensions to a 2-dimensional map. Also, the normal direction on the inner surface is calculated. The distance along this normal from the inner surface to the outer surface is determined and baked into a displacement map with the UV-coordinate system. Specific markers are mapped into UV-coordinates to track the tyre position. To predict the outside deformation of a deformed tyre, the deformation on the inner surface needs to be measured. The markers are detected and then the UV-coordinate system fit through the point cloud. Now each point in the point cloud can be offset along its normal direction by its corresponding value in the displacement map. This provides a point cloud that provides a reasonable estimate of the outside surface of the tyre.

2. Proof of Concept (2D)

A 2D proof of concept was developed to explore the feasibility of the proposed method and to explore the impact of the assumptions on the prediction accuracy. This simplifies the measurement side of the problem significantly and allows a better understanding of any issues with the approach.

A flatbed scanner was used to record the cross section of a Michelin LTX AT2 tyre. The edge of the tyre was painted white for better contrast. The points were then manually sampled on the images. Figure 2 shows an example of the scan with sampled points for both an undeformed and a deformed state. A marker was placed at the centre of the contact patch to function as a reference point from which the surface can be resampled later.

Manual sampling introduced some uncertainty in the geometry of the cross section. Figure 3 shows a worst-case region for point sampling, illustrating a 0.7mm accuracy.

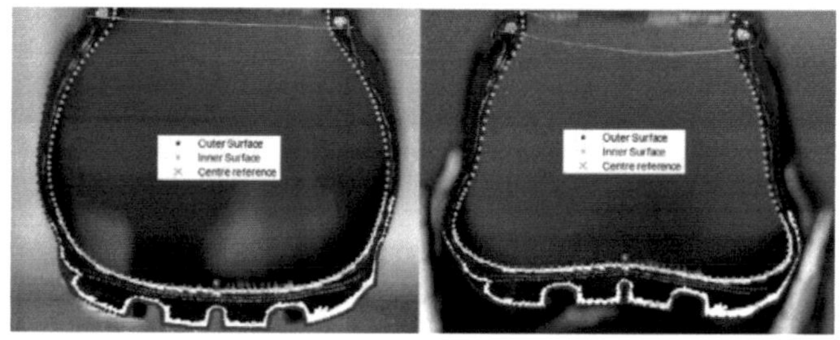

Fig. 2. Flatbed Tyre scans

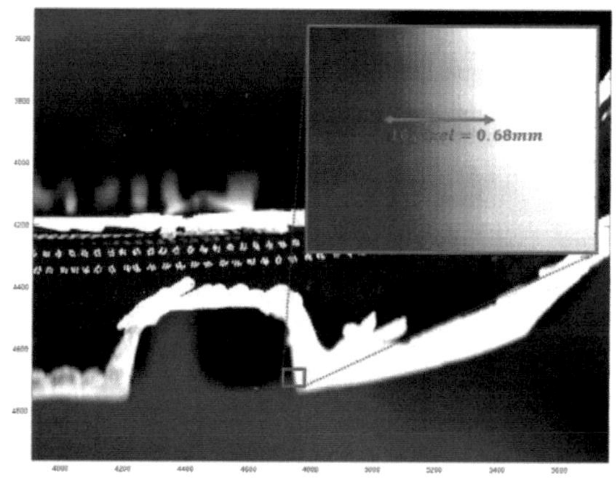

Fig 3. Region of uncertainty of sampled data

2.1. Baking displacement maps

A parametric cubic spline was fitted through the sampled inner surface. The surface is then resampled to a predefined even arc-length spacing. The points were resampled from the reference centre point on the contact patch. The node spacing can also be modified so that the regions of interest (i.e., the tread) have denser spaced nodes than the side walls. It was assumed that the area of the inner surface would remain constant. In the various tests the length of the inner surface of the deformed states remained within 0.1% of the reference state.

The vector between two consecutive points gives the tangential vector. Consecutive tangential vectors are averaged to provide the nodal vector. Cross multiplying this with a unit vector in the z direction gives the normal vector. Figure 4 shows the normal vectors and their intersection with the outer surface.

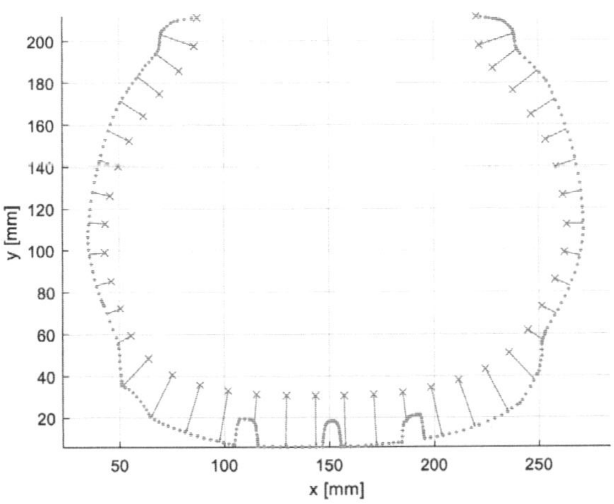

Fig 4. Calculated Normals and intersection with the outer surface

The unit normal vector can then be used to calculate the tread thickness. The intersection of each normal with the points of the outer surface is calculated by translating the outer surface points so that the current node is at the origin and then rotated so that the current normal is aligned with the x-axis. The x-intercept of the line between the two points that are nearest above and below the x-axis is the tread thickness. Figure 4 illustrates this process and shows the resulting tread displacement map.

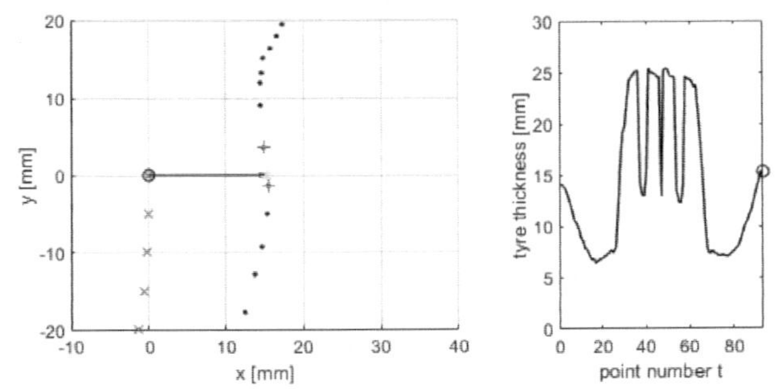

Fig 5. Left - Tread thickness calculation, Right - Tread displacement map

2.2. Applying displacement map

The inner surface of a deformed tyre is sampled with the same spacing as the reference, and the resulting normal directions for the nodes are calculated. The displacement map provides the information on how far each node needs to be displaced along its normal to represent the outer surface of the tyre.

Figure 5 shows the predicted tread against the measured tread. The prediction looks excellent with some small errors. Figure 6 shows the error of the measured surface to the nearest predicted point for various step sizes. The mean error decreases significantly with smaller step sizes. For a 0.5mm step size the mean error is between 0.5mm and 0.6mm. This is quite large, however, further tests with three other bent configurations all resulted in a similar error. Using any of these as the reference state (to derive the map) and then predict also results in this error. This suggests that the error present is largely due to the uncertainty in the measurement accuracy and not due to the underlying assumptions of the prediction. To determine the error of a prediction a total of 4 different surfaces need to be sampled (two for the map, one to predict and one to verify). Since the measurement accuracy is within 0.7mm, the expected worst-case error is up to 2.8mm. The errors present are very representative of this measurement error. Thus, the 2D model clearly illustrated the feasibility of using the proposed method.

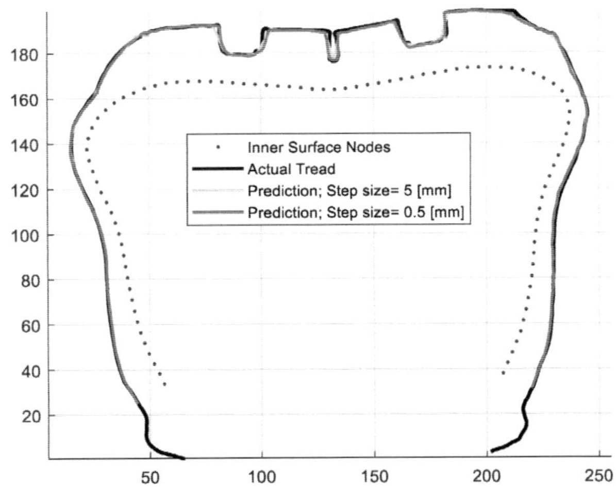

Fig 6. Tread prediction and measured tread for two different node spacings

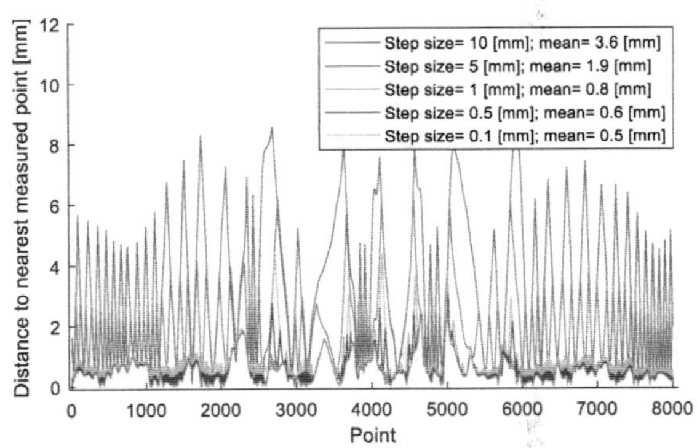

Fig 7. Error between prediction and actual tread for various step sizes

3. Reference Tyre Model

For the proposed method to work in three dimensions, a full tyre scan is needed. What makes this difficult is that both the inside and the outside of the tyre need to be captured in relation to each other, as this is what defines the displacement map. Two different approaches were attempted to scan the tyre, handheld laser scanning and photogrammetry.

3.1. Laser Scanning

Handheld laser scanners are a popular non-contact method to digitally acquire the shape of a 3D object. As demonstrated by Farhadi et al. (2018) and Phakdee and Suvanjumrat (2023) handheld laser scanners can successfully be used to measure rut profiles. Various commercial solutions exist that use laser scanners to measure tyres. These solutions are commonly intended to measure tyre wear and are set up in a way that only captures the tread. For the proposed method, the inside of the tyre needs to be captured in relation to the outside. No existing solutions to this problem were found.

Different handheld laser scanners were tried to capture the tyre, with all resulting in similar issues. The scanners were able to capture sections of the tyre very well. However, when it came to scanning the full tyre, the problems started to show. Tracking was a big issue, as the scanners kept losing their position in space. This is especially a problem on the bead when transitioning from the outside to the inside. Since the features of the tread lugs repeat, the geometry of the scan is not reliable for tracking. The scans suddenly jumped from one section to somewhere else on the tyre based on the repeating tread pattern. Some of the tracking issue could be solved with tracking markers. However, there was also the issue of drift. Over the full circumference of the tyre, the scans tended to drift off from their position. The start and end points of the scan would not line up and the circularity of the tyre would be wrong. The scanners only cover a small region at a time, so there is no way to capture the entire tyre or a large portion of it at once to get the shape and then go closer for the detail. The handheld laser scanners proved too unreliable to accurately measure the full tyre. Thus, a different measurement method is required.

3.2. Photogrammetry

In recent years, the photogrammetry approach to scan objects as an alternative to laser scanning has become quite popular. Photogrammetry can easily result in decent accuracy with an affordable setup. The object is photographed from different sides. The photos are taken in sequence moving around the object with 60-80% forward and sideways overlap. Some typical algorithms for 3D image reconstruction using structure-from-motion are described in Toldo et al. (2015). Features are detected in each image, matching features between images found, the 3D position of the camera for each image estimated and then the features are aligned in 3D space to give the structure. Stereo camera setups such as the one used in T2Cam can also be used. The advantage with this is that the extrinsic properties between two images is known from calibration. However, stereo camera setups are often cumbersome and inconvenient to use, whilst only adding small improvements.

The key challenge to getting good models is the quality of the image. Clear high-resolution images are desirable that show all the detail. Unsharp images and inconsistent lighting cause issues in the model quality. This is because the features in overlapping areas of the images will be different. Overlapping regions need to look as consistent as possible between images.

Scanning a full tyre poses several difficulties. An uninflated tyre without a rim is used. The entire tyre needs to be visually accessible for the scan at once. Normally objects can be scanned in sections and then merged afterwards. However, due to flexible nature of the tyre, repositioning the tyre between scans to observe different sections will deform the tyre differently. Thus, different sections will not necessarily line up correctly for merging.

It is necessary to scan in the entire tyre at once. For this, lighting becomes an issue on the underside and on the inside of the tyre. These regions are too dark for the camera to capture enough detail when the outside is set correctly. Shining external lights at the tyre can be difficult without proper studio lighting equipment. Using a normal camera flash also is inconvenient as each image will have the bright spot of the flash on some other part of the tyre, thus resulting in different features between images.

It was found that the best way to capture the entire tyre at once was to use a DSLR camera with a wide-angle lens and a ring flashlight, with both having a polarizing filter (shown in Fig 8). Shining only polarized light at the object and then filtering this again with a polarizing filter on the camera side removes virtually all the bright reflections and gives a neutrally lit picture (See Fig 9). Thus, pictures of different parts of the tyre will all result in similarly lit images where the detected features stay constant. This helps immensely with the reconstruction of the object from the images in photogrammetry software.

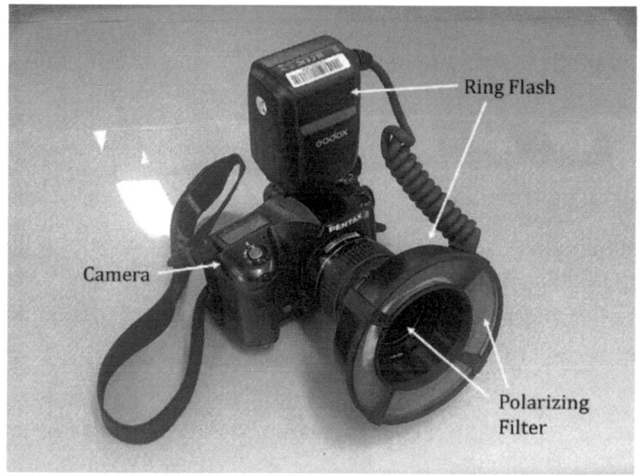

Fig 8. DSLR camera with polarizing filters and ring flashlight

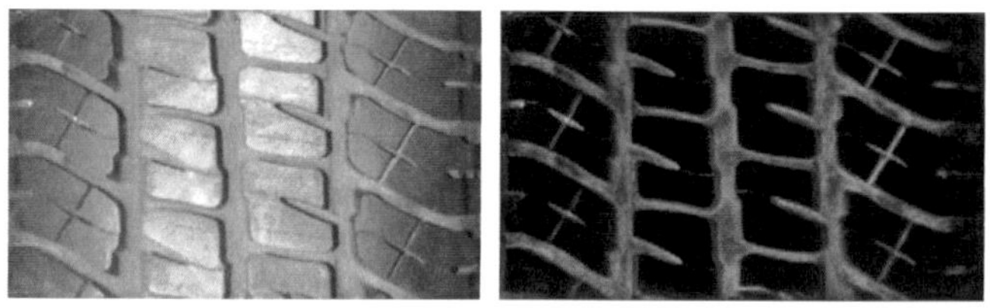

Fig 9. Effect of polarizing filters on reflections (Left - misaligned filters; Right - aligned filters)

The images were processed with the commercial software RealityCapture (Epic Games, 2024). Since April 2024, this software has become free to use. Compared to other available free software, such as Meshroom (Carsten Griwodz and Castan, 2021), RealityCapture 1.4 uses proprietary algorithms that makes it capable to process the same images much faster and tends to result in a more accurate mesh. Figure 9 shows an example of the photogrammetry scan of a tyre, where the inside and outside have already been separated for further processing.

The photogrammetry model was compared against physical measurements of the tyre. Using a notched story stick (as used in traditional woodworking), various points on a plane of the tyre were sampled on a piece of paper and then scanned into digital format. Based on some tests this technique is accurate to at least 1.5mm. The captured points and the corresponding cross section of the model are shown in Fig 11. Unfortunately, no more precise measurement equipment was available at the time of testing. The concern with the photogrammetry was that the inner surface of the tyre might be significantly offset and not in the correct position. The simple story stick measurements proved that this was not the case. As shown in Fig 11, the model is within the 1.5mm measurement accuracy of the measured points. This suggests that the inner surface of the tyre can be reconstructed with reasonable accuracy. The absolute accuracy of the approach is not known and should be investigated further if a more precise tread prediction is needed.

Fig 10. Photogrammetry scan of tyre (Top – Outside, Bottom – Inside)

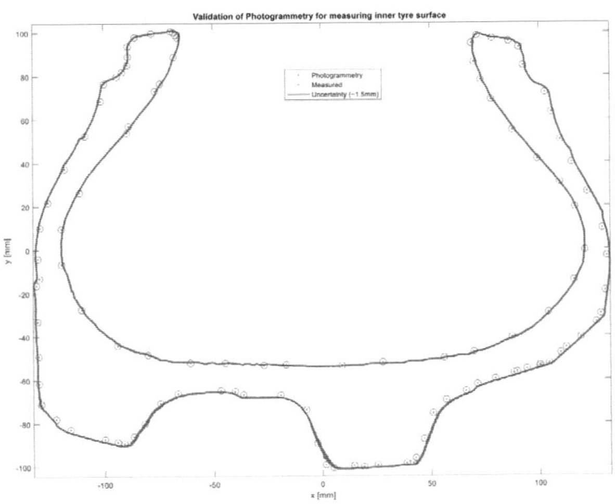

Fig 11. Validation of Photogrammetry model of tyre

The photogrammetry approach proved to be useful in scanning the full tyre. The approach was able to capture the inside and outside of a tyre in one take. It only had some issues with poor quality images, but other than that it performed well. The accuracy proved to be within measurement accuracy of the validation. This proves that this can be used as an affordable way of obtaining full 3D models of tyres.

4. 3D model and tread displacement map

A 3-dimensional model is being developed to predict the tread deformation of a tyre over the full contact patch region. The model makes use of the 2D version and rotates this around for three dimensions. The full tyre model obtained with photogrammetry allows the baking of the displacement map. T2Cam is used to measure the deformation of the inner surface. The displacement map is then used to displace the deformed inner surface to obtain the estimate of the deformed tread.

4.1. Generating displacement maps

The full reference tyre model is needed to bake the displacement map. A row of April tags (Olson, 2011) is positioned at the centre line of the tyre as shown in Fig 10. The location of each of these tags is marked in the model. The tags are later used to track the position of the tyre when captured with T2Cam and map the correct part of the tread onto the surface. The UV-coordinate system is defined so that the U-direction is longitudinally along the circumference of the tyre and the V-direction laterally from bead-to-bead.

The reference tyre is divided into evenly spaced cross-sections along V. These cross sections are then passed into a modified 2D model to determine the displacement map (illustrated in Fig 12). A reference centre plane is fit through the tags. This is the zero V coordinate. The relevant zero V coordinate is passed as the marker to the 2D model. The model then determines the tread displacement for each specific U-cross-section. Repeating this for the entire tyre results in a tread displacement map as shown in Fig 13.

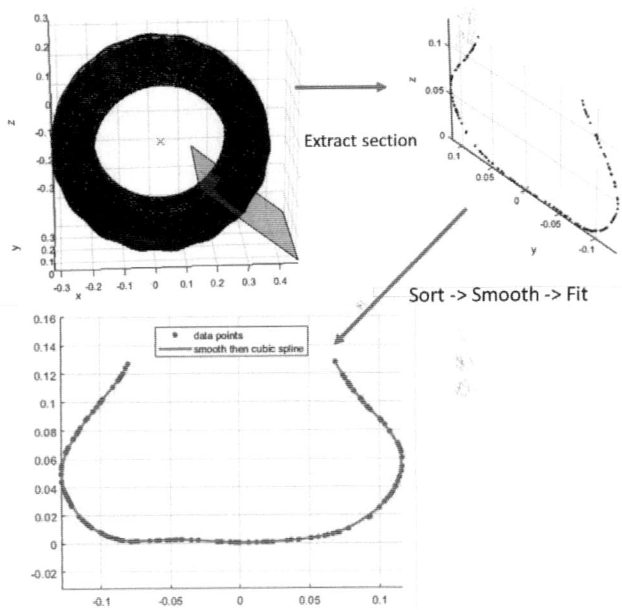

Fig 12. Cross sections extracted from the full model

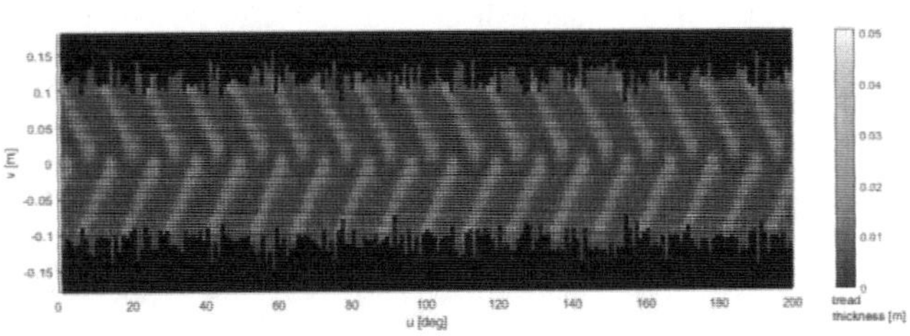

Fig 13. UV Tread displacement map

4.2. Applying tread displacement map

A speckle pattern is used on the inner surface of the tyre to add features so that the feature detection algorithm can reconstruct the inner tyre surface. April tags (36h11) were chosen as markers. These tags were chosen because they can robustly be detected. T2Cam can see multiple tags at once (see Fig 14). Thus, once measured, we have a point cloud of the deformed inner tyre surface and the locations of various tracking markers.

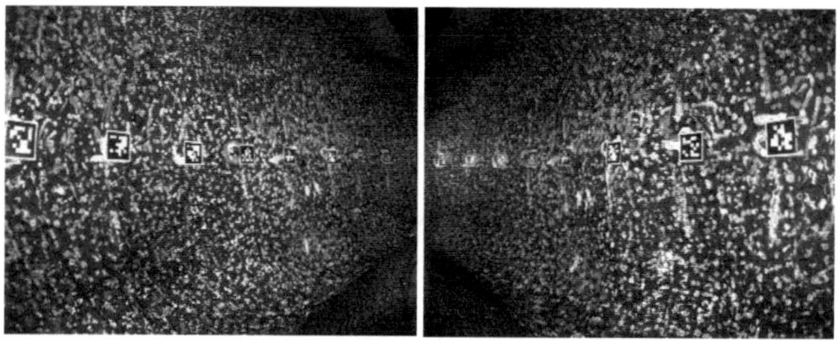

Fig 14. Images obtained from T2Cam

With the use of the tag locations, the UV-coordinate system used for the displacement map can be fitted through the point cloud. Similarly to before, the points can be displaced along their normal direction by the tread thickness. This should provide a reasonable estimate of the deformed tread of the tyre.

5. Conclusions and future work

This paper proposes a measurement technique to estimate outside tyre deformations in the contact patch region of a tyre. A 2D model was developed to investigate the feasibility of the proposed solution. The solution involves measuring the deformation of the inside of a tyre, and then offsetting this along its normal direction by the tread thickness of each point. The 2D model successfully showed that decent estimates of the deformation of the outside of a tyre can be obtained. Compared to Finite Element Models, this simple geometry offset is quick to compute and provides results of decent accuracy. The error introduced by the assumptions and simplifications is smaller than the measurement error of the surfaces.

Full 3D models of the tyre were created with a photogrammetry approach. The method uses a DSLR camera with a ring flashlight and polarising filters to provide evenly light images. This allowed the capture and reconstruction of the entire tyre, the inside and outside, in a single scan.

A 3D model to predict the outside surface of the tyre for the full contact patch region is being developed. The method works similarly to the 2D version. The solution still needs to be tested and validated experimentally.

Once the prediction of the tread deformation is available, total soil deformations can be estimated. This can be combined with measurements of the soil deformation and used to determine elastic and plastic deformation under moving vehicles. The entire system needs to be implemented on a vehicle. The following step is to investigate if soil parameters can be estimated from this.

6. Acknowledgements

We express our gratitude to A. Glenn Guthrie for assisting with T2Cam. A special thanks of gratitude is expressed to the Vehicle Dynamics Group at the University of Pretoria.

7. Declaration of competing interest

The authors declare that they have no known competing financial interests or personal relationships that could have appeared to influence the work reported in this paper.

8. References

Botha, T. R., Johnson, D., Els, P. S. & Shoop, S. 2019. Real time rut profile measurement in varying terrain types using digital image correlation. *Journal of Terramechanics,* 82, 53-61.

Carsten Griwodz, S. G., Lilian Calvet, Pierre Gurdjos, Fabien & Castan, B. M., Yann Lanthony, Gregoire De Lillo. AliceVision Meshroom: An open-source 3D reconstruction pipeline. 12th ACM Multimedia Systems Conference (MMSys 2021), Sep 2021 Istanbul, Turkey. pp.241-247.

Epic Games Inc. 2024. RealityCapture. 1.4 v. Available: https://www.capturingreality.com/realitycapture

Farhadi, P., Golmohammadi, A., Sharifi, A. & Shahgholi, G. 2018. Potential of three-dimensional footprint mold in investigating the effect of tractor tire contact volume changes on rolling resistance. *Journal of Terramechanics,* 78, 63-72.

Farhadi, P., Golmohammadi, A., Sharifi Malvajerdi, A. & Shahgholi, G. 2019. Finite element modeling of the interaction of a treaded tire with clay-loam soil. *Computers and Electronics in Agriculture,* 162, 793-806.

Feldesi, F., Botha, T. R. & Els, P. S. 2020. Full-field strain measurements of the inside of an agricultural tyre using digital image correlation. *Journal of Terramechanics,* 91, 309-318.

Gast, R. G. 2022. *Tyre Force Prediction from Deformation Measurements.* Master of Engineering (Mechanical Engineering), University of Pretoria. Available: https://repository.up.ac.za/handle/2263/89669

Guthrie, A. G. 2016. *3D computer vision contact patch measurements inside off-road vehicle tyres.* Master of Engineering (Mechanical Engineering), University of Pretoria. Available: https://repository.up.ac.za/handle/2263/56080

Guthrie, A. G., Botha, T. R. & Els, P. S. 2017a. 3D contact patch measurement inside rolling tyres. *Journal of Terramechanics,* 69, 13-21.

Guthrie, A. G., Botha, T. R., Jimenez, E., Els, P. S. & Sandu, C. 2017b. Dynamic 3d measurement of tyre-terrain interaction. *19th International & 14th European-African Regional Conference of the ISTVS.* Budapest, Hungary.

Kenarsari, A. E., Vitton, S. J. & Beard, J. E. 2017. Creating 3D models of tractor tire footprints using close-range digital photogrammetry. *Journal of Terramechanics,* 74, 1-11.

Longoria, R. G., Brushaber, R. & Simms, A. 2019. An in-wheel sensor for monitoring tire-terrain interaction: Development and laboratory testing. *Journal of Terramechanics,* 82, 43-52.

Medeiros Jr, M., Babadopulos, L., Maia, R. & Castelo Branco, V. 2023. 3D pavement macrotexture parameters from close range photogrammetry. *International Journal of Pavement Engineering,* 24, 2020784.

Olson, E. AprilTag: A robust and flexible visual fiducial system. IEEE International Conference on Robotics and Automation (ICRA), May 2011. 3400-3407.

Pegram, M. S., Botha, T. R. & Els, P. S. 2021. Full-field and point strain measurement via the inner surface of a rolling large lug tyre. *Journal of Terramechanics,* 96, 11-22.

Phakdee, S. & Suvanjumrat, C. 2023. Development of a tire testing machine for evaluating the performance of tractor tires based on the soil compaction. *Journal of Terramechanics,* 110, 13-25.

Pytka, J., Śliczniak, T., Kasprzak, P. & Gnapowski, E. 2018. Tyre-Soil Interface Determination by Photogrammetric Method. *IOP Conference Series: Materials Science and Engineering,* 421, 022030.

Toldo, R., Gherardi, R., Farenzena, M. & Fusiello, A. 2015. Hierarchical structure-and-motion recovery from uncalibrated images. *Computer Vision and Image Understanding,* 140, 127-143.

1882 | Spatio-temporal analysis of sand-density distribution beneath traveling wheel based on particle image velocimetry

Kenji Nagaoka [a,*], Seigo Sakai [a]

[a] *Department of Mechanical Engineering, Kyushu Institute of Technology, Japan*

* *Corresponding author: nagaoka.kenji572@mail.kyutech.jp*

ABSTRACT

This paper proposes a spatio-temporal analysis of sand-density distribution beneath a traveling wheel based on a particle image velocimetry (PIV) method. With the advancement of image processing technology and the higher resolution of commercially available digital cameras, the PIV methods have been widely used to directly visualize the sand flow field without tracer particles. They can output the numerous sand velocity vectors at a moment in time. By continuously connecting sand flow fields obtained, we can surmise their spatio-temporal changes, though, each flow field is obtained as a spatio-temporally independent feature. Thus, we attempt to propose a PIV-based spatio-temporal analysis accumulating the sand flow vectors and thereby show a spatio-temporal change of the sand-density distributions beneath the traveling wheel, where the sand density is represented by the relative density. Although the method requires some assumptions, this paper estimates and shows the unknown spatio-temporal sand-density distributions of the sandy terrain beneath the traveling wheel by an acumulative approach.

Keywords
Sand-density distribution
Spatio-temporal analysis
PIV
Wheel-soil interaction

1. Introduction

Small, lightweight robots have been widely used for terramechanics applications on the ground and planetary applications. For such robots, the design, development, and operation of reliable mobility systems on loose sand are required. It is well-known that wheels with fin-like grousers, so-called grouser wheels, are one of the possible solutions for their locomotion gears. On the other hand, unknown interaction physics remains to be comprehended for the more appropriate use of the grouser wheel, such as dynamic soil behaviors mechanically interacting with the grouser wheel.

Thus far, the stress distributions acting on the grouser wheel traveling on the loose sand have been analyzed based on direct measurement (Higa et al., 2019). As another major approach, visual observation of sand flow during wheel traveling has been addressed using particle image velocimetry (PIV), which can two-dimensionally measure the velocity of sand particles without contact. To date, PIV-based analyses have shown the displacement field of soil beneath a traveling wheel (Senatore et al., 2013; Skonieczny et al., 2014; Yamakawa and Goto, 2015). Also, PIV-based soil flow results have been compared with the finite element method (FEM) (Moreland et al., 2012; Saengprachatanarug and Okayasu, 2013). Recently, the PIV-based soil flow field beneath a grouser wheel has also been analyzed (Nakamura et al., 2020). Although these related works can mainly show a spatial wheel-soil interaction, a dynamic analysis of spatio-temporal sand behaviors has not been addressed.

In this paper, we propose a spatio-temporal analysis of sand-density distribution beneath a traveling wheel based on the PIV method. By continuously connecting sand flow fields obtained, we can surmise their spatio-temporal changes, though, each flow field is obtained as a spatio-temporally independent feature. Thus, we attempt to propose a PIV-based spatio-temporal analysis accumulating the sand flow vectors and thereby show a spatio-temporal change of the sand-density distributions beneath the traveling wheel, where the sand density is represented by the relative density.

2. PIV-based analysis approach for spatio-temporal sand-density distributions

We designed and manufactured wheels with different grouser lengths to discuss the effects of the wheel design on the spatio-temporal sand-density distributions. Focusing on the sand filling rate which represents the relative density of the sand, we analyze the relative change in the inter-pixel sand density in space and time. In the single-wheel traveling experiments, we used the Toyoura standard sand.

2.1. Single wheel apparatus

Fig. 1 shows the representative grouser wheel and the single-wheel test apparatus used in the subsequent experiments. As the wheel design parameters commonly used, the wheel's outer radius R was 200 mm and the number of grousers was 12, respectively. The wheel's inner radius r was 175, 180, 185, and 190 mm, and the grouser length l_G is expressed as $l_G = R - r$, where $l_G = 25, 20, 15, 10$ [mm], respectively. Throughout the experiments, the sand grains were illuminated with LED lights, and a digital camera capable of capturing 1920x1080 pixel images at 60 fps was used to acquire video of the cross-sectional sand particles' motion beneath

https://doi.org/10.56884/CHS1JPYP

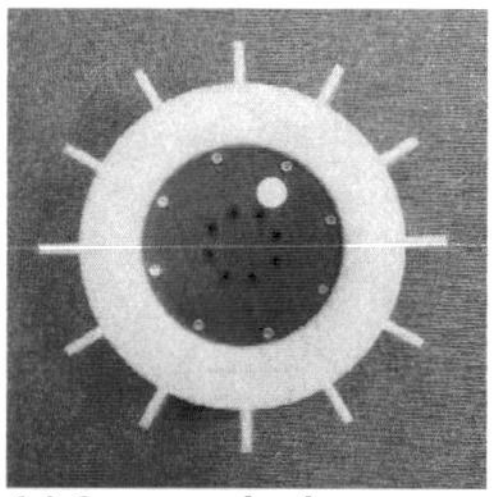

(a) Grouser wheel

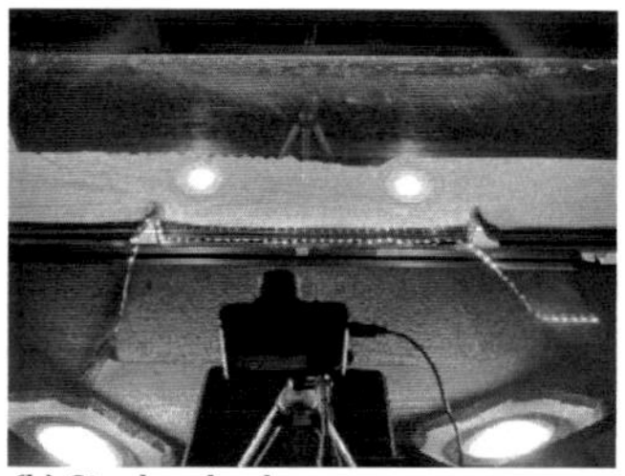

(b) Single-wheel apparatus

Fig. 1. PIV-based single-wheel apparatus

(a) Resulting PIV-based soil flow vectors

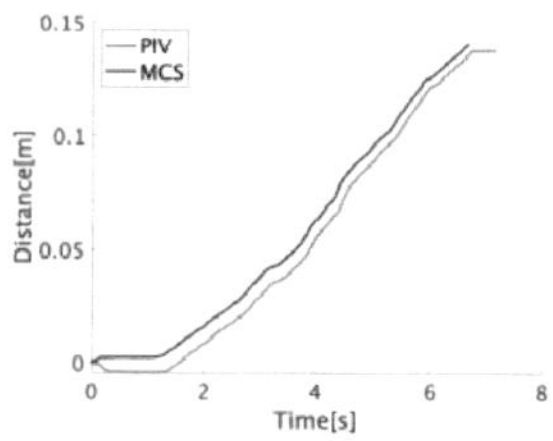

(b) Distance comparison between PIV and motion capture camera

Fig. 2. Camera calibration for PIV-based soil flow analysis

the traveling wheel for the PIV analysis, where the wheel's rotational velocity was controlled to be 5 rpm.

For the PIV method, the preliminary calibration was conducted, as shown in Fig. 2. Based on this, sufficient accuracy was achieved by aperture (F value) 8.0 in camera settings.

2.2. Image compression

In the experimental analysis, the periodic change of the sand density was difficult to visualize when all the pixels in the frame were analyzed individually due to the image resolution and processing accuracy. Hence, the total number of pixels in each video image was reduced to 1/9 of the original image by compressing as 3 × 3 pixels. Fig. 3 shows the image compression result. From the compressed image, we can observe the key features of relative, spatio-temporal sand-density change.

2.3. Spatio-temporal change in sand-density

We have analyzed three patterns, A, B, and C to represent the spatio-temporal sand-density distributions. All patterns will be compared in terms of the following conditions.

- (i) Initial sand-density value
- (ii) Sand-density increase/decrease value due to sand flow
- (iii) Process when the sand-density falls below 0
- (iv) Process when the sand is extremely density or sparse

2.3.1. Analysis pattern A

The analysis pattern A gives the following conditions.

- (i) Set all pixels to 1, which represents the sand in a pixel is considered to be one grain.
- (ii) When moving sand, the source pixel is set to −1 and the destination pixel to +1. This is set to figure out the relative density.
- (iii) If the value of a pixel is 0, it is assumed that there is an inflow of sand from the surrounding area, thus it was to set 0 → 1. With this setting, the value will never be negative and the minimum value will be 1.
- (iv) This condition is unprocessed.

(a) 1x1 pixel

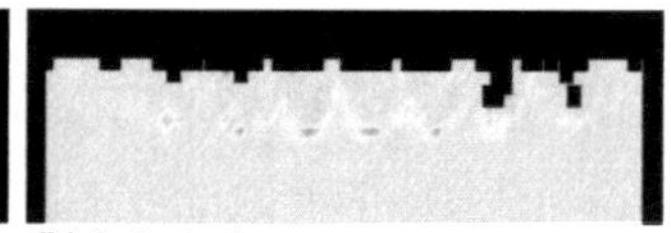

(b) 3x3 pixel

Fig. 3. Image compressed for sand-density analysis

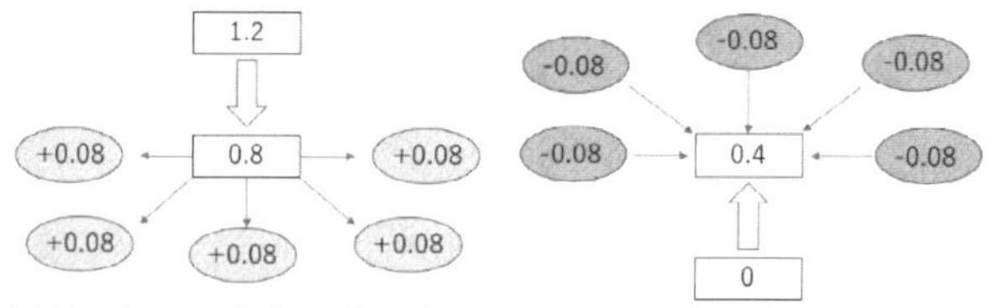

(a) Fundamental idea of sand-density distribution

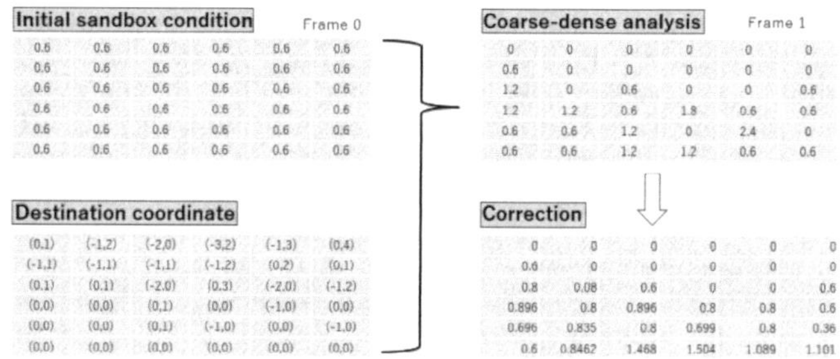

(b) Exampled sand-density change in pixels

Fig. 4. Procedure for spatio-temporal change in sand density: Pattern C

2.3.2. Analysis pattern B

- (i) Set all pixels to 1. (same as Pattern A)
- (ii) When moving sand, the source pixel is added to the destination pixel and then it is set to 0. This is because the density of the source vector is expected to decrease to zero as the sand moves.
- (iii) If the source value of the move vector is 0, the source pixel is set to −1 and the destination pixel is set to +1.
- (iv) This condition is unprocessed.

2.3.3. Analysis pattern C

- (i) Set all pixels to 0.6, which represents relative density and is expected to be reasonably evaluated and discussed.
- (ii) The source pixel of the move vector is set to 0 and it is added to the destination pixel. (same as Pattern B)
- (iii) When moving sand, the source pixel is added to the destination pixel and then it is set to 0. The density of the source pixel is expected to decrease to 0 due to the sand flow.
- (iv) The sand-density analysis as a relative density becomes an extremely dense or sparse state of the sand. Hence, as reference of the minimum and maximum relative density are assumed to be set to 0.4 and 0.8, respectively. When the density value becomes out of this range, its exceeded amount is distributed to the surrounding pixels, as shown in Fig. 4, where this distribution shall be conducted only once for simplicity of computation.

3. Results of spatio-temporal sand-density analysis

In this section, we commonly used the grouser wheel which has l_G = 25 [mm] to compare Patterns A, B, and C.

3.1. Pattern A

Fig. 5 shows the results of the spatio-temporal sand-density analysis using Pattern A. From this results, sand flow was observed toward the front and downward during grouser penetration. This increases the inter-grouser sand-density and provides traction force by pushing the grousers against the sand. In addition, the sand that entered between the grouser moved backward when the grouser wheel traveled. During this process, dense areas were generated in external regions of the wheel's outer radius. This is expected to curb wheel sinkage. Moreover, when the grouser came off contact with the sand, the dense state was generated outside the wheel's outer radius at the same time as the sand was pulled up by the grouser. From Fig. 5(b), the coarse and dense conditions after the grouser wheel traveled are related to the wheel rut on the sand surface. The deeper areas in which the wheel ruts were generated were densely packed in the sand. Periodicity can be seen in the sand similar to the wheel rut.

On the other hand, the coarse areas could not be identified, which is due to the analysis condition that the sand inflow is considered and the pixel value is set $0 \rightarrow 1$ when the pixel value reaches 0. Furthermore, the sand-density change between pixels is just 1, which denotes that even if a pixel in a dense state moves, it remains in a dense state and does not appear to have changed. This is another reason why this analysis method lacks validity. The total value in the frame differs by about 3,000 between the minimum and maximum values, This is considered reasonable when considering the inflow of sand from the surrounding areas, but lacks validity from the viewpoint of a cross-sectional sand-density analysis. For these reasons, Pattern A is not necessarily reasonable, however, this can visualize the areas where the sand flows reasonably provide dense areas.

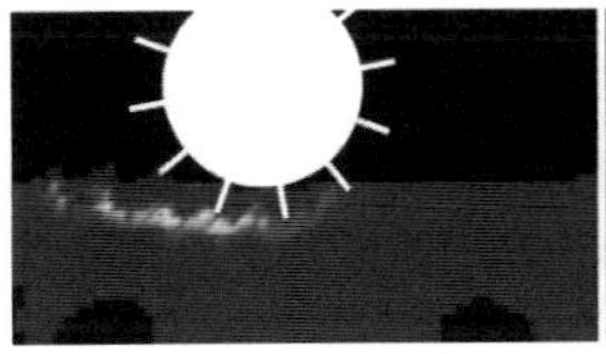

(a) Frame 112

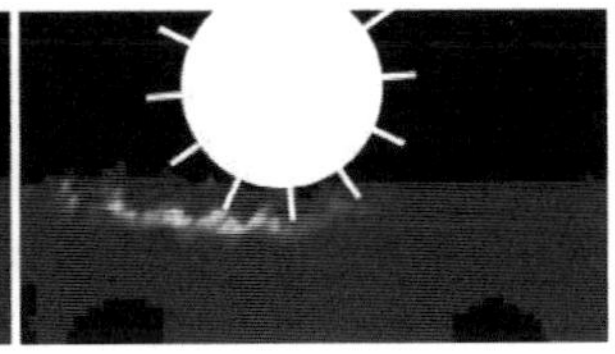

(b) Frame 117

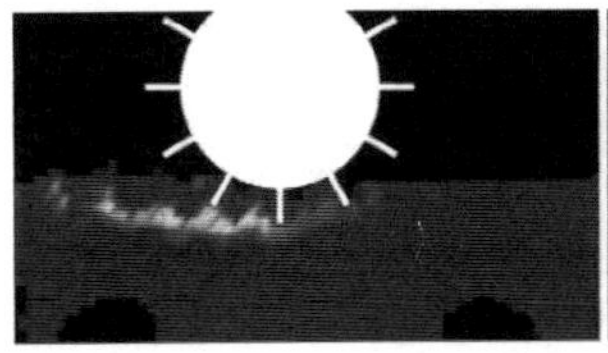

(c) Frame 122

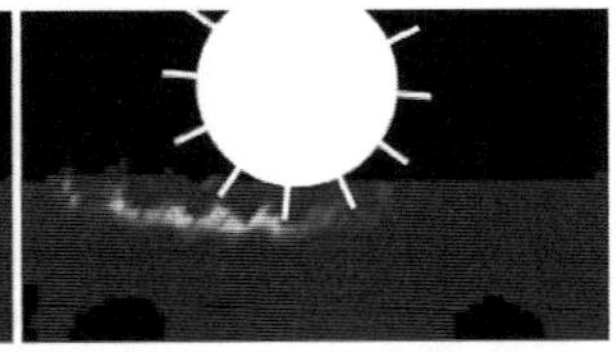

(d) Frame 127

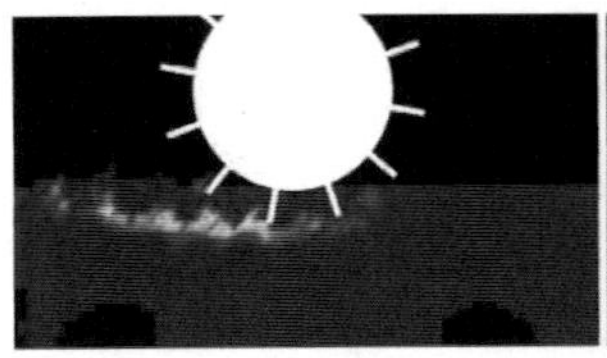

(e) Frame 132

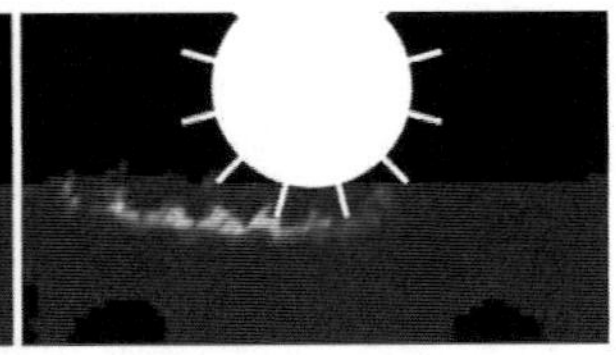

(f) Frame 137

(a) Frame-by-frame sand-density distribution change

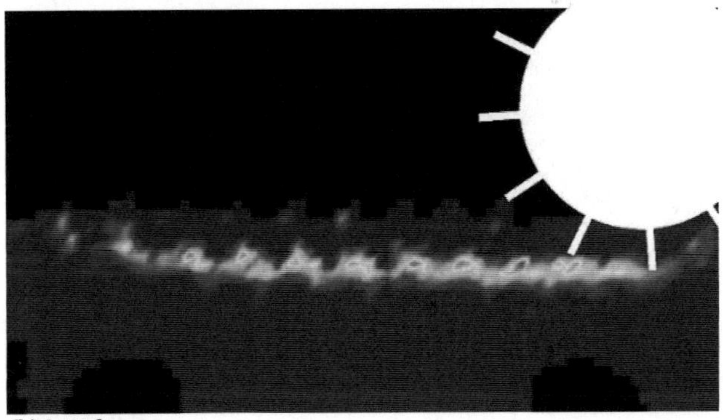

(b) Last frame

Fig. 5. Result of spatio-temporal sand-density analysis: Pattern A

3.2. Pattern B

Fig. 6 shows the results of the spatio-temporal sand-density analysis using Pattern B. This result shows that the area around the grouser became coarse at the time of grouser penetration. Compared to Pattern A, this is a result of avoiding the compensation of the coarse areas by changing the pixel values to −1 for the vector source and +1 for the vector destination when the pixel values below 0 became the movement vector source, instead of replacing them with 1 where the pixel values were 0. To focus on following the behavior of the sand, all pixel values from the source of the move vector were added to the destination of the move vector. This allows one to show a coarse and dense analysis that focuses on the spatio-temporal motion of the sand. In addition, the sand-density distribution when the grouser came off contact with the sand was different from that of Pattern A. The inter-grouser areas were coarse, whereas the sand flow outward from the wheel's outer radius and was not coarse. The wheel ruts were also not corse, indicating that the sand behavior was well captured.

However, even in Pattern B, there was a difference of approximately 3,000 between the minimum and maximum total values in the analyzed frames. The maximum value in the first frame was 1,654 and the minimum value in the last frame was −1,383, indicating a decrease in the total number. This comes from the total number of frames decreased when the grouser moved above the sand surface, which is out of the analysis area. Although the sand inevitably moves above the analysis area, it is concluded that such a change in the total dense value of each frame lacks validity for the sand-dense analysis. The calculated density values represent the ones relative to the reference value and therefore do not always correlate with actual density values, such as bulk density or relative density.

3.3. Pattern C

Fig. 7 shows the results of the spatio-temporal sand-density analysis using Pattern C. As the initial condition, all pixels corresponding to the sand were set to 0.6, which is the average filling ratio of the Toyoura standard sand used in the experiment. Compared to the other patterns, the coarse and dense areas were separated, with the density forming outside the wheel's outer radius. During grouser penetration, there was a wide dense area in the sand. This is due to the analysis condition shown in Fig. 4(b). Since the difference in the sand stress is regarded as proportional to the sand-density, it is reasonable that dense areas increase when the the grouser penetrates. The sand-density also recovered around the wheel's outer radius when the grouser came off contact with the sand, as in Pattern B. Unlike the other patterns, this analysis did not show a large difference in the total density, which

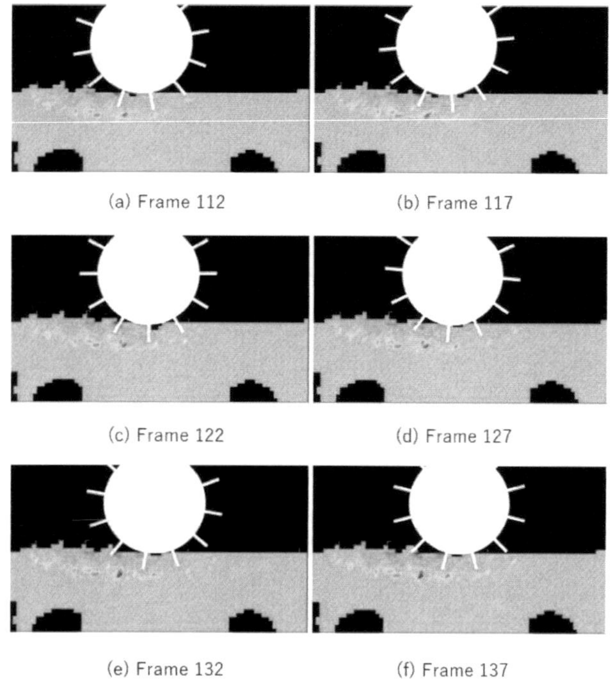

(a) Frame-by-frame sand-density distribution change

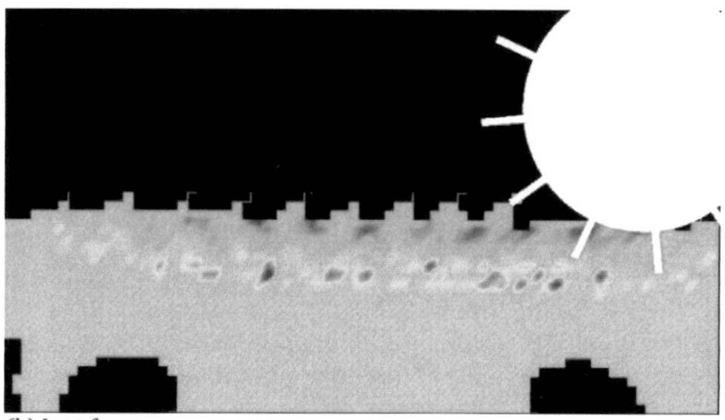

(b) Last frame

Fig. 6. Result of spatio-temporal sand-density analysis: Pattern B

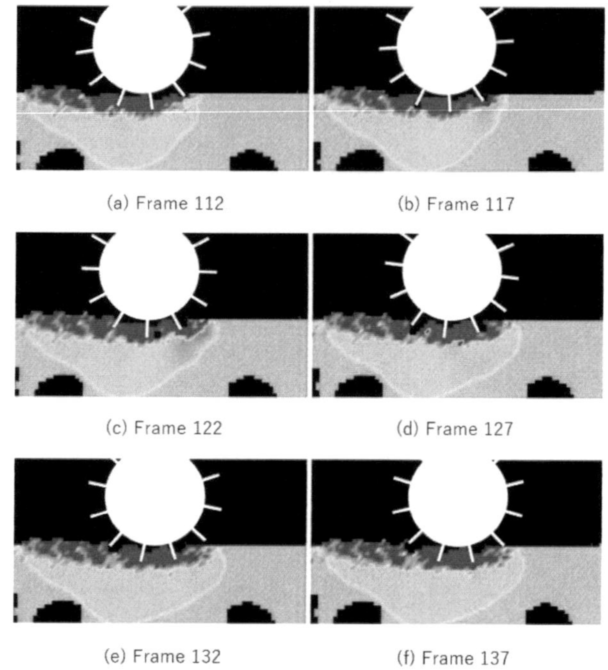

(a) Frame-by-frame sand-density distribution change

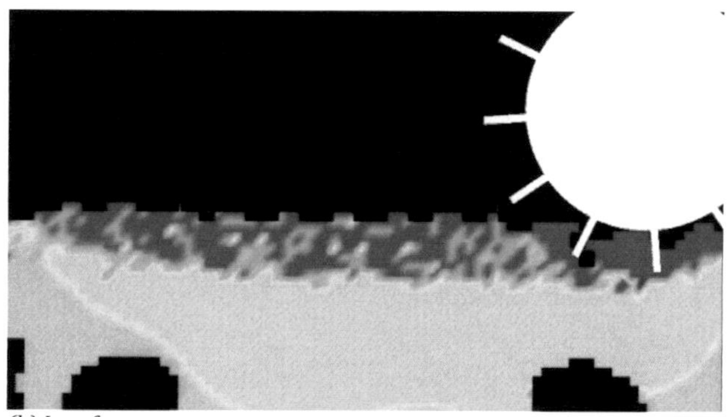

(b) Last frame

Fig. 7. Result of spatio-temporal sand-density analysis: Pattern C

was just about 100. This is also due to the analysis condition of (iv) and results in the total value being balanced.

One point where Pattern C lacks validity is that the decentralization and concentration result in the spreading out of the dense areas. This is thought to be due to the fact that Pattern C uses sand-density analysis as the relative density, and thereby concentration below 0.4 and decentralization above 0.8 lead to continued propagation.

Based on the analysis results, we can conclude that Pattern C, which did not differ in total density, shows the most reasonable results.

4. Discussions

We discuss the effects of the grouser length and the wheel slip on the sand-density distribution analysis, using Pattern C. Fig. 8 shows the grouser wheels. Here, F is the traction load to simulate different slip conditions. In this study, experiments were conducted with $F = 0, 22{,}1$ [N], where $F = 0$ [N] represents a lower slip condition and $F = 22.1$ [N] does a larger one. Here, as representative results, Fig. 9 shows the analysis with $l_G = 10$ [mm] and Fig. 10 does the one with $l_G = 25$ [mm], respectively.

4.1. Comparison by grouser length

It can be seen that the shortest grouser $l_G = 10$ [mm] at the low slip condition generated the dense area in front of the wheel's outer radius. On the other hand, for the longest grouser $l_G = 25$ [mm], the dense area was generated separately in front of and underneath the wheel's outer radius. It can also be seen that for $l_G = 25$ [mm], the sand-density behind the wheel's outer radius were smaller than that in front of the one. The reason for this is that the amount of sand flow changes depending on the grouser length, with the wheel with $l_G = 10$ [mm] that cannot flow a large amount of sand easily compacting sand in the front. Further, the wheel with $l_G = 25$ [mm] generated a dense state directly below the lowest grouser as it began to travel, forming a dense state in the divided area. In addition to this, at $l_G = 25$ [mm], the dense state formed underneath the wheel's outer radius is understood to be divided into forward and backward dense states, as the sand failure caused by the long grouser causes the dense state to flow backward behind the wheel.

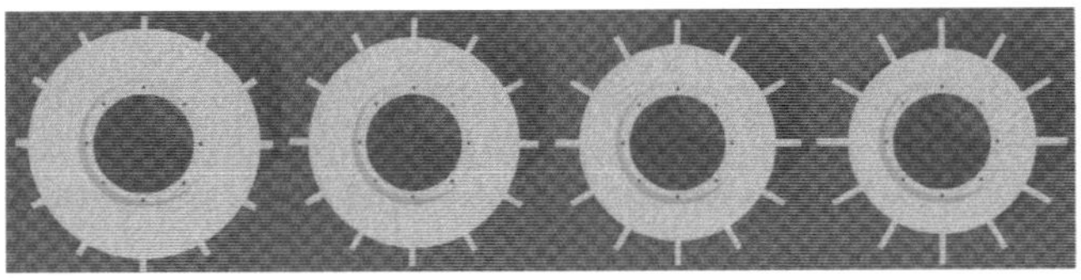

Fig. 8. Various grouser wheels

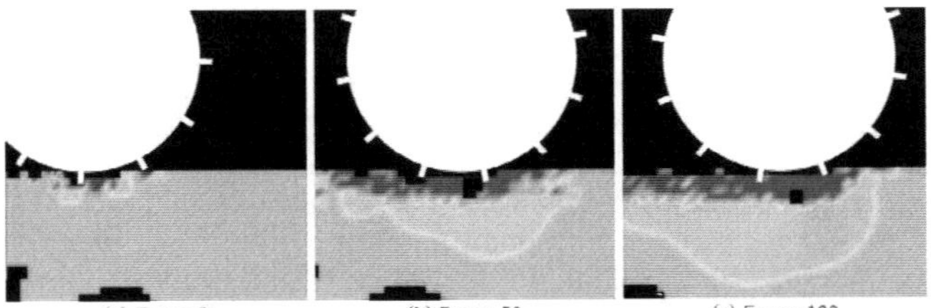

(a) F = 0 [N]: Smaller slip condition (the average slip ratio was 0.17)

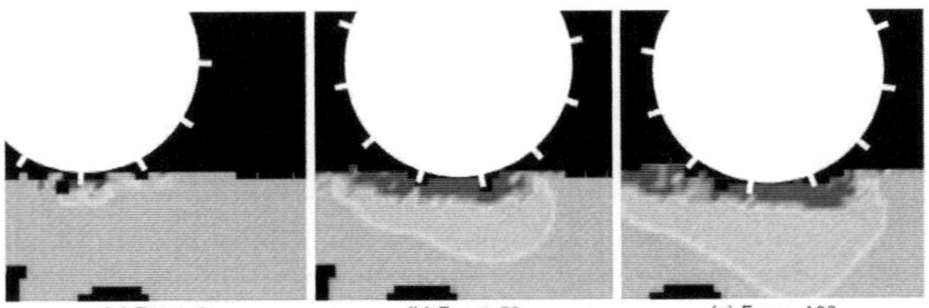

(b) F = 22.1 [N]: Larger slip condition (the average slip ratio was 0.70)

Fig. 9. Analysis of soil density distribution at l_G =10 [mm]

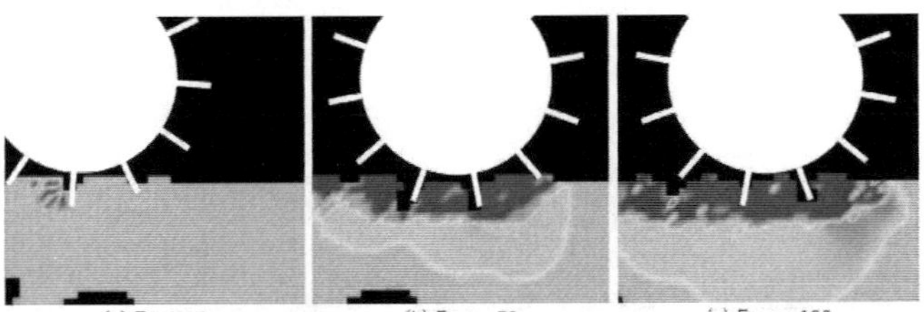

(a) F = 0 [N]: Smaller slip condition (the average slip ratio was 0.09)

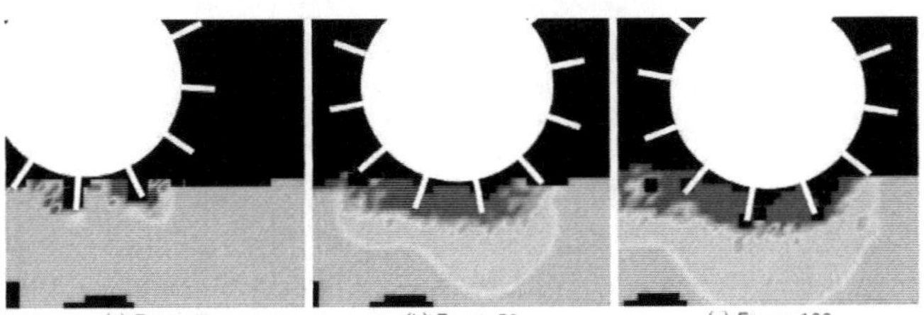

(b) F = 22.1 [N]: Larger slip condition (the average slip ratio was 0.68)

Fig. 10. Analysis of soil-density distribution at l_G =25 [mm]

4.2. Comparison by wheel slip

As the wheel slip increased, the depth of the final dense state also tended to increase. This can be interpreted as a result of the increased wheel sinkage with increasing slip, forming a dense area in deeper regions. The increase in slip was independent of the grouser length and continued to scratch the area of the coarse state. In the case of the lower slip, this was the result of a dense condition forming in front of the wheel's outer radius, which compresses the sand and allows the wheel to move forward without slipping. On the other hand, in the larger slip, the dense area in front of the wheel was relatively smaller, suggesting that the coarse area in front of the wheel was scraped backward, resulting in increased sinkage, leading to a larger slip that makes it difficult to move forward. Furthermore, in the lower slip, a dense area was formed on the lower side of the wheel's outer radius from the beginning of wheel travel, whereas in the larger slip, a dense area was not formed on the lower side, regardless of the grouser length. This result indicates that the grouser has a greater effect on the backward flow of sand in larger slip conditions. This may be because the application of a traction load causes the wheels to slip at the start of wheel travel, creating a large dense area in front of the wheels, which results in less sand being scraped by the grouser. Alternatively, it comes from that the coarse sand is scraped and moved backward, resulting in in-situ sinkage.

5. Conclusion

In this paper, we proposed a method for analyzing the spatio-temporal sand-density distributions beneath the traveling wheel by applying experimental data from a sand flow field based on PIV. We compared and evaluated three analysis patterns and the most reasonable analysis method was shown. The analysis results confirmed the effects of the grouser lengths and the wheel slip on the spatio-temporal distributions and then allowed one to discuss the wheel-soil interaction. The results obtained in this paper provide a new approach for PIV-based sand flow analysis for wheeled mobility and extend the conventional instantaneous spatial PIV data to a spatio-temporal analysis as sand-density distributions. As a key future work, to ensure the reliability of this method, a validation evaluation of the calculated density change is required. Furthermore, it is expected that a more reasonable sand flow pattern will be established and experimentally verified.

References

Higa, S., Nagaoka, K., Yoshida, K., 2019. Stress distributions of a grouser wheel on loose soil. J. Terramech. 85, 15–26.

Moreland, S., Skonieczny, K., Inotsume, H., and Wettergreen, D. 2012. Soil behavior of wheels with grousers for planetary rovers. Proc. 2012 IEEE Aerospace Conf., doi:10.1109/AERO.2012.6187040.

Nakamura, H., Nagaoka, K., Yoshida, K., 2020. Soil flow analysis beneath a grouser wheel based on a particle image velocimetry method. J. Terramech. 91, 233–241.

Saengprachatanarug, K., and Okayasu, T. 2013. Modeling of soil displacement and soil strain distribution under a traveling wheel. J. Terramech. 50(1), 5–16.

Senatore, C., Wulfmeier, M., Vlahnic, I., Andrade, J., Iagnemma, K., 2013. Design and implementation of a particle image velocimetry method for analysis of running gear-soil interaction. J. Terramech. 50 (5-6), 311–326.

Skonieczny, K., Moreland, S.J., Asnani, V.M., Creager, C.M., Inotsume, H., Wettergreen, D.S., 2014. Visualizing and analyzing machine-soil interactions using computer vision. J. Field Robot. 31(5), 820–836.

Yamakawa, J., Goto, M., 2015. Experiment and numerical analysis of driving wheel on soft sandy soil. Proc. 13th ISTVS European Regional Conf., 500–508.

Yoshida Y., Sakada, H., Yuasa, S., Nagaoka, K., 2023. A photoelastic method for stress analysis of granular terrain beneath a wheel, J. Trramech. 110, 69–78.

1902 | The combination of exhaust gas recirculation and water injection in a modern diesel engine

Daniel Szőllősi [a, b] *, Peter Kiss [c]

[a] *Doctoral School of Mechanical Engineering, MATE, Gödöllő, Hungary*

[b] *IBIDEN Hungary Technical Center, Exhaust system evaluation, IBIDEN Hungary Ltd., Dunavarsány, Hungary*

[c] *Department of Vehicle Technology, Institute of Technology, MATE, Gödöllő, Hungary*

* *Corresponding author: szollosi.daniel1@phd.uni-mate.hu*

ABSTRACT

Diesel engines are commonly found in off-road, agricultural and forestry equipment because of giving the best option in terms of weight, range and fuel storage. Given the current potential of electric powertrains, in this area, electrification does not seem to be competitive in the near future.

Regulations are tightening for on road (EURO7) and non-road vehicles (Tier 4, Stage V) in many parts of the world and tend increasingly focusing to ensure as many as possible of the vehicles with low or no emission standard, can be retrofitted to reach decreasing emissions. The overwhelming majority of vehicles fall into this category, around the world. Retrofitting should be as simple and cost-effective as possible. One such option could be the retrofitting with water injection systems.

The exhaust gas recirculation (EGR) achieves the lower temperature required for reduced nitrogen-oxide (NO_x) emissions by reducing the oxygen content and thus impairing combustion. Water injection achieves this essentially by enriching the charge air and thus increasing its heat capacity and by the heat dissipation effect of the liquid-vapor phase change.

Contrary to common belief, the two techniques are not only able to reduce NO_x separately, but also together they cause further emission reductions on NO_x and on other emission components. This has been shown through the use of practical applications, where we can see that the EGR working in low to medium load and the water injection working optimally in medium to high load conditions. As a result, the emission reductions from the two techniques add up.

This is also an important achievement, as it means, with the easily applicable engine intake water injection, we can reach further emission reduction without even modifying the engine control. However, the full potential can be reached with engine control modifications.

Keywords
NO_x
Water injection
Exhaust gas recirculation (EGR)
Retrofit emission control
Tier 4
Stage IV

1. Introduction

The main arena for human and terrestrial life is the hemisphere, which extends up to an altitude of about 80 kilometers, where the gas composition is constant. Within this, our narrowest habitat is the troposphere, where strong vertical air currents cause the mixing of surface pollutants.

Although air pollution is nowadays attributed to anthropogenic pollutants, there are also natural sources of a significant proportion of air pollutants from human activities. Since 2008, the Air Quality Directive has taken this factor into account (European Environment Agency 2012.).

Anywhere that combustion (oxidation) occurs, it is the primary cause of air pollution from human and natural sources. Since the beginning of life on Earth, these processes can have been occurring in plant and animals in the atmosphere that contains oxygen. The emergence of man and the industrial revolution marked the beginning of a new source: the human activities. Among them, there are:

- industry and chemical factories,
- power plant production,
- agriculture,
- households,
- transport.

Polluting substances are released into the air of a very small area, and their environmentally harmless dilution only occur well outside the city limits. This is one of the main characteristics of man-made pollution sources, which are typically located in large cities or industrial estates. Transport can be further broken down into water, land and air elements. Figure 1. shows the typical pollutants in the air, from which the highest proportion can relate to transportation is the nitrogen-oxide (NO_x) emissions.

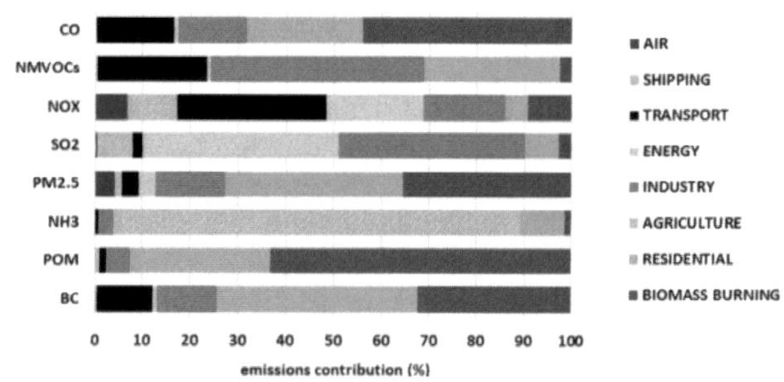

Fig. 1. Anthropogenic components of global emissions in 2010 (Lelieveld et al. 2015)

The sources of pollutants emitted by land transport are the following;

- - emissions from internal combustion engines,
- - dust from brakes and tires,
- - road abrasion and dusting of particulate matter by transport (Radnainé 2011)

The most visible and palpable kind of transportation is land transportation, and internal combustion engines are currently the most well-known propulsion technology. The most popular technology for heavy/medium duty, off-road applications is the diesel engine (Kalghatgi 2019; Ning et al. 2020.).

This research focuses on the potential concurrent operation of the exhaust gas recirculation and water injection. This is examined

https://doi.org/10.56884/XAK8652F

through the effects on the emission of nitrogen oxides as the most problematic component for the removal of pollutants. The motivation background of this paper is the retrofitting, which will be explained in the chapter 1.2.. As EGR is the most basic nitrogen oxide removal technology found in almost all vehicles with emission control systems, the impact of water injection should be considered first along with this technique. As this is about retrofitting, we also need to keep practicality in mind and this is the reason why intake water injection was chosen for the experiment. These will be explained in detail in the next chapters followed by the experimental layouts and results.

1.1. Nitrogen oxide (in the diesel engine)

Modern diesel cars have become increasingly popular in Europe and around the world in recent decades due to their outstanding efficiency, low fuel consumption, increasingly refined operation and engine noise. Behind these achievements have been significant technological advances such as common rail, turbocharging, cooled Exhaust Gas Recirculation (EGR) and continuously refining engine management systems. As a result, their low-rev torque also makes them easy to drive.

During diesel engine operation, fuel is injected into the combustion chamber, where it is atomized and vaporized in the compressed high temperature air. It forms a locally rich mixture and premixed combustion occurs after the ignition delay. Combustion is then mixing controlled until all the injected fuel is burnt, involving continuous atomization, vaporization, fuel vapor-air mixing and combustion. During this inhomogeneous combustion, temperatures can reach as high as 1600 °C and above. In such circumstances and in an overall lean air fuel ratio of higher than 25:1, nitrogen starts to react with oxygen. We can describe the formation of NO_x with 3 main mechanisms, each can bound to special circumstances. In the premixed flames, the temperature is relatively low at the start of the burning, which is not favorable for the "prompt NO" to take place (Fenimore 1971). During this low temperature combustion, the dominant reaction is the "N_2O intermediate" mechanism (Bowman 1992; Turns 2000):

$$O + N_2 \leftrightarrow N_2O \tag{1}$$
$$H + N_2O \leftrightarrow NO + NH \tag{2}$$
$$O + N_2O \leftrightarrow NO + NO \tag{3}$$

The third and most important mechanism is the thermal (or Zeldovich) mechanism. It occurs at temperatures even higher than 1300 °C in the typical mixing controlled burning (Rente 2004, Glarborg 2019):

$$N + O_2 \leftrightarrow NO + O \tag{4}$$
$$O + N_2 \leftrightarrow NO + N \tag{5}$$
$$N + OH \leftrightarrow NO + H \tag{6}$$

The NO and NO_2 are the two primary elements of NO_x emissions. Under typical circumstances, the NO, which predominates in NO_x inside the combustion chamber, at temperatures less than 200 °C, combines with oxygen in the air. Therefore, all nitrogen oxides produced during combustion can be considered as nitrogen dioxide (Paul et al. 2008).

NO_2 is acidic in the presence of moisture and therefore corrodes metals and building materials. It forms secondary particles, nitrates, which can form fog, impairing visibility. Nitrogen oxides also promote the formation of acid rain. They are also a precursor for the formation of ozone and thus play a significant role in the formation of smog (photochemical) (Sher 1998).

The mechanism of action of NO_2 on humans is twofold. On one hand, it binds to the airway mucosa and is converted into nitric acid, causing local tissue damage. On the other hand, it can be absorbed into the bloodstream, where it oxidizes the hemoglobin molecule into methemoglobin, which is unable to transport oxygen to the organs (Chiodi et al. 1983).

Because of the other main diesel emission component, the particulates can be effectively removed by the DOC (Diesel Oxidation Catalyst), DPF (Diesel Particulate Filter) (Ladommatos et al. 1998a, Rounce et al. 2012, Rossomando et al. 2021, Biró and Kiss 2023) and also by the water injection (Bedford et al. 2000, Tauzia et al. 2010) the main focus here is the NO_x emission.

1.2. Emission regulations

As has been acknowledged since the 1960s, that land transport is a key contribution to the rise in atmospheric pollutants that might cause significant climate change. This is the foundation of emission regulation, which, in reaction to health concerns, has imposed more stringent emission restrictions on new diesel engines around the world. Current emission levels require ongoing improvements and system design, and they account for a major portion of the cost of engine development. Approximately 50% reduction can be seen in nitrogen oxide emissions between each of the EURO standards for passenger cars (PC), light duty (LD), and heavy duty (HD) vehicles and also between the stages of non-road regulations. As a result, the NO_x limits in each category have been tightened by 95% or more since the regulations were introduced. From here, one of the significant ways forward is to tighten the measurement conditions, to create more severe test procedures and test cycles. The other option is to retrofit existing equipment to further reduce emissions.

As stated in the opening paragraphs of the European Commission's original 2022/0365 (COD) EURO 7 proposal (European commission 2022), *"Appropriate incentives could be put in place for older cars to be retrofitted to meet Euro 7 requirements for tailpipe emissions but also for tire and brake emissions."* In order to achieve this objective, simple and affordable ways should be put force in order to prolong vehicle life, lower emissions, and perhaps improve fuel economy. Water injection is a potential fit for each of these specifications.

However, in the non-road sector, both regulation and framework are much more incomplete. The estimated volume of these vehicles and equipment is much larger and the possibility of quantifying and regulating them is struggling with difficulties. In some more developed regions, various inventories are being tried or have been set up to estimate this. These models predict the emissions across a selected area and depend on several parameters such as fuel mix, geographic features, exhaust emission data, and others (Hagan et al. 2023). With such emission inventories available in several European nations, we may compare each country's non-road mobile machinery (NRMM) and road emissions. This indicates that the NRMM's emissions are 1.5–3.5 times greater when it comes to NO_x (Lončarević et al. 2022). In less technologically developed areas this ratio can increase significantly. We can see from that the possible high potential of retrofitting.

Many of these machines are industrial or shipping, mining machines with much smaller depreciation compared to passenger cars for example, in which case retrofitting also has a much better return on investment.

It can be concluded that much more emission reductions can be achieved through retrofitting than through further tightening of emission standards.

1.3. NO_x reduction possibilities

By the end of the 1960s, efforts to reduce emissions had already begun. Besides burning optimization with in-engine techniques, another option for reducing combustion temperatures is the use of exhaust gas recirculation (EGR), which started to be used in the early 1970s. In addition to oxidizing diesel pollutants such as carbon monoxide (CO), hydrocarbons (HC) and diesel soluble organic fraction (SOF) into harmless combustion products (Al-Aqtash et al. 2023), the DOC plays an important role in the treatment of nitrogen oxides in the exhaust system. The oxidation of NO to NO_2 is essential for the operation of modern diesel emission control systems (Wang et al. 2015). Here, DOC also functions as an auxiliary catalyst that supports the performance of other types of catalysts - placed downstream of the oxidation catalyst - that require high NO_2/NO ratios (Boutikos et al. 2019). Nitrogen dioxide is needed to enhance the performance of different types of Selective Catalytic Reductor (SCR) catalysts and to promote passive regeneration of DPFs. DOCs used in DPF/SCR applications are generally optimized for high NO_2 production. If there are equal or excess amounts of reducing substances in the gas compared to oxidizing substances, NO_x can be catalytically reduced with the non-oxidized components (CO, CH). This is the non-selective catalytic reduction. On the other hand, if there is more oxygen in the gas than necessary for the complete oxidation of the reducing components present – like in case of diesel engine operating with excess air -, the reduction of nitrogen oxides by the latter can only be achieved on a selective catalyst. The reduction process uses ammonia, which is produced by thermal decomposition and injected into the hot exhaust gases at the point of use. For this purpose, a 32,5 % mixture of urea ($[NH_2]2CO$) in demineralized water is used. Equation 7-9 shows the reaction mechanism:

$$4NO + 4NH_3 + O_2 \rightarrow 4N_2 + 6H_2O \qquad (7)$$
$$2NO + 2NO_2 + 4NH_3 \rightarrow 4N_2 + 6H_2O \qquad (8)$$
$$6NO_2 + 8NH_3 \rightarrow 7N_2 + 12H_2O \qquad (9)$$

With the use of SCR, 75-90% of the NO_x can be removed (Zhang et al. 2018). Less commonly used and mature technique is the Lean NO_x catalyst (LNC) where the hydrocarbons existing in the exhaust fumes help the reduction of the NO_x on the catalyst surface. Problems associated with this technology impede its commercial adoption. During lean engine running, nitrogen oxides are adsorbed by barium oxide, which is utilized in the lean NO_x trap (LNT). During cold start, a close-coupled, actively regenerated LNT is frequently utilized (Theiss and Lambert 2015). As exhaust temperatures rise, urea is employed to decrease NO_x over the SCR catalyst.

Another possible candidate is water injection, the background and advantages of which are described in the next chapter.

1.4. Water

Emission control systems of the internal combustion engines can be categorized into the following typical parts:

- Pre-engine processes (different fuel(s), additives);
- In-engine processes (EGR, optimized engine control);
- Post-engine processes (catalysts, filters, after-treatment systems) (Naik and Dharmadhikari 2023).

Water injection has typically been used to improve performance, and its practical applications can be seen. Although emission reduction effects have been discovered and used in marine applications since the 1990s (Turns 2000, Imahashi et al. 1995), it has never received much attention. This is largely due to the fact that there have always been techniques that have performed better in removing a single harmful emission component. The fact, that it is virtually the only significant technique that can remove particulate matter and nitrogen oxides in parallel during diesel combustion, has been overshadowed by this.

A study was conducted using a wide range of humidity conditions (Szőllősi and Kiss 2023) on a EURO V but no EGR type engine, to clearly see only the water impacts. With relatively low amount of water as humidity, we already can detect a decrease in NO_x emissions (Fig. 2).

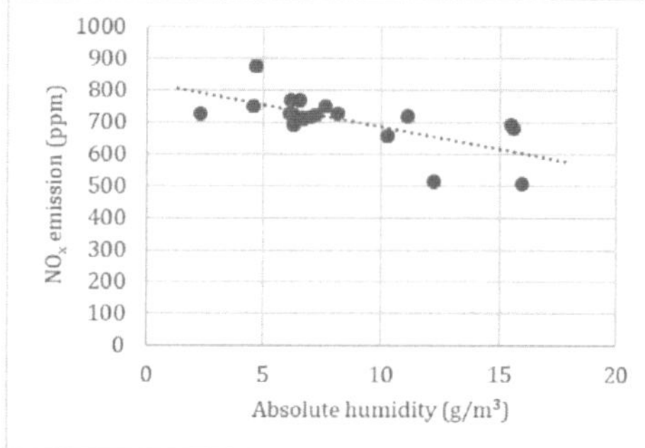

Fig. 2. NO_x emission on different Absolute humidity values (Szőllősi and Kiss 2023)

The injected water absorbs heat from the surrounding heated compressed air, evaporate, and mixes with the surrounding high-temperature high-pressure air. The primary mechanisms for NO_x reduction are the physical effects of water:

- dilution effect
- specific heat (or thermal effect)
- phase change (or charge cooling effect)

The diluting effect of water is proportional to the larger mass of the air/fuel mixture. It lowers the concentration of fuel in the cylinder charge as well as the concentration of oxygen, reducing the energy released during combustion (Prasad et al. 2015, Sun et al. 2022, Quader 1971). Another significant physical mechanism is that water has a larger specific heat capacity (2,03 J/ (g K) in liquid form and 4,18 J/ (g K) in gas form than air (1,01 J/ (g K) or fuel (2,09 J/ (g K)). The combustion temperatures are significantly lowered by the larger mass of air/fuel mixture with higher specific heat (Andrews et al. 1988, Ishida and Chan 1994, Wan et al. 2021, Prasad et al. 2015). Water vaporization has an enthalpy change of (2256 J/g) approximately 8 times larger than diesel fuel (277 J/g). Chemical effects are regularly reported in the relevant literature, however, it has no significant effect on diesel combustion in case of water injection (Neshat et al. 2019). Injection generally reduces NO_x emissions by 10% for every 20% of water to fuel ratio (Holtbecker and Geist 1998, Odaka et al. 1991). Water injection shows increased NO_x reductions at higher loads of the engine (Neshat et al. 2019, Hountalas et al. 2006, Sun et al. 2017).

The following methods are used to introduce water and its various emulsions into the engine combustion chamber:

- Intake water injection

1. Water, methanol, ethanol ...
2. Steam
3. Emulsions

- Direct injection
  1. Separate injectors
     - Water, methanol, ethanol ...
     - Emulsions
  2. Engine own injectors
     - Emulsions

Since the different designs strongly limit the types of water injections that can be used for retrofitting, at first it was evaluated depending on this. In terms of retrofitting, direct injection with additional injectors is not economical, also, due to the difficulty of implementation, it is not the domain of the subsequent conversion, but of the vehicle manufacturers. If the emulsions are to be used without engine modifications (e.g. to replace normal fuel in existing engines and use common injectors), the maximum water volume and also the NO_x reduction will be limited to about 10-20% and the engine will not be able to reach its rated power (Holtbecker and Geist 1998). It is possible to install larger injectors or to improve the fuel system to achieve further emission reduction performance, but this would require complex and significant modifications that isn't either economical. In the case of emulsions, we should also count with the instability of these mixes, and the infrastructure of the components of emulsions is not always self-evident. In case of intake port injection's or fumigation's clear advantages can be one of the cheapest investments, can be adapted and installed to any diesel engine with relative ease. However, it has the disadvantage of providing little control over injection parameters, such as timing or spatial coordinates. For this reason, the observed NO_x reductions are generally lower than for a properly built emulsion or direct injection system (Naik and Dharmadhikari 2023). Intake injection has the advantage of being a robust system with little potential for failure. It allows the system to be switched off at any time, and the engine can run even after the water has run out without a positive emission effect. Additional water tank construction is required, but the distilled water required is available at all service stations without the need for additional infrastructure.

1.5. EGR

The use of the Exhaust Gas Recirculation starts with its first applications in the 1970s. It is operating in a way that defined proportion of the exhaust gases circulated back to the engine cylinder depending on engine load, to reduce the NO_x emissions. NO_x is produced in the high-temperature diffusion flame area at the interface between the fuel-rich plume and the surrounding fresh charge. NO_x generation is an exponential function of flame temperature.

The EGR's temperature-lowering physical process, - the air/fuel dilution effect - is similar to the water injection process. By reducing the oxygen concentration, this is the strongest effect of EGR in reducing NO_x emissions. Some literature separates the thermal throttling effect from the dilution effect. This effect applies hot EGR, raises the temperature of the inlet charge, replacing comparatively dense cold air with less dense hot EGR, which lowers the mass flow rate of the input charge. It causes volumetric efficiency changes also (Ladommatos et al. 1998a).

In the combustion the O and NO radicals are responsible for the NO_x formation. With the recirculation of the exhaust gases, the overall composition of the intake changes, water and CO_2 appear. Water already existing as humidity in the burning process, but the CO_2 is a new participant. At high temperatures that occur during combustion, both of these species are capable of dissociating, and the byproducts of this process may take part in the combustion process (Wang et al. 2013). This is the chemical effect of the EGR. The direct reaction effect and the indirect catalytic reaction effect are the two categories, into which the chemical effects of water on combustion can be separated. Competition exists between the two reaction paths. The water's "third body indirect effects" depending on its 10 times larger molecule compared to the N_2. As the collision efficiency increases, the three body reactions also increase. These reactions have negative effects on the direct process (Eq. 10.), where the H_2O directly increases the rate at which OH radicals are produced (Fig. 3.).

$$H_2O + O \leftrightarrow OH + OH \tag{10}$$

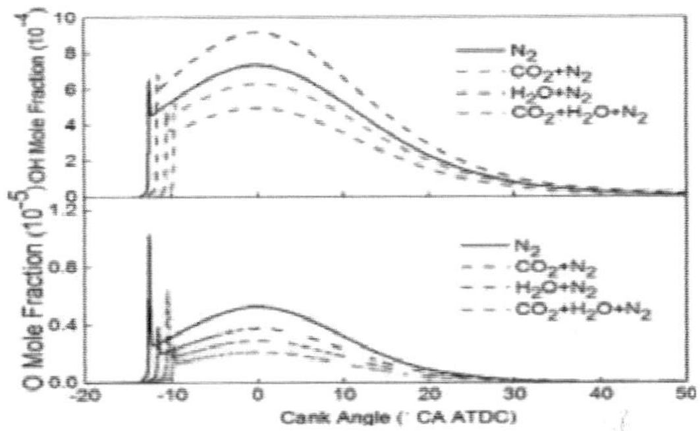

Fig. 3. H_2O and CO_2 effect on radicals formation (Zhang et al. 2021)

The increasing in the three body reactions suppress this reaction with such an alternate reaction way, that finally stops the production of the radicals (Meng et al. 2016). As can be seen from Fig. 3. the other exhaust component from recirculation, the CO_2 inhibits the O and OH radical production also. With a maximum of 30-40 % EGR rate (Brady 2013) this can lead to a significant reduction in the generation of the radicals responsible for NO_x.

The inlet temperature effect of EGR apart from the thermal throttling, also can have negative effect on the emissions, with increasing the cylinder temperatures. Although it is lesser in case of cooled EGR, with higher EGR rates it is still in effect (Ladommatos et al. 1998b).

In a study by Ladommatos, the thermal effect and the thermal throttling virtually has no effect on the NO_x emissions. The inlet temperature causes substantial increase of NO_x with changing temperature from 40 °C to 120 °C and it is responsible for the 40% of the emitted sum. NO_x at the end of the exhaust system. Dilution effect causes 75% decrease in NO_x emissions while the chemical effect is responsible for approximately 4.5% decrease. (Ladommatos et al. 2000).

There are two types of exhaust gas recirculation based on the physical design of the exhaust gas recirculation system: high pressure or low pressure ("short route" or "long route" based on other sources) (Reifarth and Ångström 2009). There is also a so-called in-engine type, which uses valve timing to return exhaust gases to the combustion chamber, but this is not discussed here. In case of the high-pressure system, the exhaust gas is tapped before the turbine side of the turbocharger and is returned after the compressor on the intake side. Its advantage is the high temperature of the exhaust gas recirculated, which is a significant advantage during the warm-up phase of the engine. By using the EGR cooler and a valve to control it, the cooled air is less damaging to the engine's efficiency, but can keep the positive effect during the warm-up (Gumus and Otkur 2023). The disadvantage of the inherently high operating temperature and the fact, that unpurified exhaust gas is recirculated, which fouls the EGR passages and the radiator. The exhaust for the low-pressure system is routed back to the compressor wheel after being tapped after the turbocharger's turbine wheel. In most situations, the drain is located downstream the turbine and the emission control system also, and it returns inert gas (free of CO and HC) to the compressor wheel. Once the operational temperature is reached, the low-pressure system is the better option since low pressure exhaust gas is colder than high pressure, thoroughly mixed, and filtered (Lapuerta et al. 2018). Therefore, in modern systems, the two types are used side by side, with the high-pressure system operating during engine warm-up and the low-pressure system or even selective catalytic reduction (SCR) afterwards.

However, at high engine loads, the EGR causes a deterioration in engine performance and efficiency, mainly because of low air fuel ratio and pumping losses, so EGR operation is suspended at these conditions (Jacobs et al. 2003, De Serio et al. 2017).

2. Experimental

2.1. Measurement system and layout

The complete system used for the measurements is shown on Fig. 4. A PSA HDI (high-pressure direct injection) DV6 1.6 L (1,560 cc) engine was used to perform the measurements. The engine is equipped with a high-pressure EGR and EGR cooler.

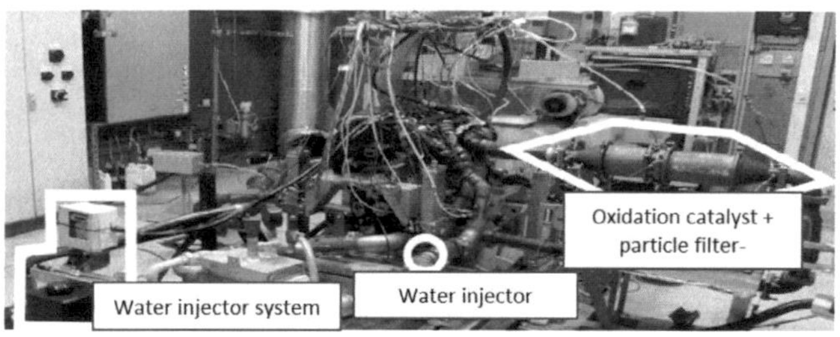

Fig. 4. Test cell with the complete measuring system

The commercially available water injection system (Snow Performance Stage 3 TD MPG-MAX) is used in the measurement, controlled by the engine's boost pressure. The main elements of the water injection system are the following:

- three chamber, high performance (~20 bar) diaphragm pump,
- nozzles of different flow rates (60 ml/min was used in my measurements),
- control unit with integrated pressure gauge.

The water injection and the boost pressure sampling point for the controller are also located after the charge air cooler (Fig. 5.).

Fig. 5. Installed water injection system

2.2. Measurements with water injection and exhaust gas recirculation

The intake manifold injection - as a technique that can be easily installed afterwards - can provide an opportunity to further reduce emissions by retrofitting the current vehicle fleet. Water injection has the effect of reducing the temperature in the cylinder chamber by means of specific heat and the heat dissipation effect of the phase change (liquid - vapor), which in turn results in the inhibition of the formation of nitrogen oxides. Exhaust gas recirculation achieves the temperature reduction by incomplete combustion (reducing the oxygen content) as a main mechanism opposed to water injection.

Based on the relevant literature, improved NO_x emission results can be expected by EGR process and also by water injection process. However, the system becomes overregulated if water injection and EGR are used together (Ogawa et al. 2013). According to the general opinion, positive effects do not add up, no improved emissions can be achieved compared to the emissions achieved by the two techniques separately, in the case of nitrogen oxides.

In order to be able to calculate the amount of water entering the combustion chamber and the water fuel ratio, a nozzle flow curve had to be created. The boost pressure at the start of injection was set to 0.2 bar and the maximum injection was set to 1.4 bar. For the measurements without EGR, its deactivation was done by software via the engine controller. Accordingly, the points of intervention (Fig. 6.) are the water injection, the EGR valve and the engine control unit required to modify it.

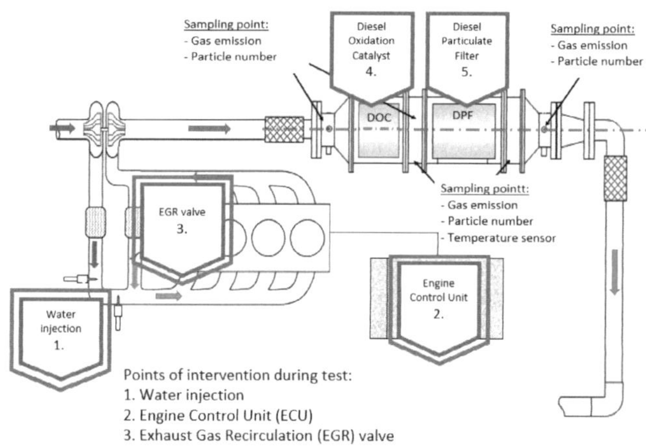

Fig. 6. Points of intervention during the EGR and water injection tests (in red bracket)

The tests were run with official emission cycles. For this purpose, the NEDC (New European Driving Cycle) cycle was selected, which consists of 3 ECE 15 cycles followed by an EUDC (Extra Urban Driving Cycle) cycle. The ECE 15 cycle represents the urban driving followed by the EUDC as a more aggressive highway driving.

Previous research has demonstrated that changes in air humidity can induce considerable differences in nitrogen oxide levels (Szőllősi and Kiss 2023). In this approach, the amount of water entering the combustion chamber was factored into the measurements.

3. Results and discussion

The actual water injection system does not operate continuously as the EGR does, but injects only from the moment the 0.2 bar (20 kPa) boost pressure is reached, as shown in Fig. 7. Therefore, lower emission reductions are seen compared to the EGR, because the water injection is not operating almost continuously.

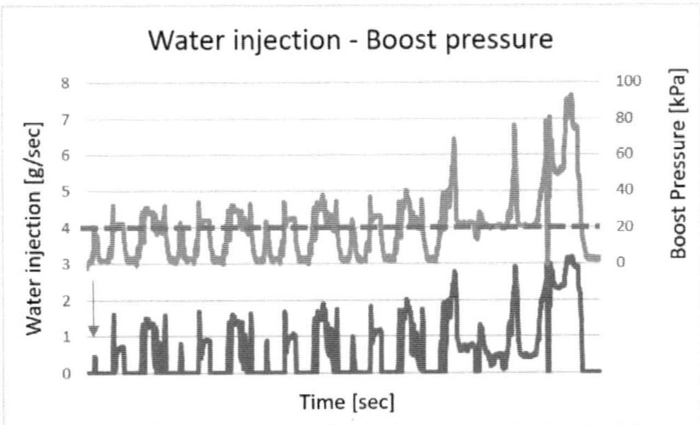

Fig. 7. Water injection during the NEDC cycle

From the results we can see that (Fig. 8.) the NO_x content was reduced by more than 25% by using water injection. However, using the two techniques in parallel, results in more than an additional 10% reduction in NO_x compared to EGR alone.

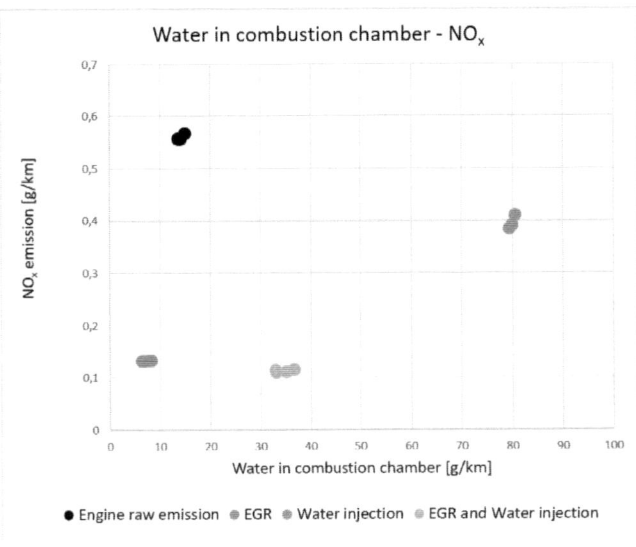

Fig. 8. Measurement results for nitrogen oxides

During the measurements, when the EGR is shut off, the high pressure EGR system of the engine causes the boost pressure to increase, resulting in higher injections from the boost pressure-controlled water injection system. This is why the diagrams show that when using water injection only, the water content in the cylinder chamber is higher.

The publications suggest that the system should become over-regulated, with water injection and EGR together leading to a reduction in efficiency on the NO_x removal (Ogawa et al. 2013). In case of the publications in the related literature, we are in an experimental environment, typically on fixed conditions, running the two devices together (fixed EGR and water injection) using experimental water injection systems or simulations. In the real systems, EGR typically operates at lower loads, reducing temperature by impairing combustion, and water injection typically operates at high loads by removing heat through specific heat and enthalpy of the phase change. An example of this can be seen on Fig. 9., where the last ~200 seconds of an EUDC cycle is shown.

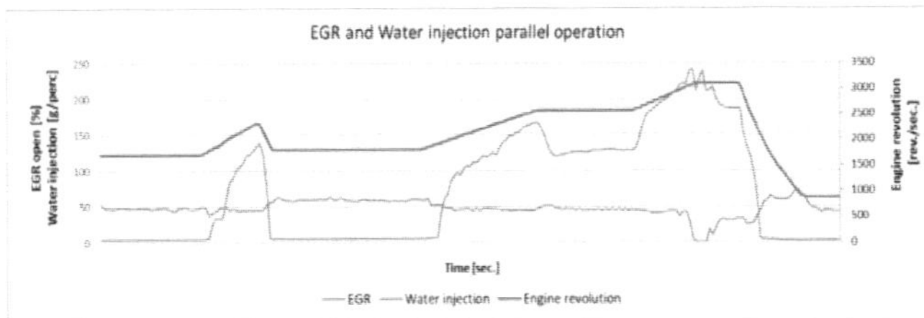

Fig. 9. EGR and water injection control on EUDC emission cycle highway section

As the vehicle accelerates, the water injection is activated and the EGR starts to close. On further acceleration, the water injection reaches its maximum by this phase of the above shown test, with the vehicle travelling at 120 km/h. Typically, at 3000 rpm, the EGR shuts off, also observed here. When the vehicle is driven at higher loads, the amount of water injection increases and the EGR opening decreases.

We can see that the two techniques could work very well in real world vehicle conditions. This is certainly a new finding for the literature. Our current results also show that improvements of NO_x emission can be seen by using the two techniques together, which could be expected to be further increased by optimizing the system.

4. Conclusions and Future work

4.1. Conclusions

The commonly applied emission treatment procedures (EGR, DOC, DPF, SCR, Water Injection) of the internal combustion engines and their effects should be considered together for the properly and effectively operating whole emission treatment system of the engine. Thinking about retrofitting it is even more crucial. The scientific materials of these individual systems often lack of this overview and the practical applications, which resulted in the findings detailed in this material. Here not just the water injection effects checked on the diesel engine emissions, but it is checked together with EGR as a most basic emission treatment system.

By using EGR and the most easily applied intake water injection separately, emissions consistent with the literature were obtained. However, a significant difference - and the most important for retrofit - is that the results were obtained with practical applications, a true EURO-rated engine, a commercially available high-performance injection system and an official (EURO) emission cycle, none of which are typical of water-injection publications. Key findings of the paper include:

- EGR typically operates at lower loads, reducing temperature by impairing combustion, and water injection typically operates at high loads by removing heat through specific heat and enthalpy of the phase change.

- Emission values improved using EGR and water injection in parallel with around 10%. The research conducted in this study was simply a proof-of-concept and by no means an optimization.
- In case of retrofitting, water injection has an increasing potential in nowadays accelerating traffic conditions, with increasing loads on the diesel engines.
- It can be concluded, that much more emission reductions can be achieved through retrofitting than through further tightening of NO_x and particulate emission standards.

4.2. Future work

There are many areas of investigation needed in order to further develop an optimally working system with EGR and water injection:

- Software coordination of EGR and water injection to exploit the potential of parallel use.
- It can be seen that during the NEDC, the water injection effect was relatively limited because of low load conditions. To further utilize the water injection and to see its effects better, further experiments with additional higher load cycles are worthwhile. From 2016, the NEDC emission cycle replaced by the WLTC (Worldwide Harmonized Light Vehicles Test Cycle) which is representing the higher loads of the current traffic conditions. Which means water injection fit increasingly into mobility trends. Also, in the non-road sector, currently used emission cycles (NRTC, NRSC) represents higher load range.
- The water injection system used in this study, was designed to be easily transferable to other engine test cells. In this way, additional data can be collected with further engines and the results obtained so far can be verified.
- In this study, only the NO_x results are presented. It is critical to research additional emission components in order to maximize the technique's utility.
- In this paper, we are checking water effect with EGR. We should assess the impact on the other emission-regulating components in further studies (DOC, DPF, SCR), because retrofitting involves adding water injection to an existing emission system.

The installation of a water injection system requires no special skills or tools. Skilled workshops can complete its installation in a vehicle in as little as a day. However, no data are available for a complete emission system. This may be due to the fact that it requires considerable preparation, sophisticated and expensive measuring equipment and the knowledge to design and control a complete emission aftertreatment system, which are altogether mostly not available. This may be the reason why such practical results are missing from the literature. PEMS (Portable emission measurement system) systems, which are now becoming increasingly available, can be a great help in such studies.

5. Nomenclature

c	Specific heat	[J/(g K)]
ΔH_{vap}	Enthalpy of vaporization	[J/g]
P_{boost}	Boost pressure	[kPa]
rpm	Engine revolution	[1/min]
AH	Absolute humidity	[g/m^3]
NO_x	Gas emission	[g/km]
$CA\ ATDC$	Crank angle after Top Dead Cent.	[°]
-	Water injection	[g/sec]
-	Pump pressure	[bar]
-	Nozzle flow rate	[ml/min]
-	Mole fraction	[mol/mol]
CO	Carbon-monoxide	[-]
CO_2	Carbon-dioxide	[-]
DOC	Diesel Oxidation Catalyst	[-]
DPF	Diesel Particulate Filter	[-]
ECE	Economic Committee for Europe Regulation 15	[-]
ECU	Engine Control Unit	[-]
EGR	Exhaust Gas Recirculation	[-]
$EUDC$	Extra Urban Driving Cycle	[-]
HC	Hydrocarbon	[-]
HD	Heavy duty	[-]
LD	Light duty	[-]
LNC	Lean NO_x Catalyst	[-]
LNT	Lean NO_x trap	[-]
$NEDC$	New European Driving Cycle	[-]
$NMVOCS$	Non-methane Volatile Organic Compounds	[-]
$NRMM$	Non-road Mobile Machinery	[-]
$NRSC$	Non-road Steady-State Cycle	[-]
$NRTC$	Non-road Transient Cycle	[-]
NO_x	Nitrogen Oxides	[-]
O	Oxygen	[-]
OH	Hydroxyls	[-]
$PEMS$	Portable Emission Meas. syst.	[-]
$PM_{2.5}$	Particulate mass	[-]
PC	Personal car	[-]
RDE	Real Driving Emission	[-]
SCR	Selective Catalytic Reduction	[-]
SOF	Soluble Organic Fraction	[-]
$WLTC$	Worldwide Harmonized Light Vehicles Test Cycle	[-]

6. Acknowledgements

Project no. C1767774 has been implemented with the support provided by the Ministry of Culture and Innovation of Hungary from the National Research, Development and Innovation Fund, financed under the KDP-2021 funding scheme.

7. Declaration of competing interest

The authors declare that they have no known competing financial interests or personal relationships that could have appeared to influence the work reported in this paper.

8. References

Al-Aqtash O., Farkas F., Sápi A. et al. 2023 Differently shaped Al_2O_3-based Pd catalysts loaded catalytic converter for novel non-road mobile machinery exhaust systems. Reac Kinet Mech Cat 136, 149–161, https://doi.org/10.1007/s11144-022-02291-x

Andrews G.E., Bartle K., Pang S., Nurein A. Williams P.T. 1988 The Reduction in Diesel Particulate Emissions Using Emulsified Fuels, SAE Technical Paper, 880348, https://doi.org/10.4271/880348

Bedford F., Rutland C., Dittrich P., Raab A. Wirbeleit, F., 2000 Effects of Direct Water Injection on DI Diesel Engine Combustion. SAE Technical Paper, 2000-01-2938, https://doi.org/10.4271/2000-01-2938.

Biró N., Kiss, P. 2023 Euro VI-d Compliant Diesel Engine's Sub-23 nm Particle Emission. Sensors, 23, 590. https://doi.org/10.3390/s23020590

Boutikos P., Březina J., Arvajová A.B., Kočí P. 2019 Comparison of O2 and NO2 impact on PtOx and PdOx formation in diesel oxidation catalysts and their reduction by CO and C3H6 pulses, Chemical Engineering Journal, Volume 377, 119654, ISSN 1385-8947, https://doi.org/10.1016/j.cej.2018.08.040.

Bowman C.T. 1992 Control of combustion-generated nitrogen oxide emissions: Technology driven by regulation. Symposium (International) on Combustion, Volume 24, Issue 1, Pages 859-878, ISSN 0082-0784, https://doi.org/10.1016/S0082-0784(06)80104-9.

Brady R.N. 2013 Internal Combustion (Gasoline and Diesel) Engines, Reference Module in Earth Systems and Environmental Sciences, Elsevier, ISBN 9780124095489, https://doi.org/10.1016/B978-0-12-409548-9.01056-3.

Chiodi H., Collier C.R., Mohler J.G. 1983 In vitro methemoglobin formation in human blood exposed to NO_2, Environmental Research, Volume 30, Issue 1, Pages 9-15, ISSN 0013-9351, https://doi.org/10.1016/0013-9351(83)90160-3.

De Serio D., De Oliveira A., Sodré J.R. 2017 Effects of EGR rate on performance and emissions of a diesel power generator fueled by B7. Journal of the Brazilian Society of Mechanical Sciences and Engineering 39, 1919–1927, https://doi.org/10.1007/s40430-017-0777-x

Europian Comission, 2022 Proposal for a REGULATION OF THE EUROPEAN PARLIAMENT AND OF THE COUNCIL on type-approval of motor vehicles and engines and of systems, components and separate technical units intended for such vehicles, with respect to their emissions and battery durability (Euro 7) and repealing Regulations (EC) No 715/2007 and (EC) No 595/2009, COM_COM(2022)0586_EN.pdf (europa.eu), (accessed on 30.04.2024)

European Environment Agency 2012. Particulate matter from natural sources and related reporting under the EU Air Quality Directive in 2008 and 2009, EEa Technical report No 10/2012, Luxembourg: Publications Office of the European Union, ISBN 978-92-9213-325-2, ISSN 1725-2237, doi:10.2800/55574

Fenimore C.P. 1971 Formation of nitric oxide in premixed hydrocarbon, Symposium (International) on Combustion Volume 13, Issue 1, Pages 373-380, https://doi.org/10.1016/S0082-0784(71)80040-1.

Glarborg P. 2019 Detailed Kinetic Mechanisms of Pollutant Formation in Combustion Processes, Computer Aided Chemical Engineering, Elsevier, Volume 45, Chapter 11, Pages 603-645,ISSN 1570-7946, ISBN 9780128195796, https://doi.org/10.1016/B978-0-444-64087-1.00011-5.

Gumus E.; Otkur M. 2023 Design of an Optimum Compact EGR Cooler in a Heavy-Duty Diesel Engine towards Meeting Euro 7 Emission Regulations. Sustainability 2023, 15, 12361. https://doi.org/10.3390/su151612361

Hagan R., Markey E., Clancy J., Keating M., Donnelly A., O'Connor D.J., Morrison L., McGillicuddy E.J. 2023 Non-Road Mobile Machinery Emissions and Regulations: A Review. Air 2023, 1, 14-36. https://doi.org/10.3390/air1010002

Holtbecker R., Geist M., 1998 Exhaust emissions reduction technology for Sulzer marine diesel engines: General aspects, Wärtsilä NSD Switzerland Ltd., Winterthur, Switzerland, July 1998

Hountalas D., Mavropoulos G., Zannis T., Mamalis, S. 2006 Use of Water Emulsion and Intake Water Injection as NOx Reduction Techniques for Heavy Duty Diesel Engines, SAE Technical Paper, 2006-01-1414, https://doi.org/10.4271/2006-01-1414.

Ishida M.c, Chen, Z., 1994 An Analysis of the Added Water Effect on NO Formation in D.I. Diesel Engines, SAE Technical Paper, 941691, https://doi.org/10.4271/941691

Imahashi T., Hashimoto K., Hayashi J.I., Yamada T. 1995 Research on NOx reduction for large marine diesel engines. ISME Yokohama;

Jacobs T., Assanis D., Filipi Z., 2003 The Impact of Exhaust Gas Recirculation on Performance and Emissions of a Heavy-Duty Diesel Engine, SAE Technical Paper 2003-01-1068, 2003, https://doi.org/10.4271/2003-01-1068.

Kalghatgi G. 2019 Development of Fuel/Engine Systems—The Way Forward to Sustainable Transport, Engineering, 5, pp. 510–518. doi: 10.1016/j.eng.2019.01.009.

Ladommatos N., Abdelhalim S.M., Zhao H., Hu Z. 1998a The effects of diesel combustion and emissions of reducing the inlet charge mass due to thermal throttling with hot EGR. SAE paper 980185,

Ladommatos N., Adelhalim S.M., Zhao H., Hu Z. 1998b The effects of carbon dioxide in exhaust gas recirculation on diesel engine emissions. Proceedings of the Institution of Mechanical Engineers, Part D: Journal of Automobile Engineering. 1998;212(1):25-42. doi:10.1243/0954407981525777

Ladommatos N., Abdelhalim S.M., Zhao H., 2000 The Effects of Exhaust Gas Recirculation on Diesel Combustion and Emissions, International Journal of Engine Research, Vol 1, pp 107-126, doi: 10.1243/1468087001545290

Lapuerta M, Ramos Á, Fernández-Rodríguez D, González-García I. 2019 High-pressure versus low-pressure exhaust gas recirculation in a Euro 6 diesel engine with lean-NOx trap: Effectiveness to reduce NOx emissions. International Journal of Engine Research. 2019;20(1):155-163. doi:10.1177/1468087418817447

Lelieveld J. et al. 2015 The contribution of outdoor air pollution sources to premature mortality on a global scale, Nature 525, pp. 367–371. https://doi.org/10.1038/ nature15371

Lončarević Š., Ilinčić P., Šagi G., Lulić Z. 2022 Problems and Directions in Creating a National Non-Road Mobile Machinery Emission Inventory: A Critical Review. Sustainability 2022, 14, 3471. https://doi.org/10.3390/su14063471

Meng S., Sun S., Xu H., Guo Y., Feng D., Zhao Y., Wang P., Qin Y. 2016 The effects of water addition on the laminar flame speeds of CO/H2/O2/H2O mixtures, International Journal of Hydrogen Energy, Volume 41, Issue 25, Pages 10976-10985, ISSN 0360-3199, https://doi.org/10.1016/j.ijhydene.2016.04.251.

Naik G.G., Dharmadhikari H.M. 2023 Methods for reducing NOx and PM emissions in compression ignition engine: A review, Materials Today: Proceedings, Volume 72, Part 3, Pages 1406-1412, ISSN 2214-7853, https://doi.org/10.1016/j.matpr.2022.09.339.

Neshat E., Bajestani A.V., Honnery D. 2019 Advanced numerical analyses on thermal, chemical and dilution effects of water addition on diesel engine performance and emissions utilizing artificial inert species, Fuel, Volume 242, Pages 596-606, ISSN 0016-2361, https://doi.org/10.1016/j.fuel.2019.01.080.

Ning L., Duan Q., Chen Z., Kou H., Liu B., Yang B., Zeng K. 2020 A comparative study on the combustion and emissions of a non-road common rail diesel engine fueled with primary alcohol fuels (methanol, ethanol, and n-butanol)/diesel dual fuel, Fuel Volume 266, 117034, doi: 10.1016/j.fuel.2020.117034

Odaka M., Koike N., Tsukamoto Y., Narusawa K., Yoshida K. 1991 Effects of EGR with a Supplemental Water Injection to Control Exhaust Emissions from Heavy-Duty Diesel Powered Vehicles, SAE Technical Paper 910739, doi:10.4271/910739

Ogawa H., Shibata G., Kato T., Setiapraja H. Hara, K. 2013 "Combustion Characteristics of Emulsified Blends of Water and Diesel Fuel in a Diesel Engine with Cooled EGR and Pilot Injection," SAE Int. J. Fuels Lubr. 6(3):977-985, https://doi.org/10.4271/2013-32-9022.

Paul K.T, Hull T.R., Lebek K., et al. 2008 Fire smoke toxicity: The effect of nitrogen oxides, Fire Safety Journal, Volume 43, Issue 4, Pages 243-251, ISSN 0379-7112, https://doi.org/10.1016/j.firesaf.2007.10.003.

Prasad H.S., Gonsalvis, J., Vijay V. S. 2015 Effect of Introduction of Water into Combustion Chamber of Diesel Engines – A Review. Energy and Power, Vol. 5 No. 1A, pp. 28-33. doi: 10.5923/c.ep.201501.06.

Quader A. A. 1971 Why Intake Charge Dilution Decreases Nitric Oxide Emission from Spark Ignition Engines. SAE Transac-tions, vol. 80, pp.

20–30. JSTOR, http://www.jstor.org/stable/44731349. (accessed on 03.05.2024.)

Radnainé Zs., 2011 Levegőtisztaság-védelem, Pannon Egyetem – Környezetmérnöki Intézet, Veszprém, ISBN: 978-615-5044-36-6

Reifarth S.,Ångström H.E. 2009 Transient EGR in a Long-Route and Short-Route EGR-System. 761-770. 10.1115/ICES2009-76107.

Rente T. 2004 Injection strategies for Heavy Duty DI Diesel Engines. Doctoral thesis, Chalmers University of Technology

Rossomando B., Meloni E., De Falco G., Sirignano M., Arsie I., Palma V. 2021 Experimental characterization of ultrafine particle emissions from a light-duty diesel engine equipped with a standard DPF, Proceedings of the Combustion Institute, Volume 38, Issue 4, Pages 5695-5702, https://doi.org/10.1016/j.proci.2020.09.011.

Rounce P., Tsolakis A., York A.P.E. 2012 Speciation of particulate matter and hydrocarbon emissions from biodiesel combustion and its reduction by aftertreatment, Fuel, Volume 96, Pages 90-99, https://doi.org/10.1016/j.fuel.2011.12.071.

Sher E. 1998 Environmental Aspects of Air Pollution, Handbook of Air Pollution From Internal Combustion Engines, Chapter 2, Academic Press, Pages 27-41, ISBN 9780126398557, https://doi.org/10.1016/B978-012639855-7/50041-7.

Sun, X., Ning, J., Liang, X,, Jing, G., Chen, Y., Chen, G. 2022 Effect of direct water injection on combustion and emissions char-acteristics of marine diesel engines. Fuel, Volume 309, 122213, ISSN 0016-2361, https://doi.org/10.1016/j.fuel.2021.122213.

Sun Z., Wang X., Wang X., Zhou J. 2017 Combustion and emission analysis of heavy-duty vehicle diesel engine, AIP Conference Proceedings, Volume 1820, Issue 1, Doi: 10.1063/1.4977307

Szőllősi D., Kiss P. 2023. Effects of humidity on the emissions of the diesel engines, International Society for Terrain-Vehicle Systems (ISTVS), 16th European-African Regional Conference of the ISTVS, October 2023, Lublin, Poland, doi: 10.56884/THMG6024

Tauzia X., Maiboom A., Shah S.R. 2010 Experimental study of inlet manifold water injection on combustion and emissions of an automotive direct injection Diesel engine, Energy, Volume 35, Issue 9, Pages 3628-3639, https://doi.org/10.1016/j.energy.2010.05.007

Theis J.R., Lambert C.K. 2015 An assessment of low temperature NOx adsorbers for cold-start NOx control on diesel engines, Catalysis Today, Volume 258, Part 2, Pages 367-377, ISSN 0920-5861, https://doi.org/10.1016/j.cattod.2015.01.031.

Turns S.R. 2000 An Introduction to Combustion. Concepts and Applications. 2nd Edition, Holman, J.P.; Lloyd, J., McGraw Hill, New York, pp 168-171

Wang, L., Liu, Z., Chen, S., Zheng, C., & Li, J. 2013 Physical and chemical effects of CO2 and H2O additives on counterflow diffusion flame burning methane. Energy and Fuels, 27(12), 7602-7611. https://doi.org/10.1021/ef401559r

Wang W., Herreros J.M., Tsolakis A., York A.P.E. 2015 Increased NO2 concentration in the diesel engine exhaust for improved Ag/Al2O3 catalyst NH3-SCR activity, Chemical Engineering Journal, Volume 270, Pages 582-589, ISSN 1385-8947, https://doi.org/10.1016/j.cej.2015.02.067.

Wan J., Zhuang Y., Huang Y., Qian Y., Qian L., 2021 A review of water injection application on spark-ignition engines. Fuel Processing Technology, Volume 221, 106956, ISSN 0378-3820, https://doi.org/10.1016/j.fuproc.2021.106956.

Zhang W., Feng T., Li Z., Chen Z., Zhao J. 2021 EGR thermal and chemical effects on combustion and emission of diesel/natural gas dual-fuel engine, Fuel, Volume 302, 121161, ISSN 0016-2361, https://doi.org/10.1016/j.fuel.2021.121161.

Zhang Y., Lou D., Tan P., Hu Z. 2018 Experimental study on the particulate matter and nitrogenous compounds from diesel engine retrofitted with DOC+CDPF+SCR, Atmospheric Environment, Volume 177, Pages 45-53, ISSN 1352-2310, https://doi.org/10.1016/j.atmosenv.2018.01.010.

1928 | A novel soil stress estimation method of wheel-soil interaction using photoelasticity

Kenji Nagaoka [a,*], Yuto Yoshida [a]

[a] *Department of Mechanical and Control Engineering, Kyushu Institute of Technology, Japan*

** Corresponding author: nagaoka.kenji572@mail.kyutech.jp*

ABSTRACT

This paper proposes a new approach to understanding the wheel-soil interaction, which is an indirect estimation method of soil stress distributions beneath a traveling wheel soil using a photoelastic method. Thus far, in the conventional studies applying the photoelastic method to the wheel-soil terramechanics, the terrain has been emulated by photoelastic disks or plates, which enable visualization of internal stresses in the simulated terrain. In particular, the photoelastic disks have performed visualization of two-dimensional dynamic stress distributions of the simulated granular terrain. With this method, we have visualized and analyzed the dynamic force chain structure of the terrain under different wheel slip conditions. Still, it is difficult for the previous configuration to simulate the dynamic behaviors of natural soil, e.g., compaction, failure, or wheel ruts. Accordingly, achieving both the stress visualization and the dynamic behaviors of soil is a significant challenge to make the photoelastic method more practical. To cope with this challenging issue, we have developed a novel experimental setup consisting of a photoelastic wheel (top layer), soil (middle layer), and a photoelastic plate (bottom layer). By vertically sandwiching the soil between the photoelastic wheel and plate, the soil stresses can be indirectly estimated to satisfy the boundary stress conditions. To achieve this approach, we have conducted calibration tests of the photoelastic wheel and plate, and then identified the force vector and contact patch corresponding to the visualized stresses. In this paper, we demonstrate that it is possible to indirectly estimate how the stress propagates and attenuates in the soil by the proposed method.

Keywords
Photoelasticity
Stress distributions
Wheel-soil interaction

1. Introduction

Field robotics has addressed mobility and manipulation in natural rough terrain for various applications, such as agriculture, construction, and planetary exploration. In recent years, the practical applications of small mobile robot systems to field operations have accelerated. In particular, to enhance the mobility performance of small machinery on rough terrain, wheel-soil interaction has been widely analyzed.

In the wheel-soil interaction, understanding high-dimensional and dynamic soil behavior is a key challenge. To cope with this, visual observation of soil flow during wheel traveling has been addressed using particle image velocimetry (PIV), which can two-dimensionally measure the velocity of soil particles without contact. So far, PIV-based analyses have shown the displacement field of soil beneath a traveling wheel (Senatore et al., 2013; Yamakawa and Goto, 2015). Also, PIV-based soil flow results have been compared with the finite element method (FEM) (Moreland et al., 2012; Saengprachatanarug and Okayasu, 2013). Recently, the PIV-based soil flow field beneath a grouser wheel has also been analyzed (Nakamura et al., 2020). Although these visual approaches can provide the soil motion, dynamic stress distributions in the soil cannot be explained.

Previously, the photoelastic method has been applied to analyses of the internal stress distributions and stress chain structures of granular particle layers. The evaluation of disorder-induced metastable relaxation based on the analyses of orientational order, which indicates the anisotropy of force chains when tapping is applied to a container filled with photoelastic disks (Iikawa et al., 2015; Iikawa et al., 2016) has been researched. Moreover, with respect to the visualization of arch effects inside the simulated powder, the analysis of arch formation processes and the regularity of the arch shape have been analyzed (Fang et al., 2020). Some studies have shown the contact force of photoelastic disks in a quasi-static flow by image processing (Zadeh et al., 2019). There have also been studies that have analyzed similar quasi-static states for changes in the contact force network due to the deformation of granular materials and compared them with the discrete element method (DEM), assuming soil pressure conditions (Leśniewska et al., 2020). However, most of these studies have been limited to experiments on static or quasi-static phenomena. In contrast, we have proposed a photoelastic method to examine the dynamic soil stress beneath a traveling wheel (Yoshida et al., 2023). This is an experimental method for two-dimensionally visualizing and measuring the dynamic stress chain structures in the granular terrain beneath a traveling wheel, using photoelastic disks as simulated granular terrain. With an orientation order of the force chain structures, we have demonstrated the feasibility of dynamic force chain analysis in the granular terrain which mechanically interacts with the traveling wheel. Furthermore, we have quantitatively investigated the effects of the wheel slip and non-photoelastic disks mixing in the photoelastic layer. Although this is an effective means for measuring dynamic stress distributions in granular terrain beneath a traveling wheel, it is difficult to simulate the dynamic behaviors of natural soil, e.g., compaction, failure, or wheel ruts, due to discrete stress propagation and physical properties of the photoelastic disks.

Achieving both the stress visualization and the dynamic behaviors of terrain particles is a significant challenge to make the photoelastic method more practical. In this paper, we propose a novel photoelastic approach that employs a multilayer structure of photoelastic materials and soil. This paper delivers the proposed approach and discusses its initial result.

https://doi.org/10.56884/15YUPU08

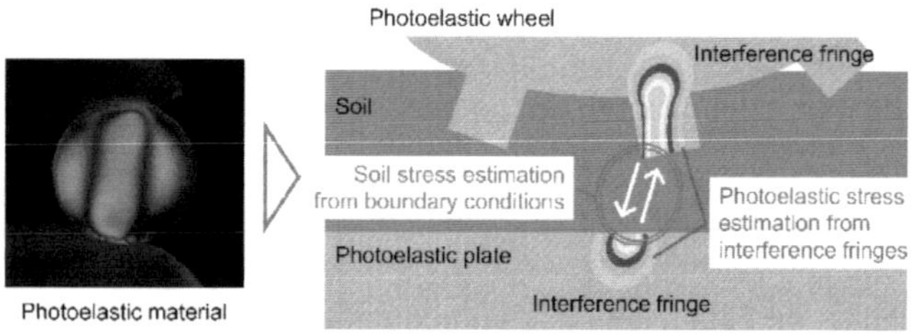

Fig. 1. Conceptual diagram of photoelasticity-based indirect estimation of soil stress distributions

2. Photoelastic method for wheel-soil system

The proposed photoelastic approach has a multilayer consisting of a photoelastic wheel (top layer), soil (middle layer), and a photoelastic plate (bottom layer). By vertically sandwiching the soil between the photoelastic wheel and plate, the directly unobservable soil stresses can be indirectly estimated to satisfy the boundary stress conditions that the photoelastic materials show. Fig. 1 illustrates the fundamental idea of the proposed approach.

2.1. Single-wheel apparatus

The photoelastic method is a two-dimensional experimental technique that uses the optical anisotropy of photoelastic materials. Photoelasticity can visualize internal stresses as interference fringes by observing the birefringence phenomenon that occurs when stress is applied to the photoelastic body through polarizing filters on both sides of the photoelastic one. The photoelastic plate was processed into the shape of a grouser wheel (ϕ120 mm in diameter, Fig. 2(a)). Also, the photoelastic plate (250×250 mm, Fig. 2(b)) was installed in the soil. The thicknesses of the photoelastic materials are all 9 mm.

Fig. 2(c) shows the overview of the single-wheel apparatus. single-wheel test apparatus which consists of the photoelastic wheel, the photoelastic plate, soil, and a translational conveyor unit including a wheel drive unit. The photoelastic materials and soil are filled between the glass plates that have a length of 1000 mm and a height of 300 mm. The translational drive unit, which serves as a linear guide, consists of a plate with a passive wheel placed in a groove on an aluminum frame and is connected to the wheel drive unit via a shaft. The wheel drive unit uses a DYNAMIXEL AX-12A motor manufactured by ROBOTIS and incorporates a pulley mechanism with a tooth ratio of 1: 2. The vertical load acting on the wheel drive unit was 375 g. The sand used in the experiment was Kanuma soil. This soil is used for horticultural purposes and was selected because its relatively large grain size prevents it from inserting into the gap between the photoelastic material and the glass plate.

2.2. Calibration of stress distributions

Prior to the wheel traveling tests, calibration of the stress distributions applied on the photoelastic wheel and plate is required. In this work, loading tests were conducted as shown in Fig. 3. The angle θ from the length of the grouser or from the vertical direction of the photoelastic plate and the contact length L, which corresponds to the contact area, were varied in the loading tests. The correlation between the applied load F and interference fringes was clarified. From the obtained stress distributions, we can calculate their macro-scale features by quantitative values

(a) Photoelastic wheel

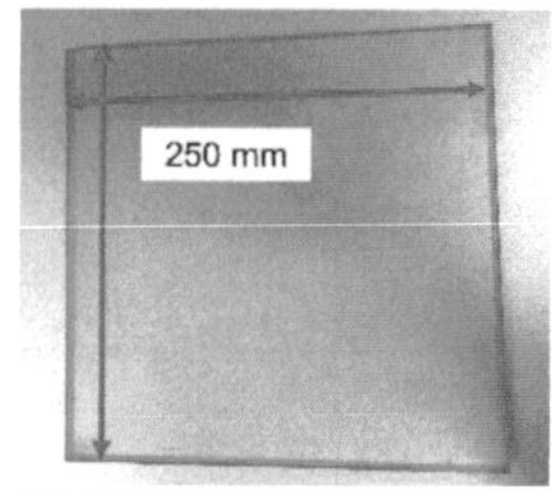

(b) Photoelastic plate

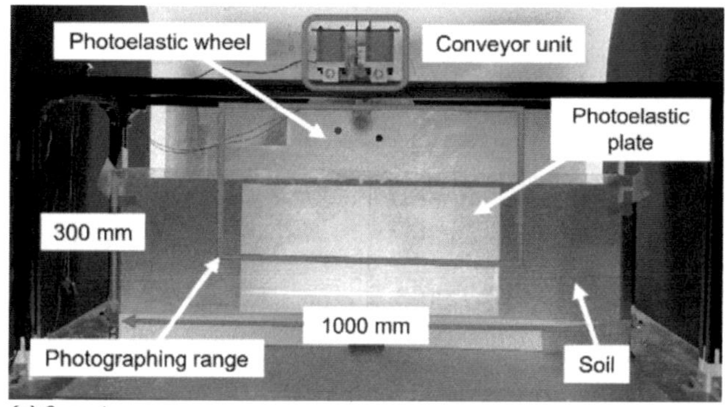

(c) Overview

Fig. 2. Single-wheel apparatus using photoelastic method

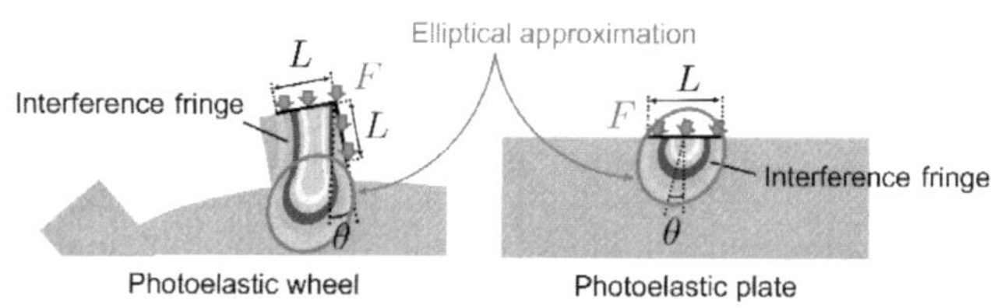

(a) Schematic view of load test

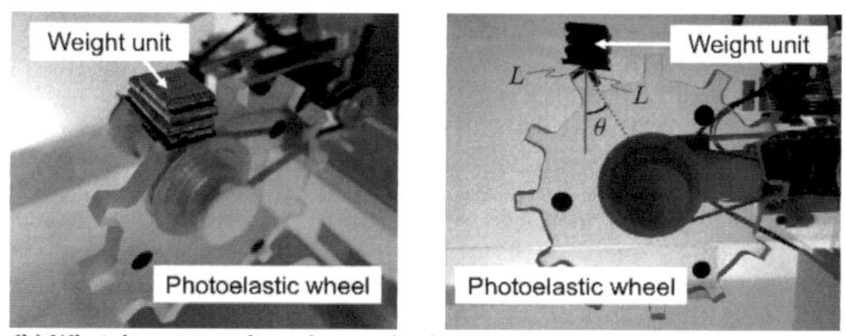

(b) Wheight unit on photoelastic wheel

Fig. 3. Load test calibration for photoelastic wheel and plate

based on elliptical approximation (see Section 2.4). The calibration results are discussed in Section 3.1.

2.3. Soil thickness

Under conditions in which the photoelastic material is installed inside the soil, the visibility of the interference fringes of the soil varies depending on the thickness of the soil layer between the photoelastic wheel and plate. Hence, a preliminary experiment was conducted to investigate the appropriate soil thickness for the single-wheel tests.

The experimental conditions and results are shown in Fig. 4, where z [mm] shows varied from 10 to 50 mm in 10 mm increments. The results show that for $z = 10, 20$ [mm], the stress distributions in the soil were observed, but for $z = 30$ [mm] and more, no significant difference was obtained in the interference

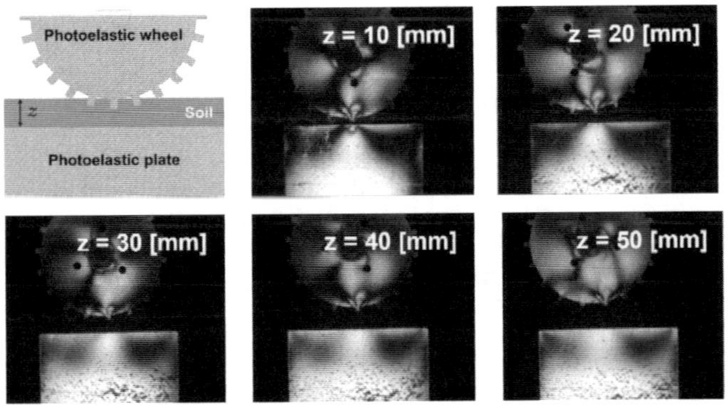

Fig. 4. Calibration test with various sand thickness

Table 1
HSV scope of photoelastic wheel and plate for hue detection

Hue	Saturation	Luminosity
0.15 ~ 0.25	0.43 ~	~ 0.9

fringes visualized. This came from what an effective stress is not propagated to the plate. Therefore, the single-wheel tests were conducted at $z = 10$ [mm]. On the other hand, the experiments in which traction load is applied will be suitable at a thickness of $z = 20$ [mm] or more to account for wheel sinkage with slippage.

2.4. Hue detection and ellipsoid approximation of interference fringes

This section introduces hue detection, which extracts features of the interference fringes. In hue detection, the RGB color values contained in an image are converted to HSV color space to specify a detailed hue range for detection. The values of the specified hue range are shown in Table 1. In the following experiment, hue detection for interference fringes was performed with a hue range of 0.15 to 0.25, a lower saturation limit of 0.43, and an upper luminosity limit of 0.9. The detected areas were drawn in white-colored dots in the images.

In previous studies, contact forces have been measured using the photoelastic method by estimating the force per disk using the brightness gradient (Howell and Behringer, 1999; Iikawa et al., 2015) or by image matching on a computer (Majmudar and Behringer, 2005; Puckett and Daniels, 2013). As a method of quantifying the results of hue detection of interference fringes, we provide the ellipse closest to all points for each detected pixel value using the least-squares method. Stresses are estimated by quantifying the interference fringes from the three parameters: major axis radius, minor axis radius, and major axis angle of the approximated ellipses and comparing them with load tests. This initial analysis focuses on identifying the boundary conditions which are the applied stresses to the soil layer, as a preliminary step leading to the contribution of quantitative estimation of the soil stress distributions. This uses the relationship that the size and orientation of the ellipse of a particular color area in the photoelastic material correspond to the stress and orientation applied.

2.5. Procedure of soil stress estimation

The experimental procedure of the soil stress estimation is shown below. The load test calibration and its process were independently conducted before the single-wheel tests.

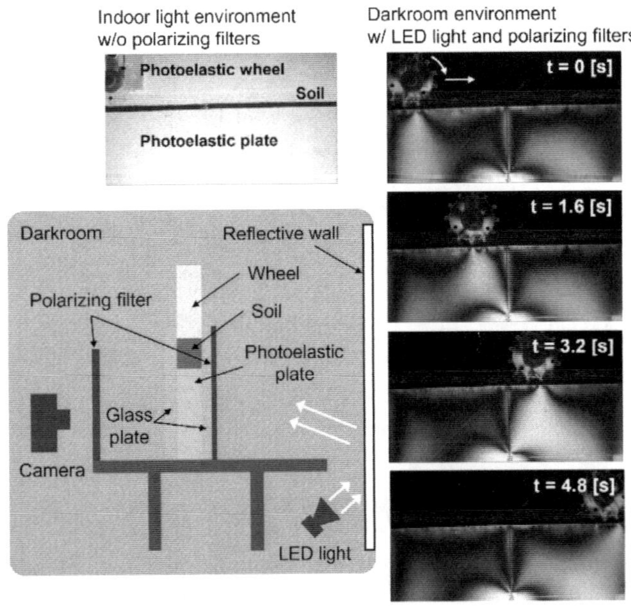

Fig. 5. Experimental overview

1. Load test calibration on photoelastic wheel and plate
2. Single-wheel traveling tests
3. Hue detection on interference fringes of the two tests
4. Ellipsoid approximation of the detection area in the image of the two tests
5. Quantification of the ellipsoid shape (size and slope) of the interference fringes
6. Identification of the most consistent stresses applied to the photoelastic wheel and plate by the comparison of the two ellipses of the load test and single wheel test
7. Estimate of the soil stress in the middle of the photoelastic wheel and plate by identified boundary stress conditions

3. Experiments

An overview of the single-wheel tests is shown in Fig. 5. The digital camera took videos with polarized filters throughout the wheel traveling experiments.

3.1. Calibration of stress distributions

Fig. 6 shows the exampled calibration results of the correlation between applied load and photoelastic stress distributions with several conditions of θ, L and F. These relationships were used for comparison with the single-wheel experiments.

3.2. Single-wheel traveling

The following procedure was used as the experimental method for wheel traveling using the single-wheel apparatus. First, tapping was applied to the terrain container. Next, the soil surface (the top layer) was leveled and the height was adjusted to a constant level. The experimental conditions were set to 25 rpm for the wheel rotational velocity and 0.1 m/s for the wheel circumferential velocity.

Fig. 7 shows the time history of the slip ratio of single-wheel traveling experiments, as shown in Fig. 5, without traction load. This result confirmed that the wheel travels on soil with a smaller

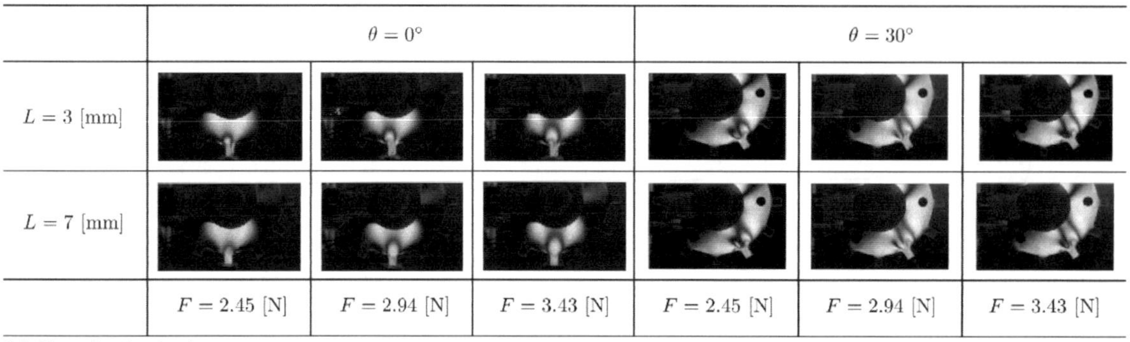

(a) Photoelastic wheel

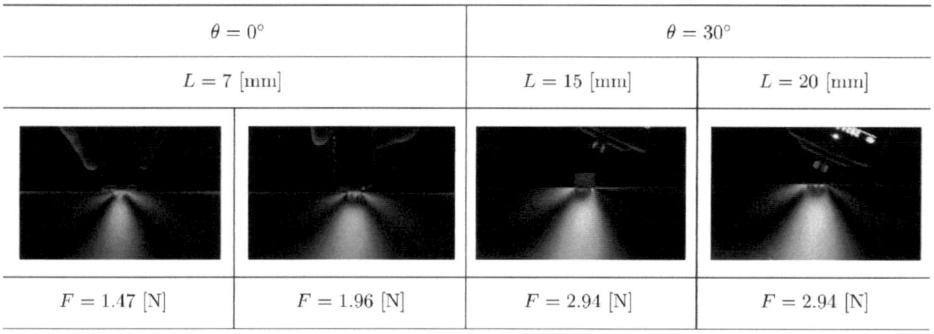

(b) Photoelastic plate

Fig. 6. Exampled calibration results of correlation between applied load and photoelastic stress distributions

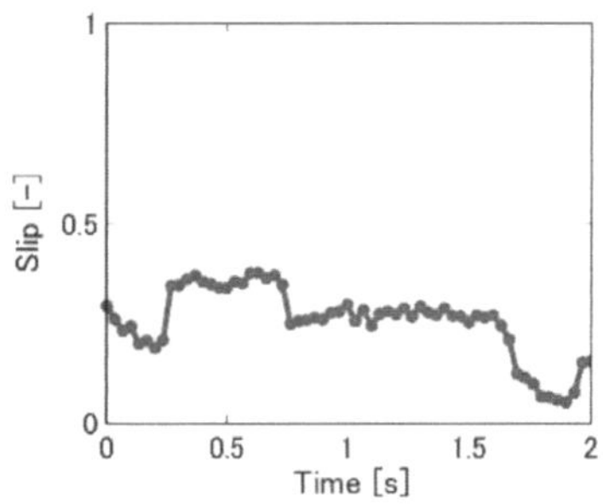

Fig. 7. Time history of slip ratio of single-wheel traveling

slip ratio to the wheel traveling on the photoelastic disks (Yoshida et al., 2023). These slip states are within practical conditions.

4. Result of soil stress estimation

4.1. Results of stresses on photoelastic wheel and plate

Figs. 8-11 shows the results of the hue detection and the ellipse approximation of the detection area, in addition to the parameters of each ellipse. For this analysis, we selected a typical case of the wheel traveling experiments. Figs. 8 and 10 show the results obtained by the single-wheel tests. Figs. 9 and 11 also show the results of the load calibration test, which is most consistent with the results of the single-wheel test.

The interference fringes of the photoelastic plate in the soil were quantified. The comparison of the results showed that the force of 2.94 N applied to the photoelastic plate at $\theta = 30^{\circ}$ and a contact patch of $L = 15$ [mm] was the closest value, as shown in Figs. 10 and 11. Thus, it was found that this stress is propagated from the traveling wheel and is another boundary condition.

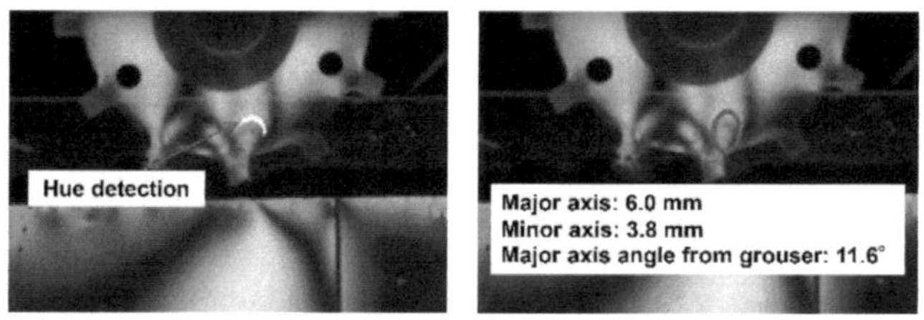

(a) Hue detection (b) Ellipsoid approximation

Fig. 8. Quantification result of photoelastic wheel traveling

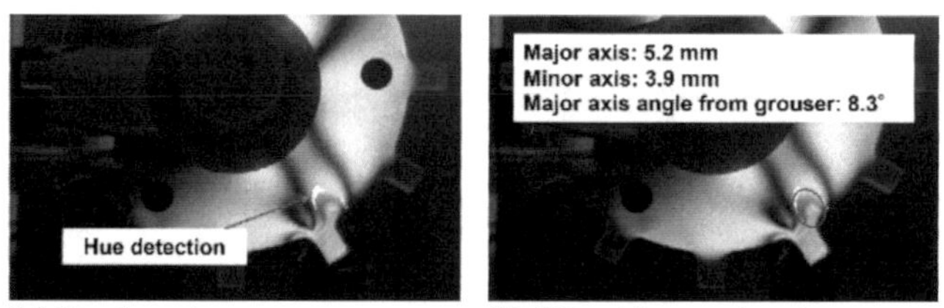

(a) Hue detection (b) Ellipsoid approximation

Fig. 9. Quantification result of photoelastic wheel load calibration test: θ=30°, L=7 [mm], F=3.43 [N]

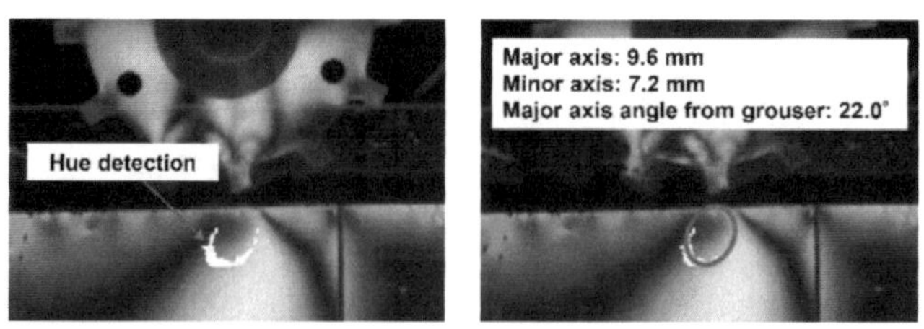

(a) Hue detection (b) Ellipsoid approximation

Fig. 10. Quantification result of photoelastic plate in soil

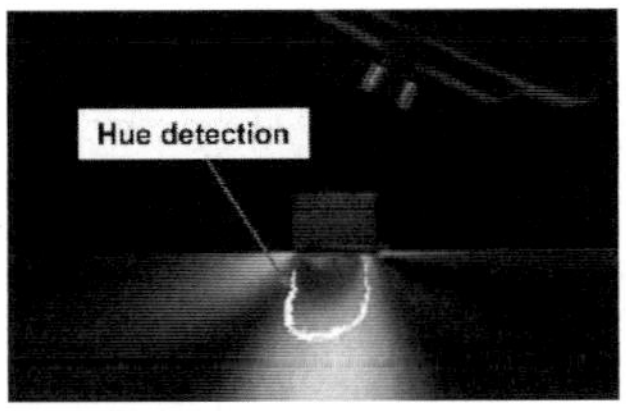

(a) Hue detection

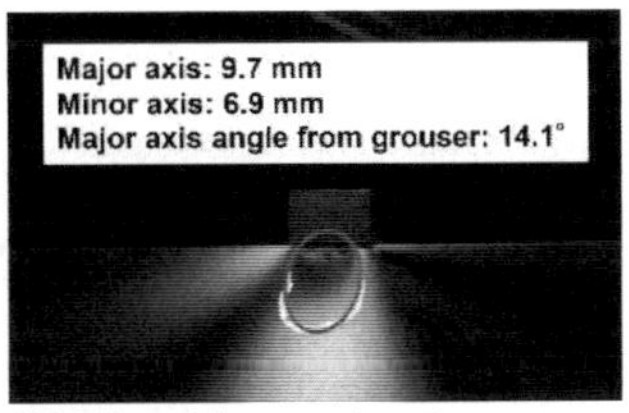

(b) Ellipsoid approximation

Fig. 11. Quantification result of photoelastic plate load calibration test: θ=30°, L=15 [mm], F=2.94 [N]

4.2. Result of soil stress estimation

Fig. 12 shows a schematic diagram of the soil stress estimation. As a result, the force of 3.43 N propagated from the grouser wheel to the photoelastic plate at an angle of $\theta = 30^{\circ}$, and the stress decayed to approximately 86% at 4.6 mm of the shortest distance. From these, the attenuation rate per unit propagation distance was calculated to 3.1%/mm. The stress propagated to the photoelastic plate with a width of 15 mm. In the front region of the direction of the wheel travel, the stress distribution was expanded by18°, and in the rear region, it was done by 9°, respectively.

5. Conclusion

In this paper, we have proposed a novel soil stress estimation method based on photoelasticity. The proposed approach indirectly estimates the soil stresses with boundary stress conditions that are applied to the photoelastic wheel and plate. With the image processing technique and ellipsoid approximation, the boundary stress conditions are quantitatively estimated, and thereby we validated that the unobservable soil stress beneath the traveling wheel can be estimated. The result confirmed that the stress propagation is decayed and expanded in the soil.

Future work looks into the quantitative correlation between photoelastic interference fringes and stresses, and high-resolution image acquisition. In the current experiments, the angle of view of the camera was wide in order to take wheel motion over a wide area, but we are currently working on focusing on the analysis area to be imaged or using multiple cameras. In addition, we plan to work on stress estimation according to slip ratio and apply the knowledge of pressure bulbs in soil mechanics to refine the distribution of stresses in the soil.

References

Fang, Y., Guo, L., and Hou, M. 2020. Arching effect analysis of granular media based on force chain visualization. Pow. Tech. 363, 621–628.

Howell, D., and Behringer, R. P. 1999. Stress fluctuations in a 2D granular Couette experiment: a continuous transition. Phys. Rev. Lett., 82, 5241, doi:10.1103/PhysRevLett.82.5241.

Iikawa, N., Bandi, M. M., and Katsuragi, H., 2015. Structural evolution of a granular pack under manual tapping. J. Phys. Soc. Japan 84, 094401, doi:10.7566/JPSJ.84.094401.

Iikawa, N., Bandi, M. M., and Katsuragi, H., 2016. Sensitivity of granular force chain orientation to disorder-induced metastable relaxation. Phys. Rev. Lett. 116, 128001 doi:10.1103/PhysRevLett.116.128001.

Leśniewska, D., Nitka, M., Tejchman, J., Pietrzak, M., 2020. Contact force network evolution in active earth pressure state of granular materials: photo-elastic tests and DEM. Granular Matter. 22 (3), 71, doi:10.1007/s10035-020-01033-x.

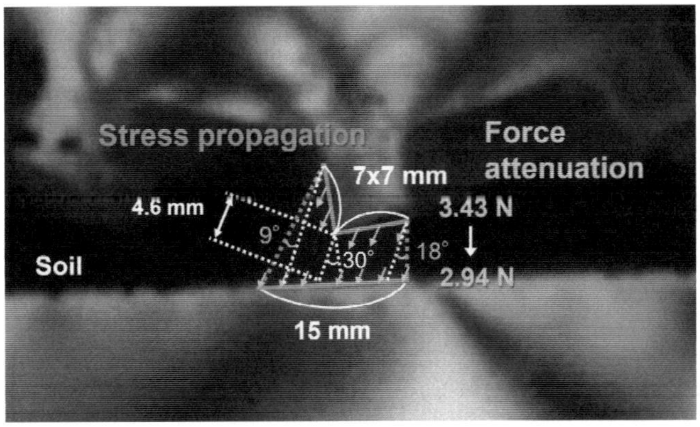

Fig. 12. Result of indirect estimation of soil stress

Majmudar, T. S., and Behringer, R. P. 2005. Contact force measurements and stress-induced anisotropy in granular materials. Nature 435, 1079–1082.

Moreland, S., Skonieczny, K., Inotsume, H., and Wettergreen, D. 2012. Soil behavior of wheels with grousers for planetary rovers. Proc. 2012 IEEE Aerospace Conf., doi:10.1109/AERO.2012.6187040.

Nakamura, H., Nagaoka, K., Yoshida, K., 2020. Soil flow analysis beneath a grouser wheel based on a particle image velocimetry method. J. Terramech. 91, 233–241.

Puckett, J. G., and Daniels, K. E. 2013. Equilibrating temperaturelike variables in jammed granular subsystems. Phys. Rev. Lett. 110, 058001, doi:10.1103/PhysRevLett.110.058001.

Saengprachatanarug, K., and Okayasu, T. 2013. Modeling of soil displacement and soil strain distribution under a traveling wheel. J. Terramech. 50(1), 5–16.

Senatore, C., Wulfmeier, M., Vlahnic, I., Andrade, J., Iagnemma, K., 2013. Design and implementation of a particle image velocimetry method for analysis of running gear-soil interaction. J. Terramech. 50 (5-6), 311–326.

Yamakawa, J., Goto, M., 2015. Experiment and numerical analysis of driving wheel on soft sandy soil. Proc. 13th ISTVS European Regional Conf., 500–508.

Yoshida Y., Sakada, H., Yuasa, S., Nagaoka, K., 2023. A photoelastic method for stress analysis of granular terrain beneath a wheel, J. Trramech. 110, 69–78.

Zadeh, A. A. et al. 2019. Enlightening force chains: a review of photoelasticimetry in granular matter, Granul. Matt. 21(83), doi:10.1007/s10035-019-0942-2.

2369 | Experimental verification of particle behavior during crushing and mixing of deteriorated asphalt pavement layers by stabilizer

Takatomo Fujii [a,*], Takashi Kurosu [a], Osamu Oikawa [a], Takashi Okayasu [b], Toyohiro Katou [b], Yuta Hirayama [b], Yukinori Tsukimoto [a]

[a] *Sakai Heavy Industries, Ltd.*
[b] *Kyushu University*

* *Corresponding author: takatomo.fujii@sakainet.co.jp, t-kurosu@sakainet.co.jp*

ABSTRACT

In the road-subbase reclamation method, a stabilizer's rotor bits (teeth) are rotated to crush deteriorated asphalt pavement in-situ while mixing it with additives such as cement and asphalt emulsion, which are then compacted with rollers to create a new, stable subbase. The uniformity and the quality of crushing and mixing is affected by the shape and arrangement of the rotor bits, rotational speed, and working speed. Therefore, it is important to understand the behavior of the crushed materials. However, since those actual behaviors inside the rotor hood cannot be observed, an attempt was made to analyze and visualize particle behaviors using the three-dimensional discrete element method (3D-DEM). In this study, an actual stabilizer was used to crush and mix the deteriorated pavement layer in a test section to compare with the analysis results. The surface of the test section was painted with spray paint in six different colors, square grid lines were drawn with white chalk, and a total of 780 numbered stickers were pasted within each grid. After the mixing test, all particles with the numbered stickers were recovered (these particles are called as "marking particles" hereafter), and the moving distance in the front-back direction were measured and a number of particles distributed in the left-right direction was counted. Test results show that about 80% of the crushed marking particles are blown away once and deposited behind the rotor, never to be mixed again. The number of marking particles counted in the left-right direction was similar and uniform. Since the test results were generally similar to the 3D-DEM results, it was concluded that this analytical method is capable of simulating the behavior of crushed material in the rotor hood.

Keywords
Road stabilizer
Experimental verification
Reclamation
Crushing and mixing behavior
Three-dimensional discrete element method (3D-DEM)
Particle size distribution

1. Introduction

In recent years, the percentage of roads maintained and repaired has increased more than new construction in Japan. Low-cost, high-quality road maintenance and repair methods are being developed to reduce maintenance and repair costs, reduce traffic control time associated with work, and reduce environmental impact. Among these, the additive-based road-subbase reclamation method (Reeder et al., 2019) has the advantage of not only crushing and mixing the deteriorated road surface and subbase in situ to rebuild a stable subbase, but also to increase the load-bearing capacity of the existing pavement structure. In addition, the thickness of the pavement structure can be reduced because of its increased strength compared to the pre-reclaimed subbase. Furthermore, this method reduces the amount of construction materials mined and transported, and the amount of subbases cut and disposed of, thus contributing to a reduction in environmental impact in terms of carbon dioxide emissions.

On the other hand, the quality of the reconstructed roadbed is highly dependent on the crushing and mixing characteristics of the road stabilizer (JCMA, 2013), which can be very costly and time consuming to verify through experimentation. Therefore, the authors attempted to analyze and visualize particle behavior during crushing and mixing using a three-dimensional discrete element method (3D-DEM) to observe the behavior of crushed materials in the rotor hood (Katou et al., 2024). The analysis model assumes a surface course, base course, and subbase course, similar to the actual pavement cross-section, with each layer consisting of spherical particles of two different particle sizes. However, the stiffness between the particles was set to represent the fracturing behavior specific to surface course asphalt mixtures.

In this study, an actual road stabilizer was used to crush and mix the deteriorated pavement layers to verify the analysis results. The surface of the test section was divided into a grid and marked with numbers and colors, and a comparison of test and analysis results regarding the distance in the front-back direction and position in the left-right direction of the marked particles before and after crushing and mixing is reported.

2. Conditions of test and 3D-DEM analysis

2.1. Specifications of test machine

Fig. 1 shows a general view of the road stabilizer used in the test, and *Table 1* shows its main specifications. This machine is a PM600 model manufactured by Sakai Heavy Industries and is self-propelled with two tires on each of the front and rear axles (Sakai, 2024). A rotor equipped with 108 bits for crushing and mixing and a rotor hood covering it are installed in the center of the frame.

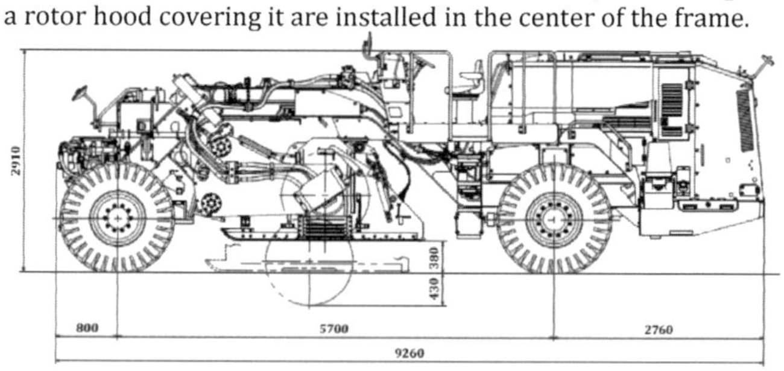

Fig. 1. General view of the road stabilizer SAKAI PM600 model

https://doi.org/10.56884/2VEBGCMK

Table 1
Main specifications of the road stabilizer

Maximum mass	22,780	kg
Operating speed	0~4,800	mm/min
Rotor revolution	130(High),100(Low)	rpm
Rotor width	2,000	mm
Rotor depth (max.)	430	mm
Number of bit	108	pieces
Engine rated output	370	kW

Fig. 2 shows a schematic diagram of the rotor and inside the rotor hood during crushing and mixing. The rotor hood and breaker bars contribute to improved crushing and mixing. The front space surrounded by the front breaker bar, bits, and road surfaces in the forward lower space inside the rotor hood is important for holding and remixing material immediately after crushing.

The structure inside the hood allows the laying height after mixing to be adjusted by opening angle of the rear gate at the rear. The dimension of 1,080 mm in *Fig. 2* represents the distance from the surface course cutting point to the gap between the rear breaker bar and the rotor bit, referred to as the "one-time mixed distance" described in *Fig. 10* in Section 3.1.

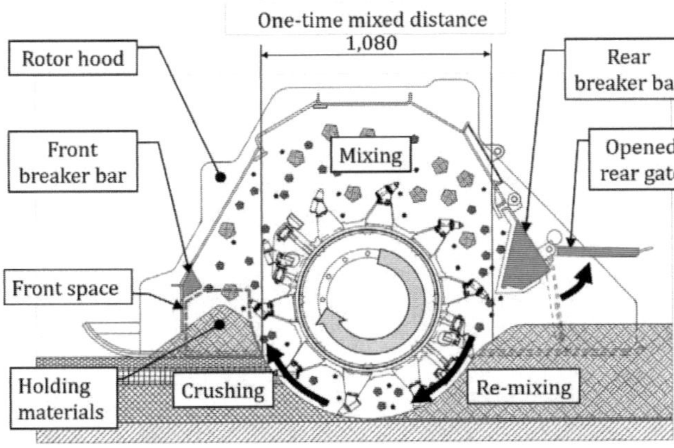

Fig .2. Schematic diagram of the rotor and inside the rotor hood during crushing and mixing

2.2. The layers and grain size distribution curves of test section

The test section consisted of three layers (surface course: deteriorated asphalt mixture 50 mm thick, base course: graded crushed stone M30 80 mm thick, and subbase course: recycled crusher run RC40 180 mm thick). The grain size distribution curves for each material are shown in *Fig. 3*.

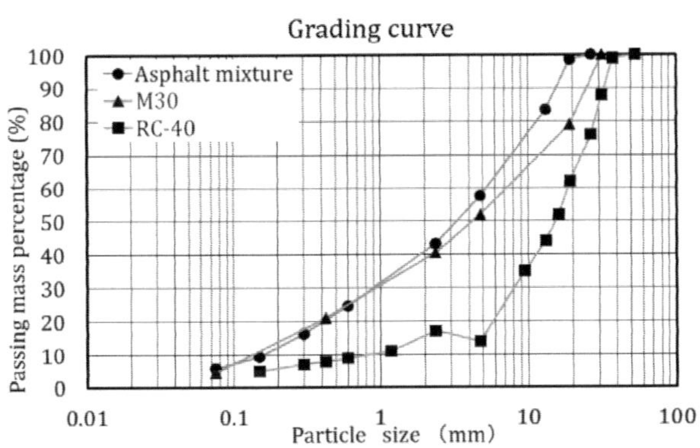

Fig. 3. Grain size distribution curve for three layers in test section

2.3. Markings by color, grid line and numbered stickers

The test section is 2,000 mm wide and 6,000 mm long, with a mixing depth of 300 mm and a total mixing volume of 3.6 m^3. To measure the particle moving distance after mixing, the surface of test section was delineated every 1,000 mm with six colors of spray paint, and a 100 mm square grid line was chalked across the central 1,300 m wide area as shown in *Fig. 4*. A total of 780 numbered stickers "markers" were pasted on the surface of each grid (*Fig. 5*). As stated above, particles with the markers recovered after mixing are called as "marking particles" hereafter. In the test section, the mixing start position of 0 is used as the reference point, the longitudinal direction of the section is the front-back, and the transverse direction of the section is the left-right. Crushing and mixing ranged from 0 to 6,000 mm, and marking particles after mixing were recovered in the range of 2,000 to 4,000 mm.

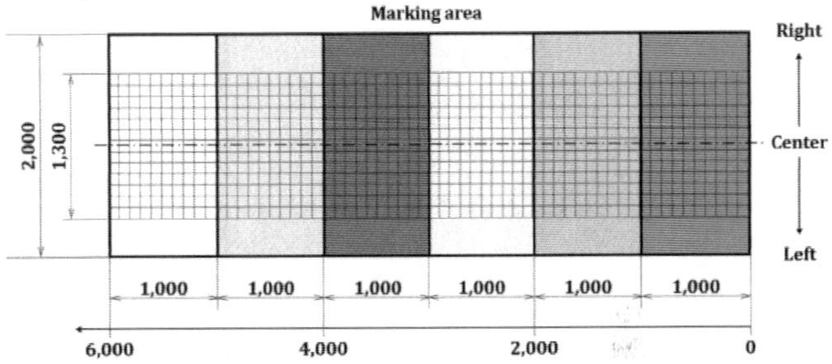

Fig. 4. Markings by color and grid line over the test sections

Fig. 5. Actual test section by markings: Color, grid line and markers

2.4. Operating conditions of the test machine

The rotational speed of the rotor was 130 rpm, pushed it to a depth of 300 mm first, and then forwarded at a speed of 1,500 mm/min. No additives were used to identify the marking particles easily after mixing. The rear gate was opened to suppress the effect of material deposited behind the rotor hood on the working speed and to keep the working speed of the test machine constant.

2.5. Recovery of marking particles and coordinate system

The marking particles were recovered by hand and with a small shovel. The surface height before mixing was used as the reference surface. *Fig. 6* shows the coordinate system and reference points of the test section after mixing. The marking particles were recovered at all four layers, i.e., the top layer (+100 to 130 mm) and every thickness of 100 mm in depth from 0 to -300 mm. The coordinates of the marking particles after recovered were recorded, and the maximum length of the particle was recorded as the maximum particle size.

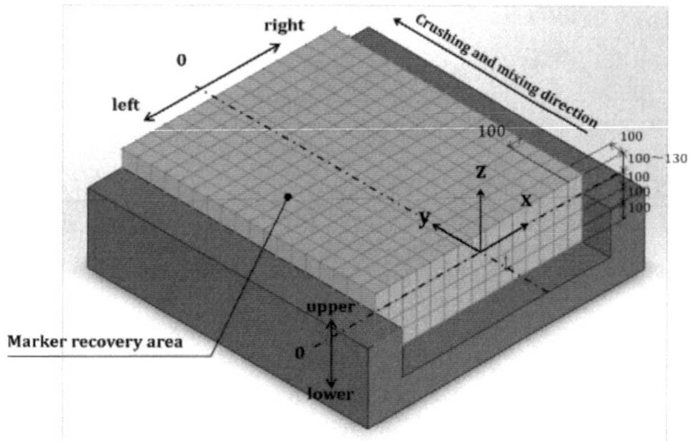

Fig. 6. Coordinate system and reference point of the test section after mixed

2.6. Analysis conditions by 3D-DEM (Katou et al., 2024)

The test results were compared with the analysis ones by 3D-DEM, which the conditions are outlined below. *Fig. 7* shows the 3D models. The rotor was placed on top of the layer model and allowed to sink vertically to a predetermined depth (300 mm) in the layer model while rotating the rotor, then moved horizontally at the speed of 1,500 mm/min as same as the test machine. The rotor equips 35 bits, the bit tip diameter is 1,150 mm, the rotor width is 700 mm, and the entire rotor is covered by a rotor hood.

The size of layer model is 7,000 mm long, 300 mm wide, and 430 mm deep. The particle size and number of particles in each layer are 50 mm and about 70,000 pieces for the surface course, 30 mm and about 80,000 pieces for the base course, and 15 mm and about 100,000 pieces for the subbase course, respectively. These conditions are the same as the test conditions except for the width of the layer model and the rotor.

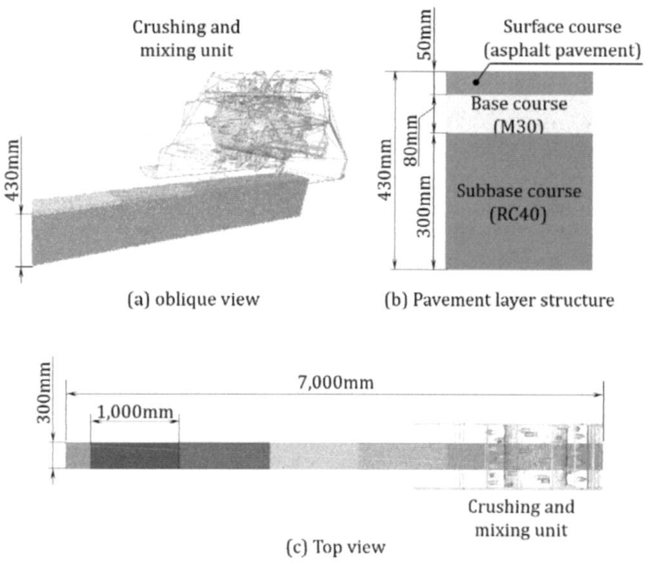

Fig. 7. 3D-DEM analysis model

Table 2 shows the particle size and passing mass percentage of the spherical particles used in the 3D-DEM analysis.

Table 2
Particle size and passing mass percentage of spherical particles used in 3D-DEM analysis

	Particle size [mm]	Passing mass percentage [%]
Surface course	10	50
(asphalt pavement)	18	100
Base course	11	50
(M30)	20	100
Subbase course	20.5	50
(RC40)	24.0	100

For the inter-particle properties, Altair's EDEM bonding model was applied to represent the interlocking phenomena between crushed stones and their bonding strength to the asphalt binder in the deteriorated asphalt mixture in the surface course.

3. Results and Discussions

In this section, the test results for the moving distance measured in the front-back direction and a number of particles counted in the left-right direction are discussed in order to evaluate the crushing and mixing conditions. These results were compared to the 3D-DEM analysis results.

3.1. Moving distance and percent of recovered marking particles in the front-back direction

The number of marking particles recovered was 153 pieces compared to the default of 780 pieces. *Fig. 8* shows the maximum particle size of the marking particles and the position of the particles after mixing. Each color indicates the section color of the initial locations of the marking particles, and the size of the circle indicates the maximum particle size. The positions before and after the move were connected by a straight line.

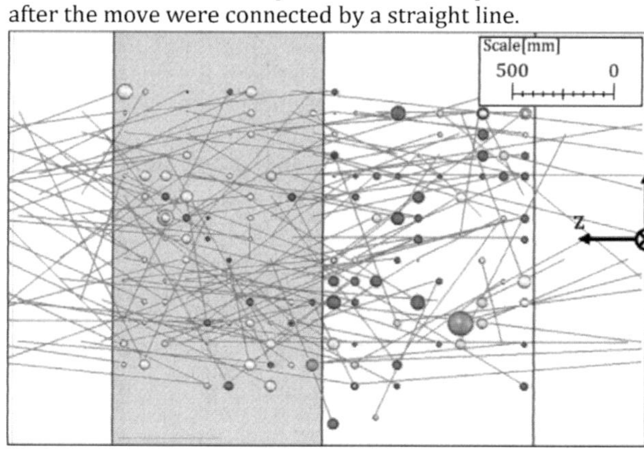

Fig. 8. Maximum particle size of marking particles and position of particles after mixing

Figure 9 is a schematic diagram showing the method of measuring the moving distance in the front-back direction of the marking particles before and after mixing. The dotted white circle with an "R" mark indicates the original position in the red zone before mixing, and the solid red circle with an "R" mark indicates the position after mixing. Similarly, the dotted white circle in the "B" mark indicates the original position in the blue zone before mixing, and the solid blue circle in the "B" mark indicates the position after mixing. The size of the circle indicates the maximum particle diameter of the marking particle as indicated in *Fig.* 8. Both the circles (R & B) are connected by a straight line. The direction of moving distance was marked with a plus (+) if the marked particles moved backward from the initial position, minus (-) if they moved forward from the initial position, and 0 if they did not move from the initial position, and the number of particles was counted for each moving distance.

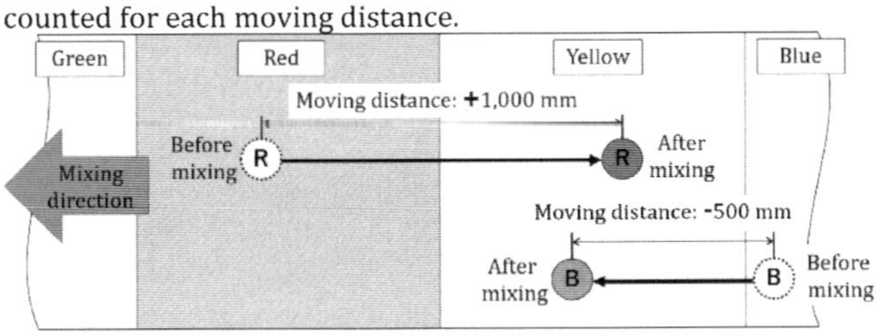

Fig. 9. Schematic diagram to measure the moving distance of marking particles

Fig. 10 shows the relationship between the moving distance in the front-back direction and the percent of recovered marking particles (called as PRMP hereafter) that is a percent of the number of particles at each moving distance in the front-back direction among the total number of particles recovered (153 pieces). Marking particles that moved backward from their initial position (0-2,000mm) were accounted for 120 pieces that is 78% of a total of 153 pieces recovered. Here, the evaluation is based on the sum of the number of marking particles counted in the three moving distance ranges. For example, the sum of the distance with the highest number of moving particles (21 pieces at 1,000 mm) and the number of moving particles in that ±100 mm range (8 pieces at 900 mm and 18 pieces at 1,100 mm) yields 47 pieces (31 %) between 900 and 1,100 mm. As described in 2.1 and in *Fig.* 2 above, this range approximates the one-time mixed distance of 1,080 mm. This implies that after the initial crushing process, most particles are blown backward and deposited through the gap between the rotor and the front and rear breaker bars, never to be mixed again. Particles distributed between 0 and 1,000 mm are considered to have been mixed multiple times or dragged along while being held in the front space as shown in *Fig. 2*.

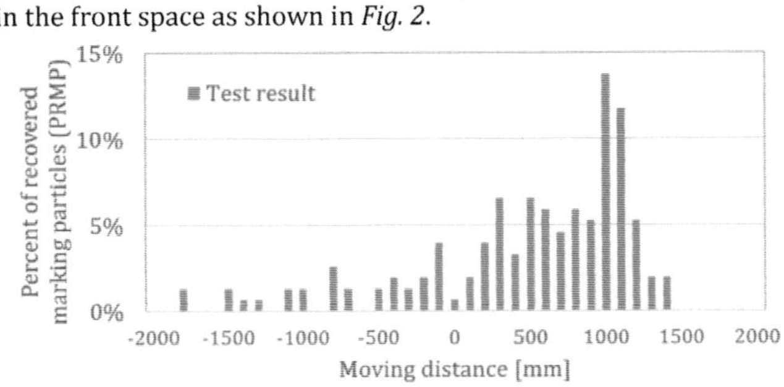

Fig. 10. Moving distance in front-back direction and PRMP

3.2. Normalized coordinate in the left-right direction and percent of recovered marking particles (PRMP)

Fig. 11 is the relationship between the normalized coordinate in the left-right direction and PRMP of the test section, with the center of the section as 0, the left end as -1, and the right end as 1. The average distribution percentages for the left and right sections are 6.3% and 7.3%, respectively. The crushed material was distributed almost uniformly on both sides of the test section.

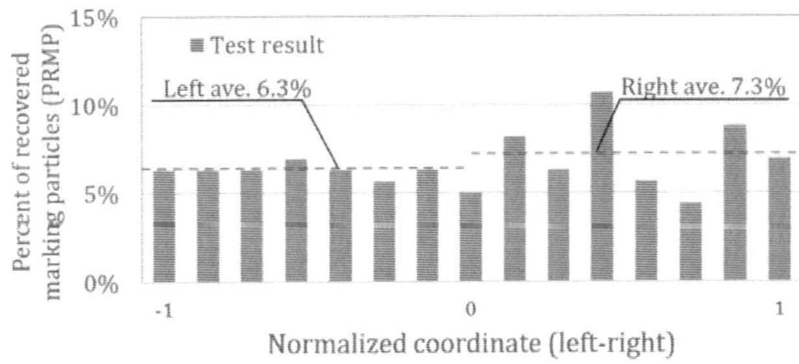

Fig. 11. Normalized coordinate in the left-right direction and PRMP

3.3. Comparative evaluation of test and 3D-DEM analysis results

Fig. 12 is the relationship between the moving distance in the front-back direction and PRMP (percent of recovered marking particles) for test and analysis results. The analysis results showed that 7,487 pieces (80%) of the crushed particles moved in the 0-2,000 mm range behind the rotor. Similar to the method of evaluating test results above, based on the total number of marking particles counted in the three moving distance ranges. For example, the sum of the distance with the highest number of moving particles (786 pieces at 900 mm) and the number of moving particles in that ±100 mm range (693 pieces at 800 mm and 738 pieces at 1,000 mm) yields 2,217 pieces (23%) between 800 and 1,000 mm.

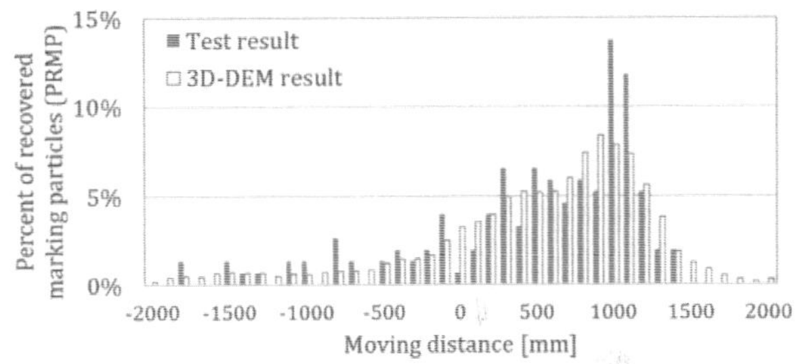

Fig. 12. Comparison of the moving distance of crushed particles in the front-back direction and PRMP (Test vs. 3D-DEM analysis results)

Fig. 13 is the relationship of normalized coordinate in the left-right direction and PRMP for test and analysis results. The analysis results show almost uniform distribution, similar to the test results except the percent of crushed particle at the left and right ends. This is simply due to the layer model width that is narrower than the rotor one as described above. Excluding the results in left and right ends, the analysis results showed that the mean distribution percentage was 5.8% for the left side and 5.8% for the right side, which was almost the similar trend as the test results (6.3% for the left side and 7.3% for the right side).

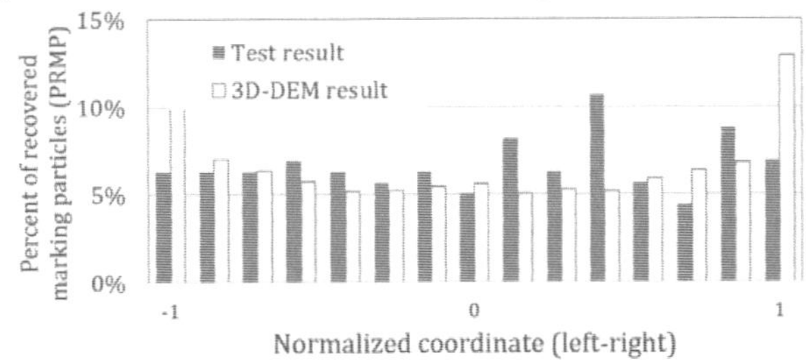

Fig. 13. Comparison of normalized coordinate in left-right direction and PRMP of crushed particles (Test vs. 3D-DEM analysis results)

4. Conclusions

A test was conducted on existing deteriorated pavement using a road stabilizer. The test results were evaluated and compared a with a 3D-DEM analysis. Results are listed as below.

- Regarding the test results for moving distance and the percent of recovered marking particles (PRMP) in the front-back direction, the number of marking particles that moved backward from the initial position was 120 pieces that is 78% of a total of 153 pieces recovered. Evaluated as the sum of the number of marking particles counted in the three moving distance ranges, the sum of the distance with the highest number of moving particles (21 pieces at 1,000 mm) and the number of moving particles in that ±100 mm range (8 pieces at 900 mm and 18 pieces at 1,100 mm) yields 47 pieces (31 %) between 900 and 1,100 mm. Since this distance range approximates the one-time mixed distance of 1,080 mm behind the rotor, it is concluded that most particles are blown backward through the gap between the rotor and the front and rear breaker bars and deposited on initial crushing, never to be mixed again.
- Regarding the 3D-DEM analysis results of the moving distance of the marking particles in the front-back direction and the percent of recovered marking particles (PRMP) also showed the similar trend as the test results. 7,487 pieces (80%) of the crushed particles moved in the 0-2,000 mm range behind the rotor. In addition, the total number of marking particles counted in the three moving distance ranges, the sum of the distance with the highest number of moving particles (786 pieces at 900 mm) and the number of moving particles in that ±100 mm range (693 pieces at 800 mm and 738 pieces at 1,000 mm) yields 2,217 pieces (23%) between 800 and 1,000 mm, which is similar range to one of the test results as above.
- The normalized coordinate in left-right direction and PRMP in the test, the average distribution ratio for the left and right sections is 6.3% and 7.3%, respectively. The crushed material was mixed almost uniformly on both sides of the section.
- The analysis results for the normalized coordinate in left-right direction and PRMP of crushed particles showed also the similar trend except the distribution rate of crushed particles at both the left and right ends. This is simply due to the width of the model layer that is narrower than the rotor width.
- Based on the similar trend between test and analysis results above, it was concluded that this analytical method is capable of simulating the behavior of crushed material in the rotor hood.
- This 3D-DEM analysis method enabled us to visualize for the first time the dynamic behavior of crushed particles in the rotor hood. Further research will be conducted to improve the crushing and mixing process by optimizing mechanical elements such as the shape and arrangement of rotor bits, rotor hoods, and breaker bars, as well as their operating conditions under various mixing conditions.

5. References

Japan Construction Machinery and Construction Association (JCMA), 2013.Japan Construction Equipment Handbook

Reeder GD, Harrington DS, Ayers ME, Adaska W., 2019. Guide to Full-Depth Reclamation (FDR) with Cement, No.SR1006P, National Concrete Pavement Technology Center

Katou T., Hirayama Y., Fujii T., Kurosu T., Oikawa O., Okayasu T., 2024. Three-dimensional Discrete Element Analysis of Crushing and Mixing Behavior of Existing Pavement Layer by Stabilizer

Sakai, Soil stabilizer and cold recycler. https://www.sakainet.co.jp/en/products/road_maintenance/ (accessed 01 May 2024).

2411 | Proposing Turning Motion of The Small and Lightweight Push-Rolling Rover with Minimal Configuration

Daisuke Fujiwara[a,*], Kojiro Iizuka[b]

[a] Suwa University of Science, Department of Mechanical Engineering and Electrical Engineering, Nagano, Japan

[b] Shibaura Institute of Technology, Department of Machinery and Control Systems, Saitama, Japan

* Corresponding author: fujiwara@rs.sus.ac.jp

ABSTRACT

Some organizations in Japan have planned planetary explorations using a small rover. However, the rovers with cylindrical wheels have a risk of failing to move in the loose soil area, such as the surface of the lunar/planet. The movement using the supporting force of the locked wheel, for example, a wheeled walking, or push-pull locomotion, can reduce the risk. Previous studies have mainly developed relatively large rovers over a mass of 10 kg. Meanwhile, our group has developed a small and lightweight (under a mass of a few kg) push-rolling rover with minimal configuration, and our previous studies indicated that the rover could climb steep slopes over 30 degrees with low slip. Planetary exploration rovers are required to move in an arbitrary direction; therefore, the developed rover also needs to have a function of a turning motion. This paper proposes the turning motion while inching for the push-rolling rover with minimal configuration and evaluates its performance by the traveling experiments at different conditions (driving conditions, and wheel shapes). The experimental results indicated that the proposed turning motion using high wheel slip and lug's wheel realized turning with low slip on loose surfaces with steep slopes. Additionally, the experimental results also showed that the slip amount of the rotational center is lower than that of a normal skid-steering rover.

Keywords
Turning motion
Wheel walking robot
Push-Pull Locomotion
Loose soil
Mobile robot
Planetary exploration

1. Introduction

JAXA and some organizations, such as ispace, are planning to a direct planetary exploration using a rover (Ispace, 2024, JAXA, 2024). Recently, a private company Dymon also has planned a direct exploration of the Moon using a small-wheeled rover (Dymon, 2024,). In these missions, the wheeled rovers are often considered. Furthermore, a small rover has been focused on due to rocket payload (JAXA, 2024, Dymon, 2024). However, the planet's surface consists of loose soil, called regolith. The wheeled type rovers are easy to become stuck on the planet's surface. To address this problem, previous studies have developed a wheel-walking rover (Andrage et al., 1998, Creager et al., 2015). The method can use the shared force between a locked wheel and soil. Furthermore, the shared force is larger than that of the driving force of the rotational wheel. Then, the rover extends or shrinks its wheelbase using the force of the locked wheel. The previous study confirmed that this method suppresses the slip on loose ground (Creager et al., 2015). Furthermore, our previous study confirmed that a lightweight wheel-walking rover with a minimal configuration can climb the loose soil with a high slope (32 degrees) (Fujiwara and Iizuka). Although the minimal-wheeled walking rover indicated high traveling performance on the loose soil, the turning motion was not developed thus far. In previous studies, some wheel-walking rovers has been developed, however, these rovers uses skid-steering motion (Andrage et al., 1998, Creager et al., 2015). As one of the few examples, a previous study has proposed an omnidirectional wheel-walking rover but the rover used many actuators (Cordes and Babu, 2016).

Therefore, this paper proposes a turning motion by a minimal-configuration rover with an articulated chassis. In the proposed motion, the rover expands its wheelbase while steering wheel. Then, this paper validates its traveling performance at different conditions on the loose soil with a high slope angle. Furthermore, this paper compares its performance to the conventional skid-steering motion.

2. Method

2.1. Wheel-walking

Figure 1 shows the conventional wheel walking mechanism, called push-pull locomotion or push-rolling locomotion. As mentioned in the introduction part, the rover locked a one-wheel unit relative to the ground. Then, the rover expands or contracts its wheelbase and move another wheel unit while driving the wheel.

In our previous study, the rover can climb the loose surface with a slope angle 32 degrees (Fujiwara and Iizuka).

2.2. Proposed turning motion

Figure 2 shows a proposed turning motion. First, the rover steers the front wheel unit δ, and the rover expands its wheelbase while rotating the joint mechanism of the rear unit. Then, the rover adjusts each steering angle of the front and rear wheel. Finally, the rover contracts its wheelbase. This motion implements rotating and expansion/contraction motion simultaneously. Therefore, the rover can move directly in a diagonal direction.

https://doi.org/10.56884/EI2EH2SM

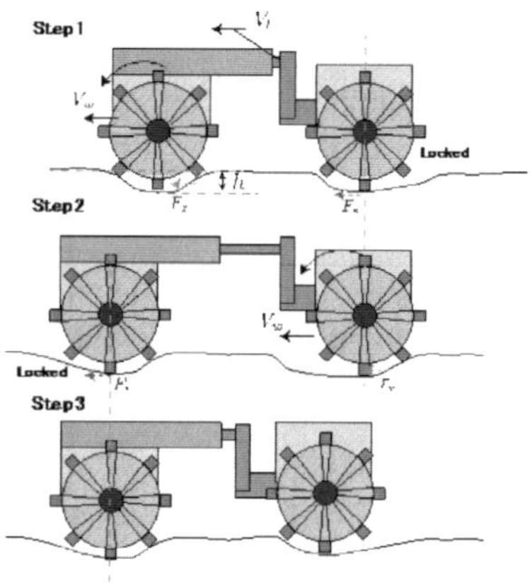

Fig. 1. Wheel-walking motion

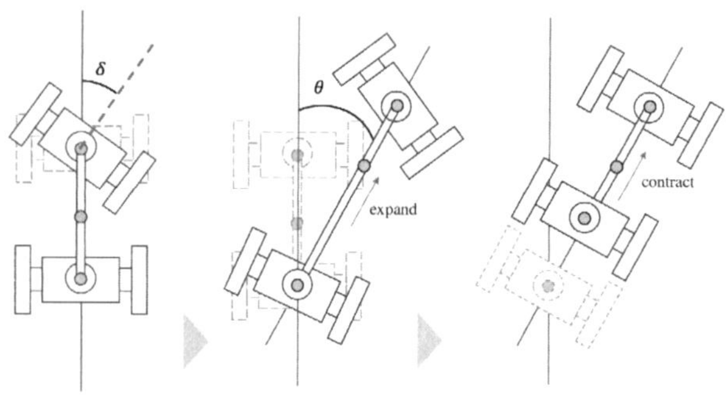

Fig. 2 Proposed turning motion

3. Experiments

3.1. Experimental settings

Figure 3 shows experimental environments. The loose soil is Silica sand No. 5, which is installed in the soil bin. The motion capture system measures marker positions attached on the rover testbed. The leveling plate smooths the soil ground before the experiment. Then, the rover testbed is set on the loose soil and performs turning motion or skid-steering motion.

To realize the proposed motion, this paper develops a testbed rover (Fig. 4, 5). The front and rear wheel unit have a steering function and a passive joint mechanism, respectively. Each wheel unit is connected by a linear actuator. Each wheel can rotate arbitrarily speed by PID control, and the linear actuator is driven on 12 V with enough torque.

Table 1 summarizes the experimental conditions and specifications of the testbed rover. This paper uses a two-wheel type: a simple circular wheel with sandpaper and a wheel with 18 lugs. The rover turns in two modes: conventional skid-steering and proposed motion. In the case of the skid-steering mode, the rotational wheel speed is 5 rpm. Meanwhile, the driving wheel speeds are high (26 rpm) and low (5 rpm) in the case of the proposed motion, while the linear actuator speed is constant (20 mm/s). Furthermore, the steering angle δ is from approximately 12° to 35°. The slope angle is set at 32°.

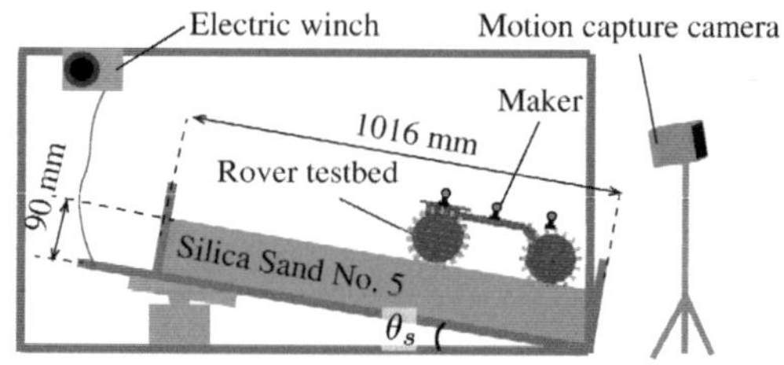

Fig. 3. Experimental environment

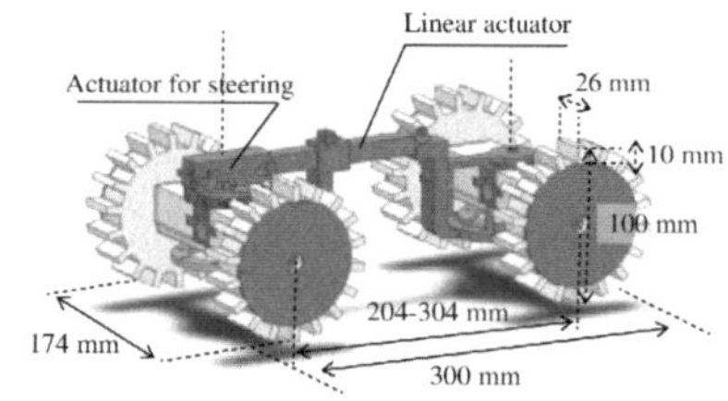

Fig. 4. Schematic view of the rover test bed

Fig. 5. Overview of the rover test bed

Table 1
Experimental conditions and rover specifications

Description [Units]	Value
Wheel width [mm]	26
Wheel diameter including lug's height [mm]	100
Number of lugs [-]	18
Mass [kg]	0.84
Rotational wheel speed (shrink and extend) [rpm]	5, 26
Rotational wheel speed (steering) [rpm]	5
Linear actuator speed [mm/s]	20
Range of wheelbase length [mm]	204-304
Soil [-]	Silica sand No. 5
Soil density [g/cm^3]	1.43
Angle of repose [°]^a	32.5
Soil cohesion [N/m^2]^b	761.84
Slope angle [°]	32

a: measured, b: reference (Mizukami, 2013)

3.2. Evaluation index

This paper mainly focuses on the slip distance of the rotation center position d_c in both turning motions (Fig. 6). Furthermore, this paper evaluates the difference in the front wheel position d_f between ideal and actual after contraction in the case of the proposed turning motion (Fig. 7).

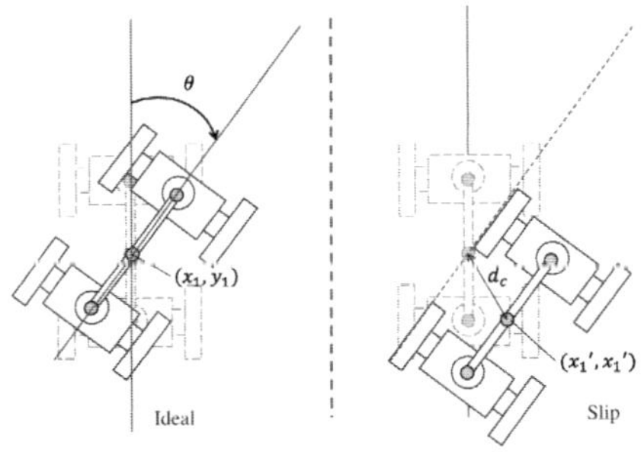

(a) Skid-steering motion

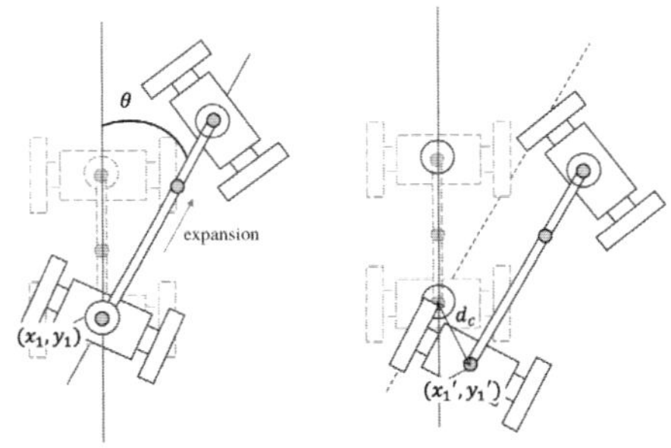

(b) Proposed turning motion
Fig. 6. Slip distance of each motion

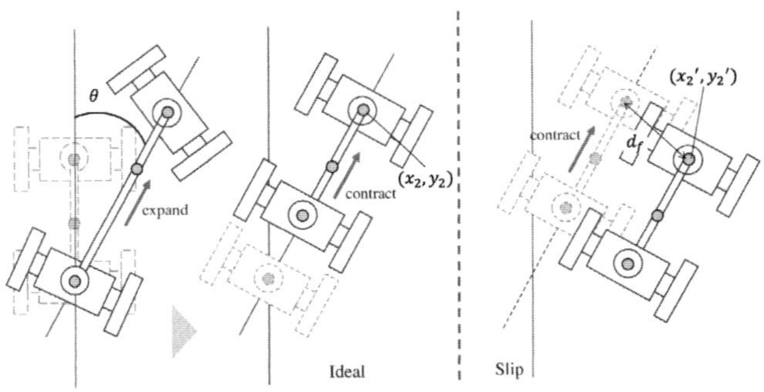

Fig. 7. Slip distance of the front position

4. Result

4.1. Skid-steering motion

Figure 8 shows the relationship between the rotation angle θ and slip displacement d_c in the case of the skid-steering motion. The indicated data is the typical data of the experiment. The displacement increases with an increasing rotational angle. Also, the average values of d_c at the angle 40°is 374.6 mm (lug's wheel) and 372.9 mm (circular wheel), respectively. $d_c$of the lug's wheel was larger than that of the circular wheel in most of the experiment but the rover of the lug's wheel indicated large d_c when the condition was worse. Therefore, the average values at 40°became a close value.

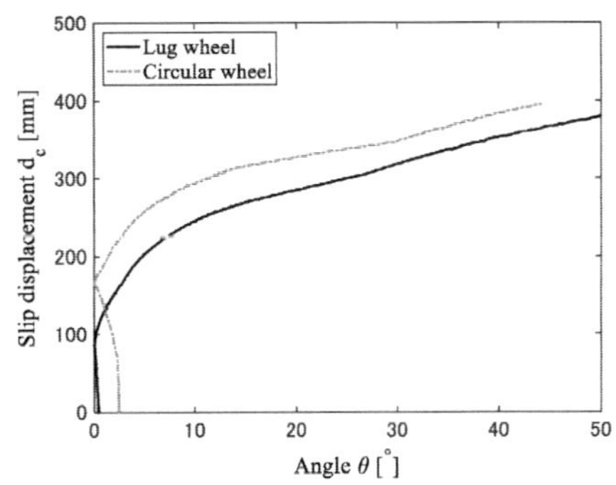

Fig. 8. Slip distance of the center position

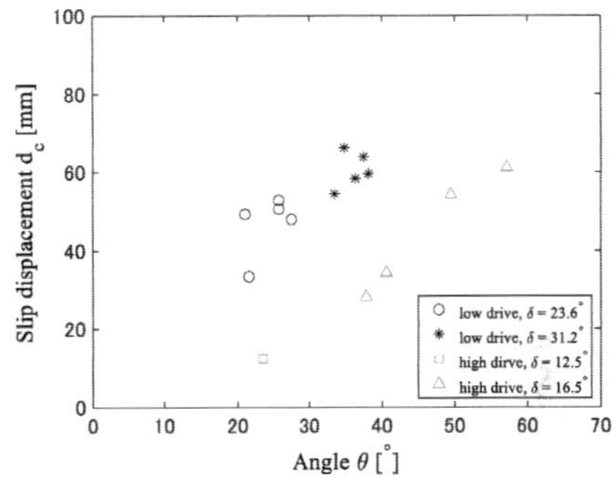

(a) Lug's wheel

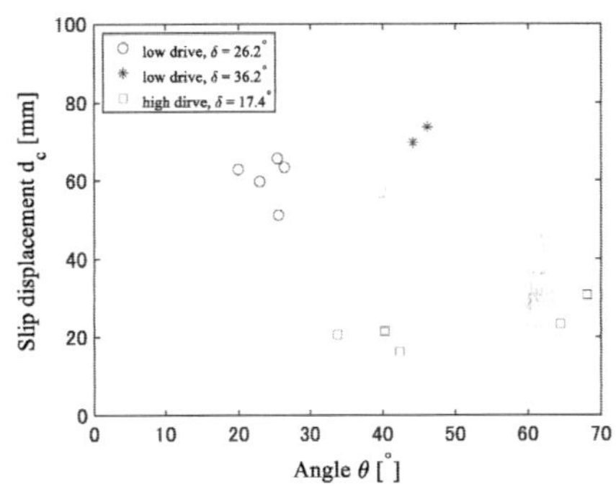

(b) Circular wheel
Fig. 9. Slip distance of the center position in the case of the proposed motion slope angle 32°

4.2. Proposed turning motion

4.2.1. Slip displacement d_c

Figure 9 shows the values d_c vs rotational angle θ at various steering angles. In the graph, a low and high drive means the rotational wheel speed is 5 and 26 rpm during expanding or contracting, respectively. When the rotational speed is high, the slip ratio of the wheel is high but the driving force and dynamic sinkage increases. These can contribute to increasing the supporting force of the locked wheel and decrease the load to the other wheel unit.

The results indicates that slip distances d_c under all conditions are smaller than that of the skid-steering. In the case of the proposed motion, d_c increases with an increasing angle θ. However, when the drive condition is high, d_c is suppressed. This trend is the same in both wheel types. As the differences between the lug's wheel and circular wheel, d_c of the circular wheel at the high drive condition and at large angle θ is small compared to the lug's wheel. When the rover uses the lug's wheel, the rotating wheel easily destroys the soil on the lower side. Therefore, d_c of the lug's wheel is large compared to the circular wheel.

4.2.2. Slip displacement d_f

According to Fig. 10, d_f increases with an increasing rotational angle θ. Furthermore, when the drive condition is low, the results indicated the increasing trend of d_f. This trend is the same in each wheel.

When the rover's drive condition is low, the slip distance d_f of the circular wheel indicates large values compared to the lug's wheel. This means that the circular wheel induces slip during contracting its wheelbase.

In both graphs, d_f is suppressed under $\simeq$ 70 mm in the case of the high driving condition. Therefore, the results confirmed that the rover can climb in a diagonal direction over 100 mm on the loose soil with a high slope angle.

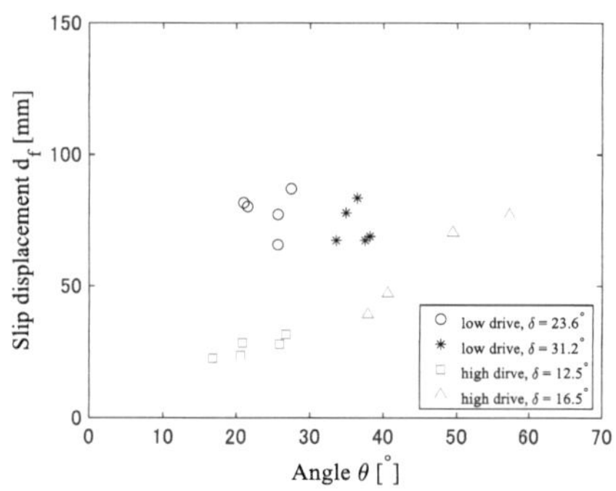

(a) Lug's wheel

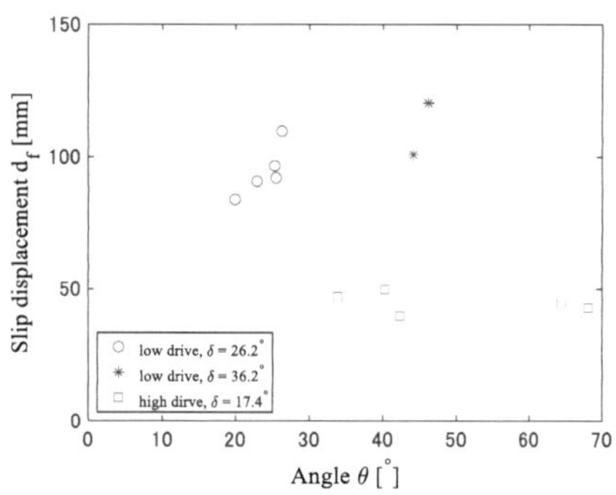

(b) Circular wheel
Fig. 10. Slip distance of the front position in the case of the proposed motion slope angle 32°

4.2.3. Relationship between angle θ and steering angle δ

According to Fig. 11, the angle θ increases with an increasing the steering angle δ. Furthermore, the angle θ depends on the driving condition, for example, the angle θ becomes large when the driving condition is high even under the small steering angle δ. Additionally, the behavior becomes unstable when the driving condition is high and the range of angle θ becomes high. This means that the front unit rapidly turns. These results suggested that there is an available range of turning angles.

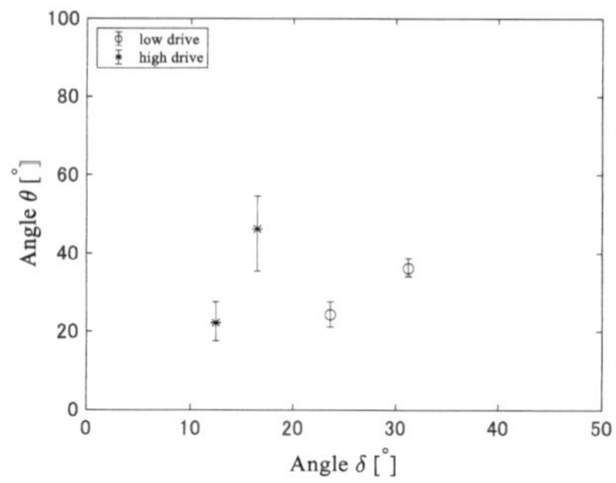

(a) Lug's wheel

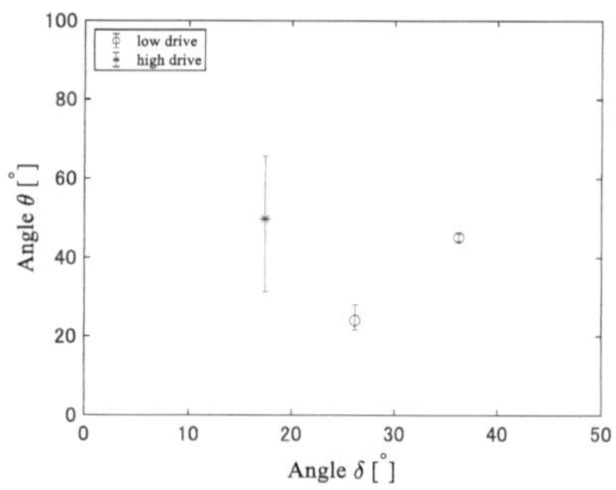

(b) Circular wheel
Fig. 11. Relationship between the angle δ and angle θ

4.3. Overview of each actual experiment

Figure 12 shows the overview of the actual experiments when the wheel is circular. When this paper focuses on the rotation center, the proposed motion suppresses the slip compared to the skid-steering motion. The results also suggested that the proposed motion could move in a diagonal direction directly.

(a) Skid-steering motion

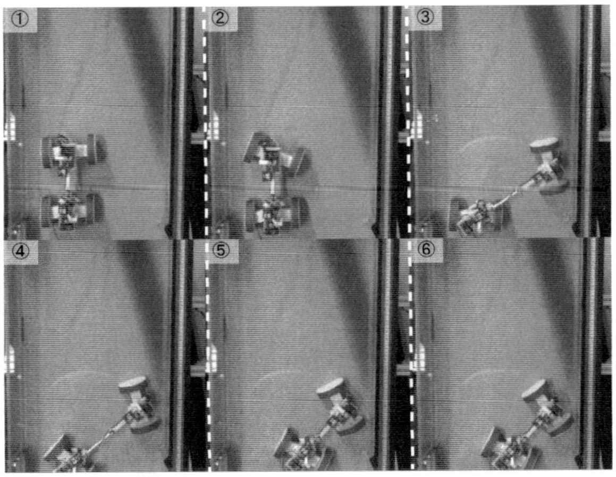

(a) Proposed turning motion: $\delta = 17.4^{\circ}, \theta = 64^{\circ}$
Fig. 12. Overview of the typical experiment: slope angle 32°, circular wheel

5. Conclusion and future works

This paper proposes a turning motion for a small wheel-working rover and confirms its traveling performance on loose soil with a steep slope at different driving conditions and wheel types. The traveling test suggested that the proposed turning motion suppresses the slip distance of the rotational center position during turning. Specifically, the high driving condition of the rover could greatly reduce the slip. Furthermore, the results confirmed that the rover could climb in a diagonal direction at least 100 mm (half of the ideal traveling distance). However, the high driving condition induces unstable conditions in the high turning angle θ. This result suggests that there is a limitation of the rotational angle. Therefore, future works have to consider the relationship between the driving condition and steering angle, and this work includes consideration of the longitudinal and lateral slip based on a mathematical simulation. Furthermore, future works confirm the turning performance in the case of another initial position, for example, the rover is located diagonally to the slopes. Additionally, validation in terms of energy efficiency should be addressed.

6. Nomenclature

d_c	Slip distance of the rotational center	[mm]
d_f	Slip distance of the front wheel	[mm]
θ	Rotational angle	[°]
δ	Steering angle	[°]

7. Acknowledgements

This work was supported by JSPS KAKENHI Grant Number JP21K14105.

8. References

Andrade, G., Ben Amar, F., Bidaud, P., Chatila, R., 1998, Modeling Robot-Soil Interaction for Planetary Rover Motion Control, Proc. 1998 IEEE/RSJ International Conference on Intelligent Robots and Systems. Innovations in Theory, Practice and Applications.

Creager C., Johnson, K., Plant, M., Moreland S., Skonieczny, K., 2015, Push–pull locomotion for vehicle extrication, J. Terramechanics, 57, 71-81.

Codes, F., Babu, A., 2016, SherpaTT: A Versatile Hybrid Wheeled-Leg Rover, Proc. of the 13th Int. Symposium on Artificial Intelligence, Robotics and Automation in Space (i-SAIRAS 2016), 2016.

Dymon Yaoki, https://dymon.co.jp/yaoki/ (accessed 3 June 2024).

Fujiwara D., Iizuka K., 2023, Suppressing the Reduction of the Traveling Displacement on Loose Soil for Rovers with Function of a Wheel-Walking, Proc. of the 16th European-African Regional Conference of the ISTVS | Lublin, Poland, October 11-13, 1453, 2023.

Ispace, https://www.ispace-inc.com/jpn/ (accessed 3 June 2024).

JAXA SLIM mission, https://www.jaxa.jp/projects/sas/slim/index_j.html (accessed 3 June 2024).

Mizukami N. 2013, Study on Development Locomotion Performance Based on Terramechanics for Wheeled Rover on Soft Ground, Doctor thesis.

2651 | Rapid assessment tools for estimating trafficability of low volume roads

John F. Rushing, [a,*] Lulu Edwards, [a] Haley Bell, [a] Margarita Ordaz [a]

[a] *U.S. Army Engineer Research and Development Center, Vicksburg, Mississippi, United States*

* *Corresponding author: john.f.rushing@usace.army.mil*

ABSTRACT

Rapid assessment of low-volume road surfaces remains a challenge when attempting to forecast allowable vehicle crossings. Variations in soil type, compaction effort, and moisture content of the soil can greatly affect trafficability, and predictive equations for soil deformation under vehicle loads often have reduced reliability for low-strength materials. Portable tools to characterize soil stiffness and corresponding relationships to load-induced deformation are needed. In this effort, researchers performed comparative testing of multiple rapid assessment tools as potential devices for giving estimations of vehicle trafficability. The test devices included a Clegg hammer and light weight deflectometer as instruments that measure response from impulse loading. Silty sand with and without chemical stabilization (using cement) at varying moisture content were used for testing. These soil states represented very weak conditions capable of supporting fewer than 50 vehicle passes to moderate strength conditions capable of supporting several thousand vehicle passes. Data from full-scale tests were used to correlate allowable traffic with data obtained from the rapid assessment tools. Recommendations from the effort include ranges of response data to categorize low-volume road surfaces based on their ability to handle ranges of vehicle loadings.

Keywords
Rapid assessment
Mobility
Soil characterization

1. Introduction

Transportation networks throughout the world are comprised predominantly of low-volume roads. These road systems provide access to rural areas and enable transit of people and goods to support basic economic activity. Often, the roads are comprised only from soil with no waterproofing surface layer such as asphalt or concrete. Limited engineering design or construction effort is typically employed when these routes are established, but their serviceability remains adequate due to limited lightweight traffic and proper maintenance. Scenarios arise when predictions of vehicle-road interaction are needed to understand trafficability on the low-volume roads. One example is the high number of relief vehicles supporting disaster recovery where traffic levels and environmental conditions could be very different. Traditional pavement evaluation techniques cannot accurately predict when significant damage to these structures will occur. Simple, rapid techniques are needed to evaluate whether low-volume roads can handle heavy traffic for short periods.

2. Previous work

The first light weight deflectometer (LWD) was developed in Germany in 1981 (Yzenas and Cook, 2019). Since that time, several studies have explored its capabilities for construction quality control as well as for assessing pavement structures (Livneh and Goldgerg, 2001; Icenogle and Kabir, 2013; and Schwartz et al., 2017). Ordaz et al. (2023) describes LWD testing for low-volume roads and suggests that impulse stiffness modulus (ISM) values below 29 kN/mm correlate to road structures capable of supporting fewer than 1,000 load repetitions when trafficked by heavy trucks (4 axle; 26,308 kg).

The first Clegg Impact Soil Tester, or Clegg hammer, was developed in Australia in the 1970s by Dr. Baden Clegg (Lafayette Instrument, 2020). The Clegg hammer has been used to test soil and aggregate surfaces by measuring the soil response to the impact of a falling mass, and it is available with hammer masses ranging from 0.5 kg to 20 kg. Freeman et al. (2008) recommended using the Clegg hammer to evaluate the strength of cement-stabilized layers for contingency airfield construction. In Freeman's study, the researchers utilized the Clegg hammer results to correlate to unconfined compressive strength and California Bearing Ratio (CBR) measurements.

3. Purpose and scope

The current study explored the use of two nondestructive portable testing devices, the LWD and Clegg hammer, as candidates for providing indices of route trafficability. The objective was to determine if either test device could accurately and repeatably provide an index that could identify soil surfaces capable of withstanding heavy truck traffic. The two test devices were a Zorn ZFG 3.0 LWD with a 300-mm-diameter loading plate and 10-kg drop weight along with a Lafayette Instrument Clegg Impact Soil Tester (Clegg hammer) with a 4.5-kg drop weight. Soil trafficability testing was performed using a four-axle heavy truck weighing approximately 26000 kg.

Test areas used to evaluate these rapid assessment tools were prepared using local materials to meet desired characteristics of a deformable soil. In the first series of tests, soils were placed in an area formed by removing a 4-m by 5-m concrete slab from a paved area and excavating to a depth of 1 m. This series of experiments

https://doi.org/10.56884/3VZXIQNF

varied by stabilizer type and dosage and resulted in a range of soil stiffness values that were verified through traffic testing. The traffic tests consisted of repetitive crossings over the soil and periodic measurements of permanent deformation.

The next series of tests were like the first but were performed at a larger scale on a simulated road lane constructed at an outdoor test site. This second set of experiments also maintained a consistent soil type with varying dosage and type of modifier used to enhance soil stiffness.

For each set of experiments, the measurements provided by the rapid assessment tools were compared to rut depth with accumulating traffic up to 1000 passes. Data were analyzed to determine if the test devices provided results that could be used to rank the areas in terms of trafficability and to create general categories for classifying suitability for withstanding heavy truck traffic.

4. Field tests

The two test series consisted of a soil surface classified as silty sand (SM) in accordance with standards for the Unified Soil Classification System (*Fig. 1)*. The soil was prepared by blending three parts sand with one part silt using local sources. Silty sand is the prevailing global soil type and is prone to rutting through densification and shear behavior (Robinson and Rabalais, 1993). In these tests, soil stiffness was modified by including soil stabilizers designed to improve strength and trafficability.

The additives were placed at 5% of the total dry weight of the silty sand. The moisture content of each silty sand stockpile before placement was approximately 10%, which was wetter than the optimum moisture content of 8%. Testing with a soil moisture content wet of the optimum represents a conservative performance result and is also more likely to be encountered. Soils drier than optimum are easier to manage since water can be added.

Minimal equipment was used to construct the test areas. This was due to the study's objective of decreasing labor, equipment, and material resources for rapid road stabilization. The equipment included a compact track loader (CTL) with a standard bucket work tool attachment, a tiller work tool attachment, and a vibratory drum compactor work tool attachment.

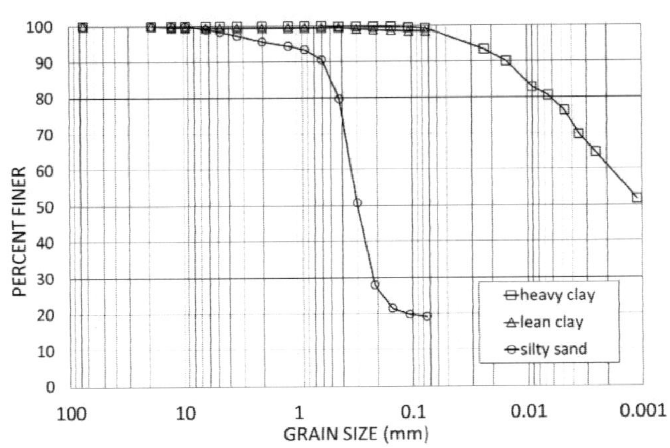

Fig. 1. Particle-size analysis for soils used in field tests

4.1. Rapid assessment tools

4.1.1. Clegg hammer

The 4.5-kg Clegg hammer was used in this study to estimate the soil strength, in accordance with ASTM D5874. The Clegg hammer consisted of a flat-ended cylindrical mass and a guide tube (Lafayette Instrument, 2020). The mass is dropped from a predetermined height marked with a line, and an accelerometer on the mass measures the peak deceleration. This measurement is converted to a Clegg Impact Value (CIV), which is reported on the tablet associated with the Clegg hammer. Each test consisted of a total of four drops, and the maximum of the four drops was used as the representative CIV for each test. Higher CIVs indicate a stronger soil layer. The CIV can be used to evaluate the relative strength of a layer of soil up to a 15-cm depth (ASTM D5874). CIV values were used as indicators to compare soil stiffness to trafficability.

4.1.2. LWD

LWDs can be used to assess materials with a range of stiffness by varying the mass of the falling weight and the diameter of the contact plate. For this study, a 10-kg weight and 300-mm-diameter plate were selected because compacted soils have stiffness values that are relatively low compared to conventional high traffic pavements. Testing was performed in accordance with ASTM E2835. The measured deflections were used to calculate ISM, an index property used to describe the overall stiffness of a pavement section, by dividing the impulse load (7.07 kN) by the deflection (Zorn Instruments, 2016). Higher ISM values represent stiffer materials.

4.2. Test Series 1

Test Series 1 consisted of neat and stabilized soil compacted within the interior area of a 38-cm-thick concrete pavement. The individual test locations consisted of one slab area, which was 4 m wide by 5 m long. After removing the pavement surface, the void was filled with a heavy clay layer *(Fig. 1)* and surfaced with a target of 18 cm of silty sand. This process was conducted with and without mixing with soil stabilizers. The soil surfaces were compacted flush with the surrounding pavement.

The natural subgrade was classified as a lean clay *(Fig. 1)* and had an average CBR of 9% with a dry density of 1730 kg/m^3 and a moisture content of 18%. Geotextile was placed on top of the natural lean clay subgrade to prevent migration of the subgrade into the base.

The heavy clay base was placed and compacted in three lifts for a total compacted thickness of 38 cm. The base had an in-situ moisture content of approximately 35% with a CBR of 3% to 5%, as estimated with the dynamic cone penetrometer (DCP). The heavy clay base was chosen to ensure a consistent base for each of the test items and was carefully controlled when placed in the test area. The dry density was approximately 1345 kg/m^3. The heavy clay base material remained in place and was reused after each test.

Preparation of each stabilizer-soil mixture occurred in batches. The CTL bucket was used to place the loose silty sand in a marked 3-m by 7-m area adjacent to the test area. Laying the soil out in this rectangle at the specified height of 20 cm ensured that the ratio of soil to stabilizer was accurate and could be repeated.

The bags containing the stabilizers were placed evenly on top of the silty sand. The bags were opened and spread on top of the silty sand pile by hand, and the material was raked over the top of the pile. The CTL tiller attachment was used to till the stabilizer product into the loose silty sand. The CTL bucket was then used to flip the mixed pile repeatedly, taking care to scoop the bottom of the pile to the top. When a visual assessment showed the mixture was a consistent color, it was ready to be placed into the test area.

The silty sand mixtures were placed and compacted in two lifts approximately 20 cm to 25 cm thick. Compaction was completed with a vibratory drum compactor CTL attachment. The stabilized surface material was cured for 24 hours after compaction before LWD and Clegg hammer measurements were taken, and traffic was applied. Test locations for the measurements are shown in *Fig. 2*. Traffic for the control item was also not applied until 24 hours after placement.

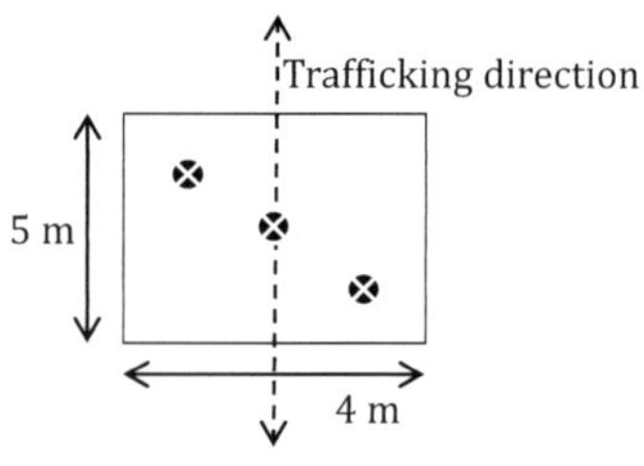

Fig. 2. Test location for LWD and Clegg hammer measurements for Test Series 1

4.3. Test Series 2

The second test series explored scaling the soil stabilization methods to provide more realistic performance data. The test areas varied in size (4.3 m wide by 12 m to 30.5 m long) and were larger than the Test Series 1 items. These test items were constructed over native soil classified as a lean clay overlain with approximately 18 cm of silty sand prepared from the same materials as Test Series 1; the silty sand surface, placed in three lifts, was 30 cm thick.

Test Series 2 included neat soil along with soil stabilized using either ordinary portland cement or a rapid-setting cementitious material. The test area was not confined with concrete as it was in Test Series 1. Replicates of test areas were prepared to evaluate the construction process of soil stabilization and provide additional data for comparison.

In contrast with Test Series 1, preparation of each stabilizer-soil mixture occurred in place to simplify the construction process. A CTL bucket was used to place the loose silty sand in the test area. The bags containing the stabilizers were placed evenly on top of the silty sand. The bags were opened and spread on top of the silty sand pile by hand, and the material was raked over the top of the loose lift. The CTL tiller attachment was used to till the stabilizer product into the silty sand. Each lift was tilled to ensure the stabilizer was evenly mixed into the test area. The silty sand mixtures were placed and compacted in one to three lifts, where each lift was approximately 8 cm to 10 cm thick when compacted. Compaction was completed with a vibratory drum compactor CTL attachment. The stabilized surface material was cured uncovered for 24 hours after compaction before LWD and Clegg hammer measurements were taken, and traffic was applied. Test locations for the measurements are shown in *Fig. 3*. Traffic for the control item was also not applied until 24 hours after placement.

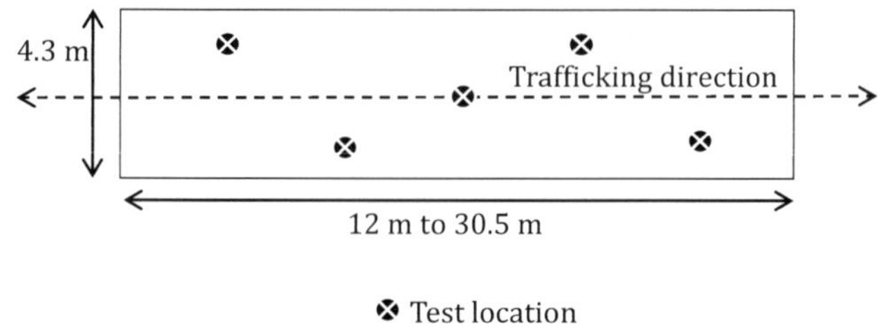

Fig. 3. Test locations for LWD and Clegg hammer measurements for Test Series 2

4.4. Traffic and performance measurements

Performance data were recorded in terms of accumulated rutting versus number of truck passes over the test area. Rut depths were measured using a straight edge, ruler, and with longitudinal and cross-section surveys. Channelized traffic was applied using a heavy truck loaded to approximately 26000 kg with tire pressures ranging from 410 to 620 kPa.

5. Results and discussion

5.1. Test Series 1

The first test series included one control test item with unstabilized silty sand that was expected to show rapid deformation under the heavy truck traffic. The remaining stabilized test items were expected to show varying levels of improvement for traffic stability depending on the dosage and type of chemical soil stabilizer used and the method for preparing and placing the material. This range of performance with a consistent soil type afforded the opportunity to characterize the soils using the rapid assessment tools to determine if they could accurately capture increase in soil stiffness that led to improved rutting performance.

Traffic on the test items proceeded until 1000 vehicle repetitions over the area were achieved or until at least 20 cm of rutting occurred. To provide comparable rut depth data for all test items, the measured deformation at 100 vehicle passes was selected for analysis. The control test item had excessive deformation exceeding 30 cm, which was nearly equivalent to the ground clearance of the vehicle (38 cm) between the axle and pavement. The strongest soils experienced fewer than 1 cm of rutting at 100 vehicle passes. Measurements from both rapid assessment tools were analyzed to determine the relative ability of the device to provide measurements relating to trafficability. Three tests were taken with each device for each test item, and the average of these data are provided in *Table 1* and plotted without the control data, for clarity, in *Fig. 4*.

Table 1
Test Series 1 soil measurements and traffic results

Test Item	CIV	LWD ISM [kN/mm]	Rut Depth at 100 Passes [cm]
1) Control	5.37	1.92	31.12
2) 3% Type I/II cement	28.97	24.84	3.18
3) 5% Type I/II cement	--[a]	43.82	0.64
4) 5% CTS Rapid Set (with citric acid)	15.68	18.98	5.08
5) 5% StabilFast	44.40	37.21	1.27
6) 2% Type I/II cement, 3% xanthan gum	13.60	6.80	3.18

7) 5% CTS Rapid Set	23.10	19.14	6.35
8) 5% Type I/II cement	40.97	28.90	0.64
9) 5% Type I/II cement (one lift)	46.37	29.30	0.64

[a]Did not test because device was in repair

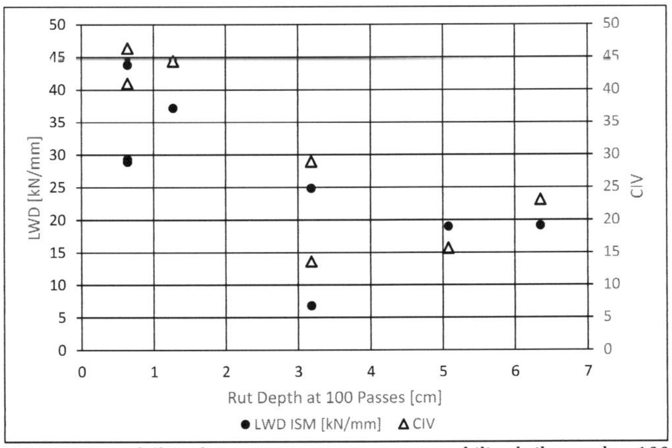

Fig. 4. LWD and Clegg hammer measurements on stabilized silty sand at 100 passes for Test Series 1

Both the Clegg hammer and LWD ranked Test Item 1, the control soil, as the weakest with test values well below those of any stabilized soils. This was expected since performance test results demonstrated that the unstabilized soil had rut depths five times greater than the worst performing stabilized material. CIVs exceeded a value of 40 for Test Items 5, 8 and 9. These values have been reported in Freeman et al. (2008) to correspond to relatively stiff materials. The stabilized soils were found to be the best performers in terms of limited rutting. Test Item 3, not tested using the Clegg hammer since it was unavailable, also showed minimal rutting. Test Items 1, 4, and 6 had CIVs less than 20 and performed the worst in terms of rut resistance. Overall, the Clegg hammer provided index values that accurately tracked the rut depth performance at 100 passes for all nine test items.

ISM values calculated from LWD measurements ranged from 6.8 kN/mm to 43.8 kN/mm for the stabilized test items. Test Items 3, 5, 8, and 9 met the threshold recommended by Ordaz (2023) of ISM values greater than or equal to 29 kN/mm to achieve 1,000 vehicle repetitions. In fact, these four items were the highest performers in terms of rutting and the only four items on which 1000 passes were achieved before the rutting threshold had been met. Test Items 2 and 7 ranked next with ISMs ranging from 19 kN/mm to 25 kN/mm. These items also ranked next in terms of rutting performance and achieved 400 to 500 total vehicle passes before failure by rutting. Like the results from the Clegg hammer, Test Item 6, which consisted of a combination of cement and xanthan gum, had the lowest ISM (6.80 kN/mm) of the stabilized test items. This result was expected since the cement content was lessened and thus resulted in a lower material stiffness. Overall, the LWD provided similar and appropriate rankings to the Clegg hammer when evaluating rut depths at 100 passes for all nine test items.

5.2. Test Series 2

Five tests were taken with each device within each test item for Test Series 2. This series had three control items with unstabilized silty sand, as presented in *Table 2*. These items varied in the way they were placed. Control 1 was placed and compacted in a single lift. Controls 2 and 3 were placed and compacted in two equal lifts. The moisture contents of Controls 1, 2, and 3 were approximately 9.6%, 8.8%, and 10.4%, respectively. From a compaction perspective, Control 1 was expected to have lower overall strength because the layer was placed as one thick lift and would likely be inconsistently compacted. Control 2 was expected to have higher stiffness because the total moisture content was closer to the optimum value of 8%. The Clegg hammer and LWD both ranked the control items in the same manner, in which Control 2 was the stiffest in terms of CIV and ISM values of 11.04 and 6.48 kN/mm, respectively, while Control 1 was the weakest test item with CIV and ISM values of 4.07 and 1.60 kN/mm, respectively. These measurements also correlated well with field performance since Control 2 exhibited the lowest amount of rutting as opposed to Controls 1 and 3.

Table 2
Test Series 2 soil measurements and traffic results

Test Item	CIV	LWD ISM [kN/mm]	Rut Depth at 100 Passes [cm]
1) Control 1 (1 lift)	4.07	1.60	19.05
2) Control 2 (2 lifts)	11.04	6.48	13.97
3) Control 3 (2 lifts)	6.86	2.24	20.32
4) 5% Type I/II cement (1 lift)	47.97	9.25	43.82
5) 5% Type I/II cement (2 lifts)	32.37	21.87	3.81
6) 5% Type I/II cement (2 lifts)	43.10	39.94	1.91
7) 5% StabilFast (2 lifts)	45.20	30.00	1.91

Test Series 2 showed variability in data measurements for both the Clegg hammer and the LWD. The difference in the Clegg hammer CIV and LWD ISM measurements could be partially due to the Clegg hammer having a limitation of measuring depths up to only 15 cm, which is not sufficient to measure some of the depths that were obtained in this investigation. The LWD ISM has been proven in other literature to successfully evaluate rut depths up to 40 cm (Ordaz et al., 2024). Overall, the point being made is that the LWD, in terms of ISM, was a better indicator of trafficability on low-volume roads than the Clegg hammer. *Fig. 5* plots the data up to a rut depth of 7 cm to provide an easy comparison to Test Series 1 results plotted in *Fig. 1*.

The CIVs for test items stabilized with 5% cement ranged from 32 to 48 (Test Items 4 through 7), with an average value of 42.2. These values were like those from Test Series 1 where the CIV for similar stabilized soil (Test Series 1, Test Items 8 and 9) averaged 43.7.

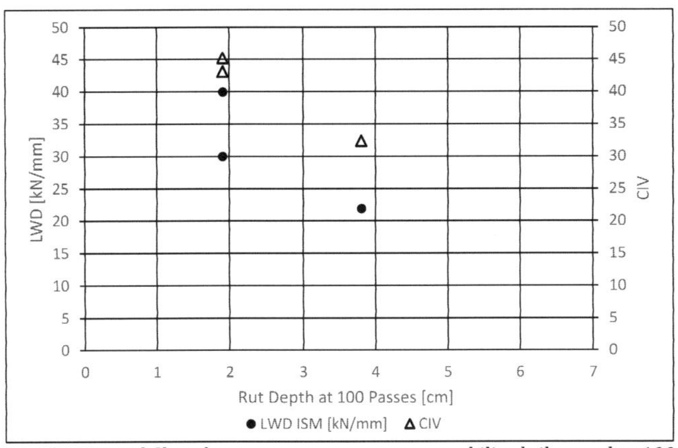

Fig. 5. LWD and Clegg hammer measurements on stabilized silty sand at 100 passes for Test Series 2

The ISM values calculated from LWD data on the stabilized test items ranged from 9 kN/mm to 40 kN/mm. Test Items 6 and 7 had ISM values exceeding 29 kN/mm that served as the indicator of suitable trafficability.

Test Items 6 and 7 both achieved 1000 vehicle passes before the target rut depth was met, and the results of the Clegg hammer and LWD were both above the threshold values of 40 and 29 kN/mm, respectively.

Test Items 4 and 5, constructed on the same day, had wet subgrade material (12% moisture content) because it had rained prior to the test. The Clegg hammer and LWD results for Test Items 4 and 5 were also inconsistent with the findings from Test Series 1. Test Item 4, placed in one lift which typically results in inconsistent stabilization, had a high rut depth at 100 passes (43.82). This was not predicted by the Clegg hammer, which measured a high CIV (47.97) but was predicted by the LWD, which showed a low ISM value of 9.25 kN/mm. Test Item 5, with a slightly lower CIV (32.37) and lower ISM (21.87 kN/mm), also had a moderately higher rut depth of 3.81 cm. This shows that the LWD and Clegg hammer may be able to predict a range of performance.

6. Conclusions and future work

The Clegg hammer and LWD show promise in indicating if a soil or stabilized soil is susceptible to rutting. These two devices are portable and can be utilized to determine route trafficability in remote areas. The Clegg hammer is limited to a 15-cm measurement depth, which means thicker depths are not applicable. Additionally, it is recommended that an evaluation of the layers under the stabilized layer, the subbase or subgrade, be measured with the Clegg hammer as well for better representation.

Like Schwartz et al. (2017), who stated that a target modulus should be set for quality assurance, the results for the LWD in this study showed that a threshold value of 29 kN/mm could be used for predicting a stable surface that allows up to 1,000 traffic passes.

Overall, the study showed that the LWD was a better indicator of trafficability on low-volume roads than the Clegg hammer. Additional testing of varying soil types and strengths is needed to validate this claim and to determine if the devices can be used to predict traffic pass number, rather than just a threshold value to determine if the roadway is suitable for up to 1,000 passes.

7. Acknowledgements

The use of trade, product, or firm names in this document is for descriptive purposes only and does not imply endorsement by the U.S. Government. The tests described and the resulting data presented herein, unless otherwise noted, are based upon work conducted by the U.S. Army Engineer Research and Development Center supported under PE 0633119, Project BL8, "Power Projection in A2/AD Environments." Permission was granted by the Director, Geotechnical and Structures Laboratory to publish this information. The findings of this document are not to be construed as an official Department of the Army position unless so designated by other authorized documents.

8. Declaration of competing interest

The authors declare that they have no known competing financial interests or personal relationships that could have appeared to influence the work reported in this paper.

9. References

Freeman, R.B., Mann, T.A. Mason, L.W., Gartrell, C.A., Moore, V.M., 2008. Quality Assurance for Rapid Airfield Construction. Technical Report ERDC/GSL TR-08-7. Vicksburg, MS: U.S. Army Engineer Research and Development Center.

Icenogle P., Kabir S., 2013. Evaluation of Non-Destructive Technologies for Construction Quality Control of HMA and PCC Pavements in Louisiana (No. FHWA/LA. 12/493). Louisiana Department of Transportation and Development.

Lafayette Instrument. 2020. Clegg Impact Soil Tester with Clegg Control App/Setup and Users Guide. Lafayette, IN: Lafayette Instrument Co.

Livneh M., Goldberg Y., 2001. Quality Assessment During Road Formation and Foundation Construction: Use of Falling-Weight Deflectometer and Light Drop Weight. Transportation Research Record: Journal of the Transportation Research Board. 1755(1): 69-77.

Ordaz, M., Doyle, J.D., Howard, I.L., 2023. Light Weight Deflectometer Evaluation of Low-Volume Road Structural Deterioration under Rapidly Increased Traffic Patterns. Transportation Research Record, 2678(4), 948-966.

Robinson, J.H., Rabalais, C.P., 1993. Performance of the Combat Engineer Vehicle with Mineplow Operating Worldwide and in Theaters of Operations. Technical Report GL-93-23. Vicksburg, MS: U.S. Army Engineer Waterways Experiment Station.

Schwartz, C.W., Afsharikia, Z., Khosravifar, S., 2017. Standardizing Lightweight Deflectometer Modulus Measurements for Compaction Quality Assurance (No. MD-17- SHA-UM-3-20). Maryland State Highway Administration.

Yzenas, J.J., Cook, K., 2019. Evaluation of Full/Partial Depth Reclamation of Secondary Roads Utilizing Sustainable Steel Slag Materials and the Light Weight Deflectometer (LWD). Proc., International Airfield and Highway Pavements Conference 2019, July 21-24, Chicago, IL, pp. 243-254.

Zorn Instruments. 2016. User Manual for the Light Weight Deflectometer (LWD) ZFG 3.0, Stendal, Germany: Zorn Instruments.

2812 | An Interaction-Aware Two-Level Robotic Planning and Control System for Vegetation Override

Charles Noren [a,*], Burhanuddin Shirose [b], Bhaskar Vundurthy [a], Sebastian Scherer [a], Matthew Travers [a]

[a] *The Robotics Institute, School of Computer Science, Carnegie Mellon University, Pennsylvania, United States of America*
* *Corresponding author: cnoren@andrew.cmu.edu*

ABSTRACT

During off-road operations, mobile robotic platforms often encounter objects that influence the platform's route. As determining the outcome of an interaction with an object (e.g., overriding) is difficult, many robotic planners are designed to avoid interactions with all environmental objects. Yet, this object-adverse planning behavior is not reflected in the actions of expert human operators, who may interact with objects in order to find a viable path towards their goal. This work intends to emulate that human operator intuition. Our objective is to improve the performance of robotic traversals in off-road terrains through the development of a planning paradigm that allows certain safe contact with environmental objects. Specifically, we design a two-level hierarchical planning and control system which couples a contact-informed regional motion planner with contact-constrained local nonlinear trajectory optimization techniques. The approach is demonstrated for classes of vegetative objects which are characterized by existing parameterized collision models from the terramechanics community. The top level of the hierarchy combines sampling-based planning techniques with a set of override (velocity) constraints derived from these collision models during the search for a minimum-time trajectory towards the platform's goal. This minimum-time trajectory is then passed to the lower-level of the architecture, which utilizes direct collocation and model predictive control techniques to ensure that the velocity constraints are enforced during trajectory execution. The capabilities of the architecture are shown both in simulation and onboard a robotic platform, where vegetation is reasoned about and then overridden depending on the environment, platform, and object geometry.

Keywords
Trajectory Optimization
Vegetation Override
Off-road Driving
Robotics

1. Introduction

When operating in the field, off-road vehicles encounter many complex terrains. These terrains can vary greatly, demanding specific driving techniques and intelligent maneuvers for each situation. Autonomous systems are increasingly used to complement the decision-making of an expert driver and determine the "best" route (e.g., fastest) through the terrain. One class of terrain that poses a unique challenge for off-road vehicle drivers is forest-like terrain, where vegetation is often encountered during operation. Rather than navigating around such vegetation, expert drivers often make the decision to override or "strike" specific pieces of vegetation to minimize the travel time to a destination.

However, determining when to strike objects in an off-road terrain is not a straightforward decision, even for expert human operators. To provide guidance to operators and logisticians, the off-road mobility and cross-country movement communities have produced models that consider both the geometric and inertial properties of a collision to determine time-optimal routes through complex environments (Bradbury et al., 2018). Of particular note are the early experiments conducted at the U.S. Army Engineer Waterways Experiment Station (WES) (later, the U.S. Army Engineer Research and Development Center (ERDC)), which developed both data-driven and first-principles models of collision (Blackmon and Randolph, 1968; Mason et al., 2012). These models were developed for both wheeled and tracked platforms and were characterized by extensive field tests (Blackmon and Randolph, 1968; Rybansky, 2020). While the vehicles considered in these studies are often human-crewed vehicles, the wider adoption of off-road-capable robotic platforms has driven interest in utilizing these off-road mobility models developed for the design of robotic platforms (Vong et al., 1999; Noren et al., 2023). Vong et al. (1999) considered these mobility models for the purpose of designing and analyzing the capabilities of robotic systems, while Noren et al. (2023) utilized these models in an online capacity to enable wheeled robotic platforms to override vegetative objects during off-road operations.

Unfortunately, the work in (Noren et al., 2023) is limited to interactions in which the collision is determined *a priori* necessary to reach the target goal. The limitation arises from the mathematical structure of the posed trajectory optimization problem, as both the point of collision and the goal state are provided as constraints in the underlying trajectory optimization problem. While the inclusion of such constraints is not uncommon in trajectory optimization works (Hargraves and Paris, 1987; Posa et al., 2013; Kelly, 2017), utilizing this framework imposes a limited *myopic* view of the environment from the perspective of the robotic platform. Specifically, the trajectory optimization problem specifies *where* in the environment the platform must travel regardless of whether there exists a more efficient route.

To provide a holistic consideration of the environment, we propose to re-frame the problem space in a manner more consistent with that considered in robotic planning algorithms (LaValle, 2006). Many robotic planning problems focus on developing a plan that connects a start state and a goal state while avoiding objects in the operating environment. Classically, this is performed by first separating the operating environment into "free" and "occupied" space. In these classical approaches, the system designer ascribes responsibility to the platform's perception system to determine what space is "occupied" and what space is "free". For problems where there is a clear separation between the "occupied" and "free" spaces, deterministic or sampling-based planning methods can often find a path between a

https://doi.org/10.56884/W3EKJYJX

start and an end state (configuration) (Karaman and Frazzoli, 2011). However, in complex cluttered environments (e.g., off-road environments), the free space may not contain a path connecting the start and end configurations (Urmson, 2005).

The planning problem where no object-interaction-free path exists is known as the "navigation among moveable obstacles" (NAMO) problem. In this particular problem, the goal is to enable the planning entity to restructure or rearrange the environment to meet the objective of the planning entity (Wilfong, 1991). Early approaches to the NAMO problem performed a state-space decomposition to find manipulation points, and then performed a heuristic search over the free-space configuration components to avoid the complexity of multi-object planning (Stilman and Kuffner, 2004). Recently, Saxena has shown that connecting the problem to the multi-agent pathfinding domain allows for a decomposition of the problem into a configuration-search step and a physics simulation step (Saxena and Likhachev, 2021; Saxena and Likhachev 2023). These planning approaches provide a framework of the planning, but do not directly consider the kinematics of the planning agent, which were shown in Noren et al. (2023) to be extremely important in order to account for the effect of the contact on the agent's motion.

Approaches which account for the kinematics commonly combine planning and boundary value control problems together. An example combination includes combining the Linear Quadratic Regulator (LQR) optimal control strategy with Rapidly Exploring Random Trees (RRT) (Karaman and Frazzoli, 2011). Recent work in this area has looked at combining LQR controllers with Control Barrier Functions (CBF) and RRT planners, but these works do not consider planning problems without free-space paths (i.e., consider only object-avoidance objectives) (Yang et al., 2019).

The primary contribution of this work is the development of a two-level interaction-aware robotic planning and control system for vegetation override that combines the work presented in (Noren et al., 2023) with classical sampling-based motion planning algorithms to develop interaction-aware plans in complex environments. This two-level planning system incorporates the collision models developed by the terramechanics community during the search-process by modeling the collision explicitly during the connections between two states of the configuration space. Given the result of this collision, a further determination may be developed on whether or not the in-collision path should be used during operation. This work begins with a discussion on the proposed two-level planning and control system, specifically covering sampling-based planning methods the controller presented in (Noren et al., 2023). A select set of simulations are provided in the third section and hardware results on a small robotic platform are presented in the fourth section. The paper concludes with a discussion of the experiments and recommended future work.

2. Method (Algorithm) Overview

A brief definition of the problem is provided to provide context for the off-road navigation problem considered in this work. Following this problem definition, the specific implementation of a sampling-based planner, RRT*, is described where the connecting mechanism between points in the configuration space is described at a high level. Finally, the last section presents an attenuated discussion of the controller and collision models used in (Noren et al., 2023) before concretizing this controller as the specific connector used in the proposed two-level planning hierarchy.

2.1. Problem Definition

Consider a robotic platform represented by a state space, $X \in \mathbb{R}^n$, a control space, $U \in \mathbb{R}^m$, and a motion profile that is described by the C^2 smooth continuous-time dynamics expression

$$\dot{\boldsymbol{x}}(t) = \boldsymbol{f}(t, \boldsymbol{x}(t), \boldsymbol{u}(t)) \tag{1}$$

In this work, the variable t will be used to represent time. Specific instants in time will be indicated via a subscript (e.g., (t_0, t_g) may represent the start and end (goal) times of the trajectory). Elements of the state and control spaces, $\boldsymbol{x}(t) \in X$ and $\boldsymbol{u}(t) \in U$, respectively, represent the state and control of a robotic platform at a specific time.

The goal of this work is to determine a set of states $\boldsymbol{x}: [t_0, t_g] \to X$ and control inputs $\boldsymbol{u}: [t_0, t_g) \to U$ to minimize the total cost associated with a trajectory. This cost, $J(t, \boldsymbol{x}(t), \boldsymbol{u}(t))$, may be evaluated along a specific trajectory and normally takes the form

$$\mathrm{J}(\mathrm{t}, \mathrm{x}(\mathrm{t}), \mathrm{u}(\mathrm{t})) = \mathrm{J_f}\left(\boldsymbol{x}(\mathrm{t_0}), \boldsymbol{x}(\mathrm{t_g})\right) + \int_{\mathrm{t_0}}^{\mathrm{t_g}} \mathrm{w}(\mathrm{x}(\tau), \mathrm{u}(\tau)) d\tau . \tag{2}$$

As described in (Kelly, 2017), the boundary cost term, $J_f(\cdot)$, and the integral cost term, $\mathrm{w}(\cdot)$, may be selected by a designer to optimize specific behaviors of a system. This work adopts a common task found in many off-road applications: minimum-time traversals. As such, the cost function described in (2) can be collapsed to $\mathrm{J}(\mathrm{t}, \mathrm{x}(\mathrm{t}), \mathrm{u}(\mathrm{t})) = t_g$.

Given a configuration space X, consider a decomposition of this configuration space into an *object-free* space, X_{free}, and *object-occupied* space, X_{obj}. Assume that the space is completely decomposable into these subspaces: $X = X_{free} \cup X_{obj}$. This decomposition is commonly used in robotic planning to describe the states that a robot may occupy during operation (Karaman and Frazzoli, 2011) without colliding with an object.

However, this common approach yields a paradigm that neglects one of the unique properties of off-road operations discussed in (Noren et al., 2023): *that off-road operations often require interactions with objects.* While "interaction" may be defined in many different ways, given the mechanical design of most off-road vehicles, "interacting" with the environment often implies that the vehicle's chassis physically strikes the object. A further decomposition of X_{obj} is required to capture this aspect of off-road driving. That is, by decomposing X_{obj} into two additional subspaces: the space of "interactable" objects, X_{io} and "noninteractable" objects (obstacles), X_{nio}, additional collision logic may be incorporated in the planning and control layer. This work assumes that, given sufficient characterization, all elements of X_{obj} are determinable members of X_{io} or X_{nio}, thus implying that $X_{obj} = X_{io} \cup X_{nio}$. In reality, such a determination may be challenging given the perception systems available onboard a robotic platform, and thus a further *"unknown-interactable" object* class, X_{uio}, may also be required. A simple but conservative approach to address this problem would be to incorporate objects of this class into the X_{nio} class.

2.2. Asymptotically Optimal Global Planning

Given the natural complexity of outdoor environments, few environments can be represented through an explicit enumeration of objects found in the environment (X_{obj}). An approach that avoids this complication is sampling-based planning

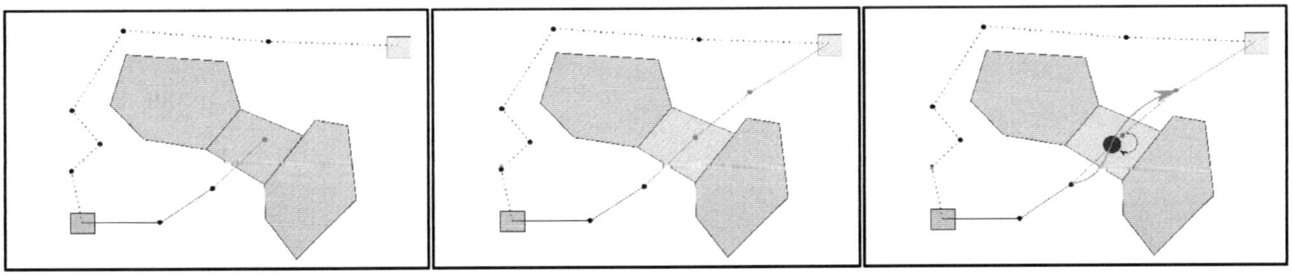

Fig. 1a.; 1b.;1c. (left-to-right): (1a.) Exploration with only noninteractable objects yielding long traversals; (1b.) Additional shorter path options are available due to interactable objects; (1c.) A depiction of an interaction-aware connector changing the original connection between nodes in the graph.

Methods (Karaman and Frazzoli, 2011), which instead rely on a collision-checking module to characterize whether a trajectory is in collision with an environmental object (see Fig. 1.). In traditional sampling-based planning approaches, points are drawn pseudo-randomly from the configuration space of a robotic platform in order to build a graph connecting an initial (starting) configuration to a desired (goal) configuration. A simplified description of this process has the algorithm drawing different points from the configuration space, checking to ensure membership in the obstacle-free subspace X_{free}, and then connecting this newly drawn point to a subset of prior-sampled points depending on the output of the collision-checking module. A depiction of this simplified approach is shown in Fig. 1a., where it can be seen that any point sampled from the space X_{obj} is rejected (in red).

The traversal distance and traversal time (assuming a constant velocity motion) may then be approximated by computing the cost-to-go from each point in the graph to the goal state.

2.3. Collision-Checking Module

Instead of the behaviors shown in Fig. 1a., this work seeks to take advantage of environmental information to perform the object decomposition described in Section 2.1. This behavior is incorporated into the collision-checking module.

2.3.1. Definition of "Interactable" Objects

Unfortunately, even with the above problem decompositions, it is not clear *how* the object-occupied space X_{obj} should be decomposed for an individual environment and terrain. Expectantly, interacting with certain objects (e.g., a concrete barrier) incorrectly may have disastrous consequences for a robotic platform. In this context, "incorrectly" is defined on a per-object basis (e.g., for the aforementioned concrete barrier, approaching at near-zero speeds may be acceptable) via interaction rules. Thus, for specific object classifications (e.g., a single-standing tree, an array of trees, etc.), an object may exist in either the X_{io} subspace or the X_{nio} subspace depending on the rules of interaction. While rules can be defined as a series of logical statements (e.g., *A robot may never interact with a concrete barrier*), such approaches do not scale with complex worlds. Rather, we utilize physics-inspired empirical models to define the interaction rules on a procedural basis. For this work, these empirical models take the form of vegetation-override models.

2.3.2. 1968 Blackmon Vegetation Override Model

The model presented in Blackmon and Randolph (1968) was regressed from a series of vegetation override tests of different vegetation types in different environments. Blackmon and Randolph (1968) provide unique regressions for force and energy expenditure for these different types of vegetation, including singular coniferous and hardwood trees, arrays of multiple trees struck in unison, and "clumps" of bamboo. From continuous measurements of pushbar force, drivetrain metrics, distance traveled, time, measurements of the impacted trees, and characterizations of the aftermath of the collision, Blackmon and Randolph (1968) construct a model primarily parameterized by the geometry of individual or multiple trees (for example, the radius of a tree) or the clump diameter.

In this study, the authors consider only experiments which require the override of a single tree or post. While Blackmon and Randolph (1968) provide additional override models, this simplification to a single class of vegetation was drawn from limitations in the perception of the necessary characterizing features for arrays of trees. Thus, the methodologies presented herein are not limited in scope to single-standing trees, aside from the limits discussed in the original manuscript itself. Given this simplification, equations B10-B12 from Blackmon and Randolph (1968), which describe the force and work required to override a single standing tree, are of particular interest. These equations largely take the form

$$F_h = K_f d_s^3 , \tag{3}$$

and

$$W = K_w d_s^3, \tag{4}$$

where F_h is the horizontal pushbar force, W is the work required to fail a single standing tree, d_s is the stem diameter, and K_f, K_w are constants that are dependent on vehicle geometry (e.g. pushbar height).

The measure of work produced from Blackmon and Randolph's (1968) models may then be combined with additional vehicle information (e.g., the operating mass) in order to generate a suitable override velocity, v_{over}, for a sensed piece of vegetation. This override velocity is then incorporated as a velocity target for a trajectory optimization problem and is further detailed in (Noren et al., 2023). The equivalent relations to determine this v_{over} may be calculated in the manner discussed in Mason et al. (2012).

The vegetation override models developed by the off-road mobility and cross-country movement communities, including those by Blackmon and Randolph (1968) and Mason et al. (2012) thus abstract complex collision interactions into useful low computational cost approximations. These approximations generally characterize the required force, work, or velocity required for a vehicle to override a subset of vegetation given some parameterization of the vehicle and environment. Furthermore, the models form an actionable criterion as the basis of an interaction rule given the reachability of the override velocity from the robotic platform's current state. By evaluating whether or not the override velocity may be reached, a robotic platform can determine if a specific vegetative object exists in the class of objects X_{io} (i.e., the necessary velocity is reachable) or X_{nio} (i.e., the necessary velocity is not reachable).

2.4. Interaction-Aware Connectors

Given the above decomposition, incorporating this collision information into the connector is necessary to ensure a physically realizable interaction. Ensuring such an interaction is realizable is possible by enforcing differential constraints between nodes in the search graph. Unfortunately, this particular requirement yields a planning problem of high computational complexity (LaValle, 2006), especially given the hybrid nature of the modeled contact dynamics.

In our prior work (Noren et al., 2023), a two-point boundary value problem was specified that uses contact models to account for hybrid dynamics experienced during collisions for robotic platform control. By running Noren et al.'s controller, which can solve a two-point boundary value problem between two given sampled states as a low-level connector into our proposed search algorithm, we can enforce a set of differential constraints on the search graph. Unfortunately, such an approach would quickly become computationally intractable and thus we move towards an approach where we selectively call the controller from Noren et al., (2023) to evaluate only on specific segments of the connected graph. The specifically evaluated segments are those where an object from the class X_{io} is encountered, such as in Fig 1c. This "second-layer" evaluation updates both the trajectory taken between two points and the cost of the traversal.

2.5. Vehicle Model

The proposed architecture is extensible to different dynamical systems, but to demonstrate its application, this paper considers a platform similar to the one shown in Fig. 2. The platform is a modified Traxxas X-Maxx with a top speed of 10 [m/s] in complex environments. While the vehicle is capable of reaching speeds of 10 [m/s], the nominal operating speed is limited to only 3 [m/s] for safety reasons. The vehicle is equipped with a set of cameras and a single Light Detection and Ranging (LiDAR) sensor for onboard localization and mapping. The vehicle dynamics (1) were modeled as an idealized 2D nonlinear bicycle model with the state: $\boldsymbol{x} = [p_x, p_y, \psi, v]$. The elements of this state vector are the vehicle's x-position (p_x), y-position (p_y), heading (ψ), and velocity (v). A diagram of the vehicle kinematics is provided in Fig. 2. The available vehicle controls included velocity and steering angle: $\boldsymbol{u} = [v, \delta]$. As such, the continuous-time vehicle dynamics may be modeled as

$$\dot{p_x} = v * \cos(\psi + \operatorname{atan}\left(\frac{L_f}{L} * \delta\right)), \tag{5}$$

$$\dot{p_y} = v * \sin(\psi + \operatorname{atan}\left(\frac{L_f}{L} * \delta\right)), \tag{6}$$

$$\dot{\psi} = \frac{v}{L} * \cos\left(\operatorname{atan}\left(\frac{L_f}{L} * \delta\right)\right) * \tan(\delta)\,, \tag{7}$$

and the characterizing platform parameters are listed in Table 1. The platform's length is L_T, wheelbase length is L, and the front axle to center of mass distance is L_f. The height of the middle of the attached bumper is h_{bumper}, the platform mass is m, and the platform width is w.

Table 1
Platform Parameters for the modified X-maxx

Property	Unit	Magnitude
L_T	[m]	0.77
L	[m]	0.48
L_f	[m]	0.23
h_{nose}	[m]	0.18
w	[m]	0.54
m	[kg]	14.7

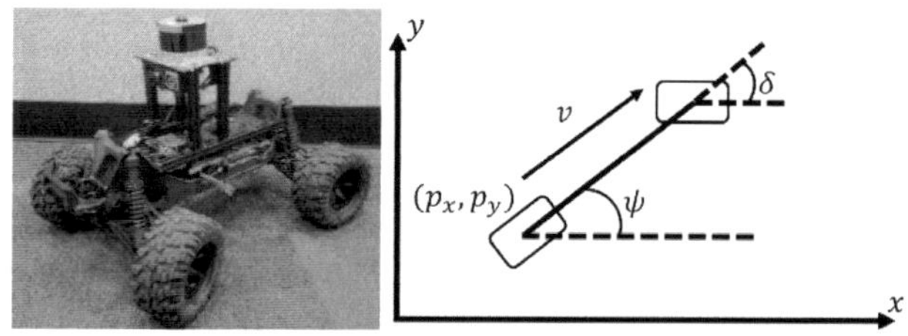

Figure 2: A modified small uncrewed ground vehicle with a similar configuration to the one utilized in this work (left). A vehicle kinematic diagram is also provided (right).

3. Analytical Analysis and Numerical Simulation

One key limitation observed in (Noren et al., 2023) was that an interactable-object needed to be selected for override prior to understanding the overall effect the object would have on the robotic platform's full path. This posed challenge arises from the associated "myopia" of local control methods, which utilize only near-field object and a limited preview horizon to determine the control of the platform. Such controllers include the controller proposed in (Noren et al., 2023). This myopia appears in at least two scenarios: 1) where a vehicle must select between striking multiple objects; and 2) where a vehicle must account for multiple object interactions. We will explore these cases in simulation in this section, but we will first present an analytical discussion of the method for a simple environmental object configuration.

In all simulated trials, a vehicle was simulated that was described by the dynamics presented in Sec. 2.5 and characterized by parameters in Table 1.

3.1. Scenario 1 - Obstacle Avoidance or Override

The first scenario considers the situation where a vehicle must determine whether to override a piece of vegetation. Such a determination in most circumstances may be difficult to compute, but in certain simple scenarios, a pure kinematic analysis may be used to determine whether to override a piece of vegetation.

Consider the scenario depicted in Fig 3., where a robotic platform must either choose to override a piece of vegetation (in brown) or avoid the vegetation and an adjacent obstacle (in red). Assuming that the vehicle initially travels at maximum velocity, v_{max}, and has a maximum constant acceleration that it can sustain, a_{max}, it is the goal of the combined planning and control system to reach the position $d_0 + d_1$ [units] directly ahead of the platform in the shortest time.

$$t_{avoid} = \frac{1}{v_{max}}(\sqrt{d_0^2 + h^2} + \sqrt{d_1^2 + h^2}). \tag{8}$$

As this travel time represents the fastest time the platform can travel between the initial position and the desired ending location, the determination of whether or not the vehicle should override the post relies solely on whether or not the time required to reach the end location, which travels through the post, t_{strike}, is less than t_{avoid}. Note that, according to the discussion in (Noren et al., 2023), the velocity needed to override the post must still be reachable from the initial configuration to have a valid trajectory. For this example, we assume this to be true (e.g., $v_{max} > v_{over}$).

Given the geometry of the above problem and the collision models discussed in Sec. 2.3, we can bifurcate the trajectory into two distinct sub-segments. The first segment represents the pre-collision motion, and the second segment represents the post-collision motion. As such, the total travel time needed to override the post with respect to these two sub-segments is

$$t_{strike} = t_0 + t_1, \tag{9}$$

where t_0 represents the travel time on sub-segment 1 and t_1 represents the travel time on sub-segment 2.

Per the arguments above, the travel time on sub-segment 0 is determined to be

$$t_0 = \frac{d_0}{v_{max}}, \tag{10}$$

which means that to yield a faster time than t_{avoid}, the constraint that

$$t_1 < \frac{1}{v_{max}}\left(\sqrt{d_0^2 + h^2} + \sqrt{d_1^2 + h^2}\right) - \frac{d_0}{v_{max}}, \tag{11}$$

must hold.

The act of overriding the post results in a velocity drop in the platform to a value given by $v_{drop} \geq 0$. Depending on how low the value of v_{drop} is, the platform can either continue to accelerate until the end of sub-segment 2 or accelerate all the way to v_{max} and cruise at this maximum speed until the end of sub-segment 2. This disparity between behaviors can be captured by computing the distance d_1' needed to reach the maximum speed.

$$d_1' = \frac{v_{max}^2 - v_{drop}^2}{2a_{max}}, \tag{12}$$

With the help of d_1', the travel time t_1 on sub-segment 2 and, subsequently, the constraint to decide between avoiding and striking the piece of vegetation via (11) can be determined as follows:

$$t_1 = \begin{cases} \dfrac{v_{max} - v_{drop}}{a_{max}} + \dfrac{d_1 - d_1'}{v_{max}}, & if\ d_1' \leq d_1, \\ \dfrac{2d_1}{v_{drop} + \sqrt{v_{drop}^2 + 2a_{max}d_1}}, & otherwise \end{cases} \tag{13}$$

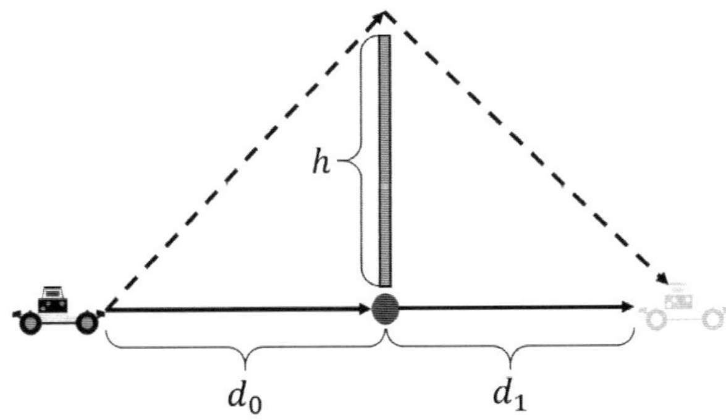

Fig. 3. Top view of a simplified scenario in which a robotic platform must determine whether to strike a piece of vegetation (in brown) to travel a more direct route, or avoid interacting with any objects in the scene (brown or red) to reach a goal location.

3.2. Scenario 2 - Obstacle Selection Tests

The second scenario considered in this work was motivated by the first identified limitation of (Noren et al., 2023). The scenario was constructed such that the robotic agent must choose between two paths, each striking a single vegetative object, to reach a goal. The first path would require the platform to interact with a 38.1 [mm] (1.5 [in]) diameter vegetative object and the second path would require the platform to interact with a 25.4 [mm] (1.0 [in]) diameter vegetative object. For both vegetative objects, Blackmon and Randolph's (1968) model was used.

The example override for this scenario is shown in Fig. 4. The starting configuration of the robotic platform is at the center of the blue marker (position: (1, 10) [m]), its initial heading is along the abscissa, and it starts with zero velocity. The desired ending configuration (shown as a gold square) is placed 34 [m] in front of the robot (position: (35, 10) [m]), with a goal heading aligned with the abscissa and a goal velocity of 1.5 [m/s]. Note that a line between the initial and desired spatial configuration generates a symmetric object field outside of the vegetative objects. Such a configuration yields a choice of trajectory in two distinct homotopy classes. Both objects appear 19 [m] ahead of the vehicle, offset exactly 2.5 [m] above (38.1 [mm]) and below (25.4 [mm]) the starting configuration.

While a biased sampling schema does exist that could lead to selecting a route through the top object, using an unbiased sampler of the configuration space yielded paths which consistently passed through the lower object. Given the analysis provided in Sec 3.1, it is clear that even with a perfect sampling schema, the geometric symmetry of the problem should bias the robotic platform to travel through the lower object. Furthermore, that same analysis indicates that a geometric configuration of vegetative objects exists in this schema, which would lead the platform to override the 38.1 [mm] object because a 25.4 [mm] object would require too much time to deviate towards (i.e., t_0 may be higher).

Finally, considering the velocity diagrams, the vehicle initially maneuvers to align with the vegetative object and then greatly increases the velocity to override the post. The depicted initial speed up (around 2.5 [sec]) and slowdown (around 7 [sec]) correlate to large changes in heading as the vehicle bounds through different curves at (x position) 2.5 [m] and (x position) 10.0 [m], respectively. The vehicle additionally accelerates into the vegetative object before maneuvering towards the goal. This acceleration was achieved by biasing the trajectory optimization to include a reference velocity (in this case, a travel speed of 1.5

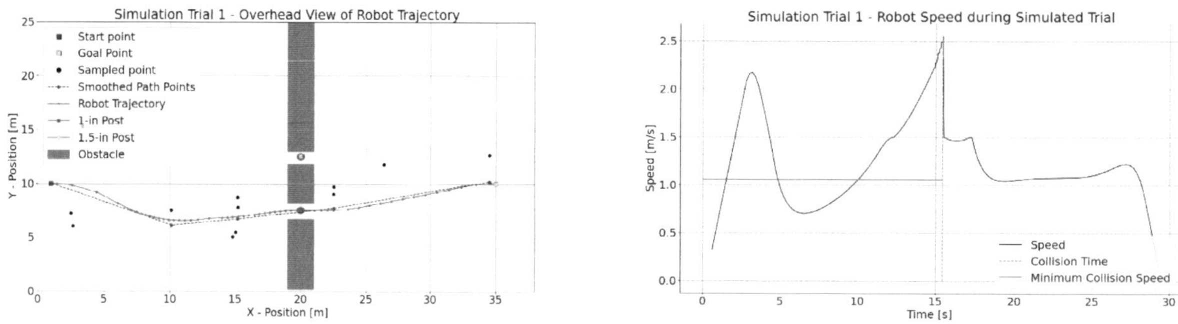

Fig. 4. The first simulated trial demonstrating a selection between two objects. The platform's spatial (left) and speed (right) trajectories reflect the collision

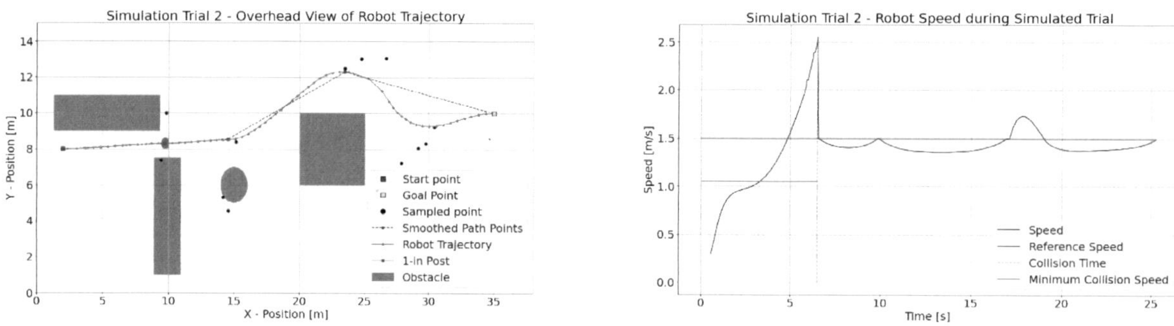

Fig. 5. The second simulated trail demonstrating multiple object interactions. The platform's spatial (left) and speed (right) trajectories reflect the collision

[m] set as a boundary condition at the end of the segment) with the expected required override velocity computed with Blackmon's model.

3.3. Scenario 2 – Multiple Object Interaction Tests

The final simulated experiment demonstrates another extension of the work presented in (Noren et al., 2023): the ability to enable an obstacle override alongside other obstacle avoidance tasks. In (Noren et al., 2023), a single obstacle was overridden in order to reach a desired goal state. However, most real-world environments are much more complicated, and a single override may affect the entire resulting path. For this scenario depicted in this section, the vehicle starts at the blue square (position: (2, 8)) with the same initial heading and velocity constraints as in the previous simulation and heads towards a goal configuration at position (35, 10). A single 25.4 [mm] object is paced in front of the robotic platform at position (10, 8), and Blackmon's model is again used to determine the override velocity.

The scenario presented in Fig. 5. involves a set of complex decisions for the platform. The platform must first choose whether to travel out of an initial confined area via an open corridor, which requires the platform to travel away from the goal (in the y-direction), or to strike the piece of vegetation. Following this set of obstacles is another larger object (starting at the x position 20 [m]) that requires the platform to deviate from its initial heading. As shown in Fig. 5., the platform biases its search away from sampled points that would take it away from the desired location or require large deviations from the goal point. Note that in this scenario, the vehicle's speed is penalized if it deviates away from the provided reference speed (again, 1.5 [m/s]). The velocity graph depicts a behavior in which the vehicle speeds up to strike the object (around 7 [sec]) and then maintains this reference velocity to the goal state.

4. Vegetation Override Hardware Trials

This section details two hardware trials, both utilizing the Blackmon and Randolph (1968) model in Sec. 2 and a platform similar to the one in Fig. 2. The vehicle was commanded to travel at a nominal 1.5 [m/s] in both experiments. A model predictive control algorithm, specifically: iterative Linear Quadratic Regulator (iLQR), was implemented to track the generated trajectory. The results of these tests are shown in Fig. 6. and Fig. 7. The platform's odometry was captured using Super Odometry (Zhao, et al., 2021), a LiDAR and IMU-centric Simultaneous Localization and Mapping (SLAM) method that estimates the platform's position, heading, and velocity during the trial.

4.1. Trial 1 – Obstacle Selection Test

The first trial scenario demonstrated on robotic hardware was an obstacle override selection test. This test was conducted with an environment configuration similar to the simulation test shown in Sec. 3.2. The object the robotic platform interacted with was a 25.4 [mm] pine dowel rod that was embedded into the soil by hand.

The executed trajectory is shown in Fig. 6. The vehicle started at a position of (1, 10) [m] and reached a target position at (25, 10) [m]. The robotic platform avoided all obstacles (in red) while overriding the dowel rod placed at position (20, 9.25) [m]. The vehicle reached a collision velocity of around 1.492 [m/s] at the point of collision, which is higher than the required override

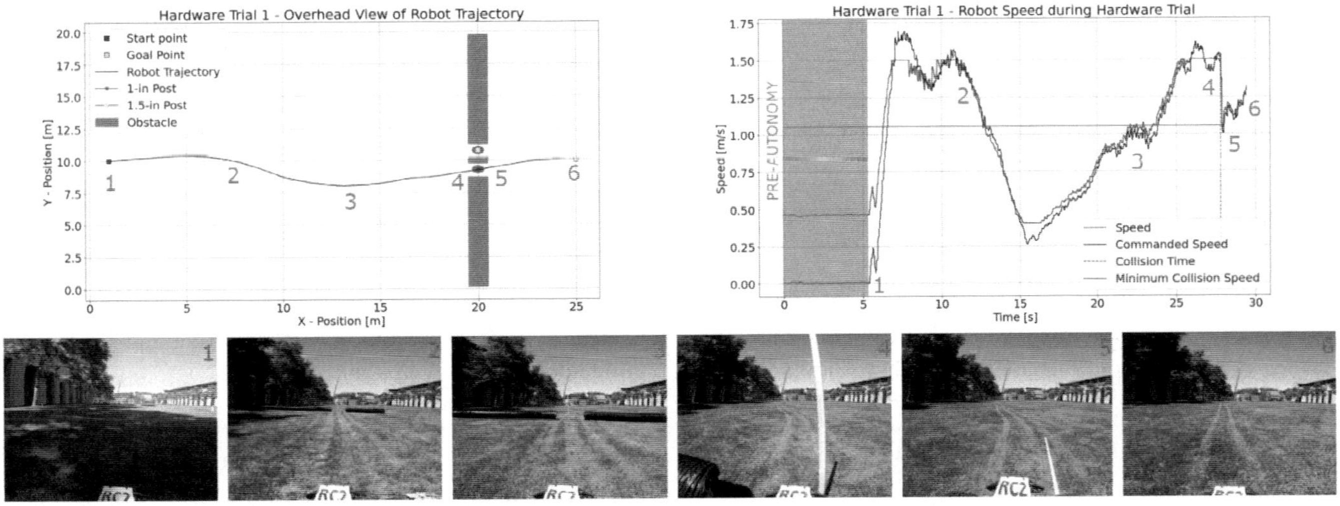

Fig. 6. Hardware Trial 1: Obstacle Selection Test. The post the platform strikes is shown in viewpoint 4. The platform's spatial trajectory (left) and speed trajectory (right) both demonstrate a collision with the dowel rod. Key frames extracted from on-board imaging shown are shown in the lower photo tableau. The position and time at which the photo was taken is denoted with red numbering on the key frame. Note that a deviation between the speed and commanded speed at the beginning of the trial exists due to a controller dead-band around commanded speeds of near-zero magnitude.

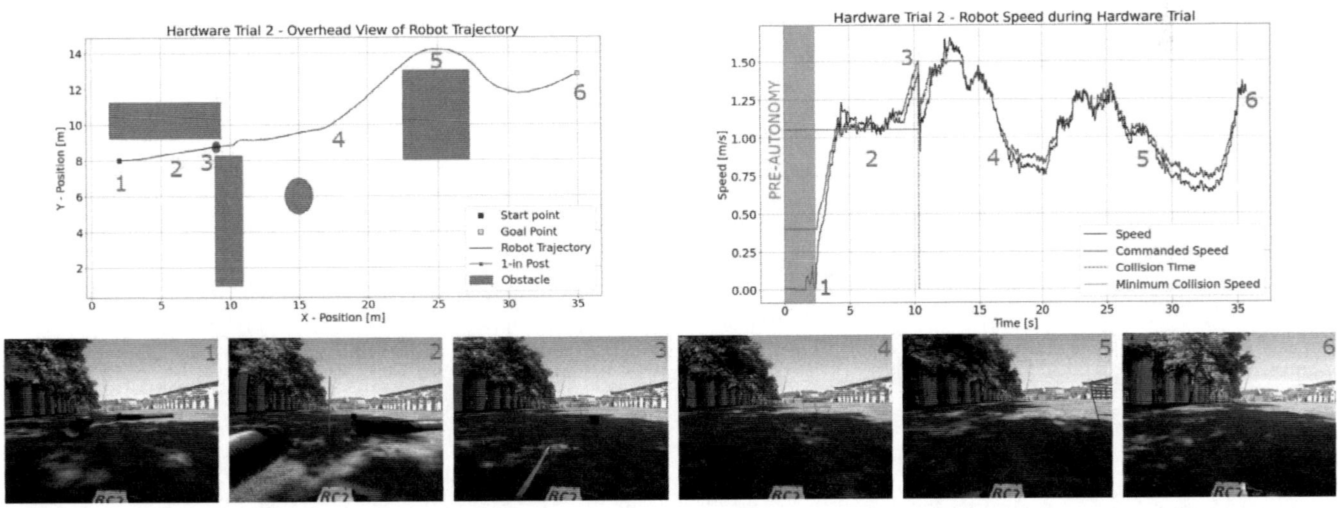

Fig. 7. Hardware Trial 2: Multiple Object Interaction Test. The post the platform strikes is shown in viewpoints 2 and 3. The platform's spatial trajectory (Right) and speed trajectory (left) both demonstrate a collision with the dowel rod, with a drop in velocity being very noticeable at around 10 [sec]. The position and time at which the lower key frames were taken are denoted with red numbering. Note that a deviation between the speed and commanded speed at the beginning of the trial exists due to a controller dead-band around commanded speeds of near-zero magnitude.

velocity calculated in Blackmon and Randolph's model. After the collision, the post was pulled from the ground and upended onto the surface. The robotic platform was not damaged during the collision. Several viewpoints taken by the robotic platform along the trajectory are also shown in Fig. 6.

4.2. Trial 2 – Multiple Object Interaction Test

The second trial scenario demonstrated on a robotic platform was a multiple obstacle interaction test. In this test, the platform would not only need to perform an override similar to the test shown in the simulations demonstrated in Sec. 3.3, but also to avoid additional zones which contained objects the platform could not interact with. These zones are depicted in red in Fig. 7. These obstacles contained objects such as tubing and chairs that the platform could not move through.

In this particular test, a 25.4 [mm] diameter post was embedded in the ground. The robotic platform started at position (2, 8) [m], the post was placed at position (9.0, 8.75) [m], and the desired location (goal) was determined to be (35, 12.8) [m]. As in the first hardware experiment, the vehicle reached a speed of 1.490 [m/s] at the point of collision. After the collision, the post was again pulled from ground and came to a rest on the surface. As in the first trial, the platform was not damaged.

5. Conclusions and Future Work

This work details a two-level motion planning architecture that combined information from existing parameterized vegetation models from the off-road mobility and cross-country mobility community, controls from hybrid trajectory optimization, and sampling-based planning methods to support complex robotic decision making in off-road environments. The method extends the current state-of-the-art by incorporating collision-aware hybrid dynamics as an intelligent connector during the planner's graph construction.

A work that is of relevance in the context of our proposed work is presented in (Xie et al., 2015). Xie combines Batch Informed Trees (BIT*) search with a two-point boundary value problem solver to perform optimal planning in time. However, Xie's work focused exclusively on obstacle avoidance and thus does not account for the hybrid-dynamics encountered during environmental interactions. An interesting extension to our work can draw inspiration from the work in Xie et al. (2015) to investigate other sampling-based planning methods and comparing their computational performance on a robotic system. Additionally, investigating how to incorporate uncertainty in the object override classes (referenced in Sec. 2.1) should be of prime importance, given the likelihood of encountering an object outside the class of identified models.

In order to further minimize connector solve times, future work can identify additional collision models (e.g., for distributed vegetation such as dense foliage) and investigate adaptive hyperparameter tuning (e.g., for the collision index).

6. Acknowledgements

This research is based upon work supported by the United States Department of Defense Science Mathematics and Research for Transformation (SMART) Scholarship program.

The SMART Scholarship program is funded through OUSD/R\&E (The Undersecretary of Defense-Research and Engineering), National Defense Education Program (NDEP) / BA-1, Basic Research.

7. Declaration of competing interest

The authors declare that they have no known competing financial interests or personal relationships that could have appeared to influence the work reported in this paper.

8. References

Blackmon, C. A., Randolph, D.D., 1968. An analytical model for predicting cross-country vehicle performance, vol. II: longitudinal obstacles. Appendix B: vehicle performance in lateral and longitudinal obstacles (vegetation). TR No. 3-783. U.S. Army Engineer Waterways Experiment Station, Vicksburg, MS.

Bradbury, M., Dasch, J., Gonzalez-Sanchez, R., Hodges, H., Iagnemma, K., Jain, A., Jayakumar, P., Letherwood, M., McCullough, M., Priddy, J., Wojtysiak, B., 2018. Next Generation NATO Reference Mobility Model (NRMM) Development. AC/323(AVT-ET-148)TP/728. North Atlantic Treaty Organization Science and Technology Organization.

Hargraves, C. R., Paris, S. W., 1987. Direct trajectory optimization using nonlinear programming and collocation. Journal of Guidance, Control, and Dynamics. 10 (4), pp. 338-342.

Karaman, S., Frazzoli, E., 2011. Sampling-based Algorithms for Optimal Motion Planning. The International Journal of Robotics Research. 30(7) pp. 846-894.

Kelly, M., 2017. An introduction to trajectory optimization: How to do your own direct collocation," SIAM Review. 59, pp. 849–904.

LaValle, S. M., 2006. Planning Algorithms, University of Illinois, Cambridge University Press (Online, accessed 04/27/2024).

Mason, G. L., Gates, B. Q., Moore, V. D., 2012. Determining forces required to override obstacles for ground vehicles. Journal of Terramechanics. 49 (3-4), pp. 191-196.

Noren, C., Vundurthy, B., Scherer, S, Travers, M., 2023. Trajectory Optimization for Vegetation Override In Off-road Driving. Proceedings of the 16th European-African Regional Conference of the ISTVS. Lublin, Poland.

Posa, M., Cantu, C., Tedrake, R., 2013. A direct method for trajectory optimization of rigid bodies through contact. The International Journal of Robotics Research. 33(1) pp. 69-81.

Rybansky, M., 2020. Determination the ability of military vehicles to override vegetation. Journal of Terramechanics. 91, pp. 129-138.

Saxena, D., Likhachev, M., 2021. Manipulation Planning Among Movable Obstacles Using Physics-Based Adaptive Motion Primitives. Proceedings of the 2021 International Conference on Robotics and Automation (ICRA). Xi'an, China. pp. 6570-6576

Saxena, D., Likhachev, M., 2023. Planning for Complex Non-prehensile Manipulation among Moveable Objects by Interleaving Multi-Agent Pathfinding and Physics-Based Simulation. Proceedings of the 2023 International Conference on Robotics and Automation (ICRA). London, United Kingdom. pp. 8141-8147.

Stilman, M., Kuffner, J., 2004. Navigation among moveable obstacles: real-time reasoning in complex environments. 4th IEEE/RAS International Conference on Humanoid Robots. 1, pp. 322-341.

Urmson, C., et al. 2009. Autonomous Driving in Urban Environments: Boss and the Urban Challenge. The DARPA Urban Challenge. Springer Tracts in Advanced Robotics, Springer, Berlin, Heidelberg. 56, pp. 1-59 https://doi.org/10.1007/978-3-642-03991-1_1.

Vong, T. T., Hass, G. A., Henry, C.L., 1999. NATO Reference Mobility Model (NRMM) Modeling of the DEMO III Experimental Unmanned Ground Vehicle (XUV). Final Report. Army Research Lab, Aberdeen Proving Ground, Maryland. pp. 1-69.

Wilfong, G., 1991. Motion planning in the presence of movable obstacles. Annals of Mathematics and Artificial Intelligence.

Xie, C., Van Den Berg, J., Patil, S., Abbeel, P., 2015. Toward Asymptotically Optimal Motion Planning for Kinodynamic Systems Using a Two-Point Boundary Value Problem Solver. Proceedings of the 2015 IEEE International Conference on Robotics and Automation (ICRA). Seattle, United States of America. pp. 4187-4194.

Yang, G., Vang, B., Serlin, Z., Belta, C., Tron, R., 2019. Sampling-based Motion Planning via Control Barrier Functions. Proceedings to the 3rd International Conference on Automation, Control, and Robots (ICACR). Prague, Czech Republic. pp. 22-29.

Zhao, S., Zhang, H., Wang, P., Nogueira, L., and Scherer, S. (2021). Super Odometry: IMU-centric Lidar-visual Inertial Estimator For Challenging Environments. In 2021 IEEE/RSJ International Conference on Intelligent Robots and Systems (IROS), 8729–8736.

2919 | Accurate Rover Mobility Analysis Using HILS-DRFT with Real-Time Parameter Tuning Approach

Takahiro Fuke [a,*], Sora Ishikawa [a], Genya Ishigami [a]

[a] Department of Mechanical Engineering, Keio university, Yokohama, Japan

* Corresponding author: taka162uke@keio.jp

ABSTRACT

Offroad mobility in extraterrestrial exploration can be significantly degraded by vehicle slippage on soft terrain covered with fine powdery sand known as regolith. Comprehensive design analysis and verification of the rover's mobility have been widely investigated using numerical and experimental approaches. Numerical simulations for the mobility analysis require accurate modeling of the contact forces between the wheel and sand. While several wheel-sand interaction models have been proposed based on the Bekker-Wong-Reece theory, the Dynamic Resistive Force Theory (DRFT) proposed in the early 2020s is a notable approach for its low computational cost and adaptability to quasi-dynamic motion. The scaling factor used in DRFT is the only parameter for representing sand-dependent parameters, and its value needs to be empirically tuned to achieve accurate and reliable mobility analysis. Therefore, the authors have integrated a single-wheel test bed into a closed-loop Hardware-In-the-Loop Simulation (HILS) associated with DRFT. This HILS-DRFT experimentally observes the characteristics of the wheel sinking phenomenon in the wheel test bed, and then, the value of the scaling factor is tuned in real-time by the difference between the observed wheel sinkage and the DRFT simulation. This real-time parameter tuning method accurately reproduces wheel mobility, particularly in its transient states, which were previously unachievable with DRFT alone. The proposed HILS-DRFT with the real-time tuning method will contribute to the efficient and reliable development of rovers.

Keywords
Terramechanics
Resistive Force Theory
Hardware-In-the-Loop Simulation

1. Introduction

Extraterrestrial exploration missions hold immense scientific significance in unraveling the processes involved in the evolution of the solar system. Furthermore, the valuable data obtained through the exploration plays a crucial role in establishing bases for expanding domain of human activities. The use of rovers capable of freely traversing the planetary surface enables the collection of more detailed sample data. However, the terrain surface is composed of a soft terrain covered with fine powdery sand known as regolith that often degrade the rover mobility performance due to slippage. Therefore, it is necessary to thoroughly test the rover's traversal performance before sending to other planets.

Comprehensive design analysis and verification of the rover's mobility have been widely investigated using both experimental and numerical approaches. Notably, numerical methods using semiempirical terramechanics model have long been researched for their potential to reduce the cost of empirical validation. The Bekker-Wong-Reece theory (Bekker, 1956,1960, 1969; Wong, 2008; Wong and Reece, 1967) forms the foundation upon which various wheel-sand interaction models have been developed. However, these models face challenges such as the necessity to consider for numerous sand parameters, their inability to accommodate complex wheel shapes and high-speed motion. In response to these challenges, the Dynamic Resistive Force Theory (DRFT) (Agarwal et al., 2021) was proposed by applying the Resistive Force Theory (RFT) (Li et al., 2013) which divides the wheel's surface into minute plates. The RFT uses the only one sand parameter called scaling factor. The RFT is advantageous as it can be applied to wheels with complex shapes and in wide range of speed. However, this method cannot accurately simulate transient states because the dynamic change of wheel-sand interaction mechanics varies the sand stiffness or the value of the scaling factor.

Therefore, we have integrated a single-wheel test bed into a closed-loop Hardware-In-the-Loop Simulation (HILS) associated with DRFT (Ishikawa et al., 2022). This HILS-DRFT experimentally observes the characteristics of the wheel sinking phenomenon in the wheel test bed, and then, the value of the scaling factor is tuned in real-time based on the difference between the observed wheel sinkage and the DRFT simulation. In this paper, we show the concept of this HILS-DRFT proposed by Ishikawa and demonstrate the validity through comparative experiments on sandy terrain.

2. Overview of HILS-DRFT

2.1. Overview of HILS

Figure 1 shows the general framework of the HILS. Compared to a full software simulation for a target system, the HILS replaces part of the numerical model with an actual experiment using a hardware testbed. The hardware testbed of the HILS measures a real phenomenon of the part of the system in a hardware environment. Subsequently, in a software environment, the numerical model calculates the system's response based on the measured values. The HILS continuously loops these experimental measurements and numerical calculations as the temporal behavior of the system progresses. The HILS is particularly useful in cases where the actual phenomenon of a target system is difficult to measure or where an accurate numerical model is barely available.

https://doi.org/10.56884/EB6EB3VY

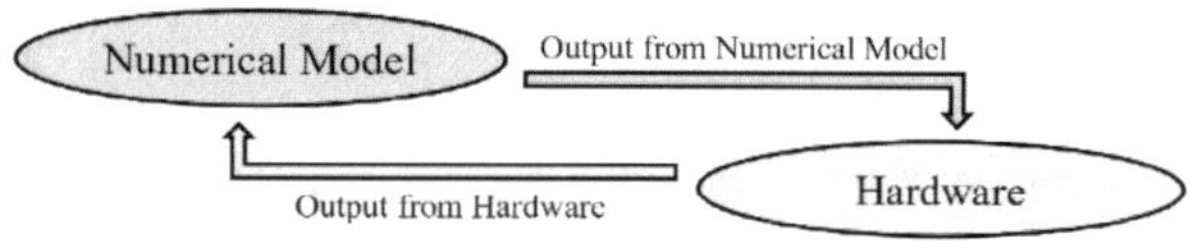

Fig. 1. Simplified flowchart of HILS

2.2. Overview of DRFT

2.2.1. Basics of RFT

The RFT is a modeling technique that divides the surface of an object into small plates, as illustrated in *Fig. 2*.

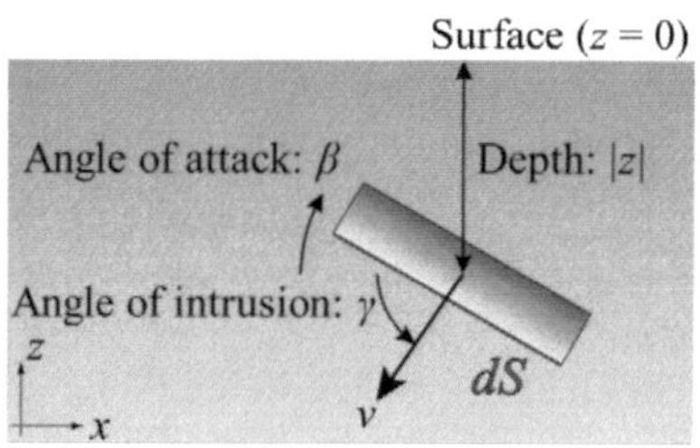

Fig. 2. Definitions of plate element's variables

Here, dS is the contact surface of each plate, $|z|$ is the depth from the sand surface, β is the attack angle, and γ is the intrusion angle. Both β and γ range from $-\pi/2$ to $\pi/2$, with their starting points and positive directions indicated by the arrows in *Fig. 2*. Under this definition, the contact force per unit area $\boldsymbol{\sigma}(\beta,\gamma,|z|)$ that each plate receives from the sand is expressed as follows:

$$\boldsymbol{\sigma}(\beta,\gamma,|z|) = \zeta\boldsymbol{\alpha}^{\text{generic}}(\beta,\gamma)|z| \tag{1}$$

Here, ζ is a dimensionless sand parameter called the scaling factor, determined by properties such as grain shape and packing density of the sand, and $\boldsymbol{\alpha}^{\text{generic}}(\beta,\gamma)$ is a function experimentally identified (Li., et al 2013). With this $\boldsymbol{\sigma}(\beta,\gamma,|z|)$ and dS, the contact force d$\boldsymbol{F}$ acting on each plate can be calculated as follows:

$$\mathrm{d}\boldsymbol{F} = \boldsymbol{\sigma}(\beta,\gamma,|z|)\mathrm{d}S \tag{2}$$

Finally, by summing up the contact force d$\boldsymbol{F}$ acting on each plate over the contact surface S, the total contact force $\boldsymbol{F}$ acting on the entire object can be calculated as follows:

$$\boldsymbol{F} = \int_S \mathrm{d}\boldsymbol{F} = \int_S \boldsymbol{\sigma}(\beta,\gamma,|z|)\mathrm{d}S = \int_S \zeta\boldsymbol{\alpha}^{\text{generic}}(\beta,\gamma)|z|\mathrm{d}S \tag{3}$$

2.2.2. DRFT for modeling wheel-sand interaction

It has been pointed out that contact force calculated from the original RFT deviates from actual results when applied to modeling wheel - sand interaction. To address this issue, (Agarwal et al., 2021) proposed DRFT that considers the following factors:

1. The momentum flux of the granular media.
2. The height decrease behind the wheel.

The momentum flux of the granular media is considered by adding $-\lambda\rho v_n^2$ to the RHS of *Eq. 1* where λ, ρ and v_n respectively represent the fitting coefficient, the bulk density of the granular media and the normal component of the plate's velocity $\boldsymbol{v}$. Introducing the outward unit normal $\boldsymbol{n}$ to each plate surface, v_n is represented by $\boldsymbol{n}\cdot\boldsymbol{v}$. The height decrease occurs behind the wheel is considered by substituting $|z|$ in the RHS of *Eq. 1* with the effective depth $|\tilde{z}|$, defined as follows:

$$|\tilde{z}| = \begin{cases} |z| - \delta h \text{ (for plates in } \theta_r) \\ |z| \text{ (for plates in } \theta_f) \end{cases} \tag{4}$$

$$\delta h = r\left(\frac{r\omega^2}{g}\right) \tag{5}$$

Here, δh is the height decrease behind the wheel, r is the wheel radius, ω is the angular velocity of the wheel, and g is the gravitational acceleration. Additionally, as illustrated in *Fig. 3*, θ_r and θ_f represent the angles that bisect the contact surface S..

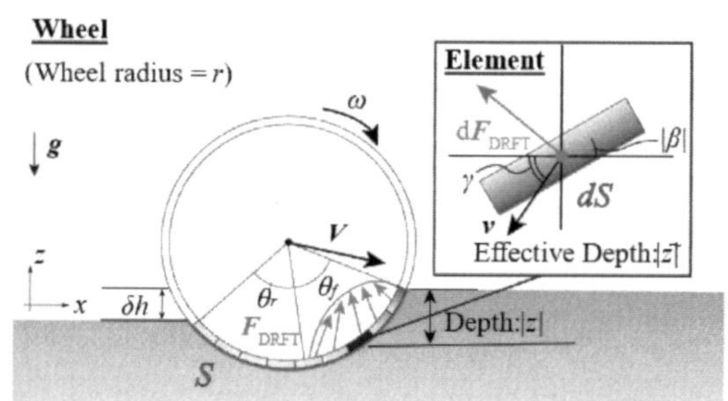

Fig. 3. Modeling of the wheel – sand interaction according to DRFT

Considering these two, the net force $\boldsymbol{F}_{\text{DRFT}}$ is given as follows:

$$\boldsymbol{F}_{\text{DRFT}} = \int_S (\zeta\boldsymbol{\alpha}^{\text{generic}}(\beta,\gamma)|z| - \lambda\rho v_n^2\boldsymbol{n})\mathrm{d}S \tag{6}$$

2.3. Hardware test bed of HILS-DRFT

The single-wheel test bed illustrated in *Fig. 4* served as the hardware test bed. This apparatus, measuring 3500 mm in total length (2500 mm travelable length) and 600 mm in width, features horizontal, vertical, and rotational axes for the wheel. The horizontal axis includes a linear motion mechanism consisting of a motor, ball screw, and linear guide. The vertical axis comprises a passive-sliding mechanism with a linear guide, while the wheel's rotational axis incorporates a motor and a gearhead. The wheel, along with the moving frame, can travel both horizontally and vertically. The moving frame connects to the ball screw via a ball screw nut, allowing for the recreation of various wheel slipping conditions. Additionally, by disengaging the ball screw nut, it is possible to observe the travel controlled solely by the wheel's rotation. Magnetic scales attached to the linear guides in both directions measure displacement from the initial position horizontally and vertically. The apparatus uses Toyoura standard sand (average grain size 0.11 mm), ensuring high reproducibility due to its uniform grain size and shape.

Fig. 4 Overview of single-wheel test bed

2.4. Framework of HILS-DRFT

Figure 5 illustrates the HILS-DRFT framework. This framework is almost identical to our previous works (Ishikawa et al., 2022), except the current framework omits the force measurements. In general, the scaling factor representing terrain stiffness is treated as a constant during numerical simulations; however, the HILS-DRFT framework assumes that it is dynamically varied. Here, $\boldsymbol{x}_{\mathrm{tw}}$, $\boldsymbol{v}_{\mathrm{tw}}$, and $\boldsymbol{F}_{\mathrm{tw}}$ are vectors with components in the horizontal x and vertical z directions, representing the position vector of the wheel's center of mass, the velocity vector, and the contact force exerted by the sand on the wheel, respectively. The initial conditions required at the start of the simulation include the wheel's static sinkage H_{static}, the initial value of the scaling factor $\zeta(0)$, the wheel's initial position $\boldsymbol{x}_{\mathrm{tw}}(0)$, and the wheel's initial velocity $\boldsymbol{v}_{\mathrm{tw}}(0)$. Only the angular velocity ω is controlled externally. Notably, the wheel operates simultaneously in both the computer environment and the wheel test bed. In this setup, only the wheel's angular velocity ω and the horizontal component of the translational velocity $v_{\mathrm{tw},x}$ are sequentially provided to the single-wheel test bed, with no direct involvement in the vertical motion. This enables real-time tuning of the scaling factor $\zeta(t)$ based on the difference between the dynamic sinkage values H_{sim} and $\boldsymbol{H}_{\mathrm{real}}$ obtained from each environment.

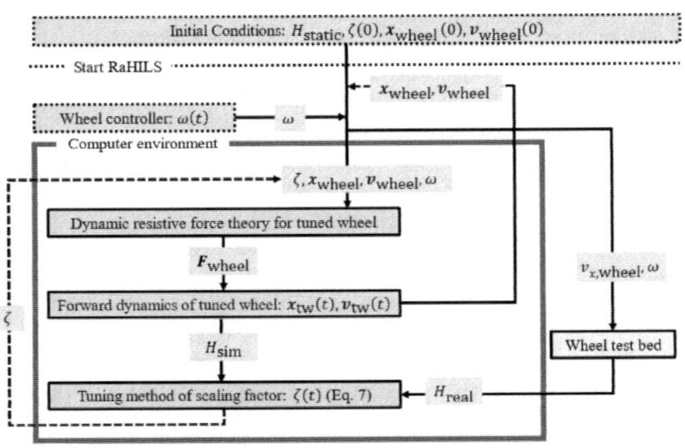

Fig. 5. Framework of HILS-DRFT

Specifically, tuning method of scaling factor in *Fig. 5* follows the equation outlined below:

$$\zeta(t+1) = \begin{cases} \zeta(t) + \Delta\zeta & (H_{\mathrm{real}} - H_{\mathrm{sim}} < -\epsilon) \\ \zeta(t) - \Delta\zeta & (H_{\mathrm{real}} - H_{\mathrm{sim}} > \epsilon) \\ \zeta(t) & (|(H_{\mathrm{real}} - H_{\mathrm{sim}})| \le \epsilon) \end{cases} \tag{7}$$

Here, $\Delta\zeta$ represents the incremental change in the scaling factor per tuning, and ϵ represents the allowable error. By appropriately setting these values, stable and precise behavior analysis becomes possible. In this study, $\Delta\zeta$ was set to 0.001 and ϵ was set to 0.1 mm.

3. Experiment

3.1. Experimental setup

As previously mentioned, the single-wheel test bed can be operated in a state where the horizontal moving frame is disconnected from the ball screw, allowing the wheel to move solely through its own rotation. This type of single-wheel driving experiment is referred as *Experiment driving* or ground truth data. In contrast, a single-wheel driving experiment using the HILS-DRFT framework will be referred as *HILS-DRFT driving*. Additionally, as shown in *Fig. 5*, two different trajectories of wheels can be observed in HILS-DRFT: one from the numerical model in a computer environment and the other from the real single-wheel testbed. We refer to the former as *HILS-DRFT (sim)* and the latter as *HILS-DRFT (real)*.

To ensure accurate comparisons, unified conditions are presented in *Tables 1 and 2*. Here, we differentiate between the horizontal and vertical masses, denoted as m_x and m_z , respectively. This distinction is necessary because the horizontal moving frame in the single-wheel test bed does not influence vertical motion. Properly setting these parameters is essential to ensure the accuracy of the forward dynamics solved in the computer environment. The initial conditions used for the HILS-DRFT framework are shown in *Table 3*, and they were configured based on the observed initial conditions from the *Experiment driving*.

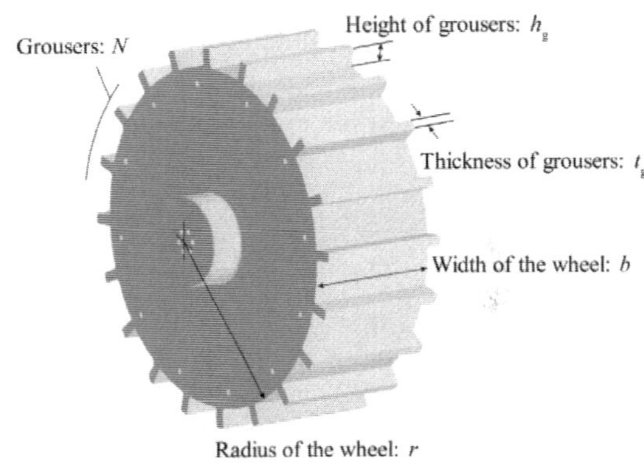

Fig. 6. Illustration of the wheel with grousers

Table 1
Typical unified parameters between HILS-DRFT and experiment

Parameters	Values
Radius of wheel r [m]	0.150
Width of the wheel b [m]	0.120
Number of grousers N [-]	12
Height of grousers h_g [m]	0.018
Thickness of the grousers t_g [m]	0.009
Horizontal mass m_x [kg]	13.902
Vertical mass m_z [kg]	18.837
Angular velocity profile of the wheel control	(Shown in *Table 2*)

Table 2
Angular velocity profile of the wheel control

Time [s]	Wheel state	ω [deg/s]
0 → 2	*Stand by*	0
2 → 7	*Constant acceleration rotation*	0→30
7 → 17	*Constant rotational speed*	30
17 → 22	*Constant deceleration rotation*	30→0
22 → 24	*Stand still*	0

Table 3
Initial conditions used for HILS-DRFT

Variables	Values
Static sinkage H_{static} [m]	0.0176
Initial scaling factor $\zeta(0)$ [-]	0.7823
Initial position $\boldsymbol{x}_{\mathrm{tw}}(0)$ [m]	$(0.0, -H_{\mathrm{static}})$
Initial velocity $\boldsymbol{v}_{\mathrm{tw}}(0)$ [m/s]	(0.0,0.0)

3.2. Validation of tuning method in HILS-DRFT

The single-wheel experiment result verifies how the real-time tuning of the scaling factor appropriately works. One typical result of the HILS-DRFT is shown in *Fig. 7.* The vertical axis represents the wheel's displacement in the wheel vertical direction Δz, and the horizontal axis represents time t. Qualitatively, the sinkage of *HILS-DRFT (sim)* matches the sinkage of *HILS-DRFT (real).* Additionally, the same experiment was executed ten times, and the root mean square error (RMSE) of Δz was less than 1 mm in all cases, which is remarkably small to the wheel radius. Also, the plot shows oscillations. To investigate the cause of these oscillations, we conducted an FFT analysis on the time variation of the settlement during *Constant rotational speed* running, and the result showed a peak at about 1.68 Hz. Considering that the wheel has 12 grousers and a rotational angular velocity of 30 degrees per second, this frequency characteristic is consistent with the influence of the wheel grousers, which is also observed in *Experiment driving.* These findings indicate that the oscillations are not due to the discreteness of the parameter tuning method proposed in this paper but rather the periodicity of the wheel grousers. Thus, the real-time tuning of the scaling factor functions appropriately.

Observing the changes in the scaling factor, we can see that it decreases from the beginning to the end of the wheel's rotation. As mentioned earlier, the scaling factor represents the terrain stiffness, so this indicates that the sand around the wheel becomes looser compared to its static state. This loosening is likely caused by the wheel's rotation, which excavates the sand. This result also implies that the value of the scaling factor is not constant and that should be tuned for static/dynamic states accordingly.

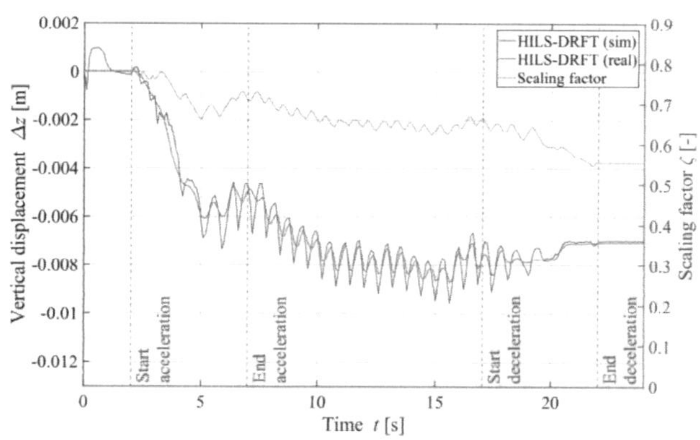

Fig. 7. Result of HILS-DRFT focused on real time tuning method

3.3. Compare the result of HILS-DRFT and DRFT

This subsection evaluates the accuracy of *HILS-DRFT driving* in replicating *Experiment driving.* Ten runs were conducted for both *Experiment* and *HILS-DRFT driving. Figure 8* illustrates the wheel trajectories for *Experiment driving, HILS-DRFT driving,* and original DRFT for comparison. For *HILS-DRFT driving,* two sets of results are available: *HILS-DRFT (sim)* and *HILS-DRFT (real).* However, only the *HILS-DRFT (sim)* results are presented here. Each plot shows the mean values of the ten runs with solid lines, and the standard deviations are indicated with light shading. The vertical axis represents the wheel's vertical displacement, Δz, and the horizontal axis represents the wheel's horizontal displacement, Δx. The results indicate that HILS-DRFT can simulate transient states more accurately than DRFT. Additionally, when comparing the RMSE of the average trajectory between *Experiment driving* and both *HILS-DRFT driving* and original DRFT, the values are 5.0 mm and 8.4mm, respectively. Therefore, HILS-DRFT demonstrates superior accuracy over DRFT not only in transient states but also in other conditions.

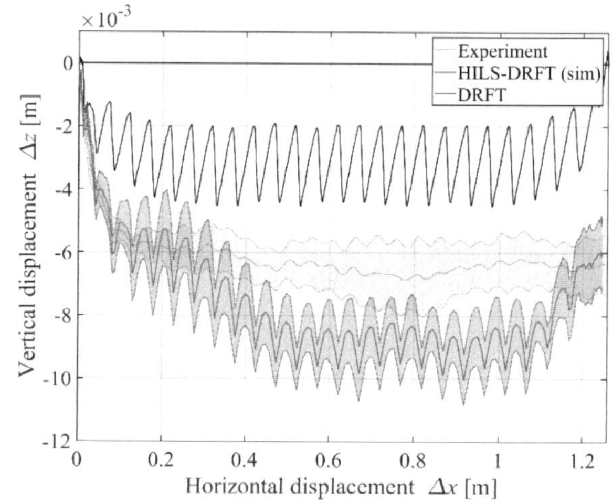

Fig. 8. Wheel trajectories of experiment, HILS-DRFT and original DRFT

However, a small error remains. By tuning the coefficients of the damping terms added to the forward dynamics for stabilizing HILS, more stable calculations and reduced errors can be achieved.

4. Conclusions and future work

This study evaluated the HILS-DRFT framework's effectiveness. The framework employs a real-time tuning method for sand parameters based on sinkage observed in a single-wheel test bed. During verification, results from the original DRFT and HILS-DRFT were compared with actual driving results. The comparison demonstrated that HILS-DRFT behavior closely matched actual wheel performance, surpassing the original DRFT. It was also notably accurate in simulating the transient state.

The tuning of the scaling factor does not merely involve fitting arbitrary parameters to align experiments with simulations. Instead, it sequentially identifies terrain stiffness caused by sand flow, compaction, and compression due to the wheel's movement. This method is applicable to analyzing vehicles with multiple wheels. Future research should extend the HILS-DRFT framework to analyze rovers' actual driving behavior, achieving more precise results.

5. Declaration of competing interest

The authors declare that they have no known competing financial interests or personal relationships that could have appeared to influence the work reported in this paper.

6. References

Agarwal, S., Karsai, A., Goldman, D.I., Kamrin, K., 2021. Surprising simplicity in the modeling of dynamic granular intrusion. Sci. Adv. 7, eabe0631.

Bekker, M.G., 1956. Theory of land locomotion. University of Michigan Press, Ann Arbor.

Bekker, M.G., 1960. Off-the-road locomotion University of Michigan Press, Ann Arbor.

Bekker, M.G., 1969. Introduction to terrain-vehicle systems. University of Michigan Press, Ann Arbor.
Ishikawa, S., Ishigami, G., 2022. Development of Hardware-in-the-loop Simulator for Rough Terrain Mobile Robot and Its Application to Real-time Tuning of Wheel Dynamics Model. Journal of the Robotics Society of Japan. 40, 542-545.
Li, C., Zhang, T., Goldman, D.I., 2013. A terradynamics of legged locomotion on granular media. Science 339, 1408–1412.
Wong, J.Y., Reece, A., 1967. Prediction of rigid wheel performance based on the analysis of soil-wheel stresses part I. performance of driven rigid wheels. J. Terramech. 4, 81-98.
Wong, J.Y., Reece, A., 1967. Prediction of rigid wheel performance based on the analysis of soil-wheel stresses part II. performance of driven rigid wheels. J. Terramech. 4, 7-25.
Wong, J.Y., 2008. Theory of ground vehicles. Hohn Wiley & Sons Inc., Hoboken, New Jersey.

3181 | Assessing sensitivities of off-road pneumatic tire on clay: A finite element investigation on tire operational and design factors

Destiny Mason[a,*], Varsha S Swamy[a], Corina Sandu[a], Alba Yerro[b], Katie Sebeck[c], and David Gorsich[c]

[a] *Terramechanics, Multibody and Vehicle Systems (TMVS) Laboratory, Department of Mechanical Engineering, Virginia Polytechnic Institute and State University, Blacksburg, VA, United States*

[b] *Department of Civil and Environmental Engineering, Virginia Polytechnic Institute and State University, Blacksburg, VA, United States*

[c] *U.S. Army DEVCOM Ground Vehicle Systems Center, Warren, MI 48397*

** Corresponding author: destinym22@vt.edu*

ABSTRACT

Military and agriculture vehicles alike often experience uncertainties regarding sinkage and traction while traversing non-linear deformable soils. The off-road tires of these vehicles are equipped with stiffer sidewalls to reduce puncture risk, increase load bearing capacity, and improve stability and traction. Since physically testing tires is time-consuming and costly, enhancing the virtual physics-based modeling capabilities can be beneficial for design decisions. Identification of key design parameters that control tire traction and mobility in such soils is essential to ensure that modelers prioritize the study and accuracy of these parameters to improve the model's reliability. To accomplish this, a tire sensitivity analysis is proposed to evaluate the tire model's robustness.

The tire used for this study is a lugged bias-ply tire with improved self-cleaning characteristics for muddy terrains. An advanced FE tire model is developed with detailed modeling of the distinct tire parts, including the bead, apex, inner layer, 2-ply carcass, sidewalls, under tread, and lugs. Two types of parameters are of interest: (a) tire operational parameters and (b) tire design parameters. The influence of tire operational parameters, i.e., tire normal load, inflation pressure, and velocity are first conducted and verified with existing studies in the literature. As a novel contribution, the tire design parameters, i.e., tire width and tire diameter, are also considered. The advanced FE tire is made to negotiate on the rigid ground first and then on low plasticity CL clay to analyze the trends in sensitivities. The results of this sensitivity analysis provide valuable insights into the reliability of the FE tire, enabling more informed conclusions to be drawn from simulated data.

Keywords
Off-road mobility
Tire modeling
Agricultural tires

1. Introduction

The performance of military and agricultural vehicles traversing non-linear deformable soils is contingent upon the complex interplay between tire characteristics, vertical loading, and soil properties. In these off-roading environments, uncertainties surrounding sinkage and traction influence vehicle mobility. To address these challenges, this paper presents an investigation into the sensitivity of tire parameters on sinkage and traction dynamics, focusing on a specific scenario characterized by a lugged biased-ply BKT TR 319 29X12.50-15 tire and non-linear deformable clay. Leveraging Finite Element Analysis (FEA) modeling, this study encompasses variations of both operational parameters including normal load, inflation pressure, and tire velocity as well as design parameters including diameter and width.

Existing literature describing the drawbar pull and sinkage of lugged tires is limited, as other research focuses on the lateral forces generated at the tire under steering conditions (F. Anstett-Collin and T. Mara et al., 2014; H Ragheb and M. El-Gindy et al., 2013; Olatunbosun O.A. and O. Bolarinwa, 2004). Ansett-Collin shared that the nonlinear model depends on parameters such as lateral friction coefficient (μ), vertical load (F_Z), cornering stiffness (K), and side slip angle (α), where it was further determined that α is the most influential parameter on the lateral force followed by FZ, and K, while C did not have an impact (F. Anstett-Collin and T. Mara et al., 2014). Other publications take a terramechanics approach and examine the interaction between an FEA tire and soil or clay (M. Marjani, 2016; P. Farhadi and A. Golmohammadi et al., 2019). These studies dive into a comparison of soil modeling, tire validation, and parameters affecting the tire-soil contact patch. Farhadi in particular examines the contact area, contact volume, and rolling resistance of a lugged agricultural tire on clay-loam soil as a result of altering vertical load, inflation pressure, and soil moisture (P. Farhadi and A. Golmohammadi et al., 2019). However, a sensitivity study of the tire parameters is not accomplished on this terrain, leaving out various parameters aspects that could have a significant impact on sinkage and traction. Lastly, an existing tire sensitivity study identifies numerous parameters affecting longitudinal and lateral forces, self-aligning torque, and rolling radius and carries out an extensive study on a rigid, dry surface in TMeasy (F. Mohammadi, 2012). In this study, modifications were applied to tire design and operational parameters by adjusting each parameter by 10-30%. This systematic approach allows for a clear comparison of each parameter's effects on the outputs of the model.

Building on existing literature, this study aims to address the research gap by conducting a sensitivity analysis on an advanced FE lugged tire model and examining its behavior over rigid and clay surfaces. The parameters of interest include tire width, tire diameter, inflation pressure, normal load, and velocity. Each of these parameters are changed independently to determine the effect it has on tire deflection, vertical sinkage, and drawbar pull.

The paper is organized as follows. Section 2 presents the methodology behind the FE model including the tire and the soil. Section 3 outlines the parameter changes made for the sensitivity analysis. Finally, section 4 provides a summary of the study's results.

DISTRIBUTION STATEMENT A. Approved for public release; distribution is unlimited. OPSEC #8714.
https://doi.org/10.56884/0AXMP2K2

2. Numerical model setup

2.1. FE tire model

The tire used in this study is a lugged bias-ply agricultural tire, BKT TR 319 29X12.50-15. As depicted in Figure 1a, the tire has a deep lug pattern at 2cm in depth for traversing soft soils, with an overall diameter and width of 75cm and 31cm, respectively. The tire was 3D scanned and subsequently cut in half to determine its parts and dimensions to construct the FEA model in Fig. 1b. The tire employs a Mooney-Rivlin hyperelastic material model for most of the components, with the exception of the carcass and is made up of the following parts: lugs, crown, tire sidewall, inner liner, inner tire, bead filler, bead, carcass and it is mounted to a rigid rim. The carcass is a composite component as it is made of both rubber and steel cords to provide structure to the tire. This combination of material and orientation of the cords results in anisotropic behavior, where the mechanical properties vary with direction. An orthotropic elastic material model is used to capture the orientation of the cords and resulting mechanical behavior. The tread patterns and other dimensions were obtained from the disassembled tire and implemented into a CAD (Computer Aided Design) model, which was then meshed, and material properties assigned within LS-Dyna. The material models used for this tire were obtained from previous studies on similar agricultural tires (H.H.R. Ghoreishy, 2001; Z. El-Sayegh, 2020).

Four element types were utilized in the model to balance accuracy and computational efficiency: constant stress elements, shell elements, fully integrated elements, and triangular elements. Constant stress elements were applied to the inner tire, bead, bead filler, and crown to accurately capture stress distributions while maintaining computational efficiency. Shell elements were used to model the thin layers of the inner and outer carcass as well as the inner liner. Full integration is implemented on the sidewall and lugs to ensure precise stress and strain calculations in these regions. Triangular elements are used on the rim to allow for simplified meshing techniques. In total, the model is made up of 11 parts and 30,480 elements.

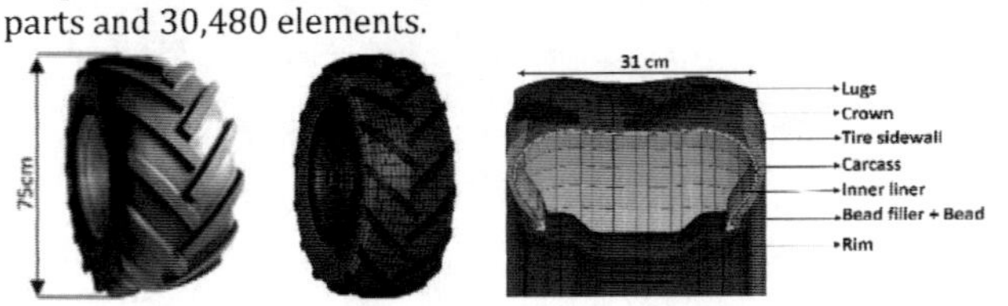

Fig. 1. (a) Lugged tire (b) Meshed FE model (c) FE labeled model

2.2. Clay material model

The soil model used in this study is representative of nonlinear deformable clay employed in previous research, as shown in Fig. 2. A homogenous rectangular prism with dimensions 0.8m (width) x 0.3m (depth) x 3.5m (length) and fixed boundaries is implemented to restrict movement, while minimizing boundary effects. The clay is represented using a material geological cap model, which effectively simulates the plastic deformation and the cohesive behavior of clay. The following parameters are defined within the material card: density, initial bulk modulus, initial shear modulus, failure envelope parameter and linear coefficient, cap, hardening law exponents and coefficient, and tension cut off as defined in previous studies (Pandit, 2023; Swamy, 2023b).

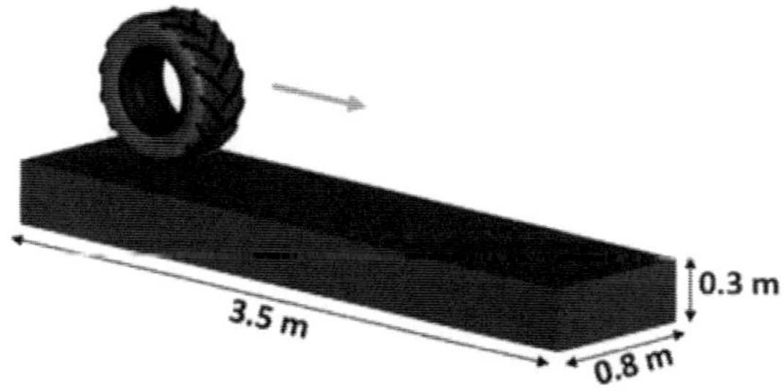

Fig. 2. Meshed FE tire-clay model

2.3. Tire-Clay FE model

The FE simulation involves the integration of the tire and soil models to accurately capture the interactions between the tire and soil, as illustrated in Fig. 2. Each of these models are created independently, as discussed in previous sections, but when imported into the same simulation, the tire is center aligned over the soil to minimize boundary affects. Table 1 below lists the initially defined parameters. Although the tire parameters are defined according to the manufacturer's guidelines, the friction coefficient is determined through physical shear testing to accurately assess the interaction between the tire and terrain. The tire undergoes inflation, followed by the application of gravity, normal load, and finally, the rolling motion is applied (implemented through angular velocity, ω and linear velocity, V). The same type of rolling tire simulation is also completed on a rigid boundary, which is simply done by removing the soil and replacing it with a rigid shell element of the same length and width dimensions with no depth, as illustrated in Fig. 3.

Table 1
Baseline operational and design parameters

Parameter	Value
Normal Load	5000 N
Inflation Pressure	140 kPa
Tire Velocity	5 km/h
Slip ratio	10%
Tire diameter	75 cm
Tire width	31 cm
Number of lugs	30
Friction coefficient	0.189

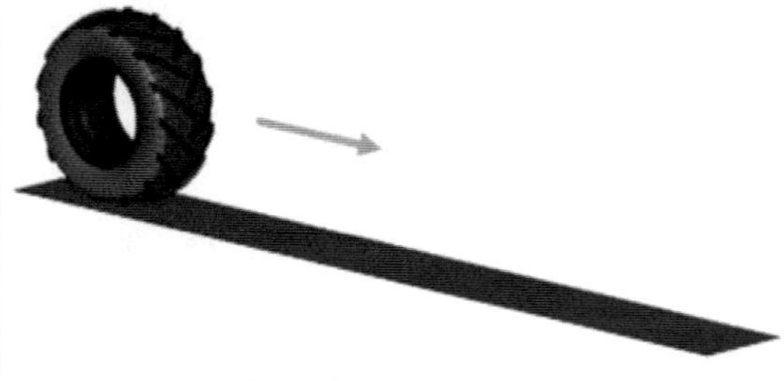

Fig. 3. Meshed FE tire-rigid model

The above simulations, both on a rigid surface and on clay, are used as a baseline for a tire sensitivity analysis. The simulation on a rigid boundary does not allow sinkage; however, traction and tire deflection can be derived from the model. Further, the same tire is simulated rolling over the clay substrate, which allows for a more advanced investigation into the interactions between a lugged tire and clay. Tire sinkage, drawbar pull, and various stresses and strains are among the various outputs that can be analyzed in this model.

3. Tire sensitivity analysis

The objective of this tire sensitivity analysis is to understand how variations in tire parameters affect the tire's performance and its interaction with soil, particularly in terms of sinkage and drawbar pull. The parameters analyzed in this study can be divided into operational and design parameters, as shown in Table 2. The operational parameters include tire normal load, tire inflation pressure, and tire velocity. The design parameters include tire diameter and tire width. Each of these parameters have a baseline value associated with the original tire, listed in Table 1; however, for the purpose of this analysis each of these parameters has been independently both increased and decreased by 10%, table 2. The purpose of this methodology is to uniformly change a single parameter in each simulation, allowing for a direct comparison between results. Further, this reveals which parameter has the largest effect on performance and how significant each parameter is in relation to the others.

Table 2
Baseline operational and design

Operational Parameters	Values
Normal load	4500 N, 5000 N, 5500 N
Inflation pressure	126 kPa, 140 kPa, 154 kPa
Tire velocity	4.5 km/hr, 5 km/hr, 5.5 km/hr
Design Parameters	Values
Tire diameter	67.5 cm, 75 cm, 82.5 cm
Tire width	27.9 cm, 31 cm, 34.1 cm

4. Results and Discussion

The study examines the impact of various operational and design parameters on tire-clay interactions, specifically focusing on sinkage and drawbar pull. Additionally, a brief analysis on how these parameters affect tire deflection and traction on rigid surfaces is presented. This analysis explores the influence various parameters including inflation pressure, normal load, diameter, velocity, and width have on the tire's traction and overall performance. Understanding this relationship is integral in creating a robust FE tire model and further optimizing tire design for mobility.

The baseline simulations were conducted under the following conditions: a normal load of 5000N, 140kPa inflation pressure, 10% slip ratio, and a constant velocity of 5 km/hr. The tire is an agricultural bias ply tire with diameter and width dimensions of 75 and 31cm, respectively. The terrains are as follows: a rigid surface with a coefficient of friction of 0.189, and a nonlinear cohesive clay modeled using a soil effective stress constitutive model.

Figure 4 illustrates the contours resulting from the tire-clay interaction simulations and serves as a reference point for the remainder of the study. From the left, Fig. 4a depicts the vertical displacement experienced by the clay beneath the tire as well as residual plastic effects after the tire has passed. Similarly, Fig. 4b and 4c depict the pressure and ZX shear stress distribution in the soil, respectively. In addition, tire sinkage and drawbar pull coefficient are plotted alongside the extended analysis in Fig. 7 and 8.

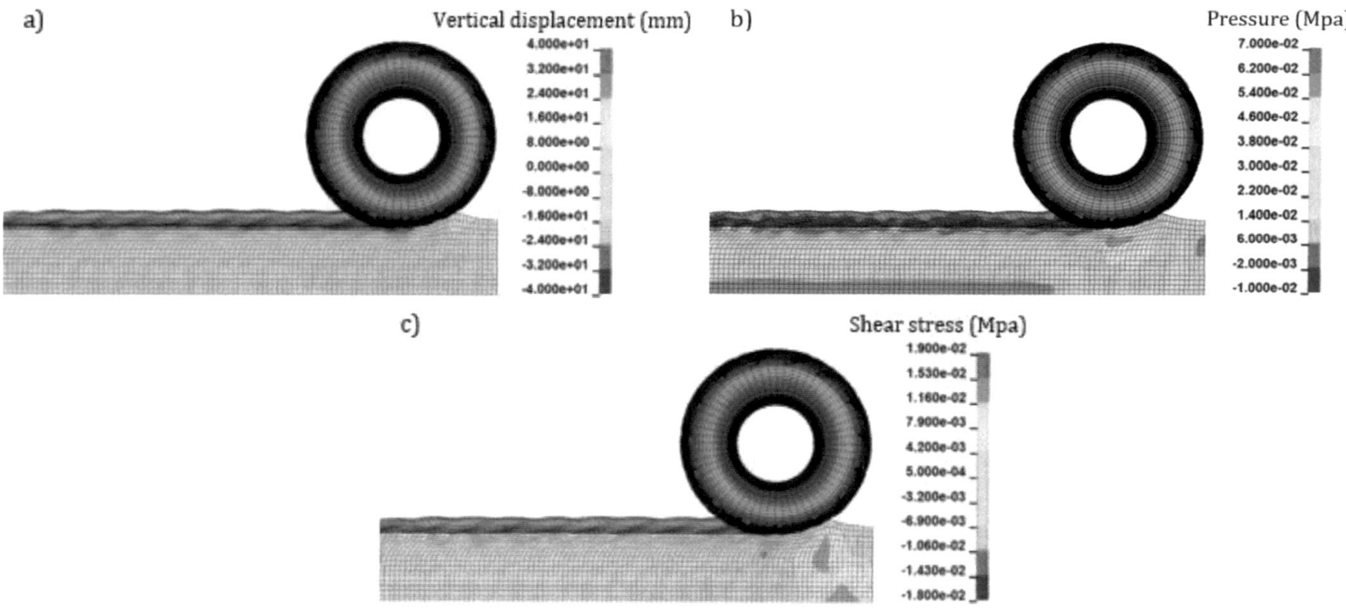

Fig. 4. Contour plots of the baseline simulation results under conditions of 10% slip, 5000 N normal load, 140 kPa inflation pressure, and 5 km/hr velocity: (a) vertical displacement, (b) pressure distribution, and (c) ZX shear stress distribution.

4.1. Tire-rigid interactions

Through simulation, each of the five parameters were individually altered to 90% and 110% of the nominal design values. The resulting FE tire model was then run on a rigid shell part yielding the results shown below in sections 4.1.1 and 4.1.2. Generally, tire deflection was positively correlated to normal load and negatively correlated to inflation pressure, tire velocity, diameter, and width. Tractive force is positively correlated with normal load and tire diameter while tire width, velocity, and inflation pressure do not have a significant affect.

4.1.1 *Variation with operational parameters*

Figure 5 presents variations in longitudinal force and tire deflection under varying operational parameters. Increasing the normal load is found to result in an increase in both deflection and longitudinal force, as can be seen in Fig. 5a. Both variables are strongly correlated with the parameter normal load. This relationship is expected as, in general, higher loads improve traction but also increase tire deflection, which in turn enlarges the contact patch, further improving traction.

Conversely, increasing the inflation pressure of the tire from 126 kPa to 154 kPa results in a relatively unchanged tractive force; however, tire deflection appears to be negatively correlated, as shown in Fig. 5b. These results indicate that increasing the tire's inflation pressure reduces tire deflection. It is possible that increasing or decreasing the inflation pressure more significantly could affect the tractive force, but the change of 28 kPa did not yield a notable effect.

Tire velocity is the last operational parameter, which was tested in this study, ranging from 4.5 – 5.5 km/hr, shown in Fig. 5c. Similar to inflation pressure, the change in velocity did not produce a notable change in tractive force. The deflection of the tire appears to be negatively correlated with inflation pressure, indicating that increasing velocity decreases the resulting tire deflection.

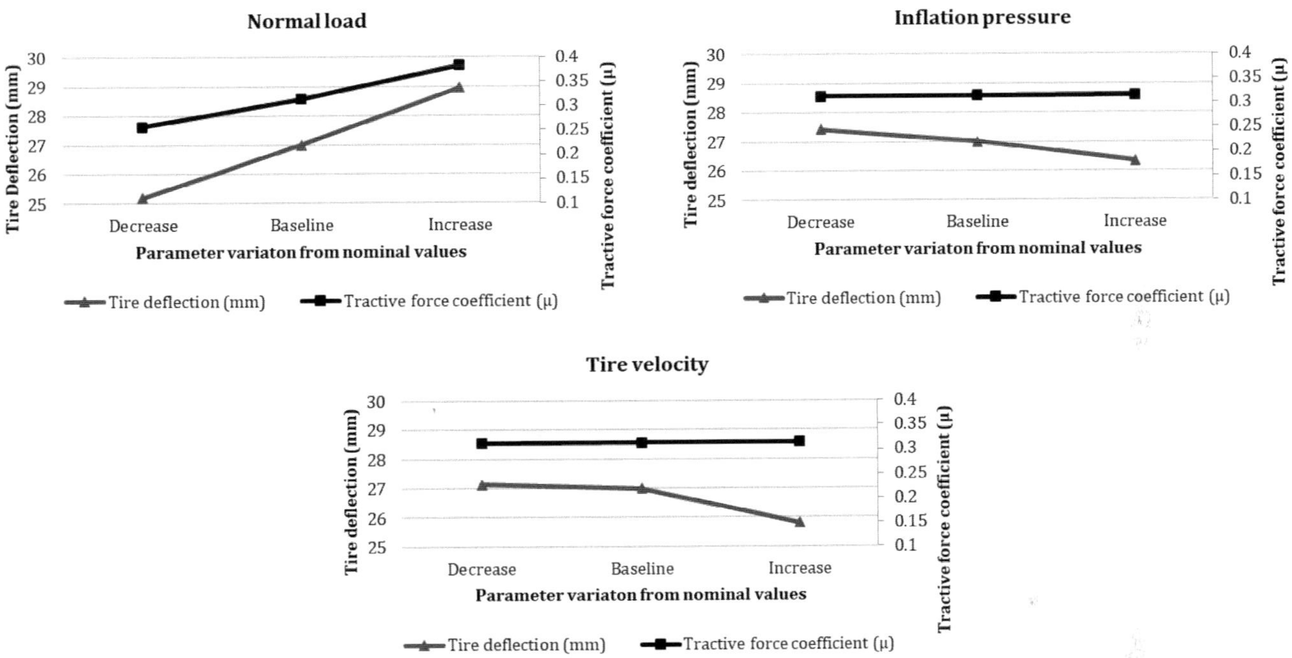

Fig. 5. The variation of tire deflection and tractive force coefficient with varying operational parameters including: (a) inflation pressure, (b) normal load, and (c) tire velocity.

4.1.2 *Variation with design parameters*

The variability of tire deflection and tractive force under varying design parameters is depicted in Fig. 6. On the left a comparison of the affect tire diameter has on these parameters is shown. Increasing the tire's diameter results in a negative correlation with tire deflection and a positive correlation with tractive force. As the tire's diameter is increased it experiences a more distributed load resulting in less deflection of the sidewall. Similarly, since the increased diameter yields a larger contact patch, the tire experiences a larger tractive force coefficient. The traction results have a similar effect when decreasing diameter, but the tire deflection appears to be nonlinear. A more extensive study will need to be done to address this.

Tire width on the other hand experiences a negative correlation with tire deflection, but tractive force does not experience a significant change from 27.9 to 34.1 cm, shown in Fig. 6b. The deflection is less with increased tire width given the increased load distribution. While the traction is not significantly changed as a result of changing the width 6.2 cm, it is likely that increasing the width further would yield an increase in traction as the contact patch increases.

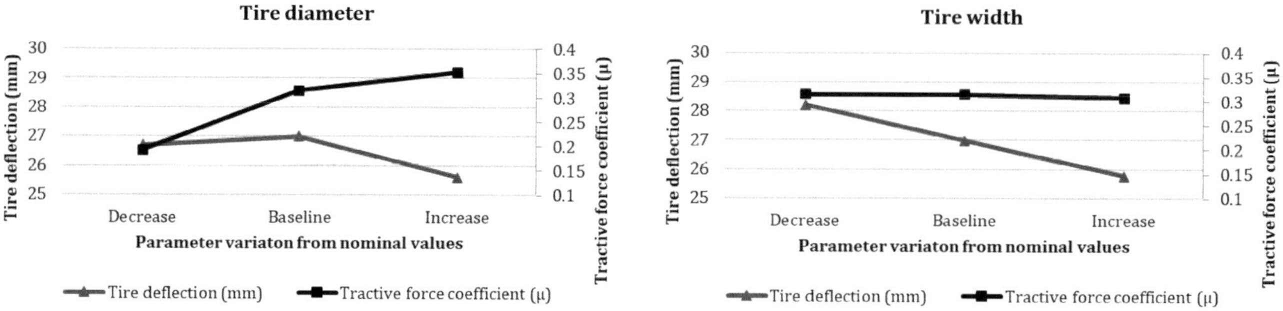

Fig. 6 The variation of tire deflection and tractive force coefficient with varying design parameters including: (a) tire diameter and (b) tire width.

4.2 Tire-clay interactions

Similar to the rigid simulations, each of the parameters were systematically altered from the nominal design values. The model was then run on FE soil yielding the results shown below in sections 4.2.1 and 4.2.2. Instantaneous sinkage is positively correlated with normal load, negatively correlated with tire velocity, width, and diameter, while inflation pressure did not make a significant impact. Drawbar pull is positively correlated with normal load and inflation pressure and negatively correlated with tire velocity, diameter, and width.

4.2.1 Variation with operational parameters

Variation in the drawbar pull coefficient and tire sinkage with varying operational parameters are shown in Fig. 7. Both parameters share a positive correlation with normal load, although instantaneous sinkage shares a more significant increase in magnitude with loading, shown in Fig. 7a. These results demonstrate that increasing the vertical load on the tire at this velocity leads to greater sinkage, as well as increase in the drawbar pull coefficient. Specifically, as the load increases, the tire sinks deeper into the soil, increasing the tire's contact area. The increased contact area causes the grip and traction to increase, thereby improving the drawbar pull coefficient.

Next, variation in the parameters with different inflation pressures are shown in Fig. 7b. The results indicate that both sinkage and drawbar pull coefficient increase as inflation pressure increases. Specifically, as the inflation pressure rises, the tire becomes less compliant. The rigidity causes the tire to dig deeper into the soil causing greater sinkage. Consequently, the tire experiences a larger contact area further increasing the drawbar pull coefficient. The increased drawbar pull can be beneficial to mobility; however, when considering more than one vehicle passing an area, deeper sinkage can cause mobility issues for the following vehicles.

Conversely, velocity shares a negative correlation with both instantaneous sinkage and drawbar pull coefficient, Fig. 7c. The reduction in sinkage can be attributed to the dynamic behavior of the tire soil interaction. At higher speeds there is less time for the tire to deform or sink into the soil, leading to less sinkage. At the speeds exhibited in this study, 4.5-5.5 km/hr, the velocity is able to be increased without fully losing traction; however, if the velocity were to increase significantly further, the tire would not have the traction or contact area necessary for mobility.

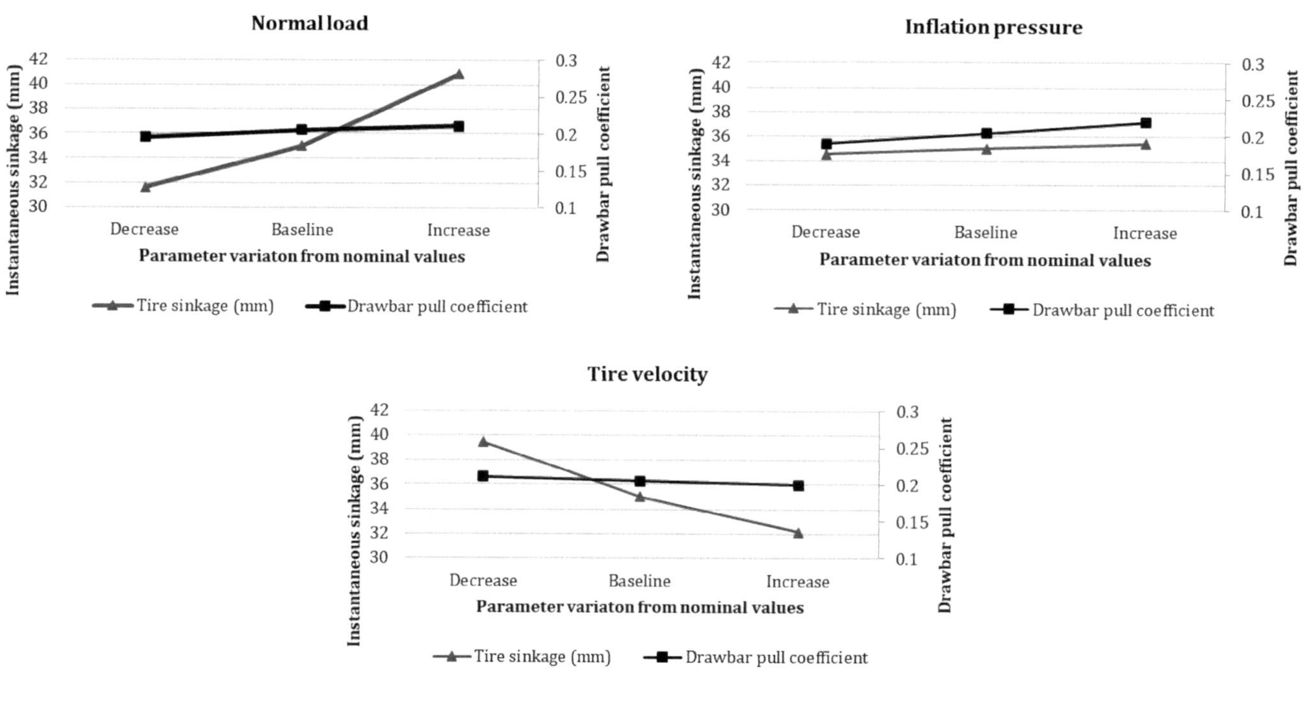

Fig. 7. The variation of tire deflection and tractive force coefficient with varying operational parameters including: (a) inflation pressure, (b) normal load, and (c) tire velocity.

4.2.2 Variation with design parameters

The result of varying design parameters can be seen in terms of instantaneous sinkage and drawbar pull coefficient in Fig. 8. Similar to the results on a rigid surface, increasing tire diameter yields a negative correlation, with both the drawbar pull coefficient and sinkage. This correlation suggests that tires larger in diameter experience less traction and sinkage due to the distribution of load over a larger surface area. The weight distribution allows the tire to sink less resulting in a smaller contact patch, thereby reducing drawbar pull coefficient. This can be beneficial in applications such as agriculture where minimizing disturbance to the soil is important.

Further investigation into these results indicate that the negative correlation may also be influenced by an increase in the slip ratio with larger tire diameters. Increasing diameter without altering angular and forward velocity results in an increased slip ratio, and vice versa, shown in equation 1. Therefore, the results show the effect of the diameter change but also that of a slip ratio change as a result. Additional setups including a change in angular velocity must be completed in order to understand this interaction.

$$Slip\ Ratio\ \% = \left(\frac{\Omega R_C}{V} - 1\right) \times 100 \quad (1)$$

Tire width experiences a negative correlation with both drawbar pull and instantaneous sinkage, as shown in Fig. 8b. The correlation indicates that wider tires distribute the vehicle's weight over a larger surface area, reducing point loads. Consequently, the tire experiences less sinkage and therefore a smaller contact area between the tire and ground. The lesser contact area causes a reduction in the drawbar pull coefficient. Similar to other parameters, this result can be beneficial for agricultural purposes as it reduces soil compaction; however, it can be disadvantageous in applications where good traction is necessary.

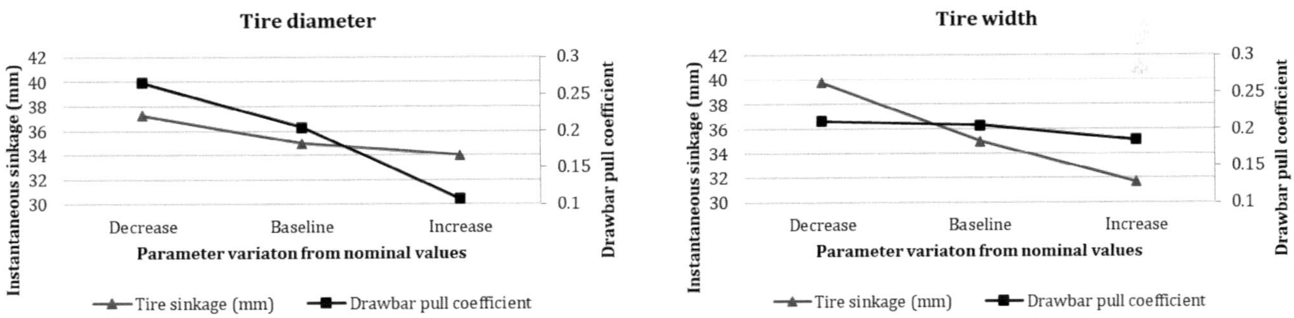

Fig. 8. The variation of tire sinkage and drawbar pull coefficient with varying design parameters including: (a) tire diameter and (b) tire width.

4.3 Summary

This study aimed to investigate the effects of various parameters on tire performance, measured by instantaneous sinkage and drawbar pull. A full summary of the tire-clay results is shared in Table 3. below. The table indicates a positive, negative, or no correlation for each parameter tested. Key findings revealed a correlation between each of the parameters. However, it was observed that certain parameters had a larger effect on sinkage while others had a greater impact on drawbar pull.

Parameters including normal load, tire velocity, and width were found to play a crucial role in sinkage, with variation in these parameters yielding notable changes in the tire's instantaneous sinkage into the soil. Drawbar pull was predominantly influenced by tire diameter. These findings emphasize the importance of the complex relationship between operational and design parameters and tire performance and mobility. Through a comprehensive understanding, a more robust FE model can be established, further influencing improvements in tire design for off-roading vehicles. Further research will incorporate the effect of additional parameters and more significant variations from the nominal values. While this study presents a simplified approach, the expanded version will incorporate more advanced techniques. Consideration will be given to methods such as principal component analysis or design of experiments to achieve a more comprehensive and detailed analysis.

Table 3
Summary of correlations between parameters and performance

	Normal Load	Inflation Pressure	Tire Velocity	Tire Diameter	Tire Width
Instantaneous Sinkage	↑	—	↓	↓	↓
Drawbar pull coefficient	↑	↑	↓	↓	↓

5. Nomenclature

R	Effective rolling radius	[m]
V	Forward velocity	[m/s]
ω	Angular velocity	[rad/s]

6. Acknowledgements

This study has been supported by the Automotive Research Center (ARC) under the Cooperative Agreement W56HZV-19-2-0001 U.S. Army DEVCOM Ground Vehicle Systems Center, as well as by the TMVS Lab and Geotechnical Engineering Lab at Virginia Tech.

7. Declaration of competing interest

The authors declare that they have no known competing financial interests or personal relationships that could have appeared to influence the work reported in this paper.

8. References

F. Anstett-Collin, T. Mara, and M. Basset, 2014. Application of global sensitivity analysis to a tire model with correlated inputs. Elsevier, 44, 54-62.

H. Ragheb, M. El-Gindy, and H. Kishawy, 2013. Development of a combat vehicle fea tire model for off-road applications. SAE Technical Paper Series. doi:10.4271/2013-01-0632.

Olatunbosun, O.A., Bolarinwa, O., 2004. Fe simulation of the effect of tire design parameters on lateral forces and moments. Tire Science and Technology 32, 146–163. doi:10.2346/1.2186779.

Marjani, M., 2016. Development of FEA wide-base truck tire and soil interaction models. eScholar. Ontario Tech University, https://ontariotechu.scholaris.ca/items/5e8662d9-d017-4879-ba37-5432b444ebff.

Farhadi, P., Golmohammadi, A., Sharifi Malvajerdi, A., Shahgholi, G., 2019. Finite element modeling of the interaction of a treaded tire with clay-loam soil. Computers and Electronics in Agriculture 162, 793–806. doi:10.1016/j.compag.2019.05.031.

Mohammadi, F., 2012. Tire Characteristics Sensitivity Study. Chalmers University of Technology.

Ghoreishy, M.H.R., 2001. Finite element analysis of a 6.45-14 bias tire under contact load. Iranian Polymer Journal 10. doi:1026-1265/2000.

El-Sayegh, Z., 2020. Modeling and analysis of truck tire-terrain interaction. University of Ontario Institute of Technology.

Review of modeling and validation techniques for tire-deformable soil interactions

Varsha S Swamy, Rashna Pandit, Alba Yerro, Corina Sandu, Denise M Rizzo, Katherine Sebeck, David Gorsich, 2023a, Review of modeling and validation techniques for tire-deformable soil interactions, Journal of Terramechanics

Swamy, V.S., Pandit, R., Yerro-Colom, A., Sandu, C., Rizzo, D., Sebeck, K., Gorsich, D., 2023b. Numerical modeling of a tire on undrained saturated clay using FEM, ALE, and SPH. In: Proc. of the 16th European-African International

Pandit R., 2023 Characterization of soft clay and clay-tire interaction for the prediction of ground mobility. Thesis, Virginia Tech.

3303 | Soil instrumentation for measuring normal stress distribution under off-road tire

Alexandru Vilsan [a,*], Nikhil Ravichandran [a], Chaitanya Sonalkar [a], Corina Sandu [a]

[a] *Department of Mechanical Engineering, Virginia Tech, Blacksburg, Virginia, United States*

* *Corresponding author: avilsan@vt.edu*

ABSTRACT

The distribution of normal stress within soil plays an important role in determining the tire-soil interaction dynamics, affecting tire traction performance, soil compaction, and overall vehicle stability. A comprehensive analysis of the normal stress patterns will allow researchers to refine tire designs, tread characteristics, and inflation pressures for enhanced off-road performance across diverse terrains. This study presents a novel soil instrumentation device designed to measure normal stress distribution in soil following a single pass of an off-road tire. The developed device employs an array of piezoresistive vertical load sensors for providing real-time, accurate data of normal stress distribution beneath the tire during operation. Several tire-soil interaction tests were performed at the Virginia Tech Terramechanics Test Rig to verify the output of the load sensors. The sensor array was embedded in GRC-1 Lunar Soil Simulant at a depth of 150 mm while an off-road tire loaded with a vertical force of 2000 N traversed over the surface of the terrain. The results obtained showed significant differences in pressure patterns under different toe angles, indicating a good correlation between the actual normal stress distribution and the readings from the sensor array.

Keywords
Off-road mobility
Soil instrumentation
Terramechanics
Soil compaction
Normal stress distribution

1. Introduction

Off-road vehicles operate in diverse and challenging terrains, from rocky trails to sandy deserts. Their ability to navigate such conditions relies heavily on tire performance. The interaction between vehicle tires and soil plays an important role in determining traction, soil compaction, and overall vehicle stability. Among the various stress components acting on the tire-soil interface, normal stress influences the tire's contact patch with the soil, affecting steering response and overall vehicle control. The objective of our research is to develop a measurement system that can comprehensively analyze the distribution of normal stress within soil under various rolling conditions. Insights from normal stress analysis will guide adjustments in tread patterns, sidewall stiffness, and contact area. Our findings will contribute to enhancing off-road vehicle technology, ensuring superior performance across diverse landscapes.

(Raper et al., 1995) conducted an experiment at the USDA-Agricultural Research Service, National Soil Dynamics Laboratory in Auburn, Alabama for investigating the impact of inflation pressure and dynamic load on tire and soil parameters. Norfolk sandy loam soil and Decatur clay loam soil were selected for the study. The experiment utilized a Goodyear 18.4 R38 agricultural tractor tire and a vehicle capable of controlling dynamic load, inflation pressure, slip, and input torque. Seven Sensotec pressure transducers were installed to measure soil-tire interface stresses, while Stress State Transducers (SST) were positioned beneath the tire center to gauge soil stress. Findings revealed that increased inflation pressure reduced rut width but had minimal effect on deformed soil cross-sectional area. Conversely, increased dynamic load heightened both rut width and deformed soil cross-sectional area.

The study presented by (Jiang et al., 2019) employed a two-wheel-drive tractor alongside piezoelectric pressure sensors to assess vertical stress variations under various tire loads and inflation pressures. Vertical stress in the topsoil was measured using sensors positioned at an initial depth of 50 mm within the soil volume. The research aimed to evaluate the impact of tire loads and inflation pressures on vertical stress.

(Gheshlaghi and Mardani, 2021) presents an evaluation of vertical stress distribution in clay-loam soil using the Smoothed-Particle Hydrodynamics-Finite Element Analysis (SPH-FEA) technique. The research aims at modeling moist soil with hydrodynamic elastic-plastic material and the Murnaghan equation of state, while the tire is simulated using Finite Element Analysis (FEA). For validating the finite element model, calibrated load cells were positioned within the soil channel at a depth of 200 mm to measure vertical stresses. Using the data collected from the measurement device, the soil model was successfully calibrated for estimating stress distribution with high correlation to experimental data for different moisture content, different levels of traffic and tire loads.

Similar to the research presented above, (Derafshpour and Mardani, 2021) introduced a real-time method for measuring and recording soil stress distribution caused by wheel pass. In the presented system, a frame with a standard dimension of 120×120 mm was utilized. Five stress sensing elements were positioned along the central line of the frame longitudinally. Each stress sensor comprised a 100 kg load cell with a measurement precision of 0.1 kg. According to the investigators, the advantages of this measurement system are related to its simplicity and geometrical flexibility of pressure gauge replacement. Other notable research articles (Nichols et al., 1987),(Van den Akker, 2004), (Arvidsson

https://doi.org/10.56884/WNKEEB0L

and Ristic, 1996) employed different methods for measuring the soil normal stress distribution.

As discussed in the brief literature review, the interaction between the tire and the soil results in three-dimensional (3D) stresses, affecting not only the soil surface but also extending into the soil body. Understanding the normal stress distribution is essential for unraveling the complex mechanics of tire-soil traction. While pressure pads are commercially available for measuring contact patch distribution, their insertion into the soil might influence the soil compaction properties therefore affecting the measurement accuracy. To address this challenge, this article presents an alternative method: embedding a grid of piezoresistive sensors on a flexible material (such as cardboard) that minimally interacts with the soil. In the present research article, for measuring the normal stress distribution caused by wheel pass, an array of 15 Tekscan FlexiForce A401 sensors was placed inside the soil at a depth of 150 mm (Ravichandran, 2024).

2. Methodology and Experimental Equipment

This section presents the methodology employed for measuring the normal stress distribution in the GRC-1 Lunar Soil Simulant after one wheel pass. Furthermore, this section describes the experimental equipment required to perform the study.

2.1. Measurement system

This research introduces an innovative soil analysis tool, specifically designed to assess the distribution of normal stress in soil after it has been traversed by an off-road tire. The novelty of this device lies in its ability to capture the changes in the soil stress profile due to the tire movement without significantly impacting the soil compaction properties. Figure 1 illustrates the main components of the measurement system.

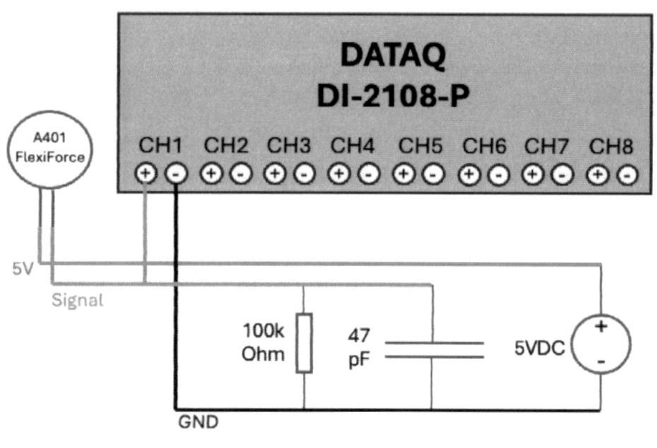

Fig. 1. Circuit diagram for one FlexiForce sensor

According to Tekscan, FlexiForce are ultra-thin, flexible force sensors designed for precise and accurate measurement of normal load. Utilizing thin-film technology, these sensors can detect force and pressure variations with high sensitivity and reliability. FlexiForce sensors provide an analog output, delivering a voltage signal proportional to the applied force or pressure. This analog output can be easily interfaced with microcontrollers, data acquisition systems, or signal processing circuits for further analysis or control.

In the experimental setup, the sensor is configured such that one side is directly connected to the 5V signal supplied by the power source. The opposite side is connected to the positive terminal of Channel 1 in the DAQ system, which corresponds to a voltage signal that varies proportionally with the applied normal load. In accordance with Tekscan recommendations, a resistor and a capacitor are connected in parallel with the Channel 1 circuit. The sensitivity of the FlexiForce sensor is influenced by the applied voltage from the power source and the values of the resistor and the capacitor. Consequently, a 47pF capacitor and a 100k Ohm resistor have been identified as the optimal components for measuring loads within the desired range. The DATAQ DI-2108-P is a versatile data acquisition system designed for precise measurement and recording of analog signals. Featuring eight input channels with a resolution of 16-bit, it enables simultaneous sampling at high speeds, allowing for real-time monitoring of various parameters. For this experiment, the measurement scale was set to 0-5V range, therefore the resolution of the DAQ system is 0.076 mV.

Figure 2 illustrates the top view of the FlexiForce A401 sensor array utilized in this study. This array configuration, selected for its suitability to the experimental setup, comprises multiple FlexiForce A401 sensors arranged in a grid pattern. Each sensor within the array is positioned to capture the variation of normal stress across the surface under examination.

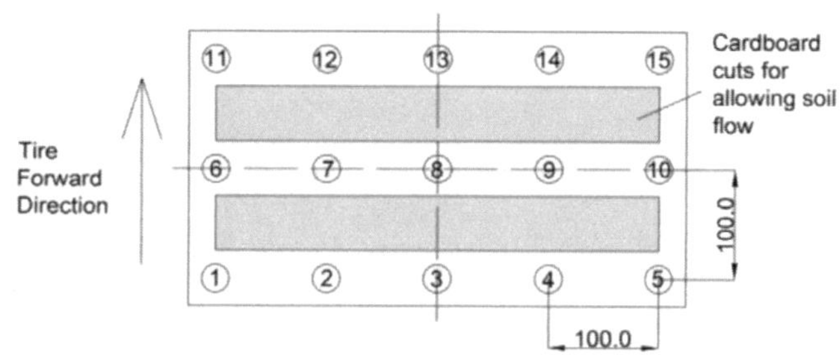

Fig. 2. Layout of FlexiForce sensors for the measurement of normal stress

A grid of 15 Tekscan A401 sensors was formed to evaluate the evolution of normal stress. Each row in the grid consists of five sensors positioned 100 mm apart. The grid, depicted in the above figure, consists of three such rows of sensors, also spaced 100 mm apart. To maintain the grid layout throughout the wheel movement and provide essential support in normal direction, the sensors are firmly attached to a slim cardboard piece. This cardboard has been specifically designed with two cuts located in the central area. The purpose of these cuts is to prevent any disruption to the natural flow and compaction of the soil. While a quantitative method for assessing the effectiveness of the cardboard cuts has not yet been developed, the presented layout intuitively helps minimize the influence of the measurement system on the natural flow of soil.

The design of this measurement device is guided by several objectives. One of the primary goals is to minimize its impact on the properties of soil compaction as much as possible. Therefore, the device can provide accurate measurements without significantly altering the soil properties. This represents an important aspect for obtaining reliable data and maintaining the soil properties during the measurement process.

As presented in Fig.1, a single data acquisition unit has the capacity to measure analog signals for eight distinct channels. However, the sensor grid developed for this study requires 15 separate data acquisition channels. The DI-2108-P is capable of expanding its measuring capabilities to a maximum of 128 channels. It incorporates a feature known as channel stretch, which facilitates synchronization across multiple units. Utilizing this feature, synchronization can be achieved across up to 16 data acquisition units, yielding a total analog measurement capacity for as many as 128 analog channels. In this particular study, synchronization was established between two distinct DATAQ DI-2108-P units to measure data across the required 15 analog channels.

2.2. Calibration procedure

Sensor calibration is a procedure that adjusts a sensor output to accurately match an established reference value. The calibration procedure in this study, presented in Fig. 3, aims at establishing a correlation between the applied normal load and voltage output of the FlexiForce A401 sensor.

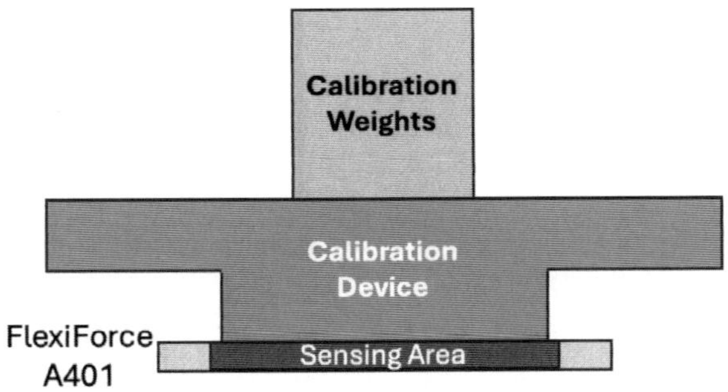

Fig. 3. Calibration procedure for FlexiForce A401 sensor

The normal load was applied using calibration weights with masses in the range of 100-500 grams. The sensing area of the sensor has a diameter of 25.4 mm, while some of the utilized calibration weights has a considerable larger diameter. Therefore, for applying a uniform normal load on the sensing area, a calibration device was created using 3D printing. Figure 4 presents the response of the first five sensors employed in the sensor grid at different applied loads.

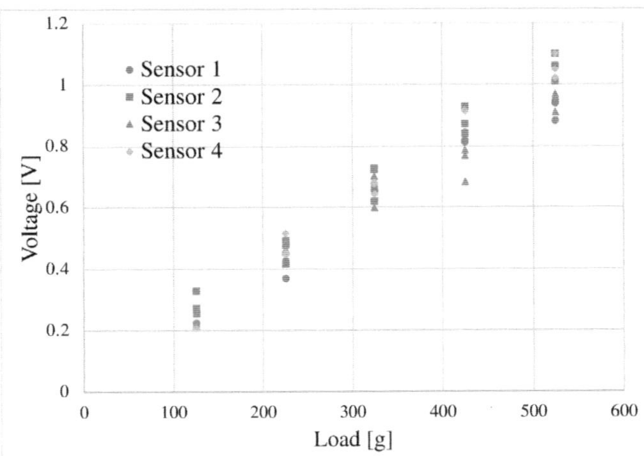

Fig. 4. Sensor response at different applied normal loads

As it can be observed from Fig. 4, for the first four sensors utilized in the measurement array, the response is significantly different. According to Tekscan recommendation, each sensor has to be calibrated individually, therefore the calibration process has been performed for all 15 FlexiForce sensors from the grid. Each sensor was subjected to five distinct vertical loads while monitoring the voltage output. The measurement process for a single vertical load was conducted three times. After the calibration process is completed for one sensor, a relation between the applied mass M and the measured voltage V is established by the calibration factor C_f according to Eq. 1.

$$M = C_f * V \tag{1}$$

Knowing the applied vertical mass M and assuming uniform load distribution over the sensing area A, the normal stress P can be calculated using Eq. 2.

$$P = \frac{M * g}{A} \tag{2}$$

The calibration factor and the coefficient of determination (R^2) are obtained by fitting a linear model through the origin of the coordinate system and the measured points. The values for each sensor are presented in Table 1.

Table 1
Calibration factor and R^2 values

Sensor number	C_f [-]	R^2 [-]
1	534.17	0.995
2	483.99	0.996
3	537.38	0.993
4	483.42	0.998
5	640.86	0.998
6	532.73	0.994
7	599.31	0.995
8	526.86	0.999
9	570	0.988
10	629.89	0.994
11	586.58	0.997
12	532.48	0.991
13	607.77	0.997
14	469.23	0.996
15	582.59	0.999

The obtained R^2 values fall within a high range of 0.988 to 0.999 for all 15 sensors, indicating that each sensor has a good linear output. Furthermore, the calibration coefficient values are significantly different among the sensors, confirming the calibration requirement for each sensor individually to ensure the accuracy and reliability of its readings.

2.3. Experimental setup

The study was conducted on the Terramechanics Test Rig at the TMVS lab of Virginia Tech. Figure 5 illustrates the positioning of the soil stress measurement device inside the test rig.

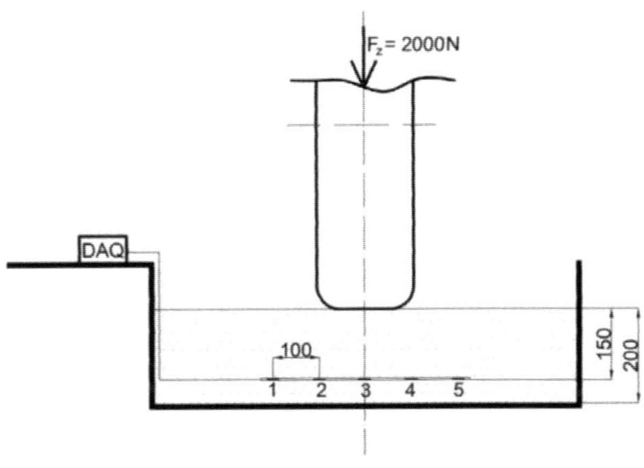

Fig. 5. Experimental setup for measuring normal stress distribution

The methodology employed for this study is similar to the methodology presented in (Sonalkar et al., 2023). The rig is 7.62 m in length, 1.82 m in width, and 1.22 m in depth. This rig has three controlled degrees of freedom: longitudinal and vertical translation, and wheel rotation; furthermore, the toe angle can be adjusted up to ±30 degrees and the camber up to 8 degrees. From the main testing area, a partition of soil with the same length but 0.9 m in width and 0.2 m in depth was created. The terrain used for the testing was the Lunar Soil Simulant GRC-1, chosen for its very dry condition with low cohesion.

The soil bed was prepared by thoroughly shoveling to remove residual stresses from the previous test run and evenly distributing the soil. After shoveling, the soil was excavated to reach a depth of 150 mm. The depth of excavation was regulated using marked pins that were embedded in the soil. Before placing the sensor grid at the depth of 150 mm, the soil had to be leveled manually without the use of a leveler, as the available soil had a height of only 50 mm. For each experiment, the sensor grid was situated in the same section of the test rig using the rig markings to ensure consistent readings. After the sensor array was placed at the depth of 150 mm, the pins were extracted, and soil was meticulously piled on top of the grid to attain a total depth of 200 mm of soil. Following this, the leveler was employed to create a smooth surface.

The selection of the sensor depth value was influenced by two key factors. Firstly, positioning the sensor at a greater depth results in minimal normal stress values, rendering the sensors incapable of detecting the impact of the wheel pass. Secondly, if the sensor is situated at a shallow depth, near the point of contact between the tire and the ground, it could potentially establish an unrealistic boundary condition, which could alter the wheel sinkage and the overall dynamics of the contact between the tire and the soil. Furthermore, to avoid any boundary effect, the total depth of the soil is chosen to be at least five times greater than the expected wheel sinkage. Therefore, the total depth of the soil is indirectly influenced by the tire size, inflation pressure, normal load and slip ratio.

The BKT AT 171, an off-road tire designed for mud boggers with an overall diameter of 785 mm, section width of 211 mm, and a load capacity of 450 kg at an inflation pressure of 140 kPa, was used for the testing. The tire was loaded with a vertical force of 2000 N. The linear speed of the test rig was set to 1 km/h for 0° slip angle case and 0.5 km/h for 6° slip angle case, with different slip ratios achieved by altering the wheel rotational speed. For this study, the data collected at 0% longitudinal slip ratio is presented.

3. Results and Discussion

3.1. Effect of slip angle

One of the studies conducted in this research article is related to the influence of tire slip angle on normal stress distribution. The tire slip angle refers to the angle between the tire longitudinal axis and the path that the tire actually follows. Figure 6 presents the positioning of the measurement device with respect to the off-road tire, for 0° and 6° slip angle settings. As presented in section 2.3, the slip angle settings can be adjusted to simulate different driving conditions, thereby providing a comprehensive understanding of how changes in the slip angle influence the normal stress distribution. The data collected from this study can be used for optimizing tire design and enhancing vehicle performance on sandy terrains.

The sensor grid schematic reveals an arrangement of sensors in a matrix of 5 columns (laterally) and 3 rows (longitudinally). As the tire traverses the grid, it initially encounters sensors 1 through 5, followed by sensors 6 through 10, and finally sensors 11 through 15 in each run. The grid is specifically designed such that the center of the tire contact patch aligns with sensors 3, 8, and 13 during the run. This layout ensures accurate data capture from the central region of the tire (Ravichandran, 2024).

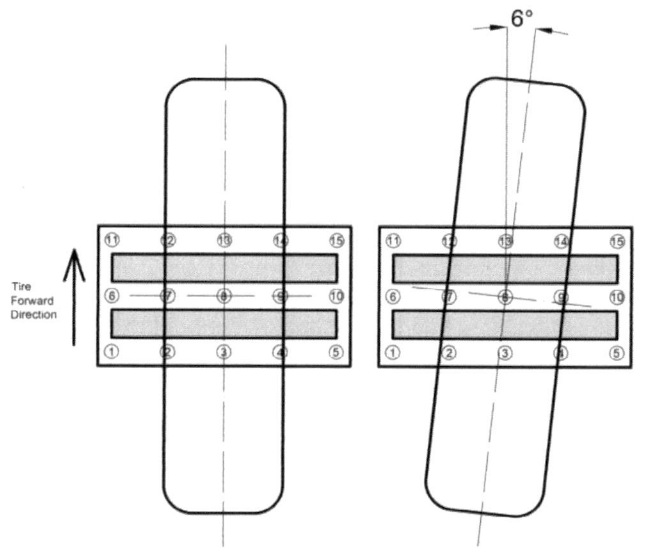

Fig. 6. Positioning of the measurement device relative to the off-road tire for 0° and 6° slip angle settings

As suggested by Fig. 6, it is expected that sensors 3, 8 and 13 to read the highest value for the normal stress, as they are located approximately at the center of the contact patch. The other sensors at the left 2, 7, 12 and at the right 4, 9, 14 are located also in the contact patch region, therefore they should record a significant increase in normal stress after one wheel pass. Sensors 1, 6, 11 and 5, 10, 15 are not in the region of the contact patch, therefore they should not read a significant increase for the normal stress values. Figure 7 presents the evolution of normal stress after one wheel pass at 0° and 6° slip angles.

The data gathered by the sensor grid, as depicted in Fig. 7, validates the hypothesis formulated from the analysis of Fig. 6. This data provides insights into the distribution of normal stress under different slip angles and the role of sensor positioning in capturing these variations. In the scenarios involving both 0° and 6° slip angles, sensor 3 consistently recorded the highest normal stress value. Specifically, it registered a stress of 28-29 kPa at a depth of 150 mm from the top of the soil. This suggests that sensor 3, due to its positioning in the center of the contact patch, is particularly sensitive to changes in normal stress. When the slip angle was 0°, sensors 2 and 4 recorded similar values for normal stress, indicating a uniform distribution of stress across the contact patch under this specific condition. Sensor 1 and sensor 5 did not measure a significant increase in normal stress after one wheel pass because they are not located in the proximity of the contact patch. When the slip angle was increased to 6°, a different pattern emerged.

As suggested by both Fig. 6 and Fig. 7, sensor 2 registered a significantly higher normal stress value compared to sensor 4. This discrepancy suggests that the slip angle plays an important role in the distribution of normal stress across the contact patch. Furthermore, for the 6° slip angle case, a time delay of approximately 0.32 s can be observed between sensor 3 and sensor 2, implying that sensor 3 comes into contact with the tire earlier than sensor 2. Furthermore, sensor 3 leaves the contact with the tire earlier than sensor 2. This behavior can be attributed to the specific geometry of the tire lugs, which interacts differently with the sensors depending on the slip angle. Although it is outside the scope of this paper, it is noteworthy that an increase in sinkage occurred when transitioning from 0° to 6° of slip angle.

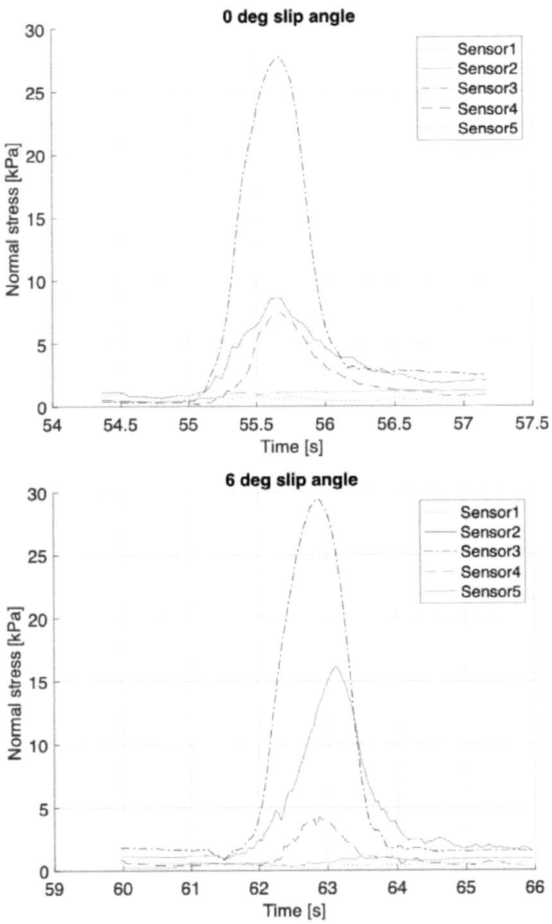

Fig. 7. Evolution of normal stress at 0° and 6° slip angles

The results discussed above pertain to the first row of sensors (1-5) in the measurement grid. It is important to note that similar patterns and observations have been made for the other sensors in the grid, specifically sensors 6-10 and 11-15. This consistency across different sensor groups further strengthens the findings and conclusions drawn from this study.

3.2. Estimation of longitudinal speed and contact length

As presented in Fig. 7, the maximum normal stress values are recorded by sensor 3 due to its location in the center of the contact patch. Similarly, the maximum normal stress values for the row 6-10 is measured by sensor 8, while the maximum normal stress value for the row 11-15 is measured by sensor 13. Figure 8 illustrates the normal stress evolution for sensors 3,8 and 13 for the 6° slip angle case. The longitudinal speed of the tire can be estimated by measuring the time difference between different normal stress peaks. For instance, sensors 3 and 8, which are 100 mm apart, recorded a time difference of 0.64 s between the normal stress peaks, leading to an estimated speed of 0.56 km/h. In a similar manner, sensors 3 and 13, which are 200 mm apart, recorded a time difference between the normal stress peaks of 1.28 s, also yielding a speed estimate of 0.56 km/h. The close proximity of these estimates to the imposed test rig speed of 0.5 km/h further validates the accuracy of the normal stress measurement system. A slight difference in peak amplitude is noticeable at sensor 13. Although not thoroughly investigated, possible factors contributing to this include differences in local soil compaction, sensor calibration, and variations in the normal load.

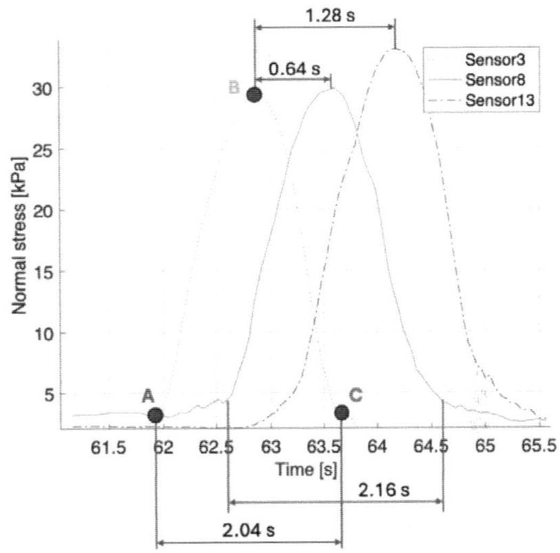

Fig. 8. Evolution of normal stress for sensors 3,8 and 13

For estimating the longitudinal contact length, Fig. 9 illustrates the three important interaction moments (A, B and C) between the tire and the sensor array. The tire dimensions and the approximate sinkage are represented accordingly to (Sonalkar et al., 2023) for the same values of 0% longitudinal slip, 2000N vertical load and 140 kPa inflation pressure.

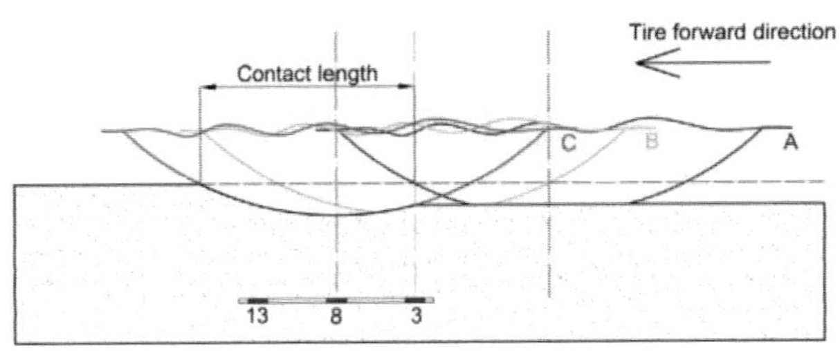

Fig. 9. The main interaction moments between the tire and the sensors

At moment A, the tire comes into contact with the soil, in the proximity of sensor 3. At this moment of time, sensor 3 begins to record an increase in normal stress, while sensors 8 and 13 do not detect any significant changes in normal stress. Moving to moment B, sensor 3 is positioned directly beneath the center of the tire-soil contact patch, therefore measuring the highest normal stress. At this moment of time, the tire is also actuating sensor 8. At moment C, the tire contact patch leaves sensor 3 region, causing the normal stress measured by this sensor to drop significantly. At this moment of time, sensor 8 is placed at the middle of the contact patch, thus measuring its maximum normal stress value, a hypothesis further supported by the data presented in Fig. 8

For sensor 3, the time difference between moment A and C is 2.04 s while the longitudinal speed is 0.5 km/h, therefore the measured contact length is 283 mm. It is worth noting that the contact length represents an estimated value, derived from the evolution of normal stress measured at a depth of 150 mm. To confirm the accuracy of the model used for determining the longitudinal contact length, further studies should be performed with the sensor grid placed closer to the soil surface.

3.3. 3D characterization of normal stress

The data presented in Fig. 7 shows the evolution of the normal stress across a single row of sensors from the measurement grid. Out of the five sensors, only three recorded a significant increase in normal stress after one wheel pass. With the sensor layout presented in Fig. 6, the total width of the sensing area is 200 mm, however only three query points are available for measuring the normal stress in the contact patch region, namely sensor 2, 3 and 4. To enhance the resolution of the normal stress data collected across the width of the contact patch, an interpolation method was employed. This involved creating an interpolated data set from the measurements taken by sensors 2, 3, and 4. The interpolation was executed using MATLAB 1D spline interpolation function, which is known for its ability to generate a smooth curve that passes through given points. Figure 10 presents a 3D plot that showcases the normal stress data for the 6° slip angle case. It includes both the original (raw) data and the interpolated data.

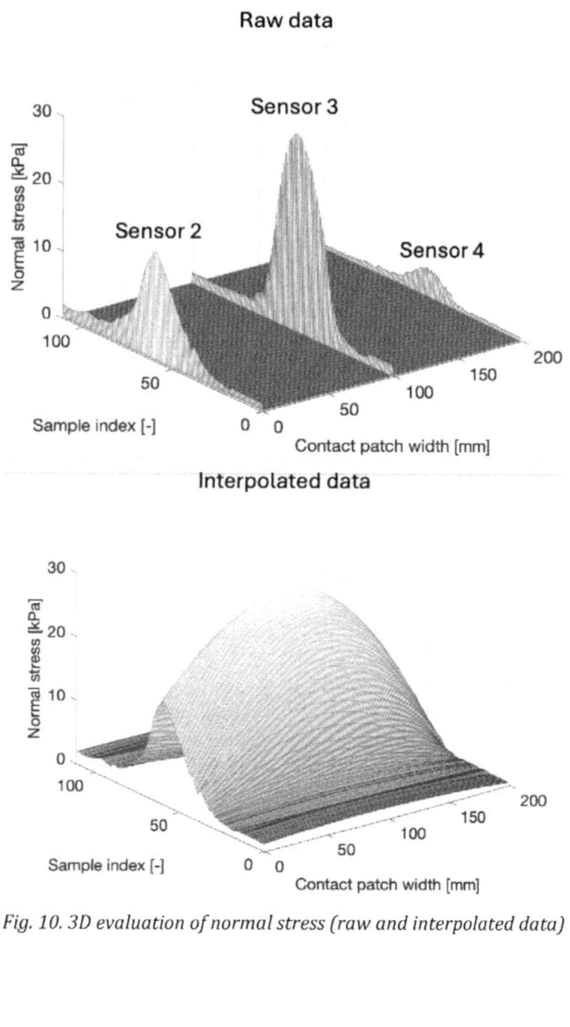

Fig. 10. 3D evaluation of normal stress (raw and interpolated data)

In this three-dimensional representation, the X-axis is used to depict the data points of normal stress that have been measured/estimated across the width of the contact patch.

The Y-axis, on the other hand, indirectly signifies the data points of the normal stress that have been measured along the length of the contact patch. The sampling was done at a rate of 0.04 seconds. For example, the time interval between the samples 31 and 83, which correspond to the entry and exit points of sensor 2 from the contact patch, is approximately 2.08 seconds. From this, we can estimate that the length of the contact patch is roughly 288 mm.

The Z-axis in this representation is used to denote the actual values of normal stress for a specific point in the X-Y contact patch plane.

The data has been interpolated to provide a more detailed view of the stress distribution across the contact patch. This gives us an estimate of the stress at any given point within the contact patch. Furthermore, the interpolation allows for a more nuanced understanding of the stress distribution under the specified conditions, offering insights that might not be immediately apparent from the raw data alone. The study of the 3D normal stress distribution has a significant contribution in evaluating the traction performance of off-road tires on deformable terrains. It should be highlighted that the stress values obtained through interpolation are only estimates, and there may be considerable deviation from the actual stress values. To validate the 3D normal stress model, it is recommended to conduct additional research with an increased number of sensors place on the width of the contact patch.

4. Conclusions and future work

In conclusion, this research presents the design, calibration, and measurement performance of a novel device capable of measuring the normal stress distribution inside the soil, after a wheel pass. This article has provided valuable insights into the distribution of normal stress within soil under various rolling conditions, particularly focusing on the interaction between an off-road tire and GRC-1 lunar soil simulant. By employing a grid of piezoresistive Tekscan FlexiForce sensors embedded into the soil, the normal stress distribution was measured without significantly altering the soil compaction properties. The findings from this study revealed the influence of tire slip angle on normal stress distribution, showing distinct patterns and behaviors under different slip angle settings. Notably, sensors positioned within the contact patch region consistently recorded high normal stress values after one wheel pass, indicating their sensitivity to changes in tire-soil interaction. The measurement system can also be used for estimating with good accuracy both the longitudinal speed of the tire and the longitudinal contact length between the tire and the terrain. Moreover, the interpolation method applied to enhance the resolution of the data has provided a more detailed understanding of stress distribution across the contact patch. This interpolated data not only improves our ability to analyze traction performance but also offers insights into optimizing tire design and vehicle performance.

For future work, it is imperative to expand testing at different depths using the measurement system to gain a deeper understanding of how normal stress distribution varies within the soil profile. It is expected that the normal stress values will increase with decreasing sensor array depth (the sensors are closer to the contact patch). Future studies should establish a correlation between the values of the normal stress and the positioning depth of the sensor array. Even though the interpolation method can improve the resolution of the measured normal stress values, there is still a need for refining the sensor grid in the lateral direction to capture more data points across the width of the contact patch. This would further enhance the resolution and accuracy of the measurements. In summary, the comprehensive analysis of normal stress distribution presented in this research contributes to advancing the understanding of tire-soil mechanics and has implications for improving off-road vehicle performance and design.

5. Nomenclature

A	Contact area	[m^2]
M	Mass	[g]
P	Normal stress	[kPa]
g	Gravitational acceleration	[m/s^2]
C_f	Calibration factor	[-]
V	Voltage	[V]
R^2	Coefficient of determination	[-]
F_z	Normal load	[N]

6. Acknowledgements

The authors would like to thank the TMVS Lab at Virginia Tech for providing the testing facilities used during this work and partially funding this work. We would also like to mention and appreciate the support of Goodyear Tire and Rubber Company for providing the soil and partially funding this work.

7. Declaration of competing interest

The authors declare that they have no known competing financial interests or personal relationships that could have appeared to influence the work reported in this paper.

8. References

Arvidsson, J., Ristic, S., 1996. Soil stress and compaction effects for four tractor tyres. Journal of Terramechanics, 33(5), 223-232.

Sonalkar, C.S., Ravichandran, N., Vilsan, A., Sandu, C., Lin, T., 2023. Assessing Performance of Light Wheeled Vehicles on GRC-1 using 3D Scanned Footprint Measurements. 16th European-African Regional Conference, Lublin, Poland.

Derafshpour, S., Mardani, A., 2021. Development of a Novel System for Measuring Soil Stress and Rut Depth Under Off-Road Vehicle Tires. *IEEE Instrumentation & Measurement Magazine*, *24*(7), 30-38.

Gheshlaghi, F., Mardani, A., 2021. Prediction of soil vertical stress under off-road tire using smoothed-particle hydrodynamics. *Journal of Terramechanics*, *95*, 7-14.

Jiang, C., Lu, Z., Upadhyaya, S. K., Memon, M. S., 2019. Measurement and Analysis of the Vertical Stress Distribution within a Cultivated Soil Volume under Static Conditions. *Transactions of the ASABE*, *62*(4), 1035-1043.

Nichols, T. A., Bailey, A. C., Johnson, C. E., Grisso, R. D., 1987. A stress state transducer for soil. Transactions of the ASAE, 30(5), 1237-1241.

Raper, R. L., Bailey, A. C., Burt, E. C., Way, T. R., Liberati, P., 1995. Inflation pressure and dynamic load effects on soil deformation and soil-tire interface stresses. *Transactions of the ASAE*, *38*(3), 685-689.

Ravichandran, N., 2024. *Tire Performance Estimation Under Combined Slip and Empirical Parametrization of the Tire Rut on Dry Sand* (Master's thesis, Virginia Tech).

Van den Akker, J. J. H. (2004). SOCOMO: a soil compaction model to calculate soil stresses and the subsoil carrying capacity. Soil and Tillage Research, 79(1), 113-127.

Wong, J. Y. 2022. Theory of ground vehicles, fifth ed., John Wiley & Sons.

3649 | Modeling the Resistive Forces on Vehicles in Deep Snow

Micah M. Borrero [a,b,*], Orian Z. Welling [b,†]

[a] *Department of Aeronautics and Astronautics, Massachusetts Institute of Technology, Cambridge, MA, United States*
[b] *US Army ERDC-Cold Regions Research and Engineering Laboratory (CRREL), Hanover, NH, United States*

* *Corresponding author:* micah.m.borrero@erdc.dren.mil
† *Corresponding author:* orian.z.welling@erdc.dren.mil

ABSTRACT

The ability to determine whether a vehicle can traverse a particular winter surface is crucial for planning and executing successful off-road operations in cold regions environments. Various works have detailed strategies and improvements to vehicle mobility models snow and ice surfaces, and others have proposed initial methods to estimate the forces required for vehicles to travel through deep snow exceeding the height of the vehicle bumper or undercarriage These methods demonstrate a strong correlation between their approach and measurements examined in the field. However, they lack sufficient field-based measurements in varied snow conditions to fully capture real-world force estimates.

In this paper, we build upon prior work to develop a preliminary model with the ability to estimate the resistive forces acting on a vehicle traveling through deep snow. We draw upon a wide range of disciplines, discussing the strength of a model based on the present knowledge of snow compression against a methodology based on fluid dynamics techniques. Both models exhibit promising results. However, these results demonstrate a stronger positive correlation using the snow compression model when compared with field-based measurements taken over a variety of winter seasons. These measurements were obtained using a purpose-built plowing apparatus, which simulated a range of bumper ride heights and profiles.

Keywords
Mobility Modeling
Deep Snow
Off-Road
Snow Compression
Drag
Plowing

1. Introduction

Having the ability to easily determine whether a vehicle is able to traverse a winter surface is pertinent when understanding the viability of a successful winter off-road operations. Existing models such as the NATO Reference Mobility Model (NRMM) have limited ability to accurately predict mobility for many winter surfaces and have no native capability to estimate forces on a vehicle driving through snow deeper than a vehicles undercarriage or bumper. To ensure the capability to plan and execute successful missions in regions with deep snow, it is important that such models are developed, and data collected to validate and fit the models.

Previous works have proposed potential models to predict the forces acting on vehicle traveling through deep snow (Welling et al., 2022), but limited data has been taken to fully understand the accuracy of such. This paper aims to fill that gap by examining the results of testing with an adjustable load-measuring plow-plate mounted on a mid-weight tracked military vehicle. These results were then used to explore the validity of two generalized methodologies – compression force and granular flow - to model the resistive forces on a vehicle traveling through deep snow and to develop improved fitting parameters to better correlate between the model and measured data.

2. Approach

In order to understand the resistive forces on a vehicle traveling through deep snow, we developed a dedicated plow plate with adjustable height and approach angle, fitted with load cells to measure the normal force on the plate, and mounted on a mid-weight military vehicle. The plow setup is shown in *Fig. 1*. The apparatus enabled us to change the orientation of the plate via translation along the vertical axis in addition to rotation about the lateral axis. Lastly, we had the ability to replace the plate allowing us to vary its size. Throughout testing, there were three primary variables that were modified: plow depth, angle, and width.

Fig. 1. Plow Apparatus

Testing was conducted at the Keweenaw Research Center (KRC) in Calumet, Michigan. For each test, a patch of fresh virgin snow was used to ensure consistency in initial ambient conditions across each test. The test vehicle was driven at a constant low

https://doi.org/10.56884/FIELAS8S

speed (below 5mph) as to mitigate the potential for skidding and maximize the accuracy of our results. Once at a constant speed data was taken for a given period of time.

3. Models

To analyze the physical interactions between the plow and snow, we developed two physically based models derived of the simplified plow force models proposed by Welling et al. (2022) along with the various theories and methods drag forces in granular media developed by Albert et al. (1999) - subsequently furthered by Ding et al.'s (2011) identification of drag induced lift in granular media. The two models are described in *Sec 3* and the results and comparison to measured data are discussed in *Sec 4*.

3.1. Snow Compression Model

The first model built upon Welling et al. (2022) stating that the forces acting on a vehicle traveling through deep snow are composed of a variety of compression and frictional forces. Welling's model proposes that at low speeds, the forces are dominated by compression and friction, and inertial forces can largely be ignored.

Welling et al. demonstrated a linear correlation between the belly drag and frontal compression forces. Although there is a linear correlation, it consistently overestimated the resulting force values (Welling et al., 2022). To compensate for this, we propose a method that takes these initial models and applies appropriate fitting parameters. These parameters are based upon experimental results taken. over the course of multiple years, including the new data collected in this effort.

Welling et al.'s snow compression model is based on the following equations which describe the plowing compression force F_{pc} and the belly compression force F_{bc}:

$$\sigma = \frac{1.11 * 10^7}{500 - \rho_0} \tag{1}$$

$$F_{pc} = A_p \sigma \tag{2}$$

$$F_{bc} = w_v \int_{z_f - d*}^{z_f} \sigma(z) dz \tag{3}$$

Where F_{pc} is a function of the plow contact area A_p and Olson et al.'s (1997) snow strength to density relationship σ. Additionally, F_{bc} is a function of the vehicle's weight w_v and the integral of the stress of the compression of the snow over the plow's depth.

Building on the findings presented by Welling et al. (2022), we postulate that *Eq. 3* presents marginal contributions to the underlying calculation of forces acting against a vehicle traveling through deep snow during the normal case. Thus, our modified model removes *Eq. 3* for this case. Notably Welling also included motion resistance forces related to the vehicles tracks or tires compressing the traversed snow - by measuring with an instrumented plow plate in this effort, we can independently measure the plowing force from the front of the vehicle and do not consider the forces on the tires - which have been studied and validated previously by Shoop et al. (2020).

Recognizing edge-effects and other limitations of this model, we also modify *Eq. 2* such that it includes an empirically fit variability factor, C_s. This variability factor is intended to encompass both underestimations and overestimations identified through direct application of this snow compression model. Furthermore, Welling et al. suggested the inclusion of a regression calibration constant to improve the model's fit to real force estimates. Building on this idea, in *Sec 4* we incorporate this constant, C_s, based on the additional data collected in this effort. An updated motion resistance formula for plowing compression that includes this constant is shown in *Eq. 4*:

$$F_{pc} = C_s A_p \frac{1.11 * 10^7}{500 - \rho_0} \tag{4}$$

3.2. Granular Force Model

As noted earlier in this paper, directly applying the compression model developed by Welling et al. demonstrated over-estimations when predicting the forces on the instrumented plow plate. With this in mind, we developed a second approach to attempt to understand the vehicle-snow interaction based on an aerodynamics-derived top-down approach.

In the field of aerodynamics, a great amount of work has been done to understand the resistive forces, commonly known as drag forces, on an object traversing through a given medium. This fundamental understanding is reflected through the simplified drag equation:

$$F_d = \frac{1}{2} \rho_\infty V_\infty^2 C_D S \tag{5}$$

The drag equation that is primarily used in the field of aerodynamics and fluid dynamics, relies on the assumption that the given medium is a Newtonian Fluid (Kyburz et al., 2022). This enables the use of factors, such as the drag coefficient C_D. In low-speed flow regimes, the drag coefficient is a function of the Reynolds number Re (Anderson & Cadou, 2024). The Reynolds number is a dimensionless and experimentally derived quantity that represents the ratio between the inertial and viscous forces within a given flow field. It is generally represented by the following equation:

$$Re = \frac{\rho_\infty * V_\infty * c}{\mu_\infty} \tag{6}$$

Although it is recognized these fundamental relations cannot be directly applied when it comes to the modeling of snow, there is precedent in avalanche research (Kyburz et al., 2022) regarding the consideration of snow as a Newtonian fluid. In order to treat snow as a Newtonian fluid, it must be slow moving relative to an object in which it impacts (Salm, 1977) - of which in the case of this study is applicable due to the speed of the test vehicle being below 5mph. Inertial-based terms are often the driving factors in previous methods modelling interactions between vehicles and snow to estimate the snow's resistive force (Olson et al., 1997). Furthermore, these methods rely on the assumption that snow is being deflected away from a vehicle due to the presence of a plow (Ravani et al., 2005).

A comprehensive method that models the resistive forces on a vehicle moving through deep snow must take into account near-zero speeds in addition to the subduction of snow beneath the vehicle. We believe that the principles ideas used in the derivation of relations within the field of aerodynamics can be applied to meet these requirements. Particularly, in this paper we derive an experimentally derived dimensionless factor C_{gran}, akin to the Reynolds number, which is proportional to the physical dimensions and orientation of the flat plate described earlier in this paper. Additionally, we produce a variant of the common drag force equation F_d based upon the coefficient C_{gran} along with the environmental factors and properties specific to snow.

Building off these initial principles, we proceed with the understanding that snow can be considered as an assembly of ice grains that are uniform in size and distribution (Nicot, 2004). We use this assumption to model our vehicle as acting within a granular medium. Through this, we can apply the concepts derived by Albert et al. (1999) which found that the depth by which an object intrudes upon a granular medium is quadratically proportional to the drag force on said object – given by *Eq. 7* where d_c represents the horizontal dimension and H the vertical dimension:

$$F_d \propto d_c H^2 \tag{7}$$

This initial relation enabled the derivation of an equivalent to a granular Stokes Law:

$$F_d = \eta g \rho H^2 d_c \tag{8}$$

We then modify this version of the granular Stokes Law - akin to the relations developed by Roth et al. (2021) - to reflect a simplified drag equation:

$$F_d = C_{gran} w z^2 \tag{9}$$

3.3. Plow Angle Adjustment

The initial 4model proposed by Welling et al. (2022) incorporated a subduction force, shown in *Eq. 3*. In the non-normal case, we find this force to be similar to a lifting force seen in models that focus on granular flow (Ding et al., 2011; Percier et al., 2011). *Equation 10* depicts how this may be incorporated to the underlying resistive force calculation.

$$F_c = \sqrt{F_{pc}^2 + F_{bc}^2} \tag{10}$$

Building off previous work, we postulate that a generalized estimation of forces acting perpendicular to the direction of travel can be captured via an angle-based percentage adjustment – given by F_{ang}.

$$F_{ang} = F C_{ang} \frac{2\theta}{\pi} \tag{11}$$

This approach is taken due to its relative simplicity and ease of general implementation. We believe that this initial simplification is critical due to the complexities of snow. As described in *Sec 3.2*, snow takes upon attributes a Newtonian fluid which allows us to rely on well-defined force models, yet at the same time we recognize that there will be interactions that these predefined models cannot capture.

4. Results

Across either model, the analysis leading to the definition of coefficients described in *Sec 3* was broken down according to cases determined by whether the angle of attack (AoA) of the plow face was normal relative to the direction of travel. This enabled ease in validation regarding the identification of C_S, C_{gran}, and C_{ang} - limiting the potential for dependent influences.

First focusing on the snow compression model described in *Sec 3.1*, in the case that the plow face is normal to the direction of travel we find a strong linear correlation between the estimated force values and the measurements obtained during testing. This is depicted in *Fig 2*.

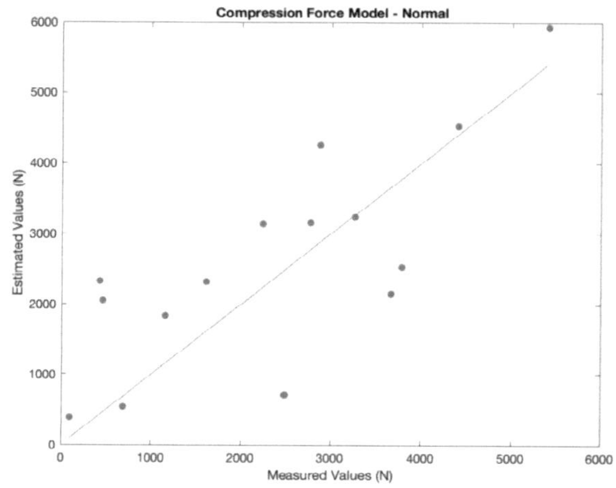

Fig. 2. Estimations based on Compression Model vs. Measured Values

Furthermore, similar results were obtained via the application of the granular force model described in *Sec 3.2*. This is depicted in *Fig 3*.

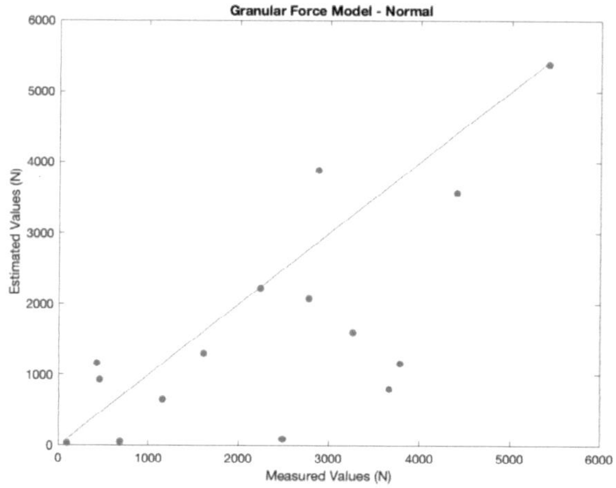

Fig. 3. Estimations based on Granular Force Model vs. Measured Values

Moving to the non-normal case, both the proposed angle-based percentage adjustment, from *Eq. 11*, and the incorporation of a lifting force, represented by *Eq. 10*, demonstrated a linear correlation – similar to what was observed in the normal case. In either set of estimations, the longitudinal force calculation was based upon the snow compression model derived in *Sec 3.1*. Although linear correlations were observed, the accuracy of these models do not demonstrate similar levels of strength as compared to the normal case. The resulting adjusted estimations are shown in *Fig. 4* and *Fig A.2* respectively.

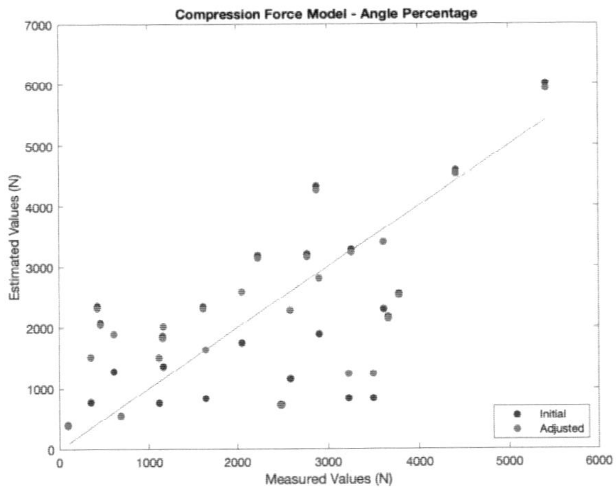

Fig. 4. Estimations based on Compression Model w/ Angle-Based Percentage Adjustment vs. Measured Values

To further explore the impacts of the plow's angle of attack on various force estimates, we conducted a brief analysis to understand the relation between AoA the empirically measured force values. *Figures 5* and A.3 depict the results of this analysis.

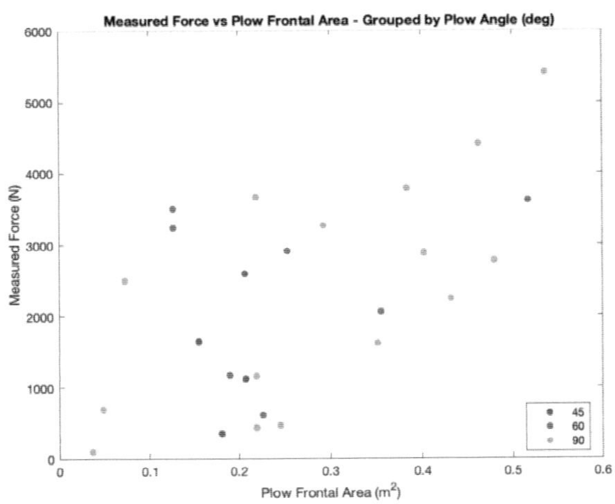

Fig. 5. Measured Force vs Plow Frontal Area grouped according to Plow Angle of Attack

5. Discussion

The general fit parameters were calculated utilizing common statistical regression methods. We assert that the initial equations derived in *Sec 3.1* and *Sec 3.2* are comprehensive, simply overshooting, or undershooting values by a given common linear factor.

To validate this assertion, we conducted a series of tests leveraging the plowing apparatus described in *Sec 2* at KRC across multiple winters. Inherently, when comparing our estimations against empirical results taken in the field, our desired result is to maximize the resultant R-squared value.

Based upon our findings shown in *Sec. 4*, we find a much stronger overall correlation in the data when leveraging the snow compression model as demonstrated by an initial R-squared value of *0.5145* – as opposed to an R-squared value of *0.4657* that is produced by the granular flow-based model. When comparing the RMSE values, we find that the compression model produces an initial value of 1.0159×10^4. Compared to the granular flow-based model with a value of 2.6916×10^3. This demonstrates a lower degree of variability produced by the granular flow-based model.

To account for variations in scaling between the estimated and empirical measurements, we introduce the regression constant, C_s, discussed in *Sec 3.1* for the compression model with a value of *0.2291*. For the granular force model, we introduce the coefficient C_{gran} with a value of *17.2177* - as discussed in *Sec 3.2*. While the underlying R-squared values remain the same, we find a reduction in both RMSE values produced by the compression and granular models – being 1.1331×10^3 and 1.4385×10^3 respectively.

To further strengthen either model, we leverage data smoothing and tightening techniques often seen in the field of control theory. Specifically, we apply a Kalman Filter to either dataset produced by the two models (Devore et al., 2012). Across multiple iterations, we did not find any viable improvements to the resulting data from the models when leveraging the smoothing technique. We believe that this may be attributed to the limited number of samples considering various independent factors. The results of this smoothing approach are shown in *Fig A.2*.

Considering the non-normal cases, we consider the dominant forces acting on the vehicle during motion to take place opposite to the direction of travel. In line with the methods used for the normal case, the base compression model described in *Sec 3.1* is used in the analysis of this case. This initial model is adjusted via an angle-based adjustment. This adjustment is represented by *Eq. 11*. Prior to the adjustment we obtain RMSE and R-squared values of *0.4044* and 1.1211×10^3 respectively. With the angle adjustment applied, we obtain RMSE and R-squared values of *0.3790* and 1.2194×10^3. By inspection, we observe a stronger model fit for the adjusted compression model when compared to the un-adjusted model – shown in *Fig. 4*. In contrast, the inclusion of this angle adjustment results in a decrease in model fit, statistically. Thus, although it is possible to apply either *Eq. 10* or *Eq. 11*, such a calculation does not improve the predictive accuracy of either model.

We included this angle-based percentage and lifting force adjustment (*Fig. 4* and *Fig. A.2* respectively) as is expected from the field of aerodynamics and studies considering Newtonian fluids. We expect this would be necessary based upon evidence from previous work. However, as noted previously we witnessed a reduction in model fit following the application of the adjustments based on *Eq. 10* and *Eq. 11*. We sought to understand this by looking at the relation between the plow AoA and measured force values. As represented in *Fig 5* and *Fig A.3*, we found that the resulting force values have virtually no dependence on the plow's AoA.

One potential source of the observed discrepancies likely stems from the significant number of independent variables impacting this analysis. It is important to acknowledge that, in empirical testing under ambient conditions, achieving consistent initial conditions is challenging. To address this, our models incorporated these initial conditions, such as plow depth and snow density, as parameters in *Eq. 4* and *Eq. 9*. However, we recognize that this approach may introduce inconsistencies into our results. Considering these inconsistencies, it is plausible that the models are optimal only under specific conditions. Specifically, the proposed compression model is designed for scenarios in which the plow is perpendicular to the direction of travel. Alternatively, the generalized parameters identified in this study (as presented in *Tab. A.1*) may be biased by experimental conditions in which our data was taken.

Ultimately, we cannot be certain as to whether the remaining noise in the data, as seen in *Fig.* 2, are attributed to experimental variability, model inconsistencies, or to novel behavior not augmented before in deep snow. Further study is required to fully understand the force interactions that may be occurring to develop an appropriate methodology to model such.

6. Conclusion and Future Work

In this paper we introduced a mechanical apparatus that simulates the varying frontal profiles of vehicles used in winter environments. This apparatus enabled for empirical data to be taken which was used to model the resistive forces acting on a vehicle traveling through deep snow. This testing setup was paired with a mathematical model that built upon previous work both within, and outside, this field.

We compared the results of a model that expanded on the methods developed by Welling et al. (2022) against one based on granular flow. Through our analysis we found the former model demonstrated a stronger fit. In studying the fitness of either model, we found that the application of appropriate fit parameters may develop sufficient results regarding the estimation of resistive forces on a vehicle traveling through deep snow.

Considering our present results for cases such that the plow angle of attack was non-normal, we found that the incorporation of an angle-based subduction force provided minimal – if not negative – improvements to the underlying models. Furthermore, we found that the empirical measurements demonstrate virtually no correlation with the plow's angle of attack. Thus, considering the ultimate simplicity of these models, we recommend using only the vertical depth of the blade and foregoing the impact of angle of attack.

We recognize that the coefficients derived may be specific to the empirical data in which we have collected and that these models are far from perfect. To enhance the validity of our assumptions, we recommend that further analysis be conducted to ensure the generality of applicability. Particularly, we believe that further calibration is needed considering the high degree of dispersion when comparing model derived estimations and empirical results for cases such that the plow's angle of attack is non-normal with respect to the horizontal.

7. Nomenclature

A_p	Plow contact area	[m^2]
c	Characteristic linear dimension	[m]
C_{ang}	Angular adjustment factor	
C_D	Drag coefficient	
C_{gran}	Granular force factor	
C_s	Compression force factor	
d_c	Transverse size	[m]
F_{ang}	AoA adjusted force	[N]
F_{bc}	Subduction force	[N]
F_c	Total force	[N]
F_d	Drag force	[N]
F_{pc}	Plow compression force	[N]
g	Gravitational acceleration	[m/s^2]
H	Plow depth	[m]
w	Plow width	[m]
w_v	Vehicle mass	[kg]
z	Plow depth	[m]
η	Granular constant	
θ	Plow angle of attack	[rad]
ρ	Snow density	[kg/m^3]

8. Acknowledgements

This work was supported by U.S. Army Engineer Research and Development Center. The authors thank Sally Shoop, Susan Frankenstein, Mike Parker, and Mark Bodie for their support throughout field testing of the plow apparatus.

9. Declaration of Competing Interest

The authors declare that they have no known competing financial interests or personal relationships that could have appeared to influence the work reported in this paper.

10. References

Albert, R., Pfeifer, M. A., Barabási, A.-L., & Schiffer, P. (1999). Slow Drag in a Granular Medium. *Physical Review Letters, 82*(1), 205–208. https://doi.org/10.1103/PhysRevLett.82.205

Anderson, J. D. (John D., & Cadou, C. (2024). *Fundamentals of aerodynamics* (Seventh edition.). McGraw Hill.

Devore, J. L., Berk, K. N., & Carlton, M. A. (2012). *Modern mathematical statistics with applications* (Vol. 285). Springer.

Ding, Y., Gravish, N., & Goldman, D. I. (2011). Drag Induced Lift in Granular Media. *Physical Review Letters, 106*(2), 028001. https://doi.org/10.1103/PhysRevLett.106.028001

Kyburz, M. L., Sovilla, B., Gaume, J., & Ancey, C. (2022). Physics-based estimates of drag coefficients for the impact pressure calculation of dense snow avalanches. *Engineering Structures, 254*, 113478. https://doi.org/10.1016/j.engstruct.2021.113478

Nicot, F. (2004). Constitutive modelling of snow as a cohesive-granular material. *Granular Matter, 6*(1), 47–60. https://doi.org/10.1007/s10035-004-0159-9

Olson, W. W., Kempainen, A., Milacic, D., & Osborne, M. (1997). *Dynamic Modeling of Forces on Snowplow Equipped Trucks.* 973193. https://doi.org/10.4271/973193

Percier, B., Manneville, S., McElwaine, J. N., Morris, S. W., & Taberlet, N. (2011). Lift and drag forces on an inclined plow moving over a granular surface. *Physical Review E, 84*(5), 051302. https://doi.org/10.1103/PhysRevE.84.051302

Ravani, B., Gabibulayev, M., & Lasky, T. A. (2005). A Snowplowing Resistance Model for Use in Vehicle Dynamic Modeling. *Mechanics Based Design of Structures and Machines, 33*(3–4), 359–372. https://doi.org/10.1080/15367730500403446

Roth, L. K., Han, E., & Jaeger, H. M. (2021). Intrusion into Granular Media Beyond the Quasistatic Regime. *Physical Review Letters, 126*(21), 218001. https://doi.org/10.1103/PhysRevLett.126.218001

Salm, B. (1977). Snow Forces. *Journal of Glaciology, 19*(81), 67–100. https://doi.org/10.3189/S0022143000029221

Shoop, S. A., Bodie, M. O., Frankenstein, S., & Bigl, M. F. (2020). *Recent improvements to snow mobility algorithms.*

Welling, O., Shoop, S., Letcher, T., Elder, B., & Bodie, M. (2022). Estimation of plowing forces on vehicles driving through deep snow. *Journal of Terramechanics, 104*, 25–29. https://doi.org/10.1016/j.jterra.2022.08.003

Appendix

Table A.1
Model Experimental Values

Variable	Value
C_s	0.2291
C_{gran}	17.2177
C_{ang}	1.0143

Table A.2
Produced Statistical Values

State	R-Squared	RMSE
Initial Compression	0.5145	1.0159 x 10^4
Initial Granular	0.4657	2.6916 x 10^3
Scaled Compression	0.5145	1.1331 x 10^3
Scaled Granular	0.4657	1.4358 x 10^3
Angle Percentage Initial	0.4044	1.1211 x 10^3
Angle Percentage Adjusted	0.3790	1. 2194 x 10^3
Lifting Force	0.1182	1.0723 x 10^4

$$F_{bc} = w_v * 1.11 * 10^7 \left[\frac{z}{500} - \frac{\rho_0 z_0}{500^2}\ln\left(1 - \frac{500z}{z_0 - 500 - p_0}\right)\right]_{z_f - d*}^{z_f} \quad \text{(A.1)}$$

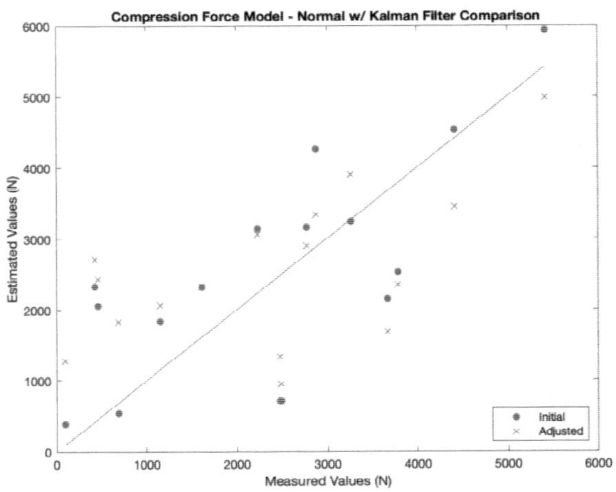

Fig. A.1. Estimations based on Compression Model w/ Kalman Filter Adjustment vs. Measured Values

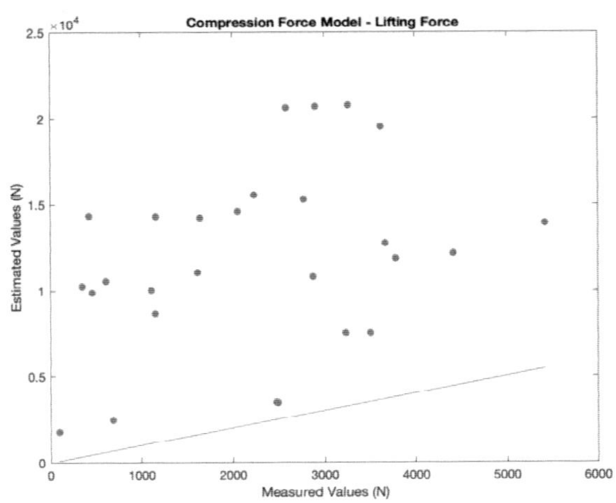

Fig. A.2. Estimations based on Compression Model w/ Lifting Force Adjustment vs. Measured Values

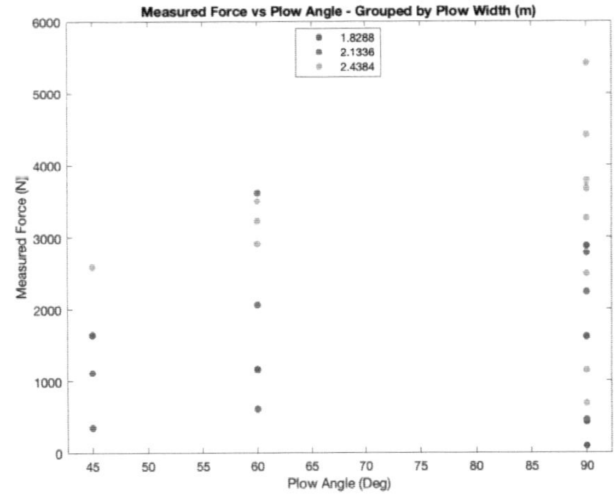

Fig. A.3. Measured Force vs Plow Angle grouped according to Plow Width

3769 | Mobility During the Transition Seasons in the Arctic

Michael Parker[a,*], Clifford Witte[a], Susan Frankenstein[a], Allan Wheeler[a]

[a] *U.S. Army Engineer Research and Development Center, Cold Regions Research and Engineering Laboratory, 72 Lyme Rd, Hanover, New Hampshire, 03755, United States*

* *Corresponding author: Michael.w.parker@erdc.dren.mil*

ABSTRACT

Mobility in the Arctic involves more than operating on just snow and ice surfaces. Weather patterns during transition seasons such as Spring, Fall, and even Summer lead to terrain of varying softness and moisture content. Ground thaw cycling is difficult to predict. In the most extreme cases the ground temperature will cycle between thawed and frozen overnight. Each of these varying ground states present unique impracticalities to vehicle mobility researchers. The surface could be blanketed by snow, insulating the surface and slowing or preventing freezing, or the experimenting team arrives to the test site a day late finding the surface already completely thawed. This research looks at reproducing freeze-thaw transitional ground surfaces in a controlled laboratory setup for testing US Army vehicles year-round. The results of this research include traction testing data of two US Army vehicles according to the same SAE standards as when conducted in Arctic field locations. U.S. Army Corps of Engineers researchers have constructed three frost susceptible soil surfaces for full scale vehicle testing at the Cold Regions Research Laboratory (CRREL) in Hanover, NH. Inside its Frost Effects Research Facility, CRREL researchers may freeze and thaw the soil as needed without relying on the difficult timing of mother nature. The facility allows the user to directly control the freeze and thaw cycles, allowing multiple vehicles to be tested on different surfaces over the course of a single winter. This speeds up the research on frost susceptible soils during these difficult transition periods and provides much needed mobility data for predictive modeling and planning purposes. Multiple soil conditions and freeze thaw states have already been studied and this paper will discuss the results for light tracked and wheeled vehicles.

Keywords
Off-road Mobility
Vehicles
Frozen
Thawed
Soil
Arctic

1. Introduction

When operating vehicles in contested environments it is not always possible or of strategic interest to drive on road networks. Additionally, road networks in Arctic environments are generally limited and constrained, providing easy choke points and targeting opportunities. However, off road mobility in Arctic environments can be extremely challenging during any season. In the summer, highly organic areas generally found in the Arctic are flooded and extremely soft making it impossible to move anywhere except on roadways, unless using highly specialized vehicles. In the winter, these areas may freeze up if the temperatures are cold enough prior to a blanket of snow insulating the soil surface. The transition seasons leading from fall into winter and winter into spring provide their own set of challenges. During an Autumn freeze it is imperative to understand how much frozen soil is required to support vehicles to maneuver in this environment. During spring thaw, the surface often gets extremely slippery with an unfrozen layer perched upon the frozen solid beneath. As this thawed layer gets deeper it creates more resistance as the vehicle sinks in deeper creating larger ruts and further decreasing mobility.

Studying mobility on the highly organic soils typically found in the Arctic is often gapped research because the ground is highly frost susceptible, and their strength varies significantly between summer and winter seasons. This makes them ill-suited for construction of roads and therefore to be avoided by infrastructure design engineers. Roads develop large sink holes or very uneven surfaces and building foundations settle, causing the structures to be condemned as they are unsafe to live in.

The ongoing war in Ukraine has renewed interest in these soils and specifically understanding mobility during the transitional freeze/thaw conditions. These conditions are extremely difficult to time in nature because the surfaces can be close to isothermal. Any solar loading or temperature swing can cause the surfaces to fully thaw or freeze back up. These changes can even change throughout the day as the sun comes up or goes down. Having the test surfaces out of the sun and in a temperature-controlled facility makes it significantly easier to plan and execute multiple tests on different vehicles in a much shorter timeline. Work conducted within this report specifically focuses on light tracked all-terrain vehicle and light tactical wheeled vehicle mobility on partially frozen, frozen, and partially thawed frost susceptible Arctic soils.

2. Background

Experimental data collection of vehicle traction on off-road surfaces during transitional seasons is challenging because the required weather conditions cannot be accurately predicted to a specific time or location within a given season.

For the best chance at capturing a transitional surface event a team of subject material experts, test personnel and instrumented test vehicles must be present on the ground at the site of interest. Therefore, key personnel must be either be staged indefinitely for weeks at the test location or on call and ready to travel. Logistically, vehicles shipped to one location which then may suffer a weather delay could impact other critical testing. These inefficiencies can ramp project costs to unsustainable levels.

Because of these issues, previous research which collects sample data of wheeled and tracked vehicles over transitional soils is limited in quantity and scope.

Previous studies which most closely identify to this research are mostly previous experiments published by CRREL This

https://doi.org/10.56884/VBHYODO9

research can be seen as a modernization and continuation of the work (Shoop 1989).

CRREL has more recently captured experimental traction data with various US Army vehicles on transitional surfaces for tractive performance updates to current models. These studies include empirical based performance results for wheels and tracked vehicles of various weight classes on the surfaces of interest. (Parker 2019).

Fig. 1. CRREL researchers surveying typical, easily traversable peat site during summer season.

There are other sectors of research being conducted on freeze/thaw cycles, but it is mostly limited in scope and targeted towards specific problems which have little to do with off-road mobility on only the top few layers of the surface and its interaction with the vehicle. Research involving transportation infrastructure in regions that experience transitional or seasons or permafrost. For example, Yu, Jing, Liu, (2022) examined the effects of freeze-thaw cycles on asphalt mixtures. However, as most of this research is focused on the construction of road surface there is little information on analytical approaches regarding the off-road trafficability of a vehicle directly over freezing or thawing soils.

Other research for the Alaska Department of Transportation and Public Facilities, CRREL, and GZR Engineering by Connor, Bjella, and McHattie (2020) does hold valuable information with updates to understanding of soils in thawing permafrost regions, however is also not focused on projections or modeling of trafficking vehicles over such surfaces.

Fig. 2. The same peat site in a questionably traversable mode several months later during a seasonal thaw.

In the latest implementation plan of artic strategy, the United States Department of defense outlines a strategic objective to develop understanding of the Arctic Operational Environment. In order to achieve this goal, it is paramount that all attempts be made to replicate the conditions in a laboratory setting in a consistent, controlled, and repeatable manner. The work conducted at the CRREL Frost Effects Research Facility will be an important step towards replicating the transitional seasonal effects for more efficient, environmentally friendly and fiscally responsible testing of US Army vehicles on some of the most difficult and elusive types of terrain.

3. Experimental testing

This study investigates vehicle mobility for a light all-terrain tracked vehicle weighing 1.9 tons, and a light tactical vehicle weighing 3.2 tons. Soils in this study included three frost susceptible soils typically found in cold regions; sand, silt and organic soil.

3.1. Test section construction

The test surface was constructed in the Frost Effects Research Facility (FERF) at the U.S. Army Engineer Research and Development Center, Cold Regions Research and Engineering Laboratory in Hanover, New Hampshire, United States. This is an enclosed facility approximately 80 wide by 300 ft long capable of cooling down to 10 degrees F with aircraft hangar style sliding doors on either end to allow for drive through testing. This end to drive through capability is required when conducting mobility testing.

A continuous soil test section was constructed inside the FERF in one of the end-to-end lanes using silt, sand and organic soil. The entire test section was 21 ft wide by 270 ft in length. This test section was constructed from the bottom up starting with the concrete base. The goal was to create a section that would allow relatively easy and natural changes in soil moisture conditions between test events to see what effect moisture has on the mobility even though this testing focuses on a single moisture content.

Pipes were laid both transverse and longitudinal in the bottom of the test sections and connected to vertical stand pipes to allow water to be added and naturally wick to the surface if required. Figure 3 shows a side view of the soil layering and piping as installed. Each test section consisted of at least one base layer, and wicking layer, and a test soil at the surface. The test lane was constructed using both the north and south bays with the north bay starting about 3 feet deeper. A foot and a half of this difference in height was filled with a poorly graded gravel with a Unified Soil Classification System (USCS) designation of GP. Then 2.5 ft of silty gravel (GM) was added to the north bay and 1.5 ft of GM to the south bay. All GP and GM layers were added in 8-10 in. lifts and compacted to 97% +/- 2%. This leveled the two test sections. Figure 4 shows soils measurements taken during construction to ensure proper compaction. Next a poorly graded sand with silt (SP-SM) was added as the wicking material in 8-10 in. lifts and compacted to 97% +/- 2%. Each test material was added on top of this in various depths in 8-10 in. lifts and compacted to 97% +/- 2%. Peat soil was mixed (25-75 by volume) with SP-SM material to create a 5% organic test surface designated as (SP) poorly graded sand with 5% organic and installed with an 8-9% moisture content. This was designed to be the thickest layer because in situ peat is typically soft and should experience the deepest ruts. The SP-SM and SM layers were installed at the depths indicated in

Figure 3 with moisture contents of 8-9% and 6-7% respectively. Once construction was finished the soils were ripped to a depth of 18 in. and tilled to approximately 9 in. to create a soft repeatable test surface.

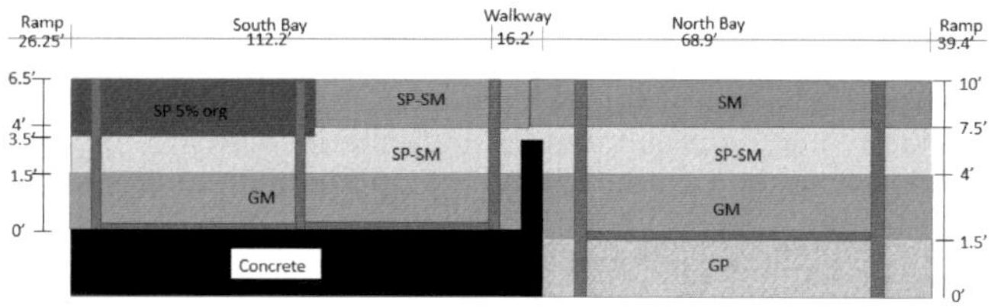

Fig. 3. Test section construction schematic (not to scale).

3.2. Soil characterization

Soil characterization was completed on the virgin (untrafficked) soil, and in the vehicle compacted motion resistance ruts using a range or instruments. Results for the rut depth, frost depth, thaw depth, density, cone index, shear strength, California bearing ratio (CBR) compaction strength for the surface and at depth.

A Troxler EGauge 4590 surface density gauge was used to measure the soil density at 2 in, 4 in, 6 in, and 8 in depths. A FieldScout SC900 Soil Compaction Meter with a 30° cone was used to continuously measure the cone index from the surface to a depth of 18 in. recording values every 1 in. (Spectrum Technologies 2022).

Fig.4. Soils measurements during construction to ensure proper compaction.

The Geonor shear vane was used to measure the soils shear strength using two different vanes, 0.79, and 1 in. in diameter. The largest set of vanes was tried first and if that did not penetrate into the soil and allow for a shear measurement then the smaller set of vanes was used. There is only one scale on the instrument so a multiplicative factor (20, 10) is applied to get the shear value. The California Bearing Ratio (CBR, 0 - 100 %) profile was measured using a dynamic cone penetrometer (DCP). Test procedure is outlined in ASTM D6951/D6951M-09 (2015). For all of the sites the 10.1 lb sliding weight was used along with the standard 60-degree, 0.79 in. diameter cone tip. A 1.1 lb. Clegg impact hammer was used to determine the strength of each soil surface. A cylinder (the hammer) is raised by hand and dropped from a height of 11.8 in. down a guide tube and the peak deceleration of the hammer on impact is measured. Four drops at a given point consist of one test (Lafayette Instrument Co 2005). Further test details can be found in ASTM D5874-16 (2016). Three replicates were taken at each grid point to allow for basic statistical analysis.

3.3. Test vehicle setup

Both test vehicles were outfitted with a compact water-resistant Aim Evo 5 data logger capable of logging multiple channels of data at once. Data recorded for each test vehicle include true vehicle speed using a fifth wheel (because we were indoors and GPS was unreliable), wheel speed using a hall effect sensor to detect cross drilled holes in the brake rotor, inline load cell force, throttle position with a string potentiometer on the gas pedal, and the IMU data in the Aim (roll, pitch, yaw, and the associated rates). Throttle pedal position was configured as zero to 100 percent with zero being no applied throttle and 100 percent being full throttle. The ground pressure of the light tactical all - terrain vehicle is 1.6 psi and the light wheeled vehicle is 12.1 psi.

For gathering tractive performance data, the vehicles were setup for testing Drawbar Pull (DBP) and Motion/Rolling Resistance (MR) according to SAE J2014. Figure 5 shows a vehicle setup in for Drawbar Pull.

Fig.5. Vehicles setup for Drawbar Pull, with 50 kN load-cell in line with tow rope.

The Drawbar Pull test consists of the test vehicle towing a holdback vehicle at a constant speed of 5 mph or 8 km/h. At constant speed the rear driver applies brakes to hold back the test vehicle as the lead driver slowly throttles up, sending more power to the driving wheels or tracks. The test driver is instructed to slowly push to full throttle as the wheels or tracks are allowed to slip at full vehicle power. If possible, the vehicle differentials are locked so that all four-wheel hubs spin at the same rate. The load-cell in the tow rope therefore acquires the pulling force of the test vehicle or Drawbar Pull.

The Motion Resistance test, also known as a rolling resistance test, switches the vehicles so the test vehicle is being towed, in neutral, at a constant speed of 5 MPH. This test measures the resistive force of the tire-surface interaction as well as internal frictions from the vehicle's drivetrain components. To acquire the true tractive power of the vehicle, the total resistance measured in the MR test is added to the Drawbar Pull test. See Parker. Et al. 2019 for more information on MR and DBP testing.

The Gross Traction (GT) of the vehicle is then defined as the following:

$$Gross\ Traction = DBP + MR_{Total}$$

3.4. Vehicle testing

Vehicle mobility testing was conducted on three separate surfaces conditions consisting of partially frozen, completely frozen and partially thawed soil. The fully frozen soil condition does not mean the soils was frozen entirely through the test section it means the soil was frozen deep enough to not have vehicle rutting or any soil deformation during the test. All results for this paper had average unfrozen moisture contents of 10.6, 9.0, and 6.1 percent for the SM, SP-SM, and SP soils respectively.

Light tracked all-terrain vehicle (LTATV) testing was conducted for partially frozen, frozen, and partially thawed soils on February 16th and March 19 and 28th respectively. Light wheeled vehicle testing was conducted for partially frozen, frozen, and partially thawed soils on February 28th and March 19 and 28th respectively. Motion resistance testing and Drawbar Pull testing were conducted for each test vehicle from north to south across the test section.

3.5. Pre and Post-test surface characterization

Table 1 and table 2 show average rut depth, frost depth, thaw depth and various other soil measurements that were taken for each test event.

Table 1. Virgin Soil and Rutting into for Tracked Light All-Terrain Vehicle

variable	2024-02-16			2024-03-19			2024-03-28		
	SM	SP-SM	SP	SM	SP-SM	SP	SM	SP-SM	SP
rut depth (in)	2.2	1.9	2.5	0.0	0.0	0.0	0.0	0.0	0.0
frost depth (in)	1.8	1.9	1.6	3.9	5.6	4.2	9.8	12.4	7.9
thaw depth (in)	0.0	0.0	0.0	0.0	0.0	0.0	1.9	1.9	1.9
dry density 0-2 in (lb/ft³)	83.0	74.3	81.8	1.41	88.0	66.8	111.7	103.0	92.4
average dry density 0-6 in (lb/ft³)	89.3	80.5	84.9	85.5	84.2	1.07	66.8	101.7	89.9
cone index 0-6 in (psi)	90.2	60.5	18.4	too hard	too hard	too hard	181.8	159.3	217.7
cone index 6-12 in (psi)	147.4	95.9	82.1	too hard	too hard	too hard	234.7	466.7	209.0
surface shear (psi)	2.0	3.2	4.6	too hard	too hard	too hard	1.46	0.6	0.7
Clegg Impact Hammer (CBR %)	6.5	10.4	6.7	14.4	18.8	12.2	5.2	4.3	4.1
depth average DCP (CBR %)	27.8	10.7	6.9	21.2	11.5	6.6	22.3	12.3	6.8

Table 2. Virgin Soil and Rutting into for Light Wheeled Vehicle

variable	2024-02-28			2024-03-19			2024-03-28		
	SM	SP-SM	SP	SM	SP-SM	SP	SM	SP-SM	SP
rut depth (in)	0.83	1.1	1.38	0.0	1.4	1.5	0.71	0.90	0.98
frost depth (in)	2.2	2.4	2.1	3.9	5.6	4.2	9.8	12.4	7.9
thaw depth (in)	0.0	0.0	0.0	0.0	0.0	0.0	1.9	1.9	1.9
dry density 0-2 in (lb/ft³)	87.4	74.9	68.7	1.41	88.0	66.8	111.7	103.0	92.4
average dry density 0-6 in (lb/ft³)	86.8	76.8	75.5	85.5	84.2	1.07	66.8	101.7	89.9
cone index 0-6 in (psi)	93.8	45.6	54.8	too hard	too hard	too hard	181.8	159.3	217.7
cone index 6-12 in (psi)	30.2	227.7	92.1	too hard	too hard	too hard	234.7	466.7	209.0
surface shear (psi)	5.2	5.1	4.3	too hard	too hard	too hard	1.46	0.6	0.7
Clegg Impact Hammer (CBR %)	8.1	10.0	4.5	14.4	18.8	12.2	5.2	4.3	4.1
depth average DCP (CBR %)	27.0	9.0	6.6	21.2	11.5	6.6	22.3	12.3	6.8

Table 1 shows that rut depth for the light tracked all-terrain vehicle is greatest for the partial thaw condition because even though the top is partially frozen the soil beneath it is completely unfrozen and providing limited bearing capacity causing the vehicle to sink in. This holds true for all soil types. Table 2 shows the rut depth for the light wheeled vehicle. Similarly, the partially thawed soil with the shallow thaw over a deep-frozen layer shows the least amount of sinkage into the soil. This indicates the deep-frozen layer is supporting the vehicles weight preventing the vehicle from sinking in further.

4. Experimental Results

Results in the following graphs show the performance of light tracked all-terrain vehicle and a light wheeled vehicle on three different soils at three different freeze/thaw states. Figure 6 shows the motion resistance results for the light track All-terrain vehicle. Results show the highest MR for the partially frozen surface which agrees with the highest rut depth also occurring on that surface.

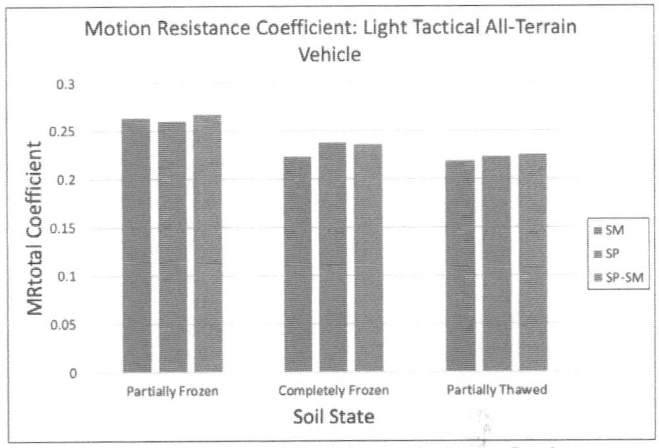

Fig. 6. Light Tracked All-terrain Vehicle Motion Resistance Results

Figure 7 shows the motion resistance results for the light wheeled vehicle. These results show a general trend of decreasing MR from partially frozen to the partially thawed soil. The SP soil shows the most linear decrease while the SM and SP-SM soils show the frozen configuration having a slightly lower MR than the shallow thawed SM and SP-SM soils. Even when frozen, the SP soil still produced the deepest ruts (highest MR) because the organic material in it made it softer and there wasn't enough moisture in the soil to cause a solid frozen layer. The SM froze solid producing no rutting and the MR was due more to surface roughness.

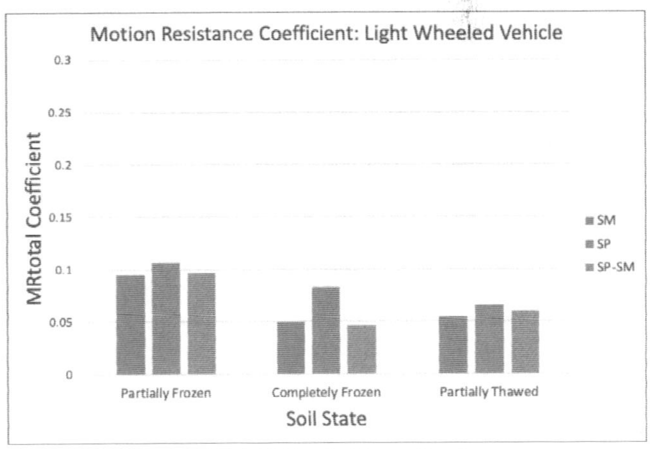

Fig. 7. Light Wheeled Vehicle Motion Resistance Results

Figure 8 shows the gross traction results for the light track All-terrain vehicle. These results show the traction values on the SP and SP-SM values reacting with soil state. The partially frozen ground having the best traction corresponding to the largest rut depth indicating the tracks are able to engage with the soil. Once the soil is frozen the rut depth goes to zero and the traction decreases as expected. As the soil starts to thaw (almost 2 inches) the traction starts to increase from the fully frozen values, but still remains less than the partially frozen soil traction values. Larger differences between the traction on different ground states were expected, based on data from previous field tests run on the same vehicles. This indicates monitoring the soil composition and its adherence to real world conditions is necessary to improve accuracy.

Figure 9 shows the gross traction results for the light wheeled vehicle. The light wheeled vehicle results show very little difference in gross traction across frozen/thawed state configurations. The partially thawed results show the SM soil type having the best traction with followed by the SP and the SP-SM having the least traction. Results in this report are for a low moisture content soil condition and it is anticipated that when moisture is added to increase to an average (more typical) moisture for the next set of tests it will have a significant impact in the results because the soil should freeze to a harder material.

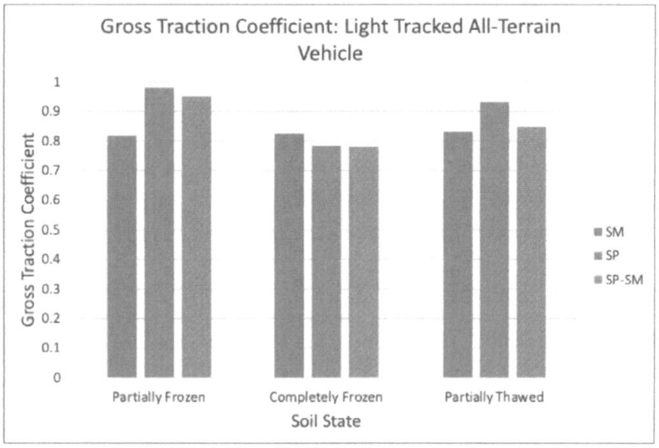

Fig. 8. Light Tracked All-terrain Vehicle Drawbar Pull Results

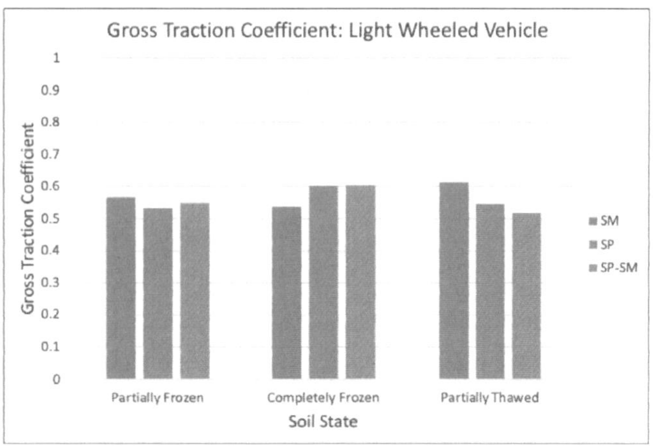

Fig. 9. Light Wheeled Vehicle Drawbar Pull Results

5. Conclusions and future work

This testing provided the initial results in a long line of testing required to fully characterize wheeled and tracked vehicles on frost susceptible soils. This work focused on the low soil moisture condition and light tracked all-terrain and light wheeled vehicle configurations. Based on previous results taken by CRREL researchers in field locations with natural surfaces of this type, both vehicle configurations showed smaller differences than expected in draw bar pull values across soil types in any freeze/thaw soil configuration. This trend is not expected to continue as the soil moisture increases in follow-on tests.

Future testing should add in a heavy wheeled vehicle and a heavy tracked over-snow vehicle on the same soil conditions before increasing the soil moisture and repeating the testing.

The research will also help to identify key improvements and modernizations to the CRREL Frost Effect Research Facility operation which will allow for increased rate of testing and consistency of the test surface.

6. Acknowledgements

All work described and data presented in this report are based upon testing conducted by the US Army Engineer Research and Development Center Cold Regions Research and Engineering Laboratory supported under 6.2 BL8 "Power Projection in A2/AD Environments, "Engineering for Battlespace Maneuvers, Task # A1030 Ground Vehicle Mobility. Permission was granted by the Director of the Cold Regions Research and Engineering Laboratory to publish this information.

7. Declaration of competing interest

The authors declare that they have no known competing financial interests or personal relationships that could have appeared to influence the work reported in this paper.

8. References

ASTM D5874-16. 2016. Standard Test Methods for Determination of the Impact Value (IV) of a Soil. ASTM International. West Conshohocken, PA.

ASTM D6951/D6951M-09. 2015. Standard Test Method for use of the Dynamic Cone Penetrometer in Shallow Pavement Applications. ASTM International. West Conshohocken, PA.

Connor, Goering, Kanevskiy, Trochim, Bjella, McHattie. 2020. Roads and Airfields Constructed on Permafrost. For the Alaska Department of Transportation and Public Facilities.

Lafayette Instrument Company. 2005. Clegg Manual. Lafayette, IN.

Spectrum Technologies, Inc. 2022. FieldScout TDR 350 Soil Moisture Meter Product Manual.

Parker, M.W., Stott, A.R., Bodie, M.O., Frankenstein, S., Shoop, S.A., "Vehicle Mobility on Highly Organic Soils", In Proceedings of the International Society for Terrain-Vehicle Systems (ISTVS), Prague, Czech Republic, September 9-11, 2019.

Shoop, S. 1989. "Vehicle Mobility on thawing soils: Interim report on CRREL's test program", US Army Corps of Engineers Special Report-89-31.

Society of Automotive Engineers (SAE), 2021, Pneumatic Tire/Wheel/Runflat Assembly Qualifications for Military Tactical Wheeled Vehicles, SAE J2014, SAE Int. Surface Vehicle Standards

Troxler Electronic Laboratories, Inc. 2016. Manual of Operation and Instruction, EGauge™ Model 4590 Surface Density Gauge. Research Triangle Park, NC.

US Office of the Under Secretary of Defense for Policy. 2023. Implementation Plan for the 2022 National Strategy for the Arctic Region.

3773 | Evaluation of the multi-pass effect of an exploration rover by single wheel testing assuming lunar gravity and soil

Tomomi Tanaka [a,†], Keisuke Takehana [a,†,*], James Hurrell [a], Kentaro Uno [a], Kazuya Yoshida [a]

[a] Department of Aerospace Engineering, Tohoku University, Sendai, Miyagi, Japan

* Corresponding author: keisuke.takehana.s1@dc.tohoku.ac.jp

† These authors contributed equally.

ABSTRACT

Exploring the lunar surface requires a wide range of mobility. Thus, wheeled mobile robots, called rovers, are crucial for traveling on the loose soil, called lunar regolith. To investigate the traction performance of the rover wheels on such a loose soil, it is essential to consider the multi-pass effect, which influences the traction performance of the wheel due to an alteration in the soil property caused by the previous wheel movement while traveling (compaction and digging).

Single-wheel testing is helpful in analyzing the driving performance of a wheel. In this study, the soil with a cohesive property and heterogeneity in its particle size and shape, which offered the nonlinear characteristics of lunar regolith, was prepared in the sandbox of a single-wheel testbed. To investigate the multi-pass effect, the passage of the rover wheel in Toyoura sand, which is a noncohesive and homogeneous sand, was compared with that in the lunar regolith simulant FJS-1. This study presents the experimental results of a repetitive running experiment conducted to measure the multi-pass effect. In the experiment, the wheel was run five times in a row over the same area in the sandbox in the same direction, and the traction coefficient, sinkage, and shear strength of the soil were measured each time. This experiment was also conducted under several conditions by varying the vertical load applied on the wheel and the slip ratio.

The wheel employed in the experiments was the same model as the grouser wheels installed on the Rashid rover, which was planned to be deployed on a lunar exploration mission.

The results revealed an increase in the shear stress due to soil compaction, particularly in the deep layer of the wheel's trace in FJS-1. However, compaction was not prominent in the shallow layer due to the soil fluidization by the grousers. On the other hand, soil compaction was not observed in either the shallow or the deep layer of Toyoura sand, where the grain sizes and shapes were almost homogeneous, and the cohesion force was nearly zero.

Keywords
Multi-pass effect
Lunar regolith
Single-wheel test
Rovers

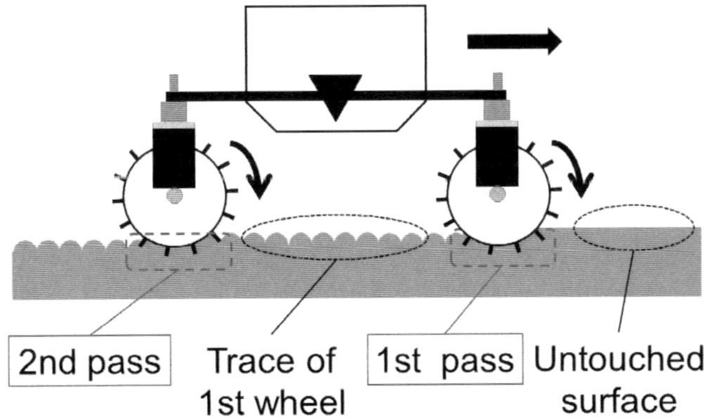

Fig. 1. Schematic showing the multi-pass effect of the rover wheels.

1. Introduction

Lunar exploration has being actively pursued in recent years. Various missions have been conducted to explore the Moon, with an increasing focus on its resource surveys and potential uses as a manned base. The rover, a wheeled robotic vehicle, is an effective tool for exploration as it can traverse the lunar surface and analyze samples on-site. At present, the use of unmanned rovers in lunar explorations is popular. These rovers require high traversability and the ability to predict their travel paths. The surface of the Moon is a low-gravity, soft terrain covered with regolith, scattered rocks, and gravel.

This study focuses on the traversability of a rover in the lunar environment. In planetary rover mobility, rigid-wheeled types are commonly used due to their simplicity and efficiency. However, circular rigid wheels tend to apply vertical stress on the soft ground, thus leading to easy disruption of the sand under the wheels. This can lead to slipping or sinking of the rover, potentially rendering it immobile. Slippage not only reduces the travel efficiency of the rover but also degrades the accuracy of odometry. Furthermore, if an unmanned rover becomes stuck due to slipping or sinking, it is difficult for it to extricate itself, forcing the mission to halt. Therefore, predicting the performance and behavior of rovers on soft ground and incorporating these into the control of the rover's movement are crucial for the success of an exploration mission. This necessitates focusing on the interaction between the wheels of the rover and terrain, and performing a wheel travel analysis based on the travel dynamics.

Terramechanics is a field of mechanics that focuses on the interaction between a terrain and machinery. It has been traditionally studied in relation to the travel dynamics of construction and agricultural machinery on unpaved grounds. In the 1960s, M.G. Bekker developed fundamental mechanical models of wheels (Bekker, 1956, 1960, 1969), which were later refined by J.Y. Wong and A.R. Reece who worked on traction models (Wong, 2008; Wong and Reece, 1965, 1967; Onafeko and Reece, 1967). However, traditional terramechanics has primarily elucidated the static behavior, whereas the dynamic behavior of wheels in motion remains unexplored. Currently, our understanding of rover travel mechanics is based on experimental evaluations, including single-wheel tests. Single-wheel tests involve attaching a wheel to a device that allows translational and rotational movements, and running it on soft ground to measure and evaluate its performance.

Grouser wheels were used in our experiment. Grousers are extensions on the surface of a wheel that enhance traction and

https://doi.org/10.56884/U5VDLHLN

reduce slippage on soft ground. The primary purpose of grousers is to increase the driving force, which they achieve by penetrating the terrain during travel and displacing the soil backward, thereby gaining greater ground reaction force and reducing slippage. We conducted stress distribution measurements of the ground reaction forces on the grousers and wheel surface, revealing that the ground reaction forces generated on the grousers are a dominant factor in the propulsion of the wheel (Nagatani et al., 2015, 2016).

Rovers typically have four or six wheels, and the soil properties change from those of the untouched surface after the passage of each previous wheel, as shown in Fig. 1. Alterations in the soil properties can affect the performance of the subsequent wheels, a phenomenon known as the multipath effect (Lyasko, 2010). However, further research is required on this effect, particularly for small and lightweight mobile robots for lunar exploration. Therefore, considering the multipath effect in tests is crucial for predicting the travel performance of a rover.

This study aimed to compare the traction performance of grouser wheels of a rover designed for a lunar mission, focusing on the multi-pass effect, to evaluate the travel performance of the rover on the lunar surface.

2. Single-wheel testbed

This section elaborates on the configuration and components of the single-wheel testbed used in the experiments, as shown in Fig. 2. The testbed was 2.5 m in length, 0.3 m in width, and 1.5 m in height, supporting a single wheel that traveled in a fixed direction across the sandbox.

The testbed consisted of the following components: a wheel drive unit, a passive vertical moving unit, a translational drive unit, a wheel, and a sandbox.

The experimental setup enabled a detailed analysis of wheel dynamics through the independent control of its rotational and translational velocities. The slip ratio, s, was defined by a formula incorporating the wheel radius, r, angular velocity, ω, and translational velocity, v_x (Eq. (1)). These variables were precisely measured by linear encoders installed on the wheel and translational drive units ensuring accurate data collection for analysis.

$$s = 1 - \frac{v_x}{r\omega} \tag{1}$$

Further, the dynamics of wheel traction were scrutinized by measuring the traction force, F_x, and the vertical load, F_z, exerted on the wheel, which were captured by force sensors. From these measurements, the traction coefficient, μ (Eq. (2)), which is a vital physical quantity that provides the traction capability, was calculated.

$$\mu = \frac{F_x}{F_z} \tag{2}$$

For the experiments, a grouser wheel, identical to that used in the Rashid rover developed by the Mohammed Bin Rashid Space Centre (MBRSC) (Almaeeni et al., 2021; Emirates Lunar Mission 2022), was chosen. This wheel was designed to enhance the traction on soft or loose surfaces. It had a diameter of 200 mm and was equipped with 14 grousers, each 20 mm in height, which contributed to its robust design.

The sandbox was 1.6 m in length, 0.3 m in width, and 0.3 m in depth, enabling a maximum wheel travel distance of 1 m. To simulate different environmental conditions, two distinct sandboxes were utilized: one filled with Toyoura standard sand, which is renowned for its uniform grain size and commonly used in scientific studies, and the other filled with a regolith simulant (FJS-1), which closely mimics the properties of lunar soil. The mechanical properties of these substrates are listed in Table 1 (Higa et al., 2019; Ryu et al., 2018; Toyoura Keiseki Kogyo Co., Ltd, 2023).

Table 1
Mechanical properties of the Toyoura standard sand and regolith simulant (FJS-1) used in this experiment.

Type	Toyoura Standard Sand	Regolith Simulant (FJS-1)
Particle size (μm)	106-300	7-4000
Specific gravity (g/cm^3)	2.63	2.94
Angle of Repose (°)	36.0	41.0
Friction Angle (°)	38.0	39.4
Cohesion (kPa)	0.0	8.0

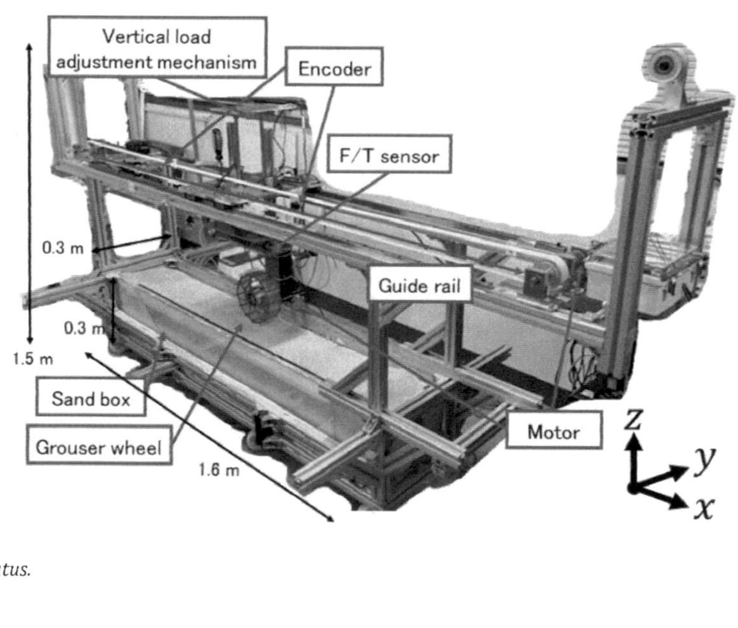

Fig. 2. Single-wheel testbed apparatus.

3. Experimental details

This section describes the experimental methods.

3.1. Experimental procedure

The experiment consisted of one set of five successive runs without leveling the ground and was performed as follows:

1. A board was used to plow the sandbox so that the ground inside it had uniform softness. The sand surface was flattened without pressing it down on the flat plate, and the sand was equalized to a uniform height.
2. The hardness of the soil was measured at depths of 15 and 30 mm, respectively, using a hand vane for a total of six points.
3. Based on the slip ratio, the rotation and translation speeds of the wheel were set and its running was started with controlled constant velocity motion.
4. During running, the reaction force acting on the axle was measured by the force sensor, and the amount of wheel sink, translational distance, and number of wheel rotations were measured by the linear encoder.
5. Rutting of the soil was avoided and the soil hardness was measured in the same manner as in step 2.
6. The wheel was returned to the starting position and steps 3 to 5 were repeated five times.
7. After five measurements, starting from step 1, the experiment was repeated three times under each condition and the data thus obtained was analyzed.

Regarding the measurement, the drawbar pull and vertical load acting on the wheels during the run were measured directly from sensors mounted on the tops of the wheels. The encoder mounted on the top of the wheel is passively displaced vertically by the movement of the wheel, and the amount of displacement is measured to determine the sinkage.

3.2. Experimental conditions

In this experiment, the vertical load, driving speed, slip ratio, and soil type were varied. The conditions for each parameter were as follows:

Vertical load: 5 N, 25 N
Driving speed: 20 mm/s
Slip ratio: 10 %, 30 %, 50 %
Soil: Toyoura standard sand, Regolith simulant (FJS-1)

The experiment on Toyoura standard sand was conducted only under the condition of 30 % slip ratio. Note that the Toyoura standard sand experiment was initially conducted as a preliminary experiment, so there are fewer slip conditions than the regolith simulant in the main experiment. On regolith simulant, experiment was conducted under all slip conditions, 10 %, 30 %, and 50 %.

Three sets of measurements were obtained for each condition, with one set consisting of five consecutive runs in one direction without leveling the ground after each run. For the Rashid rover used in this study, the vertical load applied to one wheel on Earth was 25 N and the assumed load on the Moon was approximately 5 N.

4. Data acquisition and analysis

This section describes the data captured in this test and the method used to analyze the data.

4.1. Raw data measurement

The two types of data obtained from the force sensors used in this experiment were: translational movement and wheel sinkage. Figure 3 shows the data and the analysis results obtained from the running experiments on Toyoura standard sand under the following conditions: a vertical load of 25 N, running speed of 20 mm/s, and slip ratio of 30 %. As can be seen from Fig. 3, after a certain time from the start of running, the sinkage increased from 0 and eventually its value oscillated, but its average value converged to a constant value, and the wheel running became stable. This oscillation could be attributed to the effect of the grouser. The slip ratio and vertical load were also constant while oscillating. Therefore, the section where the three points were constant was defined as the stationary section, and only data from this section, in which running became stationary, were extracted for analysis. However, as mentioned, since the values change with the vibration period, the stationary section was defined as the section in which:

-Sinkage variation was below 1.5 mm/s.
-Slip ratio fell within ±5 % of the specified experimental parameter value.
-Vertical load remained within ±10 % of the specified experimental parameter value.

4.2. Vane shear test

To ensure consistent driving conditions and account for the soil conditions before driving, a preliminary assessment of the soil hardness was conducted. The shear force of the soil was determined by performing a shear test using a hand vane equipped with blades at the end of a rod. The shear moment was measured by inserting the rod with the vane blade into the soil and rotating it. Tests were performed to ensure that the shear moment remained within the range of 0.015 to 0.020 Nm. If the shear moment deviated from the specified range, the soil was tilled again, followed by subsequent shear tests. The average shear moment values obtained from the vane shear tests were 0.015 and 0.017 Nm for the Toyoura standard sand and the regolith simulant, respectively.

The shear strength exhibited the following relationship based on the shear torque (M) at the time of measurement and the shape of the hand vane (Diameter and Height):

$$\tau = \frac{M}{\pi\left(\frac{D^2H}{2} + \frac{D^3}{6}\right)} \tag{3}$$

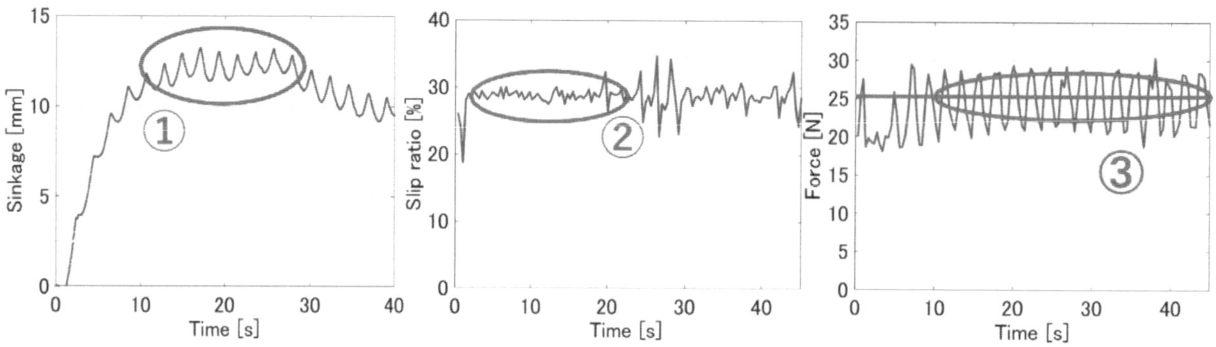

Fig. 3. Representative time-series data of a wheel traveling on Toyoura standard sand with a vertical load on it of 25 N, wheel driving velocity of 20 mm/s and slip ratio of 30 %.

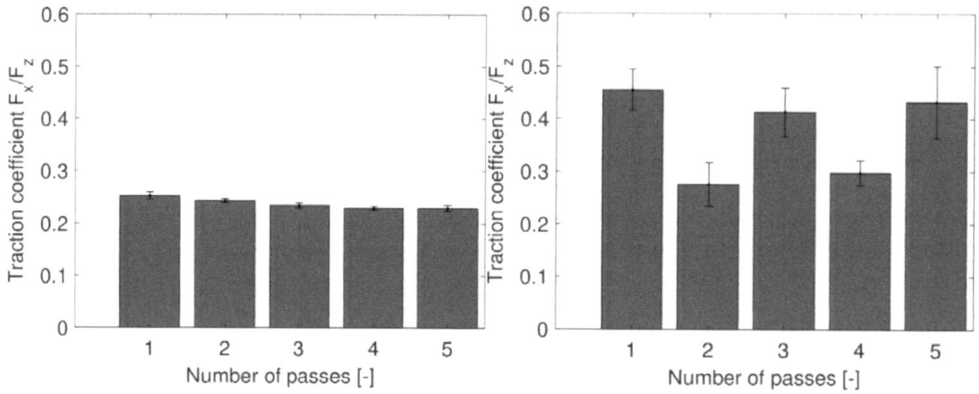

Fig. 4. Traction coefficient as a function of the number of runs of the wheel traveling on Toyoura standard sand with the vertical load of 25 N (left) and 5 N (right) on it.

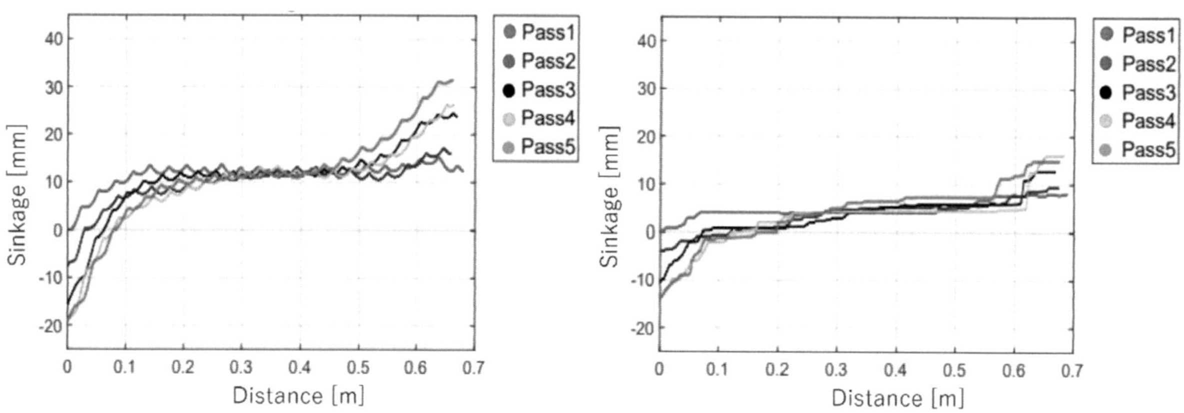

Fig. 5. Sinkage as a function of the distance traveled by the wheel on Toyoura standard sand with a vertical load of 25 N (left) and 5 N (right) on it.

5. Test results

This section presents the results of the experiment investigating the multi-pass effect. It summarizes the results of each run and discusses the impact of repeated driving on the driving performance.

5.1. Toyoura standard sand

Figure 4 shows the average of three sets of values per run on the Toyoura standard sand, which clearly indicates the change in the traction coefficient with respect to the number of runs. The slip ratio was set to 30 % in the experiment. Under a vertical load of 25 N, although the difference remained below 0.05, a slight decrease in the traction coefficient was observed. However, the values converged after the fourth run. Under a vertical load of 5 N, the traction coefficient fluctuated with each run, indicating that the effects of repeated driving are unclear.

The changes in the sinkage during driving are shown in Fig. 5. The sinkage was observed to increase gradually after each run, and the initial mounding of sand became progressively larger. This is due to the action of the grouser, which moved the sand from the front to the back as the driving is repeated. Under a 5 N vertical load on the wheel, the sinkage was less than half of that under 25 N, but a tendency for the sand to move from the front to the back due to the action of the grouser was observed with repeated runs.

Tables 2 and 3 summarize the shear strengths before and after each run. Two patterns were measured at different depths. The number of times is defined as the number of times after the run, and 0 represents the number of times before the first run. No changes due to driving were observed at depths of 15 and 30 mm, indicating that no compaction of soil occurred. This result is consistent with the characteristic lack of cohesion in the Toyoura standard sand.

Table 2
Shear strength (Toyoura standard sand, 25 N vertical load).

Depth	Driving number					
	0	1	2	3	4	5
15 mm	1.49	1.50	1.49	1.52	1.54	1.56
30 mm	2.96	2.66	2.71	2.67	2.73	2.82

Table 3
Shear strength (Toyoura standard sand, 5 N vertical load).

Depth	Driving number					
	0	1	2	3	4	5
15 mm	1.50	1.48	1.53	1.60	1.61	1.53
30 mm	2.56	2.65	2.65	2.88	2.88	2.83

5.2. Lunar regolith simulant (FJS-1)

Figure 6 shows the results for the number of runs for each condition in the regolith simulant. Compared to the Toyoura standard sand, data were obtained not only for 30 %, but also for 10 % and 50 % sand. Similar to Fig. 4, the results for the slip ratio of 30 % are shown in Fig. 6. For a vertical load of 25 N, no change in the traction coefficient was observed, a result similar to that of Toyoura standard sand. For a vertical load of 5 N, the traction coefficient fluctuated slightly from run to run, indicating that the effect of repeated runs was not pronounced.

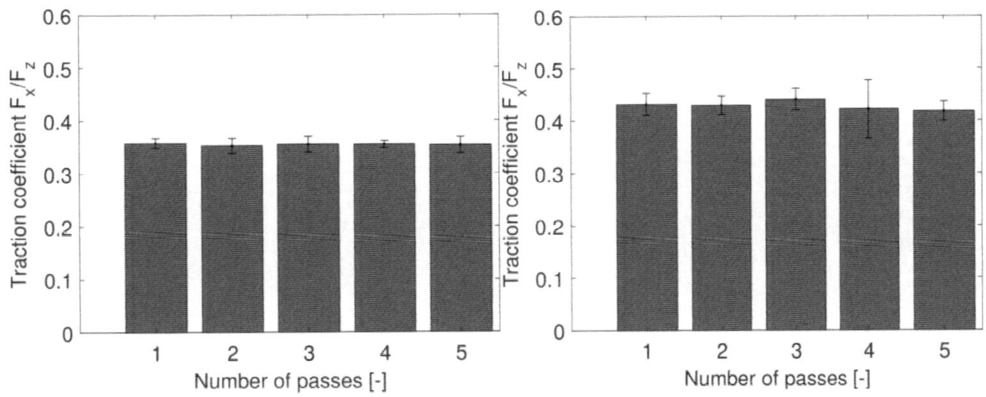

Fig. 6. Traction coefficient as a function of the number of runs of the wheel traveling on the regolith simulant with a vertical load of 25 N (left) and 5 N (right) on it.

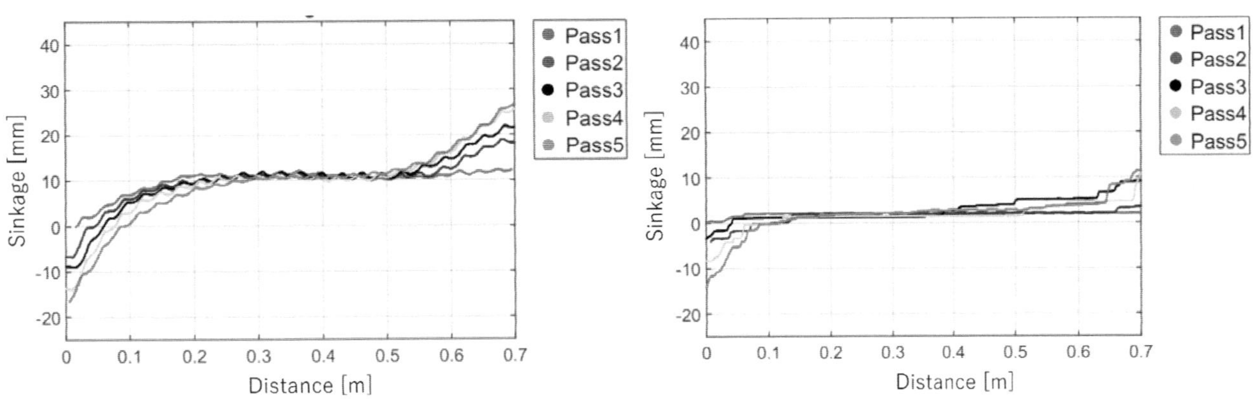

Fig. 7. Sinkage as a function of the distance traveled by the wheel on the regolith simulant with a vertical load of 25 N (left) and 5 N (right) on it.

The changes in sinkage during running are shown in Fig. 7. Similar to the results corresponding to Toyoura standard sand, the sinkage depth of the wheel traveling on the regolith simulant increased gradually after each run, with the initial mounding of the sand becoming progressively larger. This indicates that the sand moved from the front to the back with each run due to the action of the grouser.

Particularly under high-slip conditions, a large amount of sand moved, and as running continued, an incline formed on the track creation surface. As the slip ratio increased, the amount of sand moved by the grouser also increased, resulting in a greater incline.

These results are shown in Fig. 8 and have been organized according to the traction coefficient and slip ratio. The average values for each of the three instances of each run (1-5) are represented by different colored markers and indicated by error bars. Three sets of five runs were performed for each condition, and thus the data for runs 1-5 represent three runs each.

Figure 8 (left) shows the results for a vertical load of 25 N on the wheel, in which no significant differences in the traction coefficient associated with the number of runs for slip ratios of 10 and 30 % were observed. In particular, at a 30 % slip ratio, the values were almost identical, indicating that the impact of repeated runs was minimal. However, at a slip ratio of 50 %, the traction coefficient increased gradually by 0.05 from the first to the fifth run. This subtle and distinct change occurred despite the relatively stable running conditions and was attributed to the effect of repeated runs. This increase in the traction coefficient contrasts with the results obtained for the Toyoura standard sand.

Figure 8 (right) shows the results for a vertical load of 5 N. Similar to the 25 N results, no significant differences were observed with the number of runs, although the error bars were larger than at 25 N, thus showing more variability. Similar to the 25 N results, the influence of repeated runs at 30 % slip ratio was smaller than that at other slip ratios. Furthermore, across all conditions, the error bars were larger than those at 25 N, particularly noticeable at a 10 % slip ratio, where the traction coefficient slightly decreased between the second and fourth runs, most probably owing to the instability caused by the lower load. At low loads, such as the vertical load of 5 N, the effect of repeated running was insignificant. This result suggests that repeated runs on the lunar track have little effect on the running performance of the wheel.

The changes in the shear strength overruns, determined from the vane test results, are presented in Tables 4 and 5. Because no significant change in the shear strength at a depth of 15 mm was observed after one run, the shear moments at the depth of 30 mm were measured in the repeated run experiments.

Table 4
Average value of shear strength (regolith simulant, 25 N vertical load).

Slip	Depth	Run number					
		0	1	2	3	4	5
10 %	15 mm	1.35	1.92	1.90	1.99	2.06	2.12
	30 mm	1.84	3.40	3.85	3.90	4.01	4.31
30 %	15 mm	1.51	1.70	1.73	1.86	1.85	1.86
	30 mm	2.12	3.54	3.40	3.81	3.84	3.81
50 %	15 mm	1.33	1.43	1.72	1.59	1.75	1.69
	30 mm	2.11	3.18	3.37	3.04	3.57	3.37

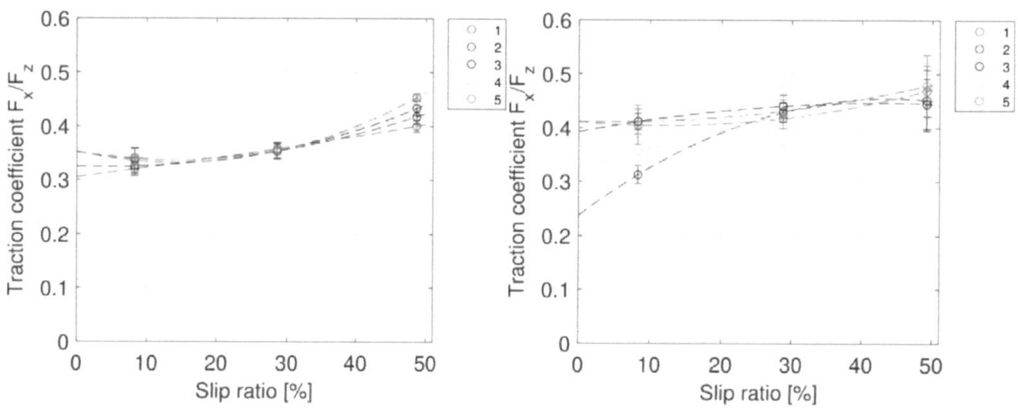

Fig. 8. Traction coefficient as a function of the slip ratio for different number of runs of a wheel traveling on the regolith simulant with a vertical load of 25 N (left) and 5 N (right) on it.

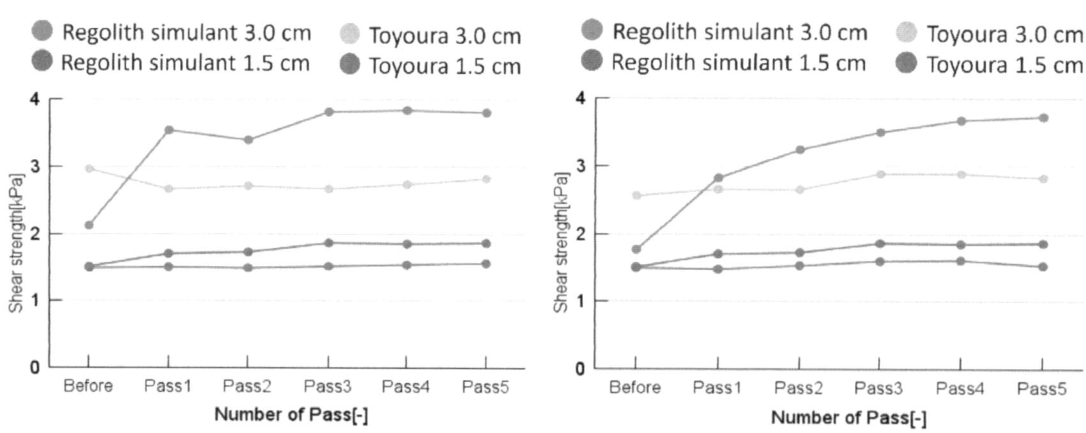

Fig. 9. Comparison of the shear strength in each the two types of soil for a vertical load of 25 N (left) and 5 N (right) on the wheel.

Table 5
Average value of shear strength (regolith simulant, 5 N vertical load).

Slip	Depth	Run number					
		0	1	2	3	4	5
10 %	15 mm	1.40	1.76	1.78	2.11	2.17	2.18
	30 mm	1.95	3.13	3.44	3.74	3.83	4.21
30 %	15 mm	1.32	1.72	1.93	2.02	1.89	2.13
	30 mm	1.77	2.82	3.25	3.50	3.68	3.73
50 %	15 mm	1.49	1.55	1.58	1.53	1.51	1.75
	30 mm	2.11	3.40	3.60	3.89	4.65	4.70

At a depth of 15 mm and 25 N (Table 4), there was little difference between pre- and post-run. Among the pre-run shear strengths ranging from 1.07 to 1.79 kPa, the largest difference between the pre-run and after the fifth run was 0.8 kPa, as read from Table 4. Considering the variation in shear strength at the time before the run, this means that the change in strength due to the run was not significant, that is, it was within the range of the variation before the run.

At a depth of 30 mm, however, the difference ranged from a minimum of 0.93 kPa to a maximum of 2.47 kPa, which is a larger change than the pre-run variation, suggesting that the shear strength changed over the five runs. Thus, it is indicated that deeper occurrence of compaction due to running than the reach of the grouser. At depths reached by the grouser, the soil was churned, preventing compaction. Further, the largest differences at depths of 15 mm as well as 30 mm occurred at a 10 % slip ratio. This is probably because slipping caused the grouser to dig more into the sand, increasing the amount of sand moved, and thus reducing the impact of compaction at higher slip ratios. For all conditions, the largest changes occurred between the first and second runs, with smaller changes occurring in the subsequent runs. At a depth of 15 mm, no significant changes were observed as the soil was consistently disrupted during running. However, at the depth of 30 mm, compaction occurred during running, but since the compaction of the soil by the wheels was nearly maximum in the first run, subsequent changes were more gradual.

From the results at the depth of 15 mm and 5 N (Table 5), almost no difference was observed before and after running. The largest difference observed was for the 30 % slip ratio condition, with a difference of 0.81 kPa, slightly greater than that observed at 25 N, but insufficient to definitively conclude that compaction occurred. On the other hand, at the depth of 30 mm, differences ranged from a minimum of 1.96 kPa to a maximum of 2.59 kPa, similar to the 25 N condition, indicating that compaction due to running occurred deeper than the reach of the grouser.

When running at a depth of 30 mm, the largest changes were observed between the first and second runs, with smaller changes in subsequent runs, similar to the results under a 25 N load. This was because the wheels compacted the soil almost maximally in the first run, leading to more gradual changes thereafter. Under 25 as well as 5 N loads, the effects of compaction were observed at a depth of 30 mm, but such phenomena were not confirmed in Toyoura standard sand. This suggests that the compaction occurred due to adhesion. At the depths reached by the grouser, as the grouser moved the sand, compaction did not occur at these depths but occurred at depths deeper than the reach of the grouser. The shear strength after compaction was expected to increase linearly with higher vertical loads, but no significant difference was observed in the results corresponding to the 25 and 5 N loads. This indicates that the compaction of regolith simulant occurs easily up to a certain level owing to its adhesive properties, but beyond that level, an additional force is required, and this threshold is considered to be above 25 N.

5.3. Comparison of each soil

The Toyoura standard sand and the regolith simulant both exhibited a tendency for sand to move from the front to the back owing to the action of the grouser during repeated runs. To compare this in detail, the changes in the shear strength for both under the same vertical load conditions are summarized in Fig. 9. The results indicate that, irrespective of the depth or vertical load, Toyoura sand showed no changes in its shear strength and no compaction occurred, which was attributed to the lack of its cohesive properties. In contrast, while the surface layers of the regolith simulant showed no significant changes in its shear strength or compaction, a notable increase in these two properties was observed in the deeper layers as the number of runs increased. This lack of surface compaction was due to the dynamic flow of sand caused by the action of the grouser. In the deeper layers, where this flow is absent, compaction occurred due to the adhesive properties of the regolith simulant. This phenomenon occurred consistently, regardless of the vertical load applied.

6. Conclusion and future work

In this study, running experiments were conducted on a rover wheel and the results were carefully examined to understand the effect of the rover wheel passage on the subsequent performance of the following wheels, termed the multi-pass effect. The experimental results were comprehensively analyzed for changes in the traction coefficient, sinkage, and soil shear strength as a function of the number of runs. The findings revealed that an increase in the shear stress was observed with soil compaction, particularly in the deep layer of the wheel trace in FJS-1. However, compaction was not prominent in the shallow layer because of the soil fluidization effect induced by the wheel grousers. On the other hand, soil compaction was not observed in either the shallow or deep layer of Toyoura sand, where the grain sizes and shapes were almost homogeneous and the cohesion force was nearly zero. In addition, the changes in sinkage demonstrated that the action of the grouser caused the sand to move from the front to the back with each run. However, neither the Toyoura standard sand nor the regolith simulant results exhibited significant changes in the traction coefficient due to repeated running.

In future work, additional experiments with different sizes and types of wheels and soils should be conducted to confirm whether the above observations are generally true. Simulations based on the discrete element method will also be useful for further investigations.

7. Acknowledgements

We are very grateful that this project was helped by Dr. Sebastian Els, Mohammed Bin Rashid Space Centre (MBRSC), who supplied important information about the wheel parameters of the Rashid rover.

8. Declaration of competing interest

The authors declare that they have no known competing financial interests or personal relationships that could have appeared to influence the work reported in this paper.

9. References

Almaeeni. S., Els, S.G., Almarzooqi, H., 2021. To a dusty Moon: Rashid's mission to observe lunar surface processes close-up. 52nd Lunar and Planetary Science Conference 2021.

Bekker, M.G., 1956. Theory of land locomotion. University of Michigan Press, Ann Arbor.

Bekker, M.G., 1960. Off-the-road locomotion. University of Michigan Press, Ann Arbor.

Bekker, M.G., 1969. Introduction to terrain-vehicle systems. University of Michigan Press, Ann Arbor.

Emirates Lunar Mission 2022 / Rover "Rashid" https://forum.nasaspaceflight.com/index.php?topic=52020.0 (accessed on 31/5/2024).

Higa, S., Nagaoka, K., Yoshida, K., 2019. Stress distributions of a grouser wheel on loose soil. J. Terramech. 85, 15–26.

Lyasko, M., 2010. Multi-pass effect on off-road vehicle tractive performance. J. Terramech. 47 (5), 275–294.

Higa, S. , Nagaoka, K., Nagatani, K., Yoshida, K., 2015. Measurement and modeling for two-dimensional normal stress distribution of wheel on loose soil. J. Terramech. 62, 63–73.

Higa, S., Sawada K., Nagaoka, K., Nagatani, K., Yoshida, K., 2016. Measurement of stress distributions of a wheel with grousers traveling on loose soil. Proceedings of the 2016 IEEE International Conference on Robotics and Automation (ICRA), 2828–2833.

Onafeko, O., Reece, A.R., 1967. Soil stresses and deformations beneath rigid wheels. J. Terramech. 4 (1), 59–80.

Ryu, B.H., Wang, C.C., Chang, I., 2018. Development and geotechnical engineering properties of kls-1 lunar simulant. J. Aerospace Eng. 31 (1), 04017083.

Toyoura Keiseki Kogyo Co., Ltd, about toyoura sand. http://www.toyourakeiseki.com/english/about.html (accessed on 31/5/2024).

Reece, A.R., 1965. Principles of soil-vehicle mechanics. Proc. Inst. Mech. Eng. Automob. Div. 180 (1), 45–66.

Wong, J.Y., Reece, A., 1967. Prediction of rigid wheel performance based on the analysis of soil-wheel stresses part I. performance of driven rigid wheels. J. Terramech. 4 (1), 81–98.

Wong, J.Y., Reece, A., 1967. Prediction of rigid wheel performance based on the analysis of soil-wheel stresses part II. performance of towed rigid wheels. J. Terramech. 4 (2), 7–25.

Wong, J.Y., 2008. Theory of ground vehicles. John Wiley & Sons Inc, Hoboken, New Jersey.

4093 | The role of tire-soil interface characteristics on performance parameters through experimental and numerical investigation

Jasleen Kaur Bheora [a,*], Varsha S Swamy [b], Destiny Mason[b], Alba Yerro[a], Corina Sandu[b], Bernardo Castellanos[a], Katherine Sebeck[c], Gorsich David[c]

[a] *Department of Civil and Environmental Engineering, Virginia Tech, Blacksburg, VA, United States*

[b] *Department of Mechanical Engineering, Virginia Tech, Blacksburg, VA, United States*

[c] *U.S. Army DEVCOM Ground Vehicle Systems Center, Warren, MI 48397, United States*

** Corresponding author: jasleenkb@vt.edu*

ABSTRACT

The capacity of off-road vehicles to navigate unprepared terrain is determined by the forces imparted by the terrain, which are directly affected by the tire-soi interface performance. The strength of the tire-soil interface is typically lesser than the soil's shear strength. In particular, the interface friction characteristics between the tire and the terrain influence the drawbar pull performance and are often overlooked in studies. As a result, accurate experimental determination of the interface friction and its influence on physics-based modeling is the focus of this paper. The tire interaction with a saturated low-plasticity clay (CL) in the plastic state is studied in this paper. A large-scale direct shear test to measure the interface friction between the off-road lugged tire and the clay at different contact pressures. The results of the tests are imported into the numerical model, which consists of a representative FE tire interacting with the characterized FE clay. The representative FE tire consists of Mooney-Rivlin rubber parts (i.e., tire body, sidewall, and lugs) on a rigid inner lining representing the inflated configuration of the tire. The soil flow and stresses as a result of a tire pass are analyzed. The results are used as a base model to conduct a sensitivity analysis of the friction coefficient on the drawbar pull and instantaneous sinkage. Comparative analysis employing different tire designs (i.e., lugged and smooth) is conducted. This research highlights how soil deformations and shear stress profiles are influenced by the friction coefficient when considering soft clayey terrain. As a conclusion, the proposed methodology shows the importance of accurate interface modeling, providing insights for better correlation between modeling and experimentation.

Keywords
Tire-terrain interaction
Interface friction
Direct shear test
Saturated soil
Tire-clay failure envelope

1. Introduction

The mobility of ground vehicles on soft fine-grained soils may pose a challenge, primarily due to the limited traction available from the soft terrain. Heavy vehicles, including military, agricultural, and earth-moving vehicles, often operate in these environments, emphasizing the significance of conducting studies in this domain. With the advancements in vehicle modeling and performance prediction capabilities, high-fidelity modeling for generating synthetic data has become a recent focus of the community. Although many studies have been performed in dry soil, the effect of vehicle mobility over soft, wet, and plastic terrains composed of fine-grained soils is still to be investigated. This study aims to advance the state-of-the-art in experimentation, parameterization, and modeling of tire-terrain interfaces in such conditions.

The physics of tire loading and rapid shearing imposed by a vehicle over wet soil is a complex phenomenon. The generation of excess pore water pressure coupled with the reorientation of soil particles may reduce the clay's shear strength causing a negative impact on vehicle mobility (Pandit, 2023). The interface friction in these cases can be low, causing the tire to lose traction. The literature reports limited experimental data, making it important to characterize the properties of the soil-tire interface for a detailed examination of the contact surface behavior.

From the modeling point of view, friction, in general, can be a sensitive parameter, and obtaining the correct value is advisable for an accurate simulation of the overall problem (Swamy et. al, 2023a). One notable effort in this area was performed by Papamichael and Vrettos, 2021 in their investigation using direct shear tests between silty sand and tire. The investigation presented herein expands the current literature to soft clay soils with a high moisture content. The tire used for this study is the BKT TR 319 29X12.50-15 agricultural tire for navigation in mud. The lugs are designed to assist stability and traction performance on wet surfaces, with self-cleaning properties.

The paper is organized as follows. First, the geotechnical lab experiments conducted to characterize the soil are described. Next, the set-up of the tire-clay direct shear testing is presented in detail, along with the results obtained from the tests. Interface friction is then compared to the soil's internal strength for more insight. The latter part of the paper introduces the numerical modeling efforts while considering the results from the soil-tire interface direct shear tests. The BKT FE tire is rolled on the characterized clay at 4000 N normal load and a 10% slip ratio. Lastly, friction sensitivity studies are conducted on the simulation to understand the effects of friction on the drawbar pull and the shear stresses in the soil.

2. Geotechnical laboratory testing

The geotechnical laboratory testing was performed to parameterize the clay and tire-clay interface model. In particular, the large-scale interface direct shear tests are novel. Parameters estimated from laboratory testing are used as inputs into the numerical model.

2.1. Characterization of Clay

As discussed by Pandit, 2023 and Swamy et. al, 2023b, clay characterization was conducted to obtain input parameters for the material constitutive law. Table 1 shows the list of laboratory testing conducted. The clay used in this project is called "craggy

DISTRIBUTION STATEMENT A. Approved for public release; distribution is unlimited. OPSEC #8701
https://doi.org/10.56884/X7EXV5XW

crunch", provided by Highwater Clays. It is a low plasticity soft clay (CL) with low swelling potential, with visual classification as pinkish, homogeneous, moist, stiff, lean clay with occasional particles of coarse sand.

To obtain a failure envelope of clay, trimmed samples from the intact specimens were tested for 34.47 kPa (5 psi), 68.95 kPa (10 psi), 137.89 kPa (20 psi), and 275.79 kPa (40 psi) pressures in consolidated undrained (CU) triaxial test conditions in accordance with ASTM D4767, 2011. The failure envelope derived from these tests indicated that the cohesion (c') of clay was 13.79 kPa (2 psi) and friction angle (ϕ') was 13.9° (Pandit, 2023).

2.2. Characterization of clay-tire interface model

Large-scale direct shear testing has been performed in this study to determine the interface properties of clay and tire contact, required for the numerical model. In particular, contact friction and cohesion between tire and clay are the most relevant parameters. Traditionally, small-scale direct shear tests are conducted to determine the drained shear strength of soils, but this technique can also be used to determine the interface properties between two materials (tire and clay in this case).

A standard test consists of a shear box with soil specimen placed into the two halves of the box. The soil is consolidated to a certain pressure value and sheared horizontally with a user-set displacement rate along a pre-defined shear plane. The test measures the amount of shear force required to instigate failure and the corresponding displacement required to cause that failure, which eventually gives us the drained strength parameters for the soil being tested. ASTM D3080, 2023 provides a standard procedure for conducting this test in soils.

To measure the strength properties of an interface, one half of the shear box is filled with one material, and the other half is filled with another material. Shearing is done along the interface. As per the authors' knowledge, there does not exist a standard procedure to test the interface between tire and soil. However, ASTM D5321, 2014 discusses the standard test procedure to determine the shear strength of a soil-geosynthetic interface. Therefore, the latest is used in this study as a guide to test the interface properties of the soil-tire system (Pandit, 2023).

2.2.1. Testing methodology

A large-scale direct shear testing apparatus with shear box dimensions of 30.48 x 30.48 x 20.32 cm (12 x 12 x 8 inches) (length x width x thickness) is used in these tests. The upper part of the box has an air bladder attached to it, a pressure regulator, and a gauge that ensures proper load application. The test equipment houses a stainless-steel water tank, which ensures submersion of the specimen throughout the test, with testing conducted in drained conditions.

To set up the testing, the tire is placed in the bottom half of the shear box, while the clay sample is placed in the upper box. Note that similar tests were performed by Pandit, 2023, with the tire on the upper box, but the results were considered unsuccessful, as discussed in Section 2.2.2. The BKT tire is cut to the required dimension, as shown in Fig. 1(a), from the original tire, which is cm 73.66 cm (29 inches) in diameter with a section width of 31.75 cm (12.5 inches). The cut tire piece is clamped on a wooden block, and its height is adjusted so that the tire-wood assembly has its tire plane along the shear plane of the apparatus. Tire grooves are filled with clay, as shown in Fig. 1(b), to ensure proper load application and appropriate interaction between clay in the upper shear box and tire placed in the bottom shear box as shown in Fig. 2(a).

Table 1. *Brief description of laboratory tests conducted on clay*

Laboratory Testing	**Parameter derived**
Atterberg Limit Test	Liquid limit and Plasticity limits
Specific Gravity Test	Specific gravity
Hydrometer Test	Particle size distribution
Proctor Compaction Test	Compaction curve
Unconsolidated Undrained (UU) Test	Undrained shear strength
Consolidated Undrained (CU) Test	Drained friction angle and drained cohesion values; ϕ' = 13.9° and c' = 13.79 kPa (2 psi)
Direct Simple Shear (DSS) Test	Undrained strength vs Consolidation pressure plot
Consolidation Test	Value of Compression index
Direct Shear Test	Friction angle and cohesion

Due to the large dimensions of the shear box, handling the soft clay and ensuring correct placement can be challenging. As a result, it is cut into smaller blocks and assembled in the shear box. Fig. 2 (b) shows clay cut into small blocks and placed in the upper shear box of the test setup, trying to minimize gaps between clay blocks. As discussed earlier, an ASTM standard does not exist for a large direct shear test, so an intact sample is used in the test because of the unavailability of a standardized method for sample preparation. A 2-inch-thick sample is used for testing to speed up the consolidation and shearing times, as it takes several months to complete the shearing of an approximately 4-inch-thick sample under drained conditions (Pandit, 2023). The thickness of the sample in drained conditions does not have an impact on the value of the friction angle. Therefore, it is possible to conduct the test on a thin sample to speed up the process.

Direct shear tests were conducted for three consolidation stresses: 34.47 kPa (5 psi), 68.95 kPa (10 psi), and 137.89 kPa (20 psi) to determine the effective stress strength parameters. To ensure drained conditions during shear, the specimens were sheared at a displacement rate of 0.000254 cm/min (0.0001 inch/min). The consolidation stage lasted about 3 days, with shearing taking between 9-12 days.

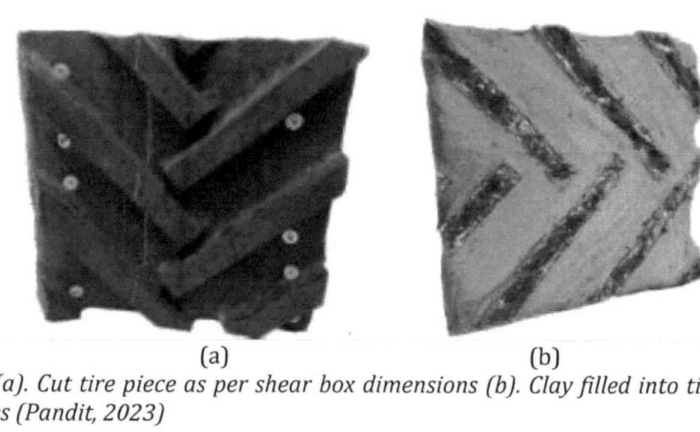

(a) (b)

Fig.1. (a). Cut tire piece as per shear box dimensions (b). Clay filled into tire grooves (Pandit, 2023)

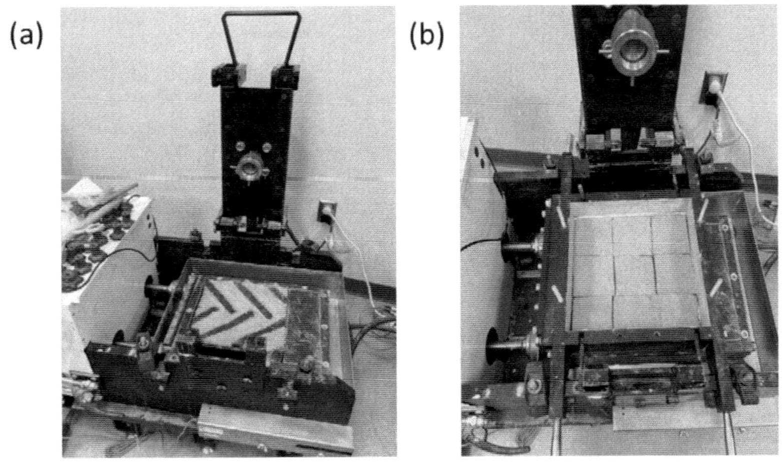

Fig. 2. (a). Tire filled with clay placed in the bottom half of the shear box (b). Clay blocks placed in the upper shear box

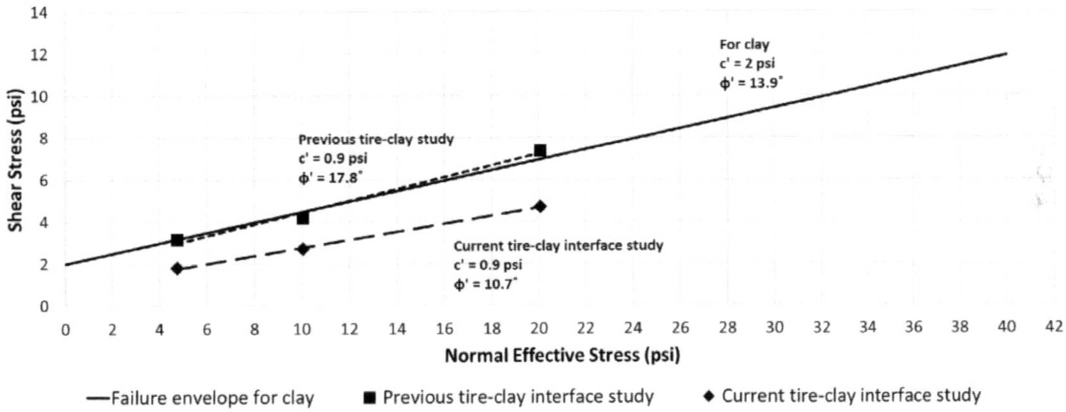

Fig. 3. Comparison of failure envelopes of clay and clay-tire; with interface tested as per previous and current study

2.2.2. Results

Three failure envelopes are presented in Fig. 3 representing three different sets of tests. The solid black envelope corresponds to the failure envelope of clay as presented by Pandit, 2023 and discussed in Section 2.1 by performing CU triaxial tests on clay samples. The dotted plot shown in Fig. 3 labeled as 'Previous tire-clay interface study' represents a failure envelope for direct shear testing done on clay and tire interface with tire placed in the upper shear box performed by Pandit, 2023. The results were considered unsuccessful as the friction angle of the tire-clay interface was found to be very similar to the friction angle of clay. A probable explanation of such behavior was attributed to the fact that the shearing of the clay itself happened instead of the tire-clay interface because of a lack of precise alignment of the shearing plane with the clay-tire interface plane (Pandit 2023). Hence, a change in methodology has been proposed in this paper with the tire piece at the bottom of the shear box, as discussed in Section 2.2.1. The latest results are presented with a dashed line labeled as 'Current tire-clay interface study'.

With the new large-scale direct shear tests, the interface failure envelope lies below the failure envelope of clay, which indicates that the shear strength of clay-tire interface is less than just clay. This behavior agrees well with the author's expectations of the clay-tire interface having lower strength than the clay-clay interface, which is consistent with, e.g., the study performed by Papamichael and Vrettos, 2021.

It can be concluded that the friction angle for the clay-tire interface surface is $\phi' = 10.7^\circ$ as compared to $\phi' = 13.9^\circ$ for the clay material. This means that a clay-tire interface will tend to fail earlier than the clay-clay interface when sheared simultaneously under similar conditions.

3. Friction law implementation into the FE model

In numerical modeling, the interactions between two disjoint parts are modeled using contacts (LSTC, 2017a.). The simplest contact definition between two bodies is modeled using the static coefficient of friction. From Section 2.2, the static coefficient of friction between the tire rubber and clay, found to be 0.189 (i.e., $\tan(10.7^\circ)$), is employed in the contact model (i.e., surface-to-surface contact). Note that the testing in Section 2.2.2 also indicates a cohesion between the tire and soil, which is not considered in this preliminary study. The tire and clay are modeled using the finite element (FE) method. The clay bed is 0.3 m deep, 1 m wide, and 2.5 m long (Fig. 4). The simulation is initialized with gravity, and the tire is vertically loaded to a force of 4000 N. It is then allowed to rotate and attain a steady state at a slip ratio of 10% and a forward velocity of 0.5 km/hr.

To assess the importance of the tire-soil interface friction in the model, sensitivity studies are conducted using friction coefficients of 0.1, 0.4, and 0.6 along with the measured value of 0.189. In particular, the drawbar pull performance and the instantaneous sinkage are investigated. Different case studies are proposed with the same simple and representative FE tire model to understand the scenarios where the interfacial friction effects are dominant: (a) the representative smooth tire on rigid ground, (b) the representative smooth tire on deformable clay, (c) the representative lugged tire on rigid ground, (d) the representative lugged tire on deformable clay. More details on each of these models are discussed in the following sections.

3.1. Representative FE tire model

The simple and representative FE tire is modeled using the dimensions of the BKT TR 319 29X12.50-15 tire (Swamy et. al, 2023), with and without the lugs. The inner lining of the tire is rigid, representing the configuration of an inflated tire. The rim is modeled as a rigid body. The various rubber parts on top of the lining, including the tire body, sidewall, and lugs, are simulated with the Mooney-Rivlin rubber material, as shown in Fig. 5. The parameters for this model are obtained from previous studies on similar agricultural tires (El-Sayegh, 2020). As recommended by Hall et al., 2004, the constant stress element is used to avoid numerical instabilities and negative volume issues. The hourglass energies are regulated using the Flanagan-Belytschko stiffness form with exact volume integration (LSTC, R., 2017a). In the smooth tire case, the lugs on the tire are removed to expose the tire body to the terrain.

3.2. Clay model

The material constitutive model for the clay is described using the "two invariant geotechnical cap material model" on LS DYNA (LSTC, R., 2017b). It is a stress–history-dependent material model with hardening effects. The cap model has the following three

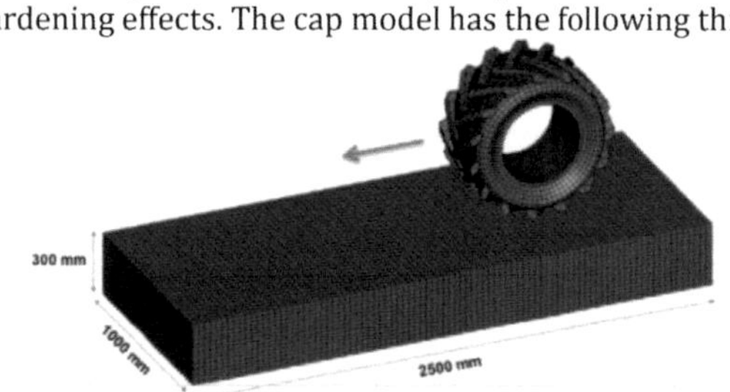

Fig. 4. Numerical simulation setup

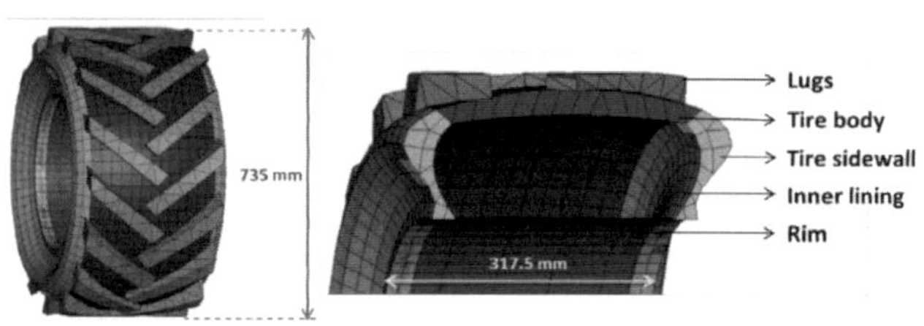

Fig. 5. FE tire mesh of simple representative tire

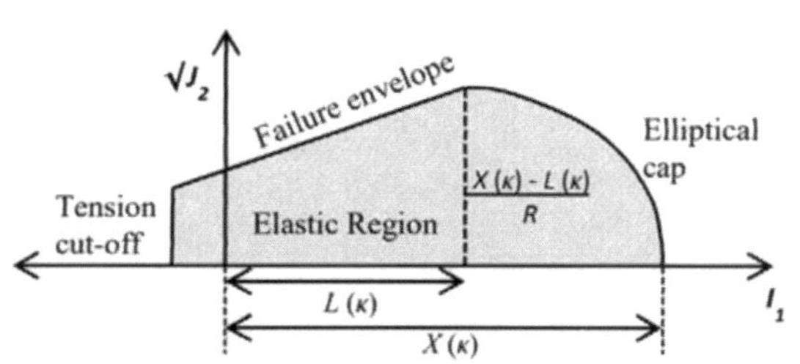

Fig. 6. Geological cap model from LS-DYNA (recreated from LSTC, R., 2017b)

failure surfaces and is represented in Fig. 6 in the I1 vs. sqrt(J2) space: (a) a non-linear failure envelope, (b) a cap surface where volumetric and shear deformations are coupled, and (c) a tension cut-off surface (Fig. 6). The intersection of the cap with the I1 axis can be associated to (three times) the highest pressure ever experienced by the soil, i.e., preconsolidation pressure. The plastic flow is associative and according to Koiter's rule. After failure, the stress path is made to reach the intersection of the failure envelope and cap to ensure no unrealistic plastic volumetric deformations at high strains. The major limitation of the model is its inability to model the softening effect, often seen at low confinement pressures.

In terms of the solid element formulation, the soil was initially modeled using full integration schemes. However, since the material is highly deformable and soft, this formulation led to negative volume issues. More robust single integration point formulations were implemented with hourglassing control (LSTC, 2017a). A mesh size of 20 mm was chosen since smaller meshes lead to instabilities.

The effect of the pore water in the simulation is considered by implementing the drainage condition; drained during gravity loading and tire placement and undrained when the tire rotates. The undrained simulation in saturated soils ensures no volumetric deformations and no dissipation of excess pore water pressure. The geological cap material model is incorporated with effective strength parameters obtained from consolidated undrained (CU) triaxial test at different confinement pressures. In general, the effective strength parameters describe the behavior of the soil skeleton individually without water. The material parameters used for this study have been tabulated in a previous work presented by the authors (Swamy et. al, 2023).

4. Numerical modeling results

The simulation results for the tire rolling on the clay with interface parameters derived from the large-scale direct shear testing are presented in Fig. 7. The total pressure experienced by the soil skeleton and the pore water along the soil's mid-cross-section is presented in Fig. 7(a). Visible soil pressure bulbs are formed due to the motion of the tire over the clay, compressing the soil up to a maximum pressure of 70 kPa. Much of the pressure from the tire is borne by the incompressible pore water (rather than the soil skeleton), resulting in a pore water pressure increase. The rest is supported by the soil skeleton. The shear stresses imparted by the tire, however, are experienced only by the soil skeleton and visualized in Fig. 7(b).

After the tire passes and the saturated soil is unloaded, it is reported that the pore water pressure does not go back to zero. This is because of the highly non-linear and plastic constitutive behavior of the clay (e.g., upon shearing, clay might tend to contract or dilate). The excess pore pressure (i.e., the difference between the total pore water pressure and the hydrostatic pore pressure in the soil) remaining in the soil is displayed in Fig. 7(c). In reality, this pore pressure is expected to dissipate with time. The dissipation rate will depend on the permeability of the soil. The dissipation mechanism is out of the scope of this paper but can be a key factor to better understand multi-pass trafficability on these complex terrains.

Finally, the rut profile and the vertical deformation of the soil are presented in Fig. 7(d). The rut shows how the soil below the tire sinks in by around 5 mm, displacing the soil in this region to the sides of the tire. The resulting soil heaps are elevated to 20 mm above the original clay surface.

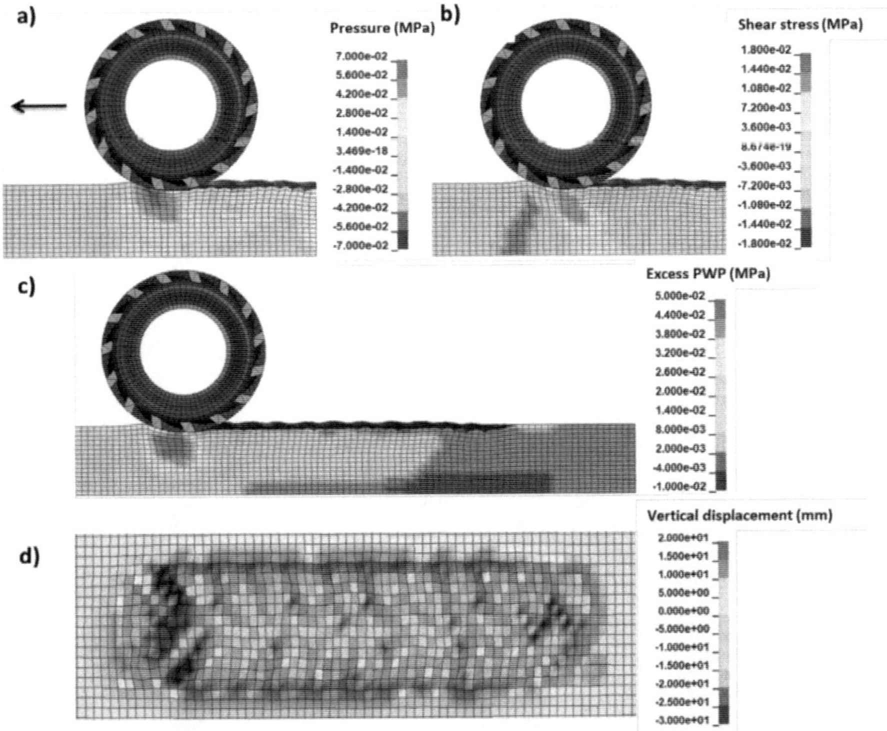

Fig. 7. Simulation results of tire rolling on saturated clay (5000 N normal load, 10% slip ratio, and friction coefficient of 0.189): (a) total pressure, (b) shear stress, (c) excess pore water pressure, (d) vertical displacement

5. Friction sensitivity studies

The simulation described in the previous section with the static friction coefficient set to 0.189 is considered the base simulation for this study. The sensitivity of the friction coefficient in different tire configurations (i.e., lugged and smooth) with different friction coefficients reasonable for tore-soil interactions (i.e., range between 0.1 and 0.6) is investigated in the following sections. The tire is run on the clay layer and on a rigid surface.

5.1. Representative lugged tire

In the representative lugged tire, the purpose of the lugs is to dig into the clay to offer considerably more traction than a smooth tire. Processing the numerical results, it was noted that one of the prominent effects of the friction coefficient is on the longitudinal backward displacement of the clay due to tire rolling. In Fig. 8, the tire moves from the right side to the left, displacing the soil in the positive x-direction according to the global coordinate system. A higher friction coefficient causes increased soil flow to the back and consequently, more shear stresses in the soil. Since soils, in general, fail in shear, the importance of selecting a realistic friction parameter becomes apparent. Investigating the shear stress profiles in Fig. 8, it is seen that with the increase in frictional coefficient, the shear stresses induced in the soil increase until a friction coefficient of 0.4. For the cases where the friction coefficient is 0.4 and higher (i.e., 0.6) the shear stress profile remains similar. This is because the interface friction coefficient is comparatively larger than the soil shear strength, which favors the failure of the soil rather than the mobilization of the interface friction and a relative displacement between the tire and the soil at the contact surface. In such a case, when subjected to the tire load, the stress path of the soil mass in direct contact with the tire-soil interface reaches the failure cap, causing failure of the soil (i.e., yielding of the soil), further causing plastic deformation. Hence, an increase in the interface friction coefficient beyond a threshold (i.e., soil shear strength) results in a similar failure pattern.

The longitudinal traction force (or drawbar pull when off-roading in clay) also experiences the influence of the friction coefficient, both on the clay and on a rigid surface. From the compiled results presented in Fig. 9, the longitudinal traction force on the clay is lesser than on the rigid surface. On a rigid surface, the tire's traction increases till the performance peaks at a friction coefficient of 0.4, as seen in Fig. 9. When operating on clay, the drawbar pull coefficient increases till it reaches a value of 0.13 at a friction coefficient of 0.4, and does not significantly increase after that. This is a direct result of the shear stress profiles. Since the shear stress profiles do not change considerably from friction coefficient 0.4 to 0.6, the drawbar pull, which can be seen as the sum of all the shear stresses integrated over the area of contact, also does not increase much. Hence, after a certain threshold friction coefficient, the increase in it will not affect the drawbar pull performance of the tire.

5.2. Representative smooth tire

The interactions discussed in this section are based on a lug-less tire, identical to Fig. 5 without the lugs. The tire used for testing and in practice is lugged; however, removing the lugs

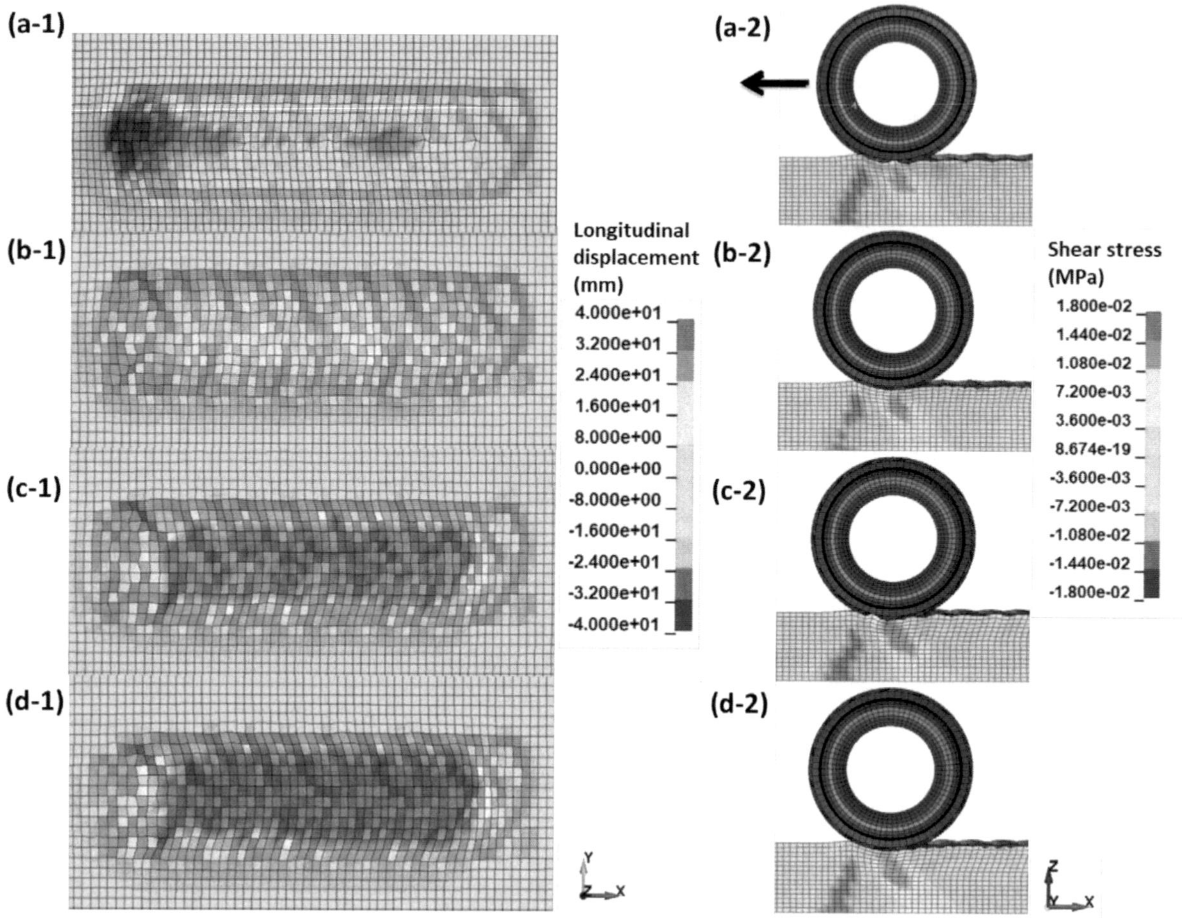

Fig. 8. Longitudinal plastic deformations (1) and shear stresses (2) in the soil after lugged tire pass with different friction coefficients (a) 0.1, (b) 0.189, (c) 0.4, (d) 0.6.

for simulation purposes, a comparative analysis can be performed, revealing the effects of parameters such as friction coefficient and tire geometry type on the results. On the rigid surface, the normalized traction force on the tire increases substantially with increasing friction (Fig. 9). In contrast, on the clay, the drawbar pull coefficient initially increases, before settling at a friction coefficient beyond 0.4. At lower friction coefficients, the tire is immobilized (negative drawbar pull), indicating how important the lugs are for mobility on this soft clay. If the lugs were not present on the BKT tire (or it got fully worn out), the tire would have been immobilized.

Additionally, similar to the lugged case, the longitudinal displacement and the shear stress plots for the smooth tire on clay are shared in Fig. 10. In the cases where the tire becomes immobilized, the soil is displaced in the direction of the tire, building up in front of the tire. The latest causes considerable bull-dozing forces that the tire is not able to overcome, resulting in negative drawbar pull (Fig. 9). In these cases, the shear stress profile shows smaller regions that aids the traction (i.e., orange and yellow regions are small). At the higher friction coefficients of 0.4 and 0.6 (this value will be problem-dependent), a considerable increase in the tractive area in contact with the tire is seen. This implies that if the soil offered more friction to the tire, the tire would have gained mobility. Although the tested clay fails to provide this support, other soils (i.e., stiffer sands) may be able to provide it, making the scenario soil-dependent. More importantly, the soil's backward displacement is crucial to mobility, aiding better tire performance.

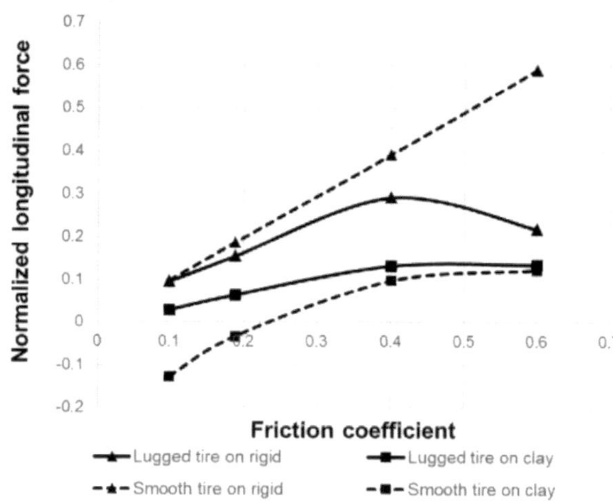

Fig. 9. The variation of normalized longitudinal traction force (or drawbar pull coefficient in the case of clay) with friction coefficient for smooth and lugged tire.

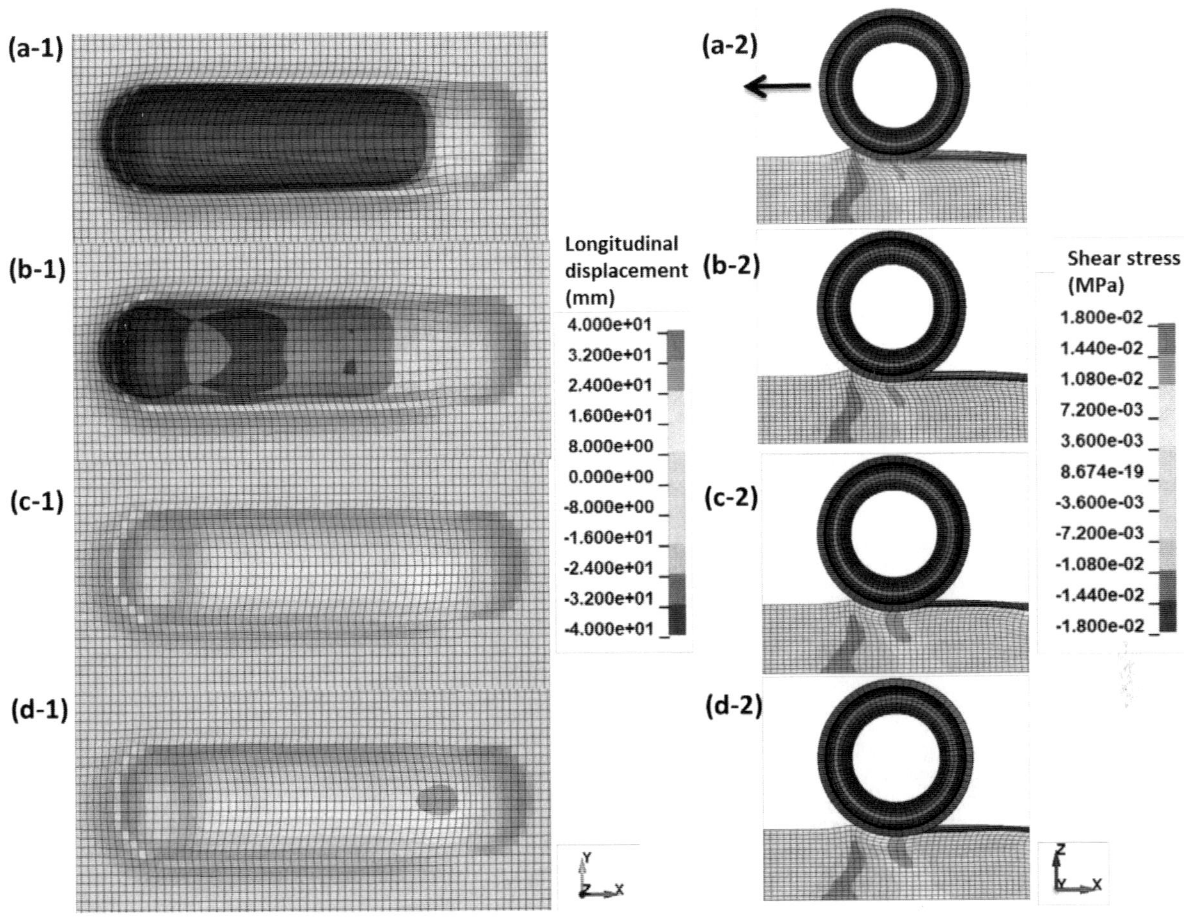

Fig. 10. Longitudinal plastic deformations (1) and shear stresses (2) in the soil after smooth tire pass with different friction coefficients: (a) 0.1, (b) 0.189, (c) 0.4, (d) 0.6

6. Conclusions

This paper presents a detailed approach to experimentally deriving the tire-clay friction properties and its subsequent incorporation into a numerical model. A methodology is proposed for conducting a large-scale direct shear test between the soft clay and tire surface. On analyzing the results and comparing the failure envelopes of the clay and the clay-tire interface, it can be concluded that the shear strength properties of the tire-clay interface are lesser as compared to the clay itself. The clay-tire interface is weaker and will tend to fail before the clay (or clay-clay interface). It is noted that the reported results tend to be conservative since using the tire cut sample directly instead of a tire rubber block sample. In the future, to increase confidence in the reported values, conducting small-scale direct shear tests with tire rubber block samples is proposed. Moreover, conducting more tests to fully represent the failure envelop is to be considered.

The result obtained from the large-scale direct shear tests is implemented into the contact model as a static friction coefficient. For this preliminary study, only the static friction coefficient is considered, however, a future scope for this work would be to also include the adhesion measured in the testing. A simulation of an FE representative tire, operating on the clay in towed mode is conducted. The results like the soil flow, stresses, and excess power water pressure are presented. The results are tabulated as a base simulation on which more sensitivity studies are carried out. Two tire geometries are considered; a lugged tire and a smooth tire.

In the lugged tire model, the results show that the friction coefficient has a prominent impact on the shear stresses induced in the soil and hence the drawbar pull. Once the soil reaches the failure envelope at a particular frictional coefficient, any increase in the friction coefficient does not considerably increase the yield stress in the cap material model. Since the shear stress profile saturates with an increase in friction coefficient, the drawbar pull, which can be represented as the integration of the shear forces over the tire contact area, also saturates. Hence, after a particular threshold interface friction value, any increase in friction minimally influences the drawbar pull, which is an important tire performance parameter. The soil flow and displacements around the tire, however, are still affected with the increase in friction. In general, the soil flows backwards due to the action of the lug giving the traction to navigate. In future works, the modeling of the soil movement using meshless techniques is proposed to allow more deformation to study post-failure rut formation more accurately. This would be very necessary in the low friction cases where hourglassing can be seen.

In the smooth tire model, the tire is immobilized at lower friction angles as a result of the soil displaced in front of the tire, causing an increase in bull-dozing forces. The effectiveness of the lug design is seen in this case when operating on soft clay. At higher friction coefficients, the tire does gain back its mobility; however, this is not realistic taking into account the interface friction measured in the laboratory for the tested soil. As a conclusion, this paper highlights the effect of friction coefficient in tire-soil modeling and why the importance of considering realistic parameters can lead to realistic scene recreations.

Declaration of competing interest

The authors declare that they have no known competing financial interests or personal relationships that could have appeared to influence the work reported in this paper.

Acknowledgements

This study has been supported by the Automotive Research Center (ARC) under the Cooperative Agreement W56HZV-19-2-0001 U.S. Army DEVCOM Ground Vehicle Systems Center, as well as by the TMVS Lab and Geotechnical Engineering Lab at Virginia Tech.

References

Allen J. M. and Kuhn J. A., 2019. Effective stress interface shear strength testing: Drained testing protocols and other associated concerns. In Geosynthetics Conference, volume 2, page 1004.

ASTM D3080-23. Standard test method for direct shear test of soils under consolidated drained conditions. ASTM International, 2023.

ASTM D4767-11. Standard test method for consolidated undrained triaxial compression test for cohesive soils. ASTM International, 2011.

ASTM D5321. Standard test method for determining the shear strength of soil-geosynthetic and geosynthetic-geosynthetic interfaces by direct shear, 2014.

El-Sayegh, Z., 2020. Modeling and Analysis of Truck Tire - Terrain Interaction. Dissertation, Ontario Tech University.

Hall, W., Mottram, J. T., & Jones, R. P., 2004. Tire Modeling Methodology with the Explicit Finite Element Code LS-DYNA. Tire Science and Technology. 32-4, 236-261. https://doi.org/10.2346/1.2186783

LSTC, 2017a. LS-DYNA Keyword User's Manual Volume I—Keyword User's Manual

LSTC, 2017b. LS-DYNA Keyword User's Manual Volume II—Material Models.

Pandit R., 2023 Characterization of soft clay and clay-tire interaction for the prediction of ground mobility. Thesis, Virginia Tech.

Papamichael S. and Vrettos C. Indentation tests and rolling simulations of a compliant wheel on soil at different consistencies. Journal of Terramechanics, 94:39–48, 2021.

Swamy, V.S., Pandit, R., Yerro-Colom, A., Sandu, C., Rizzo, D., Sebeck, K., Gorsich, D., 2023a. Review of modeling and validation techniques for tire-deformable soil interactions. Journal of Terramechanics, 109:73-92, 2023

Swamy, V.S., Pandit, R., Yerro-Colom, A., Sandu, C., Rizzo, D., Sebeck, K., Gorsich, D., 2023b. Numerical modeling of a tire on undrained saturated clay using FEM, ALE, and SPH. In: Proc. of the 16th European-African International

4142 | Root system analysis after vibratory roller compaction in dry direct seeding of rice field

Koichiro Fukami *, Naoki Matsuo, Keiko Nakano, Kimiyasu Takahashi

National Agriculture and Food Research Organization (NARO), Kyushu Okinawa Agricultural Research Center (KARC), 496 Izumi, Chikugo, Fukuoka, Japan

* *Corresponding author: kofukami@affrc.go.jp*

ABSTRACT

To improve the efficiency and optimization of compaction work using a vibratory roller to prevent water leakage in dry rice direct seeding fields, we investigated how the number of compaction passes affects soil physical properties (hardness and permeability), crop growth, and root system development. Cultivation tests showed that compaction operations effectively prevent leakage and increase the number of seedlings. We analyzed its effect on 3D X-ray CT image-confirmed vertical root elongation under all conditions. In addition, we qualitatively confirmed that the roots tended to become thinner when the number of compactions was 5 compared to other conditions. Additionally, WinRHIZO analysis revealed the following: the root dry weight per number of stems decreased with 1- and 3-pass compaction and increased with 5-pass compaction, compared to uncompacted soil. The total root length per the stems decreased after 1- and 3-pass compaction and increased after 5 passes of compaction. The composition of roots with a diameter of 0.38 mm, or more, increased after 1 and 3 compactions and decreased after 5 compactions. In addition, the composition of roots with a diameter of 0.28 mm or less increased after 5 passes of compaction. The above suggested that excessive compaction may have a negative effect on the root system of paddy rice (roots become thinner).

Keywords
Dry direct seeding of rice
Vibratory roller
Soil compaction
Prevent water leakage
Root system analysis

1. Introduction

Rice (*Oryza sativa* L.) is the most important cultivated food crop in Japan. Ensuring low rice production costs and labor-saving methods is of primary importance. Generally, expanding the management scale leads to a reduction in production costs because the efficiency of machinery used for large-scale production increases. However, in the transplanting system, which accounts for 97.3% of Japan's rice production, raising seedlings becomes a disincentive for scale expansion (MAFF, 2024). In recent years, the use of direct seeding systems in Japan has increased.

Direct seeding in Japan can be roughly classified as the direct seeding of rice on well-drained paddy fields (dry seeding) versus the direct seeding of rice on submerged soil (wet seeding) (Chosa et al., 2014). For example, a seed-shooting seeder of rice (Togashi et al., 2001a,b) combined with a tractor-mounted paddy harrow is widely used in submerged soil. In Western Japan, the requirements for the use of this method include the puddling of the field and seed coating with iron powder (Yamauchi, 2012) or calcium dioxide (CaO2) before seeding, and controlling apple snails (Wada, 2004). In contrast, dry seeding is of lower cost and enables the saving of labor compared to wet seeding because dry seeding does not have the requirements mentioned above (Farooq et al., 2011; Tasaka et al., 2013). However, with the upcoming upland cropping systems in southwestern Japan, in which rice is rotated with wheat, soybean, and barley within a two-year period, a plow pan rich in macropores has developed. Thus, the prevention of water leakage from the cultivated field is a principal challenge during dry-seeding cultivation.

Compaction using a tractor-mounted roller is crucial in preventing water leakage from a dry-seeded rice field (Kanmuri et al.,2012; Fukami et al., 2014). However, to achieve sufficient effects from the static compaction produced by the weight of the roller, both the roller and the tractor need to be upsized. Alternatively, using a vibratory roller could be expected to have a higher compaction effect through dynamic compression, even when the roller load is small. A cost reduction could also be expected due to the general-purpose use of the vibratory roller machine because the work can be done by a general-sized tractor (power 22–37 kW, 30–50 PS).

Water permeability and water retention of a field depend on the soil's structure, and changes in soil structure have a close relation to the growth of crop roots. Thus, analyses of soil structure based on X-ray computed tomography (CT) images (Munkholm et al., 2003; Peth et al., 2010; Garbout et al., 2013; Ferro et al., 2014; Tracy et al., 2015) have increased in recent years. From this perspective, Fukami et al. (2017) analyzed the structure of soil compaction by roller to prevent leakage from dry fields. However, few studies have analyzed the effects of structural changes in soil compacted by a roller to prevent leakage on root growth.

In this study, to improve the efficiency and optimization of compaction work using a vibratory roller to prevent water leakage in dry rice direct seeding fields, we investigated how the number of compaction passes affects soil physical properties (hardness and permeability), crop growth, and root system development.

2. Materials and methods

2.1. The vibratory roller

Fig. 1 and Table 1 provide a photograph and the specifications of the vibratory roller (model SV3-T, Kawabe Noken Sangyo Co., Tokyo). This machine is mainly composed of a vibration generator and a roller part. The roller's total weight (without vibrating) is 350 kg, and the roller's width and diameter are 150 cm and 40 cm, respectively. The roller compaction weight (vibrating) is 750–3125 kg (power takeoff [PTO]: 800–1100 rpm). The appropriate power of the tractor when using the vibratory roller is 22–37 kW. This machine can also be used as a vibrating wide subsoiler by changing the roller part to the curved shank (Tanaka et al., 2000).

2.2. Test conditions

We conducted field experiments in 2015 at the Kyushu Okinawa Agricultural Research Center (KARC), Chikugo, Fukuoka, Japan (33°12′N, 130°30′E, 10 m elevation). The basic soil

https://doi.org/10.56884/429UOAT3
ISBN 978-1-942112-55-6

characteristics of the 0–200 mm layer is given in Table 2. The soil texture was light clay (classification: International system). The plastic limit was 33% (gravimetric water content). Before the compaction test (May 20), we sowed (May 20) paddy rice using a slide-roll sowing machine after rotary tiller (May 18 before rainfall) in a paddy field where barley had been harvested (May 17). The conditions during the compaction test with the roller are given in Table 3. The test tractor power was 31 kW (42 PS). We set the rotational speed of the PTO at 1000 rpm, the operating speed of the roller at 2 km/h, and the pass number of the roller as 0, 1, 3, 5 (4 levels). A test section measuring 1.5 × 20.0 m with three replicates was used (Fig. 1).

Fig. 1. Photograph of the rotary tiller (a), the slide-roll sowing machine (b), the vibratory roller (c), and the test section

Table 1
Specifications of the vibratory roller

Manufacturer	KAWABE Noken Sangyo Co., Ltd.
Model	SV3-T
Roller weight/width	350 kg/150 cm
Compaction weight	750–3125 kg (PTO:800–1100 rpm)
Appropriate power of tractor	22–37 kw (30–50 PS)

Table 2
Soil characteristics of the test field

Soil texture	Light clay
Clay (<2 mm), %	37.2
Silt (2–20 mm), %	36.6
Fine sand (20–200 mm), %	16.0
Coarse sand (200–2000 mm), %	10.2
Plastic limit (% d. b.)[1)]	33

[1)] Gravimetric water content

Table 3
The conditions during the compaction test with the roller

Tractor	31 kW (42 PS)
Vibratory roller	SV3-T
PTO (rpm)	1000
Operating speed (km/h)	2
Pass number	0, 1, 3, 5 (4 levels)
Water content (% d.b.)[1)]	41[2)]

PTO: power take-off; d.b.: dry basis
[1)] Mean value of three depths (0–5, 5–10, and 10–15 cm, May 20).
[2)] 57 mm of rainfall on May 18 (Japan Meteorological Agency data).

We measured the soil gravimetric water content (% dry basis) by core sampling (core size: dia. 50 mm, height 51 mm; depth: 0–5, 5–10, and 10–15 cm). The water content before compaction was above the plastic limit. After the compaction test, we measured the penetration resistance and the saturated hydraulic conductivity. We measured the penetration resistance using a digital cone penetrometer (CP40II, Rimik, Queensland, Australia; cone top angle 30; the cross-sectional area of the cone, 2 cm^2; data acquisition intervals, 1.5 cm, three places per plot). We used a falling-head permeameter (DIK- 4050, Daiki Rika Kogyo Co., Saitama, Japan) to measure the saturated hydraulic conductivity after the core sampling. These surveys were conducted on October 13, after the rice yield survey.

2.3. Rice seedlings and yield after vibratory roller compaction

In this study, the number of seedlings and yield of each plot were investigated to analyze the effect of the leakage-preventing vibratory rollers on crop growth. The paddy rice variety used was "Nikomaru". Seeds were sown at 3.1 kg/10a, 30 cm between rows, at a sowing depth of 3 cm, and then compacted with a vibratory roller (May 20); 0 pass plots were enclosed with the corrugated plates to reduce significant leakage during the growing season (Fig. 2). Rice seedlings were surveyed on June 26 and yields on October 9. All statistical analyses (ANOVA) were performed using R version 4.4.0.

Fig. 2. Ground surface and seedlings after the vibratory roller compaction

2.4. Root system analysis of the rice stubble

Fig. 3 shows the sampling method for the root system analysis. Each sample for analysis was collected with a hand sampler (HS-25L, FUJIWARA SCIENTIFIC Co., Tokyo) and placed in a transparent cylinder made of polypropylene (PP; 95 mm dia. × 250 mm long).

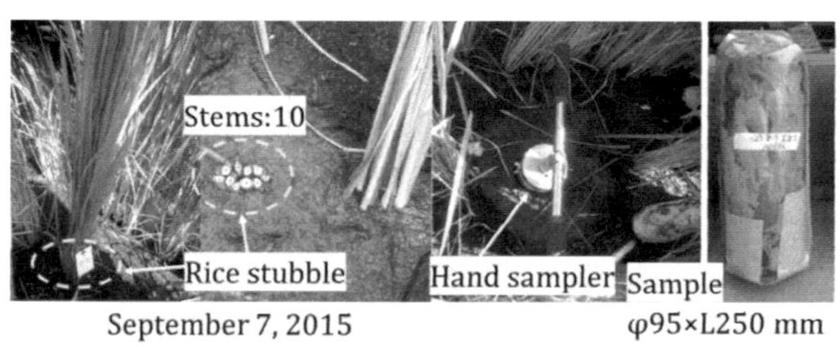

Fig. 3. Sampling method for the root system analysis

Root system sampling was conducted on September 7. We photographed the entire sample with an industrial X-ray CT scanner (TOSCANER-20000RE, Toshiba, Tokyo; voxel size 0.195 mm, slice thickness 1 mm) to obtain a three-dimensional image

(September 8). After the acquisition of the 3D images, these samples were refrigerated for quantitative analysis of the root system.

Fig. 4 shows the quantitative analysis method for the root systems. The analysis was conducted on April 21 and 22, 2018, using the following procedure. Samples were boiled in a microwave oven and then washed with water using a mesh to separate the roots from the soil and organic residues (1). The washed roots were placed in acrylic trays filled with water. Digital images (4235 × 5968-pixels, 400 dpi, TIFF) were then obtained with a scanner (Expression 11000XL, Epson America, Inc., CA, America) controlled by the WinRHIZO program (WinRHIZO Pro 2016a, Regent Instruments. Inc., Quebec, Canada). Roots found by WinRHIZO were identified by colored lines in the image, the lines were color-coded by root diameter, and the total root length and the diameter distributions were calculated accordingly. Scanning was performed several times per sample (2).

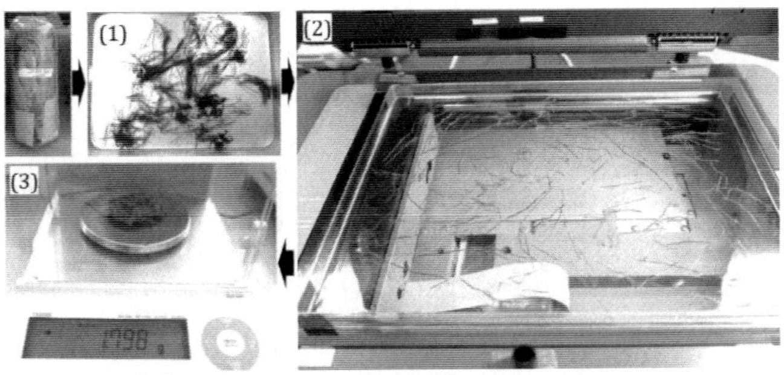

Fig. 4. Methods for quantitative analysis of the root systems

3. Results and discussion

3.1. Penetration resistance of the roller-compacted field

Fig. 5 shows the results of the comparison of the penetration resistance (PR) of the roller-compacted field at the number of passes. The PR at a range of 0–15 cm depth was increased gradually by the number of roller compactions. In addition, at the 5 pass compactions, a PR increase at approx. 22 cm depth was observed. However, no clear relationship was observed at 30–60 cm depths.

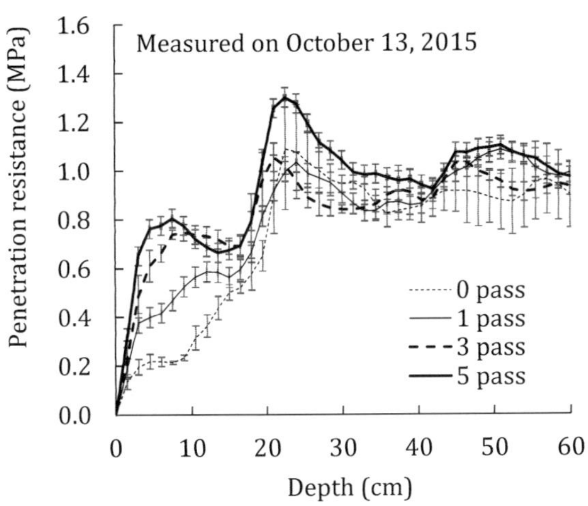

Fig. 5. Penetration resistance of the roller-compacted field

3.2. Relationship between PR and saturated hydraulic conductivity or soil density in the roller-compacted field

Fig. 6 shows the relationship between the PR and the saturated hydraulic conductivity (SHC) for each depth in the roller-compacted field. At depths of 0–5 cm and 5–10 cm, SHC decreased exponentially with increasing PR. At 10–15 cm depth, SHC was relatively small regardless of PR. A high negative correlation (coefficient of determination: 0.75) was found between PR and SHC (0–15 cm depth).

These samples with measured SHC were dried to determine dry density (DD). Fig. 7 shows the relationship between the PR and the DD for each depth in the roller-compacted field. At all depths, DD increased linearly with increasing PR. A low positive correlation (coefficient of determination: 0.39) was found between PR and DD (0–15 cm depth).

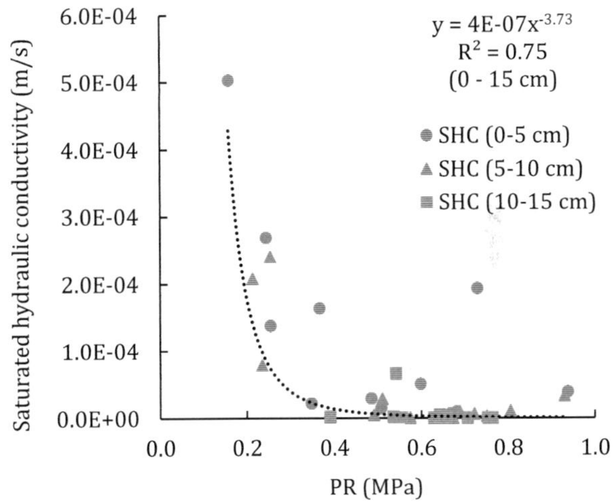

Fig. 6. Relationship between the PR and the SHC (0–15 cm depth)

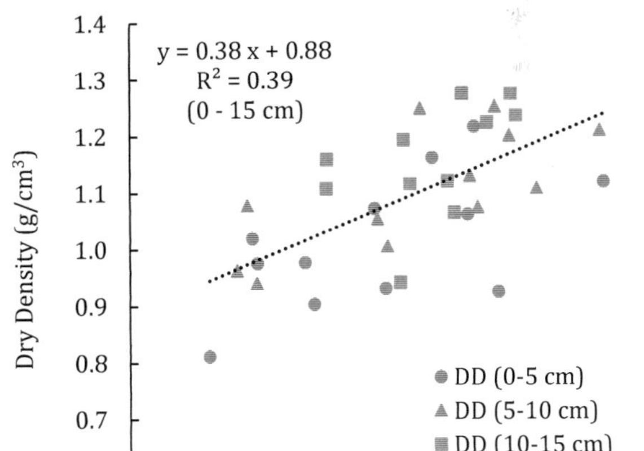

Fig. 7. Relationship between the PR and the DD (0-15 cm depth)

3.3. Effect of roller compaction on root system development

Fig. 8 shows 3D-X ray CT images of the rice stubble root systems in the roller-compacted field. The number of rice stubble stems sampled was 9, 18, 10, and 7 in the 0-, 1-, 3-, and 5-

compacting sections, respectively. The rice plants showed vertical direction root elongation under all conditions. In addition, the roots tended to be thinner in the 5 passes of compaction than in the other conditions.

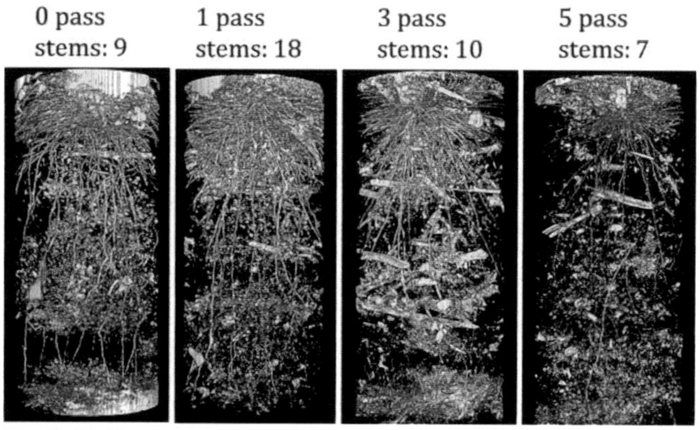

Fig. 8. 3D-X ray CT images of the root systems (Photo courtesy of Prof. Mukunoki (X-Earth Center, Kumamoto Univ.))

3.4. Quantitative analysis of roller compaction on the root system

Fig. 9 shows the root system analyzed image by the WinRHIZO program for each number of roller compactions. Roots detected by the program were color-coded by thickness, and total root length, average root diameter, and total root volume were calculated. In addition, root lengths were calculated for each root diameter as a root length distribution.

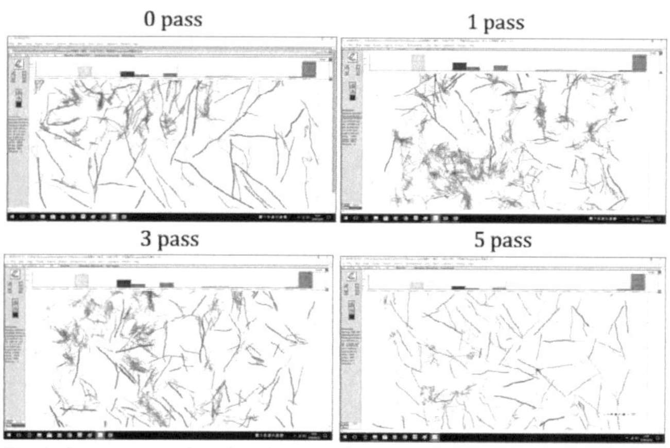

Fig. 9. Comparison of analyzed images by WinRHIZO program

Fig. 10 shows the root dry weight and that per stem relative to the number of compaction. Root weights were greatest in the 1 pass section with the highest number of rice stubble stems. The weight per number of stems decreased with 1 and 3 pass compactions, and increased with 5 pass compactions, compared to uncompacted soil.

Fig. 11 shows the root total length and that per stem relative to a number of compactions. Root total length were greatest in the 5-pass section. The total root length per number of stems decreased with 1 and 3 passes of compaction and increased with 5 passes of compaction, compared to uncompacted soil.

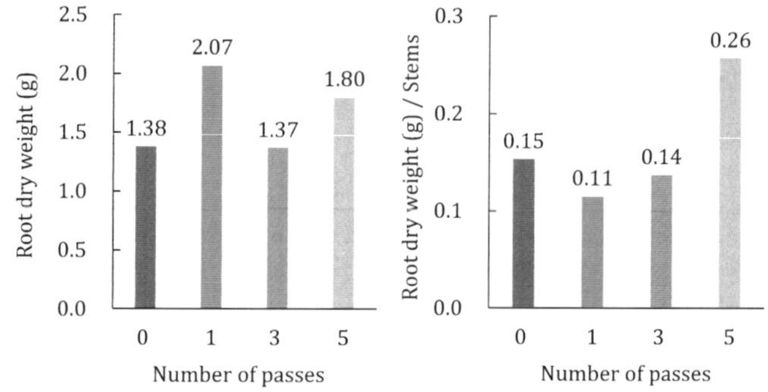

Fig. 10. Root dry weight (left) and that per stems (right) relative to the number of compactions

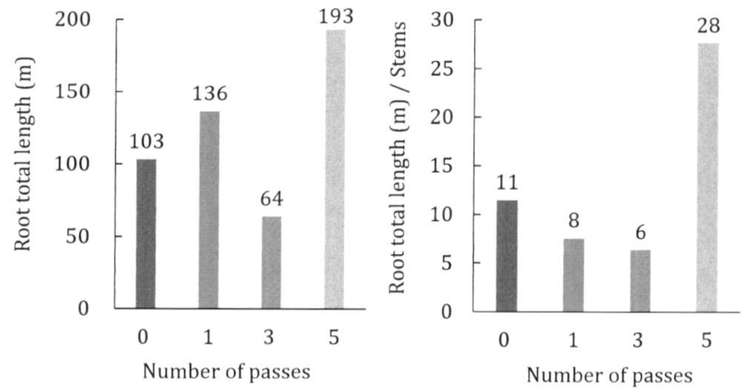

Fig. 11. Root total length (left) and that per stems (right) relative to number of compactions

Fig. 12 shows the distribution of the root length relative to the root diameter. Root lengths greater than 0.38 mm in diameter showed the same trend as the root dry weight relative to number of compactions in Fig. 10. Root lengths below 0.38 mm diameter showed the same trend as the total root length relative to number of compactions in Fig. 11.

Fig. 13 shows the distribution of the composition of root length relative to the root diameter. Root composition at root diameters greater than 0.38 mm increased with 1 and 3 passes of compaction and decreased with 5 passes of compaction, compared to uncompacted soil. However, root composition at root diameters less than 0.28 mm increased with 5 passes of compaction.

3.5. Effect of the compaction strength on soil physical properties and crop growth

Table 4 shows the ANOVA results of the soil's physical properties, rice seedlings, and yields. The roller compaction work resulted in a significant increase in PR (0–15 cm depth) and a significant decrease in SHC (0–15 cm depth). However, there was no significant difference between the number of compactions in SHC. No significant differences in compaction work were found in DD (0–15 cm depth). The number of seedlings was significantly increased by the roller compaction operation, but there was no significant difference in the number of passes. Yield was not significantly different among the compaction operations but tended to increase by approximately 20% with compaction.

These results suggest that the compaction operation was effective in preventing water leakage and increasing the number of seedlings and yields.

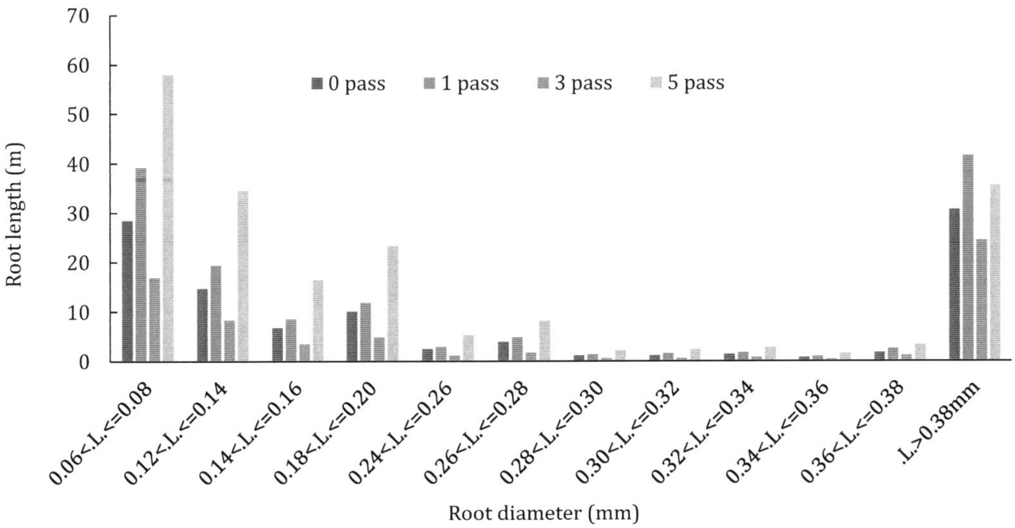

Fig. 12. Distribution of the root length relative to the root diameter

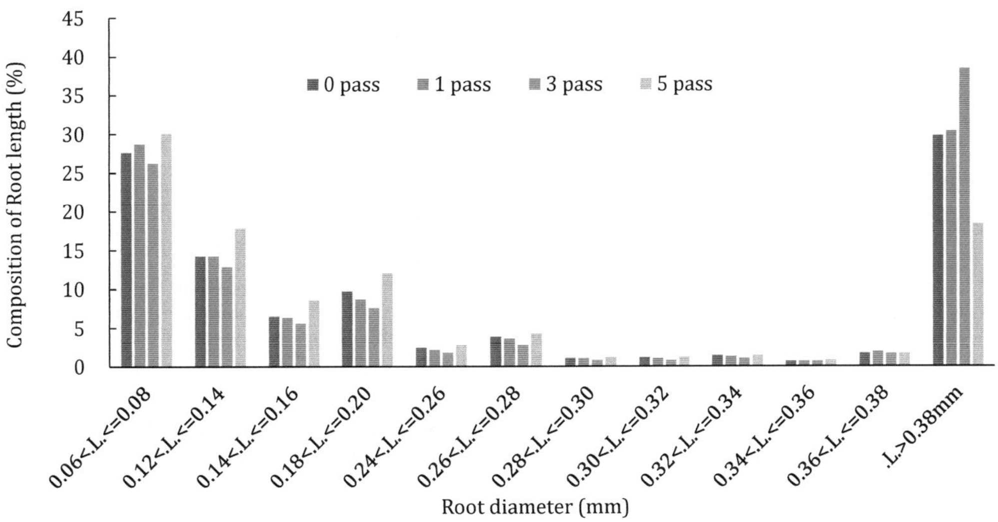

Fig. 13. Distribution of the composition (root length/total root length) relative to the root diameter

Table 4
ANOVA results of the soil physical properties, the rice seedlings and the yields

Pass	PR [MPa][2)]	DD [g/cm³][3)]	SHC [m/s][3)]	Seedlings [/m²]	Yield [kg/10a][4)]	Ratio of yield [%]
0	0.30 a[6)]	1.00	1.6E-04 a[6)]	39 a[6)]	375	100
1	0.50 b	1.09	3.5E-05 b	61 ab	443	118
3	0.69 bc	1.11	3.2E-05 b	74 b	458	122
5	0.74 c	1.18	1.0E-05 b	73 b	437	117
ANOVA[5)]	***	ns	*	*	ns	-

[1)] Section area 30 m^2, 3 replications.
[2)] Penetration resistance (PR): Average of 4.5, 9.0, and 13.5 cm depths.
[3)] Dry density (DD), Saturated hydraulic conductivity (SHC) : Average of 0–5, 5–10, and 10–15 cm depths.
[4)] Survey area 1.2 m^2, sieve 1.8 mm, moisture 15%.
[5)] Analysis of variance; significant at 5 % (*) and 0.1 %(***) risk rates.
[6)] Tukey multiple test, significant differences between different letters at 5% risk.

4. Conclusions and future work

The roller compaction work resulted in a significant increase in PR (0–15 cm depth) and a significant decrease in SHC (0–15 cm depth). However, there was no significant difference between the number of compactions in SHC.

The root weight per number of stems and the total root length per number of stems decreased with 1- and 3-pass compaction and increased with 5-pass compaction, compared to uncompacted soil. However, root composition at root diameters greater than 0.38 mm increased with 1 and 3 passes of compaction and decreased with 5 passes of compaction, compared to uncompacted soil. Root composition at root diameters less than 0.28 mm increased with 5 passes of compaction.

The number of seedlings was significantly increased by the roller compaction operation, but there was no significant difference in the number of passes. Yield was not significantly different among the compaction operations but tended to increase by approximately 20% with compaction.

These results suggest that although 1 pass of compaction was effective in preventing leakage and increasing the number of seedlings standing, excessive compaction (5 pass) may adversely affect the root system of paddy rice plants (thinning of the root diameter).

The acceleration response of a vibratory roller is also useful for the evaluation of the stiffness of compacted soil (Fujiyama and Tateyama, 2000; Fujiyama et al., 2002; Mooney and Rinehart, 2007). Therefore, in the future, based on vibration acceleration response, we will investigate the optimum work speed and PTO rotating speed for compaction work.

5. Acknowledgements

This study was supported by the Grants-in-Aid for Scientific Research (No. 25850175) from the Ministry of Education, Culture, Sports, Science and Technology. We are grateful to Prof. Toshifumi Mukunoki and Mr. Takahiro Sato of the X-Earth Center for X-ray CT analysis. Mr. Yukinari Kawahara, Mr. Teruyuki Miike, Mr. Akitoshi Honbu, Mr. Hiroyuki Itoh, Mr. Sadahiro Higashi, Ms. Tamiko Shimogawa, and Ms. Kaori Sonoda of the KARC for their field management and data collection in this experiment.

6. Declaration of competing interest

The authors declare that they have no known competing financial interests or personal relationships that could have appeared to influence the work reported in this paper.

7. References

Chosa, T., Matsumoto, T., Iwaoka, M., 2014. Machinery and information technology for biomass production. In: Tojo, S., Hirasawa, T. (Eds.), Research Approaches to Sustainable Biomass Systems. Elsevier, pp. 143–180.

Farooq, M., Siddique, M.H.K., Rehaman, H., Aziz, T., Lee, J.D., Wahid, A., 2011. Rice direct seeding: experiences, challenges and opportunities. Soil Tillage Res. 111, 87–98.

Ferro, D.N., Sartori, L., Simonetti, G., Berti, A., Morari, F., 2014. Soil macro- and microstructure as affected by different tillage systems and their effects on maize root growth. Soil Tillage Res. 140, 55–65.

Fukami, K., Nakano, K., Tsuchiya, S., Tasaka, K., Matsuo, N., Miike, T., Ito, H., Kawahara, Y., Honbu, A., 2014. Use of a leakage prevention technique on a direct-seeded dry paddy field in a warm region. J. Jpn. Soc. Agric. Mach. Food Eng. 76 (4), 341–347.

Fujiyama, T., Tateyama, K., 2000. The evaluation of compacting ground stiffness by the acceleration of vibratory roller. J. Geotechn. Eng. 115–123 (No. 652/III-51).

Fujiyama, T., Masumura, K., Tateyama, K., Ishiguro, T., Mishima, N., 2002. Applicability of roller response method to compaction control for various soil conditions. J. Geotechn. Eng. 169–179 (No. 701/III).

Fukami, K., Mukunoki, T., Nakano, K., Matsuo, N., Okayasu, T., 2017. Water leakage control by using vibratory roller on a dry-seeded rice field in southwestern Japan, Soil Tillage Res. 166, 138-146.

Kanmuri, H., Otani, R., Chiba, K., 2012. Reducing dry Andisol paddy field permeability of direct-seeding rice cultivation using plowing and compaction. Jpn. Soc. Irri\g. Drain. Rural Eng. 80 (6), 533–541.

Garbout, A., Munkholm, J.L., Hansen, B.S., 2013. Tillage effects on topsoil structural quality assessed using X-ray CT, soil cores and visual soil evaluation. Soil Tillage Res. 128, 104–109.

MAFF, 2024. Introduction of the latest direct seeding cultivation technology (2022). (Translated from Japanese).
https://www.maff.go.jp/j/syouan/keikaku/soukatu/attach/pdf/chokuha-21.pdf. (Accessed May 31, 2024)

Mooney, M.A., Rinehart, V.R., 2007. Field monitoring of roller vibration during compaction of subgrade soil. J. Geotech. Geoenviron. Eng. ASCE 133 (3), 257–265.

Munkholm, J.L., Schjonning, P., Rasmussen, J.K., Tanderup, K., 2003. Spatial and temporal effects of direct drilling on soil structure in the seedling environment. Soil Tillage Res. 71, 163–173.

Peth, S., Nellesen, J., Ficher, G., Horn, R., 2010. Non-invasive 3D analysis of local soil deformation under mechanical and hydraulic stress by mCT and digital image correlation. Soil Tillage Res. 111, 3–18.

Tanaka, H., Inooku, K., Nagasaki, Y., Miyazaki, M., Sumikawa, O., Oida, A., 2000. Measurement of soil loosening and reaction faces at subsoiling using a vibration type subsoiler. Proc. of the International Agricultural Engineering Conference Bangkok 59–65.

Tasaka, K., Wada, T., Yusa, Y., Yoshida, K., Ando, T., Tsuchiya, S., Fukami, K., Sasaki, Y., 2013. Using a repellent to suppress spawning in waterways by the apple snail pomacea canaliculata (Lamarck). Jpn. J. Farm Work Res. 48 (4), 133–141. Togashi, T., Shimotsubo, K., Yoshinaga, S., 2001a. Form of hill-seeded spot by the seed-shooting seeder of rice combined with a paddy harrow. Jpn. J. Farm Work Res. 36 (4), 187–194.

Togashi, T., Shimotsubo, K., Yoshinaga, S., 2001b. Performance of seed-shooting seeder of rice combined with a paddy harrow and the growth characteristics of rice. Jpn. J. Farm Work Res. 36 (4), 195–203.

Tracy, R.S., Black, R.C., Roberts, A.J., Dodd, C.I., Mooney, J.S., 2015. Using X-ray computed tomography to explore the role of abscisicacid in moderating the impact of soil compaction on root systemarchitecture. Environ. Exp. Bot. 110, 11–18.

Wada, T., 2004. Strategies for controlling the apple snail pomacea canaliculata (Lamarck) (Gastropoda: Ampullariidae) in Japanese direct-sown paddy fields. Jpn. Agric. Res. Q. 38 (2), 75–80.

Yamauchi, M., 2012. Direct seeding of rice crop in flooded paddy fields using ironcoated seeds. Jpn. J. Crop Sci. 81 (2), 148–159.

4213 | DEM-Based Analysis and Optimization of an Excavation Bucket Drum for In-Situ Resource Utilization

Tomoyasu Nakano [a,*], Takuya Omura [a], Genya Ishigami [a]

[a] Department of Mechanical Engineering, Keio University, Japan

* Corresponding author: nakano.tomoyasu@keio.jp

ABSTRACT

Lunar infrastructure construction involves leveling of the lunar surface and collecting lunar regolith as a building material. RASSOR 2.0 developed by NASA is one of the typical robotic vehicles as a lunar excavator. It features cylindrical rotating bucket drums for collecting regolith, positioning it as a pivotal tool for future lunar in-situ resource utilization and infrastructure development. However, the optimization of the bucket drum's shape and its motion remain as open issue. Therefore, this study aims to find an optimal design of the bucket drum through numerical simulation using the discrete element method (DEM). We introduce five key design parameters of the bucket drum: two of them are related to the bucket shape (scoop throat length, scoop inlet number) and the rest of them are to motion (bucket vertical force, horizontal velocity, and angular velocity). These five parameters were examined in accordance with two performance indices: the sand fill ratio in the drum and the power consumption of the excavation. Solving the multi-objective problem of increasing the fill ratio and reducing power consumption, we found an optimal bucket drum shape and motion. Subsequently, a bucket drum reflecting the optimal shape was fabricated, and the optimal motion was tested. The test results qualitatively matched with the results derived from the DEM analysis. This outcome highlights the validity of the relationship between the design parameters and two performance indices.

Keywords
In-situ Resource Utilization
Bucket Drum
DEM Simulation
Excavation Efficiency

1. Introduction

The necessity of lunar infrastructure development is expected to be a critical focus in the coming years. The lunar surface, characterized by its uneven terrain due to frequent meteor impacts, poses significant challenges for base construction at proposed sites. Jehn et al. (2023) lists three primary stages: clearing, grading, and compaction for effective construction site preparation. Clearing involves the removal of rocks and obstacles, grading adjusts the terrain to flatten the site, and compaction increases the soil density to provide a stable foundation for further construction. Especially, the processes of clearing and grading are crucial for removing the irregularities caused by craters, emphasizing the essential role of lunar excavators in these tasks.

To establish a sustainable human activity on the moon in the future, in-situ resource utilization (ISRU) is paramount. ISRU practices involve the direct utilization of local resources, such as extracting essential elements like oxygen, hydrogen, titanium, aluminum, and iron from lunar soil, also known as regolith (Grant et al., 1991). These resources are for various applications including rocket fuel, life support systems, and residential modules. (Mueller, 2023). Crucially, the collection of regolith is a fundamental step towards enabling sustained lunar activities, thereby making lunar excavators indispensable.

Responding to this need, a significant advancement has been made with NASA's Regolith Advanced Surface Systems Operations Robot (RASSOR 2.0), designed specifically for the low-gravity conditions of lunar environments (Mueller et al., 2016). As shown in Fig. 1, this robot incorporates rotating bucket drums equipped with digging scoops, which trap regolith inside the drum. Bucket drums offer a superior balance of efficiency, simplicity, and adaptability compared to traditional excavation methods, making them particularly well-suited to the unique challenges posed by lunar regolith and low-gravity environments. The design engages only 1-2 scoops at a time, minimizing resistive forces and reducing energy consumption (Clark, 2009). In contrast, traditional methods such as shovels and bulldozers, which rely heavily on gravity and pulling forces, face significant challenges in low-gravity conditions, including the risk of equipment overturning due to high reactive forces. Bucket drums circumvent these issues by employing lower excavation forces and maintaining minimal contact with the regolith. Thus, the RASSOR 2.0's capability to excavate and transport regolith is vital for ISRU processes, highlighting its role in infrastructure development and lunar resource extraction strategies. However, the optimization study of the shape and motion of the bucket drums has yet to be accomplished.

Fig. 1. RASSOR 2.0 by NASA

https://doi.org/10.56884/MPNVS78R

In this research, we exploit the discrete element method (DEM) to analyze the behavior of sand particles and optimize the shape and motion. Our study introduces five key design parameters: scoop throat length, scoop inlet number, bucket vertical force, horizontal velocity, and angular velocity. These parameters were evaluated based on two performance indices: the sand fill ratio in the drum and the power consumption of the excavation. By addressing this multi-objective problem, we identified an optimal bucket drum configuration that achieves a favorable balance between the fill ratio and power consumption compared to other configurations. Subsequently, experimental testing was conducted to evaluate the DEM results against the measured fill ratio and energy consumption results.

2. Selection of shape and motion parameters

2.1. Optimization function for bucket drums

To achieve the efficient excavation of lunar regolith, two primary optimization criteria were established: fill ratio, R and energy consumption, E. These criteria are essential for maximizing the amount of sand collected per excavation cycle while minimizing the energy required, thus ensuring the efficiency and sustainability of lunar operations.

The fill ratio is defined as the volume of sand collected inside the bucket drum divided by the total volume of the drum. Maximizing fill ratio is crucial for improving the efficiency of each excavation cycle. The fill ratio R is given as:

$$R = V_{\text{sand}}/V_{\text{drum}} \tag{1}$$

where V_{sand} is the volume of sand collected and V_{drum} is the total volume of the bucket drum.

Energy consumption during excavation is a critical factor due to the limited power supply available for lunar missions. In this study, energy consumption is calculated based on the energy required for three main actions: rotating the bucket drum, pushing it vertically into the sand, and advancing it horizontally. Vertical pushing energy involves the energy required to press the bucket drum into the sand. Rotational energy pertains to the energy needed to turn the bucket drum to scoop the sand. Horizontal advancement energy accounts for the energy consumed as the bucket drum moves forward through the sand.

2.2. Selection of key parameters

The shape and motion parameters of the bucket drum are critical to its performance. Five key parameters are selected for optimization: two shape parameters (scoop throat length, N_r and scoop inlet number, N_θ) and three motion parameters (vertical force, F, horizontal velocity, v, and angular velocity, ω), as listed in Table 1.

The scoop throat length, N_r, which determines the amount of sand captured in each scoop, is defined as:

$$N_r = r_{\text{in}}/r \tag{2}$$

where r_{in} is the scoop height and r is the radius of the bucket drum, as shown in Fig. 2(a).

The scoop inlet number, N_θ, which impacts the frequency of sand scooping, is given as:

$$N_\theta = 2\pi/\theta \tag{3}$$

Table 1
Five key parameters for drum design and motion

Type	Parameter	Symbol	Unit
Shape	Scoop throat length	N_r	[−]
	Scoop inlet number	N_θ	[−]
Motion	Vertical force	F	[N]
	Horizontal velocity	v	[m/s]
	Angular velocity	ω	[rad/s]

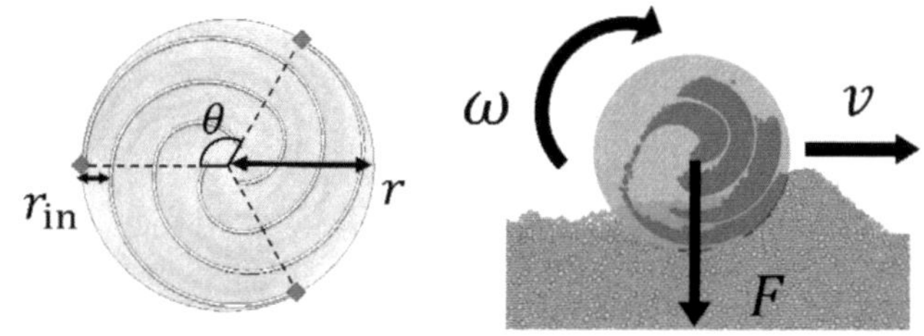

(a) Two design parameters *(b) Three motion parameters*
Fig. 2. Side view of bucket drum with five key parameters

Table 2
Discrete element method parameters for Toyoura sand

Density [g/cm^3]	2.663	
Diameter [μm]	215, 260, 311, 395, 470	
Mass ratio [%]	18.6, 32.5, 23.7, 15.8, 9.3	
Young's modulus [MPa]	20	
Poisson's ratio [−]	0.3	
Rolling resistance [−]	0.3	
	Particle-particle	Particle-bucket drum
Static friction coefficient [−]	0.5	0.3
Dynamic friction coefficient [−]	0.5	0.3
Restitution coefficient [−]	0.5	0.3

where θ is the angle between adjacent scoops. Increasing the number of scoops can enhance the fill ratio but may also increase the energy consumption.

The three motion parameters shown in Fig. 2(b) are essential for the drum's excavation function. The vertical force, F ensures that the drum maintains contact with the sand. The horizontal velocity, v corresponds to the forward movement of the excavator. The angular velocity of the drum, ω is correlated with the horizontal velocity. Optimizing these motion parameters ensures that the bucket drum operates efficiently and effectively under lunar conditions, maintaining a high fill ratio while minimizing energy consumption.

3. DEM simulations for optimization

This section focuses on optimizing the bucket drum's design and motion parameters using the DEM to maximize the fill ratio and minimize energy consumption. The optimization process involves varying two key shape parameters and three key motion parameters of the bucket drum. Initially, we optimized the two shape parameters (N_r and N_θ), followed by the optimization of the three motion parameters (F, v, and ω).

3.1. DEM excavation model setup

Commercial software Ansys Rocky DEM (Ansys Inc., 2023) was used for the analysis. In the DEM simulations, it is crucial to set the parameters for particles accurately. This study utilized the parameters for Toyoura sand provided by S. Ozaki et al. (2023), listed in Table 2. Toyoura sand is frequently employed to simulate and analyze soil behavior in the field of Terramechanics. To reproduce the particle size distribution similar to actual Toyoura sand within DEM, five representative diameters and their corresponding mass ratios were selected based on the particle size selection method by Omura et al. (2023). Details of the selection process are provided in Appendix A. Note that the particle size was scaled up by a factor of 10 in the DEM simulations to reduce computational cost. To facilitate comparison with Earth-based testing, gravity was set to 9.81 m/s^2. The bucket drum had a radius of 100 mm, a width of 50 mm, and a scoop thickness of 2 mm. The drum's mass was calculated using a density of 2000 kg/m^3, based on the carbon fiber material used in the RASSOR 2.0.

In addition to setting particle parameters, selecting suitable contact models is essential for DEM simulations. The contact forces in DEM consist of two main components: forces normal to the contact plane and forces tangential to the contact plane. This study employs the Hertzian spring-dashpot model for the normal forces and the Mindlin-Deresiewicz model for the tangential forces. These models represent the interactions between particles, as well as between particles and equipment, contributing to the simulation results.

3.2. Shape optimization via DEM

During the DEM simulation, the bucket drum undergoes three rotations to fully evaluate the relationship between the fill ratio and energy consumption. Combinations of the scoop throat length, N_r and the scoop inlet number, N_θ that satisfy the six constraints are listed in Table 3.

To ensure fair optimization conditions, constraints were imposed on the spiral shape inside the bucket drum while varying the scoop throat length, N_r and the number of scoops, N_θ. Six constraints were established:

1. The spiral structure must complete one full 360° turn around the axis of the bucket drum.
2. The distance from the inner end of the scoop to the center of the drum, r_{ctr} must be 0.2 times the radius, r.
3. The radius of curvature, r_{cvt} from the scoop tip to the adjacent scoop must be formulated with respect to the scoop throat length, N_r as:

$$r_{\mathrm{cvt}} = r - \frac{r_{\mathrm{in}} + h}{2} = r - \frac{rN_r + h}{2} \tag{4}$$

where h denotes the thickness of the scoop.

4. The spiral structure must pass through points that divide the line connecting the scoop tip and the center of the drum into $N_\theta - 1$ equal segments, ensuring consistent spacing and alignment of the scoops.
5. A tangential constraint must be used to maintain a point of tangency when connecting two curves that form the spiral structure, ensuring that the transition between curves is smooth.
6. The minimum spacing, l between the scoops must satisfy $l/r \geq 0.1$, preventing clogging and maintaining smooth operation.

Initially, nine combinations of N_r $(= 0.1, 0.2, 0.3)$ and N_θ $(= 2, 3, 4)$ were analyzed to gain insights into the relationship between the fill ratio and energy consumption. Figure 3 shows the relationship between energy consumption, E and fill ratio, R for these combinations. It was confirmed that larger values of N_r and

Table 3
Constraint-satisfying design parameters

	$N_\theta = 2$	$N_\theta = 3$	$N_\theta = 4$	$N_\theta = 5$
$N_r = 0.1$	✓	✓	✓	✓
$N_r = 0.2$	✓	✓	✓	✓
$N_r = 0.3$	✓	✓	✓	N/A
$N_r = 0.4$	✓	✓	N/A	N/A
$N_r = 0.5$	✓	N/A	N/A	N/A
$N_r = 0.6$	✓	N/A	N/A	N/A

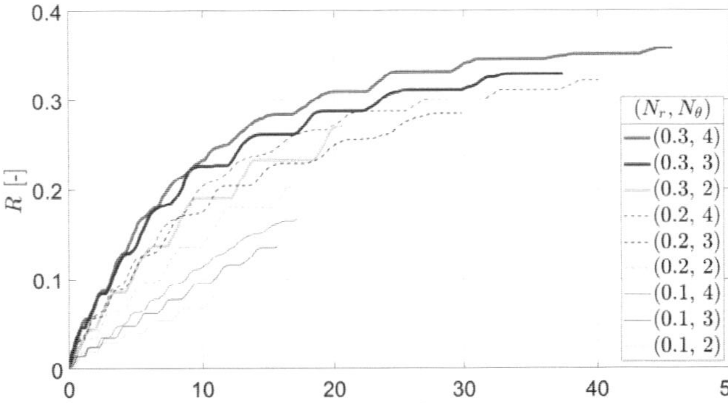

Fig. 3. Energy vs. fill ratio under different drum shapes (9 types)

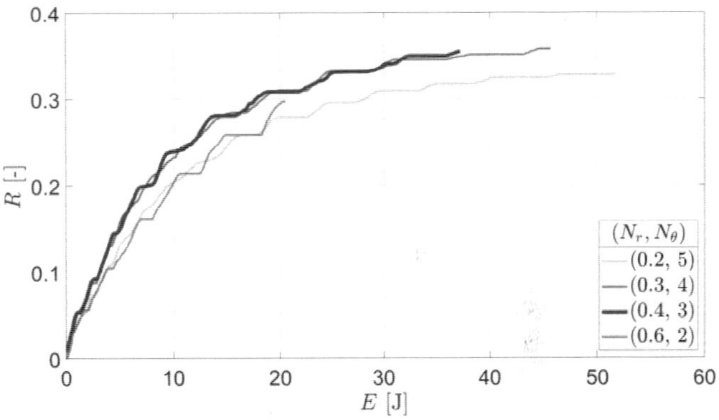

Fig. 4. Energy vs. fill ratio under different drum shapes (4 types)

N_θ result in lower energy consumption and higher fill ratio, with the influence of the scoop throat length being more pronounced.

Based on these results, the shape optimization of the bucket drum was conducted for combinations with the maximum N_r values that satisfy the constraints for each N_θ $(= 2, 3, 4, 5)$. Figure 4 illustrates the corresponding energy consumption and fill ratio. The optimal design was determined to be $(N_r, N_\theta) = (0.4, 3)$ due to its favorable balance of a high fill ratio and low energy consumption, as indicated by the blue line in Fig. 4. The combination $(N_r, N_\theta) = (0.3, 4)$, shown with the red line, has a similar result but was found less optimal due to manufacturing difficulties, higher risk of sand clogging, and increased production costs associated with more blades. Consequently, $(N_r, N_\theta) = (0.4, 3)$ was selected as the optimal set of shape parameters.

3.3. Experimental design for motion parameters

The experimental design method efficiently extracts accurate information from a limited number of experiments. This method was applied to the three motion parameters: vertical force, F, horizontal velocity, v, and angular velocity, ω. The Box-Behnken

Design (BBD) was selected to determine the optimal combination of motion parameters to minimize energy consumption and maximize fill ratio.

BBD, a statistical method introduced by Box and Behnken (1960), used for the optimization of multiple variables with minimal experimentation. This design is particularly advantageous as it requires fewer experimental runs compared to other methods, while still providing comprehensive insights into the interactions between variables (Yıldız, 2022). BBD effectively fits a second-order polynomial model to the data, allowing for accurate predictions and optimizations. Each factor is set to three levels: low (-1), medium (0), and high $(+1)$. The number of experiments, N is given by: $N = 2k(k-1) + c_p$ where k is the number of factors and c_p is the number of center points. Table 4 lists the three levels for each of the three factors. For this study, with four center points, the total number of experiments was set to 16.

3.4. Motion optimization via DEM

To analyze the impact of the motion parameters on the bucket drum's performance, specifically fill ratio and energy consumption, regression analysis was performed using MATLAB's 'fitlm' function, including second-order terms for the three motion parameters. This model accounts for the squared terms of each parameter as well as the interaction terms between parameters. This approach enables the identification of complex interactions and nonlinear effects among the parameters, providing deeper insights into their interdependencies.

Data on the fill ratio and energy consumption were collected at the 16 experimental points to construct each regression model. The regression model for the maximum fill ratio, $R_{\max}$ is given by:

$$\begin{aligned} R_{\max} = {} & 0.10 + 0.031F + 5.9v - 0.10\omega - 0.20Fv \\ & -0.0011F\omega - 2.9v\omega + 0.0013F^2 - 170v^2 \\ & -0.045\omega^2 \end{aligned} \tag{5}$$

As observed from Figs. 3 and 4, the energy consumption, E, versus the fill ratio, R, curve can be approximated by a first-order system response. The approximation equation is defined as follows:

$$R = R_{\max}\left(1 - \exp(-E/\tau)\right) \tag{6}$$

In this context, the coefficient τ is defined as the value of energy consumption at which the fill ratio reaches 63.2 % of its maximum value, $R_{\max}$. Therefore, τ evaluates the change rate of energy consumption until the fill ratio stabilizes. Thus, we conducted an optimization of the three motion parameters aimed at minimizing the time constant while simultaneously maximizing the maximum fill ratio. The regression model for the time constant, τ is given by:

$$\begin{aligned} \tau = {} & 2.7 + 1.0F + 65v - 1.2\omega + 18Fv + 0.76F\omega \\ & -69v\omega + 0.018F^2 - 363v^2 + 0.93\omega^2 \end{aligned} \tag{7}$$

Table 4
Box-Behnken design factor levels for motion parameters

Coding	Factors		
	F [N]	v [m/s]	ω [rad/s]
+1	10	0.04	−0.6
0	5	0.02	−1.2
−1	0	0	−1.4

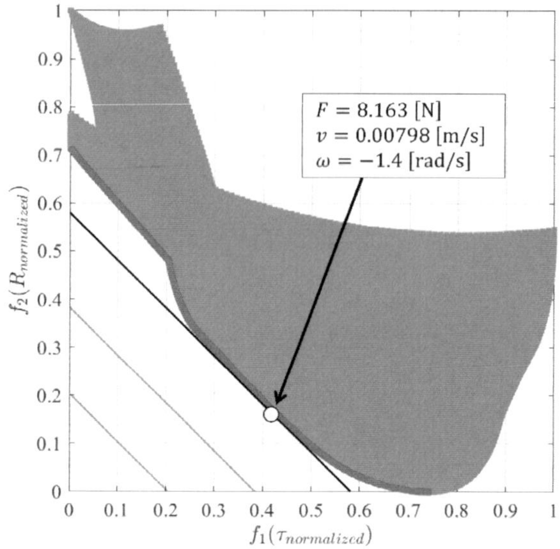

Fig. 5. Feasible region and pareto frontier with optimal solution when $w_1 = w_2 = 0.5$

To minimize the value of τ and maximize the maximum fill ratio, a multi-objective optimization was performed to find the optimal combination of vertical force, F, horizontal velocity, v, and angular velocity, ω. The motion parameters were explored within the range of upper and lower limits set in Table 4. Equation 5 was substituted into Eq. 9 as f_2 and Eq. 7 into Eq. 10 as f_1 to normalize the time constant and maximum fill ratio as follows:

$$\text{minimize} \quad w_1 f_1 + w_2 f_2 \tag{8}$$

$$\text{s.t.} \quad f_1 = \frac{\tau - \tau_{\min}}{\tau_{\max} - \tau_{\min}} \tag{9}$$

$$f_2 = 1 - \frac{R_{\max} - (R_{\max})_{\min}}{(R_{\max})_{\max} - (R_{\max})_{\min}} \tag{10}$$

Figure 5 shows the plot of the normalized time constant, f_1 versus the normalized fill ratio, f_2 as the three motion parameters are varied within the specified range. The blue region represents the set of feasible solutions, while the red curve represents the Pareto frontier. Each point on the Pareto frontier is a Pareto optimal solution (solution to the multi-objective optimization problem), having a trade-off with each other. Using the weighted sum method, the optimal solution is uniquely determined *by minimizing* the weighted sum in Eq. 8, where the weights w_1 and w_2 sum to 1. In this study, the weights for the fill ratio and energy consumption were set equally, with $w_1 = w_2 = 0.5$.

Figure 5 highlights the point on the Pareto frontier with yellow that corresponds to the minimum value that satisfies Eq. 8 under the given weights. The motion parameters at this point are $F = 8.163$ N, $v = 0.00798$ m/s, and $\omega = -1.4$ rad/s, representing the optimal solution of this study.

4. Experimental verification

This section focuses on the experimental verification of the optimal shape and motion parameters of the bucket drum determined through DEM simulations. The goal is to integrate insights obtained from excavation experiments to validate the

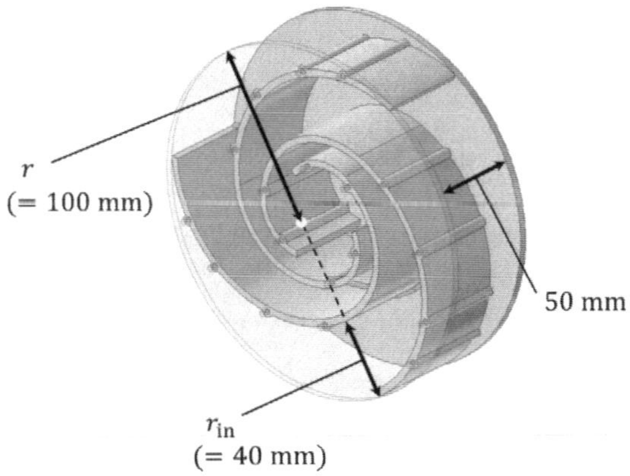

Fig. 6. CAD model with the scoop throat length and width of bucket drum for 3D printing

DEM simulation results. This section outlines the experimental setup and analysis methods used to achieve this goal.

4.1. Excavation experiment setup

Based on the optimal shape parameters, scoop throat length, $N_r = 0.4$, and number of scoops, N_θ=3, identified in Section 3, the CAD model of the bucket drum was created for 3D printing. Using the CAD model shown in Fig. 6, the bucket drum was 3D-printed with ABS-like resin due to its high bending strength, durability, and smooth surface, which reduces friction with the sand. One side of the drum was fabricated using a laser-cut acrylic plate. The transparent acrylic allowed for the observation of the behavior of the sand being trapped inside.

The experiments were conducted using an excavation test apparatus available in our laboratory. Figure 7 shows the apparatus, featuring a storage bin with dimensions of 3 m in length, 0.96 m in width, and 0.4 m in height filled uniformly with Silica sand No. 5. A camera was installed to capture the bucket drum's side view, allowing continuous monitoring of the sand fill status inside the drum during the experiment.

The optimal motion parameters obtained from DEM analysis were used for the experiments: a vertical force $F = 8.163$ N, a horizontal velocity $v = 0.00798$ m/s, and an angular velocity $\omega = -1.4$ rad/s. The bucket drum was rotated three times, and the relationship between the fill ratio and energy consumption was measured.

4.2. Analysis method

To quantify the amount of sand excavated inside the bucket drum during its rotation, a series of image processing techniques were employed. The procedure began with extracting the saturation component from the color images to remove dullness and disturbances. Next, the "imfindcircles" function in MATLAB was used to detect the bucket drum's position in the images. Following this, the detected bucket drum images were converted to grayscale and binarized using the Sobel operator to highlight edges. Subsequently, dilation processing and noise removal were applied using a median filter to enhance sand recognition. Finally, the volume of sand was estimated by counting the number of white pixels in the binarized images. Figure 8 illustrates the result of the image analysis, showing the detected area of sand inside the bucket drum.

The actuator (HEBI Robotics, X5-9) installed for the drum rotation provided the time-series data on current and voltage,

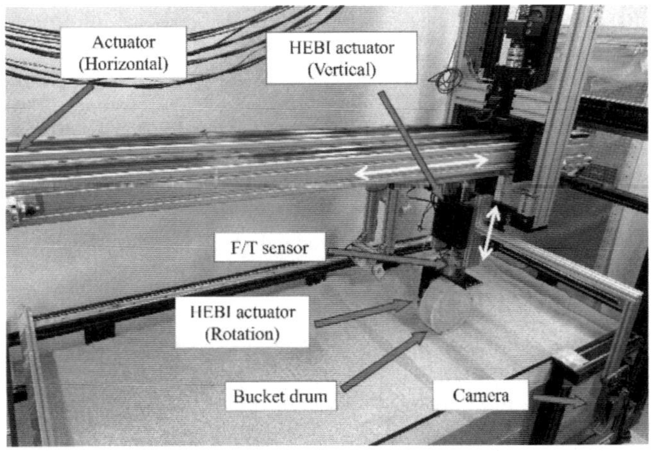

Fig. 7. Test apparatus for evaluating bucket drum excavation performance

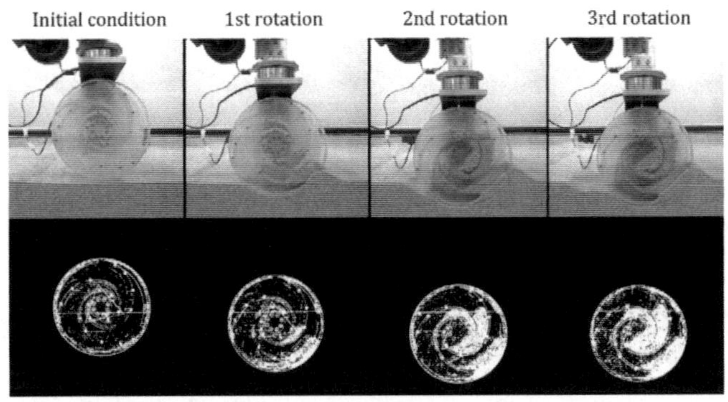

Fig. 8. Sequential visualization of the rotating bucket drum: real-time operation (top row), image processing analysis (bottom row)

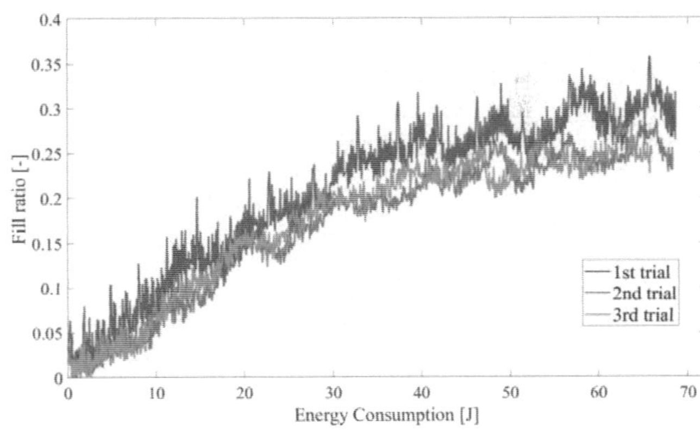

Fig. 9. Relationships between energy consumption and fill ratio in three excavation trials

which were used to calculate the energy consumption. The energy consumption was obtained by integrating the power, the product of current and voltage, over time.

4.3. Results and discussion

Figure 9 shows the relationship between energy consumption and fill ratio from the experiments three times, exhibiting a first-order system response, consistent with the DEM simulation results. This confirms a qualitative correlation between the experimental and DEM simulation results. However, the fill ratio measured from the experiments was slightly lower than that obtained from the DEM simulations. This quantitative difference could be due to several reasons. First, the image processing techniques used may not capture all the sand particles accurately, leading to an underestimation of the fill ratio. Additionally, the experiments used Silica sand No. 5, while the DEM simulations used Toyoura sand. The larger particle size of Silica sand No. 5 compared to Toyoura sand could affect the fill ratio. Overall, experimental results highlight the effectiveness of using DEM for initial design optimization. By leveraging DEM simulations, a balance between fill ratio and energy consumption can be achieved, reducing the need for extensive physical prototyping. This approach not only saves time and cost but also provides a reliable method for evaluating new design concepts and motion parameters before actual implementation.

5. Conclusions and future work

This study investigated the optimization of the excavation bucket drum's shape and motion parameters using DEM simulations, supported by experimental verification. The parameters were evaluated based on two key performance indices: the sand fill ratio in the drum and the power consumption during excavation. Handling a multitude of parameters with varying potential values poses significant challenges, particularly when striving to balance these two indices. This study found that, in simulations, the relationship between energy consumption and fill ratio follows a first-order system response, a result that was also confirmed experimentally.

Our multi-objective optimization approach, which integrated DEM simulations with statistical methods like the Box-Behnken Design, proved essential in navigating these complexities. This approach facilitated a comprehensive exploration of design possibilities, ensuring that the optimal configuration was identified without extensive physical prototyping, thereby saving both time and resources. Leveraging DEM allows researchers to explore and refine various design options before committing to physical prototypes, leading to significant time and cost savings.

Future work will involve the integration of the optimized bucket drum with a lunar rover system, assessing its performance in a fully operational scenario. Additionally, enhancements to the DEM simulation environment, such as simulating excavation of the bucket drum in lunar conditions, will be undertaken. This will provide a more comprehensive understanding of the bucket drum's performance in realistic lunar scenarios.

6. Declaration of competing interest

The authors declare that they have no known competing financial interests or personal relationships that could have appeared to influence the work reported in this paper.

7. References

ANSYS, Inc., 2023. Rocky Technical Manual, Release 2023 R2. ANSYS, Inc., Southpointe, Canonsburg, PA.

Box, G.E.P., Behnken, D.W., 1960. Some new three level designs for the study of quantitative variables. Technometrics 2(4), 455–475.

Clark, D.L., Patterson, R.R., Wurts, D.W., 2009. A novel approach to planetary regolith collection: the bucket drum soil excavator. In: AIAA Space 2009 Conference & Exposition, pp. 6430.

Grant, H.D., Vaniman, D.T., French, B.M. (Eds.), 1991. Lunar sourcebook: A user's guide to the Moon. No. 1259. Cup Archive.

Jehn, I., Dreyer, C.B., van Susante, P.J., Primeau, J., 2023. Lunar site preparation requirements for construction of infrastructure elements. ASCEND 2023.

Mueller, R.P., 2023. A review of extra-terrestrial regolith excavation concepts and prototypes. Earth and Space 2022.

Mueller, R.P., Smith, J.D., Schuler, J.M., Nick, A.J., Gelino, N.J., Leucht, K.W., Dokos, A.G., 2016. Design of an excavation robot: regolith advanced surface systems operations robot (RASSOR) 2.0. In: 15th Biennial ASCE Conference on Engineering, Science, Construction, and Operations in Challenging Environments, Reston, VA, pp. 163-174. American Society of Civil Engineers.

NASA, 2020. NASA regolith Advanced Surface Systems Operations Robot (RASSOR) bucket drum design challenge. NASA. https://www.nasa.gov/general/nasa-regolith-advanced-surface-systems-operations-robot-rassor-bucket-drum-design-challenge/ (accessed on 2024-05-16).

Omura, T., Hamasaki, R., Ishigami, G., 2023. Gravitational dependency analysis on adhesive force of granular material using a discrete element method. In: Proceedings of the Japan Society of Microgravity Application.

Ozaki, S., Ishigami, G., Otsuki, M., Miyamoto, H., Wada, K., Watanabe, Y., Kobayashi, T., 2023. Granular flow experiment using artificial gravity generator at International Space Station. npj Microgravity 9(1), 61.

Yıldız, A., Uğur, L., Parlak, İ.E., 2022. Optimization of the cutting parameters affecting the turning of AISI 52100 bearing steel using the Box-Behnken experimental design method. Appl. Sci. 13(1), 3.

Appendix A

The selection of five representative particle diameters and their corresponding mass ratios for the DEM simulations was guided by a Gaussian mixture model, as described by Omura et al. (2023). This method ensures that the simulated distribution closely aligns with the observed characteristics of Toyoura sand.

The Gaussian mixture model employed two components to capture the distinct peaks in the particle size distribution observed by Ozaki et al. (2023). The five representative diameters were derived using the mean μ_i and standard deviation σ_i of these Gaussian components:

$$\begin{aligned} d_1 &= \mu_1 - \sigma_1 \\ d_2 &= \mu_1 \\ d_3 &= \pi_1(\mu_1 + \sigma_1) + \pi_2(\mu_2 - \sigma_2) \\ d_4 &= \mu_2 \\ d_5 &= \mu_2 + \sigma_2 \end{aligned} \tag{A.1}$$

where π_1 and π_2 represent the mixture ratios of the two Gaussian components. The third representative diameter d_3 was determined by weighting the difference $\mu_1 + \sigma_1$ and the sum $\mu_2 - \sigma_2$ using the mixture ratios. This calculation was crucial for balancing the overlapping regions of the two components, thereby preventing an overemphasis on any particular range within the distribution. Figure A.1 illustrates the five representative diameters and mass ratios, as determined by this selection method. This approach reduces the number of diameters, thereby lowering computational costs while maintaining the complexity of particle interactions in the simulations.

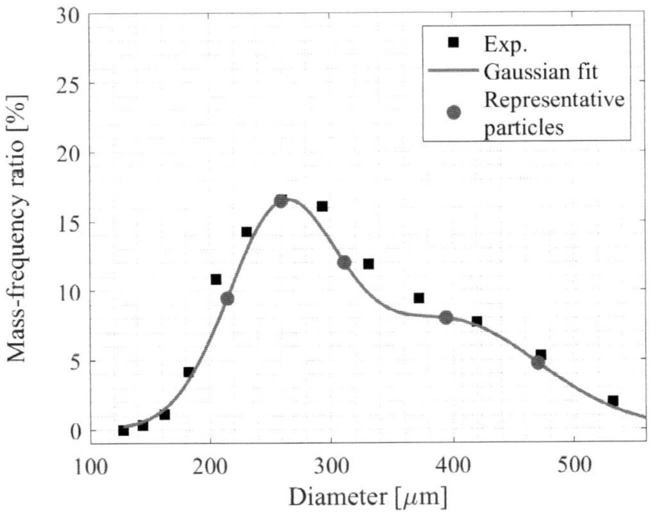

Fig. A.1. Fitting by Gaussian mixture model

4476 | Effect of initial soil color on sandy soil moisture content determination

György Pillinger [a,*], Ahmed Elawad Eltayeb Ahmed [b], Peter Kiss [a]

[a] Institute of Technology, Hungarian University of Agriculture and Life Sciences (MATE), Gödöllő, Hungary

[b] Mechanical Engineering Doctoral School, Hungarian University of Agriculture and Life Science (MATE), Gödöllő, Hungary

* Corresponding author: pillinger.gyorgy@uni-mate.hu

ABSTRACT

In the course of our previous tests, we determined the change in color due to moisture content in the case of sandy soil of three colors that can be easily distinguished with the naked eye. To characterize the color, we used the average of the 600 and 700 nm range of the reflectance curve, because in this range the curves are more linear and the points are better separated, i.e. their resolution is better. Also, this wavelength range corresponds to orange and red colors, which are better absorbed by water molecules, compared to the 400-600 nm range, which includes violet, blue, green, and yellow colors. We found that the parameters characteristic of the color of the examined sandy soils react more sensitively to the influence of soil moisture up to a moisture content of ~4-5%. The created function used the coordinates of this breaking point to determine the moisture content based on reflectance.

In our current article, we continue the further processing of the measurement results, as well as carry out new measurements to clarify the functional relationships. We examine the effect of soil porosity and grain density on the reflectance value. This means that in the future, reflectance is given as a function of saturation. The pre results predict that reflectance is more highly correlated with saturation than with moisture content alone.

Another goal is to determine the effect of the initial color of the sandy soil on the moisture content. As a reference point, we use a characteristic of the reflectance curve for 0% moisture content. We are looking for an answer to how to determine the water content of wet soil based on the data of dry soil and the reflectance of wet soil.

Keywords
Spectroscopy
Porosity
Particle density
Moisture content
Spectral signature
Sandy soil
Soil color
Reflectance
Saturation
Volumetric moisture content

1. Introduction

We are looking for the effect of the initial color (or reflectance parameter) of the sandy soil on the moisture content. The initial color is the average of the values for the 600-700 nm wavelength range of the reflectance curve for 0% moisture content. The initial color of the soil thus means the dry color of the soil. It has already been revealed from previous studies that the nature of the change in color to moisture content of sandy soils with different initial colors, i.e. the course of the curve is essentially the same, only shifted along the ordinate axis (Pillinger et al, 2023). Now we are looking for the answer to how the moisture content of the soil can be determined based on the value corresponding to 0% moisture content. Several people have already dealt with the color, mechanical properties, and saturation of sandy soils, as well as their relationships between these parameters.

Philpot (2010) analyzed the reflectance (wavelength range: 400-2500 nm) curves of soil with different moisture content. He found that the darkening of wet soil depends on several parameters, but it has not yet been clarified which of these parameters are relevant. Bedidi at al. (1992) only examined in the visible range and considered the pF value of the soil, which is a number characteristic of the soil's water-holding capacity. Thus, it provides information about the forces that hold water in the soil.

Simon et al, (2020) investigated several models of how grain size fractions and material composition of sand affect the color of Wisconsin sandy soil. The color of Wisconsin sandy soil is typically determined by the 45-250 μm and 1000-2000 μm grain sizes. The carbon and iron content of the sand is also related to the grain size.

For example, the soil carbon and iron content of the Central Sand Plains of Wisconsin, USA was highest in the fraction smaller than 45 μm. And it is the lowest in the coarse-sized fractions. Most of the carbon and iron were found in the smaller grain fractions. At least 40% of all carbon and iron in the sandy soil was present in the 250-1000 μm fractions. (Yost et. al, 2019).

The stress reduction resulting from the saturation of wet and dry sandy soils was investigated using the traditional triaxial method. That is, the effect of volumetric water content on the state of stress and how this is affected by the particle size distribution. It was found that the decrease in tension caused by saturation increases as the relative density increases. (Mahinroosta and Poorjafar, 2017)

Yoon and Go (2024) dealt with the water saturation of frozen sandy soils during melting. They show a method for determining the coefficients of the empirical water saturation model. This work also clearly shows how important the knowledge of saturation is, from the point of view of the thermal and mechanical strength of the soil.

Dry soil differs from wet soil in that it is saturated with water to some degree. Water adheres to the grains in two layers. The first layer is the adsorbed water, which is bound to the grain by adsorptive forces and approx. 1-10 molecules thick. On top of this is the solvated water shell, held together by polar, electrostatic, and ionic forces, with a maximum thickness of 200 molecules. (Kézdi, 1977) If the amount of water increases, the solvated water layer can no longer thicken, but the pores of the soil begin to fill up with water, this is called pore water. From the point of view of the color of the soil, an additional question may be, what is the relationship between the reflectance value for the established ~5% moisture content and the thickness of the water layer on the sand grain?

In several works, the dry state of the soil is usually understood as the air-dry state. The reflectance range usually covers a wider spectrum (400-2500 nm). In some places, the properties of the

https://doi.org/10.56884/GV41C6MW

water are also distinguished, for example, pore water already contains dissolved minerals compared to distilled water, so it is expected that the electromagnetic rays will behave differently in it.

2. Material and method

To investigate the relationship between saturation and reflectance, it was necessary to create different compaction states of the sand soil sample with the same moisture content. By layering the properly moistened soil into the compacting cylinder, a minimum soil volume mass was formed. The soil structure formed in this way can support its own weight without significant compression of the lower layers. After that, a hydraulic press compacted the sample in several steps (see Fig. 1). The result of the compaction steps is that, with a constant moisture content, the porosity of the soil sample decreases more and more, while its saturation and bulk density gradually increase proportionally.

Fig. 1. Soil pressure bin and plate; inner diameter of 20 cm; initial soil thickness 10 cm

Depending on the initial moisture content and the maximum pressure of the hydraulic press, the soil slowly becomes saturated until finally the water starts to squeeze out of it. The color measurements were taken after the first filling and immediately after each compression stage. The completely dry state of the soil - or its initial color - was considered to be the state created after drying at 105°C for 48 hours. The setting of the moisture content was determined by calculation, knowing the dry weight of the soil and the weight of the added water. Therefore, all moisture content data is given on a dry basis and as a percentage by weight.

Another method was used to create a fully saturated state. The soil sample was continuously saturated using the drip method in a ceramic jar until the point where the water droplet was absorbed, and the surface of the sample began to reflect. This state of the surface of the soil sample is the saturated state on which the reflectance measurement was performed.

The reflectance curve was measured with a color spectrophotometer (Konica Minolta CM700d) under normal daylight (D65), 10° observation angle and SCE mode. During the measurements, we worked with the results of diffuse reflection. The device can be used with measuring heads with a diameter of 3 and 8 mm. The reflectance range is 0-175 %, and the wavelength range is 400-700 nm. We used the MAV: Ø8 mm setting for our measurements.

The particle size distribution of the tested soils is the same as the natural state at the collection site. The soils underwent only one cleaning filter, where all particles larger than 0.5 mm were removed.

3. Results

3.1. Effect of initial soil color

To determine the effect of the initial soil color and saturation, we use the data of the reflectance curves recorded in the 600 and 700 nm wavelength range with different moisture contents of the given soil. Figure 2 shows the curves of loamy sandy soil, and Fig. 3 shows the curves of pure sandy soil at moisture contents of 0...18% and 0...16%. The water retention capacity of the loamy sand soil is a little better, so its saturation occurs at higher moisture levels, the saturation value of the tested sandy soil is 30%, and that of the loamy sand soil is 35%. The top row of dots is the reflectance curve of completely dry soil. Below this, the following curves are the reflection curves of samples with increasingly higher moisture content. The bottom one is the curve of fully saturated soil.

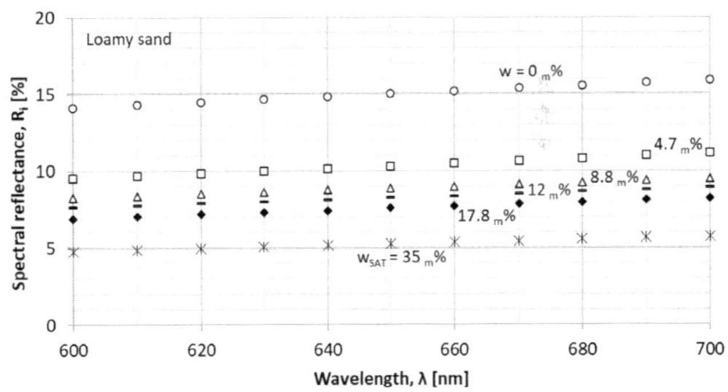

Fig. 2. Wavelength-reflectance relationship of loamy sand soil for different soil moisture contents

It can be seen from the reflectance values that the examined sandy soil starts with a slightly lighter initial value of R_{700} = 19.5%, and accordingly its wetted values take on higher values compared to the loamy sandy soil, which has a dry R_{700} value of 15.9%.

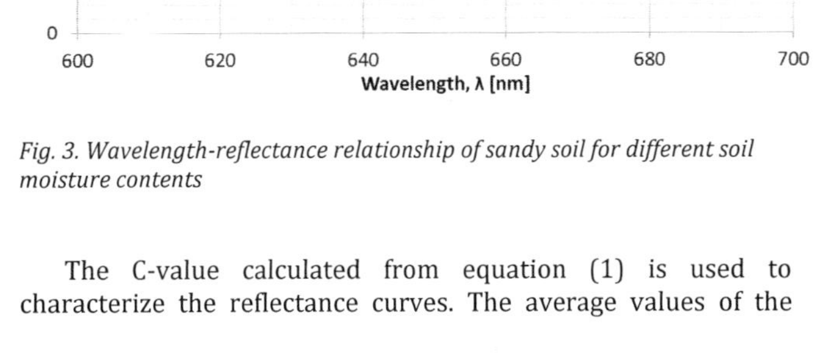

Fig. 3. Wavelength-reflectance relationship of sandy soil for different soil moisture contents

The C-value calculated from equation (1) is used to characterize the reflectance curves. The average values of the curves in the given wavelength range are plotted as a function of the moisture content (see Fig. 4). This curve is the same as the previous results (Pillinger et al, 2023), with the difference that the previous curve had a higher resolution. The reflectance values are shown on the vertical logarithmic axis, and the moisture content values on the horizontal axis. Due to the smaller number of moisture content values, the breaking point cannot be detected on it, but it is still described by an exponential function, with less accuracy. The relationships defined below are also valid for those curves, only the constants (A, n) of the equation are different.

$$C = \frac{1}{11} \cdot \left(\sum_{600}^{700} R_i \right) \tag{1}$$

If we calculate with an average constant, the accuracy will of course deteriorate. The important points of the curve are the initial "C_0" value, which is the reflectance corresponding to 0% moisture content (in the case of sand, it is 17.3%), and the end point "C_{SAT}" is the reflectance of fully saturated soil. Between the two points, the different soil moisture values and the corresponding reflectance values "C" can be seen, which decrease as the moisture content increases.

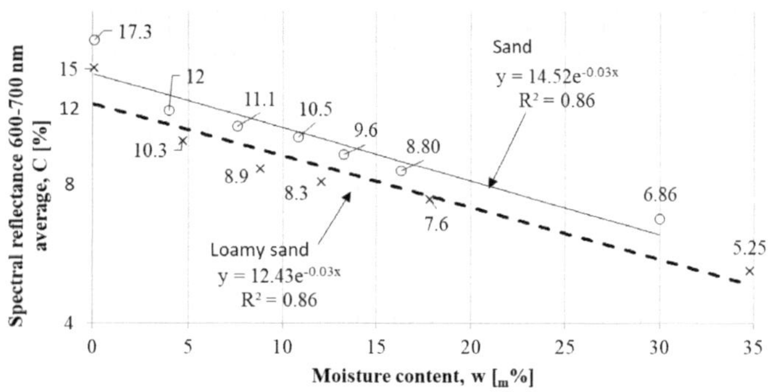

Fig. 4. The change of $C_{600-700}$ as a function of moisture content, on a semi-logarithmic scale

The next step is to calculate the deviation of each "C" value from "C_0". The deviations are determined with the relation:

$$\Delta C = C_0 - C \tag{2}$$

The results are plotted in a linear-linear diagram. It can be noted that the values of "ΔC" and the moisture content increase to a lesser extent, but they increase. The moisture content corresponding to the maximum value is the saturation moisture content. The "ΔC" value of completely dry soil is 0 [%].

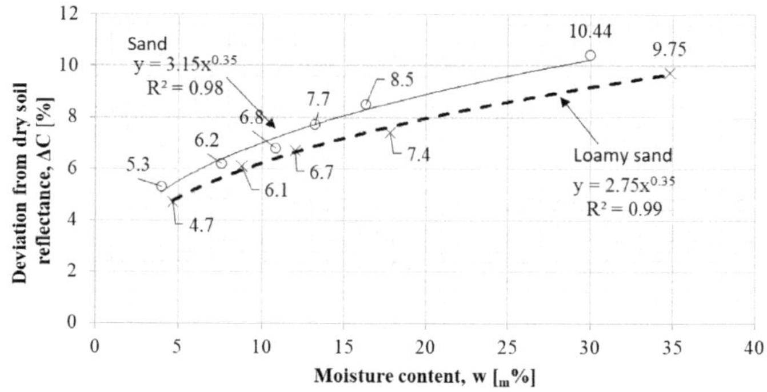

Fig. 5. Deviation of the average of the reflectance curves from the maximum value (from the reflectance value of completely dry soil)

Based on Fig. 5 it can be concluded that there is a power function relationship between the moisture content and the ΔC value:

$$\Delta C = A \cdot w^n \tag{3}$$

The equation (3) must be replaced back into the basic relationship "ΔC" (2), so we get the reflectance parameter "C" for the wet state of the soil as a function of the moisture content "w" and the reflectance parameter "C_0" of the dry soil:

$$C = C_0 - A \cdot w^n \tag{4}$$

The relationship has two constants, "A" and the exponent of moisture content "n". For accurate calculations, the values of the constants for the tested soil textures are shown in Table 1.

Table 1
The value of the constants

Soil texture	A [%]	n [-]
Sand	3.15	0.35
Loamy sand	2.75	0.35

Using equation (4), the moisture content of the soil can now also be determined using the dry reflectance parameter "C_0" of the soil and the current reflectance parameter "C" of the soil:

$$w = \sqrt[n]{\frac{C_0 - C}{A}} \tag{5}$$

3.2. Effect of saturation

It is not possible to create arbitrary starting volume masses in the compacting cylinder, and therefore arbitrary saturation values as well. The theoretical range of porosity is 0 and 1, where 0 is a completely compact soil with no pores. The soil with porosity 1 is a soil sample free of solid parts, consisting only of pores. Figure 6 shows the realized and theoretical saturation-porosity curves for soil moisture contents of 1% to 20%. The continuous thick line represents the feasible range on the theoretical dotted line. The markings on it indicate the stations where color measurements were made. The smallest possible porosity in the filled cylinder was approximately 0.65. Depending on the moisture content, different degrees of porosities were created in addition to the maximum compressive force. With a moisture content of 5%, the achieved porosity was 0.42 with a pressure of five tons, and 0.3 with 8%. A higher compression than this could not be achieved even with higher moisture contents.

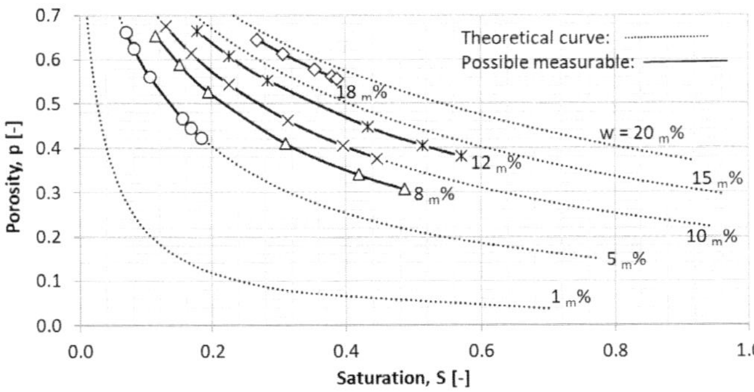

Fig. 6. The relationship between saturation and porosity during the compaction process

Saturation is the ratio of the water-filled volume to the pore volume. At 0 saturation, the pores contain no water. In the case of saturation 1, all soil pores are filled with water. Based on the set moisture content and porosity before compaction, the lowest saturation value was 0.07, and the highest value reached at the end of compaction was 0.57 (see Fig. 6).

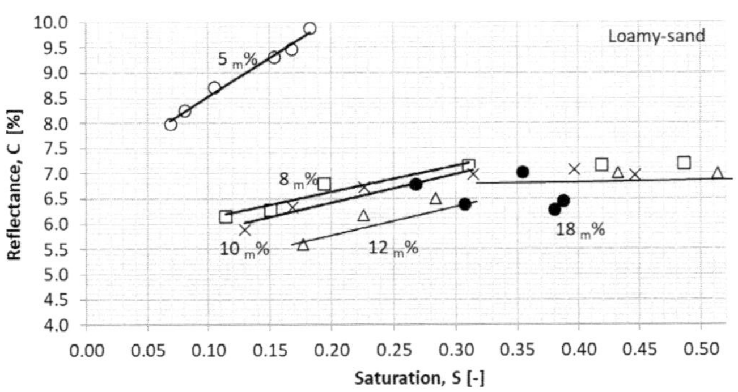

Fig. 7. Saturation-reflectance curves

A relationship can be established between the reflectance parameter taken between the compression steps and the saturation (see Fig. 7). As a result of the moisture content, the reflectance values decrease, and the soil darkens. This is the same as the results in Fig 2, 3. However, as a result of the compaction, the soil grains have become closer to each other, which increases the ratio of reflected light rays compared to the more porous soil. In this case, a more saturated state means less porosity. As the moisture content increases, the effect of the proximity of the grains (the effect of porosity) decreases and the reflectance modifying effect of water will dominate. Thus, a saturation limit can be discovered on the diagram. Where, due to the increase in water content, the adsorbed and solvated water envelope is already so thick around the grain that the effect of porosity is masked. In the case of the examined loamy sand soil, this happens approximately at a saturation level of 0.32. Above this value, the reflectance values are essentially displayed by the standard deviation of the measurement results. Therefore, beyond this saturation level, the reflectance values will fluctuate around an average. This transition occurs between 15 and 18% moisture content of the examined loamy sand soil.

Figure 2 and 3 shows the saturation reflectance curve as well. Knowing the reflectance parameters of dry soil and saturated soil, we can deduce the saturation "d" with the following equation, where "C_{SAT}" is the reflectance of fully saturated soil:

$$d = \frac{C_0 - C}{C_0 - C_{SAT}} \tag{6}$$

The "d" value does not mean the exact saturation of the soil, but is proportional to it, so it gives a value close to the real saturation level. Knowing the saturation gives important information on the consistency and thus the mechanical strength of the current soil, and there is also a relationship between the saturation and the yield strength of the soil, which is also a relevant soil property.

Table 2
Data of investigated soil textures

Soil texture	Dry	Fully saturated		Reflectance range	Soil particle distribution		
	C_0	C_{SAT}	w_{SAT} [m%]	ΔC_{max}	Clay [m%]	Silt [m%]	Sand [m%]
Sand	17.3	6.86	30	10.4	2.2	2.1	95.7
Loamy sand	15.0	5.25	35	9.75	5.8	12.1	82.1

Table 2 contains the measured parameters of the examined sand and loamy sand soil. These are the reflectance parameters for the completely dry and saturated state, the saturation moisture content "w_{SAT}", the difference between the two extreme values, i.e. the width of the reflectance parameter range that can be examined, and the composition of the individual grain fractions according to Atterberg.

Based on the current tests, it was not possible to establish any detectable effect of the grain size on the reflectance values in the range of soil grain sizes. In addition to the existing textures, by comparing dry reflectance parameters, it is possible to find the relationship between the reflectance parameters of soils with a higher clay content and sand content.

4. Conclusions and recommendations

Based on the theoretical and experimental studies, the following main conclusions can be drawn:

In the case of saturation created by compaction, it must be considered that the compaction is not completely homogeneous in the entire soil volume, so the saturation or volumetric water content will also vary in the examined volume. Larger compressions occur in the immediate vicinity of and below the pressure plate, so the saturation value is also higher in these places. As a result, the reflectance values apply only to the surface and not to the entire soil volume.

The measured reflectance curve is only valid for the ground surface. Thus, the moisture content and saturation number determined by equation (5) or (6) are only valid for the soil surface.

It was established that the results recorded with the used measuring device show no difference between the air-dried soil samples and those dried in the drying oven at 105 °C for 48 hours. The two are identical in terms of reflectance curve or color.

The measured reflectance values and calculated results are only valid at the time of measurement or for a short period of time depending on the environmental conditions. Because the surface of the soil is in direct contact with the air, where it is exposed to constantly changing solar radiation, humidity, wind and air pressure changes, which can cause rapid changes in the surface moisture content of the soil.

Near the saturation moisture content of a given soil, the measurement results are less and less separated from each other. There may be differences that cannot be detected with the instrument, or the standard deviation of the measurement errors already obscures the values. In the examined loamy sand soil, this starts already at half of the saturation moisture content value.

5. Nomenclature

A	Equation coefficient	[%]
C	Average reflectance values in the range of 600-700 nm	[%]
C_0	Average reflectance values in the range of 600-700 nm for dry soil	[%]
C_{SAT}	Average reflectance values in the range of 600-700 nm for fully saturated soil	[%]
d	Characteristic proportional to soil saturation	[-]
n	Exponent of moisture content	[-]
p	Porosity	[-]
S	Saturation	[-]
w	Soil moisture content on dry base, in mass percent	[$_m$%]
w_{SAT}	Saturated moisture content of soil	[$_m$%]
λ	Visible light wavelength	[nm]

6. Declaration of competing interest

The authors declare that they have no known competing financial interests or personal relationships that could have appeared to influence the work reported in this paper.

Project no. 2022-2.1.1-NL-2022-00012 has been implemented with the support provided by the Ministry of Culture and Innovation of Hungary from the National Research, Development and Innovation Fund, financed under the 2022-2.1.1-NL funding scheme.

7. References

Bedidi, A., B. Cervelle, J. Madeira, and M. Pouget, 1992. Moisture effects on visible spectralcharacteristics of lateritic soils. Soil Science, Vol. 153(2):129-141.

Jenifer L. Yost, Eric E. Roden and Alfred E. Hartemink, 2019. Geochemical Fingerprint and Soil Carbon of Sandy Alfisols, Soil Syst. 2019, 3(3), 59; https://doi.org/10.3390/soilsystems3030059

Kézdi Á., 1977. Talajmechanika I. Tankönyvkiadó, ISBN: 963 17 2256 2

Pillinger Gy., Ahmed E.E.A., Bessenyei K., Kiss P., 2023. Correlation between moisture content and color spectrum of sandy soils, Journal of Terramechanics, Vol. 108, page 39-45. https://doi.org/10.1016/j.jterra.2023.05.002

Reza Mahinroosta, Arash Poorjafar, 2017. Effect of stress state and particle-size distribution on the stress reduction of sandy soils during saturation; Construction and Building Materials Vol. 150; https://doi.org/10.1016/j.conbuildmat.2017.05.177

Seok Yoon, Gyu-Hyun Go, 2024. Numerical analysis study on determination of unfrozen water saturation curve for sandy soil, International Communications in Heat and Mass Transfer, Vol. 154, https://doi.org/10.1016/j.icheatmasstransfer.2024.107418

Thibaut Simon, Yakun Zhang, Alfred E. Hartemink, Jingyi Huang, Christian Walter, Jenifer L. Yost, 2020. Predicting the color of sandy soils from Wisconsin, USA; Goderma Vol. 361; https://doi.org/10.1016/j.geoderma.2019.114039

William Philpot, 2010. Spectral Reflectance of Wetted Soils, http://dx.doi.org/10.13140/2.1.2306.0169

4715 | UAV-based three-dimensional rough terrain modelling

Herman A. Hamersma[a,*], J. Christian van Aswegen[a], A. Glenn Guthrie[a], P. Schalk Els[a]

[a] *Department of Mechanical and Aeronautical Engineering, University of Pretoria, Pretoria, South Africa*

* *Corresponding author: hermanh@up.ac.za*

ABSTRACT

Accurate modelling of the interaction between a tyre and the terrain is crucial for successful vehicle dynamics simulation, particularly in challenging off-road conditions. This study introduces a novel approach for obtaining highly accurate three-dimensional terrain models for vehicle simulation. Utilizing an unmanned aerial vehicle (UAV) equipped with a high-resolution camera, we developed three-dimensional terrain models of an undulating test track with two approaches. To validate the accuracy, these models were compared with measurements obtained from a traditional mechanical road profilometer. The results demonstrate a strong correlation between the two measurement approaches. The primary advantages of the UAV method lie in its speed of data acquisition without compromising accuracy and that the measurements are immune to the terrain macro roughness. In contrast to the labour-intensive measurements required by a mechanical road profilometer and extensive subsequent post-processing, the UAV approach requires minimal time and effort. Additionally, since the UAV method is not coupled to the terrain roughness, it eliminates the integration drift associated with other ground-based approaches that are subjected to terrain excitation. This opens the door to the possibility of modelling very rough terrains and using these terrain models for vehicle dynamics simulations in extreme off-road environments.

Keywords
Terrain modelling
Road profiling
Photogrammetry
UAV surveying
Off-road mobility
Vehicle dynamics simulation

1. Introduction

The foundations of vehicle dynamics are built around the tyre-terrain interface. It is well-known that all the control forces and moments acting on a typical automobile act through four patches each the size of a man's hand. While the first tyre tester was developed in 1931 (Gillespie, 1992), the first road profilometer was developed in the 1960s by General Motors Research Laboratory (Sayers and Karamihas, 1998).

Terrain or road modelling remains relevant to this day, with three main objectives. The first objective is to better understand and hence improve a vehicle's dynamics. There are numerous examples in the literature of various approaches to accurately model a road or a terrain a vehicle model is traversing. Early approaches relied on single- or double-track models, but more modern approaches develop full three-dimensional (3D) terrain profiles. Advances in tyre modelling now allow for various contact methods between physics-based tyre models and 3D terrain models, some even capable of real-time simulation (dSPACE GmbH, 2016). There are examples of models that even include deformable terrains (Taheri et al., 2015).

The second objective is to monitor the condition of a road network for pavement management purposes. Evaluating the quality of a newly constructed road or repairs to road sections may also be lumped with this objective. Here, the focus is not on vehicle dynamics, but rather on ensuring road networks are built according to specification or to ensure efficient identification of roads requiring maintenance and monitoring of their subsequent repair (Ngwangwa et al., 2014; Medinac et al., 2020). This technique may also aid path planning applications for intelligent off-road vehicles, and may form part of the NATO Reference Mobility Model (NRMM) (McCullough et al., 2017) – e.g. a UAV may be used to survey an area prior to deploying intelligent vehicles into the area.

The third objective is terrain profile estimation. The development of autonomous vehicles and over-actuated vehicles using integrated chassis control systems to enhance vehicle performance requires detailed knowledge of what lies ahead for the ego vehicle (Hamersma and Els, 2021; Scholtz and Hamersma, 2024). This is not solely focused on potential obstacles that need avoiding; in-depth knowledge of the road condition may lead to improved performance from these modern vehicles.

Although the abovementioned objectives are dissimilar, they often make use of similar approaches to gather information and to make decisions. Recent developments, such as the proliferation of unmanned aerial vehicles (UAVs or drones) have opened up new possibilities for terrain modelling that may be beneficial to all three these objectives. This study investigates the possibility of measuring a terrain and creating a 3D model thereof with the aim simulating off-road vehicle dynamics.

1.1. Road profile characterization and classification

Various approaches exist to report road profiles and to classify roads, but for the purpose of this study, ISO 8608:2016 (International Organization for Standardization (ISO), 2016) will be used for road profile reporting.

ISO 8608:2016 specifies a uniform method of reporting road profile data from single-track and multiple-track measurements. Two methods are specified, the displacement power spectral density (PSD) and the acceleration PSD. The displacement PSD dimension is metres cubed (m^3) versus the spatial frequency (cycles/m or rad/m). A spatial frequency range is not specified, because it depends on the measurement equipment, but lower limits of 0.01 cycles/m for on-road vehicles and 0.05 cycles/m for

https://doi.org/10.56884/TK3ZX342

off-road vehicles are recommended. An upper limit is also not specified, as it depends on the application. An upper limit of 10 cycles/m is recommended for general on-road measurements, and a limit of up to 1000 cycles/m for noise purposes is mentioned.

ISO 8608:2016 also makes provision to fit a straight line to the PSD using equation (1) for the spatial frequency case (cycles/m), with a reference spatial frequency of 0.1 cycles/m. Guidance is also provided on the classification of roads, based on the degree of roughness. Assuming a constant velocity PSD with $w = 2$, different geometric means are provided at the reference spatial frequency (n_0), resulting in the various classes shown in Fig. 1.

$$G_d(n) = G_d(n_0) \cdot (n/n_0)^{-w} \tag{1}$$

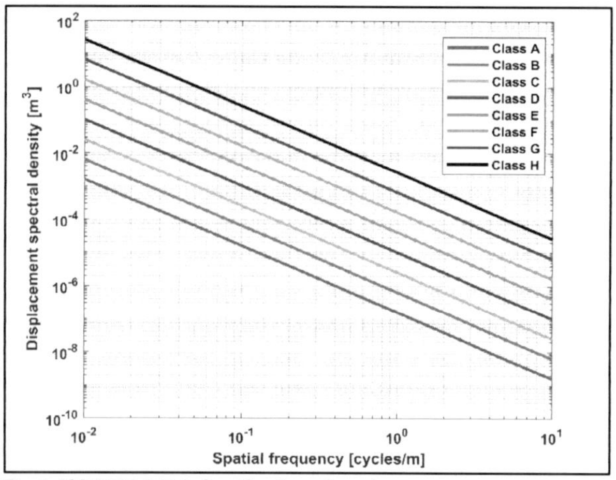

Fig. 1. ISO 8608:2016 classification of roads

1.2. Road profilometers

Various road profilometers are found in the literature, and they can generally be divided into two distinct categories. The first category is dedicated mechanical road profilometers. These instruments are typically 'walked' or towed along a road section while recording the movement of lightweight fingers or contacts to determine the vertical road profile. Typical examples are found in Becker and Els (2014) and Sayers and Karamihas (1998). More recent examples make use of laser line scanners, 3D lidar or even stereo cameras to conduct noncontact measurements of the road profile (Kern and Ferris, 2007; Varunjikar et al., 2012; Botha and Els, 2015; Lee et al., 2017; Ni et al., 2020).

The second category makes use of an instrumented vehicle, typically combining accelerometers, gyroscopes and global positioning systems with a simplified vehicle model to create a road profile (Abulizi et al., 2016). This is often referred to as an inertial profilometer.

Medinac et al. (2020) documented a case study using a UAV to monitor the haul road condition in open pit mines. Previously, the mine relied on ad hoc inspections, machine operator feedback and monthly inspections. Using a UAV, they were able to develop orthographic plots and digital elevation models to assess road smoothness and to verify road design compliance. A highly utilized road section of 1.2 km by 40 m was selected. The UAV was able to survey the road while flying at an altitude of 100 m with constant image overlap, requiring a total flight time of 24 min. The team collected 601 images and spent 1 h and 28 min in the field. This approach allowed the team to cover a very large area in a relatively short time. The results not only provided excellent information on the condition of the road, it enabled easy identification of water puddles, spillage, inappropriate road grading and road roughness. The authors anticipate improved tyre life due to reduced wear and tear, reduced maintenance costs and time savings. Some limitations were pointed out, including that UAV flights are limited to days with favourable weather conditions and its reliance on a camera, which is vulnerable to adverse lighting conditions.

1.3. Problem statement

Recent developments have opened the possibility to use UAVs for road and terrain profile measurement. The development of 3D models with UAVs has seen wide adoption in various industries, including pavement monitoring. However, for the purpose of this study, the use of a UAV to create highly accurate 3D terrain models for use with vehicle dynamics simulation will be investigated. Of particular interest is the ability to model very rough off-road terrains, where more conventional profilometers are hampered due to the severe movement of the inertial reference. For this reason, the terms terrain profile and terrain model will be used in this document, since its application is wider than just road profiling and modelling.

It is hypothesized that the methodology followed in this investigation will yield results at least comparable to that obtained with more conventional road profilometers. The expected benefit is that the measurement process will take much less time, require fewer resources and be immune to drift induced by the inertial reference, while delivering a model with comparable fidelity and accuracy.

2. Methodology

Two methods of measuring terrain profiles are used, namely surveying with a UAV and using the Can-can mechanical profilometer developed by Becker and Els (2014). The dataset measured with the Can-can mechanical profilometer is available for this investigation. The measured data obtained with both methods are then processed using the exact same method to create terrain models. The displacement PSDs of terrain models are then compared with each other.

The Can-can mechanical profilometer is a resource intensive measurement technique. The following resources are needed when conducting a measurement (typical values):

- Commissioning, requiring at least three days by one person.
- Transporting the Can-can mechanical profilometer with a large trailer, typically requiring a special driver's license.
- A measurement is conducted at less than walking pace, approximately 3 km/h.
- Once data has been collected, post-processing requires meticulous attention to detail to remove sensor noise and integration drift. For each hour of testing, an additional hour of post-processing is needed at minimum.

2.1. Surveying the terrain

Gerotek Test Facilities (GTF) is a proving ground located close to Pretoria, South Africa. It hosts test tracks ranging from relatively smooth to extremely rough. Becker and Els (2014) profiled the majority of the test tracks at GTF with the Can-can profilometer described in Section 1.2. This study was also conducted at GTF, which allows for a direct comparison of the two terrain profiling techniques. The Belgian paving at GTF is used for this study. Collecting and processing the Can-can test data of the Belgian paving at GTF with the Can-can profilometer took approximately three days.

A DJI Mavic 3 Enterprise UAV equipped with the DJI Real-Time Kinematics (RTK) module and DJI D-RTK 2 High-precision GNSS mobile station are used for this investigation. Its specifications are documented in Table 1 (DJI, 2024).

An area route is created with the DJI Pilot 2 application on the DJI RC Pro remote controller. A mapping area is drawn that encompasses the area of interest. The appropriate aircraft and camera are selected (in this case the DJI Mavic 3 Enterprise can only collect data with the wide-angle lens). Both orthographic and oblique images are collected (the oblique images at an angle of 45°). This is the shallowest angle the gimbal can rotate.

The Ground Sampling Distance (GSD) is set to the lowest possible value to ensure the highest possible resolution. The GSD is influenced by the route altitude. The flight should be conducted at the lowest safe altitude that will ensure that the UAV does not collide with objects such as trees and power lines. Finally, a 70 % side overlap ratio and 80 % frontal overlap ratio were selected – these are the default values recommended by the DJI Pilot 2 application. The DJI Pilot 2 application then automatically creates a flight route. Collecting orthographic images only requires a single route, but collecting oblique images requires flying two routes at right angles. Figure 2 shows an example of the flight routes – the bottom image is an example of an orthographic route. Both routes are needed for the oblique method. The total flight time for the orthographic area route was 46 s and the oblique area route was 6 min 26 s.

Figure 3 shows an example of an orthographic and an oblique image captured with the UAV. The photogrammetry will attempt to find matching points in all the images, resulting in numerous points outside the region of interest.

2.2. Creating point clouds

Once the images are retrieved from the UAV, photogrammetry is used to create a 3D point cloud. Reality Capture (Epic Games, 2024) is used to build this. Each image captured with the UAV contains information about the UAV position and heading and details on the camera settings in its metadata. Reality Capture uses this information to remove lens distortion and to construct the 3D point cloud using photogrammetry techniques. The user can specify the detail of the point cloud, e.g. by limiting the number of points, etc. For the purpose of this investigation, the highest detail was selected. The level of detail correlates closely with the time taken to create the point cloud and the associated file size. The computer used during this step is equipped with a 13th Gen Intel® Core™ i7-13700 processor, with 64 GB DDR5 5200 MHz RAM and a NVIDIA GeForce RTX 3060 12 GB GPU. The processing times varied, depending on the number of images used, but typically ranged between 20 min to 60 min.

2.3. Postprocessing the point cloud

The next step is to clean-up the point cloud before reorganizing it into a regular grid. CloudCompare (2021), a 3D point cloud and mesh processing software (an Open Source Project used under a GNU General Public License) is used during this step. The point cloud is rotated, translated and cropped to the region of interest. This is done in the following steps (note that this method is only applicable to straight roads):

1. Rotate the point cloud so that the x-axis is parallel to the planned direction of travel, and the negative z-axis points down.
2. Crop the point cloud to the region of interest and discard unnecessary points.
3. If necessary, shift the point cloud so that the anticipated starting point is at the origin.
4. Export the point cloud in *.pcd format.

Table 1
DJI Mavic 3E specifications (DJI, 2024)

Aircraft specifications	
Max flight speed (at sea level, no wind)	15 m/s (Normal Mode)
Max wind speed resistance	12 m/s
Max take-off altitude above sea level	6000 m (without payload)
Max flight time (no wind)	45 min
Max flight distance	32 km
GNSS	GPS + Galileo + BeiDou + GLONASS
Hovering accuracy	Vertical: ±0.1 m (with RTK) Horizontal: ±0.1 m (with RTK)
Camera specifications (wide-angle lens)	
Sensor	4/3 CMOS Effective pixels: 20 MP
Lens	FOV: 84° Format Equivalent: 24 mm Aperture: f/2.8-f/11 Focus: 1 m to ∞
ISO range	100-6400
Shutter speed	Electronic Shutter: 8-1/8000 s Mechanical Shutter: 8-1/2000 s
Max image size	5280 × 3956 pixels
Gimbal	
Stabilization	3-axis (tilt, roll, pan)
Mechanical range	Tilt: -135° to 100° Roll: -45° to 45° Pan: -27° to 27°
RTK module	
RTK Positioning Accuracy	Horizontal: 1 cm + 1 ppm Vertical: 1.5 cm + 1 ppm

Fig. 2. An example of the flight routes for a mapping area when collecting oblique images

Fig. 3. Orthographic image of the Belgian paving (top); Oblique image of the Belgian paving (bottom)

Fig. 4. Section of the Belgian paving point cloud after postprocessing in CloudCompare

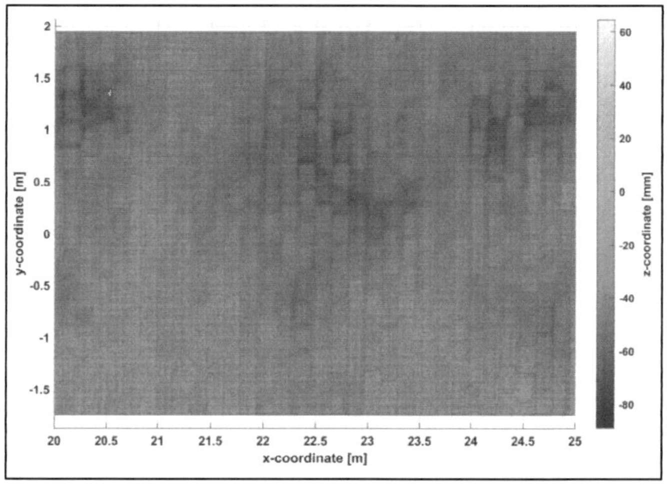

Fig. 5. Section of Belgian paving organized into a regular grid with resolution 10 mm x 10 mm with outliers removed

Figure 4 shows a section of the Belgian paving after postprocessing the point cloud with CloudCompare. Note the unstructured nature of the points comprising the point cloud – the photogrammetry has created points wherever common features/points can be identified in overlapping images.

When the 3D point cloud is aligned correctly, the point cloud is imported into MATLAB (MathWorks, 2016). The points are mean-centred and restructured into a 3D grid using 2D linear interpolation over the *x* and *y* coordinates. The 2D linear interpolation method is applied by using the `griddata` function in MATLAB, which provides the capability of organizing unstructured coordinates (such as a point cloud) into a structured, regularly spaced grid. Els and Scholtz (2017) investigated the effect of road profile measurement resolution on tyre simulations and found that a resolution of 15 mm x 15 mm is fine enough for a typical Sports-Utility-Vehicle tyre. Finer resolution is only needed for extreme high-fidelity simulations. For this reason, a resolution of 10 mm x 10 mm was used for this investigation. This ensures that the terrain models are detailed enough to provide high frequency input, but that file sizes remain reasonable.

After the interpolation has been applied there may be outliers in the coordinates because of the linear interpolation method. Outliers are removed by replacing the outliers in the data with a moving median of the data. The `filloutliers` function in MATLAB is used to identify and replace the outliers. Figure 5 shows a section of the Belgian paving at this point in the process.

With the points neatly organized into an evenly spaced grid, a displacement PSD of the terrain profile can be calculated along the longitudinal direction to classify the road according to ISO 8608:2016. A displacement PSD is calculated for each longitudinal line using the method described in Fig. 6. This results in a set of displacement PSDs over the width of the terrain. The average displacement PSD is then determined, resulting in a single displacement PSD for the terrain.

```
npts       = determine number of points along x-coordinate of
             grid
fs         = spatial sampling frequency of grid

for k = 1 to npts
    fft_x          =   Fast Fourier Transform of line k z-coordinates
    mag_fft_x      =   Single-sided, scaled amplitude of fft_x
    psd_x          =   (mag_fft_x^2)/(2*fs)

    psd_road(:,k)      = store psd_x in column k of an array
end for

psd_mean = mean of psd_road along the y-direction at each x-
           coordinate
```

Fig. 6. Determining the mean displacement PSD of the terrain profile

3. Results

The methodology described in Section 2 was first applied to data measured on the Belgian paving at GTF. The Belgian paving, images of which are shown in Fig. 3, is a 100 m long test track that is paved with cobblestones. The cobblestone dimensions are approximately 130 mm x 200 mm, with a gap of approximately 35 mm between each cobblestone. The displacement PSDs of the two measurement techniques are compared in Fig. 7. The results indicate a high degree of correlation between the two measurement techniques.

Three prominent peaks are visible at 6.1 cycles/m, 12 cycles/m and 17.5 cycles/m. These spatial frequencies correspond to wavelengths of 163.9 mm, 83 mm and 57.1 mm, respectively. The 163.9 mm wavelength is very close to the x-dimension of each cobblestone plus the gap to the next cobblestone (130 + 35). At spatial frequencies between 0.3 cycles/m to 7 cycles/m, very similar results are observed.

Generally, good correlation is seen at lower spatial frequencies between the Can-can data and the UAV data collected with orthographic images – down to approximately 0.1 cycles/m. A significant difference is visible at 0.23 cycles/m, with the orthographic UAV data showing a lower displacement PSD than the Can-can data. The oblique PSD starts to deviate from the other two techniques from approximately 0.4 cycles/m and lower. At very low spatial frequencies, a significant difference between the oblique UAV data and the Can-can and orthographic UAV data is seen.

The oblique data differs by more than an order of magnitude at 0.01 cycle/m, indicating a much rougher road than the other two methods. This is somewhat unexpected, and may simply be a numerical artefact. The total length of the test track is 100 m, thus limiting the lower spatial frequencies that can reliably be observed to approximately 0.05 cycles/m (or wavelengths of 20 m). Using a longer test track may move this point to a longer wavelength.

At higher spatial frequencies, bigger differences between the UAV and Can-can data become apparent. The UAV data tends to be smoother than the Can-can data, with significant differences visible at spatial frequencies exceeding 20 cycles/m (50 mm wavelength). It is noticeable that the peak at 17.5 cycles/m is not present on the oblique UAV data.

Various reasons for the discrepancies between the three approaches may be identified, including:

- The two UAV techniques collected images at different altitudes. The orthographic mission altitude was lower than the oblique altitude, because the UAV could safely fly along the track centreline. However, for the oblique flight mission, the altitude had to be increased to allow the UAV to fly above the vegetation next to the track.
- Even a small error in the gimbal angle, UAV attitude or roll angle may have a significant effect on the stereo photogrammetry. The gimbal angle was set to 45 ° (the shallowest angle possible), where the slightest error in the angle, position or altitude measurement may have significant consequences. The orthographic image method is less sensitive to these errors. At an angle of 90 °, the small angle assumption is generally valid due to the insensitivity of trigonometric functions at these angles.
- The frontal image overlap ratio was set to 80 % for the orthographic technique. The orthographic images captured an area of approximately 12 m x 19 m. This means that images overlapped by approximately 9.6 m, or a spatial frequency of approximately 0.1 cycles/m. This is the point where more noticeable differences between the Can-can and orthographic data become apparent.

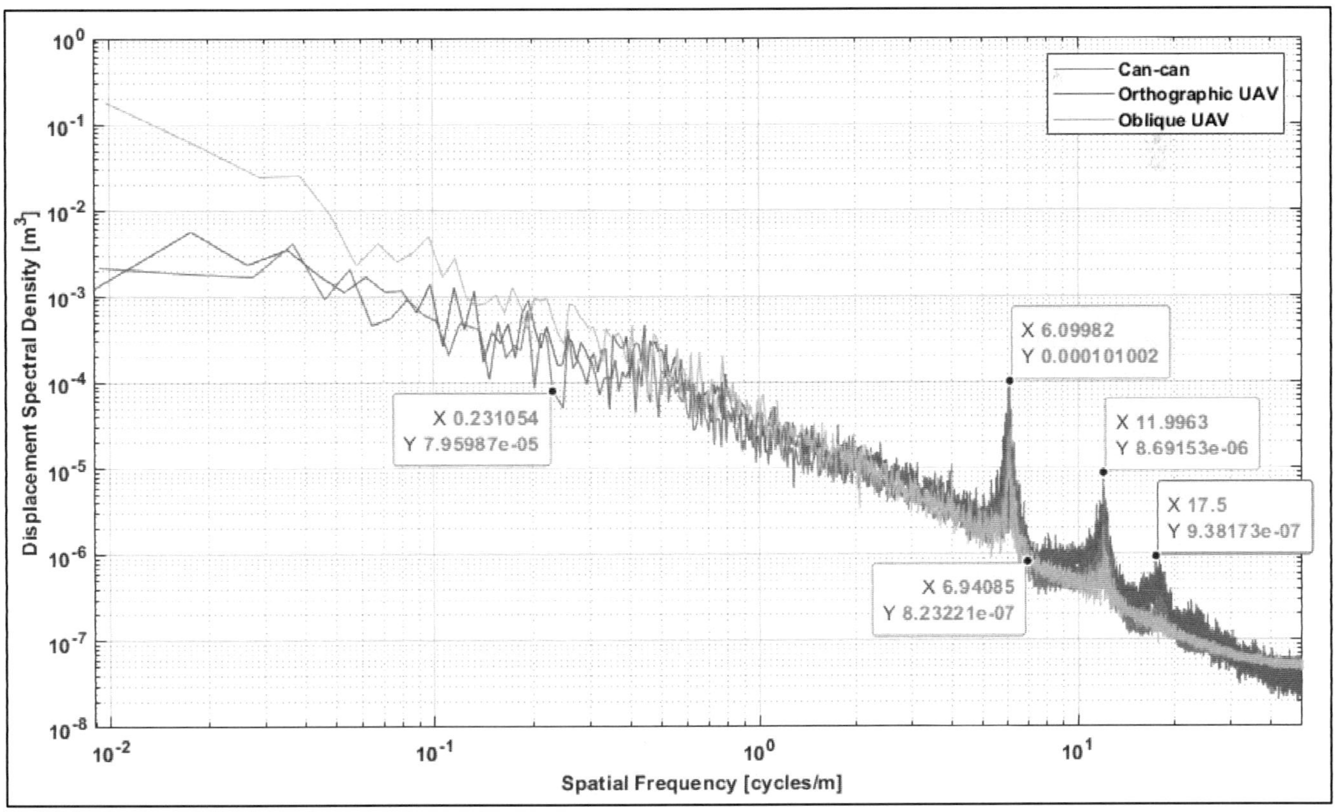

Fig. 7. Comparison of displacement PSDs of terrain profiles measured with the Can-can profilometer and the UAV

4. Conclusion and future work

4.1. Conclusion

This study set out to investigate the possibility of using an UAV to create high-accuracy 3D terrain models for use with vehicle dynamics simulations. A method to create and postprocess 3D point clouds from UAV-captured images of a terrain was developed. The resulting terrain models were analysed by calculating the average displacement PSD along the x-coordinate, and these results compared to a published displacement PSD of the same terrain, with good correlation seen at spatial frequencies of typical interest during vehicle dynamics simulations.

Two methods of collecting images of the terrain were used, the first orthographic images and the second oblique images. The terrain model created from the orthographic images correlates closely with the published displacement PSD determined from the Can-can data in the range of 0.1 to 20 cycles/m. Peaks in the displacement PSD corresponding closely to the cobblestone dimensions and spacing are visible. The terrain model created from oblique images does not match the other terrain models at the lower or higher spatial frequencies – this is a somewhat surprising observation. It was expected that the use of oblique images will give better vertical measurements. It is postulated that this observation may be due to the effect even a small error in the gimbal angle,UAV attitude or roll angle may have on the stereo photogrammetry. This has the added benefit that collecting orthographic images is significantly quicker than collecting oblique images. The presence of other obstacles, such as vegetation and infrastructure, is also of lesser consequence when collecting orthographic images.

The total time needed to create a terrain model was approximately 2 hours for a single person, including the flight, performing photogrammetry and organizing the point cloud into a grid. This is compared to the Can-can profilometer approach, which took a day to conduct tests with a team of three or four people and an additional day for post-processing. The UAV method thus reduces the time and resources required significantly, without demonstrably reducing the accuracy of the final model.

The results presented here are very encouraging and may significantly simplify and ease terrain modelling in the future, but further investigation is needed.

4.2. Future work

It is recommended that the following is considered for further investigation:

- Road profiles that can be used in a multi-body dynamics software package should be created and the simulated response compared. This will indicate if the differences observed in the displacement PSDs in Fig. 7 are significant when conducting vehicle dynamics simulations.
- The investigation should be extended to a wider array of test tracks and terrains. The Belgian paving is a very undulating road, and can be classified as Class D or E according to ISO 8606:2016. It is anticipated that the UAV approach may be less useful on very smooth roads.
- The use of different gimbal, image overlap and GSD settings should be investigated. It is anticipated that a higher overlap ratio may give even better results, but at the expense of increased flight and processing time. Collecting images at multiple altitudes may hold significant advantages to give accurate results across a larger spatial frequency range.
- Creating point clouds with Reality Capture relies on the software's built-in algorithms and a host of settings. At this point, it is unclear how these settings influence the point cloud accuracy. It is recommended that this is investigated in more detail.

5. References

Abulizi, N., Kawamura, A., Tomiyama, K. & Fujita, S. 2016. Measuring and evaluating of road roughness conditions with a compact road profiler and ArcGIS. *Journal of Traffic and Transportation Engineering (English Edition)*, 3, 398-411.

Becker, C. M. & Els, P. S. 2014. Profiling of rough terrain. *International Journal of Vehicle Design*, 64, 240-261.

Botha, T. R. & Els, P. S. 2015. Rough terrain profiling using digital image correlation. *Journal of Terramechanics*, 59, 1-17.

CloudCompare. 2021. *CloudCompare (Version 2.10.3) 3D point cloud and mesh processing software Open Source Project [GPL software]* [Online]. Available: https://www.cloudcompare.org/ [Accessed 14 May 2024].

DJI. 2024. *DJI Mavic 3 Enterprise Specs* [Online]. Available: https://enterprise.dji.com/mavic-3-enterprise/specs [Accessed 8 February 2024].

dSPACE GmbH. 2016. *FTire and ASM - Sophisticated Models for Real-Time Vehicle Dynamics Simulation* [Online]. Available: https://www.cosin.eu/wp-content/uploads/FTire_Artikel_VehicleDynamic_1601213.pdf [Accessed 22 March 2018].

Els, P. S. & Scholtz, O. 2017. Sensitivity of tyre simulation to road profile measurement resolution. *Dynamics of Vehicles on Roads and Tracks Vol 1: Proceedings of the 25th International Symposium on Dynamics of Vehicles on Roads and Tracks (IAVSD 2017), 14-18 August 2017, Rockhampton, Queensland, Australia (1st ed.).* CRC Press.

Epic Games. 2024. *Reality Capture 1.4 Data Sheet* [Online]. Available: https://www.capturingreality.com/assets/Documents/RealityCapture-DataSheet.pdf [Accessed 16 May 2024].

Gillespie, T. D. 1992. *Fundamentals of vehicle dynamics*, Warrendale, PA, SAE International.

Hamersma, H. A. & Els, P. S. 2021. Vehicle suspension force and road profile prediction on undulating roads. *Vehicle System Dynamics*, 1-27.

International Organization for Standardization (ISO) 2016. ISO 8608:2016 - Mechanical vibration - Road surface profiles — Reporting of measured data. Geneva, Switzerland.

Kern, J. V. & Ferris, J. B. Development of a 3-D vehicle-terrain measurement system Part I: equipment setup. Joint North America, Asia-Pacific ISTVS Conference and Annual Meeting of Japanese Society for Terramechanics, 2007.

Lee, J.-H., Lee, S.-H., Kang, D.-K., Na, S.-D. & Yoo, W.-S. 2017. Development of a 3D road profile measuring system for unpaved road severity analysis. *International Journal of Precision Engineering and Manufacturing*, 18, 155-162.

MathWorks. 2016. *MATLAB and Simulink* [Online]. Available: http://www.mathworks.com/ [Accessed 20 July 2016].

McCullough, M., Jayakumar, P., Dasch, J. & Gorsich, D. 2017. The Next Generation NATO Reference mobility model development. *Journal of Terramechanics*, 73, 49-60.

Medinac, F., Bamford, T., Hart, M., Kowalczyk, M. & Esmaeili, K. 2020. Haul road monitoring in open pit mines using unmanned aerial vehicles: A case study at Bald Mountain Mine Site. *Mining, Metallurgy & Exploration*, 37, 1877-1883.

Ngwangwa, H. M., Heyns, P. S., Breytenbach, H. G. A. & Els, P. S. 2014. Reconstruction of road defects and road roughness classification using Artificial Neural Networks simulation and vehicle dynamic responses: Application to experimental data. *Journal of Terramechanics*, 53, 1-18.

Ni, T., Li, W., Zhao, D. & Kong, Z. 2020. Road Profile Estimation Using a 3D Sensor and Intelligent Vehicle. *Sensors*, 20, 3676.

https://doi.org/10.56884/TK3ZX342

Sayers, M. W. & Karamihas, S. M. 1998. The little book of profiling: basic information about measuring and interpreting road profiles. University of Michigan, Ann Arbor, Transportation Research Institute.

Scholtz, S. J. & Hamersma, H. A. 2024. Investigating off-road vehicle lateral stability with integrated chassis control. *Vehicle System Dynamics*, 1-22.

Taheri, S., Sandu, C., Taheri, S., Pinto, E. & Gorsich, D. 2015. A technical survey on Terramechanics models for tire–terrain interaction used in modeling and simulation of wheeled vehicles. *Journal of Terramechanics*, 57, 1-22.

Varunjikar, T. M., Vemulapalli, P. K. & Brennan, S. N. Multi-body vehicle dynamics simulation based on measured 3D terrain data. Proc., 3rd Int. Conf. on Road Safety and Simulation, 2012.

4859 | Trafficability conditions for military wheeled trucks on cultivated fields

Kersti Vennik[a,*], Priit Põdersalu[a], Tamur Kusnets[a], Tõnu Tõnutare[b]

[a] Department of Academic Studies, Estonian Military Academy, Estonia
[b] Chair of Soil Science, Estonian University of Life Sciences, Estonia

* Corresponding author: kersti.vennik@mil.ee

ABSTRACT

Military maneuvers performed with different types of vehicles often happen outside of the road network. Essentially, areas trafficked by tracked and wheeled machinery can be divided as arable land and natural areas, depending on the impact of human activity. The latter type typically includes natural grasslands, forest and swamp areas. From the perspective of soil strength, the main difference between these land types lies in the compaction state. Compared to natural grasslands, cultivated fields are in a precompacted state. The most well-known mobility performance parameter for military vehicles is the Vehicle Cone Index (VCI). In short, this entails the minimum soil strength required for a successful passage, whereas the soil's shear strength is determined with a cone penetrometer. There are many reports and scientific papers describing trafficability experiments and validation of VCI as a Go/No-Go indicator for different soils and vehicles. However, the majority of the test results concern natural grassland areas while the applicability of the VCI calculation concept for agriculturally used areas is lacking. The aim of this study is to carry out trafficability tests with wheeled military trucks in order to validate VCI as a mobility performance parameter for cultivated fields. Moreover, changes in soil strength and moisture conditions were monitored for selected field parcels throughout the warmer half of the year. The 70 kN and 125 kN military trucks were used as test vehicles. The soils under observation included sandy loam and clay textures as well as highly organic soils. The paper provides an overview of the experiment's results and a discussion about trafficability conditions on cultivated areas.

Keywords
Vehicle cone index
Military truck
Cultivated fields
Sandy loam soil
Silt loam soil

1. Introduction

Armed forces cannot sustain or fight successfully without constantly available supplies of ammunition, fuel, water, food and, for military logistic support, trucks with different weight and payload capabilities (Gander and Biass, 2016; Hohl, 2007). This study concentrates on the trafficability prediction of military wheeled trucks.

The parameter called Vehicle Cone Index (VCI) and the trafficability prediction system connected with VCI was developed in the 1950s by the US Army Corps of Engineers. According to the International Society for Terrain-Vehicle Systems (ISTVS) standards, VCI is the "minimum soil strength in the critical soil layer, in terms of rating cone index for fine grained soils or in cone index for coarse grained soils, required for a specific number of passes of a vehicle usually one pass (VCI_1) or 50 passes (VCI_{50})." (He et al., 2020). The VCI approach is widely used, e.g. the NATO Reference Mobility Model has adopted VCI in its methodology for predicting the performance of wheeled and tracked vehicles (Wong et al., 2020). VCI can be determined with field testing and also predicted based on vehicle characteristics (Stevens et al., 2013). Cone index (CI) is determined with cone penetrometer (30° circular cone with a 3.23 cm^2 base area). For rating cone index (RCI), CI is multiplied by remolding index (RI) whereas RI is measured by special tools (Wong, 2008). In addition to mineral soils, VCI_{MK} and its calculation formula is used for peat and muskeg soils (Priddy, 1995). Calculation formulas of VCI_1 and VCI_{MK} for military vehicles were presented by Priddy (1995) and will not be reproduced here. In general, if the VCI_1 value in the critical layer is smaller than RCI for the representative area, the vehicle will not be able to conduct a successful pass on it. On peat soil, passability evaluation is based on the comparison between VCI_{MK} and CI values. The critical layer mentioned in the VCI definition depends on vehicle weight and running gear properties. Most frequently, the critical layer for military vehicles is at the depth of 15–30 cm (Collins, 1971).

There have been numerous experiments for clarifying VCI values for different vehicles and soil types. However, all the testing was carried out on military training areas, forested areas, and at the beach – thus, at natural areas (Radforth and Rush, 1964; Rush and Schreiner, 1966; Kennedy and Rush, 1968). There is a dearth of information covering agricultural soils and the applicability of the VCI approach on this type of soil. To a certain extent, the problems associated with the mechanical behavior of vehicles on agricultural soil are similar to natural grassland. Nevertheless, cultivated fields tend to be more compacted and typically the organic matter content as well as the grass cover is less tenuous compared to natural grassland soils (Vennik, 2019). Actual soil compaction as well as the shear strength state of cultivated fields depends on soil properties like texture, moisture content, organic matter content, and the cultivation state (Jabro et al., 2016), among other things. Conventional tillage, including ordinary ploughing can disrupt the soil structure and decrease the bulk density of the surface layer (Erbach et al., 1992; Osunbitan et al., 2005), i.e. up to a depth of 22–26 cm. Again, conservation tillage techniques like reduced-till farming do not include soil loosening and disturbance and, thus, soil will remain in a more compacted state (Blanco-Canqui and Ruis, 2018). Typically, the existence of farmland and natural grassland is illustrated on large-scale topographical maps.

The aim of this study was to test the applicability of the VCI approach on agriculturally managed field and, in addition, monitor the changes in trafficability conditions throughout the warmer half-year.

https://doi.org/10.56884/VOHT3ERV

2. Methods

Vehicle experiments and soil properties' measurements were carried out in 2023 in southern Estonia. As soil shear strength is, before all, influenced by moisture content (Stevens et al., 2016), all the experiments and measurements were conducted in spring and autumn when the soils were expected to have higher moisture contents. Cultivated fields for experiment and measurement sites were selected on the Estonian Basic Map (Estonian Basic Map, 2024) where arable lands and natural grasslands are depicted separately. The descriptions of vehicle experiment sites and dates are provided in Table 1. Test vehicles included the following military trucks: DAF YAD 4442 (total load 70 kN) and MAN LKV 4610 (total load 125 kN). The tire pressure of vehicles were 550 kPa and 600 kPa, respectively. Test line was at least 100 m long and machines conducted straight line movement and turns. Tracks left behind were measured for depth and sheared soil surface, or noting clearly visible tire patterns at the bottom of tracks. Cone index values up to 1 m from the tracks were measured with a half-digital handheld Humboldt cone penetrometer (30° conical tip, 3.33 cm2 cone base area, provided by Geotest Instrument Corp) for ten times. The remolding index was determined for the depth of 15–30 cm in three repetitions. Volumetric moisture content was measured at depths of 0, 15, and 30 cm using a FieldScout TDR 300 with 7.62 cm tips. Undisturbed soil samples were collected for bulk density determination with steel cylinders (5.3 cm of diameter, 4 cm of height) in three repetitions for depth of 0, 15 and 30 cm. Soil was collected from topsoil as well as subsoil for organic matter and texture analysis. The Tjurin method was applied to determine soils' organic carbon content (SOC) (Vorobyova, 1998). Soils' particle size distribution was determined with the laser diffraction method (Polakowski et al., 2023), using Malvern Mastersizer 3000. The soils' particle size distribution as well as the texture and organic matter content from the vehicles' experiment sites are presented in Table 2.

In addition to the vehicle experiments, soil moisture content was measured on different cultivated fields in southern Estonia. Volumetric moisture content was measured with FieldScout TDR 300 using 7.62 cm tips in the depth of 0, 15 and 30 cm, with five repetitions.

Table 1
Vehicle experiment sites' description

Date	Site	Description or cultivation state
04-11	SaLo1, 2	Cultivated in autumn: reduced till
04-11	SaLo3	Ploughed in autumn 2022
04-11	SaLo4	No-till after crop harvesting in autumn 2022
04-11	SiLo1, 2	Cultivated grassland, grass cover height up to 0.3 m
10-03	SaLo5	Cultivated grassland, grass cover height up to 0.5 m
10-03	SaLo6	Cultivated grassland, grass cover height up to 1 m
10-03	Pt1, 2, 3	Cultivated grassland, grass cover height up to 1 m
10-03	Pt4	No-till after crop harvesting in autumn 2023
11-07	SaLo7, 9	No-till after cabbage harvesting in autumn 2023
11-07	SaLo8	No-till after cabbage harvesting in autumn 2023, water cover up to the depth of 7 cm
11-07	SaLo10, 11	No-till after cabbage harvesting in autumn 2023
11-07	SiLo3, 4, 5, 6, 8	Cultivated grassland, grass cover height up to 0.4 m
11-07	SiLo7	No-till after crop harvesting in autumn 2023

Table 2
Physical properties of the soils from vehicle test sites

Soil	Layer [cm]	Sand [%]	Silt [%]	Clay [%]	Texture [by FAO]	SOC [% by weight]
SaLo1	0–38	53.2	40.6	6.2	Sandy loam	1.4
	38–55	53.0	40.2	6.8	Sandy loam	0.3
SaLo2	0–47	58.4	36.6	5.0	Sandy loam	1.9
	47–55	64.5	31.3	4.2	Sandy loam	1.1
SaLo3	0–28	69.0	27.1	3.9	Sandy loam	1.3
	28–55	78.2	18.7	3.1	Loamy sand	0.4
SaLo4	0–26	59.6	34.5	5.9	Sandy loam	1.2
	26–55	59.2	34.7	6.1	Sandy loam	0.6
SaLo5	0–35	73.5	24.3	2.2	Loamy sand	2.3
	35–55	71.2	25.8	3.0	Sandy loam	2.8
SaLo6	0–29	72.3	25.8	1.9	Loamy sand	1.7
	29–55	70.7	26.9	2.4	Sandy loam	0.7
SaLo7,8,9	0–38	47.4	47.2	5.4	Sandy loam	1.4
	38–55	50.0	43.1	6.9	Sandy loam	0.4
SaLo10,11	0–26	46.2	48.3	5.5	Sandy loam	2.8
	26–55	18.6	74.8	6.6	Silt loam	2.5
SiLo1	0–30	24.8	62.9	12.3	Silt loam	2.6
	30–55	18.1	72.0	9.9	Silt loam	0.4
SiLo2	0–20	22.3	66.7	11.0	Silt loam	6.0
	20–55	6.5	82.7	10.8	Silt	0.9
SiLo3,4,5	0–15	14.6	65.0	20.4	Silt loam	2.3
	15–55	8.9	70.6	20.5	Silt loam	0.6
SiLo6	0–22	13.2	69.5	17.3	Silt loam	1.3
	22–55	9.9	70.3	19.8	Silt loam	0.7
SiLo7	0–15	15.6	68.6	15.8	Silt loam	1.2
	15–55	16.7	44.5	38.8	Silty clay loam	0.4
SiLo8	0–30	14.3	63.5	22.2	Silt loam	2.1
	30–55	13.3	62.6	24.1	Silt loam	1.3
Pt1, 2	0–25	-	-	-	Peat	5.3
	25–55	-	-	-	Peat	3.8
Pt3	0–30	-	-	-	Peat	15.1
	30–55	-	-	-	Peat	17.1
Pt4	0–15	-	-	-	Peat	17.7
	15–55	-	-	-	Peat	9.8

Sand (2–0.063 mm); silt (0.063–0.002mm); clay (<0.002 mm)

3. Results

3.1. Fine-grained soils' vehicle experiment results

The calculated VCI_1 values for DAF and MAN trucks are 39 and 46, respectively. The measured RCI values for fine-grained soils' experiment sites and calculated VCI_1 values for test vehicles are presented in Fig. 1. Overall, tested soils reveal that silt loam soils tend to have lower RCI values in the critical layer in comparison to the sandy loam soils. Test vehicles got stuck for RCI values much higher than the calculated VCI_1 values. For sandy loam soil experiment sites, the highest RCI value in layer 15–30 cm where both test vehicles got stuck was 154, and the corresponding RCI value for silt loam soils was 239. Thus, soil strength as depicted by RCI was approximately 4.5 times higher at sites where vehicles experienced immobilization, as proposed VCI_1.

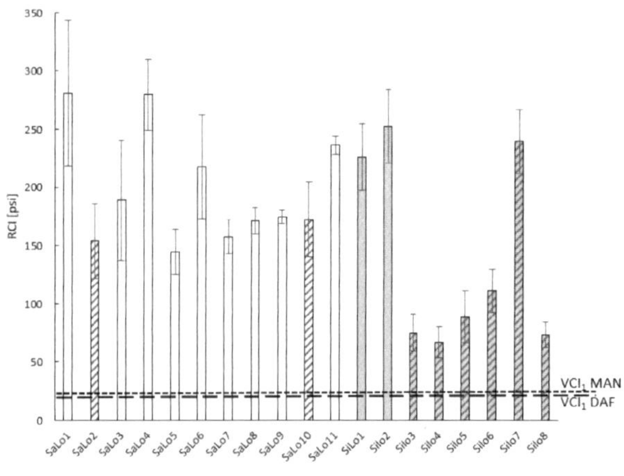

Fig. 1. RCI values for the 15–30 cm layer and VCI_1 values for MAN and DAF. Hatched columns indicating RCIs are where either DAF or MAN were immobilized. Vertical bars denote standard deviation

RI values for experiment soils in the critical layer are presented in Table 3. Overall, despite the compacted state, cultivated fields still tend to lose strength due to disturbance, i.e. RI values are lower than 1. Soils with higher clay and silt particles, i.e. silt loam soils, tend to have even lower RI values.

Table 3
RI values for fine-grained soils' experiment sites for 15–30 cm layer

Site	Average RI	Site	Average RI
SaLo1	1	SiLo1	1
SaLo2	0.7	SiLo2	1
SaLo3	1	SiLo3	0.6
SaLo4	1	SiLo4	0.6
SaLo5	1	SiLo5	0.6
SaLo6	1	SiLo6	0.6
SaLo7	0.7	SiLo7	0.9
SaLo8	0.7	SiLo8	0.6
SaLo9	0.6		
SaLo10	0.8		
SaLo11	0.8		

Fine-grained experiment soils' bulk density and volumetric moisture content data are presented in Table 4 and vehicles' rut depth as well as information on immobilization and slippage are presented in Table 5.

Table 4
Bulk density and volumetric moisture content data for fine-grained soils experiment sites

Site	Layer [cm]	Bulk density [g/cm³]	Volumetric moisture content [%]
SaLo1	0–15	1.41±0.02	21.0±4.6
	15–30	1.56±0.06	23.7±2.6
	30–45	1.57±0.07	22.1±3.5
SaLo2	0–15	1.32±0.04	43.0±3.0
	15–30	1.53±0.02	35.1±3.8
	30–45	1.57±0.06	35.0±2.8
SaLo3	0–15	1.03±0.05	12.7±3.3
	15–30	1.04±0.06	18.0±3.6
	30–45	1.59±0.03	16.3±3.9
SaLo4	0–15	1.43±0.02	20.0±4.8
	15–30	1.56±0.03	19.9±3.9
	30–45	1.57±0.03	19.9±4.3
SaLo5	0–15	1.09±0.14	26.0±2.5
	15–30	1.15±0.03	11.9±2.7
	30–45	1.31±0.02	9.3±2.6
SaLo6	0–15	1.21±0.02	40.1±4.7
	15–30	1.33±0.04	34.6±3.0
	30–45	1.56±0.03	35.3±4.0
SaLo7	0–15	1.45±0.03	29.6±4.4
	15–30	1.59±0.03	29.8±0.9
	30–45	1.67±0.05	29.5±1.0
SaLo8	0–15	1.45±0.03	40.4±4.
	15–30	1.59±0.03	38.1±6.0
	30–45	1.67±0.05	28.7±1.6
SaLo9	0–15	1.46±0.03	32.7±3.9
	15–30	1.66±0.05	27.6±1.0
	30–45	1.67±0.01	26.1±0.7
SaLo10	0–15	1.27±0.08	39.4±1.6
	15–30	1.40±0.05	35.1±2.8
	30–45	1.41±0.09	34.3±2.7
SaLo11	0–15	1.27±0.08	34.6±2.1
	15–30	1.40±0.05	34.6±4.1
	30–45	1.41±0.09	30.9±1.7
SiLo1	0–15	1.16±0.04	39.0±4.3
	15–30	1.25±0.11	39.3±3.3
	30–45	1.68±0.06	27.1±5.4
SiLo2	0–15	0.82±0.04	42.7±5.4
	15–30	0.92±0.03	43.8±5.2
	30–45	1.43±0.07	45.6±7.4
SiLo3	0–15	1.08±0.09	60.3±1.1
	15–30	1.38±0.06	62.2±1.2
	30–45	1.49±0.02	59.9±3.4
SiLo4	0–15	1.08±0.09	61.3±0.9
	15–30	1.38±0.06	57.8±3.6
	30–45	1.49±0.02	61.3±1.8
SiLo5	0–15	1.08±0.09	57.7±1.3
	15–30	1.38±0.06	55.3±2.5
	30–45	1.49±0.02	66.2±3.5
SiLo6	0–15	1.54±0.04	58.9±1.2
	15–30	1.61±0.03	60.5±3.7
	30–45	1.45±0.04	70.1±1.5
SiLo7	0–15	1.57±0.05	54.9±0.8
	15–30	1.66±0.02	52.6±5.3
	30–45	1.47±0.04	50.8±3.5
SiLo8	0–15	1.42±0.06	64.1±1.6
	15–30	1.37±0.02	66.0±1.7
	30–45	1.42±0.05	74.3±2.6

For the heavier test vehicle, i.e. MAN, the ruts left behind after traffic were slightly deeper than the ruts of DAF (Table 5). On sandy loam soils the immobilization occurred at two sites for both vehicles – at SaLo2 and SaLo10. However, vehicles' interaction with soil were different. At SaLo2 soils, test vehicles sank deep in the soil, i.e. rut depth was 0.14–0.17 m for DAF and 0.13–0.19 m for MAN (Table 5). Also, remarkably high walls surrounded the tracks, indicating plastic flow on the top soil. Although, the subsoil layer's RI value 0.7 indicated loss of strength due to the stresses (Table 3), the bulk density at 15–30 cm was higher in comparison to the 0–15 cm layer – 1.53 g/m^3 versus 1.32 g/m^3 –, and moisture content was slightly smaller in comparison to the surface layer – 35.1% versus 43.0% (Table 4). Thus, the denser frictional material laying deeper was hindering vehicle sinkage up to the belly. Vehicles' slipping was not noticed. The reason for immobilization was probably high motion resistance due to the deep ruts. All the soft and water saturated soil material from the side of the ruts was loosely displaced to the rut bottom. Again for SaLo10 the ruts were quite shallow, 0.06–0.08 m for both vehicles (Table 5). Vehicles clearly slid to both sides of the track while moving, but there was no remarkable shearing off or sinkage into the deeper layer.

Immobilization occurred due to the lack of thrust as silty soil material did not provide enough shear strength. Thus, for both described soils there were enough soil strength for 15–30 cm, but the wet and low strength conditions of the top layer seriously affected trafficability.

Table 5
Fine-grained soils' rut depth and slippage data from experiment sites

Site	DAF Rut depth [m]	DAF Tire slippage; immobilization	MAN Rut depth [m]	MAN Tire slippage; immobilization
SaLo1	0.05–0.08	CTP on TB	0.05–0.08	CTP on TB
SaLo2	0.14–0.17	Got stuck, loose soil in track	0.13–0.19	Got stuck, loose soil in track
SaLo3	0.05–0.09	CTP on TB	0.07–0.10	CTP on TB
SaLo4	0.05–0.08	CTP on TB	0.05–0.09	CTP on TB
SaLo5	0.02–0.03	CTP on TB	0.02–0.04	CTP on TB
SaLo6	0.01–0.05	CTP on TB	0.02–0.07	CTP on TB
SaLo7	0.04–0.09	CTP on TB	0.04–0.10	CTP on TB
SaLo8	0.04–0.09	SSS	0.05–0.10	SSS
SaLo9	0.03–0.08	CTP on TB	0.03–0.09	CTP on TB
SaLo10	0.06–0.08	Got stuck, high slippage of vehicle, SSS	0.06–0.08	Got stuck, high slippage of vehicle, SSS
SaLo11	0.03–0.05	Slightly SSS	0.04–0.06	Slightly SSS
SiLo1,2	0.03–0.05	CTP on TB	0.03–0.07	CTP on TB
SiLo3	0.03–0.12	Slightly SSS	0.14–0.25	Got stuck, high slippage of vehicle, SSS
SiLo4	0.23–0.25	Got stuck, high slippage of vehicle, SSS	0.16–0.29	Got stuck, high slippage of vehicle, SSS
SiLo5	0.04–0.09	SSS	0.18–0.27	Got stuck, high slippage of vehicle, SSS
SiLo6	0.05–0.08	SSS	0.14–0.27	Got stuck, high slippage of vehicle, SSS
SiLo7	0.03–0.06	Got stuck, high slippage of vehicle, SSS	0.05–0.10	Got stuck, high slippage of vehicle, SSS
SiLo8	0.09–0.22	Got stuck, high slippage of vehicle, SSS	0.12–0.25	Got stuck, high slippage of vehicle, SSS

CTP on TB - clear tire pattern on track bottom; SSS - sheared soil surface.

The relative importance of water content and also soil composition to trafficability is revealed by comparing the physical properties of SaLo2 and SaLo3, as well as SaLo10, SaLo11 and SaLo8. The field of SaLo3 was ploughed in autumn; thus, the surface soil layer was in a loose state and bulk density was quite low – 1.03 g/m³ for 0–30 cm (Table 4). While the vehicle moved, loose material was compacted, producing only relatively shallow ruts – up to 0.1 m (Table 5). The moisture content in the surface layer of SaLo2 was 43.0% and for SaLo3, it was 12.7%. Thus, although the bulk density of SaLo2 was high, its rich water content played a crucial role, making soil move to the sides instead of compaction. Again for loose soil at a dry state (SaLo3), frictional soil material was packed together without further sinkage.

At the SaLo11 site, both test vehicles carried out successful passages, but that was not the case for the nearby lying soil SaLo10. Bulk density was the same for both sites but the surface layer water content was 34.6% at SaLo11 and 39.4% at SaLo10. However, at SaLo8, although water was covering the soil surface and its topsoil moisture content value was 40% (Table 4), both vehicles could travel without a problem. A possible explanation for soil's different behavior for the same water content could lay in the difference of soil composition. At SaLo10, silt and organic matter content were higher than at SaLo8 (Table 2). Overall, it can be estimated that for sandy loam soils, the water content that reduces trafficability tends to be near 40%.

Figure 2 illustrates CI values for different depths at SaLo1, SaLo2, and SaLo3. Overall, there were no significant differences in CI values according to depth, although trafficking conditions were different (Table 5). For SaLo1 and SaLo3, rut depth was shallow – 0.05–0.10 cm, there was no slippage of soil surface, and the vehicles could move freely. At SaLo2, test trucks got stuck. Thus, a penetrometer alone cannot be used to describe soil behavior under vehicle pressure

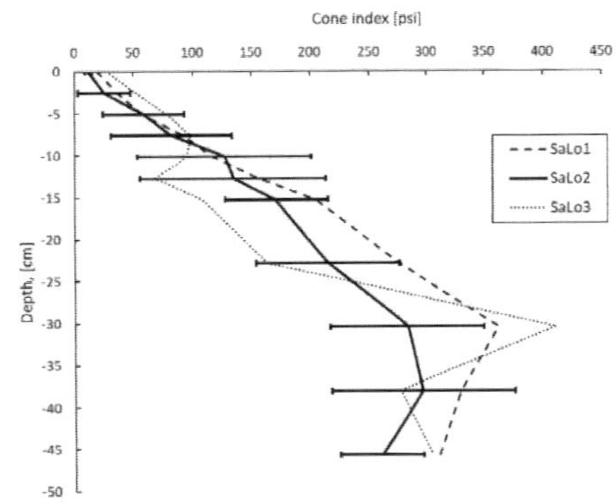

Fig. 2. Measured cone indexes as a function of depth at sites SaLo1, SaLo2 and SaLo3. Horizontal bars indicate standard deviation at SaLo2

For silty loam soils, also two different mechanisms of soil failure were observed under stress caused by wheels. At sites SiLo3, SiLo4, SiLo5, SiLo6 and SiLo8, either one or both of the test trucks sank deeply into the soil and got stuck (Table 5). At site SiLo7, test vehicles were slipping without any further sinkage on a soil layer of 0.05–0.10 cm. Slippage was so intensive that the entire surface was sheared off (Fig. 3) and vehicles could not continue traveling. A comparison of the water contents of all the mentioned soils (Table 4) reveal that the boundary value for moisture content caused immobilization because either vertical or lateral flow of soil particles is 54%.

Fig. 3. Sheared soil surface at site SiLo7

CI values for SiLo2, 4 and SiLo7 are presented in Fig. 4. CI values were smallest for all the measured depths at SiLo4 and both vehicles got stuck there (Table 5) because of deep sinkage. At SiLo2, the test vehicles could move freely, and at SiLo7, the vehicles were immobilized due to the slippage on soil surface. As can be seen from the CI profile, SiLo7 and SiLo2 had identical values for 0–10 cm, revealing that CI values alone do not indicate the risk of soil failure.

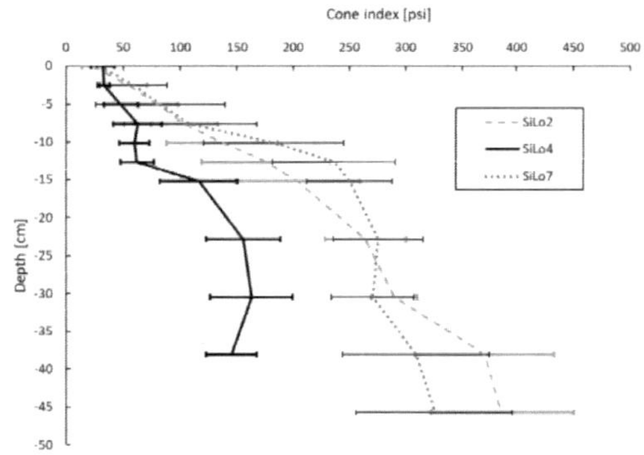

Fig. 4. Measured cone indexes as a function of depth at sites SiLo2, SILo4 and SiLo7. Horizontal bars indicate standard deviation.

3.2. Peat soils' vehicle experiment results

The calculated VCI_{MK} value is 51 for DAF and 72 for MAN. Figure 5 illustrates the measured CI and RCI values for the 15–30 cm layer. According to the standardized official VCI methodic, RI is not measured to determine trafficability on peat soils (Priddy, 1995). In this study RI was measured (Table 6), and based on the limited number of experiments it can be argued that this is also an important parameter for cultivated peat soils. First, as visible from Fig. 5, the measurement on Pt2 alone reveals that the VCI_{MK} for both vehicles matched with RCI and described the immobilization conditions correctly. In general, the immobilization of test vehicles occurred due to the deep sinkage on peat soil sites (Table 7).

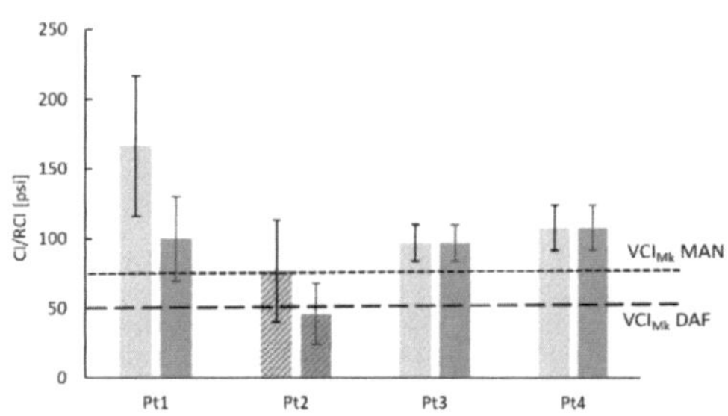

Fig. 5. RCI values for the 15–30 cm layer and VCI_{MK} values for MAN and DAF. Hatched columns indicating RCI where either DAF or MAN was immobilized. Vertical bars denote standard deviation

Table 6
RI values for peat soils' experiment sites for 15–30 cm layer

Site	Average RI
Pt1	0.6
Pt2	0.6
Pt3	1
Pt4	1

Table 7
Peat soils' rut depth and slippage data for experiment sites

	DAF		MAN	
Site	Rut depth [m]	Tire slippage; immobilization	Rut depth [m]	Tire slippage; immobilization
Pt1	0.04–0.06	No slippage	0.04–0.08	No slippage
Pt2	0.10–0.20	Got stuck, high slippage	0.13–0.22	Got stuck, high slippage
Pt3	0.04–0.08	No slippage, CBT on TB	0.08–0.13	No slippage, CBT on TB
Pt4	0.06–0.10	No slippage CBT on TB	0.07–0.13	No slippage; CBT on TB

CTP on TB - clear tire pattern on track bottom

Table 8
Bulk density and volumetric moisture content data for peat soils at experiment sites

Site	Layer, cm [m]	Bulk density, [g/cm³]	Volumetric moisture content [%]
Pt1	0–15	0.63±0.04	34.1±7.3
	15–30	1.01±0.08	37.4±1.1
	30–45	1.32±0.03	37.3±0.7
Pt2	0–15	0.63±0.04	62.7±3.1
	15–30	1.01±0.08	51.4±5.1
	30–45	1.32±0.03	54.7±6.5
Pt3	0–15	0.48±0.04	43.3±6.4
	15–30	0.50±0.03	39.0±6.9
	30–45	0.52±0.03	40.8±5.9
Pt4	0–15	0.48±0.01	44.2±13.3
	15–30	0.59±0.01	48.2±13.6
	30–45	0.53±0.04	59.4±11.1

In comparison to mineral soils, bulk density is much lower for cultivated peat soils (Table 8). Typically peat soil needs to be drained before it can be cultivated. The water table depth is increased and, as a result, organic material will decompose and mineralize (Tuohy et al., 2023). But, as can be concluded based on collected data, depending on soil and drainage conditions, bulk density can remain more or less unchanged. Test vehicles got stuck in the field of Pt2 by sinking deeply—over a depth of 0.20 meters. Thus, based on very limited number of measurements, it can generalized that the content of moisture that limits trafficability is approx. 52% for layer 0–30 cm.

3.3. Volumetric moisture content's measuring results

During the experiments, the limiting moisture contents of sandy loam, silt loam, and peat soils for cultivated fields were determined. To monitor the change in the moisture content and determine the probability of encountering untrafficable cultivated areas during spring and autumn, volumetric water content was determined of different soils in southern Estonia. Results are illustrated on Fig. 6, 7 and 8. The results revealed that in spring and autumn, the most difficult farmlands for wheeled machinery are most probably peat fields, but also a considerable number of silt loam soil areas. For sandy soils, all these moist spots are easy to notice since all five sites that exceed critical moisture value (Fig. 6) were dark brown or even blackish, or these areas were covered by water, or there were signs of excess water drying up. The moisture content of peat soils (Fig. 8) was usually highest for the deepest measured depth. As farmers probably encounter problems regarding cultivating these water saturated fields, too wet peat areas were often covered in bushes and tall grass. Although the topographic map does not capture this change yet. It is very difficult is to visually evaluate the moisture level of a field and, thus, trafficability conditions for silt loam soils (Fig. 7).

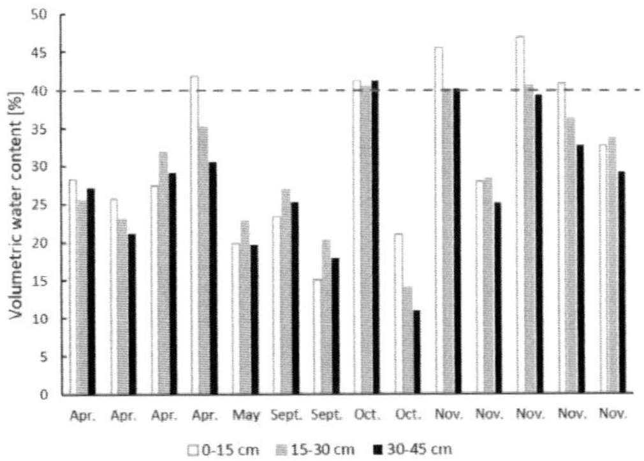

Fig. 6. Volumetric water content for sandy loam soils and critical volumetric water content value

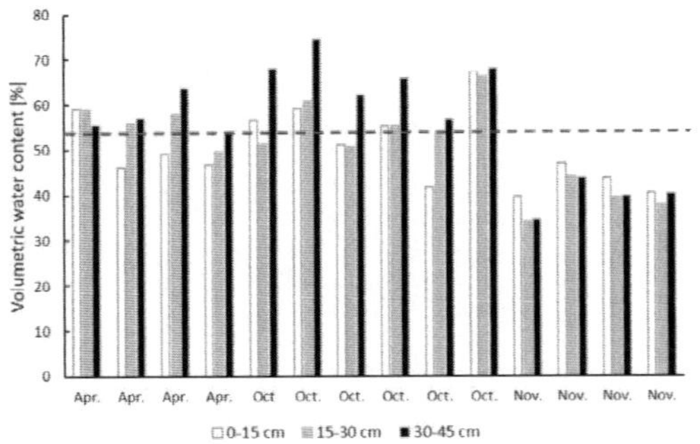

Fig. 7. Volumetric water content for silt loam soils and critical volumetric water content value

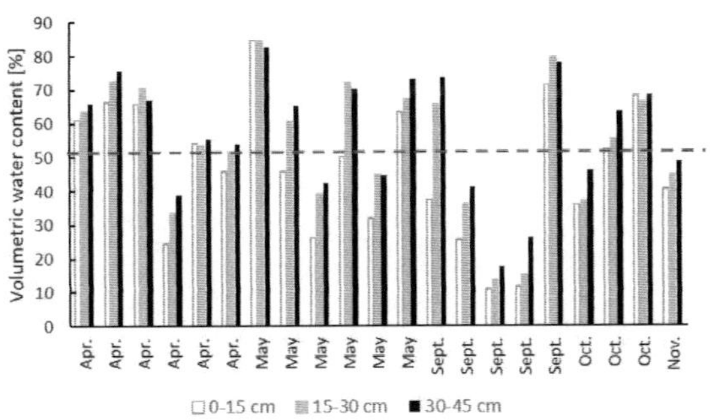

Fig. 8. Volumetric water content of peat soils and critical volumetric water content value

4. Conclusions

In this study the applicability of the VCI approach was evaluated to determine the trafficability of agricultural fields, and wheeled vehicle experiments were carried out on sandy loam, silt loam and peat soil. Altogether 27 experiments were conducted with two wheeled trucks. In general, VCI_{MK} seems to be a suitable characteristic for predicting the trafficability of wheeled tracks on peat fields when using the RCI values for the 15–30 cm layer. The well-known prediction system is less suitable for cultivated fields with mineral soils. Primarily, the conditions of the top soil layer have a significant impact on the trafficability of wheeled machinery. The limiting moisture contents for tested soils were determined and used to evaluate overall trafficability conditions. As for grasslands, the trafficability conditions are most cumbersome for peat fields. However, the strength state of silt and clay-rich soils also needs to be controlled before trafficking.

5. References

Blanco-Canqui, H., Ruis, S.J., 2018. No-tillage and soil physical environment. Geoderma 326, 164–200. DOI: 10.1016/j.geoderma.2018.03.011

Collins, J.G., 1971. Forecasting trafficability of soils. Report 10. Relations of strength to other properties of fine-grained soils and sands with fines. U.S. Army Engineer Waterways Experiment Station, Vicksburg, Mississippi.

Erbach, D.C., Benjamin, J.G., Cruse, R.M., Elamin, M.A., Mukhtar, S., Choi, C.H., 1992. Soil and corn response to tillage with paraplow. Trans. ASAE 35, 1347–1354.

Estonian Basic Map, 2024. https://geoportaal.maaamet.ee/eng/spatial-data/topographic-maps/estonian-basic-map-1-10-000-p306.html.

Gander, T.J., Biass, E-H., 2016. The Unsung heroes. Armada International, August/September 2016. Compendium Lorries & Trucks, 2–26.

He, R., Sandu, C., Mousavi, H., Shenvi,M.N., Braun, K., Kruger, R, Els P.S., 2000. Updated standards of the International Society for Terrain-Vehicle Systems. J. Terramech. 91, 185–231.

Hohl, G.H., 2007. Military terrain vehicles. J. Terramech. 44 (1), 23–34.

Jabro, J.D., Iversen, W.M., Stevens, W. B., Evans, R.G., Mikha M.M, Allen, B.L., 2016. Physical and hydraulic properties of a sandy loam soil under zero, shallow and deep tillage practices. Soil Tillage Res. 159, 67–72.

Kennedy, J.G., Rush, E.S., 1968. Trafficability of soils. Development of revised mobility index formula for šelf-propelled wheeled vehicles in fine-grained soils. TM No. 3-240. U.S. Army Engineer Waterways Experiment Station, Vicksburg, Mississippi.

Osunbitan, J.A., Oyedele, D.J., Adekalu, K.O., 2005. Tillage effects on bulk density: hydraulic conductivity and strength of a loamy sand soil in southwestern Nigeria. Soil Tillage Res. 82, 57–64.

Polakowski, C., Makó, A., Sochan, A., Ryżak, M., Zaleski, T., Beczek, M., Mazur, R., Nowiński, M., Turczański, Orzechowski, M., Smólczyński, S., Kassai, P., Bieganowski, A., 2023. Recommendations for soil sample preparation, pretreatment, and data conversion for texture classification in laser diffraction particle size analysis. Geoderma 430, 1–13.

Priddy, J., 1995. Stochastic vehicle mobility forecasts using the nato reference mobility model. Report 3. Database Development for Statistical Analysis of the NRMM II Cross-Country Traction Empirical Relationships. Technical Report GL-95-8. https://scholar.google.com/citations?view_op=view_citation&hl=en&user=9Jx-n3MAAAAJ&citation_for_view=9Jx-n3MAAAAJ:u5HHmVD_uO8C

Radforth, N.W., Rush, E.S., 1964. Trafficability tests on confined organic terrain (muskeg). Report I. Summer 1961 tests. U.S. Army Engineer Waterways Experiment Station, Vicksburg, Mississippi.

Rush, E.S., Schreiner, B.G., 1966. Trafficability tests on unconfined organic terrain (muskeg). Report I. Summer 1963 tests. U.S. Army Engineer Waterways Experiment Station, Vicksburg, Mississippi.

Stevens, M.T., Towne, B.W., Mason, G.L., Priddy, J.D., Osorio, J.E., Barela, C.A., 2013. Procedures for One-Pass Vehicle Cone Index (VCI1) Determination for Acquisition Support. Geotechnical and Structures Laboratory. ERDC/GSL SR-13-2.

Stevens, M.T., McKinley, G.B., Vahedifard, F., 2016. A comparison of ground vehicle mobility analysis based on soil moisture time series datasets from WindSat, LIS, and in situ sensors. J. Terramech. 65, 49–59.

Tuohy, P., O' Sullivan, L., Bracken C.J., Fenton, O., 2023. Drainage status of grassland peat soils in Ireland: Extent, efficacy and implications for GHG emissions and rewetting efforts. Journal of Environmental Management 344, 118391. https://doi.org/10.1016/j.jenvman.2023.118391

Vorobyova, L.A., 1998. Chemical Analysis of Soils. Moscow University Press, Moscow.

Wong, J.Y., 2008. Theory of ground vehicles (4th ed.). NJ: John Wiley and Sons, Inc.Wong, J.Y.

Jayakumar, P., Toma, E., Preston-Thomas, J., 2020. A review of mobility metrics for next generation vehicle mobility models. J. Terramech. 87, 11–20.

Vennik, K., 2019. The effect of military vehicles on rut formation on Estonian soils and natural recovery of the ruts. Doctoral Thesis of the Estonian University of Life Sciences. https://dspace.emu.ee/server/api/core/bitstreams/e487c881-ad67-4470-800a-5771883ea2d1/content.

4927 | Tip angle dependence for resistive force into dry granular materials at shallow cone penetration

Naoki Iikawa [a, b, c, *] and Hiroaki Katsuragi [c]

[a] Development Division, Komatsu Ltd, 3-25-1 Shinomiya, Hiratsuka, 254-0014, Kanagawa, Japan

[b] Komatsu MIRAI Construction Equipment Cooperative Research Center, Osaka University, 1-1 Yamadaoka, Suita, 565-0871, Osaka, Japan

[c] Department of Earth and Space Science, Osaka University, 1-1 Machikaneyama, Toyonaka, 560-0043, Osaka, Japan

* Corresponding author: naoki_iikawa@global.komatsu

ABSTRACT

In relation to the interaction of the earth's surface with machines and organisms, and its engineering applications, there has been a recent increase in interest in the penetration resistive force into granular materials at shallow depths. Previous studies have proposed various models for penetration resistive forces into dry granular materials. This study focuses on the model which has a coefficient depending on the angle of repose and the increase of resistive force in proportion to the penetration volume. In the previous studies, the model has been validated for several geometries such as cylinders, cones, and spheres. However, for cones, the model has only been validated under conditions of a tip angle close to the angle of repose. In this study, the effect of cone tip angle on penetration resistive force is investigated under several conditions with different angles of repose. This study carries out cone penetration simulations using the discrete element method. For the cone geometry, five tip angles ranging from sharp to blunt (tip angles are 15, 30, 45, 60, 75 deg) are used. The simulation results show that the penetration resistive forces for blunt cones are much higher than that computed by the model. To solve the discrepancy between the model and simulation results, this study modifies the model by assuming that the stagnant zone formed in front of the cone penetrating the granular material behaves as an effective cone. Thereby, the proposed model can calculate penetration resistive forces more accurately for cones with a wider range of tip angles than in the previous model.

Keywords
Soil characterization
Granular materials
Discrete-element modelling

1. Introduction

Recent planetary explorations have revealed that the surfaces of the Earth and several other solid planets are covered with granular materials. To understand and predict the motions of machines and organisms on these surfaces (e.g. landing (Ballouz et al. 2021), locomotion (Suzuki et al. 2023), and excavation (Miyai et al. 2019)), the interaction between objects and granular materials needs to be clarified. These phenomena vary in behavior but are in common that they include object penetration into granular materials. Thereby, it is important to estimate the penetration resistive force during object penetration into granular materials.

Previous studies have proposed various models to estimate penetration resistive force. In these models, the phenomenological models have been often used to understand the impact crater dynamics and mechanical properties of regolith in Planetary Science (Katsuragi, 2016; Okubo & Katsuragi, 2022), and to comprehend the ecology and morphology of organisms on sand in Biology (Sharpe et al., 2015). These models are generally expressed as the sum of a depth-dependent term derived from hydrostatic-like pressure and a velocity-dependent term derived from viscosity and/or inertia (Katsuragi,2016). In Terramechanics, Resistive Force Theory (RFT) which calculates the penetration resistive force for each infinitesimal surface of the intruder on each orientation relative to the ground and direction of movement, is often used (Li et al., 2013). In fact, recent study has performed simulations which reproduce the machine behaviors on granular materials by combining with RFT and Multi-Body Dynamics (Suzuki et al. 2023). While these models can accurately and quickly calculate the resistive force, it is necessary to calibrate coefficients in advance. Therefore, it is difficult to predict resistive forces when the ground material cannot be used previously.

Among these studies, a new model based on the classical theory in Geotechnical Engineering has been proposed recently (Kang et al., 2018). This model is called as modified Archimedes' law theory (MALT) and described as follows:

$$\boldsymbol{F_z}(z_p) = K_\phi \rho_g \psi \boldsymbol{g} V_p(z_p),$$
$$\text{with } K_\phi \equiv \left(2\frac{1+\sin\phi}{1-\sin\phi}e^{\pi\tan\phi}\int_0^1 \eta A(\eta,\phi)d\eta\right), \quad (1)$$

where $\boldsymbol{F_z}$ is the resistive force perpendicular to the layer; z_p is the penetration depth; ρ_g is the density of granular particles; ψ is the packing fraction (which equals 1 minus the void ratio); $\boldsymbol{g}$ is the gravitational acceleration; V_p is the penetration volume of the intruder; K_ϕ is the coefficient which depends only on the angle of repose ϕ. The details of parameters η and $A(\eta,\phi)$ in *Eq.* (1) are explained in Kang et al. (2018). MALT can estimate $\boldsymbol{F_z}$ without prior calibration (it only requires the physical properties of granular layers). However, MALT has an issue to apply them to the estimation of resistive force in actual soil ground. It is the effect of the intruder shape. Kang et al. (2018) have validated MALT with respect to several intruder geometries such as cylinders, spheres, and cones, but only for cone tip angles close to ϕ. On the other hand, Mishra et al. (2018) have reported that $\boldsymbol{F_z}$ decreases as the cone tip shape becomes sharper. Thus, it is not clear whether MALT is applicable for various cone tip shapes.

In this study, the effect of cone tip angle on penetration resistive force in dry granular layers is investigated through Discrete Element Method (DEM) simulations. Using the simulation results, this study discusses the validity of MALT for various cone tip angles and proposes the modified model which can calculate penetration resistive forces more accurately for cones with a wider range of tip angles.

https://doi.org/10.56884/N04X6LSL
ISBN 978-1-942112-55-6

2. Materials and Methods

2.1. Discrete Element Method (DEM)

This study uses an open-source DEM engine, LIGGGHTS(R)-PUBLIC Version 3.8.0 (Kloss and Goniva 2011; Berger et al. 2015) to perform the DEM simulations. In the simulations, the equations of motion for the translational and rotational directions of the particle i are respectively expressed by the following equations:

$$m_i \frac{d\boldsymbol{v_i}}{dt} = \sum (\boldsymbol{F^j_n} + \boldsymbol{F^j_t}) + m_i \boldsymbol{g},$$
$$I_i \frac{d\boldsymbol{\omega_i}}{dt} = \sum (r_i \times \boldsymbol{F^j_t} + \boldsymbol{M^j_r} + \boldsymbol{M^j_s}), \tag{2}$$

where m_i, $\boldsymbol{v_i}$, r_i, I_i, and $\boldsymbol{\omega_i}$ are mass, velocity, radius, moment of inertia, and angular velocity of particle$i$, respectively; $\boldsymbol{F^j_n}$ and $\boldsymbol{F^j_t}$ are respectively normal and tangential forces between particle i and j; $\boldsymbol{M^j_r}$ and $\boldsymbol{M^j_s}$ are respectively rolling and twisting moments from particle j to i; the symbol Σ denotes the sum of all forces or moments acting on particle i from adjacent particles.

In *Eq.* (2), $\boldsymbol{F^j_n}$ and $\boldsymbol{F^j_t}$ fully follow the Hertz model in LIGGGHTS. Each force is described as follows:

$$\boldsymbol{F^j_n} = k_n \boldsymbol{\delta_{nij}} + \gamma_n \boldsymbol{v_{nij}},$$
$$\boldsymbol{F^j_t} = \min[\mu |F^j_n|, k_t \boldsymbol{\delta_{tij}} + \gamma_t \boldsymbol{v_{tij}}], \tag{3}$$

where $\boldsymbol{\delta_{nij}}$ and $\boldsymbol{\delta_{tij}}$ are respectively normal and tangential overlap displacements between particle i and j; $\boldsymbol{v_{nij}}$ and $\boldsymbol{v_{tij}}$ are respectively normal and tangential components of the relative velocity between particle i and j; μ represents the sliding friction coefficient. Here, μ_{gg} and μ_{og} are respectively used in the case for particle-particle contact and object-particle contact. The symbol "min[ ]" denotes choice of the smaller of the two values in parentheses. The details of the elastic constants for normal contact k_n and tangential contact k_t , and the viscoelastic damping constants for normal contact γ_n and tangential contact γ_t are explained in the documentation of LIGGGHTS(R)-PUBLIC v3.X (https://www.cfdem.com/media/DEM/docu/gran_model_hertz.html, the last accessed: April 22, 2024).

In addition, $\boldsymbol{M^j_r}$ and $\boldsymbol{M^j_s}$ in *Eq.* (2) also fully follow the rolling resistance model proposed by Jiang et al. (2015). Each moment is described as follows:

$$\boldsymbol{M^j_r} = \min[0.525|\boldsymbol{F^j_n}|\chi r^*,\ 0.25(\chi r^*)^2(k_n \boldsymbol{\delta_{\theta ri}} + \gamma_n \boldsymbol{\omega_{ri}})],$$
$$\boldsymbol{M^j_s} = \min[0.65\mu|\boldsymbol{F^j_n}|\chi r^*,\ 0.5(\chi r^*)^2(k_t \boldsymbol{\delta_{\theta si}} + \gamma_t \boldsymbol{\omega_{si}})], \tag{4}$$

where χ is numerical parameter expressing rotation resistance due to particle surface irregularities; $\boldsymbol{\delta_{\theta ri}}$ and $\boldsymbol{\delta_{\theta si}}$ are respectively total relative rolling and twisting angular displacements from particle j to i; $\boldsymbol{\omega_{ri}}$ and $\boldsymbol{\omega_{si}}$ are respectively relative rolling and twisting angular velocities from particle j to i; r^* is the reduced radius. Using radii r_i and r_j of particle i and j, r^* is defined as $\frac{1}{r^*} = \frac{1}{r_i} + \frac{1}{r_j}$. The coefficient values in *Eq.* (4) are taken directly from the values determined by Jiang et al. (2015).

2.2. Simulation Setup

This study investigates the tip angle dependence of resistive force when a cone vertically penetrates a granular layer at a constant velocity. The simulation setup is shown in *Fig. 1 (a)*. The layer consists of granular particles with an average diameter of 2 mm, and the dimensions are 300 mm in width and depth and 180 mm in height, respectively. The edge of width (x-axis) and depth (y-axis) directions are periodic boundaries. Thereby, particles moving to coordinates below 0 mm or above 300 mm appear from the opposite sides. In the height (z-axis) direction, the fixed flat floor is set at the bottom of layer (z = 0 mm) to prevent particles from falling. This study set this fixed floor to the same parameters as the particles to minimize the wall effect. In contrast, the surface of a granular layer (z = 180 mm) is not set any boundary condition, thus, it allows particles move freely near the surface of a layer.

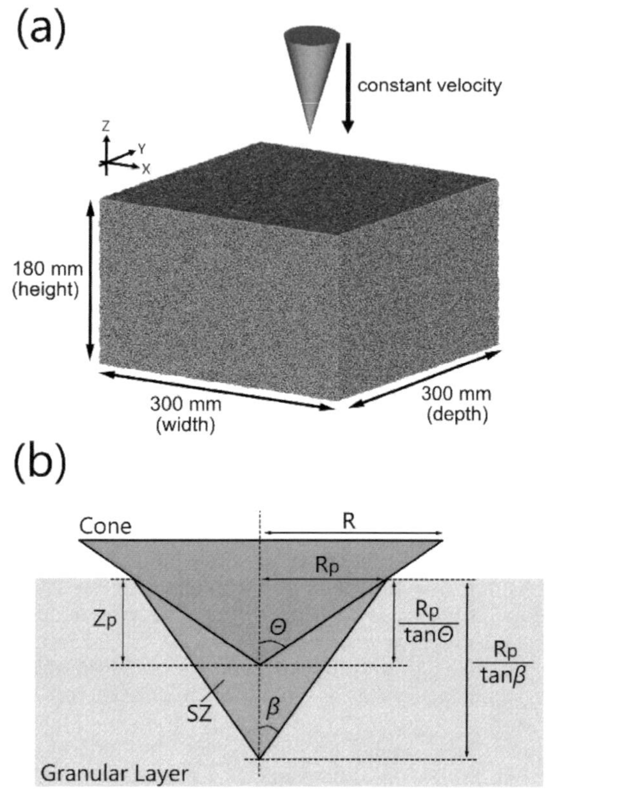

Fig. 1. (a) Schematic of simulation setup. (b) Schematic in cross-section view of the cone penetrating to granular layer.

Before the cone penetration simulation is carried out, granular layer is firstly prepared. The granular layers are adjusted to ψ = 0.60 for all conditions. After preparation, the cone positioned with the center of the surface penetrates to the layer at constant velocity 50 mm/s. The cone shape is defined by the radius R and the tip angle Θ, as shown in *Fig. 1 (b)*. In this study, five different cones are used to investigate the tip angle effect on penetration resistive forces. The specific values (Θ, R) of five cones are (15 deg, 28.6 mm), (30 deg, 28.6 mm), (45 deg, 42.9 mm), (60 deg, 42.9 mm) and (75 deg, 42.9 mm), respectively. As shown in the *Fig. 1 (b)*, z_p is defined as the depth to the cone-tip from the initial free surface level of granular layer. Cone penetrations are performed until the cone is completely buried in the layer.

In the simulation, this study uses four different parameter sets varying μ_{gg} and χ, and each condition is shown in *Table 1*. The sand piles created using these parameter sets are shown in *Fig. 2*. Calculating ϕ from the profile of sand pile using the least squares method, ϕ are obtained as 17.3 deg for Type A, 25.0 deg for Type B, 35.6 deg for Type C, and 38.6 deg for Type D, respectively.

Table 1. DEM parameter sets

Material Type	A	B	C	D
Young's modulus E [Pa]	1.0×10^9	1.0×10^9	1.0×10^9	1.0×10^9
Poisson ratio ν [-]	0.25	0.25	0.25	0.25
coefficient of restitution e [-]	0.9	0.9	0.9	0.9
particle diameter d_1, d_2, d_3 [mm]	1.7, 2.0, 2.3	1.7, 2.0, 2.3	1.7, 2.0, 2.3	1.7, 2.0, 2.3
particle mixing ratio $d_1 : d_2 : d_3$ [-]	1 : 2 : 1	1 : 2 : 1	1 : 2 : 1	1 : 2 : 1
particle density ρ_g [kg/m³]	2500	2500	2500	2500
object density ρ_o [kg/m³]	2700	2700	2700	2700
friction coefficient μ_{gg} [-] (particle - particle)	0.1	0.2	0.8	1.0
friction coefficient μ_{og} [-] (object - particle)	0.3	0.3	0.3	0.3
shape parameter χ [-]	0.05	0.2	0.6	1.0
timestep Δt [s]	4.0×10^{-6}	4.0×10^{-6}	4.0×10^{-6}	4.0×10^{-6}

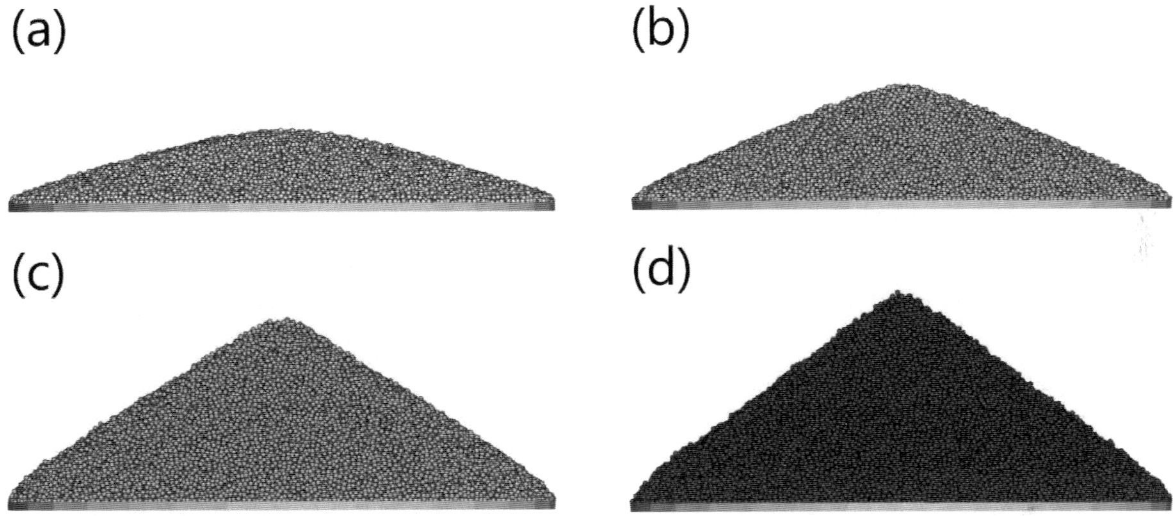

Fig. 2. Sandpiles for (a) Type A, (b) Type B, (c) Type C, and (d) Type D.

3. Results and Discussions

The results of the simulated resistive forces $\boldsymbol{F_{z,DEM}}$ are firstly compared with *Eq. (1)*. Here, V_p at the depth z_p in *Eq. (1)* is calculated as follows:

$$V_p(z_p) = \frac{1}{3}\pi\, z_p^3 \tan^2\Theta. \tag{5}$$

Moreover, K_ϕ is calculated from *Eq. (1)* using ϕ, $K_\phi = 7.3$ for Type A, $K_\phi = 20.0$ for Type B, $K_\phi = 103.7$ for Type C and $K_\phi = 177.8$ for Type D, respectively. *Figure 3* shows $\boldsymbol{F_{z,DEM}}$ with colored lines and the corresponding estimation by MALT with dashed red lines. In the simulation results, the color shades represent the difference of Θ. The horizontal axis is z_p and the vertical axis is the resistive force in log-log scale in *Fig. 3*. In *Fig. 3*, $\boldsymbol{F_{z,DEM}}$ is proportional to the cube of z_p, and qualitatively consistent with the MALT. However, for large Θ, $\boldsymbol{F_{z,DEM}}$ quantitatively deviates from MALT.

To identify and evaluate deviations between $\boldsymbol{F_{z,DEM}}$ and MALT regarding Θ, this study defines the ratio D_{MALT} as follows:

$$D_{MALT} = \frac{\boldsymbol{F_{z,DEM}}}{K_\phi \rho_g \psi \boldsymbol{g} V_p}. \tag{6}$$

The values of D_{MALT} are computed using the last 80 % of $\boldsymbol{F_{z,DEM}}$ for each Θ. The computed D_{MALT} is shown by hollow markers in *Fig. 4*. Colors and markers indicate Types presented in *Table 1*. *Figure 4* indicates that the hollow markers increase when Θ is larger than around 45 deg. The increase magnitudes depend on ϕ, with larger ϕ having a tendency towards larger values of D_{MALT} for blunt cones. If there is no effect of cone-tip shape to the resistive forces, D_{MALT} should be almost constant regardless of Θ. Therefore, this result reveals penetration resistive forces are affected by the cone-tip shapes.

The cone-tip-angle-dependent changes in penetration resistive forces are expected to be due to a stagnant zone (SZ) formed in front of the intruder penetrating to the granular layer. Here, SZ is defined as a region in which granular particles move as a rigid body in front of the intruder. In fact, Aguilar and Goldman (2016) have confirmed the formation of SZ in front of cylinder penetrating to the granular layer through experimental analysis of particle motion. Moreover, for the flat-bottom cylinder, Aguilar and Goldman (2016); Kang et al (2018) have confirmed that a sudden increase in the resistive forces occurs in the shallow penetration depth (about 0.1 times the intruder's diameter) where the SZ develops in front of cylinder.

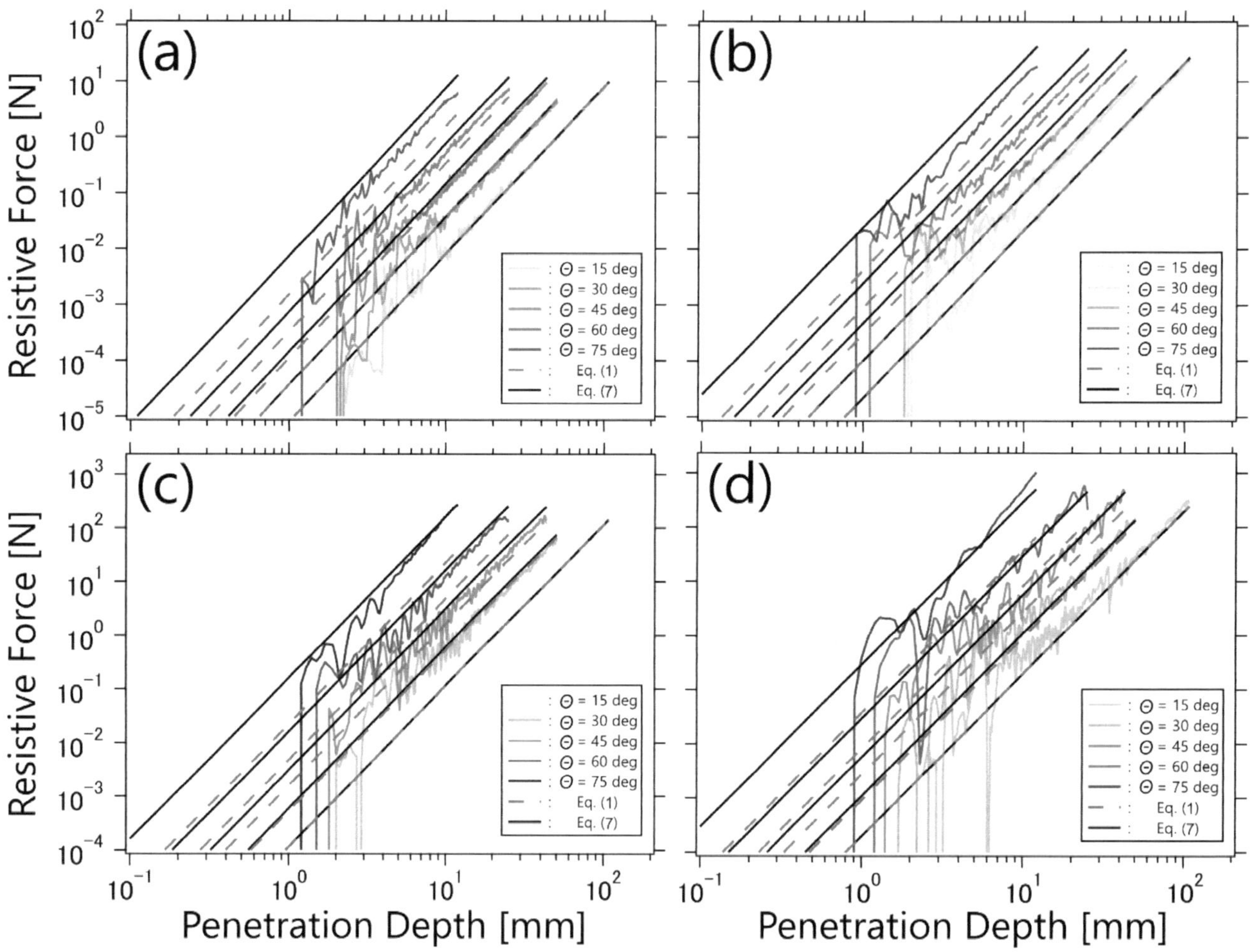

Fig. 3. Relationship between resistive force and penetration depth for (a) Type A, (b) Type B, (c) Type C, and (d) Type D.

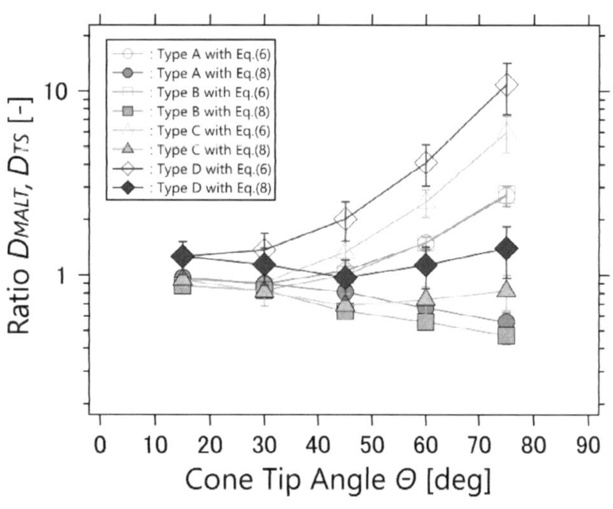

Fig. 4. Variation of D_{MALT} and D_{TS} with respect to θ.

This study investigates the particle velocity fields around a cone to clarify the formation of SZ and the effect of the tip shape on resistive forces. The particle velocity fields for Type B in the final states of cone penetration with different tip angles are shown in *Fig. 5*. Here, the colors of the particle, red and blue, indicate positive and negative particle velocities in z direction, respectively. The tip angle of SZ, β is defined as $\beta = \frac{\pi}{4} - \frac{\phi}{2}$, as in Kang et al (2018), and β = 32.5 deg for Type B. *Figure 5* shows that the distribution of the particle velocities in front of the cone varies depending on tip shape. In the case of cone tips sharper than SZ (θ = 15 and 30 deg), most of the particles in front of the cone are not moving in the penetration direction. On the other hand, in the case of a blunt cone (θ = 45, 60 and 75 deg), particles in front of the cone, especially within the SZ, are moving in the penetration direction. Thus, SZ develops in front of the cone when blunt cones penetrate.

From the above result, this study explains the discrepancy between $\boldsymbol{F_{z,DEM}}$ and MALT by considering SZ as a virtual mass. Indeed, Aguilar and Goldman (2016) have explained the large resistive force on the penetrating cylinder assuming SZ as a virtual mass. Specifically, for a blunter cone than SZ ($\theta > \beta$), this study considers the situation shown as *Fig. 1 (b)* in which SZ develops in front of the blunt cone. The penetration depths of cone and SZ are respectively $\frac{R_p}{\tan\theta}$ and $\frac{R_p}{\tan\beta}$. Accordingly, the penetration volume of blunt cone assuming a virtual mass is the product of *Eq. (5)* and $\frac{\tan\theta}{\tan\beta}$.

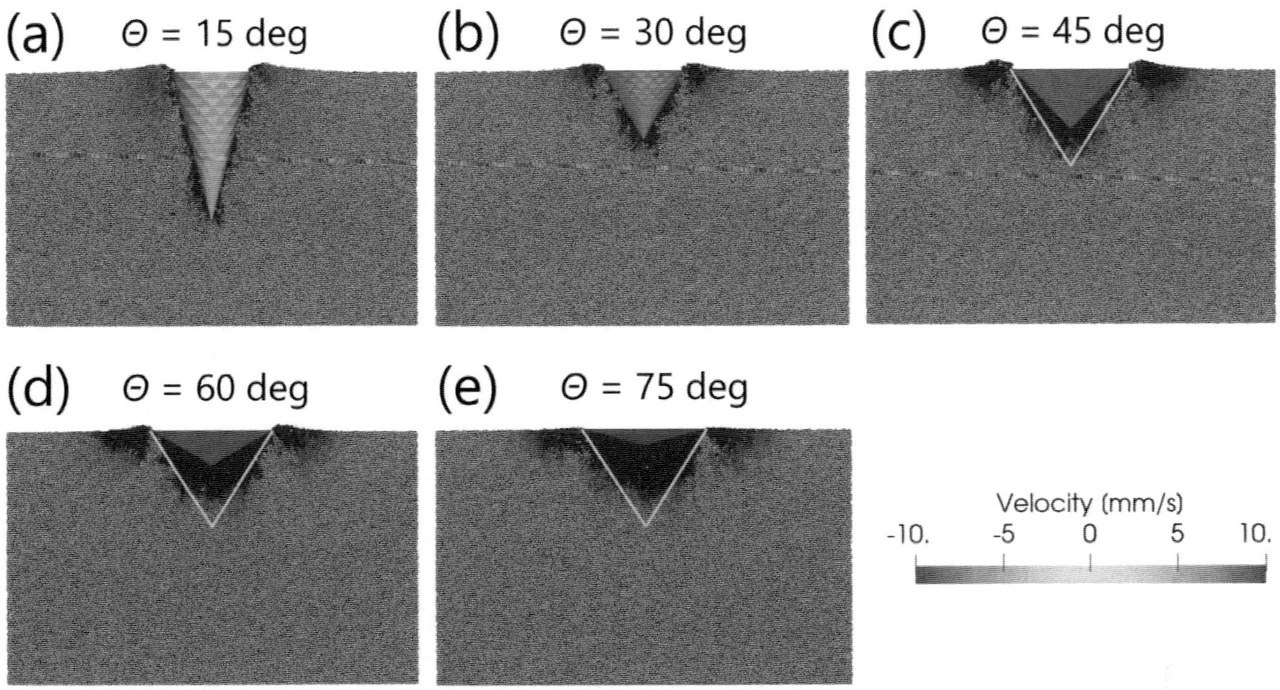

Fig. 5. Particle velocity field in z-direction on cross-section view for Type B.

Thus, *Eq. (1)* is modified by considering a virtual mass as the coefficient $f(\Theta)$, and it is as follows:

$$\boldsymbol{F_z}(z_p) = f(\Theta)\, K_\phi \rho_g \psi \boldsymbol{g} V_p(z_p)$$

with

$$f(\Theta) = \begin{cases} 1 & : \text{ if } \Theta \leq \beta \\ \frac{\tan\Theta}{\tan\beta} & : \text{ if } \Theta > \beta \end{cases} . \quad (7)$$

The estimation results of *Eq. (7)* are shown with solid black lines in *Fig. 3*. Comparing the black solid and red dashed lines in *Fig. 3*, the black lines are closer to $\boldsymbol{F_{z,DEM}}$ even for blunt cone. To identify and evaluate deviations between $\boldsymbol{F_{z,DEM}}$ and *Eq. (7)*, this study further defines the ratio D_{TS} as follows:

$$D_{TS} = \frac{\boldsymbol{F_{z,DEM}}}{f(\Theta) K_\phi \rho_g \psi \boldsymbol{g} V_p} . \quad (8)$$

D_{TS} is also shown by solid markers in *Fig. 4*. Colors and markers indicate Types of the penetrated granular layer.

From *Fig. 4*, while D_{MALT} increases with Θ, D_{TS} decreases with Θ or is almost 1 regardless of Θ. The D_{TS} behavior slightly depends on the material types. The D_{TS}, is almost 1 for Type C and D regardless of Θ, while it decreases with Θ for Type A and B. The lower limit of the estimated D_{TS} is around 0.5, which means that the difference between *Eq. (7)* and $\boldsymbol{F_{z,DEM}}$ is about twice as large. By contrast, the largest value of D_{MALT} is about 10 for blunt cones, which indicates that the ratio between *Eq. (1)* and $\boldsymbol{F_{z,DEM}}$ can be about 10 times. Therefore, the proposed model can more precisely estimate penetration resistive force to dry granular layers in most of the cases than MALT, although it may slightly overestimate the value depending on the cone tip shapes and material types.

Incidentally, the assumption for increase of penetration resistive force due to the virtual mass by SZ needs further verification. This study uses this assumption since the analysis of particle velocity field for Type B shows that particle moving zones in front of the cone match the region of the predicted SZ. However, the proposed model, which assumes an increase of penetration volume due to a virtual mass, overestimates the resistive forces on blunt cones in Type B compared to $\boldsymbol{F_{z,DEM}}$. This result means that the virtual mass produced by SZ does not fully explain $\boldsymbol{F_{z,DEM}}$. To discuss this point, more detailed studies, including an analysis of the failure mechanism of the granular layer, will be needed.

4. Conclusions and future work

This study investigated the tip angle dependence for resistive force into dry granular materials for shallow cone penetration at a constant velocity. The resistive force was calculated through DEM simulation and compared with MALT model (*Eq. (1)*). When the cone had shaper tip than the SZ, MALT can estimate the restive forces. On the other hand, when the cone had blunter tip than the SZ, MALT underestimated the resistive forces. Through the analysis of the particle velocity field around the penetrated cone, it is revealed that the increase in the resistive forces for blunt cone was caused by a virtual mass due to SZ. From this result, this study proposed a modified model considering the coefficient $f(\Theta)$ in the MALT that considered the growing volume of the SZ. In addition, comparing with the results of DEM simulation, the proposed model was validated. As a result, the proposed model can estimate the resistive force more precisely than MALT.

In this study, the resistive forces to simple intruders against a dry granular layer was investigated. However, in practical applications, it is necessary to consider more complex situations such as the intruder shape other than cones, the grain size distribution and cohesion in the ground, as well as the effects of their spatial heterogeneity. Therefore, future study should extend the proposed models for various effects and validate their effectiveness through experiments.

5. Nomenclature

$\boldsymbol{F}^j_n$, $\boldsymbol{F}^j_t$	Nomal and tangential forces between particle *i* and *j*	[kg m/s²]
$\boldsymbol{g}$	Gravitational acceleration	[m/s²]
I_i	Moment of inertia of particle *i*	[kg/m²]
k_n, k_t	Normal and tangential spring constants	[kg/s²]
K_ϕ	Coefficients of friction-derived forces	[-]
m_i	Mass of particle *i*	[kg]
$\boldsymbol{M}^j_r$, $\boldsymbol{M}^j_s$	Rolling and twisting moments from particle *j* to *i*	[kg m²/s²]
R	Radius of cone	[m]
r^*	Reduced particle's radius	[m]
r^j_i	Vector from the center of particle *i* to the contact point with particle *j*	[m]
R_p	Penetrated radius of cone	[m]
S_n, S_t	Coefficients to caluculate γ_n, and γ_t	[-]
$\boldsymbol{v}_i$	Velocity of particle *i*	[m/s]
V_p	Penetration volume of a intruder	[m³]
$\boldsymbol{v}_{nij}$, $\boldsymbol{v}_{tij}$	Normal and tangential relative velocities from particle *i* to *j*	[m/s]
z_p	Penetration depth	[mm]
β	Tip angle of stagnant zone	[-]
$\boldsymbol{\delta}_{\theta ri}$, $\boldsymbol{\delta}_{\theta si}$	Total relative rolling and twisting angular displacements from particle *j* to *i*	[-]
$\boldsymbol{\delta}_{nij}$, $\boldsymbol{\delta}_{tij}$	Normal and tangential displacements between particle *i* and *j*	[m]
γ_n, γ_t	Normal and tangential viscous dampings	[kg/s]
μ	Sliding friction coefficient	[-]
$\boldsymbol{\omega}_i$	Angular velocity of particle *i*	[1/s]
$\boldsymbol{\omega}_{ri}$, $\boldsymbol{\omega}_{si}$	Relative rolling and twisting angular velocities from particle *j* to *i*	[1/s]
ϕ	Internal friction angle	[deg]
Θ	Tip angle of cone	[deg]
ξ	Coefficient to calculate γ_n and γ_t	[-]

6. Acknowledgements

N. Iikawa would like to acknowledge the financial support from Komatsu Ltd.

7. Declaration of competing interest

The authors declare that they have no known competing financial interests or personal relationships that could have appeared to influence the work reported in this paper.

8. References

Aguilar, J., Goldman, D. I., 2015. Robophysical Study of Jumping Dynamics on Granular Media. *Nat. Phys.* 12 (3): 278–83.

Ballouz, R-L, Walsh, K. J., Sánchez, P. Holsapple, K. A., Michel, P., Scheeres, D. J., Zhang, Y., Richardson, D. C., Barnouin, O. S., Nolan, M. C., Bierhaus, E. B., Connolly Jr., H. C., Schwartz, S. R., Çelik, O., Baba, M., Lauretta, D. S., 2021. Modified Granular Impact Force Laws for the OSIRIS-REx Touchdown on the Surface of Asteroid (101955) Bennu. *MNRAS* 507 (4): 5087–5105.

Berger, R., Kloss, C., Kohlmeyer, A., Pirker, S., 2015. Hybrid Parallelization of the LIGGGHTS Open-Source DEM Code. *Powder Technol.* 278 (July): 234–47.

Kang, W., Feng, Y., Liu, C., Bluemenfeld, R., 2018. Archimedes' law explains penetration of solids into granular media. *Nat. Commun.* 9: 1101.

Katsuragi, H., 2016. Physics of Soft Impact and Cratering. *Springer*, Tokyo.

Jiang, M., Shen, Z., Wang., J., 2015. A Novel Three-Dimensional Contact Model for Granulates Incorporating Rolling and Twisting Resistances. *Comput. Geotech.* 65: 147–63.

Kloss, C., Goniva, C., 2011. LIGGGHTS–Open Source Discrete Element Simulations of Granular Materials Based on Lammps. *Supplemental Proceedings: Materials Fabrication, Properties, Characterization, and Modeling* 2.

Li, C., Zhang, T., Goldman, D. I., 2013. A Terradynamics of Legged Locomotion on Granular Media. *Science* 339 (6126): 1408–12.

Mishra, A. K., Tramacere, F., Guarino, R., Pugno, N. M., Mazzolai, B., 2018. A Study on Plant Root Apex Morphology as a Model for Soft Robots Moving in Soil. *PloS One* 13 (6): e0197411.

Miyai, S., Kobayakawa, M., Tsuji,T., Tanaka, T., 2019. Influence of Particle Size on Vertical Plate Penetration into Dense Cohesionless Granular Materials (Large-Scale DEM Simulation Using Real Particle Size). *Granular Matter Granul. Matter* 21 (4): 105.

Okubo, F., Katsuragi, H., 2022. Impact Drag Force Exerted on a Projectile Penetrating into a Hierarchical Granular Bed. *A&A* 664: A147.

Sharpe, S. S., Kuckuk, R., Goldman, D. I., 2015. Controlled Preparation of Wet Granular Media Reveals Limits to Lizard Burial Ability. *Phys. Biol.* 12 (4): 046009.

Suzuki, H., Kawakami, H., Kobayashi, T., Ozaki, S., 2023. Extended Terramechanics Model for Machine–Soil Interaction: Representation of Change in the Ground Shape and Property via Cellular Automata. *Soil and Tillage Research* 226 (February): 105578.

4966| Real-time implementation of non-linear controllers and predictors for off-road vehicle dynamics on embedded systems

Andries J. Peenze [a], A. Glenn Guthrie [a], P. Schalk Els [a,*]

[a] Department of Mechanical and Aeronautical Engineering, University of Pretoria, Pretoria, South Africa
* Corresponding author: schalk.els@up.ac.za

ABSTRACT

The vehicle dynamics of off-road vehicles tend to be non-linear concerning suspension dynamics, tyre-terrain interactions, and subsystem kinematics. These non-linearities limit the performance of control due to the need for linearization to enable real-time application. Implementing real-time non-linear controllers and prediction methods on embedded systems, such as the dSPACE MicroAutoBox 2 (MABX2), is challenging due it being computationally intensive. This challenge is particularly pronounced for controllers utilizing optimization schemes with non-linear solvers to solve optimal control problems. In this paper, we explore the possibility of embedding non-linear optimal controllers and predictors on a MABX2 for real-time applications. The implementation uses CasADi to formulate and code-generate the problems, and then executes this code using S-Functions in Simulink or off-loading the execution to a powerful external computer. We discuss the limitations and special considerations for these real-time implementations for off-road vehicle applications.

Keywords
Real-time control
Off-road vehicle dynamics
Model predictive control
Non-linear control

1. Introduction

Off-road vehicle dynamics includes many non-linear aspects that makes simple implementation of controllers, estimators and predictors challenging. Some of these non-linear dynamics includes the tyre-terrain interaction, subsystem kinematics, suspension characteristics and other actuator implementations. Conventional methods to simplify these non-linearities include linearisation, either a global linearisation of the whole system throughout the implementation or recursively linearising the system at each sample time. These methods have valid applications, and many times can be implemented in real-time on the platforms, although with varying success (Kong et al., 2015).

With implicit non-linearities the states under observation or control can change rapidly, especially with uncertain terrains and challenging obstacles. Having knowledge of these non-linearities during the solving of the states or control can be hugely advantageous (Lin et al.,2024). Solver methods that account for these non-linearities however comes at greater computational cost.

To accurately predict the future states of a non-linear system, one typically must make use of higher order integration methods. Most implementations need to make a trade-off between computational efficiency and accuracy of the solution. Linear control methods which typically have an analytical closed form, like linear quadratic regulators, can only give approximate control solutions near linear regions of the dynamical system. Optimal control problem formulations that model the non-linearities come at higher computational cost, which often make use of iterative methods to find a valid solution.

Recent advances in automatic differentiation techniques, non-linear solvers, and software, together with highly optimised hardware has led to the incorporation of more complex non-linear modelling, prediction, and control methods on various applications. Real-time implementations of these methods are becoming more prevalent, especially for systems where rapid prototyping can be used to quickly find good solutions after initial, simulation based, studies have been completed (Lin et al., 2024; Amer et al., 2024; Ricco et al., 2024; Vaskov et al., 2024).

In this paper we will discuss two implementations of non-linear methods with different levels of complexity, all realised with an embedded system that runs in real-time on a test vehicle. The first implementation uses a non-linear model of a vehicle to predict future states for a preview controller to control a semi-active suspension system. Additionally, a highly non-linear integrated chassis controller is implemented, with control on the semi-active suspension, differential braking, and active rear wheel steering. For these implementations the limitations, trade-offs and special considerations are discussed.

2. Methods and materials

For this study a Land Rover Defender 110 SUV, see Fig. 1, is used as the test vehicle. The vehicle has been modified extensively to serve as a platform for off-road vehicle dynamics research.

Fig. 1. Land Rover Defender 110 test vehicle with outriggers installed

https://doi.org/10.56884/V3EPP9IS

Some of the controllable actuators added to the vehicle for this study includes a hydro-pneumatic semi-active suspension system, a steerable rear axle, and a controllable ABS modulator. The suspension system can be switched between a hard, handling oriented spring and damper setting as well as a soft, ride-comfort oriented spring and damper setting. The steerable rear axle has a range of +/- 3 degrees of steering using a linear ball screw actuator and DC motor as depicted in Fig. 2. With the controllable ABS modulator and a brake robot actuator differential braking can be achieved for direct yaw moment control. More information of the individual controllable actuators can be found in Peenze and Els (2023).

Fig. 2. Rear wheel steering actuator

The vehicle has in strut displacement and pressure measurements, brake pressure transducers for measurement of the individual brake calliper pressures as well as steering angle sensors on the front right kingpin and rear right kingpin. These sensor measurements are captured with a dSpace MicroAutoBox 2 (MABX2) (dSpace, 2024) in real-time, which also serves as the controller for the different controllable actuators. The vehicle attitude states are measured with a Racelogic VBox 3i with dual antennas, IMU and RTK base station (Racelogic, 2024). With the MABX2 the control architecture can be developed in Matlab Simulink (MathWorks, 2024) and the code deployed to the real-time platform for experimental testing.

The controllers and predictors are developed in Matlab, by making use of the CasADi toolbox (Andersson et al., 2019) to develop the control algorithms and C-code generation to use the controllers and solvers as self-contained C-code or Simulink S-Functions. The S-Functions are deployed directly to the MABX2. The C-code can also be deployed on a high-performance standalone Intel NUC PC (Intel, 2024).

To test the controllers and their real-time ability, three different tests are performed. The performance of a heuristic controller with preview for the semi-active suspension system is tested with a non-standard single lane change at moderate speeds on a paved road in section 3. The performance of a highly non-linear integrated chassis controller is shown in section 4, where an ISO 3888-1 double lane change (International Organization for Standardization, 1999) is used. The same integrated chassis controller is tested on a gravel road with a non-standard double lane change.

3. Semi-active suspension control with preview from a non-linear model

A heuristic controller is used to change the spring and damper characteristics of a hydro pneumatic semi-active suspension system for improved handling and ride comfort. The heuristic rule compares the running root mean square (RRMS) value of the lateral acceleration and the RRMS value of the vertical acceleration (Els, 2006). If the lateral acceleration RRMS value is larger than the vertical acceleration RRMS value, the vehicle is believed to experience a handling manoeuvre and the suspension system should be switched to a stiffer handling-oriented suspension setting. If, however, the vertical acceleration RRMS value is larger than the lateral acceleration RRMS value the suspension should stay in a soft, ride-comfort oriented suspension setting. Additional criteria dictates that the vehicle will only switch to the handling mode if the lateral acceleration RRMS value exceeds 0.15g and will always switch to the handling mode if the lateral acceleration RRMS value exceeds 0.3g. For this study a previewed lateral acceleration value is used to calculate the lateral acceleration RRMS value recursively, whereas the vertical acceleration RRMS value is calculated with the currently measured vertical acceleration. The semi active suspension system's valves are switched with a bang-bang low-level control method.

Due to the inherit delays of the valves used to switch the spring and damper settings a preview of the vehicle states is necessary to overcome these delays. In a previous study (Peenze and Els, 2023) the switching delays for the semi-active suspension system was found to be 150ms.

A mathematical model is developed and used to preview the states of the vehicle through numerical integration. To capture the non-linear dynamics of the suspension system and the tyre model, a fourth order Runge-Kutta integration scheme is used. A discretisation time step of 10ms was found to give stable results with this integration scheme. Details about the mathematical model can be found in Peenze and Els (2023). To summarise, the mathematical model uses a Pacejka '89 (Bakker et al, 1989) tyre model that has been characterised for the tyre used on the test vehicle (Bosch et al. 2016). The suspension spring and damper characteristics have been experimentally determined and an ideal gas adiabatic pneumatic spring model developed, to characterise the spring, and a curve fit for the damper characteristics.

The disturbance to the mathematical model is the front wheel steer angle. A first order hold approximation for the steer angle is applied and was found to be adequate to excite the transients of the vehicle for accurate preview (Peenze and Els, 2023), given the short preview time. The gradient for the first order hold is determined with a backward difference method over 50ms of the measured front steer angle (Linström et al., 2018).

This preview method involves numerically integrating the mathematical model using a fourth order Runge-Kutta method over a 150ms preview time, with discretized time steps of 10ms. The code for this integration is generated using Matlab and CasADi. CasADi allows for the convenient creation of self-contained C-code and S-Functions, which are then used in Simulink. In Simulink, the S-Function is added and built, along with the measurements and communication, to the MABX2. For this test setup, the MABX2 operates at a sampling rate of 100 Hz.

Figure 3 shows a schematic flow diagram of the architecture used to preview and control the semi-active suspension system.

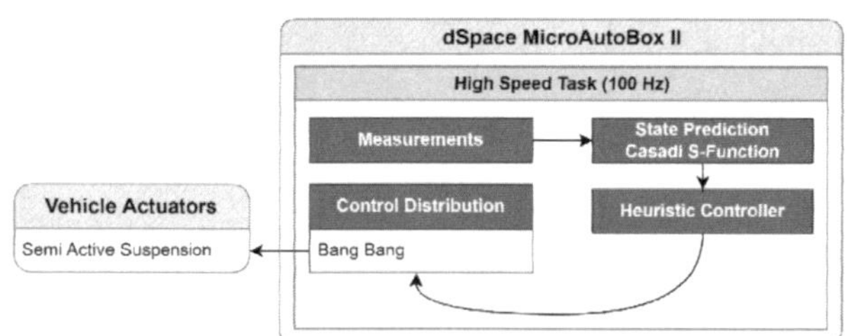

Fig. 3. Architecture for state preview and semi-active suspension control

The controlled experiment's performance is compared to a baseline experiment, where the suspension system is kept in the ride-comfort mode, and additionally with the suspension system in the handling mode. A non-standard single lane change is performed for the experiments at a moderate speed of 25 km/h. At this speed the lateral acceleration of the vehicle is adequately excited to cause significant body-roll and the effect of the semi-active suspension control can be seen. The roll angle, as measured by the VBOX 3i, is used as a metric to compare the performance.

In Fig. 4 we find that the semi-active suspension control significantly reduces the vehicle roll angle when compared to the baseline ride-comfort setting. It is also clear that the controlled suspension system follows the trend of the ride comfort experiment when the suspension setting is in the soft mode and when the suspension setting is in the hard mode, it reduces the roll angle.

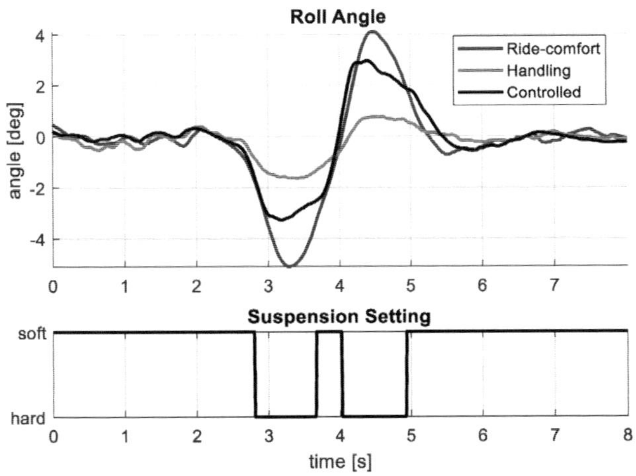

Fig. 4. Roll angle results for semi-active suspension control compared to baseline experiments

The preview states for the controlled suspension experiment are displayed in Fig. 5. We observe good correlation for the roll angle and lateral acceleration, with a phase shift which is explained by the 150ms preview time. Also, in Fig. 5 we can see that the previewed RRMS of the lateral acceleration causes the suspension system to be switched to the handling mode sooner than if the true RRMS lateral acceleration would have been used.

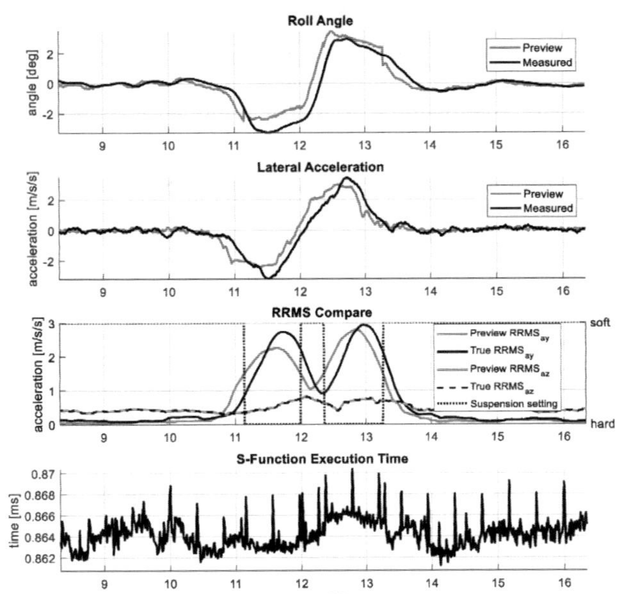

Fig. 5. Preview states and execution time

In Fig. 5 we see that the preview method execution time is between 0.86ms and 0.87ms, which is significantly less than the 10ms sampling time of the MABX2, therefore the implementation of the S-Function is executed in real time, with a lot of margin. The preview strategy successfully compensates for the actuator delay of the semi-active suspension system.

4. Integrated chassis control with semi-active suspension, rear wheel steering and differential braking

An integrated chassis controller (ICC) is presented for use on the test vehicle, with a single controller calculating the optimal distribution of the rear-wheel steering angle and differential braking of the rear wheels, given an embedded heuristic rule set for the selection of the semi-active suspension system. This controller promotes stability and improved handling by following prescribed reference states in handling manoeuvres while ensuring good ride comfort in normal or non-severe off-road driving. The external disturbance is presented in the form of a predicted driver front-steer angle input ($\delta_{preview}$). This ICC is optimal with regards to the cost function and constraints defined for the optimisation solver, and the dynamics that need to be satisfied. This optimal controller is solved with a non-linear model predictive controller (NLMPC), forming the high-level controller.

Without delving into the intricate details of the formulation, the ordinary differential equation (ODE) used in the NLMPC takes in the following states:

$$x = \{\beta, \dot{\psi}, V_x, a_y, \phi, \dot{\phi}, \theta, \dot{\theta}, CG_z, C\dot{G}_z, RRMS_{ay}, ss, ds\} \quad (1)$$

The states are side-slip angle (β), yaw velocity ($\dot{\psi}$), longitudinal velocity (V_x), lateral acceleration (a_y), roll angle (ϕ), roll velocity ($\dot{\phi}$), pitch angle (θ), pitch velocity ($\dot{\theta}$), vertical CG displacement (CG_z), vertical CG velocity ($C\dot{G}_z$), lateral acceleration RRMS ($RRMS_{ay}$), spring setting (ss) and damper setting (ds) respectively.

The control inputs to the NLMPC are:

$$u = \{\lambda_r, \delta_r\} \quad (2)$$

The control inputs are longitudinal slip ratio of the rear wheels (λ_r) and rear-wheel steer angle (δ_r) respectively. The longitudinal slip ratio between the rear left (rl) and rear right (rr) wheels are resolved as:

$$\lambda_{rl} = -\lambda_r \times (\lambda_r \geq 0) \quad (3)$$

$$\lambda_{rr} = \lambda_r \times (\lambda_r \leq 0) \quad (4)$$

To attain differential braking, the sign of the control input λ_r is used to switch between the left and right rear wheels in Eq. 3 and Eq. 4. Since the system is brake based, we require the signs of λ_{rl} and λ_{rr} to be strictly negative or zero.

The disturbance to the ODE comes from the predicted driver front steer angle input. This predicted front steer angle is determined with the method proposed by Kapania and Gerdes (2015) for steering control at the limits of handling. The steer preview is calculated recursively by numerically integrating an ODE, similar to the one used in the NLMPC, forward in time with some road curvature preview available, and updating the steer preview at each integration time step. This gives us a complete steer preview over a user selectable preview time and, as a bonus, also the hot-start states that can be used to improve the performance of the NLMPC. In Fig. 6 we can see the predicted front steer angle and the measured front steer angle for a double lane change manoeuvre.

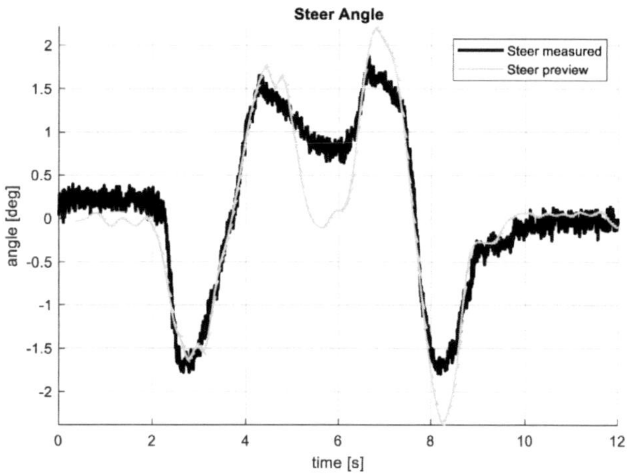

Fig. 6. Steer angle preview

The NLMPC numerical integration scheme makes use of an Euler integration method, with major time steps of 50ms (which is when the control input values are solved for) and minor integration time steps of 10ms to ensure numerical stability of the ODE. Compared to the fourth order Runge-Kutta integration method, the Euler method gave similarly good results with faster solve times. A preview and control horizon of 500ms was selected for the formulation.

The bang-bang control of the semi-active suspension system is embedded inside the ODE formulation as a heuristic rule set similar to the one described in section 3. This was done to avoid the use of a mixed integer non-linear programming (MINLP) solver. MINLPs use significantly more computational power, with orders of magnitude longer solve times than more traditional solvers.

The NLMPC is formulated with a direct multiple shooting method for solving the control inputs and satisfying the dynamic constraints of the formulation. With a direct multiple shooting method, the performance of the controller can be improved by supplying a hot start guess as an initial solution to the problem.

The NLMPC is solved with a sequential quadratic programming solver. This solver (QRQP solver from the CasADi toolbox) is one of the solvers that can be easily code generated (together with the NLMPC formulation) into self-contained C-code that can be used to generate S-Functions for Simulink, MEX-Functions for Matlab or used standalone in an external program or application. Real-time applications of these NLMPC are usually attained by assuming that one sequential quadratic iteration is sufficient to control the system (Houska et al., 2011; Diehl et al., 2009).

The side-slip angle, yaw velocity and longitudinal velocity reference states are used in the weighted quadratic cost function in the NLMPC. The side-slip angle reference state is zero throughout the preview horizon and penalises large deviations from zero. A weighting of $1/(1°)^2$ is applied to the quadratic cost of the calculated side-slip angle. With this weighting, any calculated absolute side-slip angles larger than one degree will be quadratically penalised.

The yaw velocity reference state is:

$$\dot{\psi}_{ref} = \frac{\delta_{preview} \times V_x}{L} \tag{5}$$

with L the wheelbase of the vehicle. The difference between this reference yaw velocity and the calculated yaw velocity is used in the cost function. A weighting of $1/(1°/s)^2$ is applied to the quadratic cost of the difference between the reference and calculated yaw velocity. With this weighting, an absolute value of the difference between the calculated and reference yaw velocity larger than one degree per second will be quadratically penalised.

The longitudinal velocity is added in the cost function similar to the yaw velocity, where deviations from the reference longitudinal velocity of more than 5 km/h will be quadratically penalised.

These reference states promote stability of the vehicle through the side-slip angle cost, promotes responsive turning and path following through the yaw velocity cost and reduces excessive speed scrubbing through the longitudinal velocity cost.

In the NLMPC the rear-wheel steering is constrained to operate between +/- 3 degrees. The longitudinal slip ratio is constrained to +/- 5 % slip ratio. The quadratic cost weightings for these control inputs penalises the rear wheel steering for absolute values over 3 degrees, and the longitudinal slip ratio for absolute values over 1 %.

The NLMPC solution for the rear wheel steering and longitudinal slip ratio is fed to a set of low-level controllers. These low level PID controllers are used to actuate the rear wheel steering and ABS modulators directly. The semi-active suspension system's valves are switched based on the output from the NLMPC states of the spring setting and damper setting directly.

To achieve real-time performance, the NLMPC cannot be solved on the embedded MABX2 platform, as the hardware will not provide a solution within the required time envelope. To overcome this limitation, the architecture of the ICC is designed as follows:

The MABX2 is configured to always use two tasks. The first high-priority task, executed at a rate of 200 Hz, handles all high-speed measurements and low-level control execution. The second task runs in the background, executing at a rate of 20 Hz, and is used to pre-process and transmit the hot-start states and steering preview by connecting to a User Datagram Protocol (UDP) server running on an external, small form factor, Intel NUC Linux PC.

During pre-processing, 32-bit floating point numbers are used instead of 64-bit to reduce the UDP transfer time by halving the data size, without affecting the accuracy of the states, as the difference is well within the noise band for the measurements. This also ensures that the entire hot-start vector, steer preview, reference states and controller weightings can be sent on one datagram.

The NUC runs the NLMPC using the states received from the MABX2 over UDP, with a typical solution time of 12-18ms, and outputs the required actuator control effort variables. The controller and UDP server are implemented on the NUC using Boost.Asio (2024) (UDP) and CasADi (NLMPC) in C++.

The C++ code is cross compiled on a Windows 10 PC for the Linux x86_64 architecture using Zig (2024), as the CasADi code requires a large amount of RAM during compilation.

The control effort variables are transmitted back to the MABX2 via UDP to the high-priority task, which feeds the low-level vehicle actuator controllers.

Figure 7 shows a schematic flow diagram of the architecture used for the ICC.

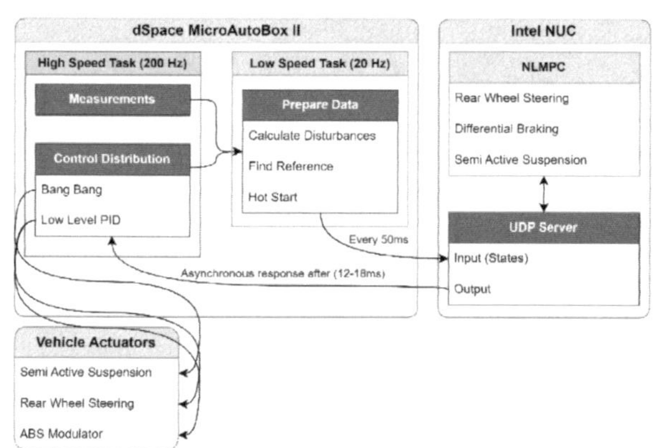

Fig. 7. Architecture of ICC

The ICC experiment's performance is compared to the baseline experiment with the suspension system in the ride-comfort mode and no rear wheel steering and no differential braking applied, and with the suspension system in the handling mode with no rear wheel steering or differential braking. An ISO 3888-1 double lane change is performed for the experiments at a moderately high speed of 50 km/h on a smooth road surface. This was the highest speed the ride-comfort baseline suspension experiment could successfully negotiate the test without the safety outriggers, installed on the vehicle (see Fig. 1), touching the ground. At this speed the lateral acceleration of the vehicle is high enough to excite the non-linear dynamics of the vehicle. The side-slip angle, yaw velocity and roll angle, as measured by the VBOX 3i, is used as the metrics to compare the performance of the ICC relative to the baseline ride-comfort test.

In Fig. 8 we find that the ICC significantly reduces the vehicle roll angle when compared to the baseline ride-comfort setting, while being particularly close to the handling setting. The side-slip angle for the ICC is larger than the other tests' results, and this can be attributed to the rear wheels steering in the opposite direction as the front wheels and generating larger lateral velocity to turn into the corner. However, the magnitude of the side-slip angles is relatively small for all the tests. The yaw velocity is in line with the other experiments' results, with some fluctuations attributed to the rear wheel steering. The similar yaw velocities indicates that the vehicle followed very similar paths between the tests.

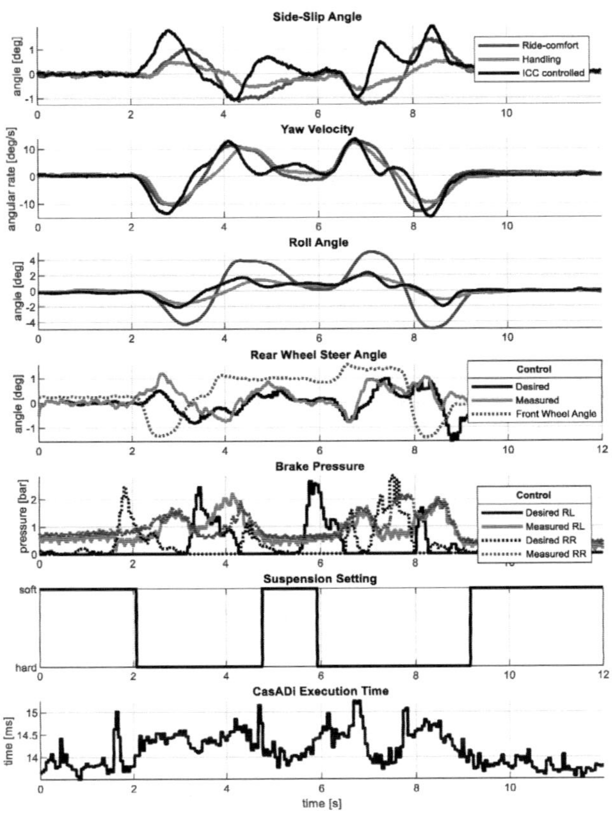

Fig. 8. Results for ISO 3888-1 double lane change with ICC

From the actuator controls we find good correlation for the measured rear wheel steering angles compared to the desired rear wheel steering angles. The desired brake pressures are extremely small and would not significantly affect the dynamics with their application. The semi active suspension control that is embedded in the ODE of the ICC formulation functioned well, with large reductions in the body-roll angle. The CasADi solve times are also presented, with the solver executing between 13ms and 16ms. This is well within the control updates of 50ms, and the system therefore has significant margins to be executed in real-time.

Additional tests were performed with the ICC on a gravel road to determine the performance on off-road scenarios. For this test, a non-standard double lane change was performed, and the ICC compared against the baseline ride-comfort suspension results with no rear wheel steering and differential braking. In Fig. 9 we see the results for the gravel road double lane change.

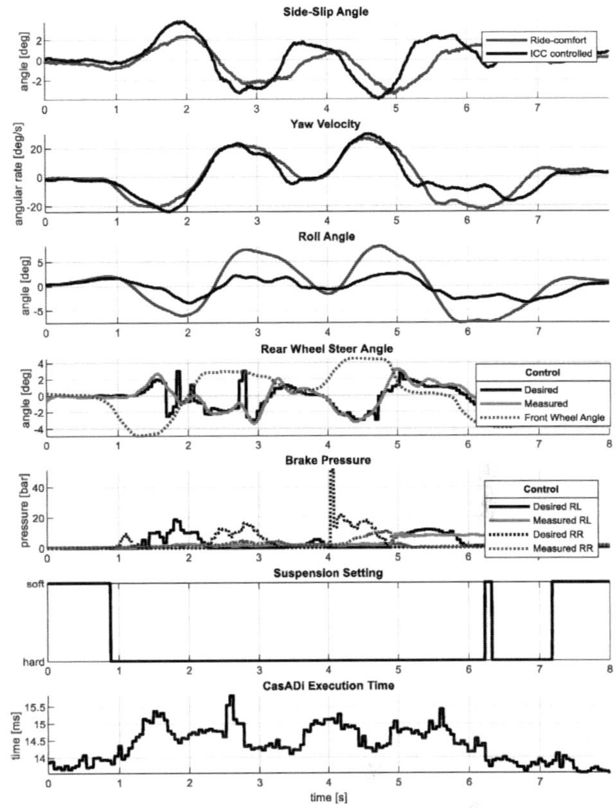

Fig. 9. Results for double lane change on gravel road with ICC

Similar to the 50 km/h ISO double lane change results the ICC significantly reduced the body-roll angle. This is evident in Fig. 10 where we overlaid images of the two tests on top of each other. The difference in roll angle can be seen from the difference in the outrigger angles. The side-slip angle for the ICC exceeds that of the baseline results. This is again due to the rear wheel steering angle acting in the opposite direction as the front wheel angle and adding more lateral velocity to the vehicle. The yaw velocity is similar to the baseline results.

Fig. 10. Image of vehicle performing a double lane change on a gravel road

The rear wheel steering angle measured and desired results correlate well, with small discrepancies that might be attributed to solver issues on the NLMPC. The differential brake performance for this test did not perform well, with large errors between desired and measured brake pressures. Initial investigations suggests that the low-level PID control method used for the ABS modulator might have been poorly tuned. Further tuning should help to reduce the errors for the differential braking. The semi-active suspension controller performed excellently and reduced the body-roll angle when compared to the baseline results. The CasADi solve times are between 13ms and 16ms, with a healthy margin to the control updates of 50ms.

From the results of the ICC, we find that the application of this controller is adequately solved in real-time. The execution time for the steer preview and hot-start states situated in the low-speed background task on the MABX2 was unfortunately not recorded but was observed to be around 8ms. To measure the full execution time of the controller this time needs to be added to the CasADi solve times to give an accurate execution time.

5. Special considerations for real-time applications

To use the S-Function previewer on the dSpace MABX2 hardware directly, like in section 3, the following need to be considered:

1. To ensure the dynamics that is solved by the integration scheme is stable, the discretisation time steps were selected to be 10ms. This value could have been reduced while keeping the 150ms preview horizon. The system should retain its real-time applicability since we observed a substantial margin in the execution times.
2. The preview horizon was selected to be 150ms to account for the semi-active suspension's valve delay times. Longer preview horizons can also be used, but caution should be taken when using the first order hold steer preview method. Previous research indicated that using the first order hold steering preview method with long preview horizons can lead to instability and poor performance (Peenze and Els, 2023).
3. The S-Function generated by CasADi, as well as the mex files and C-code necessary to compile the previewer on the MABX2 is, in total, 1.86 MB in size. Much larger files might become a problem when compiling the S-Function to the MABX2 directly.

When developing a much more complicated controller, like the ICC controller in section 4, the following points need to be considered:

1. A trade-off needs to be made between the numerical integration method used and the preview horizon or discretized time steps. If it is essential to use a higher order method like the fourth order Runge-Kutta method, shorter preview horizons need to be used. Similarly, if smaller discretized time steps are necessary to have the ODE solutions stable, an Euler method can to be used to ensure solver execution times. The control horizon can also be shortened (to values less than the preview horizon) to achieve real-time performance.
2. Hot starting of these quadratic solvers is beneficial to give quality solutions quickly. This adds additional computational cost with an integration scheme used to produce these hot-start states. Luckily this integration scheme was performed in parallel to recursively finding the front wheel steering preview. Other methods, like using the previous solution of the NLMPC shifted by one time step, can be used as the hot start. This is only advised if the controller can give guaranteed stable and robust solutions.
3. For longer preview horizons, either to overcome any actuator delays or finding more optimal control inputs, the quality of the disturbance estimation becomes imperative. In this study the disturbances, in the form of the front wheel steer preview angles, are determined based on the road curvature, and is assumed to be readily available from a source external to the ICC. If other disturbances are considered, the addition of their calculations or estimation must be considered and optimised to execute in real time.
4. One drawback of a complex implementation, like this ICC example, is the off-loading of the problem to an external PC. This adds additional overhead to send the information to the external PC and adds additional monetary cost since more computer hardware is necessary to solve the ICC. The limitation for this study was the execution time and memory requirements to implement the generated C-code for the CasADi NLMPC formulation directly to the MABX2. A more powerful dSpace system, like the MicroAutoBox 3, might solve this problem but was not explored in this study.
5. The QRQP solver used in the NLMPC has shown instances where the solution is not smooth and possibly not optimal, see Fig. 9. To ensure execution of the NLMPC at each control time step, an *"error_on_fail = false"* flag had to be added to the solver. Other solvers can be explored to find more robust solutions for the NLMPC.
6. To generate self-contained C-code for the CasADi NLMPC, the options for solvers are limited to sequential quadratic methods for CasADi version 3.5.5 used in this study.

Overall, given these special considerations, the implementation of real-time previewers and controllers are easily achievable with good results. With a good understanding of the implications of the trade-offs that need to be made, any reasonably feasible controller can be implemented in real-time and experimentally tested.

6. Conclusion and recommendations

In this study we considered two real-time implementations for control namely, a heuristic controller with preview and a highly non-linear ICC. Both controllers gave good results and operated within the real-time margin allocated to them based on their control time steps.

For the semi-active suspension system with preview control, the controller performed excellently given the assumptions that had to be made regarding the front steering angle. The body roll angle is significantly reduced during the handling manoeuvre by switching the suspension system to the handling setting, while staying in the ride-comfort setting otherwise.

The ICC gave good results for both the high speed on-road double lane change and the off-road double lane change. The tuning of the low-level controller of the ABS modulator is identified as a point for improvement, but overall, the controller performed well. The execution time of the ICC is found to be consistently bound between 13-16ms, with the hot-start and steering preview calculation observed to be 8ms. This is well within the 50ms solve time allocated for the ICC.

Implementing the ICC directly on a dSpace MicroAutoBox 3 system need to be investigated. This can simplify the architecture needed for the ICC and reduce any overhead for communication and reduce the cost of implementing the system.

Some further investigation into different solvers for the ICC can also lead to more robust control applications. Re-evaluation of the fundamental mathematics to describe the ODEs can also potentially improve accuracy and robustness.

The controllers have been tested in a structured environment using high quality measurement equipment that might not be practically feasible. Additional attention needs to be given to investigate the effects of unstructured environments with various uncertainties such as varying friction conditions or mixed terrain applications to test the robustness of the system.

7. Declaration of competing interest

The authors declare that they have no known competing financial interests or personal relationships that could have appeared to influence the work reported in this paper.

8. References

Amer, N.H., Dalboni, M.A.T.T.E.O., Georgiev, P., Caponio, C., Tavernini, D., Gruber, P., Dhaens, M. and Sorniotti, A., 2024. Integrated torque-vectoring and anti-roll moment distribution strategies based on optimal control: influence of model complexity and road curvature preview. Vehicle system dynamics, pp.1-34.

Andersson, J.A., Gillis, J., Horn, G., Rawlings, J.B. and Diehl, M., 2019. CasADi: a software framework for nonlinear optimization and optimal control. Mathematical Programming Computation, 11, pp.1-36.

Bakker, E., Pacejka, H.B. and Lidner, L., 1989. A new tire model with an application in vehicle dynamics studies. SAE transactions, pp.101-113.

Boost.Asio (2024) Boost.Asio. Boost C++ Libraries. Available at: https://www.boost.org/doc/libs/1_76_0/doc/html/boost_asio.html (Accessed: 10 June 2024).

Bosch, H.R.B., Hamersma, H.A. and Els, P.S., 2016. Parameterisation, validation and implementation of an all-terrain SUV FTire tyre model. Journal of Terramechanics, 67, pp.11-23.

Diehl, M., Ferreau, H.J. and Haverbeke, N., 2009. Efficient numerical methods for nonlinear MPC and moving horizon estimation. Nonlinear model predictive control: towards new challenging applications, pp.391-417.

dSpace (2024) MicroAutoBox II. dSpace. Available at: https://www.dspace.com/en/inc/home/products/hw/micautob/microautobox2.cfm (Accessed: 10 June 2024).

Els, P.S. (2006). The ride comfort vs. handling compromise for off-road vehicles. Ph.D. Thesis. University of Pretoria. Available at: http://hdl.handle.net/2263/26302. (Accessed: 10 June 2024)

Houska, B., Ferreau, H.J. and Diehl, M., 2011. An auto-generated real-time iteration algorithm for nonlinear MPC in the microsecond range. Automatica, 47(10), pp.2279-2285.

Intel (2024) Intel NUC Kit NUC6i7KYK. Intel ARK. Available at: https://ark.intel.com/content/www/us/en/ark/products/89187/intel-nuc-kit-nuc6i7kyk.html (Accessed: 9 June 2024).

International Organization for Standardization (1999), ISO 3888-1: Passenger cars – Test track for a severe lane change manoeuvre - Part 1: Double lane-change, ISO 3888-1. Available at: https://www.iso.org/standard/31317.html (Accessed: 27 May 2024).

Kapania, N.R. and Gerdes, J.C., 2015. Design of a feedback-feedforward steering controller for accurate path tracking and stability at the limits of handling. Vehicle System Dynamics, 53(12), pp.1687-1704.

Kong, J., Pfeiffer, M., Schildbach, G. and Borrelli, F., 2015, June. Kinematic and dynamic vehicle models for autonomous driving control design. In 2015 IEEE intelligent vehicles symposium (IV) (pp. 1094-1099). IEEE.

Lin, C., Siampis, E. and Velenis, E., 2024. Path-tracking control at the limits of handling of a prototype over-actuated autonomous vehicle. Vehicle System Dynamics, pp.1-23.

Linström, B.V., Els, P.S. and Botha, T.R., 2018. A real-time non-linear vehicle preview model. International Journal of Heavy Vehicle Systems, 25(1), pp.1-22.

MathWorks (2024), MATLAB and Simulink (Version 2020b). Available at: https://www.mathworks.com/products/matlab.html (Accessed: 27 May 2024)

Peenze, A. & Els, P.S. (2023) 'Improving predictive control methods on off-road vehicles with realistic steering preview inputs', in Pytka, J., Tomiło, P and Martelli, M. (eds.), Proceedings of the 16th European-African Regional Conference of the ISTVS, Lublin, Poland, October 11-13, 2023, pp. 72-77.

Racelogic (2024) VBOX 3i. VBOX Automotive. Available at: https://vboxautomotive.co.uk/index.php/en/products/data-loggers/vb3i (Accessed: 10 June 2024).

Ricco, M., Alshawi, A., Gruber, P., Dhaens, M. and Sorniotti, A., 2024. Nonlinear model predictive control for yaw rate and body motion control through semi-active and active suspensions. Vehicle System Dynamics, 62(6), pp.1587-1620.

Vaskov, S., Quirynen, R., Menner, M. and Berntorp, K., 2024. Friction-adaptive stochastic nonlinear model predictive control for autonomous vehicles. Vehicle System Dynamics, 62(2), pp.347-371.

Zig (2024) Zig Programming Language. Available at: https://ziglang.org/ (Accessed: 10 June 2024).

5295 | Model-based online optimal control for off-road vehicles in reduced gravity

Chaitanya Sonalkar [a, *], Adwait Verulkar [a], Corina Sandu [a]

[a] *Department of Mechanical Engineering, Virginia Tech, Blacksburg, Virginia, United States*

* *Corresponding author: csonalkar@vt.edu*

ABSTRACT

There has been an interest in the exploration of planetary bodies for more than half a century. Scientific research has been focused on understanding the origins of life, potential for habitability in extraterrestrial environments, and discovery of rare minerals and energy sources. Extraterrestrial rovers have been used previously by space agencies for the exploration of lunar and Martian surfaces. Reduced gravity on such surfaces influences the performance and mobility of extraterrestrial vehicles. Gravity can change how the soil behaves under the wheel as well as the traction force and sinkage developed by the wheel. Thus, the sinkage rate and depth of a wheel and motion resistance are also impacted by gravity. As the rovers are operated by interplanetary telemetry, autonomous real-time control is employed in these systems due to communication delays. Optimal online control with a high-fidelity terramechanics model may not be feasible and it necessitates the development of simplified models which consider the effect of reduced gravity on dynamics of a rover. The goal of this work is to study the effect of reduced gravity on Lunar terrain and its effect on vehicle performance of rovers through analytical modeling of tire-terrain interaction. We have developed an optimal control algorithm to improve vehicle stability and trajectory tracking performance in standard maneuvers. The simplified terrain models can also be extended for hardware-in-loop simulations.

Keywords
Off-road mobility
Vehicle control dynamics
Extraterrestrial rovers
Optimal control
Terramechanics

1. Introduction

As the field of extraterrestrial exploration has gained global interest, continuous efforts are being made to develop technology to aid those missions. The most common way for exploration is using unmanned planetary rovers. As humans plan to build and sustain colonies in space, it is necessary that these rovers carry heavy payload over larger distances while consuming minimal power. These rovers also encounter harsh conditions due to low gravity, uncertain solar load, and reduced tractive surface. Hence, improving efficiency and power management of onboard power systems is of utmost importance. As the rovers are controlled by interplanetary telemetry, real time control is not very feasible in all scenarios.

The need for studying the effect of reduced gravity on rover mobility can be illustrated through the case study of the Spirit Rover. Spirit was sent to Mars by NASA Jet Propulsion Laboratory (JPL) to conduct missions. Though it completed its planned missions and lasted more than 6 earth years, the rover stopped working because it got stuck in the sand. The primary reason for this was the effect of reduced gravity, as the rover was tested on earth gravity and the effect of reduced gravity on the mobility of the Spirit rover could not be accurately estimated on earth.

The mechanical properties of the terrain in low gravity conditions, which largely influence the mobility of the rovers, are not available. This coupled with the influence of different gravity and atmosphere pose a challenge in determining the exact properties of the terrain. Testing in low gravity conditions is not feasible, except using parabolic flights. Various analytical methods have been adopted to compensate for the change in gravity, such as testing at lower load (equivalent weight according to change in gravity), scaled wheel models and Granular scaling laws. However, these laws are also based on empirical relationships. Thus, for using a physics based high fidelity model it is necessary to develop an optimal control strategy.

Navigating the extraterrestrial surface is also challenging due to the presence of craters, rocks and geographic features which have formed over the surface. Extensive work has been done on path planning strategies of the rovers. Our work primarily focuses on controlling the power delivery (throttle) and vehicle maneuvering (steering) of the rover, considering the physics of the tire-terrain interaction phenomenon. A learning-based model predictive control (referred to as MPC, here on) algorithm has been developed and implemented to control the vehicle.

This paper focuses on optimal control of lunar roving vehicles using a surrogate tire model based on a physical tire-terrain interaction model. Section 2 of the paper discusses various experiment and modeling approaches for simulating tire-terrain interaction and in low gravity conditions. Section 3 focuses on the modeling approach for vehicles, tires and the MPC controller. Results and overview of our work are presented in section 4. Section 5 and Section 6 discuss the conclusions and future work.

2. Literature Review

(Kobayashi et al., 2010) studied the performance of a single rigid wheel traveling straight on levelled terrains under varying gravity conditions (1/6 g, 1/2 g, 3/4 g, 1 g and 2 g). The reduced gravity environment was generated by performing the experiment on an aircraft traveling in a parabolic path such that a partial gravity field was maintained for 20-40s in a convex path whereas 2g gravity condition was generated by the centrifugal effect via circular flight pattern.

https://doi.org/10.56884/ZXZEHLSM

The testing done on the two types of dry sand: lunar soil simulant and Toyoura sand, indicated that while the sinkage of the wheel under fixed gravity conditions decreases as the wheel load decreases, there is no meaningful change in sinkage with respect to wheel load in a low gravity environment. This is because while sinkage increases due to the confining stress reduction in low gravity, the wheel load decreases causing sinkage to decrease. Thus, the effect of wheel load and effect of soil resistance nullify each other out. Sometimes though, the soil tends to have remarkably high flowability in a low gravity environment and thus the effect of soil resistance will tend to dominate.

The experiments were also able to show that the traction of wheel-terrain interfaces decreases as the gravity level decreases. This is believed to be due to the increase in soil flowability which lessens the shear resistance to the wheel rotation thereby reducing the traction of the wheel. Overall, it is concluded that the wheel mobility deteriorates due to the relative decrease in the driving force while the wheel sinkage remains constant.

(Kovács et al., 2020) had performed experiments to understand the effect of reduced gravity on the velocity profile of the soil as well as the traction force and sinkage developed by the wheel. They managed to combine the results from the experimental study with the traditional semi-empirical terramechanics model to develop mathematical models capable of estimating drawbar pull in reduced gravity environments.

A similar process to Kobayashi was followed to generate low gravity environments. The experimental apparatus consisted of a sandbox with a robotic gantry with a linear actuator that advances the wheel horizontally, a motor that turns the wheel and pneumatic actuators that apply a downward force on the wheel thus ensuring the slip rate and normal force of the wheel are controlled. The wheel-soil interactions were in turn observed with a high-speed camera.

The dynamics model was developed by considering the pressure sinkage relationship to calculate the tangential and normal stress distribution of the wheels. Empirical formula developed by (Wong, 2022) which has gravitational acceleration factored out explicitly is used. Since flexible wheels were used in the experimental setup, the deformation of the wheels results in a large contact area meaning the contact pressure drops. This effect is modelled implicitly as the sinkage associated with a surrogate wheel with larger radius is lower. A linear relationship for the exponent component of the slip sinkage phenomenon is considered and the shear stress distribution is determined by the Janosi-Hanamoto formula (Wong, 2022). The shear deformation was calculated based on the assumption of steady-state motion.

Comparing the results of the experimental setup and simulation, it was observed that the simulated sinkage varies quite a bit when compared to experimental data. This was attributed to the simulated sinkage corresponding to steady state values which is not the case in the experimental setup. The magnitude of the experimental drawbar pulls oscillations matched well with the simulation data obtained with a surrogate wheel model considering the wheel flexibility. The applied torque values were also found to match well with simulated data, further enhancing the accuracy of the mathematical model.

(Setiawan et al., 2009) had developed a 14 DOF vehicle model capable of capturing both the ride and handling behavior of the vehicle. The 14 DOFs consisted of the spring mass movements in the longitudinal, lateral vertical, roll, pitch, and yaw motion as well as the vertical motion and wheel spin of each wheel. Lumped mass was used to represent the sprung and unsprung masses while the vehicle body was considered rigid. The outer and inner steering angles were assumed to be same, and the wheels were assumed to always be in contact with the ground. The outputs from the model, specifically the yaw rate, lateral acceleration, the roll angle of the vehicle and the slip angle at each wheel were compared to experimental results.

(Slonaker et al., 2017) proposed and validated a set of scaling laws in granular locomotion for the case of rigid arbitrarily shaped wheels. The scaling was done by assuming that intrusion forces arise from resistive force theory or by assuming that granular materials behave as a continuum obeying a frictional yield criterion. The hypothesis was experimentally confirmed using pairs of wheels of various shapes and sizes under many driving conditions in a common sand bed.

For a physics-based tire model, two analytical models were considered for the study. The first is the Chan-Sandu tire model (Chan et al., 2014) which can be used to simulate rigid and flexible tires on a deformable terrain. It is a purely analytical model which requires less parametrization. It is a 2D tire model which has uniform properties across its width. The second tire model considered was the Hybrid Soft Soil Tire Model or HSSTM (Sandu et al., 2019, 2020; Taheri et al., 2019). It is a 3D tire model and was developed using a lumped mass approach. Although this tire model has more accuracy and higher fidelity, the parameterization of this tire model is a tedious task. Also, it might not be suitable for a real time application considering the computational time required.

As a high-fidelity model proposes the challenging task of rendering the Model Predictive Control (MPC) intractable, a learning-based MPC model was adopted for the Formula Student (FSAE) vehicle by (Pinho et al., 2023), to overcome the computational challenges as well as to counter the recursive infeasibility of the optimization-based problems. The control algorithm can learn from previous training data sets, experimental results, or simulations to improve the accuracy of the model. Some common machine learning models such as neural network or Gaussian process regression models are used in place of physics-based models to augment the control algorithm. A similar approach has been adopted in our study which aims at developing and training a surrogate tire-terrain interaction model based on a physics-based model.

The optimal control of rovers has previously been studied by (Tang et al., 2022). Their work focuses on developing an optimal path planning framework based on visual information. The terrain constraints and obstacles can be encountered at the same time with high map accuracy. Ding et al. (Liang Ding, 2010), proposed a high fidelity terramechanics model which was used for dynamics simulation of wheeled robot. The model incorporated multi-body dynamics with wheel-soil interaction model to incorporate the wheel-lug effect and slip sinkage.

3. Mathematical Modeling

Mathematical modeling of a system involves using mathematical equations to describe a state of the vehicle depending on a few vehicle parameters. This system of mathematical equations can predict the behavior of a vehicle for the given set of vehicle input parameters which helps us to simulate and understand the working of a vehicle under a specific maneuver.

We have modelled a lunar roving vehicle with 7 DOFs i.e., 3 DOFs of vehicle namely longitudinal, lateral and yaw motion and a rotational degree of freedom for each wheel. In literature, usually a bicycle model is used for developing path-planning and following algorithms. For this study, a four wheeled model was developed to account for terrain influence. The rover navigates with the help of the front wheel steering mechanism with power being delivered to all four wheels. The tire-terrain interaction was modelled using Bekker-Wong parameters.

3.1. Vehicle Model

For the lunar roving vehicle, we have modeled a four-wheeled vehicle with rigid wheels. As the velocity of the rovers is relatively low as compared to other wheeled vehicles, load transfer in longitudinal and lateral direction to acceleration/braking and cornering is considered negligible. The values of all the parameters used for rover are as follows –

Table 1. Vehicle parameters (Apollo11Space)

Vehicle Parameter	Value
Mass	600 kgs
Track width	1.3 m
Wheelbase	1.8 m
Weight distribution	40:60(Front: Rear)
Tire radius	0.35 m
Tire Width	0.25 m

The torques applied on the wheels generate angular velocity which in turn generates slip ratio on the wheels. The thrust generated on the wheels is then calculated which propels the vehicle forward. The thrust value is then used to calculate acceleration in longitudinal and lateral direction. The acceleration is then used to determine the slip angles, vehicle velocity and vehicle position in the global co-ordinate system. The equations which describe the behavior of the vehicle are as follows –

$$Yaw\ acceleration\ ((\varphi))^{\cdot\cdot} = 1/I\_z\ [l\_f/2\ [(F\_fr - F\_fl\ ) \times cos(\beta) + (F\_fr + F\_fl\ ) \times sin(\beta)] + l\_r/2\ [(F\_rr - F\_rl\ ) \times cos(\beta) - (F\_rr + F\_rl\ ) \times sin(\beta)]] \quad (1)$$

$$Longitudinal\ acceleration\ (a_x) = \frac{1}{m}[F_{fl} + F_{fr} + F_{rl} + F_{rr}] \times \cos(\beta) \quad (2)$$

$$Lateral\ acceleration(a_y) = \frac{1}{m}[F_{fl} + F_{fr} + F_{rl} + F_{rr}] \times \sin(\beta) \quad (3)$$

$$Longitudinal\ velocity\ (v_x) = \int (a_x + \dot{\varphi} \times v_y).dt \quad (4)$$

$$Lateral\ velocity\ (v_y) = \int (a_y + \dot{\varphi} \times v_x).dt \quad (5)$$

$$Wheel\ velocity\ Front\ (v_{\omega f}) = \sqrt{(v_y + \sqrt{(t/2)^2 + a^2}) \times \dot{\varphi})^2 + {v_x}^2} \quad (6)$$

$$Wheel\ velocity\ Rear\ (v_{\omega r}) = \sqrt{(v_y + \sqrt{(t/2)^2 + b^2}) \times \dot{\varphi})^2 + {v_x}^2} \quad (7)$$

$$Slip\ ratio\ Front\ (\kappa_f) = \frac{v_{\omega f} - \omega_f R_w}{v_{\omega f}} \quad (8)$$

$$Slip\ ratio\ Rear\ (\kappa_r) = \frac{v_{\omega r} - \omega_r R_w}{v_{\omega r}} \quad (9)$$

$$Side\ slip\ angle\ (\beta) = \arctan(\frac{v_x}{v_y}) \quad (10)$$

$$Vehicle\ Position\ (X) = \int (v_x \times \cos(\varphi) + v_y \times \sin(\varphi)).dt \quad (11)$$

$$Vehicle\ Position\ (X) = \int (v_x \times \cos(\varphi) + v_y \times \sin(\varphi)).dt \quad (12)$$

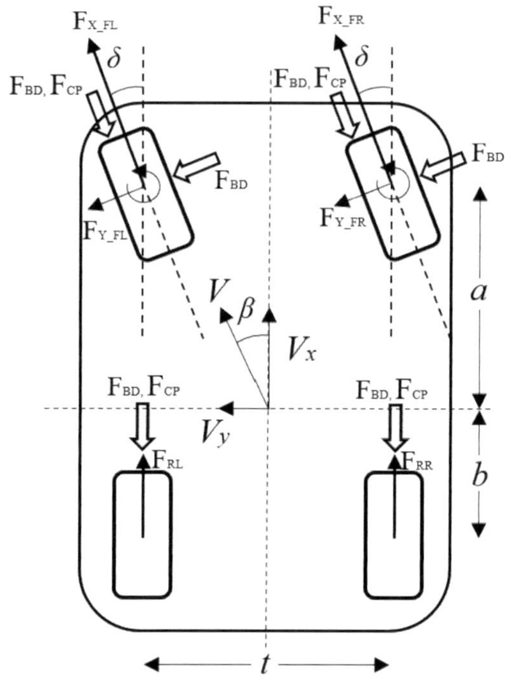

Figure 1. Free body diagram of the vehicle

3.2. Tire-Terrain Interaction Model

A rigid tire is used for lunar roving vehicles. The Chan-Sandu tire model (Chan et al., 2014) was chosen which allowed to simulate the tire-terrain interaction with rigid wheel. The tire model takes vertical load (Fz), slip ratio (κ), camber angle (γ), slip angle (α) and wheel speed (ω) as an input to calculate longitudinal force (Fx), lateral force (Fy), aligning moment (Mz) and overturning moment (Mx).

For the rigid wheel model, the angle of entry and the angle of exit is estimated for a stationary loaded wheel. The main assumption is that the soil experiences a shear failure in the longitudinal direction. Then, Meyerhof's equation (Wong, 2022) was used to find the bearing pressure of the soil mass under foundation to implement the theory of plasticity. It is then further used to compute the angle of maximum radial stress. Once this point is located, the shear stress is calculated using the Janosi-Hanamoto equation. The Drawbar pull (DP) (eq. 13), vertical load (W) (eq. 14) and the angular torque (eq. 15) are calculated by integrating the components of shear thrust, radial load and compaction resistance over angle of exit to angle of entry.

$$DP = R_u b\left(\int_{-\theta_r}^{\theta_e} \tau_x(\theta)cos\theta d\theta - \int_{-\theta_r}^{\theta_e} \sigma_n(\theta)sin\theta d\theta\right) \quad (13)$$

$$W = R_u b\left(\int_{-\theta_r}^{\theta_e} \sigma_n(\theta)sin\theta d\theta + \int_{-\theta_r}^{\theta_e} \tau_x(\theta)cos\theta d\theta\right) \quad (14)$$

$$T = R_u^2 b\int_{-\theta_r}^{\theta_e} \tau_x(\theta).d\theta \quad (15)$$

The lateral force generated in the soil is a result of the generated shear force across the contact patch and the bulldozing effect. The total lateral force (eq. 19) acting on the wheel is a sum of the shear force (eq. 18) and the bulldozing resistance (eq. 16), as shown in the following equations.

$$F_{ybd} = \left(\int_0^{2\sin(\theta_e)} \left(\gamma_s\left(R_u(cos\theta - cos\theta_e)\right)^2 N_\gamma cos\gamma_t + c\left(R_u(cos\theta - cos\theta_e)\right)N_c cos\gamma_t + q\left(R_u(cos\theta - cos\theta_e)\right)N_c cos\gamma_t\right) d_\lambda\right)\tan(\alpha) \tag{16}$$

$$q = \frac{\gamma_s\left(\int_{-\theta_e}^{\theta_e} zd\theta\right)\left(\int_0^{2sin\theta_e} \lambda\tan(\alpha)\,\mathrm{d}\lambda\right)}{mz} \tag{17}$$

$$F_{ycp} = R_u b\int_{-\theta_r}^{\theta_c} (c + \sigma_n\tan(\phi))^{(1-e^{-\frac{jy}{K_y}})} d\theta \tag{18}$$

$$F_y = -(F_{ycp} + F_{ybd}) \tag{19}$$

For combined slip conditions, the shear strength envelope of the soil is considered to be bounded by a unilateral constraint. The limit to the constraint is defined and after some simplifications, the longitudinal and lateral shear stresses are integrated into the final equation as follows –

$$F_x = R_u b\left(\int_{\theta_a}^{\theta_e} \tau_x(\theta)cos\theta d\theta + \frac{\mu_x^2.s}{\sqrt{(\mu_y\tan(\alpha))^2 + (\mu_x.s)^2}}\int_{-\theta_r}^{\theta_a} \tau_{\max}(\theta)cos\theta d\theta\right) - \int_{-\theta_r}^{\theta_a} R_u\sigma_u(\theta)sin\theta d\theta \tag{20}$$

$$F_y = \left(-R_u b\left(\int_{\theta_a}^{\theta_e} -\tau_{ycp}(\theta)d\theta + \frac{\mu_y^2.tan(\alpha)}{\sqrt{(\mu_y\tan(\alpha))^2 + (\mu_y.s)^2}}\int_{-\theta_r}^{\theta_a} \tau_{\max}(\theta)d\theta\right) - \left(\int_0^{2\sin(\theta_e)} \sigma_p d\lambda\right)\tan(\alpha)\right)\cos(\gamma_t) + F_z\sin(\gamma_t) \tag{21}$$

$$M_x = -F_z R_u\sin(\gamma_t) - F_y R_u\cos(\gamma_t) \tag{22}$$

$$M_z = F_{ycp}\sin(\theta_N) + R_u\sin(\theta_e) \tag{23}$$

The soil parameters used for this study are for the lunar soil simulant or GRC-1. The estimated values for the GRC-1 were obtained from the tests performed by (Oravec, 2009). The values of the parameters used are presented in Table 2.

Table 2. Soil parameters

Soil Parameter	Value
Soil cohesion parameter (K_c)	100 kN/m^{n+1}
Friction angle parameter (K_ϕ)	-100 kN/m^{n+2}
Friction angle (ϕ)	42 deg
Cohesion (c)	0.2 kPa
Sinkage index (n)	1
Unit weight of soil (γ)	1200 kg/m^3

Using the above equations, the Force-slip characteristics of the rigid tire were extracted from the Chan-Sandu tire model. The model cannot be directly incorporated into the MPC code, hence the data obtained was fitted using a regression model. It was made sure that the regression model is continuous across all its input dimensions. The fitted Force-slip, Figure 2 and Figure 3, and Moment-slip curve, Figure 4, are given below.

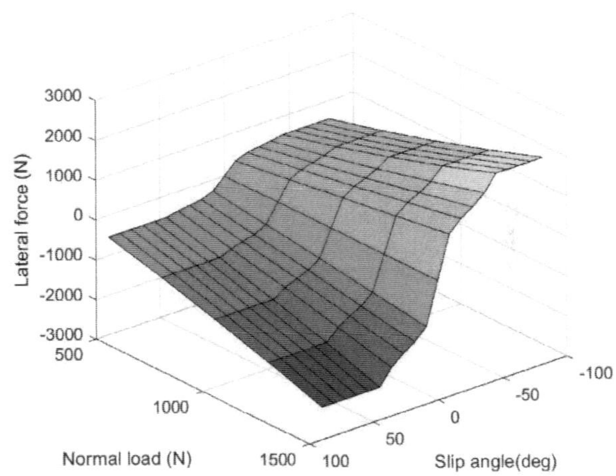

Figure 2. Lateral Force as a function of normal load and slip angle

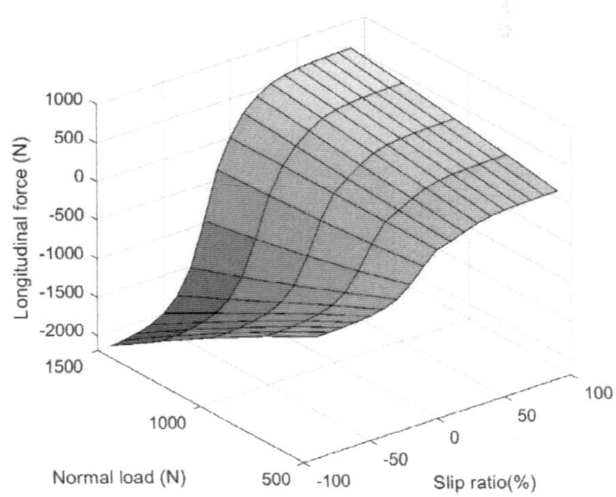

Figure 3. Longitudinal force as a function of normal load and slip ratio

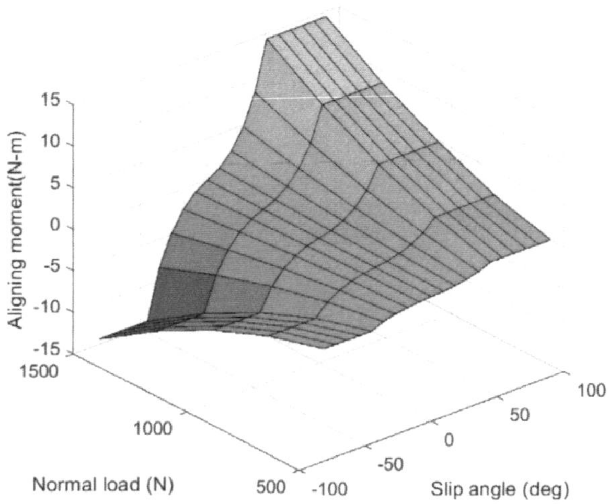

Figure 4. Aligning moment as a function of normal load and slip angle

3.3. Scaling using Granular scaling laws

The scaling factors used for considering the effect of gravity were developed using the Resistive Force Theory (Slonaker et al., 2017). Using dimensional analysis, the following equation (Eq. 24), describing the wheel's steady state driving limits, was proposed.

$$\left[\frac{P}{Mg\sqrt{Lg}}, \frac{V}{\sqrt{Lg}}\right] = \Psi\left(\sqrt{\frac{g}{L}}t, f, \frac{g}{L\omega^2}, \frac{\xi DL^2}{Mg}\right) \tag{24}$$

Here, f is a point-set denoting the dimensionless shape of the vehicle, D is the wheel width, L, is the characteristic length that scales f. The wheel is considered to have a constant rotational speed (ω). ξ is termed as 'Grain-structure coefficient' and depends on the grains, the intruder roughness, and gravity. Ψ is a four input-two output function, which has Power (P) and Longitudinal velocity (V) as output.

As we are concerned only about the effect of gravity, the above relationship can be extended by assuming that the terrain behaves as an ideal Coulomb material. The following relationship derived by comparing two tests with inputs (g', L', M', D', ω') and (qg, rL, sM, sr-2D, q1/2r1/ ω),

$$\langle P' \rangle = q^{\frac{3}{2}} r^{\frac{1}{2}} s \langle P \rangle \tag{25}$$

Now, as power can be written as a product of torque and angular velocity, the effect of gravity on driving torque can be written as following.

$$\langle T' \rangle = q^{\frac{3}{2}} r^{\frac{1}{2}} s \langle T \rangle \tag{26}$$

The above relationship was used to scale the value of driving force. As the lateral force and resistance generated by tire arises due to bulldozing and lateral shearing, its effect was directly considered by replacing the value of gravity. The limitation of using this method is that it does not take into account the changes in bulk properties of the terrain due to changes in intra-granular forces in low gravity.

3.4. Control

The problem involves optimal control of rover, which is a resource constrained system, such as path constraints, state constraints etc. Classical controllers based on robustification techniques such as phase and gain margins cannot be suitable for such systems. In addition, such systems also must control multiple parameters such as state boundaries, control thresholds etc. Since the tire-terrain interaction model is highly non-linear, classical techniques do not work on such models. All the above things can be handled by an optimization-based controller such as an MPC. An MPC model can be used to develop a multi-input multi-output controller. In MPC, we can accommodate the physics required for the vehicle and the tire model.

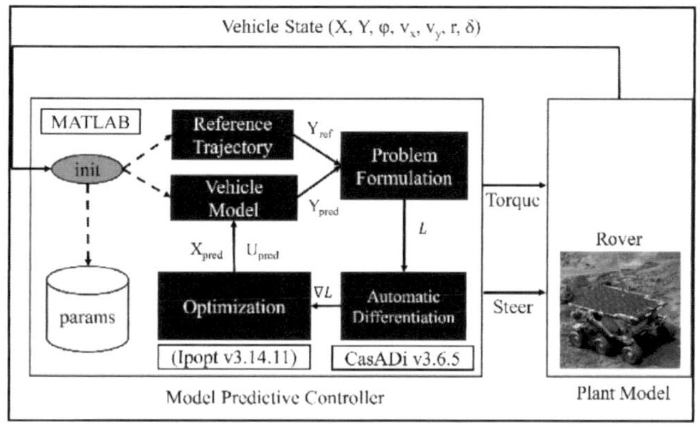

Figure 5. MPC architecture

The rover is controlled by MPC by modulating the torques on the wheels and the steering angle. The internal control model for the MPC is a four wheeled rigid suspension dynamic vehicle model with Chan-Sandu tire model for lateral and longitudinal control forces. The model consists of seven states, as shown in Eq. 27.

$$x = \left[X, Y, \varphi, V_x, V_y, \dot{\varphi}, \delta\right]^T \tag{27}$$

The control model has freedom to manipulate the rate of change of steering and the slip ratios on the four wheels independently. The torque and angular velocity at the wheels can be obtained as a function of these slip ratios, which are used to control the rover. By doing this, we avoid numerical issues, like dividing by zero.

$$u = \left[\dot{\delta}, \kappa_{fl}, \kappa_{fr}, \kappa_{rl}, \kappa_{rr}\right]^T \tag{28}$$

Since conventional models like LQR and PID are not feasible due to non-linear multi-input multi-output (MIMO) system dynamics, online MPC is a more appropriate method to approach this problem as it simultaneously optimizes all inputs.

The objective function is set such that the rover does not exceed the maximum traversable speed of 3 m/s as well as minimizing the cross-track error.

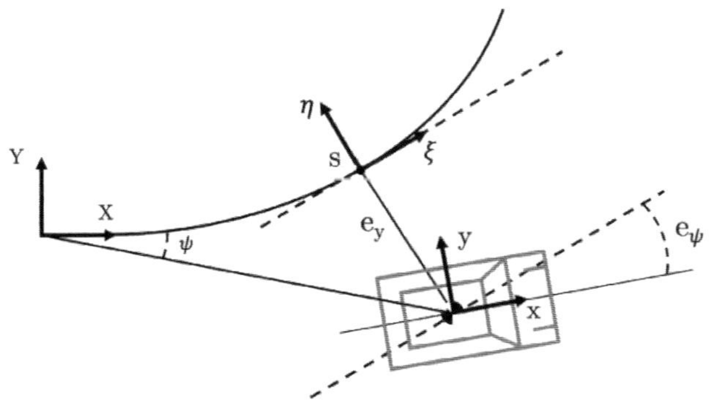

Figure 6. Local and global vehicle frames

The dynamics of the vehicle can be represented in the Eq. 1 – Eq. 12, as the following first order ODE –

$$\dot{x} = f(x,u) \tag{29a}$$

The objective function we are trying to solve is for a reference tracking problem, where the reference trajectory is given by x_{ref}. x_{ref} is given by a vehicle-on-rails model and it contains the same states calculated by the lumped-mass kinematics based on the desired cartesian path coordinates.

$$\dot{x} = \int (x - x_{ref})^T Q(x - x_{ref}) + u^T Ru$$
Such that, $\dot{x} = f(x,u)$
$$x_{min} \leq x \leq x_{max}$$
$$u_{min} \leq u \leq u_{max} \tag{29b}$$
$$g(x,u) < 0$$

Here, $g(x,u)$ represents any additional constraints such as minimum and maximum slip angles.

This cannot be solved on a computer numerically as the control signal is not discretized. We are using Multiple shooting to transcript the continuous optimal control problem into a constrained nonlinear program which can be solved using classical optimization techniques such as interior point methods and sequential quadratic programming. For this example, we are using the open-source optimizer IPOPT along with an academic license for the sparse linear solver MA57. By discretizing Eq. 29b into a N-step horizon,

$$\min_u \psi = \sum_{i=1}^{N} \left[\left(x^i - x^i_{ref}\right)^T Q(x^i - x^i_{ref}) + u^{i^T} Ru^i \right]$$
Such that, $x^{i+1} = RK4\left(x^i, u^i\right)$ $for\ all\ (i = [2, n-1])$ (30a)
$$x_{min} \leq x \leq x_{max},$$
$$x^1 = x_o$$
$$u_{min} \leq u \leq u_{max}$$

Where RK4 is a fourth order Runge-Kutta integration step.

For gradient calculations, a computational graph based automatic differentiation has been employed using the CasADi(Andersson et al., 2019). This package generates sparse matrix representations for the Jacobian of non-linear constraints and gradient and Hessian of the objective function. For this problem, they are represented in Figure 7 and Figure 8, respectively.

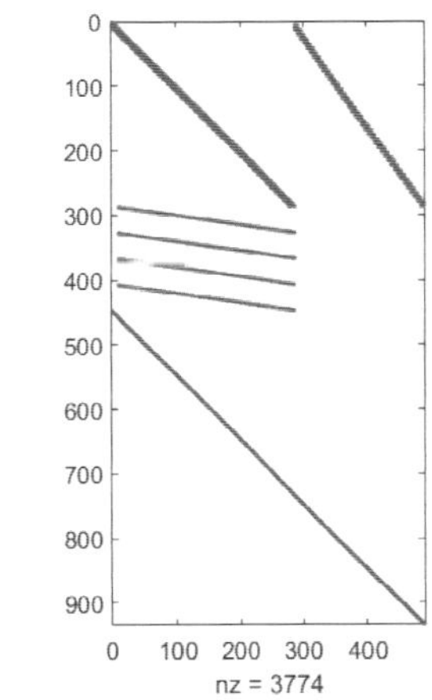

Figure 7. Jacobian of dynamic constraints w.r.t. free variables

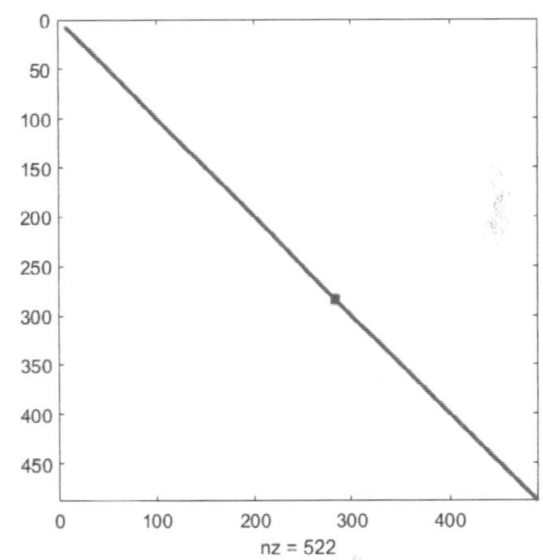

Figure 8. Hessian of Objective function w.r.t free variables

Since it is highly sparse, we are using a sparse symmetric linear solver which employs the multi-frontal method (HSLMA57).

4. Results and Discussion

The results of preliminary MPC simulation are shown below for a randomly generated trajectory. With the four wheeled model, the MPC model was not able to generate reasonable slip ratios, and further investigation is required. However, if the traction forces are calculated independently from the MPC with the steering control inputs, the vehicle was able to follow the desired path trajectory. This is because there lies a chance of discontinuity arising as the slip ratio changes from positive to negative value or vice versa.

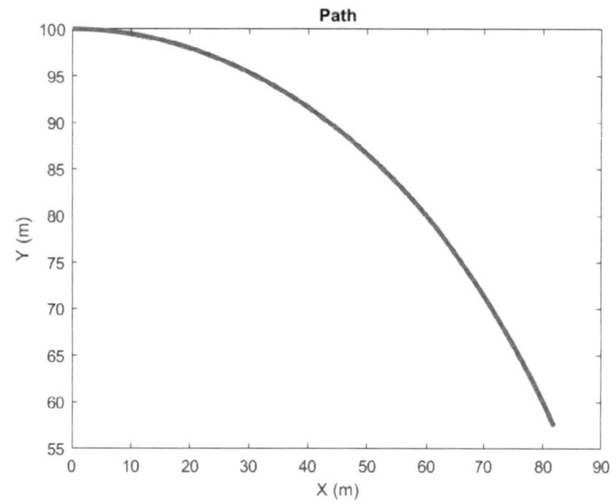

Figure 9. Vehicle path and reference path (Circular trajectory)

Figure 9 represents the vehicle trajectory against the reference trajectory, which overlaps each other. The path used as reference was a quarter circle. As can be observed the vehicle path is very close to the reference path indicating a very good convergence on the steering input.

Figure 10 depicts the steering wheel angle, as controlled by the MPC controller while Figure 11 represents the slip ratios on all four wheels.

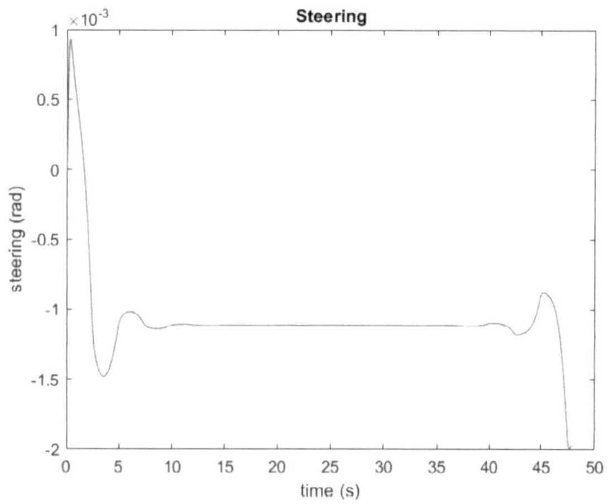

Figure 10. Steering wheel angle vs time (Circular trajectory)

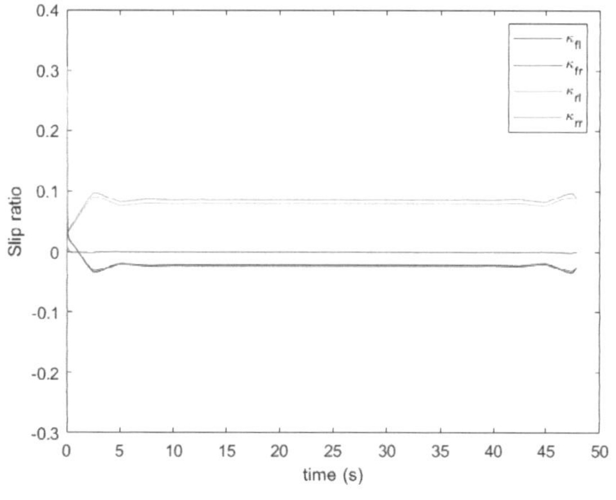

Figure 11. Slip ratios vs time (Circular trajectory)

To validate the model further, a sine wave reference trajectory was used. The reference and actual path which also overlaps for this case, Steering wheel angle, slip ratios and vehicle speed for the sine wave trajectory are represented in Figure 12, Figure 14, Figure 15 and Figure 16, respectively.

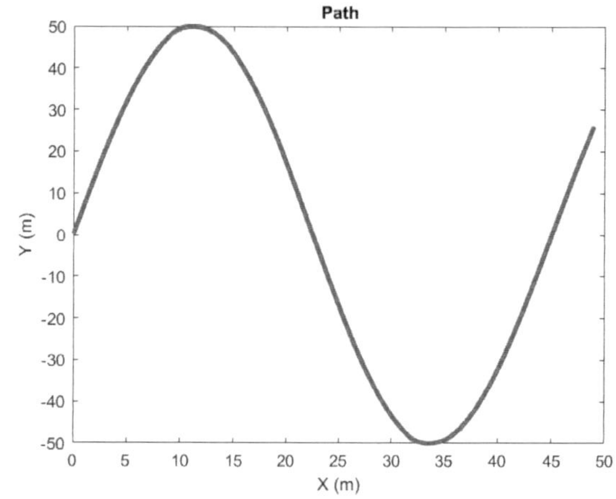

Figure 12. Vehicle path and reference path (Sine trajectory)

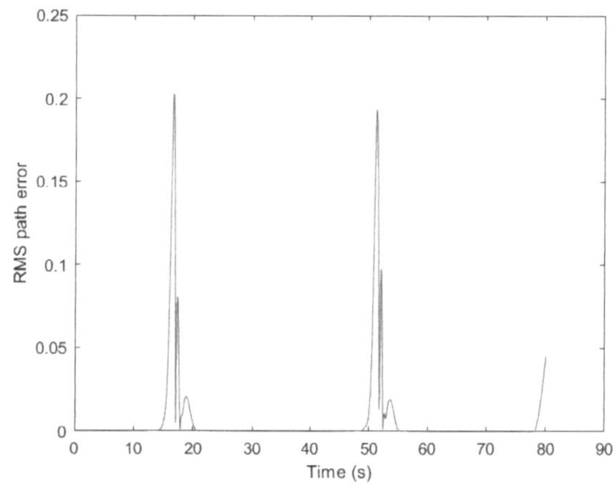

Figure 13. Cross tracking error vs time (Sine trajectory)

Figure 13 represents the cross-track error of the vehicle. Cross track error is the lateral offset or deviation of the vehicle center of gravity from the reference path. For the above simulation, a maximum cross track error of 0.02 m was obtained.

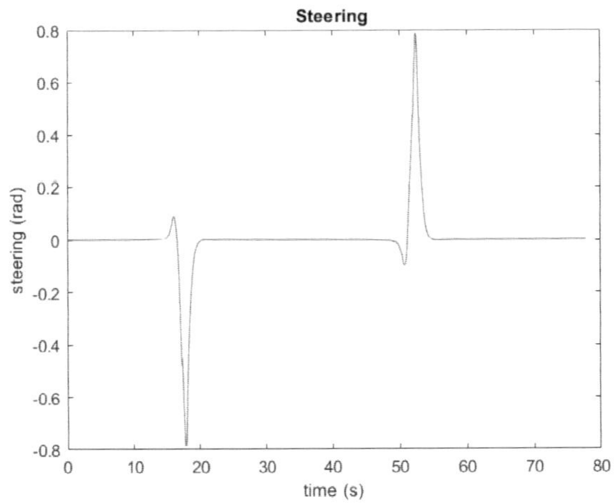

Figure 14. Steering wheel angle vs time (Sine trajectory)

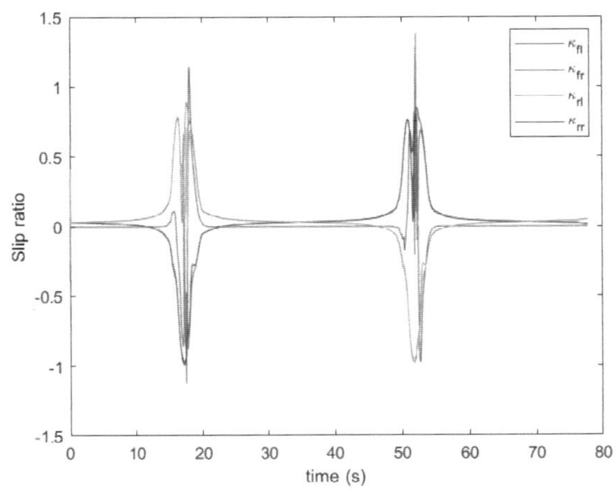

Figure 15. Slip ratios vs time (Sine trajectory)

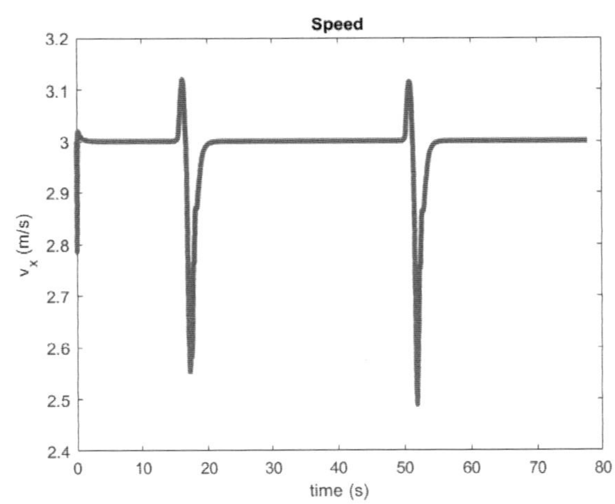

Figure 16. Vehicle speed vs time (Sine trajectory)

5. Conclusions

The aim of this research was to integrate a physics-based tire-terrain interaction model with a model predictive control algorithm. The Chan-Sandu tire model used for calculating the tire-terrain interaction parameters, such as generated longitudinal and lateral force, etc., is not continuously differentiable. The main reason for it is the conditional statements required to determine the maximum shear stress location. Hence, it cannot be directly integrated into the MPC model as it is difficult to linearize as a plant. Also, all the functions used in the Chan-Sandu model are not continuous. This results in the model being non-differential at some points making the algorithm unsolvable.

A surrogate model was developed using the above tire model. The tire model provided a continuous function for lateral force, longitudinal force and aligning torque as a function of slip and normal load.

The MPC model was able to control the vehicle and keep it on the defined path. However, it was also noticed that the algorithm sometimes converged to local infeasibility. This indicated that the control model is not robust enough to handle the complexity of the system. More work should be performed to increase the robustness of the algorithm so as to integrate a complete physics-based tire-terrain interaction model with slip-sinkage.

6. Future work

The Chan-Sandu tire model should be updated for the specific application to be used in the MPC plant. All the functions used in the above tire model should be made continuous across the range of input parameters. For this study, only the normal load and the slip were considered from the tire perspective, as an input to the Tire-terrain interaction model. An updated tire model should be developed which also considers the wheel speed to account for the strain rate and slip sinkage.

It was observed that the MPC model converged to the local infeasibility for a few times. Work should be put in to increase the robustness of the algorithm while maintaining the fidelity of the vehicle and the tire-terrain interaction model. The current model has a high complexity due to which sinkage was not considered for the control algorithm. A simplified slip-sinkage model should be added to the surrogate tire-terrain interaction model to consider sinkage as a control parameter.

It was observed that the model relied more on torque vectoring than steering input to turn the vehicle. By penalizing the slip ratios to ensure that slip ratios are always positive, i.e., the vehicle does not perform differential braking, the above condition could be avoided.

Additionally, the control model could be integrated into a Physics-informed neural network (PINN) which can increase the accuracy of the algorithm when trained using reinforcement learning.

7. Nomenclature

a	Distance from Vehicle C.G, to Front axle	[m]
b	Distance from Vehicle C.G, to Rear axle	[m]
t	Track width of vehicle	[m]
a_x	Vehicle acceleration in X direction	[m/s²]
a_y	Vehicle acceleration in Y direction	[m/s²]
V_x	Vehicle velocity in X direction	[m/s]
V_y	Vehicle velocity in Y direction	[rad/s]
X	Global x coordinate of vehicle	[m]
Y	Global y coordinate of vehicle	[m]
α	Slip angle	[rad]
β	Vehicle side slip angle	[rad]
δ	Steering input	[rad]

ω	Wheel angular velocity	[rad/s]
φ	Yaw angle	[rad]
$\dot{\varphi}$	Yaw rate	[rad/s]
$\ddot{\varphi}$	Yaw acceleration	[rad/s^2]
F_{FL}	Tire Force on front left wheel	[N]
F_{FR}	Tire Force on front right wheel	[N]
F_{RL}	Tire Force on rear left wheel	[N]
F_{RR}	Tire Force on rear right wheel	[N]
F_{CP}	Compaction resistance	[N]
F_{BD}	Bulldozing resistance	[N]
τ	Shear stress	[N/m^2]
σ	Radial stress	[N/m^2]
x	Vehicle States	
u	Model outputs	
P	Power	[N.m/s]
T	Driving Torque	[N.m]
W	Normal load on Tire	[N]
DP	Drawbar pull	[N]
R_u	Unloaded radius	[m]
q	Resistance due to Surcharge	[N]
λ	Linear distance (Front of wheel to angle of entry)	[m]
γ_s	Unit weight of soil	[kg/m^3]
N_q	Soil surcharge coefficient	
N_c	Soil cohesion coefficient	

8. Acknowledgements

The authors would like to thank the TMVS Lab at Virginia Tech for providing the resources needed for this project. The authors would also like to thank the Science and Technology Facilities Council, as part of United Kingdom Research and Innovation (UKRI) for providing Academic License for the sparse linear solvers (MA57, MA86, and MA97) from the Harwell Subroutines Library (HSL) which were used in this research.

9. Declaration of competing interest

The authors declare that they have no known competing financial interests or personal relationships that could have appeared to influence the work reported in this paper.

10. References

Andersson, J. A. E., Gillis, J., Horn, G., Rawlings, J. B., & Diehl, M. (2019). CasADi: a software framework for nonlinear optimization and optimal control. Mathematical Programming Computation, 11(1), 1–36. https://doi.org/10.1007/s12532-018-0139-4

Apollo11Space. (n.d.). Apollo11Space. Retrieved August 9, 2024, from https://apollo11space.com/apollo-11-space-about-us/

Chan, B. J., Sandu, C., Chan, B. J., & Sandu, C. (2014). Development. In Int. J. Vehicle Systems Modelling and Testing (Vol. 9, Issue 2).

Kobayashi, T., Fujiwara, Y., Yamakawa, J., Yasufuku, N., & Omine, K. (2010). Mobility performance of a rigid wheel in low gravity environments. Journal of Terramechanics, 47(4), 261–274. https://doi.org/10.1016/j.jterra.2009.12.001

Kovács, L. L., Ghotbi, B., González, F., Niksirat, P., Skonieczny, K., & Kövecses, J. (2020). Effect of gravity in wheel/terrain interaction models. Journal of Field Robotics, 37(5), 754–767. https://doi.org/10.1002/rob.21924

Liang Ding, K. N. K. S. A. M. K. Y. H. G. Z. D. (2010). Terramechanics-based High-Fidelity Dynamics Simulation for Wheeled Mobile Robot on Deformable Rough Terrain. I E E E.

Oravec, H. A. (2009). UNDERSTANDING MECHANICAL BEHAVIOR OF LUNAR SOILS FOR THE STUDY OF VEHICLE MOBILITY.

Pinho, J., Costa, G., Lima, P. U., & Ayala Botto, M. (2023). Learning-Based Model Predictive Control for Autonomous Racing. World Electric Vehicle Journal, 14(7). https://doi.org/10.3390/wevj14070163

Sandu, C., Taheri, S., Taheri, S., Els, S., & Jimenez, E. (2020). Hybrid soft soil tire model (HSSTM). Part III: Model parameterization and validation. Journal of Terramechanics, 88, 1–15. https://doi.org/10.1016/j.jterra.2019.11.002

Sandu, C., Taheri, S., Taheri, S., & Gorsich, D. (2019). Hybrid Soft Soil Tire Model (HSSTM). Part II: Tire-terrain interaction. Journal of Terramechanics, 86, 15–29. https://doi.org/10.1016/j.jterra.2019.08.004

Setiawan, J. D., Safarudin, M., & Singh, A. (2009). Modeling, simulation and validation of 14 DOF full vehicle model. International Conference on Instrumentation, Communication, Information Technology, and Biomedical Engineering 2009, ICICI-BME 2009. https://doi.org/10.1109/ICICI-BME.2009.5417285

Slonaker, J., Motley, D. C., Zhang, Q., Townsend, S., Senatore, C., Iagnemma, K., & Kamrin, K. (2017). General scaling relations for locomotion in granular media. Physical Review. E, 95(5–1), 052901. https://doi.org/10.1103/PhysRevE.95.052901

Taheri, S., Sandu, C., Taheri, S., & Gorsich, D. (2019). HYBRID SOFT SOIL TIRE MODEL (HSSTM). PART I: TIRE MATERIAL AND STRUCTURE MODELING.

Tang, H., Bai, C., & Guo, J. (2022). Optimal Path Planning of Planetary Rovers with Safety Considerable. Lecture Notes in Electrical Engineering, 861 LNEE, 3308–3320. https://doi.org/10.1007/978-981-16-9492-9_325

Wong, J. Y. (Jo Y. (2022). Theory of ground vehicles.

5334 | Enhanced open-loop control of automatic gear shifting in hydromechanical CVT for agricultural tractors

Massimo Martelli [a,*], Pietro Marani [a], Silvia Gessi [a], Damiano Chiarabelli [a]

[a] *CNR-STEMS, National Research Council, Inst. of Sci. and Tech. for Sustainable Energy and Mobility, Ferrara, Italy*

* *Corresponding author: massimo.martelli@cnr.it*

ABSTRACT

Continuously Variable Transmissions (CVTs) are typically implemented in agricultural tractors by means of a power-split configuration between mechanical transmission and infinitely-variable hydrostatic unit (IVHU).

The present work is based on an optimized output-coupled/compound architecture developed by the authors for a 200-kW reference tractor, originally equipped with a standard input-coupled CVT.

A two-stage planetary gearset – serving as the power-split device – is complemented by a four-gear mechanical gearbox, to efficiently cover the full velocity range, up to 40 km/h, with power-shift capability, to supply an uninterrupted traction torque even during gear change.

The main topic is the control of the transmission, and the backbone of the algorithm is based on an open loop ideal mapping. At any given constant engine speed and for each gear, an ideal reference condition is considered – in which the hydraulic units have ideal efficiencies – to analytically define a theoretical open-loop function, providing the requested IVHU transmission ratio as a function of vehicle velocity, and the theoretical shaft synchronization point, to shift to the previous/next gear.

Real-world efficiencies affect the synchronization dynamics and, consequently, the traction force while shifting gears. An enhanced control has then been devised, keeping the simple open-loop foundation, with an additional smoothing contribution, triggered by the shaft speed feedback when speed matching is detected.

The problem is studied via a lumped-parameter simulation model, focusing on the effect of the control on the longitudinal dynamics of the vehicle-wheel-terrain system. An exploratory investigation is carried out to tune the function, based on vehicle speed, drawbar pull and gear, for a smooth operation in a suitable range of operating conditions.

Keywords
Continuously Variable Transmission
Agricultural tractor
Power-shift
Lumped-parameter modeling

1. Introduction

CVTs are considered to be the best option among agricultural tractor transmissions, to achieve independence of the vehicle speed on the engine speed. This speed decoupling is desirable to allow the use of every engine working point (thus including the best possible fuel economy), for every power demand (Renius, 2020).

Hydrostatic transmissions are a mature, robust and reliable technology, capable of continuously variable ratios, with a very practical ratio control (via displacement control on the hydraulic units), making them the technology of choice for several decades now. A direct hydrostatic drive is not a viable option for large tractors: due to the moderate efficiency of hydrostatic units, delivering the whole engine power would be very inefficient; also, massive size units would be required.

The solution is given by power-split transmissions, where the engine power is split between a fixed, high efficiency mechanical path and a variable, lower efficiency hydraulic path (i.e. the hydrostatic transmission), so that the continuous variation is maintained, but with a higher overall efficiency. Power splitting is achieved via a planetary gearset (PGS); a classification of the main architectures (input- output- and compound-coupled) and theory of operation are provided in the landmark work (Kress, 1968).

For the already mentioned efficiency and sizing constraints, CVT transmissions for large tractors must be equipped with a discrete number of mechanical gears (ranges). Consequently, ensuring an uninterrupted power transmission to the wheels during gear shifts (i.e. a power-shift capability) becomes vital for optimal operation. This requires a setup with a two-stage PGS, proper grouping on the mechanical gears in multiple output shafts and (at least) two independent clutches.

Nowadays, power-split transmissions are widely popular in the passenger car sector because they enable hybrid (combustion engine + electric) powertrains (Chau, 2015). Following a general trend of knowledge transfer from the automotive to the agricultural industry, electric power-split CVTs are now a prominent field of research for the development of innovative agricultural driveline architectures (Rekow et al., 2017).

The present paper focuses on the control of a power-split power-shift transmission, proposing a simple yet effective method to enhance the synchronization dynamics, new to the state of art.

2. The system under investigation

The object of this study is a power-split CVT transmission, for a reference 200-kW heavy-duty agricultural tractor; it includes (for the forward direction of motion) four mechanical gears ($F1$ to $F4$), so that effective operation can (in principle) be achieved across the whole speed range (up to 40 km/h), in all relevant working conditions. The four gears are divided in two groups, *odd* ($F1$, $F3$) and *even* ($F2$, $F4$), each one with its own output shaft, in a dual-clutch layout. In any gear, the corresponding clutch (A/B for *odd/even*) is engaged while the other one is disengaged. At gear shift, the incoming gear is always of the opposite type (*odd/even*) than the outgoing one and the two clutches (i.e. their engagement state) must be swapped.

https://doi.org/10.56884/APA3YYLX

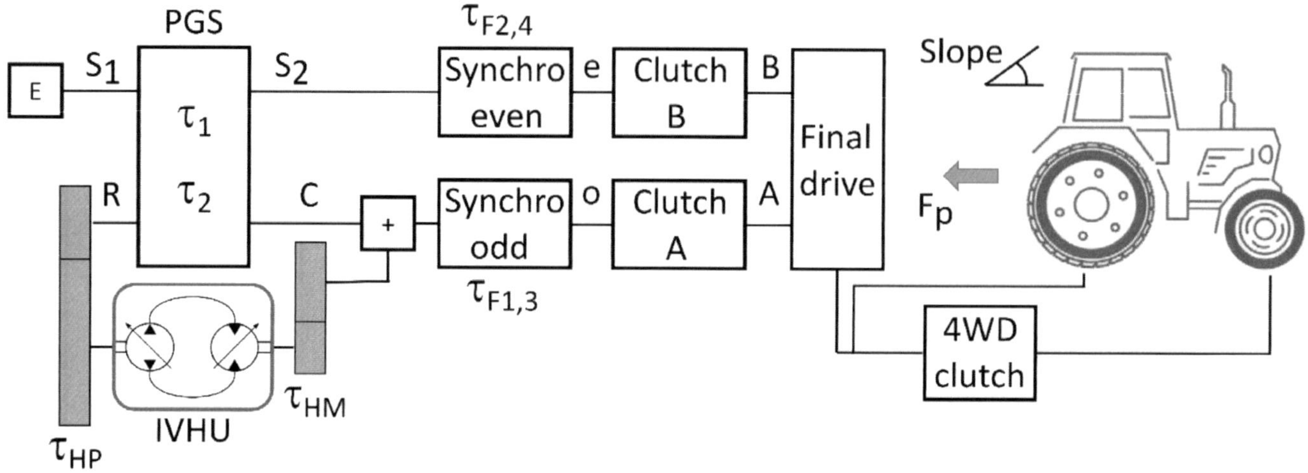

Fig. 1. Block diagram of the physical system (not including the control system)

Unlike single-clutch layouts (like the ones in traditional passenger cars), where a transient loss of traction torque at the driving wheels is inherent and unavoidable (when the clutch is disengaged), a properly tuned coordinated command ensures that at no time both clutches are completely disengaged, resulting in an uninterrupted traction torque (i.e. with *power-shift* capability).

The CVT is of the traditional hydromechanical type (currently the most common in agricultural tractors): its mechanical core is a two-stage, four-shaft planetary gearset (PGS), paired with an infinitely-variable hydrostatic unit (IVHU), the latter implementing the actual continuous variation of the transmission ratio. The four PGS shafts are: carrier (C), shared between the two stages; ring (R) and sun 1 (S1), on the first stage; sun 2 (S2), on the second stage. The ratios for the two stages are τ_1= -0.30, τ_2= 0.20.

(NOTE: Throughout the work, all mechanical ratios will have a positive/negative sign, due to the adopted sign convention for rotational velocities and, specifically to the PGS, depending on the use of internal/external gearwheels; the details are outside the scope of this paper and will not be further discussed).

The IVHU is a hydrostatic transmission, with two volumetric units, pump (HP) and motor (HM), in a typical full circuit including boost pump, pressure relief valves and flushing valve (Vacca and Franzoni, 2021). Following the general convention, the two units are identified as "pump" and "motor" but, in reality, both of them switch between pumping and motoring mode during operation; at any time, one unit is pumping and the other is motoring (with the exception of transient "singularity" points, where the operating mode is not definable in a meaningful way). HP and HM are connected to the R and C shafts of the PGS, respectively, each through its own mechanical reduction (ratios: τ_{HP} = -0.65, τ_{HM}= 0.35). Both units are of the variable displacement type, with zero-displacement capability, and they have the same size (D_{HP} = D_{HM} = 66 cm^3/rev).

The tractor is powered by a diesel engine (E), which is directly connected to the S1 shaft. C and S2 are the two output shafts of the PGS, connected respectively to the odd and even branches of the mechanical gearbox. Each branch is made of a synchronizer, selecting the proper mechanical ratio based on the active gear (odd gears: τ_{F1} = 0.59, τ_{F3} = 2.15; even gears: τ_{F2} = 0.19, τ_{F4} = 0.68), followed by a clutch (A/B) connecting/disconnecting the output shaft (o/e) to the final drive. This component applies a final mechanical ratio, to drive the tractor rear wheels; a 4WD clutch allows transferring power also to the front wheels, switching from 2WD to 4WD operation.

A block diagram representation of the system is shown in Fig. 1.

2.1. The CVT architecture

The above-described architecture is the output of a previous work by the authors (Marani et al, 2023), where alternative options were studied for the reference tractor, originally equipped with a standard input-coupled CVT, using a fixed displacement HM. The aim of the work was to improve the transmission efficiency, while achieving minimum sizing of the two volumetric units. It was required that the number of mechanical gears (four) be kept unchanged, but the use of either a fixed- or variable-displacement HM was allowed. The result is an optimized output-coupled/compound layout, with an output-coupled (compound) configuration for odd (even) gears.

The PGS imposes a set of relations between the homologous physical quantities at its four shafts: kinematic (speed transmission, Eq. 1), torque transmission (Eq. 2), torque balance (Eq. 3) and power balance (Eq. 4):

$$\omega_{S2} = \frac{\omega_{S1} + (\tau_2 - 1)\omega_C}{\tau_2} \tag{1}$$

$$T_R = -\tau_1 T_{S1} + \frac{\tau_1}{\tau_2} T_{S2} \tag{2}$$

$$T_{S1} + T_R - (T_{S2} + T_C) = 0 \tag{3}$$

$$P_{S1} - (P_R + P_{S2} + P_C) = 0 \tag{4}$$

In any gear, the typical control target for the vehicle (in a driving scenario) is the longitudinal speed requested by the operator (v_{req}). At a given engine speed (ω_E), this translates, for the transmission, into a requested value for the transmission ratio ($\tau_{T,req}$), which is defined by Eqs. 5 and 6 for odd and even gears, respectively:

$$\tau_T = \frac{\omega_{o1,3}}{\omega_E} = \frac{\tau_{F1,3}}{1 - \tau_1 + \tau_\alpha} \tag{5}$$

$$\tau_T = \frac{\omega_{e2,4}}{\omega_E} = \tau_{F2,4}\left(\frac{1}{\tau_2} - \frac{1 - \tau_2}{\tau_2}\frac{1}{1 - \tau_1 + \tau_\alpha}\right) \tag{6}$$

The auxiliary ratio τ_α is defined by Eq. 7:

$$\tau_\alpha = \frac{\tau_1}{\tau_{HP}\tau_{HM}} \frac{\alpha_{HM}}{\alpha_{HP}} \frac{D_{HM}}{D_{HP}} \tau_{\eta V} \quad (7)$$

where factor $\tau_{\eta V}$ accounts for the real (non-unitary) volumetric efficiencies in HP and HM, when they are not assumed as ideal devices (Eq. 8):

$$\tau_{\eta V} = \begin{cases} \dfrac{1}{\eta_{V,HP,p} * \eta_{V,HM,m}} & F1, F3 \\ \eta_{V,HP,m} * \eta_{V,HM,p} & F2, F4 \end{cases} \quad (8)$$

In Eqs. 5 and 6, all the mechanical ratios are fixed; therefore, τ_T can be adjusted only via τ_α, i.e. by changing the displacement settings α_{HP} and α_{HM}.

The formulae for torque transmission are provided in Eqs. 9 to 12:

$$T_{o1,3} = -\frac{\tau_1(1-\tau_\gamma)-1}{\tau_{F1,3}} T_{S1} \quad (9)$$

$$T_{e2,4} = \frac{1}{\tau_{F2,4}} \frac{\tau_2(\tau_1(1-\tau_\gamma)-1)}{\tau_1(1-\tau_\gamma)-\tau_2} T_{S1} \quad (10)$$

$$\tau_\gamma = \frac{1}{\tau_{HP}\tau_{HM}} \frac{\alpha_{HM}}{\alpha_{HP}} \frac{D_{HM}}{D_{HP}} \tau_{\eta M} \quad (11)$$

$$\tau_{\eta M} = \begin{cases} \eta_{M,HP,p} * \eta_{M,HM,m} & F1, F3 \\ \dfrac{1}{\eta_{M,HP,m} * \eta_{M,HM,p}} & F2, F4 \end{cases} \quad (12)$$

While the kinematic relations defined by Eqs. 5 to 8 are undoubtedly at the core of speed control design, Eqs. 9 to 12 don't share such a prominent role, under the assumption of constant ω_E. Nonetheless, the latter equations were used in the preliminary design phase - which is out of scope of the paper - for the sizing of the hydrostatic units and to verify the operating envelope of the diesel engine of the reference tractor.

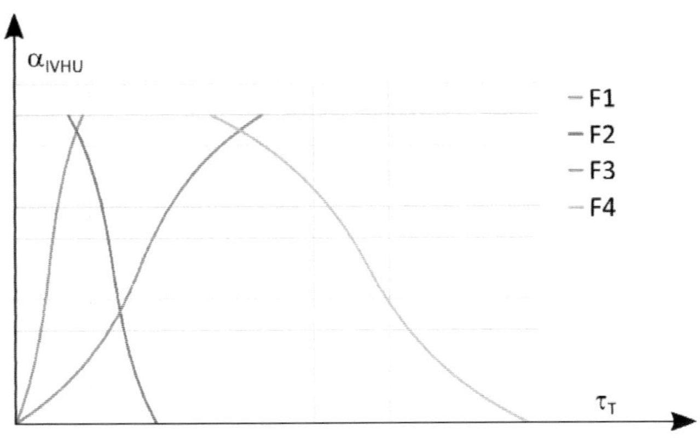

Fig. 2. IVHU control curves (forward gears F1 to F4)

2.2. The IVHU control

A given value of τ_α can be achieved by an infinite set of (α_{HP}, α_{HM}) pairs; nonetheless, keeping two independent control variables would be meaningless and impractical. A generalized single command for the IVHU (α_{IVHU}) was then defined, in the range [0, 2], with (α_{HP}, α_{HM}) following an HP-HM sequential control, as per Eq. 13:

$$\begin{cases} \alpha_{HP} = \max(min(\alpha_{IVHU}, 1), 0) \\ \alpha_{HM} = \max(\min(2-\alpha_{IVHU}, 1), 0) \end{cases} \quad (13)$$

Thanks to the two-stage PGS, τ_T increases - for odd (even) gears - as α_{IVHU} increases (decreases), as clearly shown in the control characteristic in Fig. 2. This results in three synchronism points, at the intersection of each pair of consecutive Fi curves, where the power-shift can be effectively executed, with no discontinuity on the command.

3. The model

The system is modeled, following a lumped-parameter approach, in a simulation framework previously developed by the authors (Martelli et al, 2023), suitable for both field and road operation. Relevant details about the test case and the modeling choices for the different subsystems are provided below.

3.1. Test case

Due to the focus on automatic gear shifting, a road operating condition is considered, with an operator input requesting a sequence of vehicle acceleration, constant speed and deceleration phases, covering the range from standstill to 40 km/h and back. The transmission is therefore set to 2WD. A typical working condition, where the engine speed is constant and the vehicle velocity is controlled only via the CVT (IVHU), is applied. The allowed range for ω_E is [1500, 2200] rpm.

3.2. Engine

To focus as much as possible on the inherent characteristics of the CVT transmission, the engine is modeled as an ideal angular velocity generator.

3.3. Clutches

For the two wet multi-disc clutches, a simple model is adopted, which generates a rotary friction proportional to the input command.

3.4. Mechanical components of the transmission

An ideal engagement is assumed for all gearwheel pairs (i.e. parasitic phenomena such as backlash are not considered). Mechanical losses are lumped as viscous friction losses at the output shafts, with coefficients tuned based on the results of previous research works (TASC, 2020).

3.5. IVHU

The focus is on the main losses, given by the volumetric and hydromechanical efficiencies of HP and HM. The model has the capability to use both high-fidelity maps - where the efficiency is a function of pressure, shaft speed and displacement setting - and constant efficiencies. The latter approach was adopted, to streamline the analysis; moreover, this is a sensible approach, since the optimized architecture designed by the authors does not correspond to a production vehicle, so that HP and HM cannot be reliably matched with commercially available devices. Efficiencies are either unitary, for the ideal ("reference") case, or defined as follows for the real-world case: η_V = 0.9, η_M = 0.8.

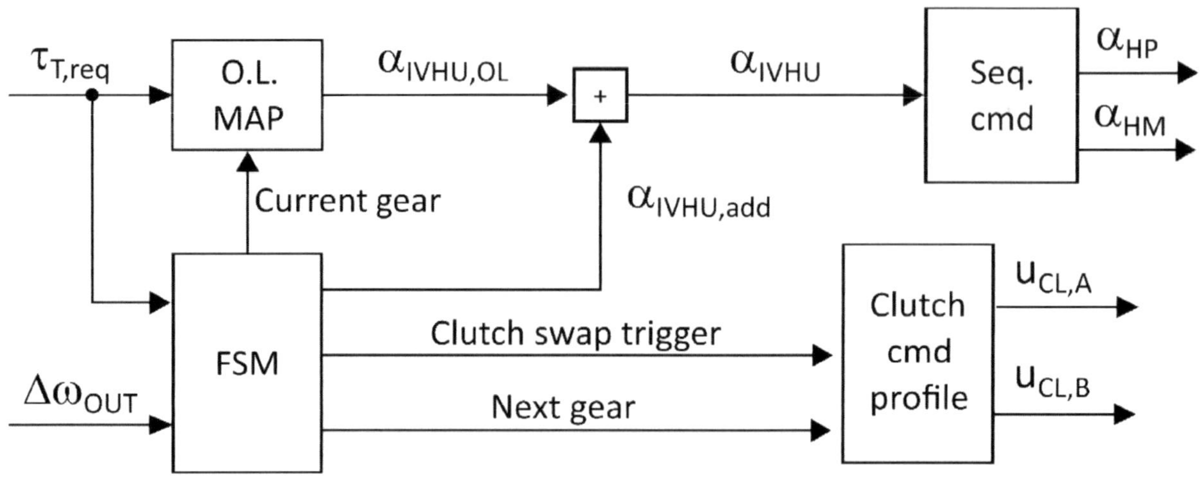

Fig. 3. Block diagram of the control system

The displacement adjustment dynamics are emulated via a simple first-order lag on the command signal.

3.6. Vehicle

The model is purely longitudinal, with a fixed weight distribution between the front and rear axles (59% on the rear) of the reference 12,000 kg tractor. Vehicle translation dynamics, modeled via the classical Newton's 2nd law, consider: traction/resistance forces from the wheels, drawbar pull, slope (gravitational) force and inertia.

3.7. Wheels and terrain

As for the vehicle, the model is purely longitudinal. The (net) traction force (F_X) is a function of the longitudinal slip ($i_{\%}$), following a "Magic Formula" approach (Wong, 2001). A "Simplified Pacejka" formulation (Siemens, 2022) is used (Eq. 14):

$$F_x = F_Z * D * \sin\left(C * \arctan\left(\frac{BCD}{C * D} * i_{\%}\right)\right) \tag{14}$$

where:

$$i_{\%} = \frac{r_W * \omega_W - v}{|v|} \cdot 100 \tag{15}$$

with constant effective rolling radii: r_W = 0.87 m (rear); 0.66 m (front). The remaining parameters are as follows: BCD = 0.065, C = 1.2132, D = 0.09. The rolling resistance depends only on the vertical load, with a constant resistance coefficient.

Angular rotation dynamics, following again Newton's 2nd law, consider: wheel inertia, driving/braking torque (from the transmission), torque generated by the tire-terrain interaction. Full details about the "1-D" vehicle-wheels-terrain model are provided in (Martelli et al., 2023).

4. The control system

The control system developed in this work is conceived as a *simple, yet effective, control logic, for the automatic gear shifting in a CVT transmission*. Its purpose is to be a robust tool, within the general simulation framework at CNR-STEMS, to allow a reliable and insightful virtual prototyping of agricultural tractors, so that innovative transmission architectures can be developed and properly characterized/compared. At this stage, it is not intended to be a finalized design, such as the extremely refined algorithms on OEM current production ECUs but it could represent a possible first layer of a much more complex logic.

As shown in the block diagram (Fig. 3), the system has two inputs: $\tau_{T,req}$ (operator input) and $\Delta\omega_{out}$ (feedback signal), and four outputs: α_{HP}, α_{HM} (IVHU commands) and $u_{CL,A}$, $u_{CL,B}$ (clutch engagement commands). An open-loop map (O.L. MAP) is defined for the ideal (unitary efficiency) case, which implements the ideal IVHU control curves (Fig. 2): given the current gear, the block provides an open-loop value for the IVHU command ($\alpha_{IVHU,OL}$), as a function of $\tau_{T,req}$ alone.

Considering e.g. a vehicle acceleration (i.e. an increasing $\tau_{T,req}$) in $F1$, with ideal hydraulic units, the $F1$-$F2$ upshift could be simply triggered at the synchronism point (the intersection point between the two curves in Fig. 2); in other words, all gear shift points could be pre-programmed. With real-world units, the non-unitary volumetric efficiencies distort the curves (see Eqs. 5 to 8), so that each synchronism point is shifted (i.e. it occurs at a different, non-predetermined τ_T). For this reason, the actual speed difference between the odd and even output shafts ($\Delta\omega_{out}$), to trigger the gear shift when $\Delta\omega_{out}$ is within a given threshold. This task is performed by a finite state machine (FSM), whose top-level state diagram is shown in Fig. 4: *Fi* is the (stable) state for the *i*-th gear, e.g. $F1$; *Switch Fo to Fi* is the transition state, from the *o*-th to the *i*-th gear, e.g. *Switch F1 to F2*; state-transition conditions (events) are indicated as *eoit*, where *o* = current (possibly outgoing) gear, *i* = (possibly) incoming gear, *t* = type of event (*a*, *c*, *d*, *m*, *s*), e.g. *e12s*.

All the transition states implement an internal (second level) FSM, as the one in Fig. 5. When $\tau_{T,req}$ exceeds a predefined threshold value (event *e12s*), the "window" for a possible F1-F2 upshift opens and the top-level state is switched from *F1* to *Switch F1 to F2*. In the *Synchronizer pre-selection* sub-state, the appropriate position is pre-selected in the *F2-F4* synchronizer (in this case *F2*, according to the possible upcoming shift to *F2*); then, a transition to sub-state *Wait for speed match* occurs when the pre-selection is done (event *e12d*).

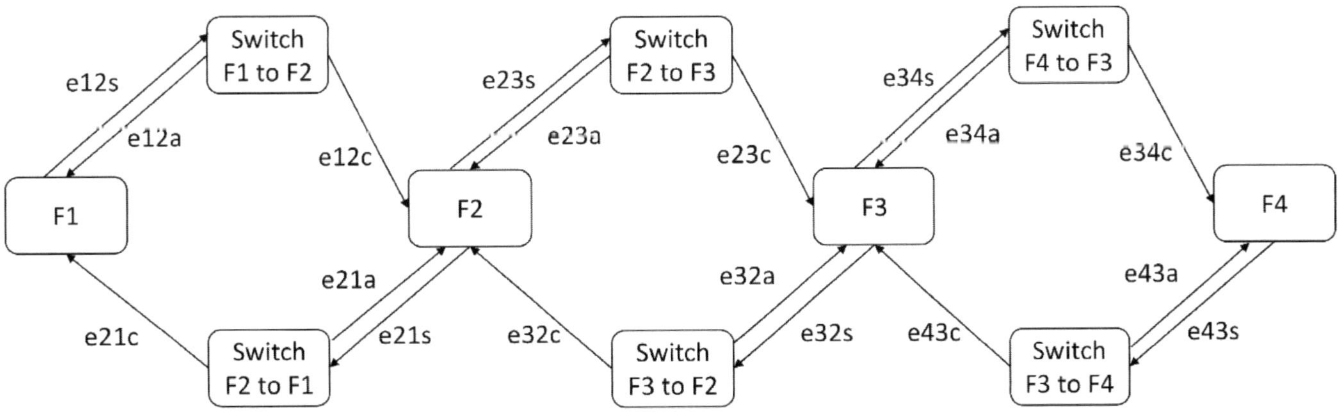

Fig. 4. State diagram of the FSM block, top-level

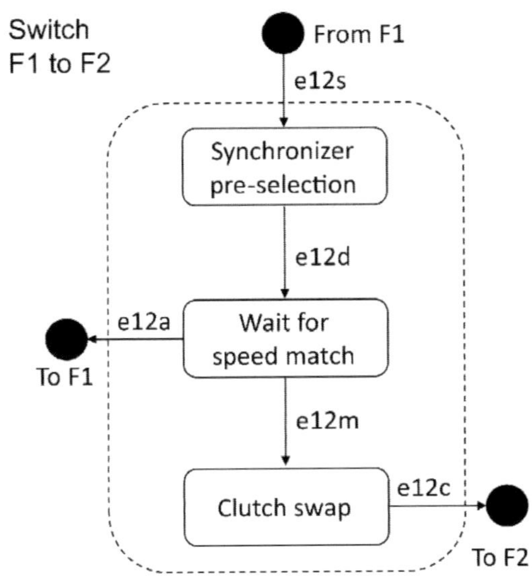

Fig. 5. FSM state diagram, second level (state Switch F1 to F2*)*

If, before the actual speed match, the operator requests a deceleration and $\tau_{T,req}$ drops below threshold (event *e12a*), the gear shift is aborted, returning to top-level state *F1*. Otherwise, when $\Delta\omega_{out}$ enters its tolerance band around zero (event *e12m*), the current time is stored (as t_0) and $\Delta\alpha_0$ is set, according to Eq. 16:

$$\Delta\alpha_0 = \alpha_{IVHU,curr}(t_0) - \alpha_{IVHU,next}(t_0) \tag{16}$$

Then, the sub-state is switched to *Clutch swap*, where the $\alpha_{IVHU,add}$ signal is active (i.e. a non-null output is generated), as per Eq. 17:

$$\alpha_{IVHU,add} = c_\alpha * \Delta\alpha_0 * e^{-(t-t_0)/k_t} \tag{17}$$

Clutches A and B are swapped, following two predefined time-dependent curves (engaging/disengaging clutch). When the swap is done (event *e12c*), top-level state *F2* is reached. All the six *Switch Fo to Fi* states operate in the same way, with the necessary adjustments between upshifts and downshifts (i.e. "acceleration" becoming "deceleration", etc.).

To understand Eqs. 16 and 17, Fig. 6 shall be considered. The *F1* (green) and *F2* (dashed blue) curves are the open-loop map output ($\alpha_{IVHU,OL}$) for the *F1* and *F2* gears, respectively, which, as already discussed, refers to the ideal efficiency case. So, with the actual speed match occurring at t_0, after the theoretical synchronism point (where the two curves intersect), a direct switch from the fist curve to the second would result in a command discontinuity (yellow curve, c_α = 0).

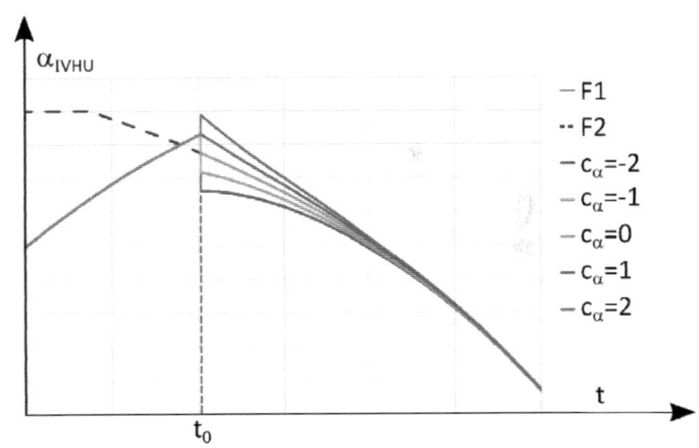

Fig. 6 IVHU control, additional contribution output for c_α = -2, -1, 0, 1, 2 (F1-F2 upshift)

$\Delta\alpha_0$ is the output gap (at t_0), to be compensated by $\alpha_{IVHU,add}$; then, the additional output must decay to zero, to go back to the open-loop control map (for the new gear), Eq. 18:

$$\alpha_{IVHU,add}(t) = \begin{cases} \Delta\alpha_0 & t = t_0 \\ \cong 0 & t \gg t_0 \end{cases} \tag{18}$$

Therefore, an exponential function as the one in Eq. 17, with c_α = 1, is a sensible choice to fulfill the requirements, in a simple way. As it will be shown in the next section, the initial intuition (i.e. to require a continuous α_{IVHU} output) does not correspond to the best system performance; consequently Eq. 17 provides a general formulation, where coefficient c_α is a tunable parameter. $c_\alpha > 1$ and $c_\alpha < 1$ correspond to "over-compensation" and "under-compensation" conditions. Any curve with $c_\alpha \neq 1$ introduces a discontinuity (at t_0).

It is worth noting that, with fixed (and known) volumetric efficiencies, an actual, 100% accurate, open-loop map could be implemented, removing the need for the above-described compensation. This is clearly not an option when the goal is designing a robust control, capable to effectively manage real-world setups, with variable efficiencies that, in general, will not be (reliably) known in the whole range of possible operating conditions.

5. Simulation results

5.1. Test case

In addition to what already specified in section 3.1, the following test conditions were applied to the model: a zero-slope terrain, with all the longitudinal load forces for the vehicle lumped into the (equivalent) drawbar pull (F_p). In each test run, F_p is constant; three load conditions are considered: no load (F_p = 0), low load (F_p = 5,000 N), high load (F_p = 10,000 N).

5.2. Results

The performance of the control system is evaluated in terms of smoothness of the vehicle longitudinal velocity during gear shifting. As an example, the results for the low load condition and ω_E = 1,800 rpm are shown in Fig. 7 (top), where the three upshifts and the three downshifts are recognizable at $t \approx$ 10.1 s (*F1-F2*), 16.1 s (*F2-F3*), 32.4 s (*F3-F4*), 80.1 s (*F4-F3*), 95.1 s (*F3-F2*) and 102.7 s (*F2-F1*).

Five tentative values are considered for the c_α coefficient: -2, -1, 0, 1, 2, and, as expected, the chosen value has an heavy impact on the system response, determining whether v will follow an overshooting, undershooting or relatively smooth profile. A zoomed view of the *F3-F4* and *F4-F3* transitions is provided in sub-plots *a* and *b* at the bottom of the figure.

Table 1 summarizes the optimal values of c_α (among the five previously selected), in terms of vehicle velocity smoothness.

Table 1
Optimal values of c_α (ω_E = 1,800 rpm)

			Load (F_p)
Gear shift	no	low	high
Upshift: F1 to F2	0	0	0
Upshift: F2 to F3	-1	-1	-1
Upshift: F3 to F4	-1	-1	-1
Downshift: F4 to F3	0	0	0
Downshift: F3 to F2	1	1	1
Downshift: F2 to F1	0	0	0

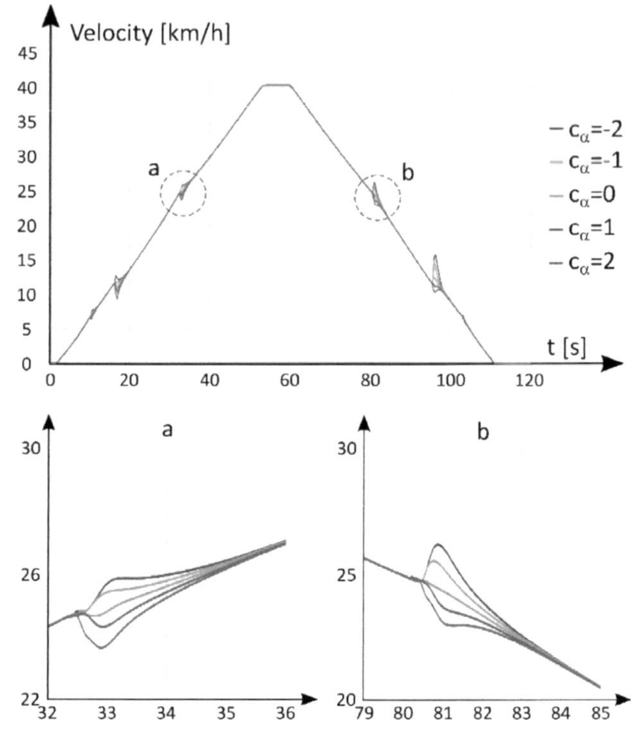

Fig. 7 (top) Resulting vehicle velocity with automatic gear shifting for the reference cycle (ω_E = 1,800 rpm; F_p = 5,000 N); (bottom) Zoomed view of the F3-F4 upshift (a) and F4-F3 downshift (b).

A first interesting result is the independence (at least at this first level of analysis) of c_α from F_p, in all six gear shifts. Then, a clear trend is visible, with null- to under-compensation required for upshifts and null- to full-compensation for downshifts.

A qualitative explanation can be achieved via graphical analysis: the upshift case is considered in Fig. 8, for the *F1-F2* transition at ω_E = 1,800 rpm and F_p = low.

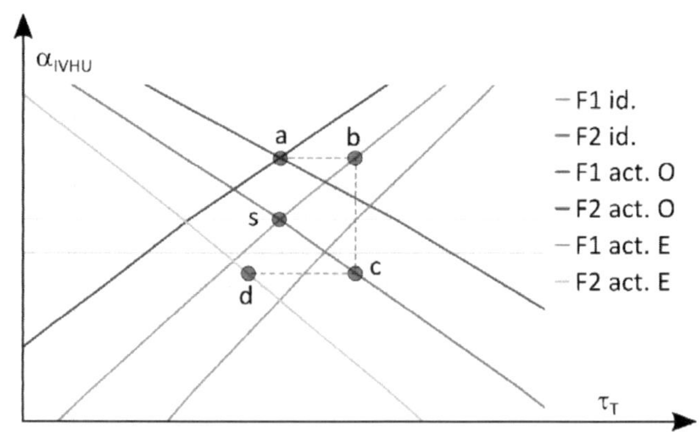

Fig. 8. Graphical analysis of the F1-F2 upshift (ω_E = 1,800 rpm; F_p = low)

As the vehicle accelerates (in the *F1* gear) the α_{IVHU} output – coinciding with the $\alpha_{IVHU,OL}$ output generated by the open-loop map – is given by the operating point moving from left to right on the *F1 id.* curve (i.e. the ideal curve for *F1*). If HP and HM were indeed ideal, speed synchronism would occur in point *s*, at the intersection of *F1 id.* with *F2 id.* (i.e. the ideal curve for *F2*). Instead, with non-unitary volumetric efficiencies, the actual control curves for *F1* and *F2* - with *F1* engaged - are *F1 act. O* and *F2 act. O*, respectively (*act. O* means *the actual curve, with odd-gear IVHU volumetric efficiency factor*, Eq. 8), so that the actual synchronism occurs at point *a*.

Since the open-loop map works with the *F1 id.* curve, the requested τ_T will have to increase to point *b*, to reach the α_{IVHU} output of point *a*. Then, after the gear shift to *F2*, the open-loop map working point jumps to *c* (on the *F2 id.* curve), while, similarly to what seen for *F1*, the actual curve is *F2 act. E*. (*E* clearly means *even-gear*). Therefore, the α_{IVHU} output in *c* generates an actual lower τ_T (point *d*), with a net decrease in the transmission ratio (given by the horizontal gap between points *d* and *a*). For a perfect transition, *d* must be horizontally aligned to *a*; this corresponds to moving c (and b) to the right, which translates to a further increase of τ_T, with respect to the ideal synchronism. This means accentuating the effect of non-ideal efficiencies, instead of compensating it, hence an "under-compensation".

The same analysis can be made for the upshift case (see Fig. 9, *F3-F2* transition at ω_E = 1,800 rpm and F_p = low).

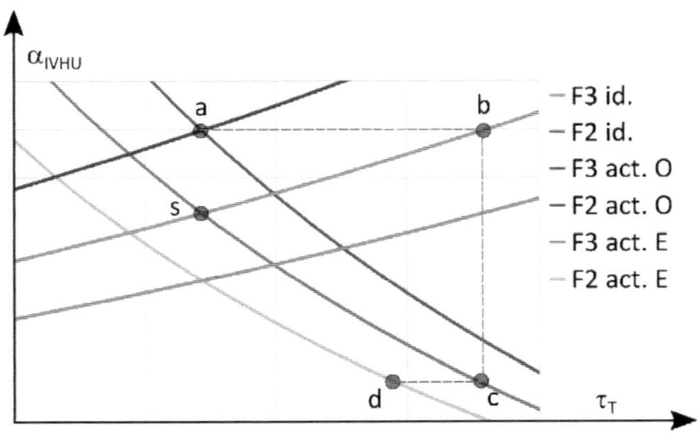

Fig. 9. Graphical analysis of the F3-F2 downshift (ω_E = 1,800 rpm; F_p = low)

In this case, the net result is an increase in τ_T from *a* to *d*, to be compensated via an opposite action than the one for the upshift transition, hence an actual ("full") compensation. The instantaneous vertical jump from point *b* to *c* (i.e. with no change in τ_T) considered so far is clearly an idealized condition, which would require an instantaneous clutch swap. In reality, the clutch swap transient generates a complex dynamic response, superimposing to the effect of the IVHU control, thus skewing the global response in terms of τ_T (and, consequently, on v). This is arguably the reason for the null compensation (c_α = 0) being the optimal option in three of the six gear shifts.

The above-mentioned superimposition is a complex dynamic phenomenon, which cannot be effectively studied qualitatively and is not further analyzed in the present work.

Finally, the results in Table 1 can be refined, using a more granular set of candidate c_α values; as an example, Fig. 10 shows the performance of c_α = -0.5 and c_α = 0.5 in the *F3-F4* and *F4-F3* shifts, respectively.

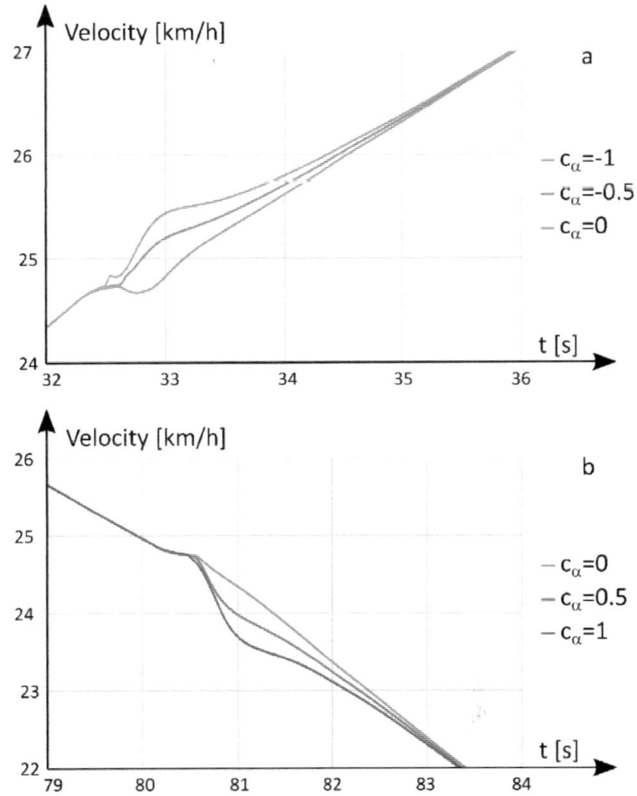

Fig. 10 Fine tuning of c_α. (a) F3-F4 upshift; (b) F3-F3 downshift (ω_E = 1,800 rpm; F_p = 5,000 N)

Conclusions and future work

A simple control logic - based on an enhanced open-loop scheme - has been successfully implemented, in the simulation framework at CNR-STEMS, to provide reliable automatic gear shifting capabilities in hydromechanical CVTs for agricultural tractors.

It has been tested on a lumped-parameter, purely-longitudinal model of a vehicle equipped with a four-gear, output-/compound-coupled architecture designed by the authors for a 200-kW, 12,000-kg reference tractor.

Non-ideal efficiencies of the hydraulic units have been taken into account, via constant non-unitary values, providing useful insights on the gear shift dynamics and the corresponding implications on the tuning of the control systems.

A method to compensate the discontinuity in the synchronization transients has been devised, to identify optimal control compensation coefficients.

Possible further developments include: implementing more realistic variable efficiency maps in the simulation model; extending the control scheme to incorporate typical functionalities of real-world production ECUs; investigating the potential for a transition to an electric hybrid architecture.

6. Nomenclature

NOTE: When two units of measurement are listed for the same physical quantity, the first one is the SI unit, used in equations (so that conversion factors are not required), while the second one is the practical unit, used when analyzing results.

add	Additional control output (to the open-loop output)	[-]
BCD	Simplified Pacejka, longitudinal stiffness	[-]
C	Simplified Pacejka, shape factor	[-]
c_α	Amplitude coefficient for the additional control output	[-]
$curr$	Current gear (outgoing gear during gear shift)	[-]
CVT	Continuously Variable Transmission	[-]
D	Simplified Pacejka, peak factor	[-]
D_{HM}	Hydraulic motor max displacement	[m³/rad], [cm³/rev]
D_{HP}	Hydraulic pump max displacement	[m³/rad], [cm³/rev]
F_p	Drawbar pull	[N]
F_x	Wheel net longitudinal traction force	[N]
F_z	Wheel vertical load	[N]
Fi	i-th forward gear	[-]
$i_{\%}$	Wheel longitudinal slip	[%]
$IVHU$	Infinitely-variable hydrostatic unit	[-]
k_t	Time constant for the additional control output	[s]
$next$	Next gear (incoming gear during gear shift)	[-]
OL	Open loop	[-]
P_C	Carrier power	[kW]
P_R	Ring power	[kW]
P_{S1}	Sun 1 power	[kW]
P_{S2}	Sun 2 power	[kW]
PGS	Planetary gearset	[-]
r_W	Wheel effective rolling radius	[m]
req	Requested (by the operator)	[-]
T_C	Carrier torque	[Nm]
$T_{e2,4}$	Even gear output torque	[Nm]
$T_{o1,3}$	Odd gear output torque	[Nm]
T_R	Ring torque	[Nm]
T_{S1}	Sun 1 torque	[Nm]
T_{S2}	Sun 2 torque	[Nm]
t	Time	[s]
t_0	Gear shift time	[s]
$u_{CL,A}$	Clutch A command	[-]
$u_{CL,B}$	Clutch B command	[-]
v	Vehicle longitudinal velocity	[m/s], [km/h]
α_{IVHU}	IVHU command	[-]
α_{HM}	Hydraulic motor displacement setting	[-]
α_{HP}	Hydraulic pump displacement setting	[-]
$\Delta\alpha_0$	Open-loop output gap at t_0	[-]
$\Delta\omega_{OUT}$	Speed difference between the odd and even gear output shafts	[rad/s], [rpm]
$\eta_{i,j,k}$	Hydraulic unit efficiency, where: • i (type): M = hydromechanical; V = volumetric • j (unit): HM = hydraulic motor; HP = hydraulic pump • k (operating mode): m = motoring; p = pumping	[-]
$\tau_{F1,3}$	Odd gear synchronizer ratio	[-]
$\tau_{F2,4}$	Even gear synchronizer ratio	[-]
τ_{HM}	Hydraulic motor ratio	[-]
τ_{HP}	Hydraulic pump ratio	[-]
τ_1	PGS stage 1 ratio	[-]
τ_2	PGS stage 2 ratio	[-]
τ_T	Transmission ratio	[-]
τ_α	Auxiliary ratio for IVHU-mediated angular velocity transmission	[-]
τ_Y	Auxiliary ratio for IVHU-mediated torque transmission	[-]
$\tau_{\eta M}$	IVHU hydromechanical efficiency factor for torque transmission	[-]
$\tau_{\eta V}$	IVHU volumetric efficiency factor for angular velocity transmission	[-]
ω_C	Carrier angular velocity	[rad/s], [rpm]
ω_E	Engine angular velocity	[rad/s], [rpm]
$\omega_{e2,4}$	Even gear output angular velocity	[rad/s], [rpm]
$\omega_{o1,3}$	Odd gear output angular velocity	[rad/s], [rpm]
ω_{S1}	Sun 1 angular velocity	[rad/s], [rpm]
ω_{S2}	Sun 2 angular velocity	[rad/s], [rpm]
ω_W	Wheel angular velocity	[rad/s], [rpm]

7. Declaration of competing interest

The authors declare that they have no known competing financial interests or personal relationships that could have appeared to influence the work reported in this paper.

8. References

Chau, K.T., 2015. Electric vehicle machines and drives – Design, analysis and application, John Wiley and Sons, Singapore, ISBN 978-1-118-75252-4, pp. 315-342

Kress, J., 1968. Hydrostatic Power-Splitting Transmissions for Wheeled Vehicles — Classification and Theory of Operation, SAE Technical Paper 680549, https://doi.org/10.4271/680549.

Marani, P., Chiarabelli, D., Gessi, S., Martelli, M., 2023. Input Coupled and Output Coupled Power Split Transmission Performance under Comprehensive Working Conditions, Proceedings of the ASME/BATH 2023 Symposium on Fluid Power and Motion Control, FPMC2023, https://doi.org/10.1115/FPMC2023-111467

Martelli, M., Chiarabelli, D., Gessi, S., Marani, P., Mucchi, E., Polastri, M., 2023. Comprehensive lumped parameter and multibody approach for the dynamic simulation of agricultural tractors with tyre–soil interaction, IET Cyber-Systems and Robotics, Volume 5, Issue 3, September 2023, https://doi.org/10.1049/csy2.12092

Rekow, A.K., Jeffries, D.L., Fritz, N., Mueller, D., Gugel, R., 2017. Method for controlling gear shift, US Patent 9,562,592, Deere & Company, Moline, IL, USA

Renius, K.T., 2020. Fundamentals of Tractor Design, Springer, Switzerland, https://doi.org/10.1007/978-3-030-32804-7, pp. 127-151

Siemens Industry Software NV, 2022. Simcenter Amesim, version 2210, https://plm.sw.siemens.com/en-US/simcenter/systems-simulation/amesim/, accessed 2024-05-23

TASC, 2020. TASC – Trattrici Agricole Smart & Clean (*Smart & Clean Agricultural Tractors*), Italian Regional Project, Region of Emilia-Romagna (2020). POR FESR 2014-2020. https://www.tascproject.eu/en/, accessed 2024-05-23

Vacca, A., Franzoni, G., 2021. Hydraulic Fluid Power – Fundamentals, Applications, and Circuit Design, John Wiley and Sons, ISBN 9781119569114, pp. 573-613

Wong, J.Y., 2001. Theory of ground vehicles – Third Edition, John Wiley & Sons, New York, ISBN 0-471-35461-9, pp. 58-65

5620 | Step-Climbing Motion Acquisition of Tracked Robot with Flippers without Using Environment Information by Reinforcement Learning

Ryosuke Eto [a,*], Hayatake Sato [a], Junya Yamakawa [a]

[a] *Department of Mechanical Engineering, National Defense Academy of Japan, Yokosuka, Japan*

* *Corresponding author: eto@nda.ac.jp*

ABSTRACT

Tracked robots, which have flippers on the front, back, left, and right sides, are expected to be used for disaster investigation because of their high performance to overcome obstacles such as debris and bumps. However, it is difficult for the operator to control the robot because of its high degree of freedom. Therefore, the system that automatically controls the flippers and crawlers is required.

In this study, we examined the acquired motion of a tracked robot with flippers to climb a step without using environment information by reinforcement learning. The learned motions are targeted to climb a step efficiently with a small amount of motion and to prevent the large impact when landing on a step. In order to reduce the amount of information required for the decision of the motion, only the information obtained from the internal sensors is used without the information of the surrounding environment. The agent Learned in a simulation environment using multi-body dynamics. First, the robot was trained to climb a step from the front, and the effectiveness of the acquired motion was confirmed from the results of a step climbing simulation and experiments using the trained agent. Then, the motion of the robot for climbing a step from an angle acquired by randomly changing the robot's initial orientation was clarified.

Keywords
Tracked robot
Flipper
Step-climbing
Reinforcement learning

1. Introduction

Disaster response robots that can traverse obstacles such as steps at disaster sites are expected to play an important role in disaster area. In particular, tracked robots with flippers are expected to play an active role because they can freely change their posture and have high traversing ability. However, because the high degree of freedom of the robot makes it difficult to maneuver, it is desirable for the robot to be semi-autonomous to assist the operator in maneuvering.

Reinforcement learning has recently been applied to control tracked robots with flippers, which have a high degree of freedom. In a previous study (Mifu Totani, et al., 2019), a step-climbing motion was learned by reinforcement learning in order to climb a high step by controlling the crawler speed and flipper angles. In other study (Hainan Pan, et al., 2023), a motion of traversing an obstacle by controlling the flippers was learned. However, these models did not take into account the amount of action. Moreover, the approach to the step was made from the front only, and the approach from an angle was not considered. They also used the environmental information from external sensors such as cameras and LiDAR for the observed value. At disaster sites, these sensors may not be able to acquire data due to rain, smoke, or mud splashes. In addition, data accuracy may be low due to the unevenness of the terrain over which the vehicle is traveling.

In this study, a tracked robot with the flippers (Dr Robot's Jaguar V6 with arm) shown in Fig. 1 was used as a target model to acquire a step-climbing motion by reinforcement learning, aiming at acquiring a motion to approach a step at an angle with a small amount of action and mitigation of impact of landing on a step without using environmental information.

Fig. 1. Tracked robot with flippers

2. Reinforcement learning

Reinforcement learning is a process in which a robot learns optimal behavior through repeated trial and error. Learning was performed by simulation because it is time-consuming and involves the risk of failure if it is performed on an actual robot. Multibody dynamics was used to calculate the robot motion in the simulation, and Digital Elevation Map with height information at grid points was used as the terrain. Penalty method was used to calculate the interaction between the robot and the terrain.

When the robot chooses an action based on a certain strategy, it acquires a scalar amount of reward if the action meets the objective. As the robot learns, it will take actions that maximize the reward. By reinforcement learning, the robot can acquire behaviors that adapt to the environment by providing rewards according to the goals. In this study, we used Proximal Policy Optimization (John Schulman, et al., 2017), which is often used for continuous values such as robot control inputs and has high learning stability. The learning stability is enhanced by clipping the policy function to prevent it from updating too much.

https://doi.org/10.56884/DWSSCCSE

3. Learning condition

In this study, the observed information received by the robot consisted of 12 variables: target speed, attitude angular velocities, attitude angles except yaw angle, peripheral velocities of the crawlers, and angles of the flippers, without information of a step. The output behavior consisted of five variables: angular velocities of the front, back, right, and left flippers and crawler peripheral velocity. The networks of actor and clitic of learning agent consisted of two intermediate layers with 400 and 300 units, respectively, and Rectified Linear Unit (ReLU) was used as the activation function. The output of the actor network is normalized to a range from -1 to 1.

3.1. Learning goals

In the first simulation, the robot is initially positioned as shown in Fig. 2 and moves straight ahead against a step. The three control targets are the crawler peripheral velocity of the main body (①) and the angular velocities of the front and rear flippers (② - ⑤). The peripheral speeds of the crawlers attached to flippers are the same as that of the main body. The speed is positive in the forward direction and the direction of rotation of the flipper is positive clockwise. The manipulator (⑥) is not moved.

The goal of this research is to obtain a step-climbing operation by controlling the crawler peripheral speed and the four flippers (front, back, left, and right) angles. Figure 3 shows the step-climbing method. Figure 3 (a) shows the method of climbing at a high speed with almost no movement of the flippers, and Fig. 3 (b) shows the method by mitigating the landing impact while moving the flippers. In this study, the learning target is to mitigate the landing impact by moving the flippers as shown in Fig. 3(b), because the step-climbing operation at the high speed shown in Fig. 3(a) may cause the robot to crash into an obstacle beyond the step, and also does not mitigate the landing impact. To avoid excessive speed, the target speed is set at 0.2 m/s.

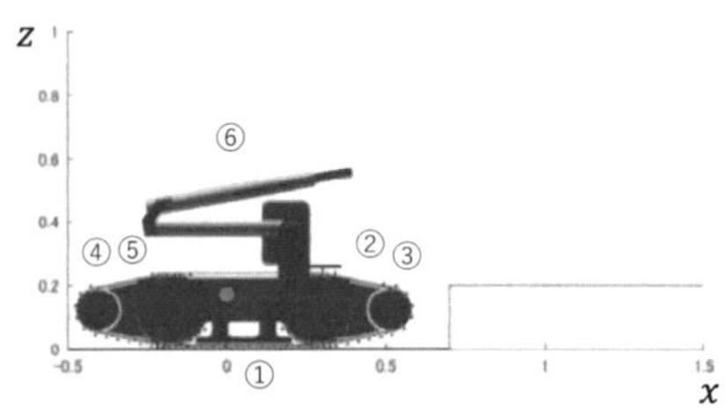

Fig. 2. Initial state in simulation

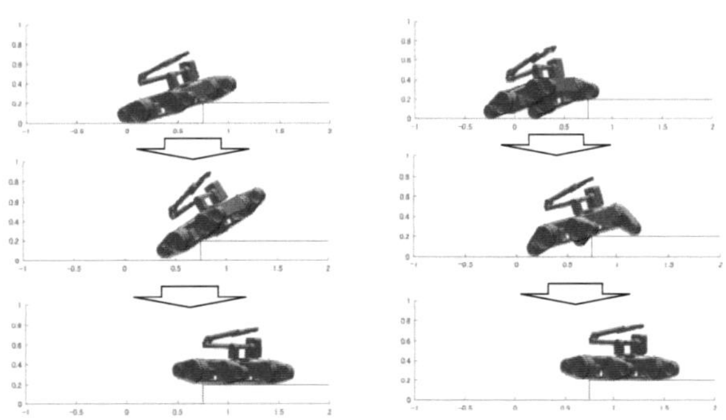

(a) without using flippers *(b) with using flippers*

Fig. 3. Sequence of step-climbing

3.2. Initial conditions

The initial conditions in each episode were random within the range shown in Table 1, and the initial velocity v_0 was set to 0.

Table 1
Initial conditions of simulation

	Min.	Max.	Unit
Initial position of a robot	-0.1	0.1	[m]
Position of a step	0.8	1.0	[m]
Friction coefficient μ	0.7	1.0	[-]
Target speed	0.15	0.25	[m/s]
Flipper angle	$-\pi/8$	$\pi/8$	[rad]

3.3. Goal conditions

The goal line was set at 1.5 m in x direction forward of the robot from the initial position. A reward of 1 was given for success in reaching the goal line, and a penalty of -1 was given for failure in the episode when either of the following was reached.

-The robot moves backward to $x = -0.3$ m
-The robot's pitch angle exceeded π /2 rad
-Robot roll angle exceeded π /2 rad
-Flipper angle exceeds $\pm\pi$ /2 rad
-Time 15 s elapsed

4. Rewards

Reward design is an important setting that determines the progress of reinforcement learning. Rewards are given at each step, and the robot learns so that the reward obtained becomes higher. In this study, we set the robot's movement reward r_x, crawler's target speed reward r_{tv}, speed reward r_v, flipper speed reward r_u, flipper movement reward r_θ, and landing impact reward r_z to make the crawler belt speed close to the target value and to allow the robot to climb a step with fewer movements. The total reward R is defined as Eq. 1.

$$R = ar_x + br_{tv} + cr_v + dr_u + er_\theta + fr_z \quad (1)$$

where, a, b, c, d, e, f are hyperparameters.

4.1. Travel Reward

The difference between the position x of the center of gravity at step k and k-1 was normalized by the distance traveled at the target speed v_{target} in time step Δt and determined as in Eq. 2 to be the travel reward.

$$r_x = \frac{x_k - x_{k-1}}{v_{target} * \Delta t} \quad (2)$$

4.2. Target Speed Reward

Equation 3 is set so that the speed difference is a negative reward when the crawler peripheral speed v_k is greater than the target speed v_{target}. The target speed reward is zero when the crawler speed does not reach the target speed v_{target}.

$$r_{tv} = -\left(\frac{v_k - v_{target}}{v_{target}}\right)^2 \quad \text{if } v_{target} < v_k \tag{3}$$

4.3. Speed Reward

To avoid severe acceleration/deceleration of the crawler peripheral speed, normalize the difference between the speed v_k one step ago and the maximum speed v_{max} as in Eq. 4, and square the difference to give a negative reward.

$$r_v = -\left(\frac{v_k - v_{k-1}}{v_{max}}\right)^2 \tag{4}$$

4.4. Flipper Speed Reward

In order to reduce the flipper movement, the flipper speed $u_k{}^{(i)}$ is normalized by the maximum speed value u_{max} as in Eq. 5 and the sum of squares is given as a negative reward. i represents the four flippers (front, back, left and right).

$$r_u = -\sum_{i=1}^{4}\left(\frac{u_k{}^{(i)}}{u_{max}}\right)^2 \tag{5}$$

4.5. Flipper Movement Reward

To keep the flipper angle close to zero, normalize the angle $\theta_k^{(i)}$ by the upper flipper angle limit $\theta_{max}^{(i)}$ as in Eq. 6 and give the sum of squares as a negative reward.

$$r_\theta = -\sum_{i=1}^{4}\left(\frac{\theta_k^{(i)}}{\theta_{max}^{(i)}}\right)^2 \tag{6}$$

4.6. Landing Impact Reward

The impact of landing on the step on the robot can cause it to malfunction. Therefore, a negative reward is given according to the magnitude of the landing impact, as in Eq. 7. where F_z is the force in the z-direction received by the body part and F_{zmax} is the maximum value of F_z observed.

$$r_z = -(F_z/F_{z\,max})^2 \tag{7}$$

5. Curriculum learning

In reinforcement learning, it has been found that learning can be done efficiently by gradually increasing the difficulty level of the learning target from a simple one according to the progress of learning (Yoshua Bengio, et al., 2009). The learning target was a step-climbing movement, so the step height was changed according to the progress of learning. The step height was initially set at 0.1 m, and was increased by 0.01 m for every 9 out of 10 successful attempts until reaching a height of 0.21 m. The step height was set at 0.1 m at the beginning of the learning process, and was increased by 0.01 m for every 10 successful attempts until reaching a height of 0.21 m at the end of the curriculum. After reaching 0.21 m, which was the end of the curriculum, the height was set randomly in the range of 0 to 0.2 m every 25 trials.

6. Results of learning

The behavior acquired by reinforcement learning was initially verified by simulation, and then verified on the actual machine. To facilitate learning, a frontal approach was initially learned with three variables (crawler peripheral speed and front and rear flipper angular velocities) as control targets, and based on the results, a diagonal approach was learned with five variables (crawler peripheral speed and four flipper angular velocities) as control targets. The learning was performed using the MATLAB Reinforcement Learning Toolbox on an Intel Core i7-12700, 2.1GHz, Windows 11 operating system with a GeForce RTX 3060 GPU.

6.1. Front approach

Initially, the flipper angle was assumed to be the same on both sides, and the five control targets were controlled by three variables: crawler peripheral velocity, angular velocity of the front flipper and rear flipper, and the effectiveness of the method was confirmed by learning a frontal approach to a step. The hyperparameters of the reward were $a = 02$, $b = 0.5$, $c = 0.05$, $d = 0.001$, $e = 0.01$, and $f = 0.5$. Parameters were determined by trial and error to approach the expected behavior.

6.1.1. Learning transition

Figure 4 and 5 show the relationship between episodes and rewards. Figure 4 shows the results of learning a 0.2 m step without curriculum learning, and Fig. 5 shows the results of learning with curriculum learning. In the case of no curriculum learning, the robot could not learn the movement to climb the step, although it moved forward, and the reward remained low. On the other hand, when curriculum learning was used, the learning ended at episode 326, and learning continued randomly after episode 327. The average reward increased after the curriculum learning, because the robot overcame the low steps. The computation time was approximately 60 minutes when curriculum learning was used, and approximately 40 minutes when curriculum learning was not used. The reason for the shorter computation time without curriculum learning is that each episode reached the failure condition and was terminated early.

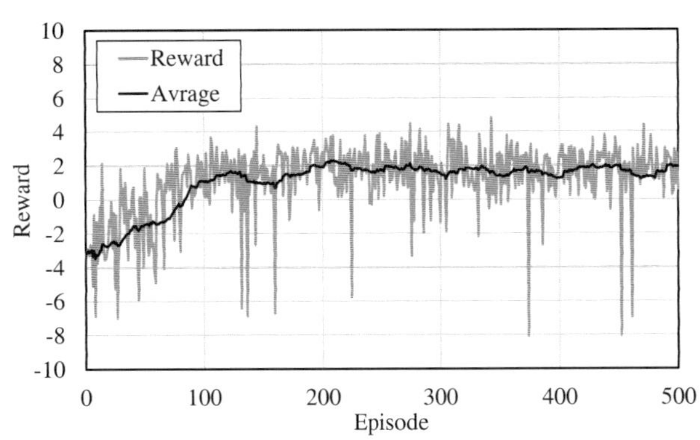

Fig. 4. Reward without curriculum learning

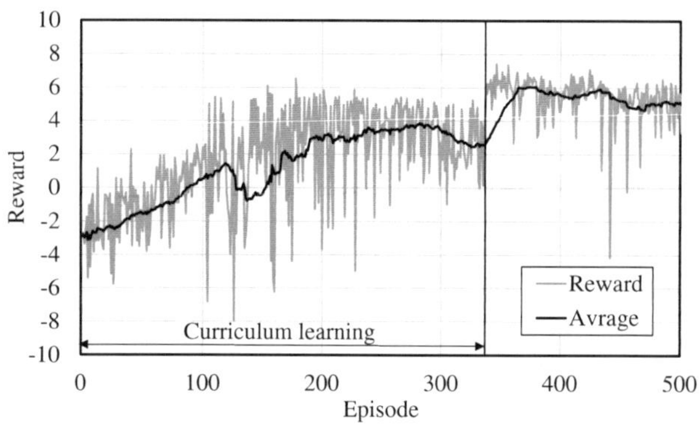

Fig. 5. Reward with curriculum learning

6.1.2. Acquired actions

The acquired behaviors were confirmed by simulation assuming the robot's initial position as 0 m, target speed as 0.2 m/s, step position as 0.8 m, step height as 0.2 m, and front-back flipper angle as 0 degree. The acquired behaviors are shown in Fig. 6. By lowering the rear flippers and making the pitch angle closer to horizontal, the robot's center of gravity can easily go over the step edge, preventing the robot from tipping over. If the robot's center of gravity is higher or further back, the center of gravity is less likely to exceed the step edge, and the robot is expected to acquire the behavior of lowering the rear flippers more. By lowering the front flippers, it was able to ascend the step by mitigating the impact of landing.

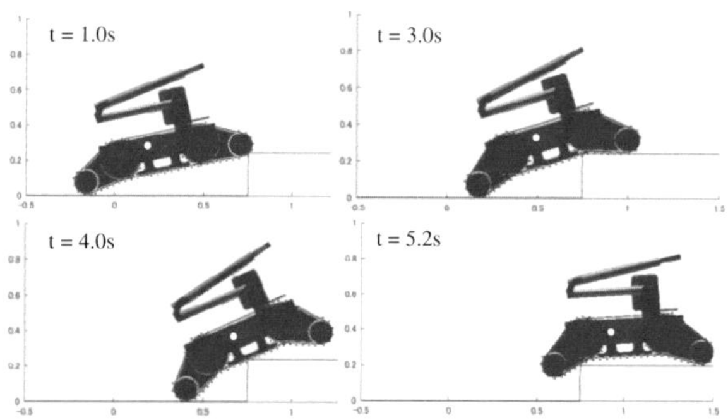

Fig. 6. Sequence of step-climbing by using acquired action

6.1.3. Impact of landing impact reward r_z

The landing impact reward r_z was set to mitigate the impact of landing on a step. Figure 7 shows the robot's behavior when the landing impact reward r_z is varied. Figure 8 shows the result of vertical forces F_z acting on the robot. Figure 7(a) shows the robot learned with landing impact reward, and Fig. 7(b) shows the robot with landing impact reward set to 0. When the landing impact reward is taken into account, the front flipper is lowered to mitigate the landing impact. The large value observed at around 1.5 seconds is the force at the time of contact with the step, and the value of over 600 N observed after contact with the step is because the robot is lifting its body up to go over the step. The large value observed around 5 and 6 seconds is the landing impact, and the peak of impact was suppressed by the landing impact reward.

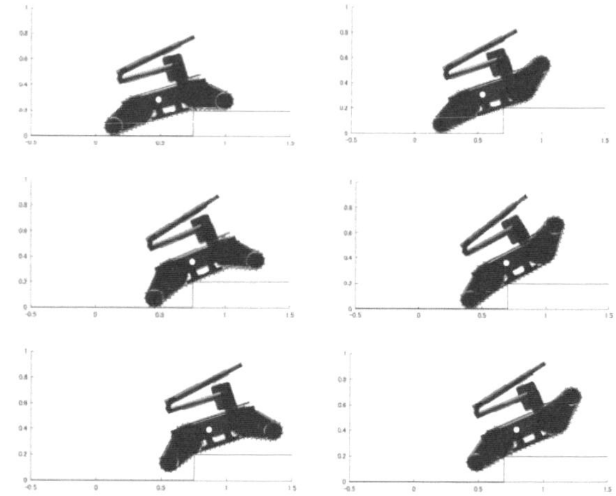

(a) f = 0.5 *(b) f = 0*

Fig. 7. Difference of actions with considering landing impact

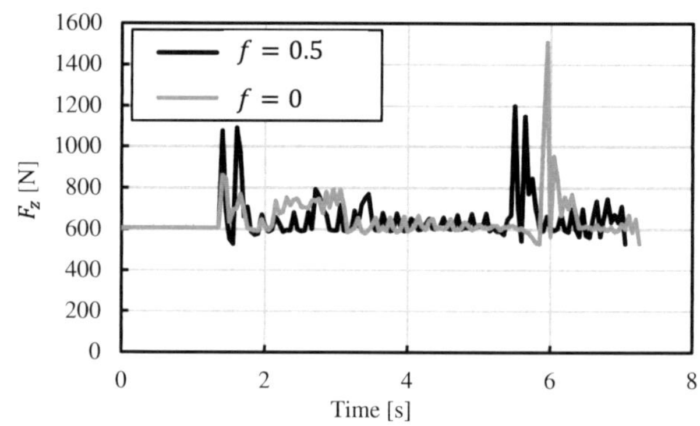

Fig. 8. Simulation results of F_z

6.2. Diagonal approach

The five variables to be controlled were the crawler peripheral speed and the front-back and left-right flippers, and transfer learning was performed based on the neural network parameters obtained in the previous section.

6.2.1. Initial conditions

The initial conditions for the diagonal approach simulation are shown in Table 2. The angle at which the robot approaches the step was reproduced by changing the yaw angle around the robot's initial center of gravity position, as shown in Fig. 9(b). Hyperparameters were set to $a = 02$, $b = 0.5$, $c = 0.05$, $d = 0.001$, $e = 0.01$, and $f = 0.5$ as in the frontal approach.

Table 2
Initial conditions of diagonal approach simulation

	Min.	Max.	Unit
Initial position	−0.1	0.1	[m]
Step position	0.8	1.0	[m]
Friction coefficient μ	0.7	1.0	[-]
Target velocity	0.15	0.25	[m/s]
Yaw angle	$-\pi/4$	$-\pi/4$	[rad]
Flipper angle	$-\pi/8$	$\pi/8$	[rad]

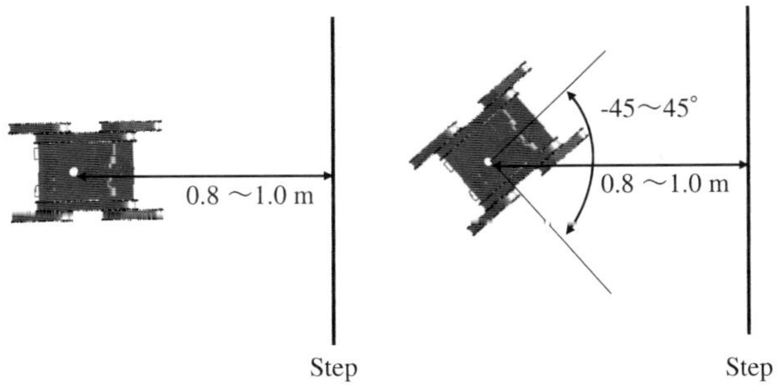

(a) frontal approach *(b) diagonal approach*

Fig. 9. Initial positions and orientations in diagonal approach simulation

6.2.2. Learning transition

The rewards per episode are shown in Fig. 10. Curriculum learning was terminated by clearing a step of 0.21 m height in episode 268, and the step height was randomly set after episode 269. The reason for the higher rewards from the beginning may be that the agent already had a neural network trained under similar conditions because of the use of transition learning. The part where the average reward decreased several times after curriculum learning was due to the fact that the neural network was learning steps with a height of around 0.25m. The computation time for 1000 episodes was approximately 100 minutes.

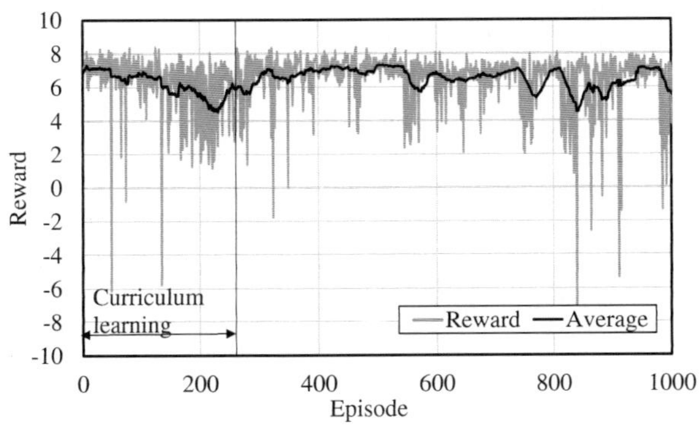

Fig. 10. Results of rewards in diagonal approach

6.2.3. Acquired actions

Figure 11 shows the attitude change when the robot started at an initial position of 0 m, a target speed of 0.2 m/s, a step position of 0.8 m, a step height of 0.2 m, a front-rear flipper angle of 0 degrees, and a yaw angle of -45 degrees. All four flippers were lowered, but the speed of lowering was adjusted so that the three points were as close to the ground as possible.

Figure 12 shows the angular change of each flipper. The right rear flipper, located farthest from the step, was lowered continuously to maintain the roll and pitch angles, while the left rear flipper was lowered halfway to maintain contact with the three points, but stopped halfway to avoid catching on the step. The left front flipper was lowered in advance to prepare for the impact after climbing a step, and the right front flipper was lowered slightly to suppress the impact by reducing the roll angle.

The change in attitude angles is shown in Fig. 13. The roll angle changed significantly, but the pitch angle was maintained. This is because the robot was not turned over by the rolling motion, so the reward was not negatively affected by the roll angle. On the other hand, the pitch angle was controlled because the robot was turned over due to pitching motion and the robot was penalized.

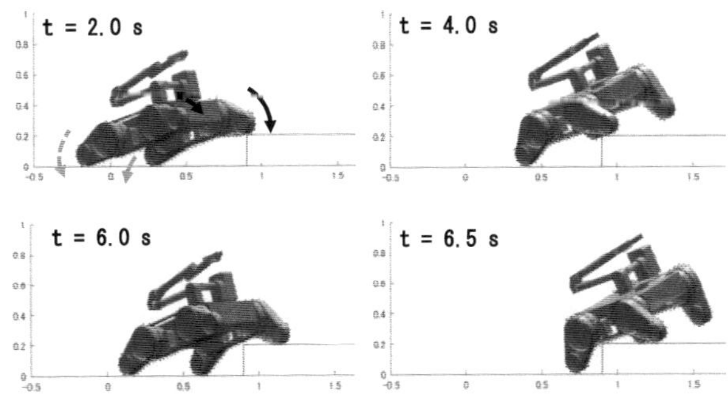

Fig. 11. Sequence of step-climbing from an angle

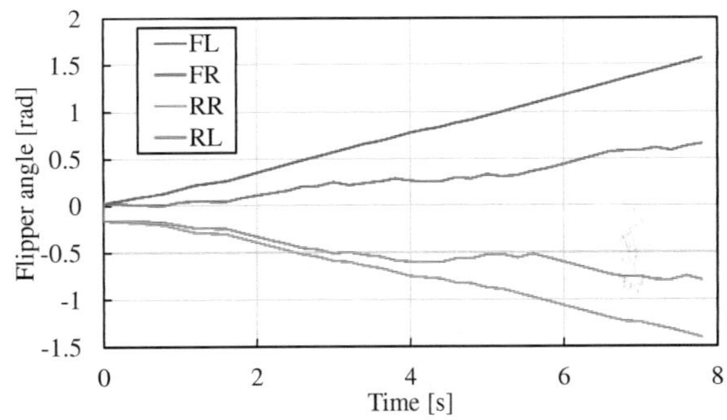

Fig. 12. Results of flipper angles

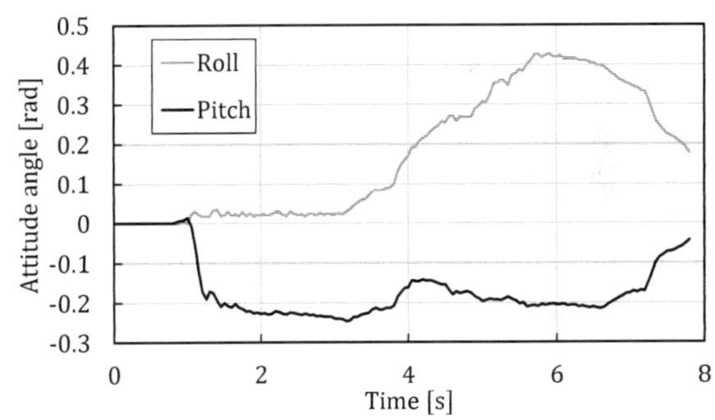

Fig. 13. Results of robot attitude angles

7. Experimental test

Experiments on the actual robot were conducted to confirm the validity of the acquired actions by reinforcement learning.

7.1. Experimental system

ROS was used to exchange signals between the robot and the computer. Fig. 14 shows the robot control system. The robot has a rotary encoder, gyro sensor, and accelerometer, and the PC converts the robot's measured data into attitude angles and attitude angular velocities, which are output by a trained agent. The agent takes 12 variables as input values and outputs 5 variables. Communication between the robot and computer was performed at 5 Hz via Wi-fi.

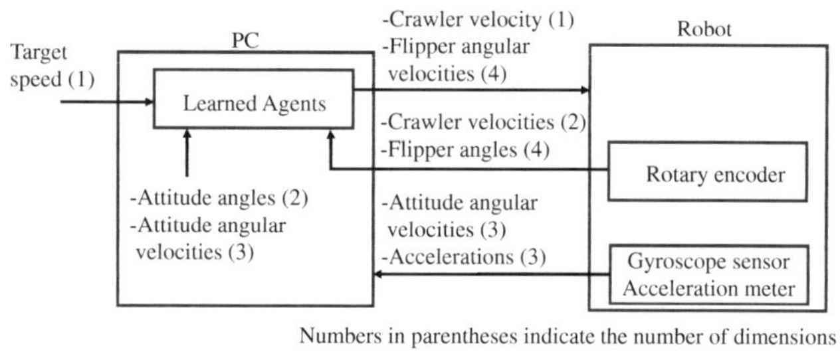

Fig. 14. Robot control system

7.2. Experimental conditions

The experiment was conducted with the arrangement shown in Fig. 15. Concrete blocks were used to reproduce a step 0.18 m high. The robot was initially positioned 1.0 m from the step, with a target speed of 0.2 m/s, a front flipper angle of -20 degrees, and a rear flipper angle of -11 degrees.

Fig. 15. Initial state of experimental test

7.3. Experimental results

Figure 16 shows the behavior of the machine when it was successfully raised in an experiment. It was confirmed that the robot could climb a step by lowering the rear flipper, as the behavior acquired through learning. Both the left and right crawlers produced almost the same speed, but they were not able to maintain the target speed, and the actual speed of the left and right crawlers decreased after contact with the step, and increased all after climbing the step. This is thought to be because a large torque was required for climbing, and the speed was not as fast as the commanded value.

In the experiment when the robot approaches the step from an angle, it could not climb the step because only one flipper contacted the step, which required a large torque on the flipper and crawler in contact with the step. Therefore, the rewards regarding the flipper torque and crawler torque should be added in reinforcement learning.

8. Conclusions

In this study, we examined the step-climbing behavior of a tracked robot with flippers using reinforcement learning without environmental information. Initially, the effectiveness of the method was verified by using a frontal approach. Based on the verified results, the flipper motion was acquired in a diagonal approach by setting rewards and learning by randomly changing the approach angle.

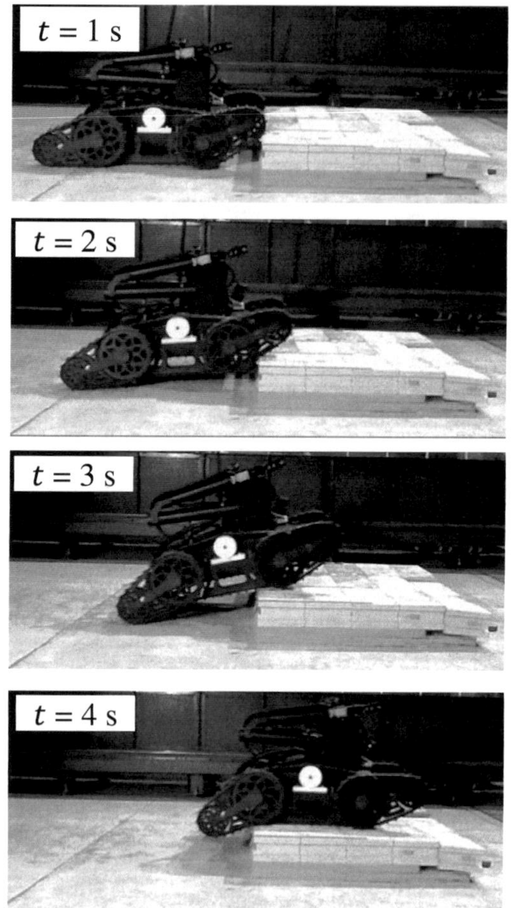

Fig. 16. Motion of step-climbing in experiment

However, the behavior of the actual machine was different from that of the simulation due to insufficient torque. Therefore, it is necessary to add additional rewards related to the driving torque of flippers and crawlers for reinforcement learning in the future.

Furthermore, since this study targets a climbing motion to a single step, the robot learned to lower its front flipper in order to reduce landing impact. In actual disaster situations, there are multiple steps and stairs. For these obstacles, if the front flippers are lowered, the robot climbing motion cannot be performed efficiently. Therefore, future studies should be conducted in fields with a continuous series of steps.

9. References

John Schulman, Filip Wolski, Prafulla Dhariwal, Alec Radford, Oleg Klimov, 2017, Proximal policy optimization algorithms, arXiv: 1707.06347.

Hainan Pan, Xieyuanli Chen, Junkai Ren, Bailiang Chen, Kaihong Huang, Hui Zhang, Huimin Lu, 2023, Deep Reinforcement Learning for Flipper Control of Tracked Robots in Urban Rescuing Environments, remote sensing, Vol. 15, 4616.

Mifu Totani, Noritaka Sato, Yoshifumi Morita, 2019, Step climbing method for crawler type rescue robot using reinforcement learning with Proximal Policy Optimization, 12th International Workshop on Robot Motion and Control.

Yoshua Bengio, Jerome Louradour, Ronan Collobert, Json Weston, 2009, Curriculum learning, Proceedings of the 26th Annual International Conference on Machine Learning, 41-48.

5720 | Suppress Slip While Crossing Loose Slopes Using Reverse Rotation Behavior of Rovers With Function of Independent Contraction/Expansion Mechanism

Tsukasa Mochizuki [a,*], Daisuke Fujiwara [b], Kojiro Iizuka [a], Takaaki Sakata [a], Tatsuhiko Suga [a]

[a] *Graduate School of Engineering and Science, Systems Engineering and Science, Shibaura Institute of Technology, Saitama, Japan*

[b] *Department of Mechanical and Electrical Engineering, Suwa University of Science, Nagano, Japan*

* *Corresponding author: bq20087@shibaura-it.ac.jp*

ABSTRACT

In recent years, many organizations have developed lunar rovers for traveling over uneven terrain. Our research team has studied a rover using independently extending/retracting left and right wheelbases. In the case of four wheels, this method allows three wheels to remain stationary while moving. The independent extending/retracting locomotion of the wheel enables the traction to increase. This method provides a greater support force and improves the climbing performance on a loose slope. However, increased load on a sloped surface causes sideslipping, and the rover tilts its posture in downward. Our previous study confirmed that intentionally increased sinkage by the large wheel slip increased side forces from the soil. However, the lateral force is still insufficient and cause sideslipping on steep slopes. To increase the side force, this study focuses on the resistance between the wheel and the ground, along with the direction of wheel rotation. Specifically, we propose reverse rotation of the upper side wheel of the rover. This method increases dynamic sinkage by increased resistance from the soil during wheel driving. To confirm the effectiveness of method, this study conducts experiments that traversing loose slope with the rover. The results suggest that the proposed method greatly reduces sideslipping. In addition, the results indicated that the wheel sinkage increased. Therefore, the reverse rotation could increase the normal force of the wheels on the upper side. In summary, proposed method increased side forces and suppressed sideslipping. Hence, the reverse rotation prevents tilting the rover's posture downward on the slope.

Keywords
Rover
Soft ground
Slip
Sideslip

1. Introduction

Many organizations have been focusing on the field of space exploration in recent years. Among manned missions, NASA's Artemis program aims to achieve manned lunar landings (NASA, 2024). While in unmanned exploration, JAXA's Smart Lander for Investigating Moon (SLIM) successfully landed on the moon (JAXA, 2023). These missions could potentially lead to discoveries about the origins of the solar system, extraterrestrial resources, and the expansion of operational areas. For unmanned exploration rovers, there are wheel-based, crawler-type, and legged rovers (Yoshida, 2011). Among them, wheel-based rovers are predominantly used due to their simpler mechanisms and lower risk of malfunction. However, the lunar surface is characterized by craters and is covered by a fine sand known as regolith. Consequently, rovers are difficult to gain sufficient force from the ground. The Mars exploration rover Spirit, used in past missions, became immobilized due to getting stuck (NASA, 2019). To address this issue, previous studies have proposed models like Scarab and Sherpa, which incorporate a discrete motion known as Push-Pull Locomotion into wheeled rovers (NASA, 2017; Roehr et al., 2014). This approach involves extending and retracting the wheelbase between the front and rear wheels. With the front wheels stationary, rotating the rear wheels and contracting the wheelbase as shown in *Fig. 1*. This driving technique ensures that stationary wheels remain in constant contact with the ground, providing greater support for driving.

Additionally, the ability to traverse slopes is essential for the direct exploration. However, sideslipping occurs during traversal. Ishigami et al. (2005) proposed effective methods for suppressing sideslipping which uses steering with normal driving. There have been many studies on Push-Pull Locomotion (NASA, 2017; Roehr et al., 2014), while slope traversal data with PPL have been insufficient. In order to clarify the performance of PPL on slope traversal, our group evaluated its performance in an experimental environment. Although the suppression ability of PPL locomotion was larger than that of normal traveling, side slipping occurred on the loose surface with slope. To solve the side slipping on high slope, our research group proposed a method called 4WAPPL. This approach lets each wheel move independently like legs and incorporates a sinking action by rotating wheels freely. This technique more effectively uses the resistance between the stationary wheels and the ground than previous studies. Additionally, this method can significantly reduce sideslipping on traversing soft ground (Iizuka et al., 2023). The detailed operation method is shown in Fig. 2. Additionally, we confirmed 4WAPPL with steering has further abilities to suppress sideslipping (Iizuka et al., 2023). Meanwhile, a rover with a simple mechanism is important due to the risk of failure. These kinds of rovers tend to have limited function and mechanism. Sideslipping causes the rover to move in unexpected directions during traversal. This issue arises because the ground under the wheels disintegrates during rover propulsion. This leads to a downward tilt in orientation and overall sideslipping as the ground under each wheel collapses.

This study considers another approach to suppress sideslipping during slope traversal in 4WAPPL without using turning the steering. Our preliminary experiment confirmed that the reverse rotation of the upper side wheel during slope traversing increased these amount of sinkage. The increased sinkage changes the attitude of the rover and suppresses the load transfer. This paper focuses on this effect and proposes a novel method for suppressing sideslipping. Additionally, to verify the effectiveness of the proposed method, we conducted experiment of driving steep slope with a test bed rover.

https://doi.org/10.56884/8KPL7W26

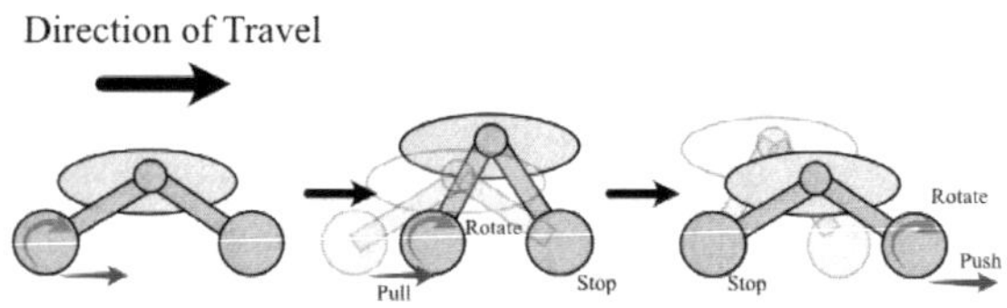

Fig. 1. Sherpa locomotion

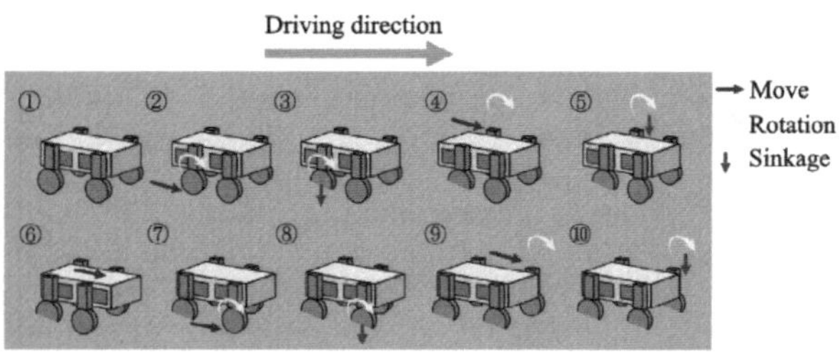

Fig. 2. Driving method of 4WAPPL

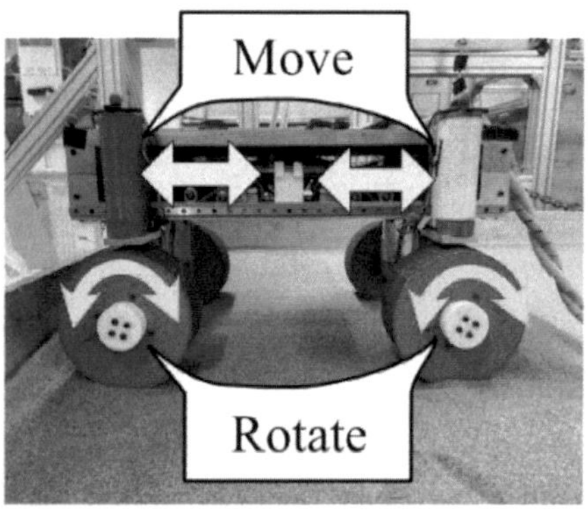

Fig. 3. Mechanism for movement

2. Method

2.1. Previous studies

Figure 2 shows the detailed operation method of the proposed 4WAPPL. The operation consists of steps 1 through 10 and repeats multiple times. *Figure 3* illustrates the operation mechanism.

1. Initial state
2. Right rear wheel rotates forward and moves forward.
3. Turn the right rear wheel and the wheel sinks.
4. Rotate the left rear wheel forward and move forward.
5. Rotate the left rear wheel and the wheel sinks.
6. Return all the wheelbase and move the body forward.
7. Rotate the right front wheel forward and move forward.
8. Rotate the right front wheel forward and the wheel sinks.
9. Rotate the left front wheel forward and move forward.
10. Rotate the right front wheel, and the wheel sinks.

2.2. Effect of side force

The relationship of forces acting on the wheels is illustrated in *Fig. 4* and *5*. The side force F_y occurs in response to the wheel's angle. R_b results from the wheel's side pushing the soil. r and b represent the wheel radius and wheel width, respectively. h represents the wheel sinkage. F_s is the force obtained by integrating R_b over the contact area. τ_y is the force on the bottom of the wheel in contact with the ground in the Y axis direction. Additionally, F_u is obtained by integrating force τ_y over the contact area. The angle θ_f and θ_r represent the wheel entry angle and the wheel exit angle respectively. The total side force, F_y is given by *Eq. 1* and is the combined result of these forces F_s and F_u(Ishigami et al., 2005).

$$F_y = F_u + F_s \tag{1}$$
$$= rb\int_{-\theta_r}^{\theta_f} \tau_y(\theta)d\theta + \int_{-\theta_r}^{\theta_f} R_b(r - hcos\theta)d\theta$$

According to *Eq. 1*, an increase in the contact area between the wheel and the ground results in an increase in the integrated forces F_s. Consequently, the side force F_s also increases. Therefore, an increase in the sinkage amount, which enlarge the contact area can mitigate sideslipping. Additionally, (Inaba and Iizuka, 2021) confirmed differences in the forces exerted on the wheels at the upper and lower sides of the slope through the traversing experiment on a sloped surface. *Figure 6* details the differences in force between the upper and lower sides. Due to the center of gravity shifting towards the lower wheels, the force W_L and side force F_{yL} on the lower wheels are greater than force W_L and side force F_{yU} on the upper wheels. This larger load on the lower wheels leads to ground collapse and is one of the causes of sideslipping. Increasing the sinking of the upper-side wheels and equalizing the side forces across all wheels can suppress sideslipping. To increase side forces, this paper focused on the wheel rotation direction during locomotion. Specifically, this paper uses reverse wheel rotation to increase sinkage.

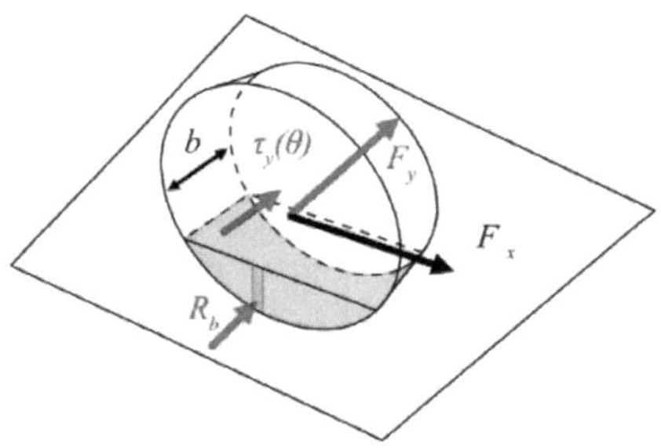

Fig. 4. Schematic view of the static sinkage model

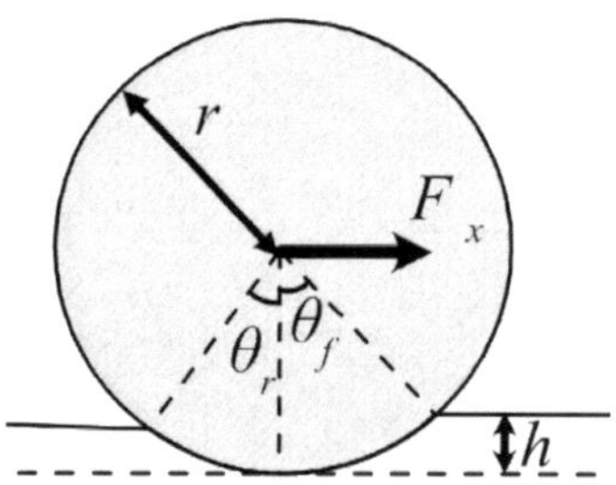

Fig. 5. Forces acting on wheels

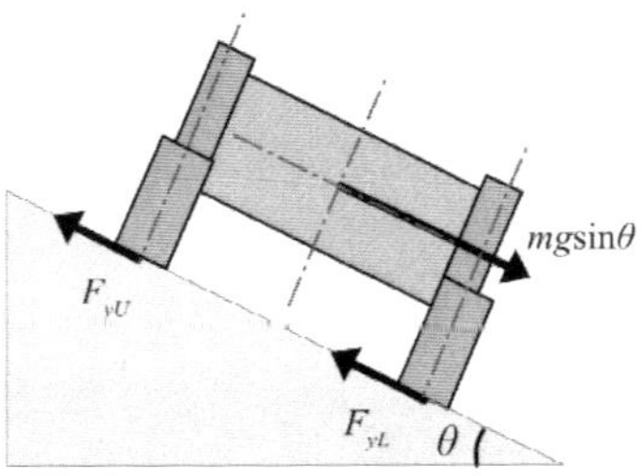

Fig. 6. Forces acting on the rover on a slope

2.3. Preliminary experiment - confirming reverse rotation of the wheel

In this chapter, the authors conduct preliminary experiments focused on wheel rotation direction on a flat soft ground. The experiments aim to compare the effects of normal and reverse rotations. The experiments involve reversing the direction of wheel rotation when wheels move forward.

2.3.1. Experimental method

Figure 7 shows the testbed rover used in the experiment. *Table 1* summarizes the rover specifications. The length, height, and width of the experimental field are 2000 mm, 150 mm, 1500 mm, respectively. To measure the movement trajectory of the rover, motion markers are placed on the rover, and The motion capture system measures the position trajectory of the rover as indicated in *Fig. 8*. The marker position is the center on the top of the rover. The experimental conditions are outlined in *Table 2*. The authors conduct two kind of motion experiments, conventional motion and reverse motion. The conventional motion sequence is as indicated in *Fig. 2*. Meanwhile, reverse motion is as indicated in *Fig. 9*. The direction of wheel rotation is reversed during operations 2, 3, 7, and 9, as indicated in *Fig. 9*. After repeating each operation three times, we measure the wheel sinkage. The rover repeats each sequence five times. To confirm the effect of the reverse rotation, this paper measures sinkage and calculates the average of the five trials.

Fig. 7. Testbed rover

Table 1 Rover's parameters

Parameters	Unit	Value
Weight	N	150
Wheelbase length	mm	325
Wheel size	mm	R 160, W 70
Rover size	mm	W 456, L 581, H 297
Length between the center and wheel	mm	162.5

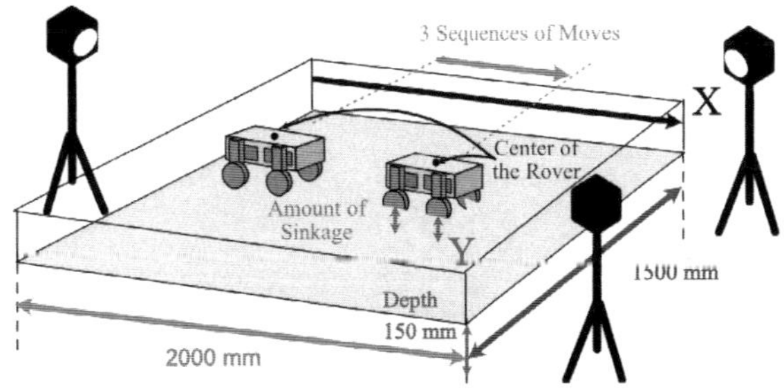

Fig. 8. Test field for preliminary experiment

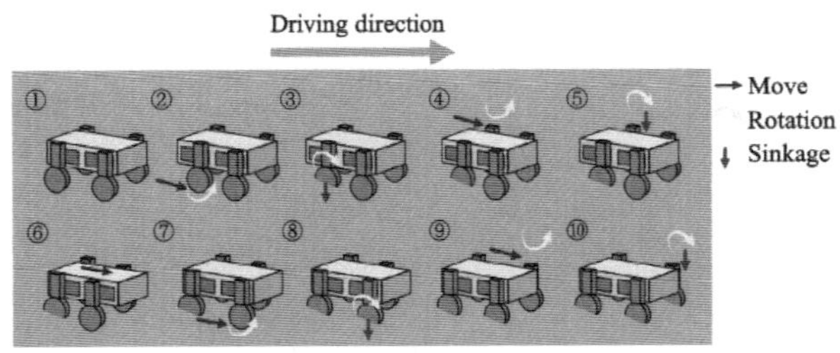

Fig. 9. Driving method of 4WAPPL in reverse rotation

Table 2 Conditions for the preliminary experiment

Item[unit]	Value
Soil	Slica sand No.5
Slope angle [°]	0
Wheelbase speed [mm/s]	8
Wheel rotating speed [mm/s]	34
Traveling cycle	3
Trial of experiment	5

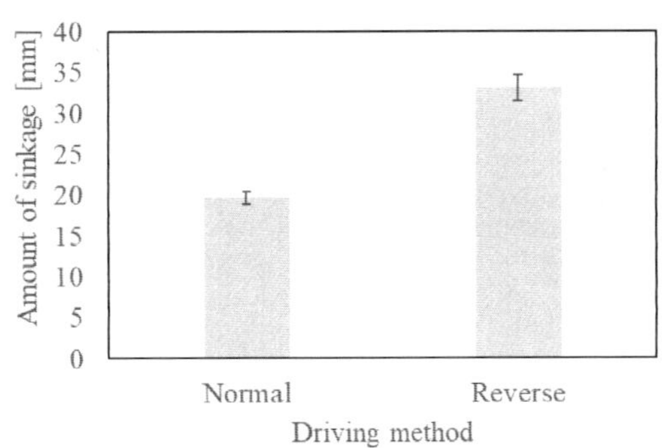

Fig. 10. Results of the preliminary experiments

2.3.2. Experiment results

Figure. 10 shows the experimental result. When comparing the amount of sinking after traversal, amount of sinkage caused by reverse rotation is significantly greater than normal rotation. Therefore, reverse rotation in 4WAPPL is efficient method to increase sinkage.

2.4. Advantage of reverse rotation

As an effective method for traversing slopes, the authors propose the reverse rotation of the wheels on the upper side of the slope shown in *Fig. 11*. The following section discusses two factors that reduce sideslipping by reversing the wheels on the upper side.

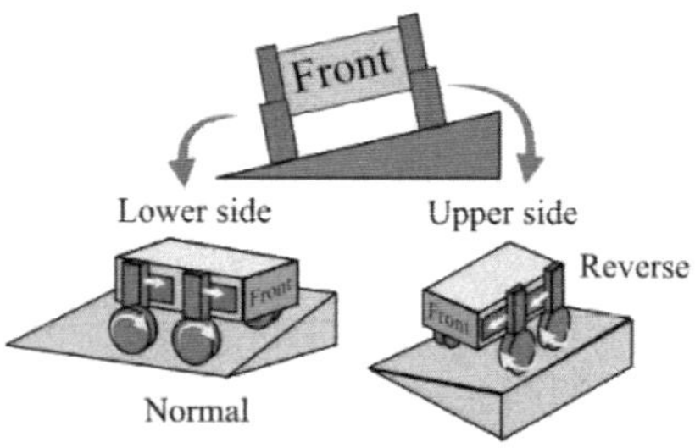

Fig. 11. Reverse rotation of the wheels on the upper side

2.4.1. Load balancing

As stated in Section 2.2, the concentration of load on the lower wheels leads to ground collapse. Varying the difference in sinkage between the lower and upper wheels addresses this problem. Preliminary experiments show that reversing the wheel direction increases sinkage. *Figures 12* and *Figure 13* show the difference in sinkage between the normal and reverse rotation of the upper wheels, respectively. As the upper wheels sink and the contact area increases, the force F_s acting on the wheel side increases, as shown in *Eq. 1*. Consequently, the side force F_{yU} and the normal force F_{zU} increase to F_{yU}' and F_{zU}' as shown in *Fig. 12* and *Fig. 13*. Reversing the upper wheels reduces the rover angle θ *to* θ'. This equalizes the load distribution as shown in *Eq. 2* and *Eq. 3*. Therefore, this driving method helps reduce the risk of ground collapse.

$$F_{zL} > F_{zL}' \tag{2}$$

$$F_{yU}' > F_{yU} \tag{3}$$

2.4.2. Moment facing uphill

One cause of sideslipping is the collapse of the ground beneath the front wheels during rover operation, causing the rover's angle to tilt downward. When the rover uses the reverse rotation of the upper side wheel, a moment acts counterclockwise as shown in *Fig. 14*. The moment M formula is given by *Eq. 4*. The rover's wheel driving force is F_x. The distance from the rover's center of gravity to the wheels is R, and the angle from the rover's center to the wheels is θ_R. This moment M can maintain rover's angle and help suppress sideslipping.

$$M = 4F_x R \sin\theta_R \tag{4}$$

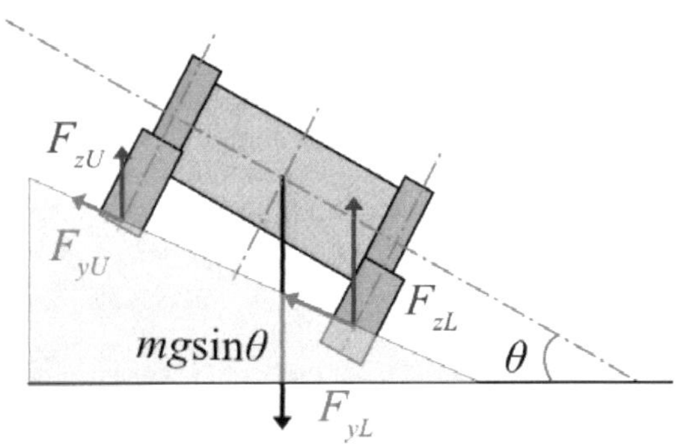

Fig. 12. Normal rotation of all wheels

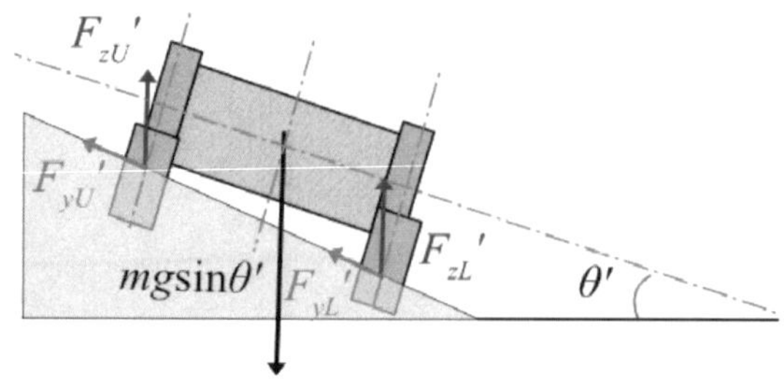

Fig. 13. Effect of reverse rotation of the wheels on the upper side

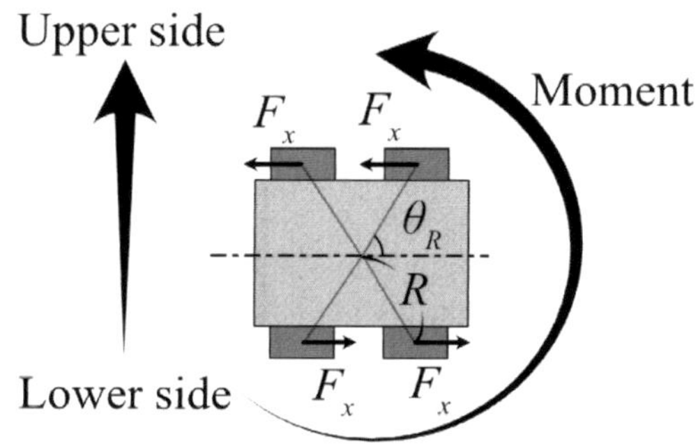

Fig. 14. Moment acting on reverse driving method

3. Driving Experiment

In this chapter, the authors conduct driving experiments focused on reverse rotation of upper side wheels. This experiment aims to confirm the difference between normal rotation and reverse rotation of upper side wheels.

3.1. Experimental conditions

We use same testbed rover as the preliminary experiment shown in *Fig. 7*. To measure the trail of the rover, motion markers and motion captures are placed, in the same as the preliminary experiment. These tests use normal wheel rotation (as shown in *Fig. 2*) and reverse rotation of only the upper slope wheels (as shown in *Fig. 15*). The slope angle is 20 degrees, and the number of trial is five times. The experimental filed is the same as shown in *Fig. 8*. *Table 3* shows experimental conditions. To evaluate driving performance, the amount of sideslipping, the rover's angle, and the amount of sinkage are measured. *Figure 16* shows a schematic diagram of the parameters. The driving distance is 500 mm in the X axis direction. The amount of the sideslipping is the rover's displacement in the Y direction. The rover's angle is measured relative to the X axis, with the upper slope as positive and the lower slope as negative. The sinkage is calculated as the average of each trial for each upper and lower wheel.

Table 3 Experimental conditions

Item[unit]	Value
Soil	Slica sand No.5
Slope angle [°]	20
Wheelbase speed [mm/s]	8
Wheel rotating speed [mm/s]	67
Traveling cycle	3
Trials of expeirment	5

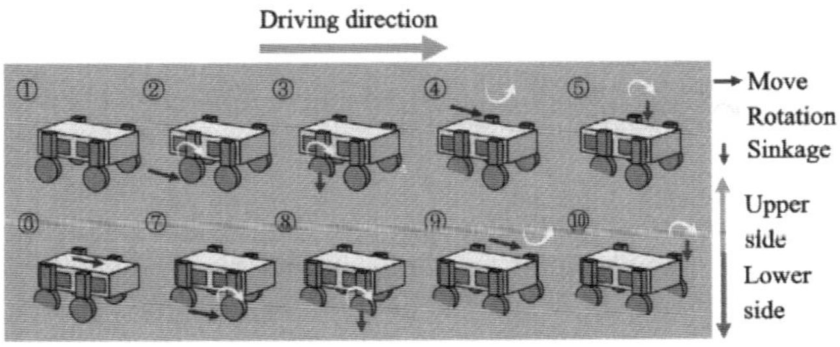

Fig. 15. Driving method of 4WAPPL in upper side reverse rotation

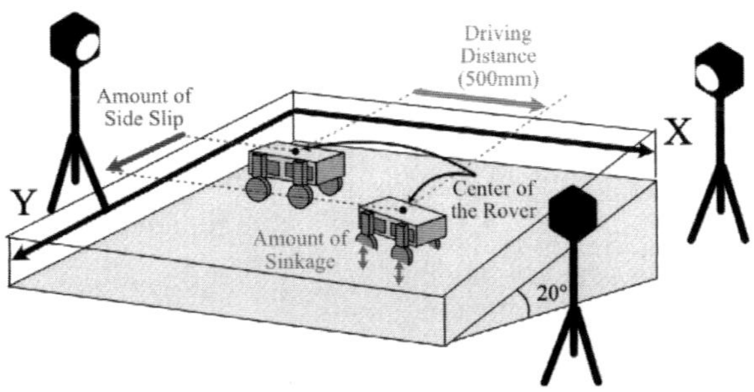

Fig. 16. Test field for experiment

3.2. Experimental Result and Discussion

Figure 17, *Figure 18*, and *Figure 19* show the amount of sideslipping, rover's angle, and subsidence amount, respectively. Additionally, *Figure. 20 (a)* and *(b)* display the results of driving with normal rotation, while *Fig. 21 (a)* and *(b)* show the results of driving with reverse rotation of the upper side wheels. First, we discuss the amount of sideslipping and sinkage. Comparing normal rotation and the reverse rotation of the upper side wheels. The experimental results confirmed a significant reduction in sideslipping across all trials with reversed rotation of the upper wheels as indicated in *Fig. 17, Fig. 20* and *Fig. 21*. This is due to the difference in sinkage compared to normal driving. In the case of the normal rotation, the lower wheels sank more, whereas in the reverse rotation, the upper wheels sank more as shown in *Fig. 18*. As a result, the normal force and side force on the upper wheels increased, while the load on the lower wheels decreased. These factors contributed to reducing ground collapse. Second, we discuss the relationship between sideslipping and rover's angle. Comparing the rover's angles in normal rotation and reverse rotation of the upper wheels, the reverse rotation consistently inclined more toward the upper slope as indicated in *Fig. 19*. This could be due to the upper slope-facing moment described in Section 2.4.2. This moment tilted the rover towards the upper slope, helping reduce sideslipping. Overall, reversing the upper slope wheels significantly reduced sideslipping and enabled stable slope traversal comparable to using steering. These results suggest the importance of sinkage on upper side wheels and tilting the rover's angle toward upper side on the slope. This experiment confirmed the effectiveness of the proposed method for crossing slopes with the specific parameters such as size and mass used. Changing the wheel rotation direction could reduce slipping, making this method applicable to small rovers without steering mechanisms and those with different mechanisms. However, its effectiveness under different conditions with varying parameters has not been verified. Therefore, further experiments are essential for practical application.

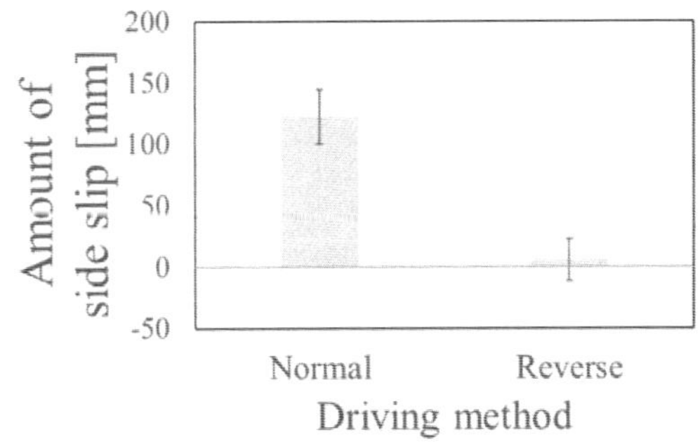

Fig. 17. Amount of side slip after driving "Reverse rotation" means only rotating upper wheels

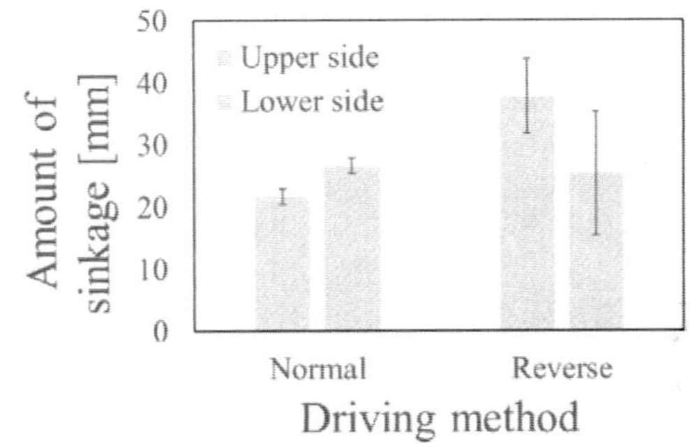

Fig. 18. Amount of sinkage after driving

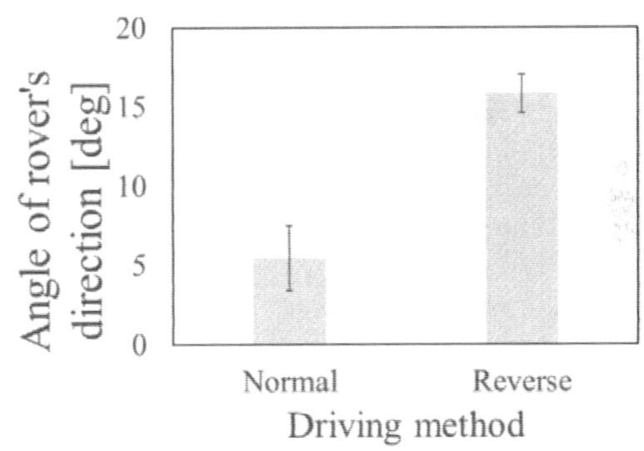

Fig. 19. Angle of rover's direction after driving

(a) Before driving in normal rotation

(b) After driving in normal rotation
Fig. 20. Experimental results for normal rotation

(a) Before driving with upper wheel in reverse rotation

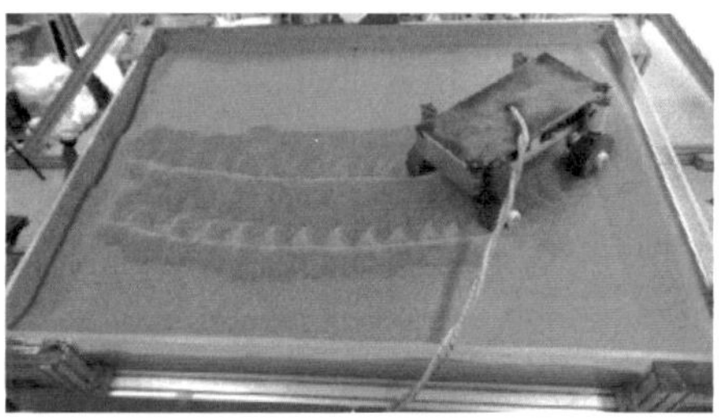

(b) After driving with upper wheel in reverse rotation
Fig. 21. Experimental results for reverse rotation

4. Conclusions and future work

This study proposed a driving method that suppresses sideslipping during slope traversal in 4WAPPL without using turning the steering. The proposed method involved reversing the direction of wheel rotation during an inching or push-pull scheme to minimize sideslipping. A driving technique that reverses only the upper slope wheels was tested to determine its effectiveness in suppressing sideslipping through practical driving experiments. As a result, reversing the upper slope's wheels led to greater sinking of the upper wheels, improving the previously insufficient support. This helped suppress ground collapse on the lower slope and enabled better control of sideslipping. This reverse rotation on upper wheel tilted the rover toward the upper slope, confirming the potential for suppressing sideslipping. As a result, the proposed method suppressed sideslipping during slope traversal in 4WAPPL without using turning the steering. At this stage, the specific principle causing the increased sinking remains unclear. Additionally, running methods that simultaneously perform front and rear extension. such as scarab rover, might not achieve sufficient sideslip suppression through reverse rotation. In future study, we plan to conduct single-wheel movement experiments to investigate the wheel's behavior during reverse rotation. Additionally, we plan to experiment with other types of rovers to determine whether the reverse rotation method is effective.

5. Nomenclature

b	Width of the wheel	[m]
F_s	Force acting on the side of the wheel	[N]
F_u	Force acting on the bottom of the wheel	[N]
F_x	Driving force	[N]
F_y	Side force	[N]
F_{yL}	Side force on the lower wheels	[N]
F_{yL}'	Side force on the lower wheels in reverse rotation method	[N]
F_{yU}	Side force on the upper wheels	[N]
F_{yU}'	Side force on the upper wheels in reverse rotation method	[N]
F_{zL}	Normal force on the lower wheels	[N]
F_{zL}'	Normal force on the lower wheels in reverse rotation method	[N]
F_{zU}	Normal force on the upper wheels	[N]
F_{zU}'	Normal force on the upper wheels in reverse rotation method	[N]
g	Gravitational acceleration	[m/s^2]
h	Amount of sinkage	[m]
m	mass	[kg]
M	Moment	[N]
r	Radius of the wheel	[m]
R	Distance from the rover's center of gravity to the wheels	[m]
θ	Rover angle	[rad]
θ_f	Wheel entry angle	[rad]
θ_r	Wheel exit angle	[rad]
θ_R	Angle from the rover's center to the wheels	[rad]

6. Acknowledgements

This work was supported by JSPS KAKENHI Grant Number 24K07409.

7. Declaration of competing interest

The authors declare that they have no known competing financial interests or personal relationships that could have appeared to influence the work reported in this paper.

8. References

Iizuka, K., Suga, T., Sakata, T., Fujiwara, D., 2023. Skidding Suppression Method using "Discrete 4-Wheel-Drive typed Rover" considering Steering Angle at traveling across Loose slope, Proc. Of the 16the Europian-African Regional Conference of the ISTVS. 889. https://doi.org/10.56884/ABCD1234.

Ishigami, G., Miwa, A., Yoshida, K., 2005. Steering trajectory analysis of planetary exploration rovers based on all-wheel dynamics model, Proc. of the 8th Int. Symp. on Artificial Intelligence, Robotics and Automation in Space, Vol. 603.

Inaba, K., Iizuka, K., 2021. Proposal of a Planetary Exploration Rover with a Wheel using Sinking Performance for Crossing Slope, Journal of the Robotics Society of Japan. Vol. 39, No.10, 958-965,

JAXA, 2023. Live Coverage: Launch of the X-Ray Imaging and Spectroscopy Mission (XRISM) and the Smart Lander for Investigating Moon (SLIM) onboard the H-IIA Launch Vehicle No. 47. https://global.jaxa.jp/countdown/slim_special_site.html (accessed 15 April).

NASA, 2017. Scarab, https://www.nasa.gov/image-article/scarab/. (accessed 15 April).

NASA, 2019. Mars Exploration Rovers: Spirit and Opportunity. https://science.nasa.gov/mission/mars-exploration-rovers-spirit-and-opportunity/imagegallery/rspirit.jpg.html (accessed 15 April).

NASA, 2019. https://www.nasa.gov/wp-content/uploads/2020/12/artemis_plan-20200921.pdf, (accessed 15 April).

Roehr, T.M., Cordes, F. and Kirchner, F., 2014. Reconfigurable integrated multirobot exploration system (RIMRES): Heterogeneous modular reconfigurable robots for space exploration, Journal of Field Robotics, Vol. 31, No. 1, 3-34.

Yoshida, K., 2011. Development of Mobile Robots for Lunar/Planetary Exploration. Journal of the Japan Society of Precision Engineering, Vol. 77, No1, 12-15.

6086 | Traveling analysis of wheel for lunar exploration rover based on extended terramechanics model: Examination of similarity law of gravity

Mai Shimizu [a,*], Shingo Ozaki [a]

[a] *Yokohama National University, 79-5 Tokiwadai, Hodogaya-ku, Yokohama, Kanagawa 240-8501, Japan*

* *Corresponding author: Shimizu-mai-fx@ynu.jp*

ABSTRACT

The development and operation of exploration rovers is important for human lunar exploration, which is considered an international space exploration goal. However, the lunar surface is a completely different environment from that on the Earth because it is under 1/6G and is covered with regolith. Hence, it is difficult to experimentally evaluate rover's traveling performance in advance. Under these circumstances, the extended terramechanics (xTerramechanics) model has been proposed that takes terrain surface deformation into account. In this study, systematic single-wheel traveling analyses assuming a lunar surface are performed based on the xTerramechanics model. First, the effect of soil parameters assuming Toyoura sand and regolith simulant (FJS-1) is examined. Then, the simple similarity law of gravity is examined by setting gravity levels as 1G and 1/6G under three different soil conditions (dense Toyoura sand, dense FJS-1, and loose FJS-1). Furthermore, an analysis is also conducted under the condition in which the wheel load is adjusted to the lunar surface (1/6M under 1G), and the significance of the simulated environment-experiments on the Earth was discussed. As a result, it is revealed that the simple similarity law of gravity might be valid only when a cohesion of soil is negligible. In addition, it is shown that when conducting experiments on the Earth ground, it is important to consistent the sinkage and wheel contact area to the 1/6G condition than to consistent the wheel load to the 1/6G condition.

Keywords
Traveling analysis
Extended terramechanics model
Terrain deformation
Model experiment
Gravity

1. Introduction

An important topic of the ongoing human lunar exploration program is the development and operation of unmanned rovers to survey the topography and resources, and manned pressurized rovers to transport supplies and astronauts. However, the lunar surface is a completely different environment from that of the Earth, with gravity about one-sixth that of the Earth, high vacuum, and large temperature fluctuations. In addition, the lunar surface is an off-road environment covered with soft soil called "regolith," which can easily cause the rover to get stuck, and elucidating the interaction between the regolith and the rover's wheels and pre-evaluating its traveling performance are issues.

The conventional terramechanics model (Bekker-Wong-Reece model) (Wong and Reece, 1967) proposed by Bekker, Wong, and Reece is widely used for wheel traveling analysis. However, although BWR model can evaluate the reaction forces acting on the wheel, it cannot evaluate the terrain surface deformation caused by the interaction between the wheel and the soil. Under these circumstances, Suzuki et al. (Suzuki et al., 2023; Watanabe et al., 2023) proposed an extended terramechanics model (xTerramechanics model) that can describe ground deformation due to the wheel-soil interaction by discretizing the ground surface with cells based on the concept of cellular automata and introducing soil moving rules.

In this study, numerical simulations implementing the xTerramechanics model were used to examine the traveling behavior of the wheels assuming a lunar exploration rover. Although previous studies (Suzuki et al., 2023; Watanabe et al., 2023) have confirmed the validity of the xTerramechanics model, all of them were comparisons with experimental results using Toyoura sand, a typical standard sand. In contrast, we adopted regolith-simulated soil (FJS-1) as an environment more similar to the lunar surface, and simulated the traveling behavior at one-sixth the gravity of the earth. First, the soil parameters were changed to those of FJS-1, and differences in traveling performance and behavior were confirmed. Then, focusing on soil cohesion, the gravity was set to 1/6G and 1G for the Toyoura sand and FJS-1 conditions, and the effect of gravity was examined.

Here, regarding a gravity scaling, several studies have already been reported (e.g., Karpman et al., 2020; Kovács et al., 2020; Yoshida et al., 2006) but uses conventional terramechanics model. Hence, comprehensive analysis with models that consider soil movement are thought to be of impotence. In this study, the similarity law of gravity is simply defined as the law that the coefficient of traction remains consistent when only the gravity on the wheel and the ground is changed without changing the soil conditions. Furthermore, since researches and developments of lunar exploration rovers have been often conducted on the earth under conditions where the wheel mass is set to 1/6M (1G) assuming the wheel load on the lunar surface, we examined the significance of "simulated-environment experiments on the ground" by conducting analysis under the same conditions with different wheel masses.

2. Theory

In this study, the interaction between the wheel and soil is analyzed using the xTerramechanics model. The xTerramechanics model is a combination of the classical terramechanics model, Bekker-Wong-Reece model (Wong and Reece, 1967) (BWR model) and terrain surface deformation model (Suzuki et al., 2023; Watanabe et al., 2023). The terrain surface deformation model

https://doi.org/10.56884/K4GQX5BD

considers soil movement due to the interaction between soil and an object. This section briefly describes each of these theories.

2.1. BWR model

BWR model is widely used to evaluate the traveling performance of off-road vehicles. The vertical stress $\sigma(\theta)$ acting on a wheel surface is defined by the contact angle θ as in Eq. 1.

$$\sigma(\theta) = \left(ck_c' + \rho g l k_\phi'\right)\left(\frac{\lambda(\theta)}{l}\right)^n \tag{1}$$

where c is the soil cohesion [Pa], k_c' is the sinkage modulus influenced by cohesion [-], ρ is the bulk density [kg/m^3], g is the acceleration of gravity [m/s^2], l is the wheel contact length (the length of the arc where the wheel is in contact with the soil) [m], k_ϕ' is the sinkage modulus influenced by friction angle [-], and n is the sinkage exponent [-]. The amount of sinkage $\lambda(\theta)$ at a contact angle θ [rad] is obtained as follows.

$$\lambda(\theta) = \begin{cases} \frac{D}{2}\left(\cos\theta - \cos\theta_f\right) \ (\theta_m < \theta < \theta_f) \\ \frac{D}{2}\left\{\cos\left[\theta_f - \frac{\theta - \theta_f}{\theta_m - \theta_r}(\theta_f - \theta_m)\right] - \cos\theta_f\right\} (\theta_r < \theta < \theta_m) \end{cases} \tag{2}$$

where D is the wheel diameter [m], θ_f is the entry angle [rad], and θ_r is the exit angle [rad]. Also, θ_m is the maximum stress angle [rad], which is the angle at which the wheel receives the maximum normal stress from the soil, and is expressed by the following equation:

$$\theta_m = (a_0 + a_1 s)\theta_f \tag{3}$$

where a_0 and a_1 are the maximum stress angle constants [-], which are parameters that depend on the wheel geometry and soil. s is the slip ratio [-] and is defined as

$$s = 1 - \frac{v}{(D/2)\omega} \tag{4}$$

where ω is the angular velocity of the wheel [rad/s] and v is the translational velocity [m/s]. The shear stress τ_x occurring in the tangential direction of the wheel surface is given by the following equation using the normal stress in Eq. 1.

$$\tau_x = (c + \sigma(\theta)\tan\phi)\left(1 - e^{-\frac{j_x(\theta)}{k_x}}\right) \tag{5}$$

where ϕ is the internal friction angle [rad] and k_x is the shear deformation modulus along x-axial [-]. $j_x(\theta)$ is the amount of sliding between the wheel and the soil [m], which is obtained by integrating the wheel-soil sliding velocity over time and is expressed by the following equation:

$$j_x(\theta) = \frac{D}{2}\left[\theta_f - \theta - (1 - s)\left(\sin\theta_f - \sin\theta\right)\right] \tag{6}$$

From the above, the drawbar-pull F_x and the vertical reaction force F_z can be evaluated as follows:

$$F_x = \frac{D}{2}B\int_{\theta_r}^{\theta_f}\left[\tau_x(\theta)\cos\theta - \sigma(\theta)\sin\theta\right]d\theta \tag{7}$$

$$F_z = \frac{D}{2}B\int_{\theta_r}^{\theta_f}\left[\tau_x(\theta)\sin\theta + \sigma(\theta)\cos\theta\right]d\theta \tag{8}$$

where B is the wheel width [m].

2.2. Terrain surface deformation model

The terrain surface deformation model adopts cellular automaton that discretizes the ground surface into cells and generates terrain surface deformation based on information such as the height of each cell and its contact with the wheel. There are three types of terrain surface deformation actions: rolling action, moving action, and compressive action. In addition, changes in soil density due to wheel compaction are also considered.

The rolling action is the action of sand rolling by gravity when the slope of the formed road surface is greater than the angle of repose (Nishimori and Ouchi, 1993; Suzuki and Ozaki, 2020; Werner, 1995), and is expressed by the following equation:

$$dh_{i,j}^{roll} = -rh_{i,j} + \frac{r}{6}H_{cross} + \frac{r}{12}H_{diag} \tag{9}$$

where the subscripts i and j are the cell numbers corresponding to the x- and y-directions, respectively. $h_{i,j}$ [m] is the cell surface height, and r is the outflow ratio [-]. Also, H_{cross} is the sand inflow term from orthogonal cells, and H_{diag} is the inflow term from diagonal cells, which are given as follows:

$$\begin{cases} H_{cross} = h_{i+1,j} + h_{i-1,j} + h_{i,j+1} + h_{i,j-1} \\ H_{diag} = h_{i+1,j+1} + h_{i+1,j-1} + h_{i-1,j-1} + h_{i-1,j+1} \end{cases} \tag{10}$$

The moving action is caused by the interaction between the object and the ground. This action is represented as

$$\begin{cases} dh_{i,j}^{rem} = -S_{i,j} \\ dh_{i+di,j+dj}^{rem} = R_1 S_{i,j} \end{cases} \tag{11}$$

Eq.11 indicates that the terrain surface height decreases by a specific height $S_{i,j}$ [m], and that soil with a height equal to $S_{i,j}$ multiplied by a specific ratio R_1 [-] moves to the adjacent cell. Note that the specific ratio R_1 is a parameter that takes into account the compaction or expansion of the soil as it moves. The direction of soil movement is in the opposite direction of the stresses that the wheel receives from the soil. In addition, the amount of movement $S_{i,j}$ is given as

$$S_{i,j} = R_2\delta_{i,j} \tag{12}$$

where $\delta_{i,j}$ is the amount of wheel penetration into the soil [m] during each numerical calculation step. The amount of penetration multiplied by a specific ratio R_2 [-] becomes the amount of movement $S_{i,j}$.

The compressive action corresponds to the terrain surface compressing as the wheels pass over it, and is expressed by the following equation:

$$dh_{i,j}^{comp} = -R_3\delta_{i,j} \tag{13}$$

As with the moving actions, the soil is reduced in height by a specific ratio R_3 [-] times the penetration $\delta_{i,j}$ [m], but no movement into other cells takes place. The compressive action is more pronounced in loose soils.

The deformation of the terrain surface is determined by applying the deformations evaluated by the above three actions to all cells as follows:

$$h'_{i,j} = h_{i,j} + dh_{i,j}^{roll} + dh_{i,j}^{rem} + dh_{i,j}^{comp} \tag{14}$$

When terrain surface deformation occurs, the soil parameters are also considered to have changed in real world phenomena. Therefore, since the normal stress generated at the wheel depends on the bulk density, the change in bulk density due to ground compaction by the wheel is considered as follows.

$$\rho = \rho_0 \exp(-\varepsilon_v) \quad (15)$$

where ρ is the density after terrain surface deformation [kg/m^3], ρ_0 is the initial density [kg/m^3]. Further, the volumetric strain ε_v [-] is given by

$$\varepsilon_v = \ln \frac{V}{V_0} \quad (16)$$

Here, V is the volume after terrain surface deformation [m^3] and V_0 is the initial volume [m^3].

3. Analysis conditions

The Simscape® environment was used to simulate the dynamic system, and the xTerramechanics model was implemented using the MATLAB®-based Simscape language. The ode4 solver was employed to solve the equations of motion, and the time increment was set to t=0.002 s. Table 1 shows the wheel specifications used in the analysis. Here, forced-slip traveling was assumed as the wheel motion condition. The ground surface was discretized into cells with $dx = dy =$ 5 mm on a side. The acceleration of gravity in a 1G field was assumed to be g = 9.8 m/s^2.

Table 1
Specification of rigid wheel.

D	: Wheel diameter	0.2	[m]
B	: Wheel width	0.1	[m]
M	: Wheel mass	15.7	[kg]
ω	: Angular velocity	7	[deg/s]

Table 2
Parameters of BWR model.

ρ	: Bulk density	1.5	[g/m^3]
k'_c	: Sinkage modulus influenced by cohesion	0	[-]
k'_ϕ	: Sinkage modulus influenced by friction	34.3	[-]
n	:Sinkage exponent	1	[-]
k_x	: Shear deformation modulus	0.035	[-]
κ	: Sinkage ratio	1.0	[-]
a_0	: Maximum stress angle constant	0.2	[-]
a_1	: Maximum stress angle constant	0.3	[-]

Table 3
Soil cohesion and Internal friction angle of Toyoura sand and FJS-1.

c_1	: Soil cohesion (Toyoura sand)	0.6	[kPa]
c_2	: Soil cohesion (FJS-1)	1.44	[kPa]
ϕ_1	: Internal friction angle (Toyoura sand)	34.8	[deg]
ϕ_2	: Internal friction angle (FJS-1)	41.6	[deg]

Table 4
Parameters of terrain surface deformation model.

r	: Rolling ratio	0.2	[-]
R_1	: Expansion and compression ratio	0.99	[-]
R_2	: Moving ratio	0.2	[-]
$k_{\max\_d}$	: Number of rolling repetitions	6000	[count/s]
d_{act}	: Interval of moving action	0.0005	[m]

3.1. Comparison by sand type

In the parametric analysis of Toyoura sand and FJS-1, among the various soil parameters, we focused on soil cohesion, internal friction angle, and angle of repose, and examined the effects of these values on traveling performance.

Table 2 shows the basic parameters of the BWR model used for stress evaluation. For simplicity, these values are independent of sand type, and the values for Toyoura sand are used (Suzuki et al., 2023). The values of soil cohesion and internal friction angle, which were varied as an effect of sand type, are shown in Table 3. Here, the soil cohesion and internal friction angle of FJS-1 were referred from the work of Kobayashi et al. (2009). The angle of repose was assumed to be approximately equal to the internal friction angle and was set to the same value.

The parameters related to terrain surface deformation are shown in Table 4. These parameters are also assumed to be independent of sand type and were set based on previous studies (Suzuki et al., 2023). Because the bulk density used in this analysis assumed a moderately dense condition, compaction and bulk density change were ignored for the terrain surface deformation.

3.2. Examination of the similarity law of gravity

In the analysis of the similarity law of gravity, the effects of changing the gravity from the earth gravity (1G) to the moon gravity (1/6G), and the effects of changing the wheel mass to 1/3M and 1/6M on the earth (1G) were examined under three different ground conditions.

Table 5 shows the values of gravitational acceleration and wheel mass. Here, 1/6G is assumed to be on the lunar surface, and the gravity similarity law was examined by comparing the results of 1G and 1/6G at the wheel mass (1M) shown in Table 5. In addition, when the wheel mass is changed to 1/6M at 1G, the wheel load is the same as that on the Moon. By comparing the results of 1/6M (1G) and 1/6G (1M), the significance of the experiments in which the loads were aligned in different gravity fields was discussed. The 1/3M wheel mass assumes the load under the Mars gravity (1/3G), and the condition was added for reference.

Three types of ground conditions were prepared: dense Toyoura sand, dense FJS-1, and loose FJS-1. The dense condition is assumed to be on the Earth, and the loose condition of FJS-1 assumes that the lunar surface has a very low bulk density. The specific parameters are shown in Table 6. In addition to the cohesion, internal friction angle, and angle of repose set for the sand type. Further, ρ, k'_c, and k'_ϕ were also set for each ground condition. ρ was determined by referring to the Lunar Sourcebook (Carrier et al., 1991) and the test results of maximum and minimum bulk densities (Kobayashi et al., 2010). The k'_c, and k'_ϕ for FJS-1 are such that the ratio of the cohesion term to the friction term in the normal stress evaluation Eq. 1 of the BWR theory is 2:8 for each condition. Other values are the same as those shown in Table 2.

Table 5
Conditions of gravity and wheel mass.

g	: Acceleration of gravity (1G)	9.80665	[m/s^2]
g	: Acceleration of gravity (1/6G)	1.63	[m/s^2]
M	: Wheel mass (1M)	15.7	[kg]
M	: Wheel mass (1/3M)	5.2	[kg]
M	: Wheel mass (1/6M)	2.6	[kg]

Table 6
Soil parameters.

	ρ [g/cm^3]	c [kPa]	ϕ [deg]	k'_c [-]	k'_ϕ [-]
Toyoura sand (Dense)	1.58	0	34.8	0	34.3
FJS-1 (Dense)	1.90	1.5	41.6	8.5	27.4
FJS-1 (Loose)	1.55	1.5	41.6	7.0	27.4

Table 7
Parameters of terrain surface deformation model.

	r [-] (Rolling)	R_1, R_2 [-] (Moving)	R_3 [-] (Compression)
Dense	0.2	0.99, 0.2	0
Loose	0.2	0.99, 0	0.01

Table 7 shows the parameters related to terrain surface deformation. The terrain surface deformation methods were set separately for dense and loose soils. For dense soil, the compressive action is neglected, while the terrain surface deformation is considered to be due to rolling deformation and moving action. On the other hand, in the case of loose soil, the compressive action was assumed to be dominant, as if traveling on snow, and the moving action were neglected.

4. Analysis Results and Discussion

4.1. Comparison by sand type

This section describes the parameter analysis of Toyoura sand and FJS-1. Figure 1 shows the relationship between drawbar-pull, sinkage, and slip ratio during steady-state traveling obtained from the analysis. It also shows the results for the same conditions using only the BWR model without considering terrain surface deformation. Figure 2 shows the deformation of the terrain surface and the stress distribution in the horizontal (x-axis) and vertical (z-axis) directions for each ground condition for a slip ratio 0.3. Contour plots of terrain surface deformation show the initial terrain surface height in green, higher in red, and lower in blue. The stress distribution in the x-axis positive direction is shown in red, the stress distribution in the x-axis negative direction in orange, and the stress distribution in the z-axis positive direction in blue.

Figure 1 shows that the xTerramechanics model analysis indicates that the drawbar-pull is greater in the case of FJS-1. The increased drawbar-pull in FJS-1 is considered to be due to the increased shear stress caused by the soil cohesion and internal friction angle, which can be confirmed from the stress distribution in Fig. 2. The sinkage of the Toyoura sand is larger than that of the FJS-1. However, as shown in Fig. 2, the sand pile pushed out in front of the wheel is higher in FJS-1, which is due to the larger angle of repose of FJS-1.

Furthermore, comparing the results of the analysis using the xTerramechanics model with the results of the analysis using only BWR model under the same conditions, the results using the xTerramechanics model show a smaller drawbar-pull for both Toyoura sand and FJS-1. This is thought to be related to the presence or absence of sand piles due to soil moving action in front of the wheel and soil removal action behind the wheel, as shown in Fig. 2, because the terrain surface deformation was not processed in the BWR model-only analysis. The sinkage of Toyoura sand and FJS-1 were almost identical in the BWR model analysis, but the xTerramechanics model results showed a difference. As shown in Fig. 2, this is thought to be due to changes in the vertical stress distribution caused by terrain surface deformation, soil cohesion, and internal friction angle.

These results confirm that the xTerramechanics model can represent more realistic behavior than BWR model with respect to terrain surface deformation, drawbar-pull, and sinkage, regardless of sand type.

4.2. Examination of the similarity law of gravity

First, a comparison is discussed between gravity conditions on the earth (1G) and the moon (1/6G). Here, the xTerramechanics model is used in the following analysis.

Fig. 3 shows the analysis results of the relationship between the steady-state values versus the slip ratio. In addition to the drawbar-pull and sinkage, coefficient of traction (drawbar-pull/wheel load) is also shown in the figure to discuss the similarity law of gravity. Fig. 4 shows the deformation of the terrain surface under each condition at a slip ratio 0.3. The color of the terrain surface deformation indicates the road surface height: green is the initial terrain surface height, red is higher, and blue is lower.

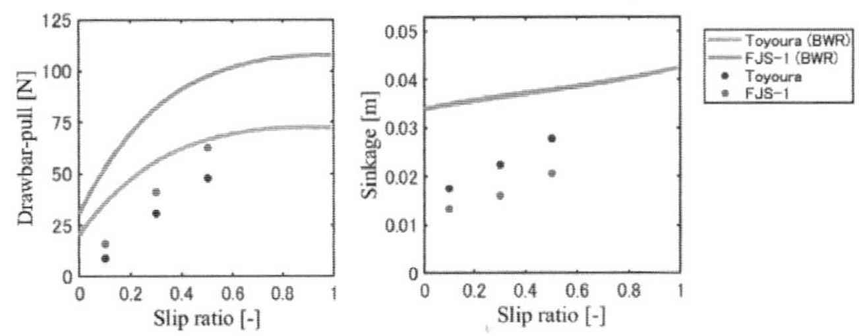

Fig. 1. Relationship between the traveling characteristics and slip ratio.

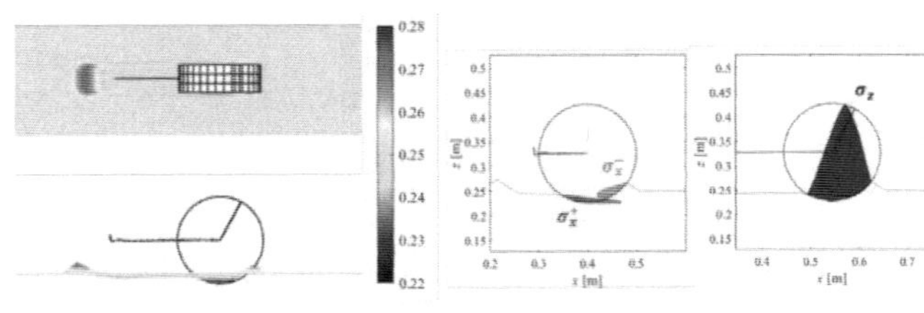

(a) Toyoura sand

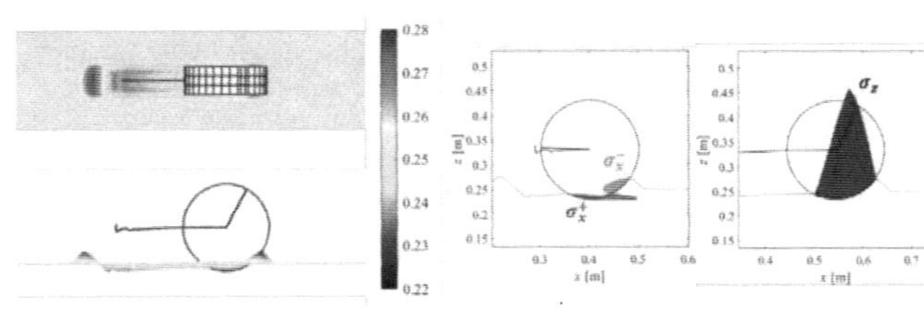

(b) FJS-1
Fig. 2. State of road deformation and stress distribution at s=0.3.

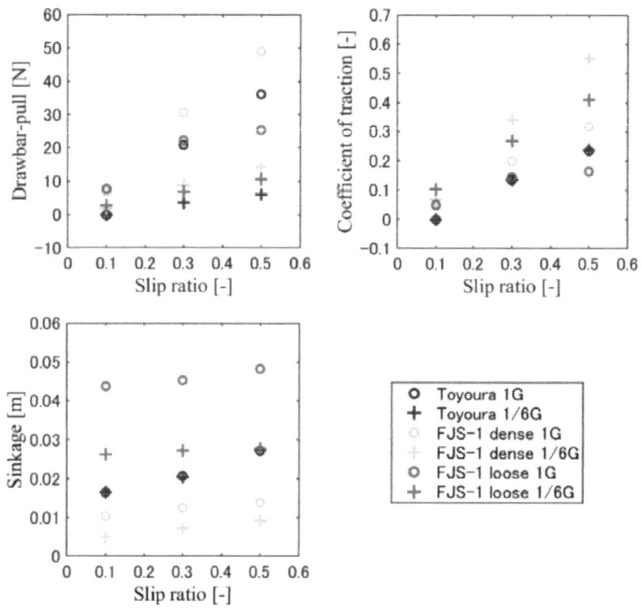

Fig. 3. Relationship between the traveling characteristics and slip ratio under 1G and 1/6G.

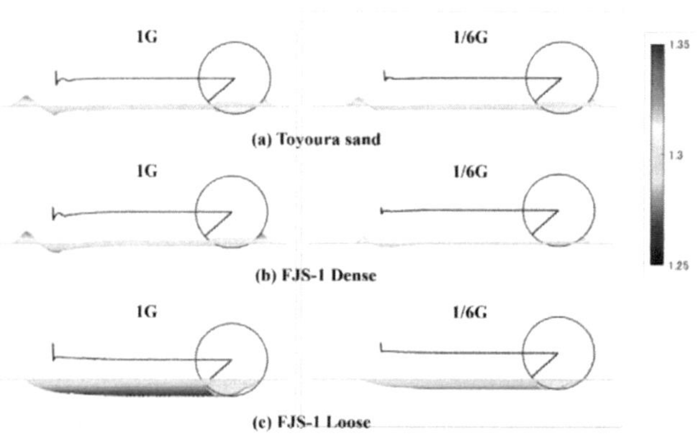

Fig. 4. State of road deformation at s=0.3 under 1G and 1/6G.

From the graph of drawbar-pull in Fig. 3, 1G and 1/6G do not coincide for any of the soil conditions. However, the coefficient of traction and sinkage are almost consistent for the Toyoura sand, while they are not consistent for the FJS-1 in both the dense and loose conditions. Therefore, it is suggested that the similarity law of gravity is basically not valid for FJS-1, but it is valid for Toyoura sand under the conditions of the present analysis. The reason for this thought be that Toyoura sand was assumed to have no soil cohesion ($c = 0$), whereas the FJS-1 sand was assumed to have soil cohesion. Note again that the analysis results are based on changing only the gravity applied to the wheel and the ground for each soil condition.

Figure 4 shows that for the dense Toyoura sand and FJS-1, a pile of sand is formed in front of the wheel due to terrain surface deformation caused by the moving action. On the other hand, in the loose FJS-1, the terrain surface deformation is caused by compressive action, and it can be confirmed that the wheel travels while compressing the ground without moving soil in front of the wheel.

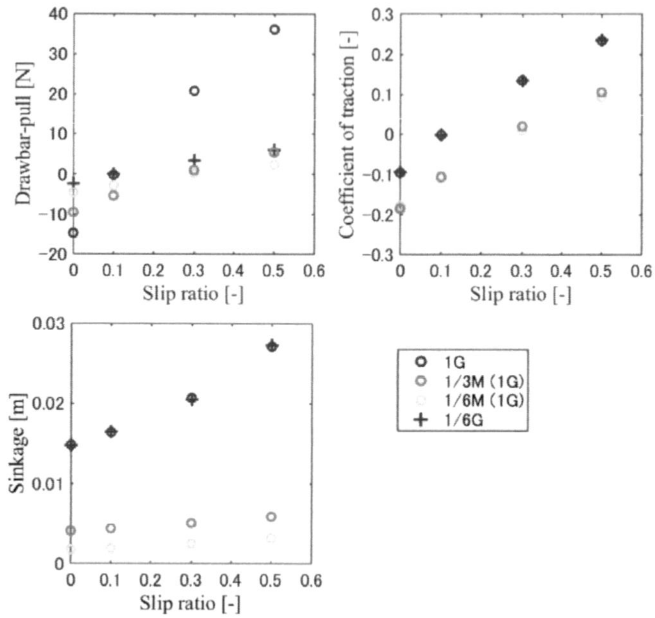

Fig. 5. Effects of gravity and wheel load on traveling characteristics (Toyoura sand).

Next, the analysis results for the case where the wheel mass is 1/6M on the earth (1G) are described.

Figure 5 shows the relationships among drawbar-pull, coefficient of traction, sinkage, and slip ratio in Toyoura sand under the 1G condition with wheel masses of 1/3M and 1/6M, and under the 1M condition with gravity conditions of 1G and 1/6G. The horizontal (x -axial) stress distributions under these conditions are also shown in Fig. 6. Here, blue represents the stress in the x-axis positive direction and red represents the stress in the x-axis negative direction.

Comparing the 1/6G and 1/6M (1G) results in Fig. 5, it can be seen that neither the drawbar-pull, coefficient of traction, nor the amount of sinkage is consistent. The drawbar-pull and coefficient of traction show similar trends, but the tendency of the amount of sinkage to increase with the slip ratio is different, with the 1/6G condition having a larger amount of sinkage and a higher dependence on the slip ratio. Fig. 6 also shows that the sinkage, terrain surface deformation, and stress distribution are different even when the wheel loads are matched. In the 1/6M condition, the sand pile in front of the wheel is smaller, but the smaller sinkage also results in a smaller traction force. On the other hand, in the 1/6G condition, the sand pile in front of the wheel was larger than in the 1/6M condition, but the larger sinkage resulted in a larger ground contact area, and a greater traction force was obtained. The reason for this difference is thought to be the greater influence of gravity acting on the ground itself under the conditions of the present analysis. The same tendency was observed for the 1/3M as for the 1/6M.

Next, the analysis results for the dense FJS-1 and the loose FJS-1 with a wheel mass of 1/6M are presented.

Figures 7 and 9 show the relationships among drawbar-pull, coefficient of traction, sinkage and slip ratio under 1/3M (1G), 1/6M (1G), 1G, and 1/6G conditions in FJS-1. The horizontal (x-axis) stress distributions under these conditions are shown in Fig. 8 and Fig. 10. Here, blue represents the stress in the positive direction along the x-axis and red represents the stress in the negative direction along the x-axis.

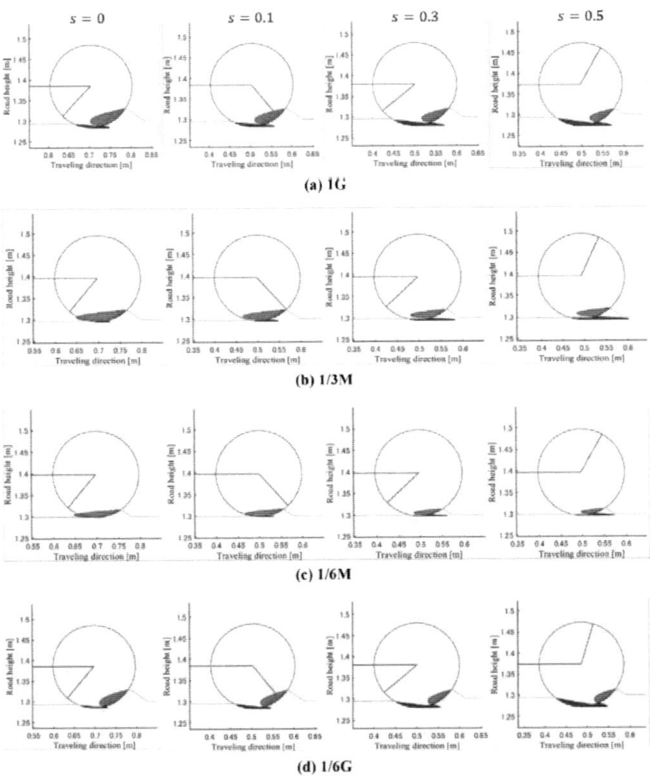

Fig. 6. Effects of gravity and wheel load on stress distribution (Toyoura sand).

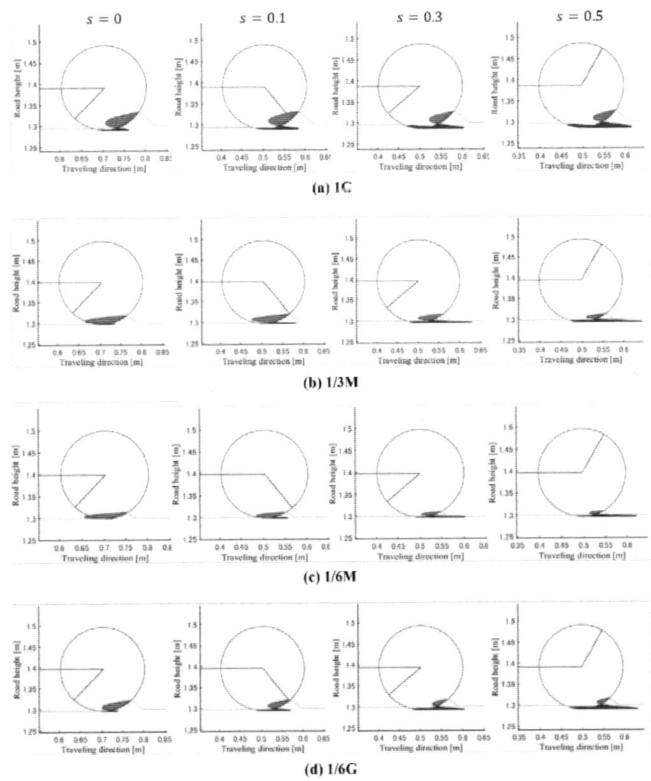

Fig. 8. Effects of gravity and wheel load on stress distribution (FJS-1, Dense).

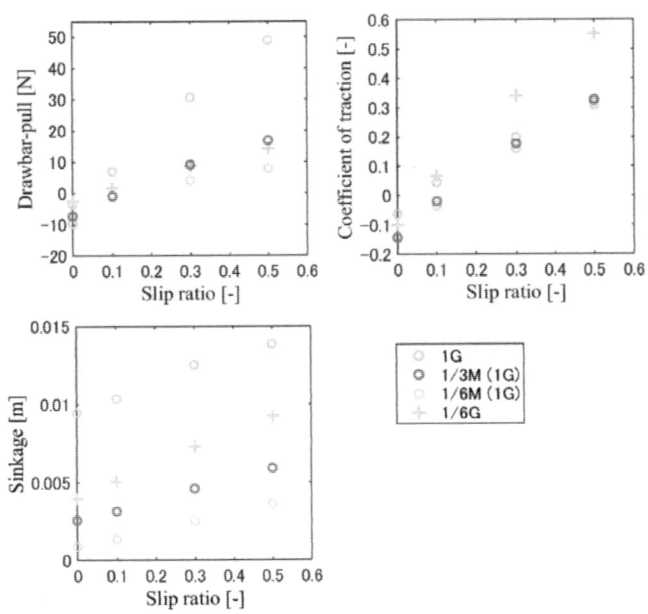

Fig. 7. Effects of gravity and wheel load on traveling characteristics (FJS-1, Dense).

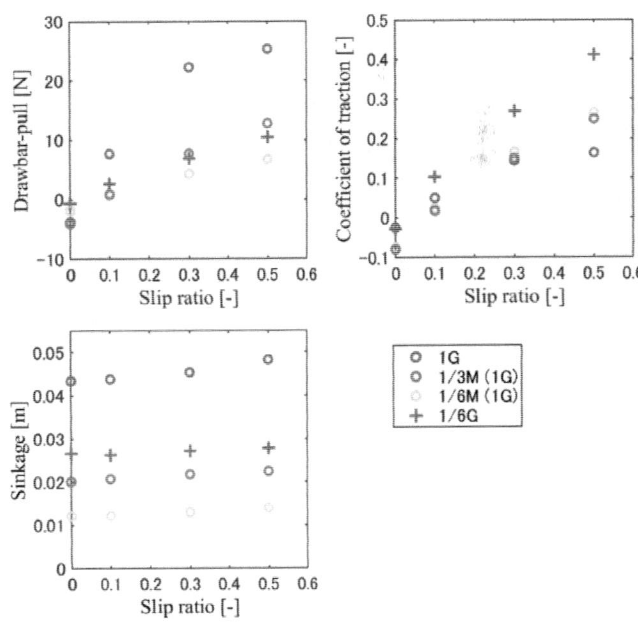

Fig. 9. Effects of gravity and wheel load on traveling characteristics (FJS-1, Loose).

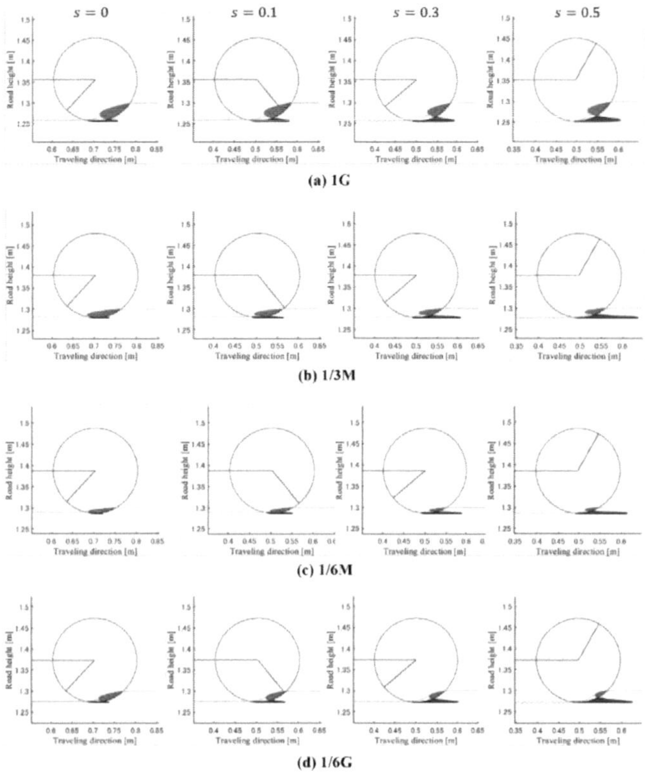

Fig. 10. Effects of gravity and wheel load on stress distribution (FJS-1, Loose).

From the dense FJS-1 results shown in Fig. 7 and Fig. 8, it can be seen that neither the drawbar-pull, sinkage, nor coefficient of traction agreed with the results for the 1/6G condition when wheel mass was set to 1/6M at 1G. The dependence of the sinkage on the slip ratio, the deformation of the terrain surface, and the trend of the wheel contact area were similar to those of the Toyoura sand case. Note that, in the case of FJS-1, the influence of cohesion at 1/6G is not negligible compared to that of Toyoura sand.

As can be confirmed from results of the loose FJS-1 in Fig. 10, the main terrain surface deformations are rolling deformation and compressive action, thus the terrain surface deformation and stress distribution are similar for the 1/6M and 1/6G conditions. However, since the amount of sinkage is different, the ground contact area of the wheel is also different. Hence, the drawbar-pull shown in Fig. 9 shows a difference, confirming that the coefficient of traction is not the same between the 1/6M (1G) and the 1/6G condition.

In summary, the simple similarity law of gravity might be valid only when the soil cohesion is zero, as in the case of Toyoura sand, and is basically not valid when the soil cohesion is present, as in the case of FJS-1. Even if only the wheel load is the same as that on the lunar surface, the coefficient of traction does not match if the gravity conditions are different. The reason for the mismatch is that the amount of sinkage and the deformation of the terrain surface, that is, the contact area between the wheel and the ground, do not match. Therefore, it is more meaningful to match the amount of sinkage and the contact area between the wheel and the ground with the 1/6G condition than to match the wheel load with the 1/6G condition when conducting the simulated environment-experiments on the Earth.

5. Conclusions

In this study, a traveling analysis was conducted under conditions assuming that the wheel travels on the lunar surface, implementing the xTerramechanics model. Specifically, the effects of sand type, gravity environment, and wheel load conditions on drawbar-pull and sinkage, which are typical indicators of traveling performance, were comprehensively confirmed. As a result, the validity of the xTerramechanics model was confirmed, the feasibility of the simple similarity law of gravity was examined. Furthermore, the effective condition setting for the simulated environment-experiments on the Earth was suggested.

References

Carrier, W.D., Olhoeft, G.R., Mendell, W., 1991. Physical properties of the lunar surface Lunar sourcebook: a user's guide to the Moon., Cambridge University Press, pp.475-594.

Karpman, E., Kövecses, J., Holz, D., Skonieczny, K., 2020. Discrete element modeling for wheel-soil interaction and the analysis of the effect of gravity, J Terramech. 91, 139-153

Kobayashi, T., Fujiwara, Y., Yamakawa, J., Yasufuku, N., Omine, K., 2010. Mobility performance of a rigid wheel in low gravity environments. J Terramech. 47, 261-274

Kobayashi, T., Ochiai, H., Suyama, Y., Aoki, S., Yasufuku, N., Omine, K., 2009. Bearing capacity of shallow foundations in a low gravity environment. Soils and Foundations. 49, 115-134

Kovács, L.L., Ghotbi, B., González, F., Niksirat, P., Skonieczny, K., Kövecses, J., 2020. Effect of gravity in wheel/terrain interaction models. J Field Robot 37, 754–767

Nishimori, H., Ouchi, N., 1993. Formation of ripple patterns and dunes by wind-blown sand. Phys. Rev. Lett. 71, 197–200.

Suzuki, H., Kawakami, H., Kobayashi, T., Ozaki, S., 2023. Extended terramechanics model for machine--soil interaction: Representation of change in the ground shape and property via cellular automata. Soil Tillage Res. 226, 105578

Suzuki, H., Ozaki, S., 2020. Multi-stage terramechanics simulation: Seamless analyses between formation of wind ripple pattern and wheel locomotion. Cogent Eng. 7. 1758289.

Watanabe, Y., Nakano, S., Suzuki, H., Ozaki, S., 2023. Numerical analysis of wheel locomotion on soft soil using the extended terramechanics model based on cellular automata. J Terramech. 109, 9-20

Werner, B.T., 1995. Eolian dunes: Computer simulations and attractor interpretation. Geology 23, 1107–1110.

Wong, J.Y., Reece, A.R., 1967. Prediction of rigid wheel performance based on the analysis of soil-wheel stresses. Part II. Performance of towed rigid wheels. J Terramech. 4, 71-125

Yoshida, K., Watanabe, T., Mizuno, N. and Ishigami, G., 2006. Terramechanics-based analysis and traction control of a lunar/planetary rover. Springer Tracks in Advansed Robotics, vol 24, Springer, Berlin, pp.225-234.

6412 | Traversing abilities simulation of a biomimetic robot on granular soil terrain

Zhenwen Zhou [a], Xiang Lei [a], Dehai Guan [a], Yan Zhang [a], Gabriel Lodewijks [b,*], Guangming Chen [a,*]

[a] *College of Mechanical and Electrical Engineering, Nanjing University of Aeronautics and Astronautics, Nanjing 210016, China*

[b] *School of Engineering, University of Newcastle, Callaghan, NSW 2308, Australia*

** Corresponding author: guangming2017@nuaa.edu.cn*

ABSTRACT

Biomimetic mobile robots that adopt animal features have proven superior traversing abilities on granular terrains over traditional vehicles. To analyze traversing properties of the biomimetic robots, numerical simulations to predict interaction behaviors between robots and soil are developed. Nonetheless, available simulation methods are not yet capable of predicting interactions between robot feet and soil subjected to arbitrary motions. This work illustrates a co-simulation method Simulink-Recurdyn-EDEM to model robot traversing abilities on granular soil. Using soil simulation parameters determined by experimental tests, the traversing behaviors for the robot to move on granular surfaces are modeled. Meanwhile, the dynamic reaction forces on robot feet are numerically estimated. The simulated robot behaviors and dynamics are in agreement with experimental tests. Therefore, an effective simulation method to assess the traversing abilities of robot interacting granular soils is demonstrated.

Keywords
Biomimetic robot
Dynamics
Discrete element method
Foot/soil interaction
Co-simulation

1. Introduction

B The explorations of soft soil terrains, such as lunar and Martian surfaces require specific vehicle or robot rovers. The wheeled or C-shape leg vehicles have shown serious sinking due to the weak strength of soils, which can be impossible to extricate and thus resulting in abandonment (Rankin et al., 2021). Inspired by desert animals that freely walk on sand/soil surfaces without sinking, biomimetic robots that adopt the structure and locomotion characteristics of these desert animals have been designed (Li et al., 2012; Dang et al., 2019). As a result, various biomimetic robots for adapting to granular surfaces have been proposed (Roberts et al., 2021; Arm et al., 2023; Chen et al., 2024).

To test the traversing properties of the biomimetic robots, experimental testbeds are built. Roberts et al. constructed a soft granular platform to test the single-legged robot leaping performance (Roberts et al., 2021). Arm et al. built a testing facility using ESA's ES-4 Martian soil analogue and evaluated robot mobile characteristics (Lee et al., 2020; Arm et al., 2023). Chen et al. built soft soil terrains resembling the surfaces of Mars to test robot locomotion performance (Chen et al., 2023; Chen et al., 2024). These researches show that the behaviors and traversing abilities of robots can be assessed using physical experiments. However, creating large amount of soil materials to build testbeds is of high expense, and the optimization of equipment structures is time-consuming by using experimental tests. Moreover, in experiments, it is difficult to study microscopic behaviors of foot-ground interaction on granular soils, which is not conducive to adjust structure and control parameters.

To effectively evaluate interaction behaviors that mechanical parts engage with granular soils, numerical approaches are developed (Y.-S. Kim et al., 2022; Rodriguez et al., 2022). For example, using the discrete element method (DEM) coupling with multibody dynamics (MBD), Lu et al. modeled wheel-sand interactions and predicted wheel reactions in a sandy environment(Lu et al., 2023). Ucgul (Ucgul, 2023) simulated the cutting behaviors for an agricultural plow engaging soils and assessed the rotational speeds of several plows. Wang et al. (Wang et al., 2019) modeled the flow and passage of soybean particles on sieve plates. The current MBD-DEM co-simulation method can model simple motions when machines interact with particles. It is not yet capable of modeling the complex motions of the machines that engage with granular particles.

This paper contributes a co-simulation approach (Simulink-Recurdyn-EDEM) to model interaction behaviors when complex movements of machines engage with granular soils. The simulation parameters for granular soil were determined based on calibration experiments using the JLU Mars1 analogue (Liu et al., 2023). Referring to experiments, the effectivity of the established simulation model was verified, and the traversing characteristics for a biomimetic robot walking on granular soil were predicted. In addition, the dynamic behaviors between robot feet and soil at micro scales were analyzed.

2. Simulation setup

To predict the traversing abilities of the robot on granular soils, this section illustrates the data communication method for the co-simulation platform using Simulink-Recurdyn-EDEM. Besides, the configuration for simulating a lizard-like robot (Chen et al., 2024) moving on a granular soil surface is elaborated.

https://doi.org/10.56884/2OT1ETM6

2.1. Data communication method

To provide precise control of the robot's complicated motion, a motion control program needs to be established using Simulink(Ji et al., 2022). To simulate the traversing behaviors of a lizard-like robot (Chen et al., 2024), the robot model is imported to Recurdyn, where the robot's basic motions are evaluated. To build the granular soil surfaces, the discrete element program EDEM(Teslenko et al., 2021) is selected. Based on the data transfer protocol (Zhao et al., 2021), a communication interaction method for co-simulation of Simulink, Recurdyn, and EDEM is established. This procedure consists of five steps (a-e) by virtue of a transfer station 'Shared Memory', which is given in Fig. 1. The solid lines denote the commands transmitted from the sender to the 'Shared Memory'. The dashed lines represent the corresponding commands transmitted from 'Shared Memory' to receivers. The detailed five steps (a)-(e)are given below.

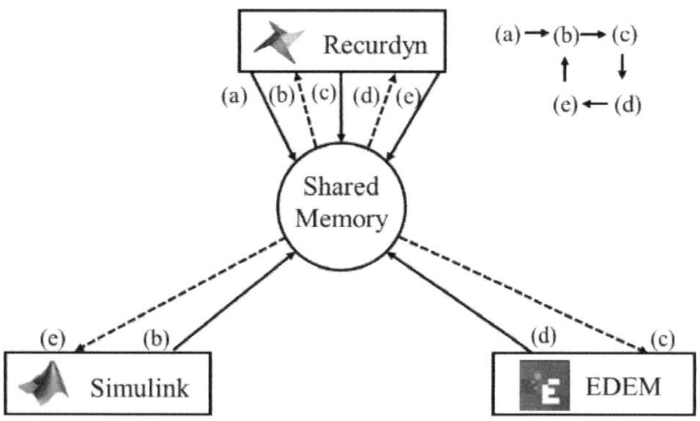

Fig. 1. Co-simulation information interaction method

(a) Recurdyn establishes 'Shared Memory', calling Simulink and EDEM, and waiting for control data output from Simulink.

(b) Simulink outputs control data, which is to control the motion of the Recurdyn model through 'Shared Memory', and ensure that the control time step is consistent with that in Recurdyn.

(c) Recurdyn transmits the state information of the simulation model (including simulation parameters and motion constraints) to EDEM through 'Shared Memory', then the geometry state information between Recurdyn and EDEM is associated.

(d) EDEM transmits the particle state information to Recurdyn through 'Shared Memory', so that the particle information between EDEM and Recurdyn is linked. Meanwhile, the structure and environmental parameters are also unified.

(e) Recurdyn integrates the status information of the structure and particle, and transmits it to Simulink through 'Shared Memory', then waits for new data from Simulink.

2.2. Co-simulation configuration

Fig. 2 illustrates the procedure to simulate the traversing abilities of the lizard-like robot referring to the date communication method. First, the robot model is imported to Recurdyn, where the material and motion parameters are configured, and the robot's basic motions are evaluated. Coordinated motion planning for the spine and leg is performed with a composite non-impact curve (S. Yuan et al., 2019). The obtained trajectories are for the inverse kinematic solution to determine robot joint movements and provide information for motion control in Recurdyn. To establish a particle ground, accurate and effective intrinsic particle parameters and contact parameters of the selected soil (Yang et al., 2021) are required. The determination of these parameters is discussed in the next section.

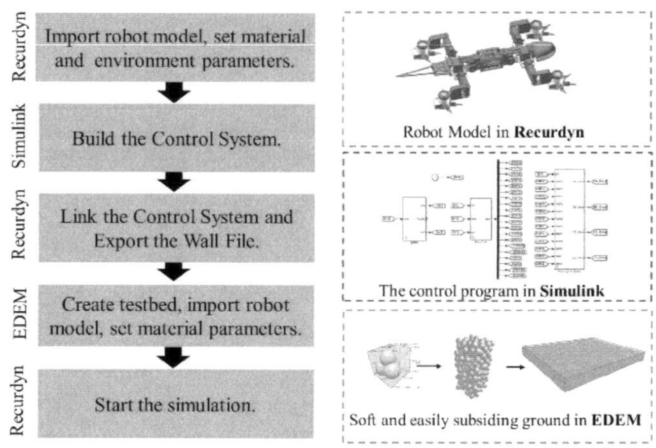

Fig. 2. Co-simulation method of traversing behavior of the robot

3. DEM parameter calibration

In this work, JLU Mars1 is selected as the granular material for simulating the traversing abilities of a lizard-like robot. Considering the properties of low moisture content, high dispersibility, and low cohesion characteristics of the granular material, the Hertz-Mindlin (no slip) contact model in the discrete element method (Zhou et al., 2021) is adopted (Liu et al., 2023). To produce accurate and reasonable simulation results, the DEM particle parameters and contact parameters are determined.

3.1. Particle parameters

To determine DEM parameters of particle diameter and density, a rectangular container with sizes of 100×100×100mm is created in the software EDEM to simulate the free fall and accumulation of spherical particles. After the container is fully filled, the mass of particles in the container is measured to estimate particle density. Due to the larger number of particles, enormous computational time is required. To reduce calculations by reducing the number of particles, the particle diameter can be enlarged (Zhou et al., 2021), while accounting for the angle of repose of single spherical particles (Yamane et al., 1998). Therefore, based on the results that particle diameter enlargement in the range of up to five times in DEM still yields acceptable simulation results (Du et al., 2017; Fei et al., 2019), spherical particles with a radius of 1.75mm are used referring to the particle size range of 0.068-2mm (Liu et al., 2023). In addition, tetrahedral model consisting of four spherical particles is selected to model all irregular particles of JLU Mars1. The experimental measured bulk density is 1206.5 kg/m^3 (Liu et al., 2023). For approaching such value, the simulated single particle density is determined to be ρ=2100 kg/m^3, with deviation of about 1.5%.

Based on that Poisson's ratio (v) play a minor impact on the mechanical characteristics of granular soil flowability (Hu et al., 2021), v=0.25 is selected for the modelled particles (J. Sun et al., 2020; Jin et al., 2023). To determine shear modulus, settlement test is carried out by pressing a plate (55×55×10mm) on a granular terrain created in a box (100×100×100mm). The contact area of the plate with soil is approximately equivalent to that between the robot foot and the ground. A force of 10N is applied to the plate corresponding to the load at the robot foot. The subsidence is

5.24mm subjected to particle shear modulus 10MPa in the simulated settlement test. The subsidence is close to the actual sinking of 5.5mm under static equilibrium conditions for the robot traversing on JLU Mars1. Resultantly, a shear modulus of 10Mpa is selected as the simulated particle shear modulus for JLU Mars1(Nguyen et al., 2020).

3.2. Contact parameters

The test of the angle of repose can be used to estimate interaction parameters of particle-particle static friction coefficient (μ_s), particle-particle rolling friction coefficient (μ_r), particle-equipment static friction coefficient (μ_s'), particle-equipment rolling friction coefficient (μ_r'), particle-particle collision restitution coefficient (e), and particle-equipment collision restitution coefficient (e') (Zhu et al., 2022). Fig. 3(a) illustrates the experimental measurement of the heap angle for JLU Mars1. The angles from the left and right of the front view of the pile were recorded. The same experiments were conducted five times, and the average angle of 32.45° was used as the final angle of repose for JLU Mars1.

Fig. 3. Heap angle measurement of JLU Mars1 in simulation and experiment

The basic ranges for contact parameters for JLU Mars1 are selected based on the literature (Zhou et al., 2021; Fang et al., 2022) and the GEMM database in EDEM (H. Yuan et al., 2023), namely: $e \in (0.15,0.25)$, $\mu_s \in (0.35,0.65)$, $\mu_r \in (0.16,0.24)$, $e' \in (0.15,0.25)$, $\mu_s' \in (0.18,0.54)$, $\mu_r' \in (0.3,0.5)$. Using Design-Expert software (S. Sun et al., 2024), Plackett-Burman experiments are conducted, and the sensitive parameters (μ_s, μ_r and μ_s') of angle of repose for JLU Mars1 are identified. The non-sensitive parameter values (Zhu et al., 2022) are given as $e = 0.2$, $e' = 0.2$ and $\mu_r' = 0.4$. Through steepest ascent trials, the ranges of the sensitive parameter values (Jiang et al., 2023) have been narrowed down, which are: $\mu_s \in (0.35,0.55)$, $\mu_r \in (0.16,0.24)$ and $\mu_s' \in (0.18, 0.42)$. Subsequently, utilizing the Box-Behnken experiment (Lenaerts et al., 2014), regression equations were established for calculating the sensitive parameters μ_s, μ_r and μ_s'. In combination with the experimentally measured pile angle of 32.45°, the values of sensitive parameters were determined to be $\mu_s = 0.51$, $\mu_r = 0.16$ and $\mu_s' = 0.25$.

With the determined values of contact parameters and particle models, simulations of the angle of repose test for JLU Mars1 are conducted. The simulation results are displayed in Fig. 3(b), and the angles are measured at different surrounding sites. The contour fitting using Matlab yields a simulated angle of repose 30.65°. The angle is lower than the experimental results, and the reason for this is that the used particle models are uniformly tetrahedrons that result in lower contact friction. In reality, various shapes of irregular particles exist. The deviation from the actual physical experiment angle of 32.45° is around 5.5%, indicating that the obtained values for contact parameters can be used to model characteristics of the real soil. The final determined DEM parameters and values for simulating the Mars soil analog JLU Mars1 are listed in Table 1.

Table 1
Calibration results of modelled particle parameters

Parameters	Values
Particle radius/mm, R	1.75
Particle density/kg/m^3, ρ	2100
Particle Poisson's ratio, ν	0.25
Particle shear modulus/Mpa, G	10
Particle-particle collision restitution coefficient, e	0.2
Particle-particle static friction coefficient, μ_s	0.51
Particle-particle rolling friction coefficient, μ_r	0.16
Particle-structure collision restitution coefficient, e'	0.2
Particle-structure static friction coefficient, μ_s'	0.25
Particle-structure rolling friction coefficient, μ_r'	0.4

4. Simulation results

In this section, co-simulations are performed to model robot traversing behaviors on granular surface terrain. To investigate robot foot-ground interactions, the dragging tests for a foot with grasping abilities and without grasping are conducted in experiments and simulations. By applying tripod gait to the robot, the dynamic reaction forces on robot feet are numerically estimated.

4.1. Robot foot/ground interaction

Fig. 4 illustrates experiments and simulations of the grasping abilities of a robot foot, in which Fig. 4(a-b) displays dragging without foot grasping under a quarter of robot weight at a constant speed 2 mm/s (Ohzono et al., 2020; J.-Y. Kim et al., 2021). The stable dragging forces are 0.85N in experiment and 0.92N in simulation. Fig. 4(c-d) depicts dragging under the motion of foot grasping. It is measured that the dragging forces are approximately 4.27N in experiments and 4.56N in simulation. It is demonstrated that foot using foot grasping the ground results in a considerable increase in driving force (R. Zhang et al., 2020). To investigate the reasons for the increase in dragging force subject to grasping by biomimetic toes, particle velocities during the interactions are examined (L. Zhang et al., 2022).

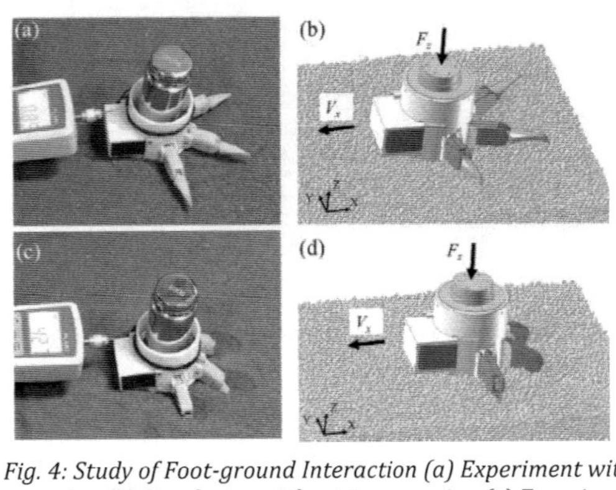

Fig. 4: Study of Foot-ground Interaction (a) Experiment without toe grasping (b) Simulation without toe grasping (c) Experiment with toe grasping (d) Simulation with toe grasping

Fig. 5(a-b) compares particle velocities during robot foot-ground interaction with and without toe grasping. Using the EDEM tool Geometry Bin, it was calculated that 2376 particles are driven by the foot during foot-ground interaction without toe grasping, generating velocities range 0.6 to 3.0 mm/s. Instead, 9077 particles are aroused when toes grasp soils, and the velocity ranges are much wider. This suggests that biomimetic toes can grab more particles, so the contact area of granular soils is solidified and thus the foot driving force is enhanced.

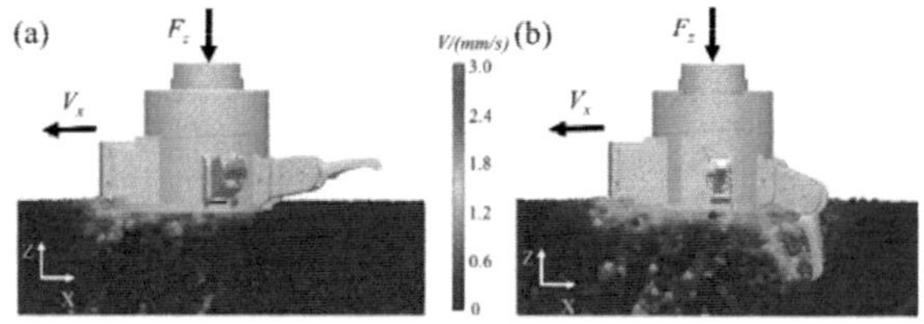

Fig. 5: Particle velocity cloud maps during foot-ground interaction (a) No toe grasping (b) Toe grasping.

4.2. Robot traversing test

Fig. 6(a-b) shows simulations and experiments for the robot traversing a granular surface without toe grasping. It is seen that the robot performs identical postures as experimental results. Because of the soft granular properties, the robot experienced sinking, measuring roughly 9mm in experiments and 11.2mm in simulation. This discrepancy can be ascribed to the lower angle of repose in the simulation calibration process. The robot walked five cycles in experiments, resulting in a total displacement 289mm and an average single-cycle displacement 57.8m. In the simulation, the soil material, motions of robot, and calculation settings are unchanged. Therefore, to obtain the slip ratio one cycle is applied, and it turns out approximately two weeks of physical time were elapsed with the lab computer (Logic processor 128 cores, 256GB RAM, and base speed 2.7Ghz), leading to simulation displacement 61.3mm. The displacements in both simulation and experiment are smaller than theoretical displacement for a single cycle of the tripod gait 68mm, which is attributed to slippage on the soft ground (Su et al., 2023). The slippages from the simulation deviated by 5.2% from those of the experiment, which is due to the accuracy of calibrations of particle parameters and numerical calculation settings.

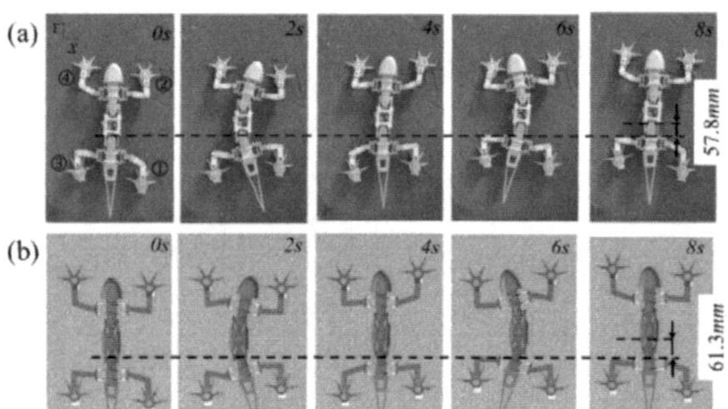

Fig. 6: Analysis of tripod gait without toe grasping: (a) Experiment (b) Simulation

with toe grasping in one cycle, which demonstrated that consistent results of simulation and experiments are obtained. The subsidence of the robot is about 9.5mm in the experiment and 11.76mm in the simulation. The increase in the subsidence is due to the fact that foot grasping introduces impulse on the toes and causes higher contact force on the granular surface. In experiments, the robot walked five cycles, resulting in total displacement of around 363mm and an average single-cycle displacement of approximately 72.6mm, whereas, in simulations, the single-cycle displacement is predicted 77.6mm based on repetition tests. In both simulation and practical scenarios including toe grasping, the robot displacements exceed the theoretical value (68mm). This is apparently attributed to the toe-grasping motion in comparison to the results without toe-grasping. Table 2 compares the results for robot traversing with toe grasping and without grasping. The slippages from the simulation deviated by 5.3% from those of the experiment and the discrepancy is also affected by the accuracy of the simulator.

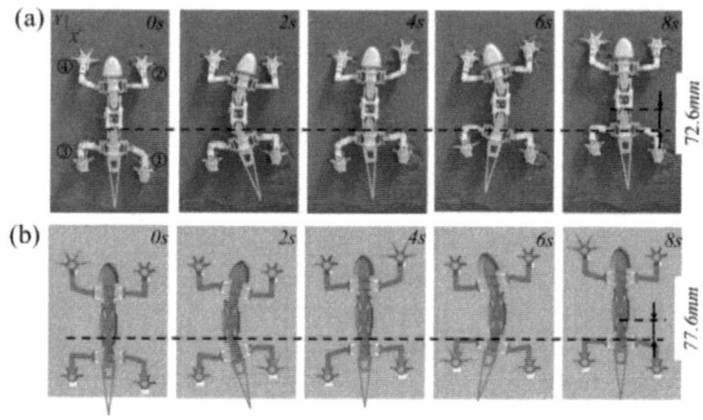

Fig. 7: Analysis of Tripod Gait with Toe Grasping. (a) Experiment., (b) Simulation

Table 2
Performances of the robot with and without toe grasping for one cycle

Tripod Gait	Verification Platform	Subsidence (mm)	Displacement (mm) by toe	Displacement (mm) by leg	Slip rate (%)
Without grasping	experiment	9	0	57.8	15
	simulation	11.2	0	61.3	9.8
With grasping	experiment	9.5	9.1	63.5	6.6
	simulation	11.76	10.5	67.1	1.3

From Table 2, it can be observed that when the toes were not used, the slippage rate in simulations is roughly 9.8%, whereas in experiments it is around 15%; when the toes grasp the ground, the slippage rate in simulations fall to 1.3%, while in experiments it is reduced to 6.6%. To investigate the mechanism of grasping in enhancing forward displacement, a study on the interaction forces F_y between the toes and the soil was conducted. Because each foot goes through the stages of grasping, holding grasping, releasing and holding releasing. Hereby foot ② is selected to illustrate the force F_y in the Y direction of foot-ground interaction, as shown in Fig. 8.

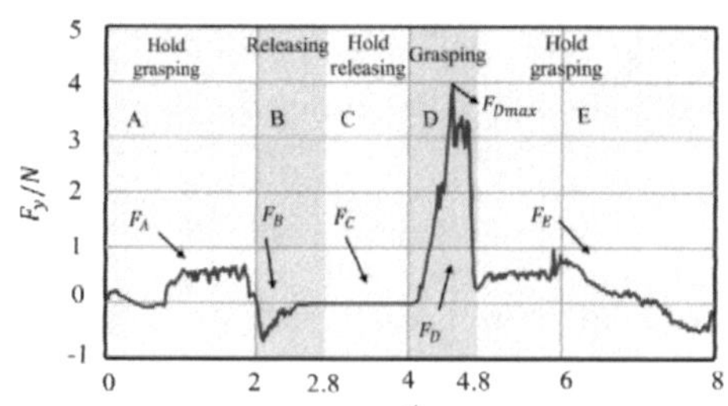

Fig. 8: Force analysis on foot ② in the Y-direction when traversing the granular surface

Figure 8 shows that when toes maintain grasping, positive forces F_A and F_E are generated to prevent the robot from slipping, as observed in both Region A and Region E (R. Zhang et al., 2017). During the toe releasing process, the robot toes sustain the load of soils and result in a negative Y-direction force, as depicted in

Region B of Fig 8. As the toe grasps soils, the robot experiences a positive Y-direction force F_y, as shown in region D of Fig. 8. For the region E, when foot ④ swings, the inertial force produces a counterclockwise rotation torque, which drives the counter foot ② to push forward the soil, resulting in negative forces. During the time interval 4s-4.8s, the driving force ($F_{D,max}$) created by one foot is larger than the resistant force ($\bar{F}r,max$) generated by the other three feet, i.e., $F_{D,max}) \geqslant Fr,max$=3max{|$F_A$ |,|F_B|}, such that toe grasping results in an increased forward displacement for the robot.

5. Discussion

This study develops an integrated simulation system, Simulink-Recurdyn-EDEM For simulating interaction for equipment of complex motion-engaging granular soils. Compared to the MBD-DEM simulation system (Y.-S. Kim et al., 2022; Rodriguez et al., 2022), the simulation system in this study accurately recreates the robot's complicated motion control. Furthermore, it is convenient to investigate the robot mobility performance (Lee et al., 2020; Arm et al., 2023; Chen et al., 2024) in the microscopic and reveal mechanisms for foot-ground contact.

This research analyzes the robot's foot-ground interface via modeling and experimentation (Ohzono et al., 2020; J.-Y. Kim et al., 2021). The paper simulates the traversing abilities of the robot with and without toe grasping on the soft and easily subsiding ground, analyzing the mechanisms of robot subsidence, slippage, and the effect of toe grasping on robot displacement, demonstrating the role of biomimetic toes in preventing slippage. There is some disparity between simulation and experimentation, which is caused by the selected particle models. Nonetheless, consistent results between experiments and simulations are obtained, which demonstrates the rationality and effectiveness of the co-simulation platform. This plate can be used to efficiently evaluate the performances of equipment of various motions engaging with granular particles.

To acquire accurate simulation analysis results, more precise calibrations for determining particle parameters are required. First, the particle shape and particle size distribution of soil particulates can be more effectively calculated by using sieve analysis (Zhu et al., 2022). Then the determinations of shear modulus and Poisson's ratio of soil particles can be improved by a triaxial compression test (Zhou et al., 2021). Using the modeled particles and their properties, the particle contact parameters can be further refined based on the angle of repose tests (Zhu et al., 2022). Moreover, by using high-performance computing clusters, the time step of the simulation system can be reduced, which is beneficial to enhancing numerical precision (Siegmann et al., 2021).

6. Conclusions

This paper introduced a co-simulation system, namely, Simulink-Recurdyn-EDEM, to investigate the performances of robots with complex motions engaging with granular solids. The experiments of traversing tests on granular terrain demonstrate the effectiveness of the simulation approach. It is identified that toe grasping by quadruped robots solidifies granular terrains and increases traction forces. As a result, robot forward distance is increased. This work improves simulation technologies for investigating interactions when complex motions of machines imposing granular soils.

7. Acknowledgements

The authors gratefully acknowledge financial support from the Ministry of Science and Technology of China (G2023181009L).

8. Declaration of competing interest

The authors declare that they have no known competing financial interests or personal relationships that could have appeared to influence the work reported in this paper.

9. References

Arm, P., Waibel, G., Preisig, J., Tuna, T., et al. 2023. Scientific exploration of challenging planetary analog environments with a team of legged robots. Sci. Robot, 8(80), eade9548.

Chen, G., Qiao, L., Zhou, Z., Lei, X., et al. 2024. Biomimetic lizard robot for adapting to Martian surface terrain. Bioinspiration Biomim.19, 036005

Chen, G., Qiao, L., Zhou, Z., Richter, L., et al. 2023. Development of a Lizard-Inspired Robot for Mars Surface Exploration. Biomimetics, 8(1), 44.

Dang, R., Jiang, L., Yao, Q., Xu, P., et al. (2019). The research status on foot design and motion control of the legged robot in soft terrain. IEEE International Conference on Unmanned Systems (ICUS), Beijing, China

Du, Y., Gao, J., Jiang, L., & Zhang, Y. 2017. Numerical analysis on tractive performance of off-road wheel steering on sand using discrete element method. J. Terramechanics. 71, 25-43.

Fang, M., Yu, Z., Zhang, W., Cao, J., et al. 2022. Friction coefficient calibration of corn stalk particle mixtures using Plackett-Burman design and response surface methodology. Powder Technol. 396, 731-742.

Fei, D., Xuefeng, S., Wuyun, Z., Fengwei, Z., et al. 2019. Simulative calibration on contact parameters of discrete elements for covering soil on whole plastic film mulching on double ridges. Nongye Jixie Xuebao/ Trans. Chin. Soc. Agric. Mach. 50(2).

Hu, C., Gao, J., Diao, J., & Song, X. 2021. Numerical simulation of tire steering on sandy soil based on discrete element method. AIP Adv. 11(1).

Ji, K., & Liu, X. (2022). Co-simulation technology of arc and mechatronics System. The 2022 IEEE 3rd China International Youth Conference on Electrical Engineering (CIYCEE), Wuhan, China.

Jiang, D., Chen, X., Yan, L., Gou, H., et al. 2023. Parameter calibration of discrete element model for cotton rootstalk–soil mixture at harvest stage in Xinjiang cotton field. Agriculture, 13(7), 1344.

Jin, H., Lu, Y., Wu, W., & Han, F. 2024. Influence of screw blades on the performance of a screw - drive granary robot. J. Field Robot. 41:2133–2146

Kim, J.-Y., Hwang, Y.-T., Baek, J.-H., Song, W.-Y., et al. 2021. Study on inter-ply friction between woven and unidirectional prepregs and its effect on the composite forming process. Compos. Struct. 267, 113888.

Kim, Y.-S., Lee, S.-D., Baek, S.-M., Baek, S.-Y., et al. 2022. Development of DEM-MBD coupling model for draft force prediction of agricultural tractor with plowing depth. Comput. Electron. Agric. 202, 107405.

Lee, J., Hwangbo, J., Wellhausen, L., Koltun, V., et al. 2020. Learning quadrupedal locomotion over challenging terrain. Sci. Robot. 5(47), eabc5986.

Lenaerts, B., Aertsen, T., Tijskens, E., De Ketelaere, B., et al. 2014. Simulation of grain–straw separation by discrete element modeling with bendable straw particles. Comput. Electron. Agric. 101, 24-33.

Li, C., Hsieh, S. T., & Goldman, D. I. 2012. Multi-functional foot use during running in the zebra-tailed lizard (Callisaurus draconoides). J. exp. Biol. 215(18), 3293-3308.

Liu, Y., Qi, Y., Pan, D., Chen, Z., et al. 2023. Gradeability of 'Zhu Rong'Mars rover based on the simulated Martian terrain. J. Terramechanics. 106, 57-73.

Lu, S., Xu, X., & Wang, W. 2023. Coupling dynamic model of vehicle-wheel-ground for all-terrain distributed driving unmanned ground vehicle. Simul. Model. Pract. Theory. 128, 102817.

Nguyen, N. H., Bui, H. H., & Nguyen, G. D. 2020. Effects of material properties on the mobility of granular flow. Granul. Matter. 22, 1-17.
Ohzono, T., Saed, M. O., Yue, Y., Norikane, Y., et al. 2020. Dynamic manipulation of friction in smart textile composites of liquid - crystal elastomers. Adv. Mater. Interfaces. 7(7), 1901996.
Rankin, A., Maimone, M., Biesiadecki, J., Patel, N., et al. 2021. Mars curiosity rover mobility trends during the first 7 years. J. Field Robot. 38(5), 759-800.
Roberts, S. F., & Koditschek, D. E. 2021. Virtual energy management for physical energy savings in a legged robot hopping on granular media. Front. Robot. AI, 8, 740927.
Rodriguez, V. A., Barrios, G. K., Bueno, G., & Tavares, L. M. 2022. Coupled DEM-MBD-PRM simulations of high-pressure grinding rolls. Part 1: Calibration and validation in pilot-scale. Miner. Eng. 177, 107389.
Siegmann, E., Enzinger, S., Toson, P., Doshi, P., et al. 2021. Massively speeding up DEM simulations of continuous processes using a DEM extrapolation. Powder Technol. 390, 442-455.
Su, Y., Yang, H., Ding, L., Xu, C., et al. 2023. A unified foot–terrain interaction model for legged robots contacting with diverse terrains. IEEE/ ASME Trans. Mechatron. 29(4), 2661-2672.
Sun, J., Chen, H., Duan, J., Liu, Z., et al. 2020. Mechanical properties of the grooved-wheel drilling particles under multivariate interaction influenced based on 3D printing and EDEM simulation. Comput. Electron. Agric. 172, 105329.
Sun, S., Hu, B., Wu, X., Luo, X., et al. 2024. Research on a vibrationally tuned directional seed supply method based on ADAMS-EDEM coupling and the optimization of system parameters. Agriculture, 14(3), 433.
Teslenko, D., & Belyakov, V. (2021). Simulation of the process of interaction between the vehicle and snow in the environment of DEM and MBD-systems. IOP Conf. Ser.: Mater. Sci. Eng. 1086 012036.
Ucgul, M. 2023. Simulating Soil–disc plough interaction using discrete element method–multi-body dynamic coupling. Agriculture, 13(2), 305.
Wang, Y., Yu, J., Yu, Y., & Zhang, Q. 2019. Validation of a coupled model of discrete element method with multibody kinematics to simulate the screening process of a swing-bar sieve. Powder Technol. 346.
Yamane, K., Nakagawa, M., Altobelli, S., Tanaka, T., et al. 1998. Steady particulate flows in a horizontal rotating cylinder. Phys. Fluids. 10(6), 1419-1427.
Yang, Y., Wen, B., Ding, L., Li, L., et al. 2021. Soil particle modeling and parameter calibration for use with discrete element method. Transactions of the ASABE, 64(6), 2011-2023.
Yuan, H., Cai, Y., Liang, S., Ku, J., et al. 2023. Numerical simulation and analysis of feeding uniformity of viscous miscellaneous fish bait based on EDEM software. Agriculture, 13(2), 356.
Yuan, S., Zhou, Y., & Luo, C. (2019). Crawling gait planning based on foot trajectory optimization for quadruped robot. IEEE International Conference on Mechatronics and Automation (ICMA), Tianjin, China
Zhang, L., Zhai, Y., Chen, J., Zhang, Z., et al. 2022. Optimization design and performance study of a subsoiler underlying the tea garden subsoiling mechanism based on bionics and EDEM. Soil Tillage Res. 220, 105375.
Zhang, R., Han, D., Ma, S., Luo, G., et al. 2017. Plantar pressure distribution of ostrich during locomotion on loose sand and solid ground. PeerJ, 5, e3613.
Zhang, R., Pang, H., Wan, H., Han, D., et al. 2020. Design and analysis of the bionic mechanical foot with high trafficability on sand. Appl. Bionics Biomech. 2020, 3489142
Zhao, L., Zhang, M., & Jin, X. 2021. Construction and application of a high precision 3D simulation model for geomechanics of the complex coal seam. Sci. Rep. 11(1), 21374.
Zhou, L., Gao, J., Hu, C., & Li, Q. 2021. Numerical simulation and testing verification of the interaction between track and sandy ground based on discrete element method. J. Terramechanics., 95, 73-88.
Zhu, J., Zou, M., Liu, Y., Gao, K., et al. 2022. Measurement and calibration of DEM parameters of lunar soil simulant. Acta Astronaut. 191, 169-177.

6422 | Sinkage Study in Granular Material for Space Exploration Legged Robot Gripper

Arthur Candalot [a,*,†], James Hurrell [a,†], Malik-Manel Hashim [b], Brigid Hickey [c], Mickael Laine[a], Kazuya Yoshida [a]

[a] *Department of Aerospace Engineering, Tohoku University, Sendai, Miyagi, Japan*
[b] *Technische Universität Berlin, 10623 Berlin, Germany*
[c] *University at Buffalo, NY 14068, United States*

* *Corresponding author: acandalot@dc.tohoku.ac.jp*

†*These authors contributed equally*

ABSTRACT

Wheeled rovers have been the primary choice for lunar exploration due to their speed and efficiency. However, deeper areas, such as lunar caves and craters, require the mobility of legged robots. To do so, appropriate end effectors must be designed to enable climbing and walking on the granular surface of the Moon.

This paper investigates the behavior of an underactuated soft gripper on deformable granular material when a legged robot is walking in soft soil. A modular test bench and a simulation model were developed to observe the gripper sinkage behavior under load.

The gripper uses tendon-driven fingers to match its target shape and grasp on the target surface using multiple micro-spines. The sinkage of the gripper in silica sand was measured by comparing the axial displacement of the gripper with the nominal load of the robot mass. Multiple experiments were performed to observe the sinkage of the gripper over a range of slope angles. A simulation model accounting for the degrees of compliance of the gripper fingers was created using Altair MotionSolve software and coupled to Altair EDEM to compute the gripper interaction with particles utilizing the discrete element method. After validation of the model, complementary simulations using Lunar gravity and a regolith particle model were performed.

The results show that a satisfactory gripper model with accurate freedom of motion can be created in simulation using the Altair simulation packages and expected sinkage under load in a particle-filled environment can be estimated using this model. By computing the sinkage of the end effector of legged robots, the results can be directly integrated into the motion control algorithm and improve the accuracy of mobility in a granular material environment.

Keywords
Pressure-sinkage
Toyoura sand
Lunar regolith
Soft gripper
MotionSolve EDEM
Legged robot
Space exploration

Fig. 1 SCAR-E hexapod robot walking in dry soil.

1. Introduction

Since the beginning of space exploration, observation of the surface of multiple celestial bodies has been achieved in our solar system. Thanks to pioneer missions, valuable data and samples were collected, and the feasibility of space exploration was confirmed. In recent years, the discovery of high value resources such as water on the Moon (Li and Milliken, 2017) has motivated the development of more advanced robots. These resources will become a key factor in the creation of extraterrestrial infrastructures and potentially support deeper space exploration.

1.1. Wheeled and legged space robots

The first planetary missions focused mainly on surface exploration using wheeled rovers (Rodríguez-Martínez et al., 2024; Welch et al., 2013). Such robots allow for reliable locomotion on flat ground and can overcome minor obstacles. However, the sandy surface of the Moon or Mars creates challenging environments with the risk of causing wheels to slip and sink, especially on steeper terrain. New areas of interest, such as craters, caves, and steeper terrain, cannot be accessed with conventional wheeled rovers. For this reason, alternative approaches have been considered to improve mobility in uneven terrain. Currently, research on snake-like locomotion in granular environment (Marvi et al., 2014), crab-shaped robots (Graf Nicole M. and Behr, 2019), or leg-wheel hybrid robots like RHEX (Saranli et al., 2001) is being conducted for Earth-based applications, although the technology may produce successful results in space.

Legged robots are also becoming a viable solution for the exploration of extreme environments due to their increased ability to climb and overcome obstacles. Recent quadrupeds (Di Carlo et al., 2018) and biped robots (Komizunai et al., 2010) usually employ spherical or flat appendices as feet, which provides stability even on softer terrain (Yang et al., 2024).
Planetary exploration robots (Parness et al., 2017; Uno et al., 2021) are often equipped with spine grippers to grasp the rough surface of rocks, allowing them to descend into lunar caves. Current studies focus on the climbing performances of gripping systems,

https://doi.org/10.56884/FLURUDA3

Fig. 2 Gripper side view

Fig. 3 Gripper bottom view

yet the gripper will be used as a foot during walking. In a granular environment, the foot is expected to sink at each step under the load of the robot, potentially hindering their locomotion. During field testing of the SCAR-E hexapod robot platform (Fig. 1), currently under development at the Space Robotics Laboratory of Tohoku University, it has been observed that if the depth of the sinkage is not taken qualitatively into account in the autonomous GAIT algorithm, discrepancies occur between the computed position and the real position, causing locomotion failure. Therefore, the terramechanic behavior of the equipped end effector must be assessed to ensure safe autonomous navigation in a granular environment.

1.2. Pressure-sinkage model

The wheel-soil interaction has been an important topic of interest in the development of wheeled vehicles for off-road applications to improve mobility. One of the foci resides on the quantification of expected sinkage during locomotion. Bekker et al. (1962) first introduced a model linking the sinkage to pressure applied and the granular material properties. The model was further refined by Wong & Reece (1967) to include the dynamics of wheel rotation. Grouser wheels (Nagaoka et al., 2020) have been used to improve contact with the soil, but slipping and dynamic sinking still pose a challenge due to shearing of the material, especially on inclined terrain (Ishigami et al., 2006; Oettershagen et al., 2019).

In the case of legged robots, the end effector compresses the granular material and then applies a tangential force as a step is made. The sinkage is therefore induced by the dynamic entry of the end effector into the soil, as well as the normal and tangential pressure applied (Yang et al., 2024). For an equivalent vehicle weight, a stepping motion may generate less shearing than a conventional grouser wheel, allowing a legged robot to overcome steeper slopes.

1.3. Gripping system

The proposed gripper (Fig. 2 and Fig. 3) for this study is a soft gripping system with compliant fingers to match the shape of its target and maximize the contact area for improved performances

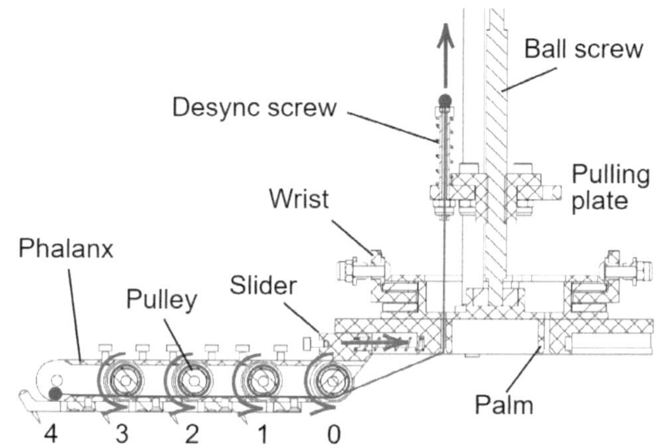

Fig. 4 Section view of the gripping system

(Candalot et al., 2024). It is constituted of 8 tendon-driven fingers, each divided into 4 rotating phalanges. A tether is attached to the extremity of the finger and is guided through a series of pulleys linked to each phalanx, as shown in Fig. 4. This architecture was previously used in the softgripper developed by Hirose & Umetani (1978), which shows great results in matching irregular shapes. Each phalanx is identical in length and shape to provide a modular option for longer or shorter fingers.

The gripper is actuated by a DC brushless motor linked to a fine pitched ball screw, creating a linear motion to pull the tethers and providing a self-locking feature to reduce power consumption during grasping. The gripper opening is achieved passively by relaxing the tethers thanks to tension springs located between the phalanges. Each tether is attached to the pulling plate through a spring-loaded screw that allows compliance for the finger to individually match the local curvature of the target.

The gripper phalanges are equipped with multiple micro-spines to hook onto the asperities of a rocky surface and generate gripping. The fingers are also attached to the gripper using a prismatic joint to create a tangential motion over the surface of the target. These features increase the probability of the micro-spine to hook a suitable asperity while also creating a clamping action, therefore increasing the gripping performances.

2. Method

In this section, the experimental and simulation method to determine the sinkage behavior of the gripper in soft soil will be described.

The challenge of replicating an extraterrestrial environment on Earth commands us to rely on simulation models to determine the terramechanic behavior of regolith in reduced gravity. Therefore, models of the gripper and granular material were created using the Altair HyperWorks and EDEM suite, a coupling of multi-body dynamics and particle modelling. To validate the simulation model, the gripper sinkage is experimentally quantified in Toyoura sand and compared to the simulation results.

2.1. Gripper model

The gripper model was created using multibody simulation software, Altair MotionSolve. Due to the highly compliant geometry of the fingers, a single solid body would not accurately demonstrate the sinkage behavior of the gripper. Therefore, a total of 41 bodies including the gripper palm, eight finger sliders, and 32 phalanges (Fig. 5) were assembled and constrained to replicate the gripper compliant properties.

Fig. 5 MotionSolve gripper model

Unnecessary hardware, such as screws, bearings, axles, and clips were removed to simplify the model. The tension springs in charge of the passive opening were replaced by torsional springs centered on the phalanx joint axis. The finger slider springs were also included in the model. The tether and pulley actuation were simulated by directly applying torque onto the phalanx's joints, and motion angle limits were added to accurately replicate the compliance range of the fingers.

To recreate the dynamic motion of a step, gravity is applied to the model, as well as a load on the gripper palm that represents the robot weight using the function $\vec{L}$ detailed in Eq. 1 and Eq. 2.

$$\vec{L} = \begin{Bmatrix} L_x \\ L_y \\ L_z \end{Bmatrix} = \begin{Bmatrix} \sin(\theta)\sigma(t) \\ 0 \\ \cos(\theta)\sigma(t) \end{Bmatrix} \quad (1)$$

With the function $\sigma(t)$ defined as:

$$\sigma(t) = STEP(t, t_1, mg, t_2, -5) + STEP(t, t_3, 0, t_4, -61) \quad (2)$$

The angle θ represents the slope inclination of the granular material, t the simulation time value, and STEP the third derivation interpolation function provided by MotionSolve.

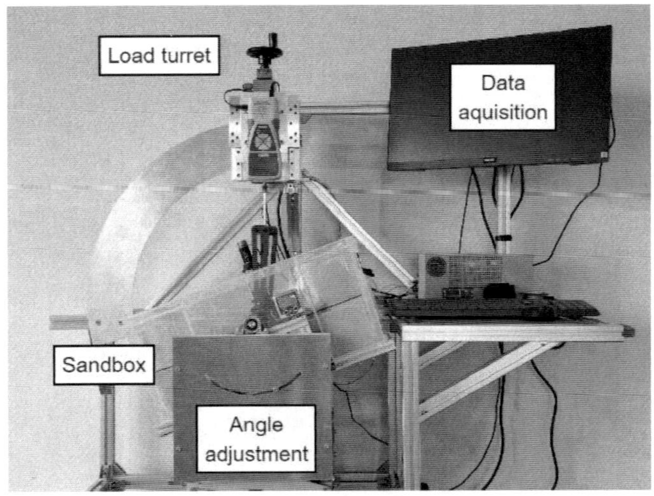

Fig. 6 Gripper testing bench

To avoid unwanted dynamic behavior due to the fast approach of the gripper in the granular material, the gripper weight *mg* is compensated for by the load from t0 to t1 to allow the fingers to settle in their resting position. Then the gripper contacts with the medium and stabilizes under a load of 5N from t2 to t3. Finally, the load gradually increases from t3 to t4 to the maximum of 66N, corresponding to the weight that a robot of 20kg under Earth gravity would apply to the gripper while standing on 3 legs.

2.2. Particle model

The interactions between the gripper model and the particles are computed using the Discrete Element Method (DEM) provided by Altair EDEM software. The particle model was previously developed to simulate the single-wheel performance of a micro-rover, with soil parameters independently calibrated based on an angle of repose experiment and validated against single-wheel experiments before being used to study lunar behavior. The parameters for Toyoura sand, which closely mirror the silica sand used in the experiment, were validated in the prior study and are now integral to this model.

Interaction between particles or a collection of particles is simulated as soft spheres, with interactions between them defined by contact models and associated particle parameters. Two key parameters are rolling friction and static friction, accounting for the non-sphericity of actual particles. The standard Hertz-Mindlin contact model and a Type C rolling model were used (Ai et al., 2011; Coetzee and Scheffler, 2023).

Table 1 Particle model properties for Toyoura sand and Regolith simulant

Parameter	Unit	Toyoura sand	Regolith
Particle size	mm	1	1
Particle number		1249000	1249000
Density	kg/m³	2650	2857
Poisson's ratio		0.25	0.3
Young's Modulus	N/m²	$5e^7$	1e8
Coefficient of Restitution		0.3	0.4
Static friction		0.65	0.81
Rolling friction		0.15	0.42
Cohesion	N/m²	0	0
Angle of repose	°	34	39

Fig. 7 Gripper sinkage at a 35 °slope angle

2.3. Testing environment

The sinkage behavior of the gripper can be characterized by the vertical displacement required for the granular material to generate a reaction force equal to the applied load. The vertical displacement may vary depending on the properties of the granular material, the load and the angle of slope, as the material will slip and collapse. For assessing the gripper sinkage under a wide range of parameters, a testing bench, presented in Fig. 6 was developed.

The test bench is composed of a manually actuated linear turret on which is attached a force meter with a resolution of 0.01N and an acquisition frequency of 0.1s. The gripping system is connected to the force meter using a rigid assembly. The turret is mounted on a curved rail to adjust the load angle from a range of 0° to 90° that can be dialed with a precision of 0.1° using an absolute digital inclinometer. The axial displacement of the turret is measured through a graduated analog display with a resolution of 0.01mm. A sandbox is mounted on the bench using an inclination-adjustable system to simulate a ground inclination from a range of -45° to 45°. The wrist of the gripper is a universal joint with roll and yaw freedom, allowing the load to be applied at the center of the gripper regardless of inclination or orientation.

The sandbox is a 446 x 332 x 218mm container filled with silica sand to a height of 100mm from the bottom. Silica sand has been chosen for its low cohesion and the uniformity of the size of its particles. It is closely related to Toyoura sand in terms of mechanical properties and is widely available.

3. Results

In this section, a comparison of the experimental and simulated results of the gripper sinkage behavior will be presented.

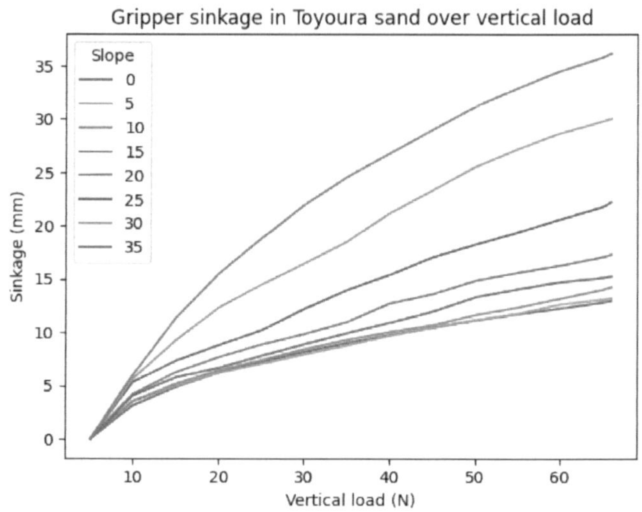

Fig. 8 Gripper sinkage at different ground inclinations

3.1. Experimental results

The sinkage of the gripper was demonstrated by first applying an initial load of 5N and recording the corresponding position of the gripper palm using the linear turret indicator. The load was then gradually increased, and the relative displacement was recorded in increments of 5N until a maximum load of 66N. To accurately represent the resting state of the gripper during walking, no tension is applied to the tethers and the fingers are free to take a more advantageous shape.

To observe the impact of the soil inclination on the sinkage, the sandbox angle was adjusted from 0° to 35° at 5° increments. Experiments above an inclination of 35° were classified as unsatisfactory as the slope angle exceeded the angle of repose of the silica sand, inducing high slipping and collapsing of the surface.

As presented in Fig. 8, the results of the experiment show that the sinkage increases while a heavier load is applied to the gripper. At a maximum of 66N, the gripper sank into the sand at an average of 12.98mm on flat ground, to 36.19mm with the ground inclined at 35°. The sinkage depth is comparable for a slope angle from 0° to 10°, but it increases greatly as the angle increases beyond 20°. At a steeper angle, the pressure generated by the gripper is not perpendicular to the ground surface, making the granular material prone to shearing and collapsing. Especially at inclinations of 30° to 35°, the granular material is close to its angle of repose and amounts of material are sheared away from the surface as the gripper digs into the surface (Fig. 7). The maximum sinkage values are presented in Table 2 and compared in percentage with the gripper diameter D = 250mm.

Table 2 Gripper maximal sinkage

Slope angle (°)	Sinkage (mm)	Percentage
0	12.98	5.2
5	13.23	5.3
10	14.26	5.7
15	15.23	6.1
20	17.31	6.9
25	22.26	8.9
30	30.05	12
35	36.19	14.5

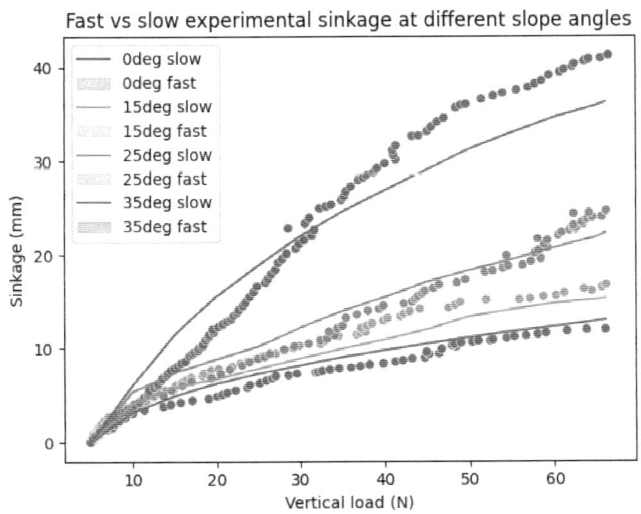

Fig. 9 Impact of entry speed on the gripper sinkage

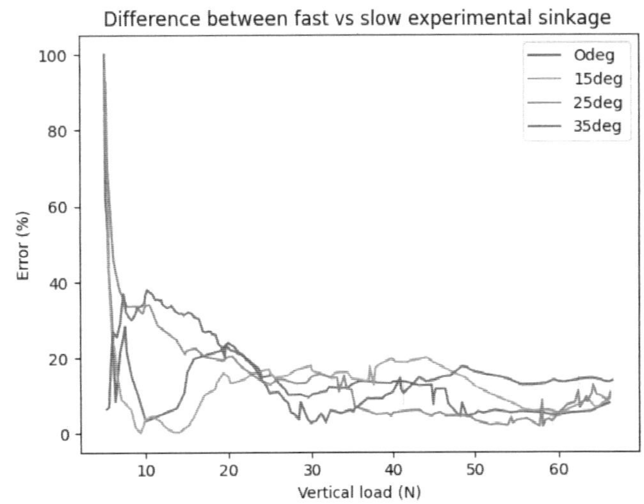

Fig. 10 Error percentage between slow and fast entry

Fig. 9 Gripper sinkage under 66N at 0°

Fig. 1 Gripper sinkage calculated in EDEM.

In the presented experiment, the manual increase of the load and recording of the displacement allows the granular material to settle and compact. On average, a single test takes approximately one minute to perform. Therefore, a second set of experiments was conducted to observe the impact of the entry speed of the gripper on the granular material. The load is aimed to continuously increase to 66N in around 10s. This represents 6 times increase in entry speed, from a range of 0.22mm/s to 0.6mm/s to a range of 1.2mm/s to 3.6mm/s.

The load was recorded using the force meter feed at a frequency of 0.1s and the vertical displacement was taken using a video camera. The sinkage comparison is presented in for a slope angle of 0°, 15°, 25°, and 35°.

The results in Fig. 9 show that the speed of entry has a moderate impact on the sinkage of the gripper. Although a faster entry does not seem to generate a deeper or shallower sinkage, an average difference of 20% can be expected. It is noted that the final depth of the sink at a load of 66N is consistent by around 10% between fast and slow entry. The highest difference is observed at a higher slope angle, especially at 35°. During experimentation, it has been remarked that a higher entry speed generated more granular material collapse, therefore increasing the sinkage depth.

3.2. Simulation comparison

The gripper pressure-sinkage simulation was performed using Altair MotionSolve and EDEM software. The simulation parameters introduced in Eq. 2 were set to compute the sinkage over a period of 10s, in a manner similar to faster experimentation. The simulation required 9h of computation with a timestep of 0.01s, and the simulated sinkage can be observed against the experimental results in Fig. 13. The simulated sinkage is consistent with the experimental data, with a difference percentage mean of 13.49 % and a standard deviation of 9.15%. The most drastic difference can be observed starting at a load of 40N. Before 40N, the difference percentage mean averages 9.25% with a standard deviation of 5.61%.

Table 3 Simulation parameters

Parameter	Symbol	Value
Simulation duration	t_{max}	10s
Time step	f	0.01s
Settling start	t_1	0
Settling end	t_2	0.5
Load start	t_3	0.5
Load end	t_4	10
Slope angle	θ	0°, 15°, 25°, 35°

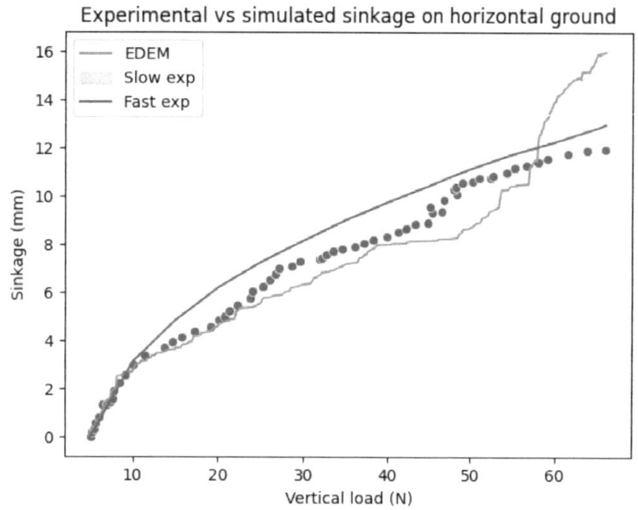

Fig. 13 Experimental vs simulated sinkage on horizontal ground

A visual appreciation of the simulated results reveals that some particles clip through the gripper model at higher load. The sharp increases in sinkage after 50N are evidence that particle clipping occurs and voids in the granular material are quickly filled by other particles, resulting in deeper sinkage. An increase in the particle density of the simulated sandbox is considered, although this would exponentially lengthen the computation time. The gripper model may also be simplified further by removing unnecessary geometries.

3.3. Space environment simulation

Figure 14 introduces the sinkage computed in lunar gravity using the regolith particle model introduced in 2.2. As a comparison, the data collected from the experiments on Earth were correlated to lunar gravity by adjusting the force values by the factor $s = \frac{1.62}{9.81}$. It can be observed that the simulated sinkage values are similar to the experimental values, as shown by the difference estimation presented in Table 4.

Table 4 Difference between lunar simulation and experimental results

Experiment	Mean (%)	SD (%)	Min (mm)	Max (mm)
Slow entry	13.71	12.53	0.01073	2.7396
Fast entry	13.35	7.32	0.0019	3.7531

Although the granular material properties are different between Toyoura sand and regolith, it is possible to consider that gravity does not hold a high impact on the sinkage behavior of the gripper. In lower gravity environments, the robot weight applied to the gripper is greatly reduced, which would result in a lower sinkage than on Earth. However, gravity also impacts the granular material particle interactions, causing the soil to be less dense and compact, therefore inducing higher sinkage. These two factors are believed to compensate for each other, explaining the similarity in sinkage between terrestrial and lunar environments.

Kobayashi et al. (2010)present similar conjectures regarding the effect of gravity on wheel sinkage, although reduced gravity appears to have a negative impact on traction performance, reinforcing the argument that alternative means of locomotion need to be developed.

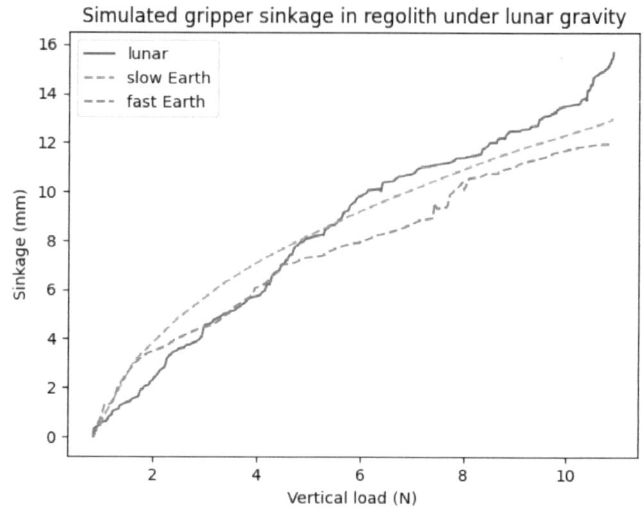

Fig. 14 Gripper sinkage simulated in lunar environment.

Nonetheless, it can be concluded that sinkage experiments in Toyoura sand or regolith under Earth gravity can offer a rough estimate of the expected sinkage in lunar environment.

4. Discussion

This study demonstrates that a legged robot end effector is subject to significant sinkage while walking in a granular environment. The sinkage is directly affected by the properties of the granular material, the load, the inclination of the ground, and gravity.

The velocity of entry in the granular material moderately impacts the sinkage behavior of the gripper and must be taken into consideration for qualitative estimations, particularly on steeper ground. For the case of wheeled vehicles and robots, multiple pressure-sinkage models such as Bekker and Reece propose a logarithmic correlation between pressure and sinkage based on material properties. Then it can be discussed that to reduce the sinkage, the gripper palm could be modified to increase the contact surface, thus reducing the pressure applied to the granular material. In addition, flexible webbing between the fingers could also be integrated if the grasping performances are not hindered.

Although the experiments yield satisfactory and repeatable results, improvements can be made by integrating an automated linear actuator to control the sinkage rate and displacement of the gripper. Such a device can be directly coupled with the force meter to generate more precise results, but also reproduce more complex situations such as exponentially increasing load, variable velocity entry, stomping and so on.

The gripper sinkage simulation holds promising results, and accuracy of the model can be confirmed by comparison with the experimental data. The issue of particle clipping needs to be addressed to completely validate the simulation model, and comparisons with different slope angles are to be conducted.

Finally, the method can also be tested using granular materials more appropriate to observe the sinkage of the gripper in space environment, such as regolith simulant. Based on the simulation results in lunar conditions, the sinkage is seemingly close to the experimental values on Earth gravity, and further experiments are required to confirm the similarity. Tests can also be conducted in Martian conditions to observe the extent of this correlation.

Regolith simulant is extremely thin compared to silica sand and requires measures to safely conduct experiments and protect the gripper from infiltration and clogging. An origami-based protective cover made of polyvinyl fluoride is currently being developed to seal the gripper from its environment while still offering full mobility and compliance of the fingers.

5. Conclusion

This research introduces the terramechanic behavior of an adaptive gripping system for legged robots to navigate and operate on extraterrestrial surfaces, addressing the limitations of wheeled rovers in accessing deeper lunar areas such as caves and craters.

Experimental observations of the gripper's sinkage in Toyoura sand, coupled with simulations using Altair MotionSolve and EDEM, provided valuable insights into to the gripper terramechanic performance on granular terrain. The simulation model extended the analysis to lunar gravity conditions, enhancing relevance to actual lunar missions.

The findings indicate that the gripper's sinkage can be predicted and integrated into the motion control algorithms of legged robots, potentially improving their mobility and operational accuracy on granular surfaces. This study lays the groundwork for enhancing the functionality of robotic systems in extraterrestrial environments, with future research aimed at optimizing the gripper design and refining the simulation models for broader lunar conditions.

6. Acknowledgement

This work was supported by the Japan Science and Technology Agency "Support for Pioneering Research Initiated by the Next Generation", Asteroid Mining Corporation Ltd., GP-Mech Program of Tohoku University, Japan and the Space Robotics Laboratory of Tohoku University, Japan.

7. Declaration of competing interest

The authors declare that they have no known competing financial interests or personal relationships that could have appeared to influence the work reported in this paper.

8. References

Ai, J., Chen, J.-F., Rotter, J.M., Ooi, J.Y., 2011. Assessment of rolling resistance models in discrete element simulations. Powder Technol 206, 269–282. https://doi.org/https://doi.org/10.1016/j.powtec.2010.09.030

Bekker, M.G., Kitano, M., Kuma, A., 1962. Theory of Land Locomotion: The Mechanics of Vehicle Mobility. University of Michigan Press.

Candalot, A., Hashim, M.-M., Hickey, B., Laine, M., Hunter-Scullion, M., Yoshida, K., 2024. Soft Gripping System for Space Exploration Legged Robots, in: Walking Robots into Real World (CLAWAR 2024). Springer Nature Switzerland.

Coetzee, C., Scheffler, O.C., 2023. Comparing particle shape representations and contact models for DEM simulation of bulk cohesive behaviour. Comput Geotech 159, 105449. https://doi.org/https://doi.org/10.1016/j.compgeo.2023.105449

Di Carlo, J., Wensing, P.M., Katz, B., Bledt, G., Kim, S., 2018. Dynamic Locomotion in the MIT Cheetah 3 Through Convex Model-Predictive Control, in: 2018 IEEE/RSJ International Conference on Intelligent Robots and Systems (IROS). pp. 1–9. https://doi.org/10.1109/IROS.2018.8594448

Graf Nicole M. and Behr, A.M. and D.K.A., 2019. Crab-Like Hexapod Feet for Amphibious Walking in Sand and Waves, in: Martinez-Hernandez Uriel and Vouloutsi, V. and M.A. and M.M. and A.M. and P.T.J. and V.P.F.M.J. (Ed.), Biomimetic and Biohybrid Systems. Springer International Publishing, Cham, pp. 158–170.

Hirose, S., Umetani, Y., 1978. The development of soft gripper for the versatile robot hand. Mech Mach Theory 13, 351–359. https://doi.org/https://doi.org/10.1016/0094-114X(78)90059-9

Ishigami, G., Miwa, A., Nagatani, K., Yoshida, K., 2006. Terramechanics-based analysis on slope traversability for a planetary exploration rover. Proceedings of the Twenty-Fifth International Symposium on Space Technology and Science, Kanazawa 2006.

Kobayashi, T., Fujiwara, Y., Yamakawa, J., Yasufuku, N., Omine, K., 2010. Mobility performance of a rigid wheel in low gravity environments. J Terramech 47, 261–274. https://doi.org/https://doi.org/10.1016/j.jterra.2009.12.001

Komizunai, S., Konno, A., Abiko, S., Uchiyama, M., 2010. Development of a static sinkage model for a biped robot on loose soil, in: 2010 IEEE/SICE International Symposium on System Integration. pp. 61–66. https://doi.org/10.1109/SII.2010.5708302

Li, S., Milliken, R.E., 2017. Water on the surface of the Moon as seen by the Moon Mineralogy Mapper: Distribution, abundance, and origins. Sci Adv 3, e1701471. https://doi.org/10.1126/sciadv.1701471

Marvi, H., Gong, C., Gravish, N., Astley, H., Travers, M., Hatton, R.L., Mendelson, J.R., Choset, H., Hu, D.L., Goldman, D.I., 2014. Sidewinding with minimal slip: Snake and robot ascent of sandy slopes. Science (1979) 346, 224–229. https://doi.org/10.1126/science.1255718

Nagaoka, K., Sawada, K., Yoshida, K., 2020. Shape effects of wheel grousers on traction performance on sandy terrain. J Terramech 90, 23–30. https://doi.org/https://doi.org/10.1016/j.jterra.2019.08.001

Oettershagen, P., Lew, T., Tardy, A., Michaud, S., 2019. Investigation of specific wheel-terrain interaction aspects using an advanced single wheel test facility. ASTRA 2019.

Parness, A., Abcouwer, N., Fuller, C., Wiltsie, N., Nash, J., Kennedy, B., 2017. LEMUR 3: A limbed climbing robot for extreme terrain mobility in space, in: 2017 IEEE International Conference on Robotics and Automation (ICRA). pp. 5467–5473. https://doi.org/10.1109/ICRA.2017.7989643

Rodríguez-Martínez, D., Uno, K., Sawa, K., Uda, M., Kudo, G., Diaz, G.H., Umemura, A., Santra, S., Yoshida, K., 2024. Enabling Faster Locomotion of Planetary Rovers With a Mechanically-Hybrid Suspension. IEEE Robot Autom Lett 9, 619–626. https://doi.org/10.1109/LRA.2023.3335769

Saranli, U., Buehler, M., Koditschek, D., 2001. RHex: A Simple and Highly Mobile Hexapod Robot. Departmental Papers (ESE) 20.

Uno, K., Takada, N., Okawara, T., Haji, K., Candalot, A., Ribeiro, W.F.R., Nagaoka, K., Yoshida, K., 2021. HubRobo: A Lightweight Multi-Limbed Climbing Robot for Exploration in Challenging Terrain, in: 2020 IEEE-RAS 20th International Conference on Humanoid Robots (Humanoids). pp. 209–215. https://doi.org/10.1109/HUMANOIDS47582.2021.9555799

Welch, R., Limonadi, D., Manning, R., 2013. Systems engineering the Curiosity Rover: A retrospective, in: 2013 8th International Conference on System of Systems Engineering. pp. 70–75. https://doi.org/10.1109/SYSoSE.2013.6575245

Wong, J.-Y., Reece, A.R., 1967. Prediction of rigid wheel performance based on the analysis of soil-wheel stresses part I. Performance of driven rigid wheels. J Terramech 4, 81–98. https://doi.org/https://doi.org/10.1016/0022-4898(67)90105-X

Yang, H., Zhang, C., Ding, L., Wei, Q., Gao, H., Liu, G., Ge, L., Deng, Z., 2024. Comparative study of terramechanics properties of spherical and cylindrical feet for planetary legged robots on deformable terrain. J Terramech 113–114, 100968. https://doi.org/https://doi.org/10.1016/j.jterra.2024.100968

6476 | Improved trafficability over soft soils using ground matting

John F. Rushing, [a,*] Brad Hansen, [a] and Mike Parker [b]

[a] *U.S. Army Engineer Research and Development Center, Vicksburg, Mississippi, United States*
[b] *U.S. Army Engineer Research and Development Center, Hanover, New Hampshire, United States*

* *Corresponding author: john.f.rushing@usace.army.mil*

ABSTRACT

Soft soils pose mobility challenges, even for vehicles designed with superior off-road capabilities. When numerous vehicles travel the same path, permanent deformation of the soil can result in rut depths that exceed vehicle ground clearance. These challenges can be overcome by modifying ground conditions to improve bearing capacity or spreading wheel loads over a greater area. Researchers at the U.S. Army Engineer Research and Development Center conducted field tests to quantify the performance benefits of a ground matting system made of connected fiberglass panels designed to improve vehicle mobility on soft soils. Soil conditions included silt, sand, and highly organic soil with varying strength. Test vehicles included wheeled trucks with gross weights of approximately 6350 kg per axle. Performance of the matting system was assessed by the number of allowable vehicle crossings with and without matting present. Results from testing showed that allowable number of vehicles increased by at least a factor of ten on the weakest soils. Data presented herein includes geotechnical site characterization, soil deformation as a function of traffic, and material characteristics for the fiberglass matting system.

Keywords
Terrain surfacing
Off-road mobility
Site stability
Matting

1. Introduction

Paved or improved road networks are not always available to allow access to remote terrain for industrial or military applications. Off-road mobility for vehicles operating in these areas can be limited when weak soils are present. Engineering effort for improving ground conditions is not always practical from a resource or time perspective. In these cases, temporary roadways provide an optimal solution for accessing desired locations and offer the added benefits of being recoverable and reusable.

Many products have been introduced to civilian and military markets to provide temporary road access. Interconnected wood beams are often used by oil and gas exploration firms to support heavy equipment. Lightweight geotextiles are often used across sandy soil to enhance traction and limit movement of soil grains. Interlocking metal or polymeric planks provide spools of matting that can be rolled across weak soils to spread loads and improve mobility. Each system has desirable characteristics and is selected based on factors such as cost, weight, and time to install depending on the intended use. To provide a solution for a variety of applications, the U.S. Army Corps of Engineers designed a matting system comprised of fiberglass panels that can be joined to emplace temporary roads or staging areas. The system, termed ACE-mat, provides a solution that can be manually installed without the need for material handling equipment and can greatly improve operations over soft soils for heavy vehicles.

2. Background

The ACE-mat system was originally developed in 1999 and patented in the United States in 2004. Preliminary testing using this design was performed by Santoni et. al (2001). They evaluated a variety of approaches for improving mobility over an extremely weak (California Bearing Ratio (CBR) = 0.1-0.5) soil. The ACE-mat was used atop geofoam blocks and stabilized sand to provide additional load carrying capacity. While performance in terms of vehicle pass count was lower than other approaches, they did note that the system had benefits of lower logistics and cost compared to other methods.

Santoni (2003) investigated ACE-mat (described as multi-purpose mat) for enhanced coastal trafficability over sandy soils. The system was compared to hexagonal plastic mats and geofiber-reinforcement. The ACE-mat performed well in this test and withstood the target 500 vehicle crossings from a 5-ton transport truck. ACE-mat was recommended as a potential solution to increase mobility over loose sands as a result of this study. Further testing of mat surfacings for beach roads by Rushing (2007) compared ACE-mat to eight other products and recommended it as a top performer based on its ability to limit soil deformation under traffic and for its ease of installation and relatively low cost. Rushing and Harder (2019) also included ACE-mat in a study comparing multiple matting systems and found it capable of supporting up to 1,000 vehicle crossings over loose sand and 300 crossings over a medium-strength soil.

3. Research approach

This study sought to quantify the benefit of ACE-mat when used as a temporary road surface over a range of soil conditions. The design traffic was a heavy transport truck weighing 29000 kg. Soils included silt, sand, and a highly organic soil intended to

https://doi.org/10.56884/BPMDMI7R

represent a range of soft surface conditions in order to develop guidance on best practices to employ ACE-mat.

3.1. Test conditions

Testing was performed under shelter at ambient conditions in the frost effects research facility (FERF) at the U.S. Army Engineer Research and Development Center, Cold Regions Research and Engineering Laboratory in Hanover, New Hampshire, United States. Site preparation took place from July to September, and traffic testing was completed in October 2023.

The soil test section was constructed inside the FERF which was designed to be a drive-through test facility with two 6.4 m wide lanes. This testing used the one entire lane in order to allow for adequate lengths of each test soil and the drive-through capability needed for continuous trafficking of the soils. Three 19.8 m soil test sections were constructed one after the other with a 10.7-12.2 m transition in and out of the lane to form one continuous surface approximately 82.3 m long.

Test sections were constructed in a concrete basin from the bottom up with each section being at least 2.4 m deep to ensure proper construction and to allow for natural moisture wicking from the bottom if required. The base layer consisted of a series of pipes running both transverse and longitudinal in the lane with stand pipes coming to the surface to allow water to be added, if required. A silty gravel with a Unified Soil Classification System (USCS) designation of GM was then added in 20.3-25.4 cm lifts and compacted to 97% (+/- 2%) compaction. This was followed by a poorly graded sand with silt (SP-SM) also compacted in 20.3-25.4 cm lifts at 97% (+/- 2%) at 8-9% moisture content followed by each of the test surface materials which included silt (ML), poorly graded sand with silt (SP-SM), and peat mixed with SP-SM to create a 26.5% organic content organic silt material (OL). Since organic soils are typically softer and experience higher vehicle rutting, that test soil had higher total layer thickness (0.9 m) compare to the SP-SM and ML test sections (0.75 m). Each of the test soils were also placed in 20.3-25.4 cm lifts at 97% (+/- 2%) compaction at 6-7, 8-9%, and 8-9% moisture contents, respectively. Once construction was finished the soils were ripped to a depth of 45.7 cm then tilled to approximately 22.9 cm to create a soft surface.

3.2. Soil characterization

Post-construction site characterization included a suite of portable tools for providing index values correlating to soil trafficability. Results are presented herein for cone penetrometer, Clegg impact tester, and lightweight deflectometer tests.

A Field Scout SC900 cone penetrometer with a 30° cone having a 322 mm^2 area was used to provide a vehicle trafficability index for the upper 150 mm of the soil layer. A Clegg soil impact tester was used to measure Clegg light impact values from a 0.5kg falling mass in accordance with ASTM D5874. A Zorn Laboratory Lightweight Deflectometer was used in accordance with ASTM E2835 to provide deflection measurements from a 10 kg falling mass and 300 mm plate area. These deflection data were then used to calculate the impulse stiffness modulus (ISM).

3.3. Fiberglass matting

The ACE-mat is a matting system which uses square fiberglass panels arranged in a grid to create a road surface. The current version of the mat contains holes placed in corners to allow connection to adjoining mats using a mat connector or by using 1.3 m screw anchors that also secure the matting to the ground.

Each mat has a useable surface of 1.83 m by 1.83 m, with 20.3 cm of overlap or underlap on each side, providing a logistics footprint of 2.03 m by 2.03 m. Holes are located in the corners to allow anchors to be placed and to connect the mats. Depending on the application, not all holes may be used. Elongated handholes on two sides of each mat allow for easy carrying. Each panel weighs approximately 54.4 kg and can be carried and emplaced by two people. The top surface of the mat has an aggressive diamond-shaped embossed pattern for traction. The mats were placed by hand, and mat connectors were installed with an electric impact drill and fabricated socket. The entire length of mat was emplaced within approximately 5 hours.

For the purpose of this study, mats were used to create a road surface, two panels wide, across the entire test area. *Fig. 1* shows mat panels with connectors after installation on the three soil types.

Fig. 1. ACE-mat installed over soft soil, insert to show the connection method

3.4. Performance testing

Performance testing involved cyclic traffic application across the test lane to identify the number of vehicle passes achieved before soils deformed to a depth of 150 mm or until the desired pass level was reached. Trafficking used two test vehicles to include a light wheeled truck around 2700 kg and a heavy wheeled truck around 29000 kg. Direct comparison among traffic vehicles was not a specific objective of the study. Rather, the goal of the effort was to determine relative performance of the matting system under various soil conditions and the benefit obtained when heavy truck traffic was expected.

Traffic was applied initially on bare soil using a light transport truck for a total of 250 passes. At intervals of 1, 5, 10, 25, 50, 100, 200, and 250 passes, traffic was paused to measure rut depth by means of a straight edge and ruler. This allowed for a comparision to show how well the matting performed throughout trafficking. Soils were ripped and tilled after bare soils testing was finished but before the matting was installed on 19-20 October 2023.

Initial trafficking on the matting began on 21 October. Traffic was applied initially using a lightweight transport truck for a total of 200 passes. At intervals of 1, 5, 10, 25, 50, 100, and 200 passes, traffic was paused to measure rut depth. The test vehicle was stopped on the matting to deflect the mat and cause contact with the soil. The light truck was used to ensure rapid soil deformation did not occur. After the materials demonstrated the ability to withstand the load, a heavy truck was used to continue traffic. Soils were ripped and tilled again at the completion of the light truck mat testing before starting the bare soils heavy truck trafficking. Rut depth measurements were taken at traffic intervals of 1, 5, 10, 15, 20, 25, 30, and 35 on the bare soil to capture the rapid rate of soil deformation. Soils were ripped and tilled at the completion of the heavy truck bare soils testing before installing the matting. On the soil covered with matting, traffic intervals selected for measuring rut depth were spaced further apart because rutting did not progress as quickly. Traffic was paused at intervals of 1, 5, 10, 25, 50, 100, 200, 250, and 350 for these measurements.

4. Results

Capturing soil trafficability improvements resulting from the ground matting system provided relative indices to quantify benefits for three different soils and two types of truck traffic. Soil characterization data informs the range of conditions for which the matting is beneficial and continues to build a database for establishing usage guidance. Soil tests were performed before and after traffic was applied. When testing the matting, measurements were taken immediately prior to mat placement and immediately after it was removed. Data are categorized as "bare soil" associated with control tests and "matting area" referring to areas which received the mat. No soil tests (i.e. Clegg, DCP, LWD) were performed on the mat surface. Pertinent results from testing are described below.

4.1. Soil characterization

Cone index test results are given in *Table 1*. These results are the average of three readings. Because the same soil area was used for "control" and mat testing, data was collected prior to each event to ensure conditions were similar. Although the same tilling procedure was used to prepare the test site, some variation in surface strength is expected due to equipment operators.

The peat soil had consistently lower cone index values compared to other soil types while the silt showed highest resistance to penetration. The values obtained before traffic ranged from 41 to 752 kPa, all of which suggest loose soil susceptible to deformation from ground vehicle traffic.

After traffic, all items showed an increase in cone index value. The magnitude of the increase was correlated to the vehicle type, with areas trafficked by the heavy truck showing greater increases. This observation was expected since the heavier vehicle has the capability to further compact the soil. The ML soil had the highest cone index value prior to traffic and afterwards for the light truck area; the SP-SM soil, however, had higher post-traffic reading in the heavy truck area.

Table 1. Cone Index (0-150 mm, kPa)

		bare soil			matting area		
		ML	SP-SM	OL	ML	SP-SM	OL
Light truck	before traffic	752	476	110	324	103	41
	after traffic	3013	2199	1620	1862	1358	758
Heavy truck	before traffic	352	83	145	393	159	117
	after traffic	2144	2510	2013	3130	3578	2523

The Clegg light impact value is selected as an index for comparing soil strength in this study based on the soil types. The test standard is applicable when using a lighter 0.5 kg hammer to evaluate lower strength soils such as fine grained cohesionless, highly organic, saturated, or highly plastic soils having a maximum particle size less than 9.5 mm. According to Clegg impact tester results (*Table 2*), the peat soil showed the lowest deflection while the ML and SP-SM soils were similar to each other. All values ranged from 0.7 to 3.1, indicating a low-stiffness soil. Clegg impact values (CIV) values in this range correlate to very low CBR values (Al-Amoudi et. al 2002) which would indicate a limited number of vehicle passes would be expected before significant deformation.

The Clegg light impact values showed noticeable increases after trafficking for all soils. Areas trafficked with the light truck showed lower stiffness increases for the area covered with matting. This fact indicates that the matting distributed some of the load from the light truck. On the other hand, stiffness values for areas trafficked with the heavy truck increased comparably for soil with or without matting. This may be attributed to the significantly higher number of heavy truck passes on the mat covered soils versus the bare soils tests.

Table 2. Clegg Light Impact Value

		bare soil			matting area		
		ML	SP-SM	OL	ML	SP-SM	OL
Light truck	before traffic	11.6	6.7	2.7	6.5	5.6	2.0
	after traffic	23.5	36.0	8.2	17.8	8.0	4.2
Heavy truck	before traffic	3.8	6.0	1.8	6.0	5.1	1.8
	after traffic	31.1	28.5	9.3	28.7	35.1	12.5

Data from the lightweight deflectometer (LWD) were used to calculate the ISM value as an index of soil stiffness relating to trafficability (*Table 3*). Ordaz et al. previously reported that ISM values calculated from LWD measurements lower than 29 kN/mm correlate with road structures incapable of supporting more than 1000 load repetitions when trafficked by heavy trucks (4 axle; 26308 kg). All ISM values collected during this test fell far below that threshold; thus, relatively high deformation was anticipated.

Prior to traffic, the ISM values for the SP-SM soil were slightly higher than those measured for the ML soil. With the exception of the test using the light truck where matting was placed, the OL soils had the lowest ISM. The LWD results after trafficking the soils showed trends less consistent with other devices. In most cases, the ISM increased after traffic. There were cases, even with the heavy truck, that showed decreases in ISM after traffic. The trends were not consistent, regardless of whether matting was present of not. While the LWD indices were in a range that suggest that the soil should deform under the vehicle loads, the device did not do a good job of ranking them.

Table 3. ISM value (kN/mm)

		bare soil			matting area		
		ML	SP-SM	OL	ML	SP-SM	OL
Light truck	before traffic	2.2	3.1	0.7	1.2	1.6	1.3
	after traffic	5.5	12.5	1.5	6.0	3.7	0.9
Heavy truck	before traffic	2.1	2.9	0.7	1.8	2.1	0.7
	after traffic	1.3	8.5	1.2	1.8	11.2	2.1

4.2. Performance testing

The performance testing data provided the most practical assessment to quantify benefits of the matting system for improving mobility. *Fig. 2* shows the measured rut depth versus number of vehicle passes for the light truck. These rut depths were measured over the matting by deflecting the material under load from the truck. It was suspected that the mat was in contact with the soil at this time, but this assumption cannot be guaranteed.

Total number of vehicle passes applied on the bare soil was 250, whereas this number was reduced to 200 in areas with matting due to time constraints during testing and the rut depth remaining consistent. It was observed that all bare soils rapidly deformed under the loads of the light truck. However, this rate of deformation was dependent on the soil type. The OL soil deformed more quickly than the others during the first few traffic passes. The ML soil deformed at a greater rate than other soils and continued to show deformation throughout testing. Both the SP-SM and OL reached most of the deformation within the first 20 passes of the light truck.

The area covered with matting showed some immediate deformation, but this was limited throughout testing. After 200 light truck repetitions, measured rut depths were not increasing and were limited to around 4 cm, which was much lower than on the areas of bare soil.

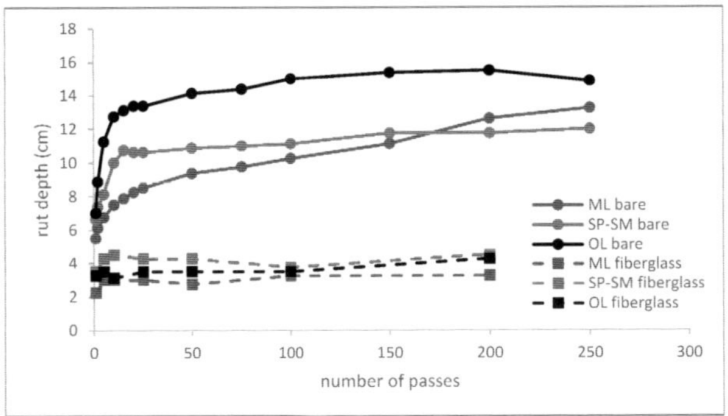

Fig. 2. Rutting versus pass count for light truck during Test 1

Fig. 3 shows the rut depth versus number of passes for the heavy truck traffic. The traffic was ceased on the areas of bare soil after 35 passes due to the fact that the measured rut depths were all over 20 cm, which was above the failure threshold. The OL soil showed similar trends to the results from light truck traffic in that it deformed at a higher rate and achieved deeper rutting than the other soils. Also, the ML soil had lower rut depths but continued to deform at a significant rate while the other soil types showed a decrease in the rate of rutting after 20 passes.

The soil covered with matting received ten times the number of heavy truck passes while showing limited rutting. Even after 350 passes, measured rut depths were below 10 cm for all soil types. Most of the rutting occurred from early densification of the soil under traffic and did not continue to accumulate. Overall, the matting provided enhanced mobility in each test area.

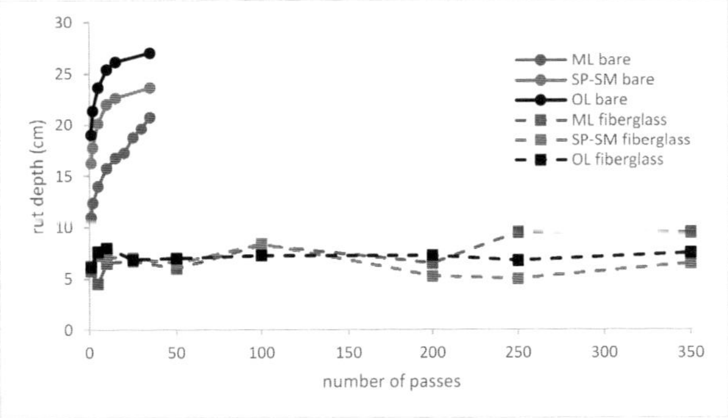

Fig. 3. Rutting versus pass count for heavy truck during Test 1

5. Conclusions and future work

This series of tests provided relative performance data to capture mobility improvements when fiberglass panels were used as temporary road matting. The observed rut depth on areas with matting was approximately 1/3 of that for bare soils. Most of the rutting in areas with matting occurred early in the traffic cycle, and the total allowable number of vehicle passes was greatly increased by the use of the matting.

While this effort identified that mobility matting provides substantial benefit over a range of soil conditions, additional work is recommended. First, a family of performance curves need to be developed that capture anticipated allowable vehicle pass count as a function of soil properties for a range of vehicle classes. Reliably creating these curves will require additional testing on other soil types and traffic with different vehicles. Further, a field soil test or measuring device is needed to provide a soil strength index that can be correlated to the expected number of vehicle passes. A range of soil tests, such as cone penetrometers and vane shear devices should be evaluated to determine which has the best correlation to matting performance.

6. Acknowledgements

The use of trade, product, or firm names in this document is for descriptive purposes only and does not imply endorsement by the U.S. Government. The tests described and the resulting data presented herein, unless otherwise noted, are based upon work conducted by the U.S. Army Engineer Research and Development Center supported under PE 0633119, Project BL8, "Power Projection in A2/AD Environments," Permission was granted by the Director, Geotechnical and Structures Laboratory to publish this information. The findings of this document are not to be construed as an official Department of the Army position unless so designated by other authorized documents.

7. Declaration of competing interest

The authors declare that they have no known competing financial interests or personal relationships that could have appeared to influence the work reported in this paper.

8. References

Al-Amoudi, O.S.B., Asi, I.M., Wahhab, H.I.A.A. and Khan, Z.A., 2002. Clegg hammer—California-bearing ratio correlations. Journal of materials in civil engineering, 14(6), pp.512-523.

Ordaz, M., Doyle, J.D., Howard, I.L., 2023. Light Weight Deflectometer Evaluation of Low-Volume Road Structural Deterioration under Rapidly Increased Traffic Patterns. Transportation Research Record, 2678(4), 948-966.

Rushing, J.F. and Harder, D.E. 2020. Improved Vehicle Mobility by Using Terrain Surfacing Systems. ERDC/GSL MP-20-1, Vicksburg, MS.

Rushing, T.S., J.S. Tingle, and Q.S. Mason. 2007. Evaluation of Expeditionary Mat Surfacings for Beach Roads. ERDC/GSL TR-07-1, Vicksburg, MS.

Santoni, R. 2003. Enhanced Coastal Trafficability: Road Construction Over Sandy Soils. ERDC/GSL TR-03-7, Vicksburg, MS.

Santoni, R., C.J. Smith, J.S. Tingle, and S.L. Webster. 2001. Expedient Road Construction Over Soft Soils. ERDC/GSL TR-01-7, Vicksburg, MS.

7322 | Lateral tyre characterization rolling tyre vs. static tyre testing

Carl M. Becker [a,*], P. Schalk Els [a]

[a] University of Pretoria, Pretoria, South Africa

* Corresponding author: carl.becker@up.ac.za

ABSTRACT

Tyre models used in soft soil simulation analysis requires tyre parameters in the form of stiffness in multiple directions. These parameters are obtained from measurements on hard terrain as these parameters are a function of the tyre carcass construction. Vehicles used on soft terrain are also used on hard terrain.

Many off-road vehicles used in construction, mining, agriculture and forestry use large tyres operating under heavy loads. Testing of large tyres is not a trivial or inexpensive exercise and outdoor testing has limitations on repeatability and load application. This is the case for testing on soft terrain/soil and hard terrain, thus it is preferred to conduct laboratory tests when characterizing tyres as higher loads can be applied and conditions can be controlled. This study investigates the lateral tyre characteristics measured during dynamic/rolling and static tyre tests.

Tests are conducted on the actual concrete surface of interest, typically used in field tests with the use of a Dynamic Tyre Test Trailer. Static tests are conducted on the same tyre over representative surfaces in a laboratory with the use of the Static Tyre Test Rig. Multiple tyres are tested and measured tyre characteristics compared. The data can be used to parameterize tyre models of special, large off-road tyres.

Keywords
Tyre characterization
Friction coefficient
Tyre stiffness
Field testing
Laboratory testing
Surface roughness
Large tyres

1. Introduction

Over the last two decades most of the design methodologies used in passenger car development is being increasingly implemented in the detail design and simulation of large off-road vehicles. This includes construction, mining and agricultural vehicles amongst others. Evolving vehicle design technology incorporates simulation up to the point where homologation of new vehicles is possible based on simulation results, instead of real-world testing (TüV SüD, 2021). Simulation results are however only as reliable as the input data which starts at the tyre. In vehicle ride or handling simulations the primary excitation into the vehicle is via the tyres, thus the importance of having reliable tyre data for accurate tyre modelling and vehicle simulation. In most cases tyres are the only suspension on many agricultural, mining and construction vehicles, thus the importance of accurate tyre models. With the expansion of farms, agricultural vehicles are used more on asphalt roads and needs to be safe in an evasive maneuver or during an emergency brake scenario. As a result, designers need to simulate these maneuvers to improve the safety of these vehicles. Stayner et al., (1984) and Lines (1991) indicated that poor information about the agricultural tyre's behavior over any terrain has a very large effect on the simulation errors. The advances made in tyre technology to date have placed a large question mark on the reliability on the limited historical collection of tyre data available especially in the agricultural and construction industry. Other vehicle factors that are affected by the tyre characteristics range from off-road traction and soil compaction to on-road handling, ride comfort, braking and motion resistance. All of the on-road characteristics of a tyre are used on off-road applications and both are limited to the friction coefficient between the tyre and terrain. However, most of the time the tyre is operating in the linear region of the tyre characteristic.

Obtaining tyre data is not a trivial or inexpensive exercise, as a result, tyre manufactures are reluctant to freely supply tyre data. Test equipment suitable for large agricultural and off-road vehicles are very limited due to the size of the tyres, the required operating load and logistics around testing these large tyres at the designed operating conditions.

2. Tyre models

Every tyre model used in simulations requires different sets of parameterization data. Different tyre parameterization test methods exist and vary from laboratory tests to field tests on different surfaces. Tyre parameters include longitudinal, lateral and vertical stiffness which are dependent on the inflation pressure and carcass construction. The tyre-terrain interaction parameters are dependent on the friction coefficient between the tyre and terrain, which is dependent on the material properties of the rubber compound and tread pattern, which directly effects the motion resistance and maximum traction available. Each test surface has different surface roughness and will result in different measured tyre characteristics related to maximum longitudinal or lateral forces generated by the tyre; however, the longitudinal and lateral stiffness should remain the same as this is determined by the carcass construction and inflation pressure.

Tyre models fall broadly into two types of models namely:

- Empirical or semi-empirical models, for example a Pacejka/ Magic Formula model, Hybrid Soft Soil Tire Model, etc.

https://doi.org/10.56884/AU5QX0XH

- Physics based models, for example FTire, RMOD-K, Hohenheim, etc.

The Pacejka tyre model, also known as the Magic Formula, MF, was formulated by Dr. H.B Pacejka in the 1980s (Bakker, et al., 1987). The model is widely known and used for handling simulations on smooth terrains, but has limitations when it comes to ride comfort simulations (Becker, 2009). The model is semi-empirical and completely dependent on experimental data, obtained from a rotating wheel, for parameterization. The following data sets are the minimum required to parameterize a useful Pacejka model:

- Longitudinal force vs. % longitudinal slip at three vertical loads
- Lateral force vs. slip angle at three vertical loads
- Vertical stiffness and damping at three vertical loads.

Curve fits are made to the test data and the curve fit coefficients are used in the simulation model. Changes in tyre operating conditions, for example a different inflation pressure, requires that a new MF model needs to be parameterized. If it is possible to use static tyre tests for parameterization data it will enable the agricultural, construction and mining industry to obtain multiple affordable and accurate tyre models.

The FTire model is a physics-based flexible structure tyre model developed by cosin scientific software (2017). The model can be used for handling and ride comfort simulations on rough terrain. The FTire model can be parameterized at multiple inflation pressures with the use of static tests to obtain vertical stiffness (on flat surface and cleats), lateral stiffness and longitudinal stiffness together with footprints and tyre geometry. Modal analysis will contribute to model parameterization but it is not an essential requirement. Damping over cleats and dynamic tests in the form of lateral force vs. slip angle and longitudinal force vs. % longitudinal slip are used for model validation. The required tests can be conducted on a flat track or drum tester. FTire is a very flexible/adjustable tyre model and enables the user to adjust the amount of tread on the belt and inflation pressure amongst other parameters. However, it is not always possible to obtain all of the required parameterization data on larger tyres due to the increased size and load required to test these tyre at representative operating loads.

3. Tyre parameterization tests

Tyre characteristics can be measured using multiple methods. Each method has different constraints and have their own set of positive and negative features. This is the case for passenger cars, sports cars, motorcycles, commercial trucks, heavy duty trucks, agricultural, mining and construction vehicles. Many test laboratories are spread worldwide and all have the capability to parameterize relatively small tyres. Testing larger tyres has its own challenges as one needs large amounts of driving power, logistics and infrastructure to test these tyres at the designed loads. Typical indoor laboratory test rigs include drum test rigs and flat track rigs. Some test rigs can only be used for traction characterization, where other test rigs can conduct lateral force vs. slip angle tests, motion resistance measurements or even simulate track racing conditions. These laboratory test rigs tend to have different test surfaces (corundum surfaces/sandpaper or steel belts) and on some of these test rigs the test surface can be changed or contains a soil bin.

The Vehicle Dynamics Group (VDG) in the Department of Mechanical and Aeronautical Engineering at the University of Pretoria in South Africa, has developed a Dynamic Tyre Test Trailer (DTTT) and a Static Tyre Test Rig (STTR) as described by Becker and Els (2018). The DTTT tests rolling tyres with diameters larger than 800mm and up to 2000mm, where two of the same tyres are tested in opposed lateral slip directions up to a maximum vertical load of 50kN per wheel. The DTTT can measure tyre forces and moments for lateral force vs. slip angle measurements, longitudinal brake force vs. longitudinal slip measurements as well as combined lateral and longitudinal force measurements during braking at pre-determined lateral slip angles. The STTR can measure the static stiffness of a tyre in the longitudinal or pure lateral direction at a predetermined vertical load applied to the tyre on test surfaces that are interchangeable. The STTR can test up to 250kN applied vertical loads and can accommodate 3500mm diameter tyres.

The University of Hohenheim, has multiple large tyre test rigs in the form of a flat belt test stand, single wheel test trailer and an instrumented test tractor as presented by Witzel (2018). These test rigs/platforms can be used to measure longitudinal forces, lateral forces and motion resistance on larger agricultural tyres. Other single wheel test trailers have been built for agricultural tyres, used and upgraded from early 1960's to present. These trailers can be used on any terrain. Billington (1973) introduced the NIAE MkII single wheel tester in 1973. Ambruster and Kutzbach (1989), developed a single wheel tester which parameterized a driven wheel at pre-set slip angles. Shmuleviuch et al. (1996) presented a field single wheel tester. These trailers can be used on any terrain, however the size of the test tyres and the vertical load applied to them, with the use of these test rigs are limited. Both Kawase et al., (2006) and Phakdee and Suvanjumrat (2023) has developed tire testing machines to evaluate the performance of tractor tires based on the soil compaction. These test rigs are designed for smaller agricultural vehicles used in Asia and are limited to low vertical loads. It is however possible to obtain static parameterization data on larger tyres with the use of the STTR at the University of Pretoria. This raised the question of how static parameterization data compares to dynamic parameterization data. Tests were conducted on a 280/70R16 agricultural tyre, a 235/85R16 SUV tyre and a 225/45R18 passenger car tyre. Although these are relatively small tyres, the investigated principles can be used on large diameter tyres for future research. The findings of this analysis are discussed in this paper.

4. Parameterization Test Results

In this section the different parameterization test results are presented and compared.

4.1. Static/non-rolling tyre

A large number of factors influence the characteristics of a tyre. This includes, but are not limited to, operating temperature, inflation pressure, rubber compound, terrain/test surface, vertical load, tread condition etc. Some of these factors can be controlled by the operator, designer and tyre manufacture, however none of them has control over all of these factors. As an example, Fig 1 shows the large differences in measured lateral tyre characteristics on the agricultural tyre in question, with the use of the STTR, by only changing the roughness of the test surface during static testing. It can be seen that the lateral stiffness of the tyre is not affected by the friction coefficient of the test surface. The test is conducted by applying a predetermined vertical load to the tyre, followed by a pure lateral displacement of the road plate. The tests conducted on concrete will be used during the analysis in this study as the roughness of the concrete corresponded to the surface on the outdoor test track.

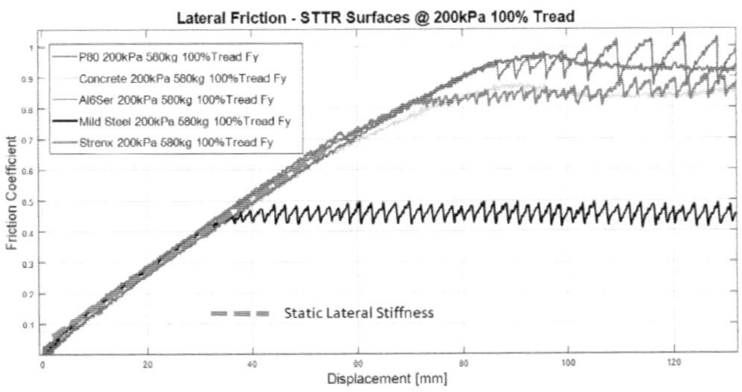

Fig 1: Lateral tyre characteristics of the agricultural tyre for different test surfaces with static lateral stiffness indicated.

4.2. Dynamic/rolling tyre

Dynamic/rolling tyre characterization tests were conducted with the use of the DTTT in order to obtain the lateral characteristics of a tyre on concrete. Test were conducted at inflation pressures of 80kPa and 200kPa and a speed of 4km/h, with a vertical load of 5.68kN. The test is conducted by driving at a constant velocity with the applied vertical load followed by changing the slip angle of the tyre. Force measurements were recorded with the use of wheel force transducers. The test conditions were representative operating conditions of the agricultural vehicle on which the tyre of interest is typically fitted to.

An example of the unfiltered lateral friction coefficient vs. slip angle test data is shown in Fig 2. The effect of the lugs in the tread pattern and radial runout are clearly seen in the vertical force oscillations in the data.

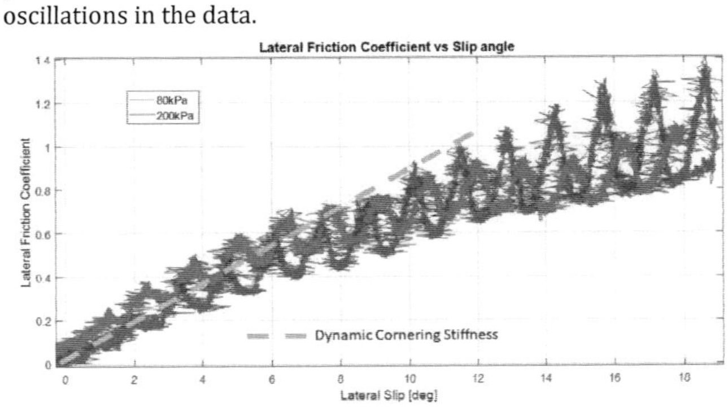

Fig 2:Unfiltered lateral friction coefficient vs slip angle measurements on Trelleborg TM700 280/70R16 at 4km/h, at 80kPa and 200kPa inflation pressure, with dynamic cornering stiffness indicated.

4.3. Static/Non-rolling vs. Dynamic/Rolling

Outdoor dynamic tyre tests are time consuming, expensive and less repeatable for testing over different times of year when compared to tests conducted in a laboratory. Also, tyre size/load limitations are a major factor when conducting dynamic/rolling tests on large tyres. If static test rig results could be used to, at least estimate, tyre parameters that represent dynamic/rolling characteristics, it can become more affordable to obtain tyre characteristics in order to parameterize more tyre models. In a previous study by Becker and Els (2022), the dynamic tyre characteristics were transformed to static characteristics with the results correlating well. The lateral friction coefficient vs. slip angle tests were translated to the lateral friction coefficient vs. pure lateral displacement with the use of trigonometry to calculate the lateral displacement component of the contact patch, $Slip_{Lat\_mm}$, at the specific measured lateral slip angle, β, and the contact patch length, L, as shown in Eq. 1:

$$Slip_{Lat\_mm} = Ltan\beta \tag{1}$$

A more practical application will be to transform the static lateral stiffness (lateral friction coefficient vs. pure lateral displacement) to dynamic/rolling tyre cornering stiffness (lateral friction coefficient vs. slip angle).

The estimated slip angle, α_{est}, is calculated with the use of the contact patch length, $Contact_x$ as a function of the applied vertical load, the measured lateral displacement during the static test, $Disp_y$ and the component, A, related to the lateral stiffness of the tyre, which is a function of the contact patch length. This transformation equation is shown in Eq. 2.

$$\alpha_{est} = A \sin^{-1}\frac{Disp_y}{Contact_x} \tag{2}$$

The result of the comparison between the transformed static tests on the STTR on a concrete surface, with the dynamic/rolling tyre tests from the DTTT on the 280/70R16 agricultural tyre is shown in Fig 3. The transformation on the 235/85R16 SUV tyre is shown in Fig 4.

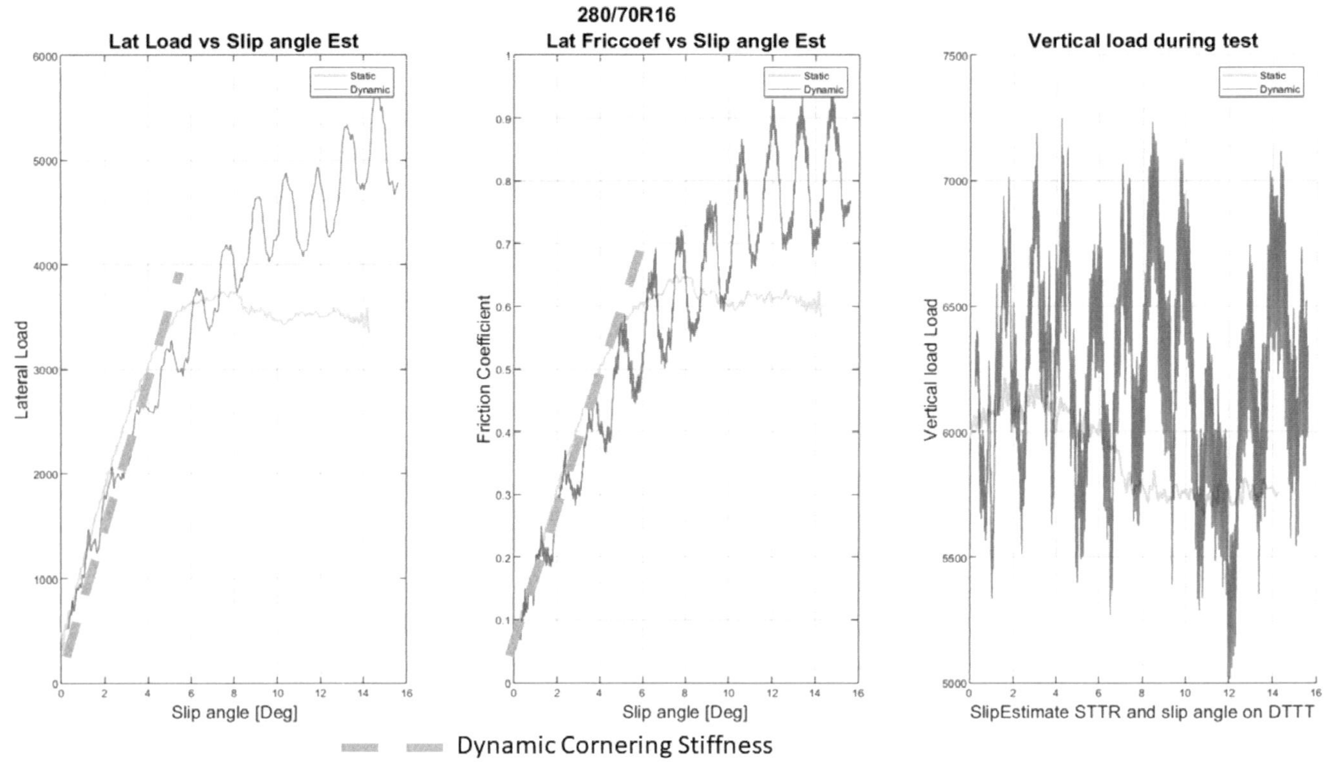

Fig 3: Static lateral stiffness transformed to dynamic cornering stiffness on 280/70R16 agricultural tyre.

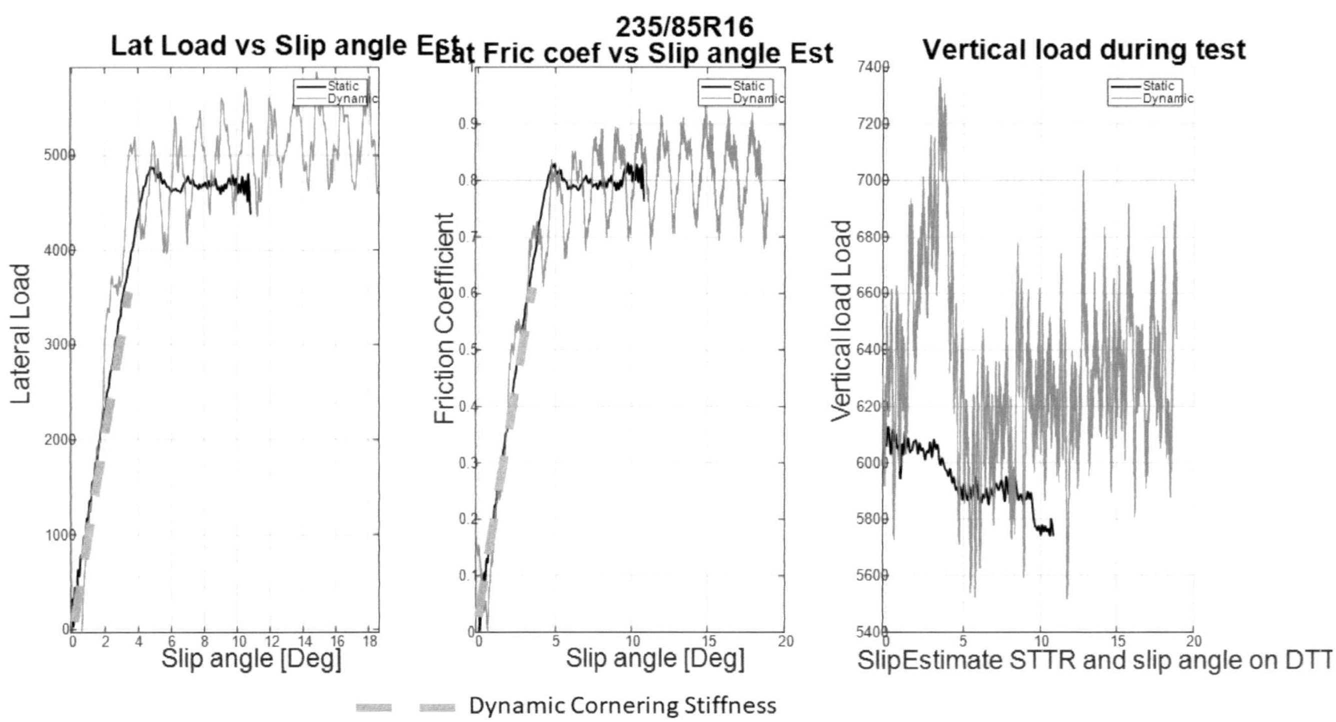

Fig 4: Static lateral stiffness transformed to dynamic cornering stiffness on 235/85R16 SUV tyre.

Very good correlation in cornering stiffness is seen in the transformation from pure lateral displacement during the static tests to slip angle in the dynamic tests. Although the maximum friction coefficient did not correlate on the agricultural tyre, as shown in Fig 3, the lateral stiffness measured with the static tests, represented by the green and black lines in Fig 3 and Fig 4, respectively, correlates very well with the dynamic measurements on both tyres. The maximum friction coefficient is dependent on the condition of the test surface, ambient temperature and temperature of the tyre, compared to the cornering stiffness which should be stable and repeatable as it is a function of the inflation pressure and tyre carcass construction. Typical dynamic cornering stiffness estimations are represented by the dashed green line in both figures. In addition, the same tests were conducted on a passenger car tyre with a low aspect ratio, 225/45R18. The results are shown in Fig 5.

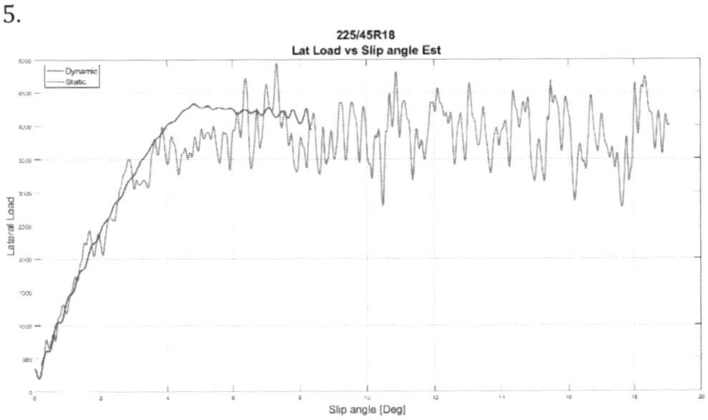

Fig 5: Dynamic vs. Static lateral tyre characterization date from 225/45R18 passenger car tyre.

The results shown in Fig 5 indicate that it is possible to use this transformation from static tests to dynamic tests on tyres with large lugs, as used on agricultural tyres, SUV tyres with high aspect ratios and passenger car tyres with low aspect ratios.

5. Conclusions

The results from the static/non-rolling tyre tests can successfully be transformed to represent dynamic/rolling tyre tests. This indicates that it is possible to obtain accurate tyre characteristics from static tests which are representative of a rolling tyre at low velocities. This is an important observation as most of the large agricultural, mining and construction vehicles operate at low speeds. This will allow engineers to obtain valuable tyre parameterization data at a lower cost with the use of static tyre testing. This data can also be used to parameterize physics-based tyre models for rolling tyre simulations, only when using carefully executed static tests on representative test surfaces. The difference in boundary conditions can be eliminated during tyre model validation as the physics-based tyre model can simulate the boundary conditions and movements of the STTR which is used to measure the tyre characteristics.

Future studies will investigate and validate this methodology on larger tyres.

6. Nomenclature

A	Lateral tyre stiffness component	[]
$Contact_x$	Contact patch length as a function of the applied Fz	[mm]
$Disp_Y$	Lateral displacement measured during static test	[mm]
L	Contact patch length	[mm]
$Slip_{Lat\_mm}$	lateral displacement component	[mm]
α_{est}	Estimated slip angle	[°]
β	Measured lateral slip angle	[°]

7. References

Ambruster, K, Kutzbach, H.D., 1989, Development of a single wheel tester for measurement on driven angled wheel. In: Proceedings of 5th European conference of the ISTVS, Wageningen, The Netherlands; p. 8–14.

Bakker, E, Nyborg, L and Pacejka, H.B., 1987, Tyre modelling for use in vehicle dynamics studies, Society of Automotive Engineers, Warrendale, PA.

Becker, C.M, 2009, Profiling of rough terrain, Master of Engineering Thesis, University of Pretoria, South Africa, http://hdl.handle.net/2263/29833.

Becker, C.M. and Els, P.S., 2018, Static and Dynamic Parameterization Test Rigs for Large Tyres, Proceedings of the 10th Asia-Pacific Conference of the ISTVS, Kyoto, Japan, July 11-13.

Becker, C.M. and Els, P.S., 2022, Tyre Parameterization Tests: Dynamic vs. Static, Proceedings of the 11th Asia-Pacific Regional Conference of the ISTVS, Harbin, China, September 26-28.

Billington, P.W., 1973 The NIAE Mk II single wheel tester. J Agric Eng Res; 18:67–70.

cosin scientific software, 2017, FTire, Modelization and Parameter Specification. 2017-1-r14630. [pdf]. Software documentation and user guide.

Kawase, Y., Nakashima, H. and Oida, A., 2006, An Indoor Traction Measurement System for Agricultural Tires. Journal of Terramechanics, Vol. 43, 317-327.

Lines, J.A., 1991, The Suspension Characteristics of Agricultural Tractor Tyres. PhD Thesis, Cranfield Institute of Technology Silsoe College.

Phakdee, S. and Suvanjumrat, C., 2023, Development of a Tire Testing Machine for Evaluating the Performance of Tractor Tires Based on the Soil Compaction. Journal of Terramechanics, Vol. 110, 13-25.

Shmuleviuch, I., Ronai, D., Wolf, D., 1996. A new field single wheel tester. Journal of Terramechanics, Vol. 33(3), p133–41.

Stayner, R.M., Collins, J.A. and Lines, J.A., 1984, Tractor Ride Vibration Simulation as an Aid to Design. Journal of Agriculture Engineering Res. 29.345-355.

TüV SüD, 2021, Virtual Simulation to Ensure the Safety of Vehicles, website: https://www.tuvsud.com/en/e-ssentials-newsletter/automotive-essentials/e-ssentials-3-2017/virtual-simulation-to-ensure-the-safety-of-vehicles, accessed 5 October 2021.

Witzel, P., 2018. The Hohenheim tyre model: a validated approach for the simulation of high-volume tyres – Part I: Model structure and parameterisation. Journal of Terramechanics. 75, 3–14.

7453 | The impact of changes in the river regime on the mobility of off-road vehicles

Marian Rybansky [a,*], Filip Dohnal [a], Martin Hubacek [a], Jaromir Capek [a], Vladimir Kovarik [a]

[a] *Department of Military Geography and Meteorology, University of Defence, Brno, Czech Republic*

* *Corresponding author: marian.rybansky@unob.cz*

ABSTRACT

The mobility of off-road vehicles, especially military vehicles, integrated rescue system vehicles and civilian vehicles in an open terrain where it is necessary to cross rivers, is dependent on the current hydrological conditions and technical parameters of the vehicles. In addition, we can also calculate the ability of drivers to manoeuvre the vehicle while crossing watercourses.

The article describes the methodology for determining the parameters of changes in the depth and speed of the water flow and includes the results of testing the ability to wade through water bodies or overcome them by swimming amphibious vehicles. The mentioned classification of changes in the characteristics of watercourses is important for a better understanding of the influence of watercourses on the mobility of vehicles in open terrain outside the road network.

Keywords
Off-road vehicle
Military vehicle
Vehicle wading
Water regime changes
Hydrological transformation
River depth
Flow velocity
River bed characteristics

1. Introduction

The mobility of off-road vehicles is dependent on a number of terrain factors that affect the movement of wheeled and tracked vehicles. Important factors include mainly terrain slopes, terrain roughness, soil types, adhesion, rolling resistance, low and high vegetation, water obstacles, settlements, communication, meteorological conditions and others.

There is a large body of literature describing the general methodologies of the cross-country movement evaluation using terrain data (e.g., Affleck et al, 2009; Melloh and Shoop, 2009; Ahlvin and Haley, 1992; Blackmon and Randolph, 1968; Bruzelius et al., 2010; Cibulova, 2017, 2017a; Creager et al., 2015; David et al., 2023; Dohnal et al., 2017; Drozd et al. 2018; Hubacek et al., 2016; Mason et al., 2012; Mazal, et al., 2016; Nohel et al., 2019; Pokonieczny et al., 2018, 2023, 2023a; Rybansky, 2007, 2009a 2009b, 2009c, 2009d, 2010, 2013, 2014, 2018; Shoop, 2006; STANAG 3992, 1990; Stodola and Mazal, 2010, 2010a; Taheri et al., 2015.

One of the most difficult terrain environment that determines the directions of rescue and military operations are watercourses. According to the European Catchment and Rivers Network System database (ECRIN), the main watercourses in Europe form a network of barriers of a total length of 1.65 × 106 km (Belletti, 2020) restricting the off-road movement of military vehicles – see Fig. 1.

Fig. 1. River network in Europe, https://milospopovic.net/map-rivers-with-sf-and-ggplot2-in-r/

The ability of a vehicle to cross a river depends on whether or not it can either ford this obstacle or cross it by swimming. When crossing the river, it is necessary to determine the depth and speed of the water flow, the nature of the bottom and the parameters of the river profile (especially the bank slopes at the entrance and exit of the river). There are not many articles available in the world literature that deal with the analysis of watershed passability and changes in watershed hydrological parameters that have a significant impact on the ability of off-road vehicles to overcome water obstacles. The most significant publications of this nature include HEC-4 Monthly Streamflow Simulation 1971 ed U.S. Army Corp of Engineers; Ahlvin and Haley, 1992; Jacobsen, 1980; Dohnal, 2018; Pokonieczny, 2018a; Murakami, 2017; Sosnowicz and Kosiuczenko, 2021; STANAG 3992, 1990; Yuta, 2017 and Yamada, 2020. The Department of Military Geography and Meteorology, University of Defence in Brno (UNOB), Czech Republic has conducted research, including modelling the impact of terrain on military vehicle mobility, for more than 20 years.

https://doi.org/10.56884/EII98HEJ

These investigations have also included vehicle-river interactions. UNOB has cooperated with the Brno University of Technology (BUT) and Czech Hyrometeorological Institute (CHMI) for many years to determine the impact of rivers on the mobility of military vehicles.

In cooperation with these institutions it is possible to obtain information on the profile characteristics of the rivers of the Czech Republic in real time - see Fig. 2. However, the information obtained from these representative profile characteristics cannot be sufficient to conduct rescue or military operations, where it is necessary to know the possibilities of crossing the river in any section suitable for or wading or swimming by amphibious off-road vehicles.

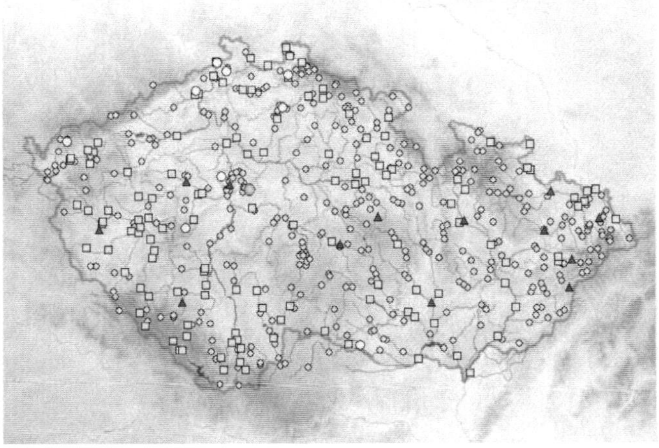

Fig. 2. Density of watercourses and reporting points in the Czech Republic (chmi.cz)

2. Factors affecting the possibility of crossing rivers by off-road vehicles

Basic factors affecting the river crossing ability are at particular:

- overall covering of terrain by drainage;
- drainage structure;
- characteristics of banks of water expenses and watercourses;
- watercourse width;
- depth of watercourses and water expenses;
- character of a bottom;
- water flow rate and flow speed;
- climatic effects;
- mutual position of a drainage and other geographical subjects.

These factors can be divided into those that are relatively stable and those that can be highly variable. Variable factors that significantly determine the ability of off-road vehicles to cross watercourses include, in particular, water depth and flow velocity.

The depth of watercourses and water expenses is very changeable (see also Rybansky et al., 2018). The depth depends on precipitations, flow pattern, i.e. on stability of water sources, on watershed profile, flow speed and character of bottom. The depths of streams, torrents and little rivers come up to 1m rarely. Depth of most rivers varies from values lower than 1 meter up to several meters. At higher water level, the depth rises, water escapes from banks and floods the circumambiency.

To determine critical values of watershed depth for purpose of its passing by military vehicles, we can in a rough way to use table 1. However, these watercourse depths do not include the effect of the nature of the bed. The sediments at the bottom of watercourses mainly affect the rolling resistance and the coefficient of adhesion, which is an important parameter of the traction force that a vehicle can achieve when crossing a watercourse. Character of a bottom depends on layers of underbed of watershed, on the bottom sediment runoff, i.e. on the material (shingle, sand, mud, bottom scour) pushed or towed by stream longways in direction of bed slope. Intensity of these dislocations and thickness of bottom scour are dependable on the bed slope, quantity of water and flow speed. If a watercourse has a muddy bottom, even a shallow river constitutes the NOGO obstacle.

Table 1
The most permissible depths for wading (Vsevojsk military directine)

Units/ Equipment	Wading depths of rivers at flow speed:		
	< 1 ms^{-1}	< 2 ms^{-1}	> 2 ms^{-1}
motorized infantry units	1.00	0.80	0.60
personal vehicle	0.60	0.50	0.40
truck 1.5t up to 2t	0.60	0.50	0.40
truck 3t up to 3.5t	0.80	0.70	0.60
truck 5t	0.90	0.80	0.70
armored personnel carriers	0.70	0.60	0.50
artillery drawn by tracks	1.00	0.90	0.80
medium tanks	1.20	1.10	1.00
heavy tanks	1.50	1.40	1.30

The values of the depths given in Table 1 are, as stated above, indicative and depend both on the specific technical parameters of the vehicles and on the parameters of the particular watercourse profile.

Flow speed (see also Rybansky et al., 2018) depends on water quantity, bed slope, watershed profile, bank roughness and quantity of a bottom scour and suspended load.

Mean speed of flow v is quantity given at meters per second and is determined as arithmetic mean of partial speeds measured at various places of cross profile of a river. Measurements are made e.g. by use of hydrometric wing rotating in water flow. The speed of flow at position of cross profile given is calculated from number of revolutions per time unit. In the hydrological practice, providing stable flow of water, we can express mean profile speed v as follows:

$$v = C.\sqrt{RI} \quad (1)$$

$$C = \frac{1}{n} R^{-\frac{1}{6}} \quad (2)$$

C	Chézy´s factor	
n	(n = 0.020 – 0.050) is the coefficient of roughness	
R	$R = F/O$ is hydraulic radius	[m]
F	flow section line surface	[m^2]
O	wetted perimeter of watershed	[m]
I	longitudinal inclination of level	[°]

By the crossing of a watercourse, we have to consider that the flow speed at different points of a river is not constant. The lines joining points with equal speed follow approximately the cross section line of the watershed bottom. The speed of flow has the highest value at direct sections upon the deepest point at the centre of flow profile just below the water level and decreases in direction to banks and bottom.

The line joining points with maximum speed is stream contour line. Mean speed on the entire flow profile is about 30% lower than the speed measured in stream contour line. The speed of flow measured in mountain gills is 5 – 7 ms^{-1}, at floods up to 12 ms^{-1}. The speed of flow of lowland watercourses varies from 0.5 up to 1.5 ms^{-1}. The flow speed rises at high water level and floods up to 3 ms^{-1} and even more.

To evaluate the pattern of watercourse, we have to determine the movement of water at given section, which we have to pass over. At steady-state motion in profile given, the speed of flow is stable and constant. At sections where the cross profile or the watershed inclination are variable, the flow speed is unequal and unstable.

To evaluate water level and its changes, there are stream gauging stations. They are placed so they can obtain the best possible realistic characteristics of watercourse pattern. Stream gauging stations at important places are fitted with recording devices (limnigraphs) to record continuous changes of water level in time.

For the determination of critical flow speed of watercourse to pass it by military vehicles we can use the data contained in table 1.

The abovementioned factors are subjects of changes (depth, width and flow speed at particular). It is very important to determine possible future change of these values from the point of view of a vehicle passability. If causal precipitation does not come after a certain period without precipitation on the drainage area, that change is very slow, water flow rates and water levels have decreasing trend. It is because of fact that only groundwater discharge of drainage area and the groundwater resources in drainage area decrease generate the water flow in watercourses gradually. For each drainage area and profile relevant, it is possible to construct a so-called depletion curve that has exponential form and, according to this curve, we can predict relatively accurately the future water flow rate in a certain period.

If we have no other information about the watercourse, we can very roughly estimate the flow rate by its long-term value *Q*a. With utilization of stage-discharge relation curve we can read the corresponding water level from the water flow in section given (depth of water over the lowest point of bottom) and, by known form of profile, to read the entire remaining attributes. In discharge section lines, we also know the relation between mean section speed and water level.

A more difficult situation occurs by the attempt to estimate mentioned attributes after precipitation or, to estimate future flow rate after predicted rainfall, since each predicted flow rate is affected by the course of both previous and future rainfalls. What part of both previous and future rainfalls will obtain, this amount depends on position of profile in drainage. Quantity of predicted flow rates can be undersized, but it can reach the dimensions of disaster. That all depends on intensity and space distribution of previous and future rainfalls and on soil infiltration ability of a drainage area.

The process of precipitation and discharge shows progressive transformation of drainage precipitation in drainage up to discharge by closing hydrographic profile. The abovementioned process is affected by a number of factors being influential in selection of instruments to solve relations of precipitation and discharge and subsequent prediction of flow rates. Water level in drainage area is affected both by climatic (meteorological) and geographical factors (Rybansky, 2009).

Considering climatic factors, storm rainfalls have decisive influence in small drainages even if it's a short duration, but impinge with high intensity the entire area of drainage given and, in reference to drainage area extent, are problematically predictable and that is why we have difficulties with subsequent protective measures. In the event of regional precipitation causing floods on large drainage areas (low intensity and large surface affected or, if you like, snow melt and a combination of both aforementioned features) the problem is solved by another way. Here we have relatively sufficient time to issue precipitation forecast and to give subsequent information concerning changes. The entire process is however very complicated.

We can to include, for small drainage areas, following geographical factors:

- the area size and form of drainage;
- altitude;
- relief;
- form of stream network;
- watercourse parameters - bank roughness, length, inclination, form of section line (profile), inundation;
- hydrogeological features of a drainage area (accumulation capability of a ground and subsoil, soaking capability of soil, sort of canopy) etc.

Mutual interaction of the abovementioned factors is subsequently projected onto size of changes of watercourse.

To make further analysis of mechanism of generation of increased water flow rate on watercourses, we have to describe and make an analysis of (Rybansky, 2009):

- causes of change of attributes of watercourse crossing ability;
- method of solution of process of precipitation and discharge;
- classification of models existing;
- questions of operative forecasts;
- possibility to obtain information relating to water levels and water flow rates;
- method of calculation of changes of crossing ability parameters;
- possibilities of completion of changes of topology information.

2.1. Causes of change of attributes on watercourse crossing

It is very important to determine possible future change of these values from the point of view of trafficability. If causal precipitation does not come after a certain period without precipitation on the drainage area, that change is very slow, water flow rates and water levels have decreasing trend. It is by reason of fact that the water flow in watercourses is generated only by groundwater discharge of drainage area and the groundwater resources in drainage area decrease gradually. For each drainage area and profile relevant, it is possible to construct a so called depletion curve which has exponential form and, according to this curve we can predict relatively accurately the future water flow rate in a certain period.

If we have no other information about the watercourse, we can very roughly estimate the flow rate by its long-term value *Q*a. With utilization of stage-discharge relation curve we can read the corresponding water level from the water flow in section given (depth of water over the lowest point of bottom) and, by known form of profile, to read the entire remaining attributes. In discharge section lines we also know the relation between mean section speed and water level.

A more difficult situation occurs by the attempt to estimate mentioned attributes after precipitation or, to estimate future flow rate after predicted rainfall, since each predicted flow rate is affected by the course of both previous and future rainfalls. What part of both previous and future rainfalls will obtain, this amount depends on position of profile in drainage. Quantity of predicted flow rates can be undersized, but it can reach the dimensions of disaster. That all depends on intensity and space distribution of previous and future rainfalls and on seeping ability of soil of drainage area given.

The process of precipitation and discharge shows progressive transformation of drainage precipitation in drainage up to discharge by closing hydrographic profile. The abovementioned process is affected by a number of factors being influential in selection of instruments to solve relations of precipitation and discharge and subsequent prediction of flow rates. Water level in drainage area is affected both by climatic (meteorological) and geographical factors.

Considering climatic factors, storm rainfalls have decisive influence in small drainages even if it's a short duration, but impinge with high intensity the entire area of drainage given and, in reference to drainage area extent, are problematically predictable and that is why we have difficulties with subsequent protective measures. In the event of regional precipitation causing floods on large drainage areas (low intensity and large surface affected or, if you like, snow melt and a combination of both aforementioned features) the problem is solved by another way. Here we have relatively sufficient time to issue precipitation forecast and to give subsequent information concerning changes. The entire process is however very complicated.

We can to include, for small drainage areas, following geographical factors:

- the area size and form of drainage;
- altitude;
- relief;
- form of stream network;
- watercourse parameters - bank roughness, length, inclination, form of section line (profile), inundation;
- hydrogeological features of a drainage area (accumulation capability of a ground and subsoil, soaking capability of soil, sort of canopy) etc.

Mutual interaction of the abovementioned factors is subsequently projected onto size of changes of watercourse.

To make further analysis of mechanism of generation of increased water flow rate on watercourses, we have to describe and make an analysis of:

- causes of change of attributes of watercourse passability;
- method of solution of process of precipitation and discharge;
- classification of models existing;
- questions of operative forecasts;
- possibility to obtain information relating to water levels and water flow rates;
- method of calculation of changes of passability attributes;
- possibilities of completion of changes of topology information.

To understand causes of changes of watercourses trafficability, we have to get acquainted with definitions:

- process of precipitation and discharging;
- hydrological transformation;
- hydraulic transformation.

The process of origin of sudden changes caused by whether storm rainfall or by regional rainfall, eventually by snow melt, which are the content of the independent chapter, can be seen as two separate problems – hydrological transformation and hydraulic transformation. Their course is evident and clarified by Fig. 3.

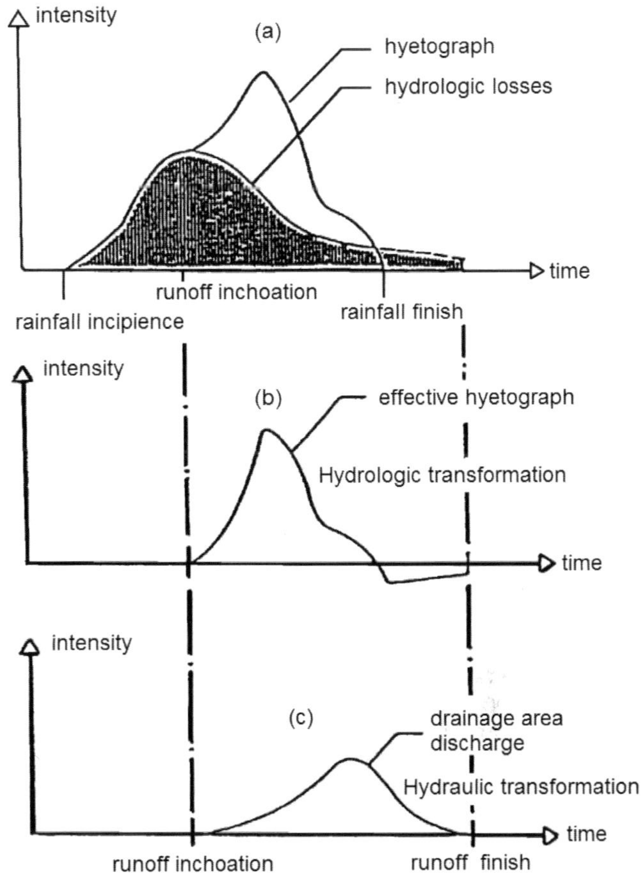

Fig. 3. Hydrologic and hydraulic transformation (by Jacobsen, 1980)

Hydrologic transformation presents process of gradual separation of hydrologic loss from rainfall intensity (rainfall totals). Losses participating in hydrologic transformation are caused by:

- interception;
- humidification;
- evaporation;
- infiltration;
- surface retention.

We can depict the occurence of losses and their share in total discharge by Fig. 4.

The size of these losses is given partly by character of drainage area represented by geographic and hydrogeologic conditions, by canopy and partly by level of groundwater resources and surface waters (reservoirs) situated at drainage area - see also Rybansky, 2009. Groundwater resources represent the level of groundwater and rate of saturation of soil profile by water depending on foregoing precipitation at drainage area given.

The losses by interception represent the impact of vegetation detaining on its surface part of precipitation. The size of retained quantity is commensurable to sort of vegetation and its value for the forests is varying at interval 40% (spruce) to 24% (pine tree).

The loss by interception is calculated according to balance relation among precipitation amount reduced by the amount fallen through foliage, the amount gradually dropping away from foliage surface, evaporation and change of retention capacity of a foliage.

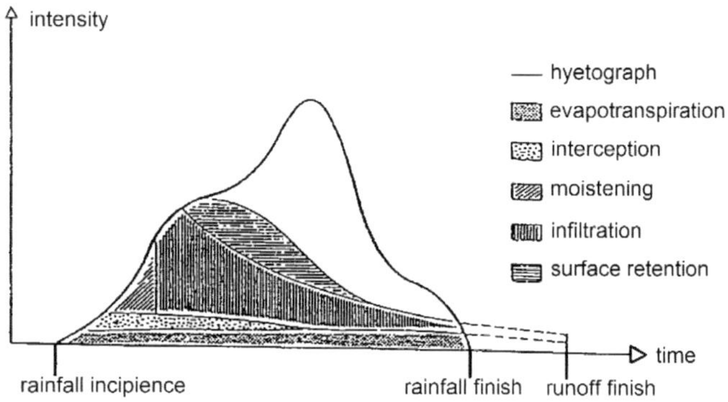

Fig. 4. Share of losses in hydrologic transformation (by Jacobsen, 1980)

The losses by humidification present process of surface humidification that precedes infiltration and surface retention. As the Fig. 4 shows, this process is under way simultaneously with evaporation and infiltration. Respective boundary between humidification and infiltration is very hardly definable.

Evaporation losses present both evaporation from soil surface and evaporation from surface of plants. On the ground of very small values (average value of potential evapotranspiration is in summer varying about 100mm per month), when we simulate the runoff in course of precipitation, we can omit this loss by reason that the air in the immediate vicinity of terrain surface is saturated by water vapour.

The losses by infiltration present the loss caused by the penetration of water into soil environment. One of the widely known methods at present frequently used is the Green-Ampt simulation models method. It is an integral method presupposing homogenous soil profile penetrated by the humidity front with exactly defined forefront on the depth of humidification.

The losses by surface retention presents primary accumulation of water in local depressions when the intensity of precipitation exceeds the intensity of infiltration. The quantity of that loss depends both on the relief (slope and roughness) and on the vegetation cover (canopy).

The hydaulic transformation presents gradual process of rise of areal and concentrated runoff from the drainage area, caused by effective casual precipitation and by underwater runoff. At first, rain gullies and erosion rills rise, concentrating the areal surface runoff, subsequently they merge into streams and larger watercourses. If the drainage area contains one or more reservoirs then their function presents point transformation of concentrated runoff, i.e. process of time distribution of water between reservoir inflow and flow-off. This is achieved by retention of the part of volume of water flowed through reservoir by functional space of reservoir given. To provide protective function of reservoir, the independent flood control basin is allotted in a reservoir as a rule.

The flood control basin is divided generally both to the controllable flood basin which is situated under the unhatched spillway dam crest (flow-out from this space is controllable by bottom outlets or by hatched spillways) and to uncontrollable flood basin which is situated over the security spillover, where the impact of transformation is given by hydraulic features e.g. of diversion weir. Resulting transformation effect is conditioned by volume of water-bearing capacity retained or shifted in time, which can cause, in the extreme event, on drainage area where the demanded water-bearing volume exceeds capability (volume of future reservoirs limited by terrain morphology), the rise of needfulness to develop the system of reservoirs. The feedback of small drainage areas to casual storm rainfall is very quick – about several hours. That is why we can simplify the problem of simulation of the course of water runoff to the simulation of surface runoff and consider the subsurface runoff as constant.

This simplification is not possible to use on medium and large drainage areas. In the course of precipitation and runoff processes, the impingement of change of underwater runoff is not negligible. That change is manifested in particular at the descent branches of hydrograms. On immediately repeating floods we can see that the abovementioned impingement is of significance even at the progressive branches of hydrograms and is presented both by shape of hydrogram and by value of culminating the runoff reached. Modelling of water runoff from a catchment - represents the modelling of two processes. The first is surface and groundwater runoff, the second is concentrated runoff in streams. Models based on physical principles (describing the problem by equations based on the laws of motion) require temporal and spatial discretization of the problem to solve. The spatial discretization represents a certain schematization of the actual watershed (determining the discrete points at which the solution is sought). The temporal discretization represents the determination of the period and time step of the solution.

Surface runoff - can be described by the differential equations of Saint-Venant. The solution of these equations is then the depth y and velocity v of the surface runoff on a slope of a given width B. The first, the continuity equation, is based on the law of conservation of mass:

$$\frac{\partial Q}{\partial x}+\frac{\partial S}{\partial t}=i_e \qquad (3)$$

Q	runoff quantity from an area of a given width	$[m^3s^{-1}]$
S	flow area	$[m^2]$
i_e	effective rainfall intensity	$[ms^{-1}]$
x	horizontal distance in the direction of flow	$[m^2]$
t	time	[s]

Presented form of continuity equation is appliable under these simplifying presuppositions:

- we presume parallel flow in direction of x axis, which enables to solve the problem as an unidimensional one. The solution is expressed by two status values, i.e. flow area S (or more precisely the height of water column on slope y) and runoff quantity Q (mean speed of profile given v);
- the solution is made on an infinite wide slope and the surface runoff forms continuous layer on the entire slope;
- the implementation of presupposition of constant specific weight of water omitting the transport of floating solids;
- precipitation intensity and hydrological losses are constant on the entire slope and hence they are the function of time.

The second equation is the momentum equation – it results from the motion conservation law. It is a partial differential equation of the hyperbolic type:

$$\frac{\partial Q}{\partial t}+\frac{\partial}{\partial x}\left(\frac{Q^2}{S}\right)=gS(I_0-I_f)-gS\frac{\partial y}{\partial x} \qquad (4)$$

Q	runoff quantity from an area of a given width	$[m^3s^{-1}]$
S	flow area	$[m^2]$
y	water depth on a slope	[m]
g	acceleration of gravity	$[ms^{-2}]$
Io	slope gradient	[%]
I_f	energy line inclination	[%]

The motion equation applies for the same simplifying presuppositions as in the event of continuity equation extended by presupposition of the small gradient α (up to 10°), when $\sin\alpha \cong \mathrm{tg}\alpha \cong I_0$.

The solution of those differential equations is possible only by setting of initial and limitative conditions.

Initial conditions are:
$Q\ (x, t = 0)$;
$S\ (x, t = 0)$.

Limitative conditions are:
$Q\ (x=0, t)$;
$S\ (x = 0, t)$;
$i_e\ (x,y,t)$.

Remark:

In practice, at the solution of the areic surface discharge, we use frequently the simplification by utilization of application of kinematic wave approximation - see also (Rybansky, 2009), resulting from presupposition of balance between the power of friction (resistance) and the power of gravity. At the presupposition mentioned, we can substitute the motion equation by simple equation describing relation between status values $Q(x,t)$ and $S(x,t)$ at stable flow, i.e. by the equation:

$$Q = 1{,}49 \frac{1}{n} R^{2/3} I_0^{1/2} S \quad (5)$$

Concentrated surface runoff (stream network) – is again describable by differential equations of Saint-Venant. The solution of those equations is then the flow $Q(x,t)$ and flow area $S(x,t)$ of the stream network. Water inflow to stream network consists of the areic surface runoff $q(x,t)$ and underwater runoff $q_z(x,t)$. The aggregate of these values than substitutes rainfalls effective intensity at the right side of continuity equation.

Initial conditions are:
$Q\ (x, t = 0)$;
$S\ (x, t = 0)$.

Limitative conditions are:
$Q\ (x=0, t)$;
$S\ (x = 0, t)$;
$q(x,t)$;
$q_z(x,t)$.

Reservoirs

If a stream network has reservoirs (discreteness points) it is necessary to extend limitative conditions by the principal equation of a reservoir (simple differential equation), having the form as follows:

$$\mathrm{d}\,V(t)/dt = Q\,(t) - O(V(t)) \quad (6)$$

$Q(t)$	total flow quantity of a reservoir in time t	$[m^3s^{-1}]$
$O(V(t))$	presents total discharge of reservoir in time t which is the function of reservoir filling and depends on hydraulic characteristics of spillway devices	$[m^3s^{-1}]$
$V(t)$	instant filling of a reservoir	$[m^3s^{-1}]$

Numerical solution is possible only providing that we set the additional initial condition in the form of $V(t=0)$. The feeding quantity $Q(t)$ is the solution of an upstream situated stream network, $O(t)$ is given by the manner of discharge control of relevant functional space of reservoir.

For purpose of a respective solution a drainage is frequently schematized i.e. substituted by an oriented evaluated graph $G(N,H,P)$ consisting of vertexes - set N (water resources and consumptions, control nods), of edges - H (watercourse sections) and surfaces hanged on them – set P whereon the entire drainage-basin area is divided.

The above-mentioned changes caused by hydrological and hydraulic transformation have a significant impact on the ability of vehicles to cross watercourses by wading or swimming. The profile parameters can be determined by physical measurements, e.g. tachymetry and hydrometric wing (manual or digital), or we can use contact methods, e.g. SONAR, or non-contact methods, e.g. SONAR suspended on a rope under a drone or RADAR. The timeliness and reliability of the data will always be important. The older the profile data, the less reliable it will be in relation to hydrological changes in the watercourse.

3. Possible methods

The passability of off-road vehicles through watercourses can be determined on the basis of previous experience and regulations - see Table 1, or on the basis of the manufacturer's technical data or verification by field tests. Verification by field tests can be carried out on test polygons (canals, fords) or directly on natural sections of rivers. The most complex testing is directly on rivers – see Fig. 5, where it is necessary to prepare the vehicles before testing (exhausts, intakes, breakwaters, etc.) to avoid damage or drowning of the vehicles.

For this testing, it is necessary to measure the profiles and depths of the watercourses to map the sections where the vehicles will wade and where they will float. Subsequently, the slopes and character (surface and roughness) of the banks for entering and exiting the watercourse must be measured, and the traction and adhesion forces required for a given vehicle or group of vehicles to exit the same track must be measured. To overcome fine-grained or incoherent bottom sediments, it is necessary to measure traction (adhesion) forces and rolling resistance. A dynamometer can be used for this measurement and the vehicle can be anchored in the direction of traction behind a fixed object (tank, large tree, etc.). A recovery vehicle is essential for this testing. Objective measurement of traction forces in the river bed using a dynamometer is possible if the river bed is wide and the tow rope is parallel to the transverse profile of the terrain – see Fig. 6.

Fig. 5. Overcoming water obstacle by SMSS (Anon. 2017 SMSS Squad Mission Support System Unmanned group vehicle Army Recognition (Ramillies: Army Recognition Group)

When measuring the traction forces at the exit of the river, a problem arises especially with short banks, where it is very problematic to anchor the traction rope so that it is parallel to the bank profile - see Fig. 6. Exit options can vary considerably depending on weather conditions, e.g. in rain or when the slopes are covered with snow or ice and the tensile force of the vehicle can decrease significantly as the coefficient of friction decreases and the rolling resistance increases. Under these adverse conditions, the descent into the river is also problematic, with the risk of the vehicle sliding down the slope - see Fig. 7.

Fig. 6. Possibility to measure the tractive force of a vehicle leaving the river - fixed vehicle is at the rear (photo M. Rybansky)

A vehicle can also collide with the river bed when descending quickly down a slope (ramp) into a river - see Fig. 9, 10.

When the vehicle descends into the water - see Fig. 9, the speed, the opposing vectors of gravitational force F_g and buoyancy force F_b and the hydrodynamic pressure force F_H on the breakwater and the front of the vehicle are decisive for a possible collision with the terrain. A vehicle - see Fig. 10 may also collide with the bottom of a watercourse, particularly if the bank slope α is greater than the angle of approach of the vehicle between the underside of the front wheel and the bumper or front of the vehicle. Vegetation can be another obstacle to wading watercourses. If we are crossing vegetation on river banks with a heavy vehicle, we can measure the strength of its resistance using dynamometry - see Fig. 6.

Snow-covered watercourse banks can be impassable obstacles, especially for wheeled vehicles - see Fig. 8.

The great variability in the nature of the profile of watercourses leads to simplification and parameterization of the main parameters. This is particularly applicable to off-road vehicle test facilities, i.e. modified fords and test channels.

Fig. 7 Armored tracked vehicle BVP-2 descending into the river (photo M. Rybansky)

Fig. 8. Also the small snow-covered watercourse banks can be an impassable obstacle (photo F. Dohnal)

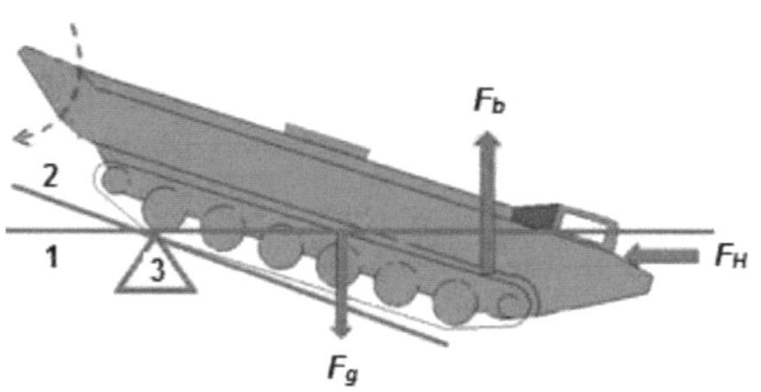

Fig. 9 The mechanism of appearance of the moment of forces causing the sink of the transporter: 1 - water line, 2 - slope, 3 - the momentary centre of rotation, F_H - hydrodynamic pressure force, Fg - gravity force, Fb - buoyancy force (by Sosnowicz and Kosiuczenk, 2021)

These facilities are often concreted or paved with stone blocks to prevent damage to these objects. However, these structural modifications lead to a partial distortion of the test results and may be reflected in the vehicle's ability to drive into and out of the water, as paved banks have different surface friction and rolling resistance values than unpaved and often unloadable and muddy banks, especially when the river is crossed by multiple vehicles on the same track.

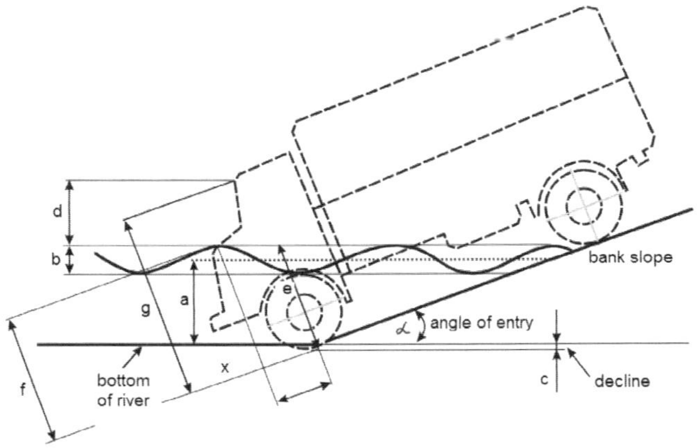

Fig. 10 Determination of wading depth

x	vehicle overhang in contact with the water	[m]
α	angle of entry	[°]
a	depth of water obstacle at still water	[m]
b	wave height (crest of wave in cross section)	[m]
c	subsidence due to bottom embedment as a function of specific pressure and bottom bearing capacity	[m]
d	splash height	[m]
e	wading depth $e = (a + b/2 + c) / \cos\alpha$	[m]
f	apparent wading depth due to vehicle displacement $f = e + x.tg\alpha$	[m]
g	apparent wading depth due to vehicle overlap increased by the proportional part of the splash height $g = f + d - \cos\alpha$	[m]

Very difficult obstacles are micro-relief shapes that can occur along the banks or in the river profile itself - see Fig. 11.

Number	Name	Form type	Evaluated parameters
1	embankment	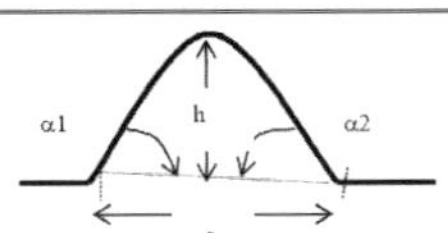	slope gradients ($\alpha1$, $\alpha2$) height (*h*) width (*s*)
2	excavation	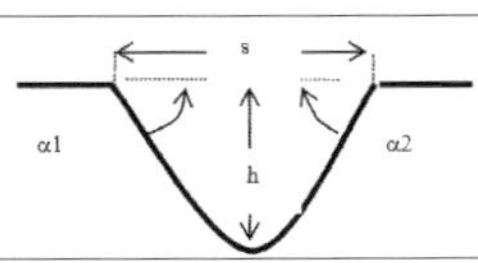	slope gradients ($\alpha1$, $\alpha2$) height (*h*) width
3	crater	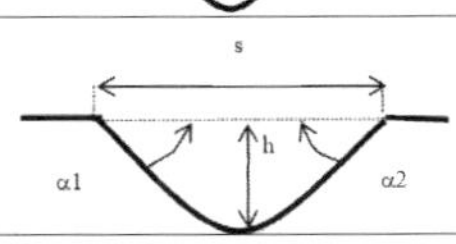	slope gradients ($\alpha1$, $\alpha2$) depth (*h*) width (*s*)
4	terrain step (climbing)	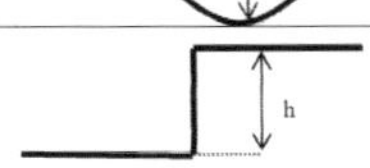	height (*h*)
5	terrain step (descent)	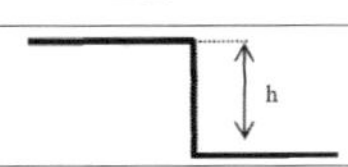	height (*h*)
6	trench, scarp (passing)	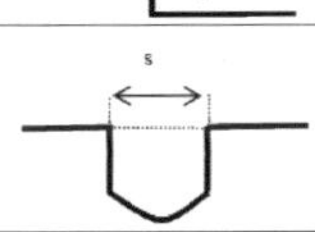	width (*s*)

Fig. 11 Micro-relief forms in overcoming water obstacles

4. Results and discussion

For the above variables of watercourse crossing, the most important is the change in river depths and current velocity due to precipitation and snow melting, which can affect both wade ability and vehicle swimming.

For the purpose of practical testing of the influence of the water parameters on the wading ability, a test channel in the military training area Podivice was used. The reason for the testing was to investigate the effect of water depth on the wading capabilities and speed of vehicle movement, as well as to find out from what water depth the vehicle starts to swim and also what speed it is able to reach during navigation. The following vehicles were tested:

- LR 110 (light passenger wheeled vehicle) - Fig. 12
- T 810 (heavy goods wheeled vehicle) - Fig. 13
- PANDUR (heavy armored wheeled vehicle) - Fig. 14
- BVP-2 (amphibious tracked infantry fighting vehicle) - Fig. 15

Fig. 12 LR 110 wading water obstacles

Fig. 13 T 810 descending into the water obstacle

Water from the adjacent water reservoir was gradually (0.2 m at a time) fed into the canal. The ford is 5 m wide, 20 m long and the maximum depth is 2.5 m. The gradients are 15° at the entrance and 20° at the exit.

The time taken by the vehicle to cross the water obstacle was measured for each passage, from the moment its front wheels touched the water surface until the rear wheels lost contact with the water surface. At the same time, the length of the water section in the ford was measured as a function of a certain depth. All the values mentioned are given in Table 2.

Fig. 14 PANDUR coming out of the water obstacle

Fig. 15 BVP-2 swimming over water obstacle

It can be seen from the table that the speed of the vehicles decreases with increasing water depth. The trend data for the above effect of water depth on vehicle movement is shown in the following graphs - see Figs. 16.

Table 2
Water obstacle crossing times with increasing water depth

Depth [m]	Time to cross the water section [s]				Section length [m]
	LR 110	T 810	PANDUR	BVP-2	
0.2	8.6	12.0	8.2	11.3	22.70
0.4	7.4	12.4	8.2	11.7	23.20
0.6	1.9	12.7	9.9	12.6	23.70
0.8		13.2	13.4	13.7	24.20
1.0		17.5	17.2	14.5	25.70
1.2		22.9	20.4	17.0	26.20
1.4			26.8	19.7	26.74
1.5			35.2	swimm	27.28
1.6			45.7		28.95
1.8			swimm		-

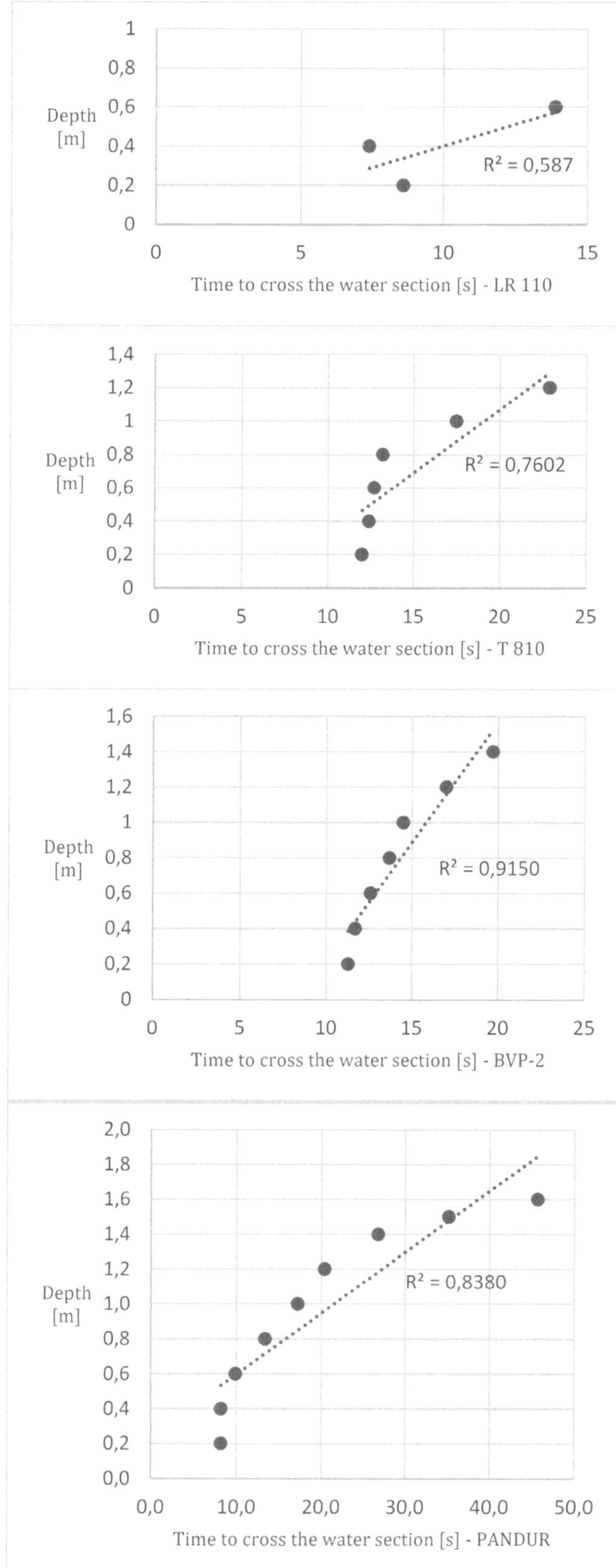

Fig. 16 Graphs of the dependence of the time required to wade through a water obstacle on the water depth

The trends shown are the result of both the resistance of the water braking the vehicle and the effect of water depth on the degree of caution of drivers.

A specific case would be the movement of an unmanned ground vehicle (UGV) or an unmanned autonomous system (UAS). The movement of this vehicle would depend on the sensor equipment (RGB camera, LIDAR, RADAR) and on the capability of navigation - Global Navigation Satellite Systems (GNSS), autonomous or inertial (IMU).

It would also be important whether this type of vehicle can swim, as is the case with the TAROS wheeled autonomous vehicle - see Fig. 17.

Fig. 17 Wading/swimming testing of UGV TAROS V2 in the water channel

The above results of testing the effect of changes in the parameters of watercourses on the speed of vehicles are important not only in terms of calculating the time needed to cross the watercourses, but also for determining the method of crossing the rivers, where the change in depth due to precipitation may lead to a situation where instead of wading through the water it will be necessary to calculate the swimming maneuver.

This fact also determines the methods of measuring the parameters of river profiles, i.e. their depth, the nature of the bed and the current speed of the rivers. The authors would like to focus on mapping these parameters in the near future.

This focus can be summarised in the following stages:

- Identify experimental test areas for rivers that are representative of the Czech Republic and the United States;
- Characterize rivers: catchment structure characteristics, depth and speed of the water flow, the nature of the bottom and the parameters of the river profile (micro-relief forms of river banks);
- Develop methods for river parameters data collecting (terrestrial and remote sensing methods), analysis of data acquisition options from hydrometeorological stations and institutions;
- Measure tractive force effort (adhesion force) of vehicles leaving the river: single and multiple passes of vehicles (wheeled and tracked), different soil types and moistures.

5. Conclusion

Watercourses are one of the main obstacles to the movement of rescue or military vehicles. The aim of this paper was to describe the parameters of watercourses that affect the possibility of crossing watercourses by wading or swimming. The possible changes that occur over time due to hydrological and hydraulic transformations on watercourse profiles as a result of precipitation were also described. In order to test the possibilities of wading or swimming, a test channel was used to replace a real watercourse. Although the unevenness and strength of the water bed on the real watercourse profile also affect the vehicle, the results obtained are an aid to both the speed of vehicle movement and to determine at what depth the vehicle starts to float, which requires considerably more demanding driving skills. The test channel also allowed the simulation of changes in water depth in a shorter time than would be the case in real water course.

6. Acknowledgements

This paper is a particular result of the defence research project DZRO VAROPS managed by the University of Defence in Brno and the specific research project 2024–2026 at Department K-210 managed by the University of Defence, Brno.

7. Declaration of competing interest

The authors declare no conflict of interest. The funders had no role in the design of the study; in the collection, analyses, or interpretation of data; in the writing of the manuscript, or in the decision to publish the results.

8. References

Affleck, R.T., Melloh, R.A. and Shoop, S.A. Cross-country mobility on various snow conditions for validation of a virtual terrain. Journal of Terramechanics 46 (2009) 203-210.

Ahlvin RB, Haley PW. NATO reference mobility model. 2nd ed. NRMM II User's Guide. TR GL-92-19. US Army Engineer Waterways Experiment Station, Vicksburg, MS; 1992.

Belletti, B. et al. "More than one million barriers fragment Europe's rivers." Nature 588.7838 (2020): 436-441.

Blackmon CA, Randolph DD. An analytical model for predicting cross-country vehicle performance, vol. II: longitudinal obstacles. Appendix B: vehicle performance in lateral and longitudinal obstacles (vegetation). TR No. 3-783. US Army Engineer Waterways Experiment Station, Vicksburg, MS; 1968.

Bruzelius, N. et al. Lärobok i militärteknik. Vol 5 Farkostteknik [Textbook in Military Technology. Vol 5: Vehicle engineering]. Stockholm: Swedish Armed Forces, 2010.

Buza, A.K. and Divos, F. Root Stability Evaluation with Non-Destructive Techniques. ActaSilv. Lign. Hung.,Vol. 12, Nr. 2 (2016) 125–134, DOI: 10.1515/aslh-2016-0011.

Cibulová K., 2017. Mobility during Crisis Situations. Key Engineering Materials, 2017, no. 755, p. 236-241. ISSN 1013-9826.

Cibulová, K., 2017a. The Mobility during Crisis Situations. In: Structural and Mechanical Engineering for Security and Prevention ICSMESP 2017. Switzerland: Trans Tech Publications, p. 236-241. ISSN 1662-9809.

Creager, C., Johnson, K., Plant, M., Moreland, S. and Skonieczny, K. Push–pull locomotion for vehicle extrication. Journal of Terramechanics 57 (2015) 71–80.

Dohnal, F., Hubacek, M, Sturcova M., Bure,s M. and Simkova, K., 2017. Identification of Microrelief Shapes Along the Line Objects Over DEM

Data and Assessing Their Impact on the Vehicle Movement 2017 International Conference on Military Technology (ICMT) (Brno: University of Defence) pp. 262–267.

Dawid, W., Pokonieczny, K., & Wyszyński, M., 2023. The application of the methodology to define terrain passability on a detailed level in various configurations. Aviation and Security Issues, 3(1). https://doi.org/10.55676/asi.v3i1.61.

DOHNAL, Filip; RYBANSKÝ, Marian; ŠIMKOVÁ, Kateřina. Water regime and prediction of its impact on river crossing in different types of zonal landscape. In: Useful Geography: Transfer From Research To Practice. Brno: Masaryk University, 2018, p. 132-142. ISBN 978-80-210-8907-5.

Drozd, J., Stodola, P., Kristálová, D. and Kozubek, J. Experiments with the UAS Reconnaissance Model in the Real Environment. In: Modeling and simulation for autonomous systems, 2018. Volume 10756, p. 340-349, DOI: 10.1007/978-3-319-76072-8_24.

Hošková-Mayerová, Š., Talhofer, V., Hofmann, A. and Kubíček, P., 2013. Spatial Database Quality and the Potential Uncertainty Sources, Advanced Dynamic Modeling of Economic and Social Systems, Studies in Computational Intelligence, 2013, Volume 448/2013, 127-142, DOI: 10.1007/978-3-642-32903-6_10.

Hubacek, M., Kovarik, V., Talhofer, V., Rybansky, M., Hofmann, A., Brenova M and Ceplova, L., 2016. Modelling of geographic and meteorological effects on vehicle movement in the open terrain Central Europe Area in View of Current Geography (Brno: Masarykova univerzita) pp. 149–159.

Jacobsen, P., 1980. Urban surface runoff simulation. Ph.D. Thesis, Dep. Of Sanitary Engineering, TU of Denmark, DK 2800 Lyngby.

Mason, G.L., Gates, B.Q., Moore, V.D., 2012. Determining forces required to override obstacles for ground vehicles. J. Terramech. 49, 191 196. DOI: 10.1016/j.terra.2012.04.001.

Mazal, J., Stodola, P. Procházka, D., Kutej, L., Scurek, R. and Prochazka, J. Modeling of the UAV Safety Manoeuvre for the Air Insertion Operations. In: *Modeling and simulation for autonomous systems, MESAS 2016.* Volume 9991, p. 337-346, DOI: 10.1007/978-3-319-47605-6_27.

Murakami, H. et al., 2017, Development of prototype 2 of the remote-controlled semi-underwater heavy transport robot, The 17th Symposium on Construction Robotics in Japan.

Nohel, J., Stodola, P. and Flasar, Z., 2019. Model of the Optimal Maneuver Route [Online First], IntechOpen, DOI: 10.5772/intechopen.85566. Available from: https://www.intechopen.com/online-first/model-of-the-optimal-maneuver-route.

Pokonieczny, K. and Rybansky, M. Method of developing the maps of passability for unmanned ground vehicles. In: 9th IGRSM International Conference and Exhibition on Geospatial & Remote Sensing (IGRSM 2018) 24-25 April 2018, Kuala Lumpur, Malaysia. Kuala Lumpur, Malaysia: IOP Conference Series: Earth and Environmental Science, 2018, p. 1-9. DOI: 10.1088/1755-1315/169/1/012027.

Pokonieczny, K. & Dawid, W., 2023. The application of artificial neural networks for the generalisation of military passability maps, International Journal of Cartography, DOI: 10.1080/23729333.2023.2231589.

Pokonieczny, K., Dawid, W. and Wyszyński, M. Methodology of Using Pathfinding Methods in Military Passability Maps, 9'th International Conference on Military Technologies (ICMT), 23-26.05.2023, Brno, Czech Republic, IEEE Proceedings.

Rybansky M., 2007. Effect of the Geographic Factors on the Cross Country Movement during Military Operations and the Natural Disasters. In: International Conference on Military Technologies 2007, University of Defence, Brno (Czech Republic), pp. 590 – 596. ISBN 978-80-7231-238-2.

Rybansky, M., 2009a. The Cross–Country Movement – The Impact and Evaluation of Geographic Factors, CERM Brno, 2009, 113 pp, ISBN 978-80-7204-661-4.

Rybansky, M., 2010. The Cross–Country Movement – Modeling. Brno, 2010, 100 pp, ISBN: 978-80-7204-717-8.

Rybansky, M., 2014. Modelling of the optimal vehicle route in terrain in emergency situations using GIS data. In: 8th International Symposium of the Digital Earth (ISDE8) 2013, Kuching, Sarawak, Malaysia 2014 IOP Conf. Series.: Earth Environmental Science 18 012071, doi:10.1088/1755-1315/18/1/012131, http://dx.doi:10.1088/1755-1315/18/1/012131.

Rybansky M. and Vala, M., 2009. Geographic Conditions of Military Transport Using Roads and Terrain. In: ICMT'09 -International conference on military technologies 2009, Brno (Czech Republic), 9 pp, ISBN 978-80-7231-649-6 (978-80-7231-648-9 CD).

Rybansky, M. and Vala, m., 2009. Relief Impact on Transport. In: ICMT'09 - International conference on military technologies 2009, Brno (Czech Republic), 2009, 9 pp, ISBN 978-80-7231-649-6 (978-80-7231-648-9 CD).

Rybansky, M. and Vala, M., 2009. Analysis of relief impact on transport during crisis situations. In: Moravian Geographical Reports, 2009, vol. 17, No. 3, p. 36 - 43, ISSN 1210-8812.

Rybansky, M., Dohnal, F., Hošková-Mayerová, Š., Svatoňová, H. The impact of drainage on terrain UGV movement. In: 9th IGRSM International Conference and Exhibition on Geospatial & Remote Sensing (IGRSM 2018) 24-25 April 2018, Kuala Lumpur, Malaysia. Kuala Lumpur, Malaysia: IOP Conference Series: Earth and Environmental Science, 2018, p. 1-10. DOI: 10.1088/1755-1315/169/1/012027.

Shoop, S.A., Richmond, P.W. and Lacombe, J. Overview of cold regions mobility modeling at CRREL. Journal of Terramechanics 43 (2006) 1–26.

Sosnowicz, R. and Kosiuczenko, K. Numerical Analysis of the Ability of a Floating Vehicle to Overcome a Water Obstacle. Advances in Military Technology Vol. 16, No. 1, 2021, pp. 149-157, ISSN 1802-2308, eISSN 2533-4123DOI 10.3849/aimt.01476.

STANAG 3992 - AGeoP-1 Terrain Analyses, Field Manual No. 5-33. Headquarters Department of the Army", Washington, DC, 1990.

Stodola, P. and Mazal, J., 2010. Optimal Location and Motion of Autonomous Unmanned Ground Vehicles WSEAS Transactions on signal processing 6 68–77.

Stodola, P. and Mazal, J., 2010. Autonomous Motion of Unmanned Ground Vehicles in General Environment. In: 9th WSEAS International Conference Recent Advances in Signal Processing, Robotics and Automation (ISPRA 2010) (Cambridge: University of Cambridge) pp 226–231.

Taheri, Sh., Sandu, C., Taheri, E.P., Pinto, E. and Gorsich, D. A technical survey on Terramechanics models for tire–terrain interaction used in modeling and simulation of wheeled vehicles. Journal of Terramechanics 57 (2015) 1–22.

Yamada, M., Yamauchi, G. and Hashimoto, T. Fundamental study on underwater trafficability for tracked vehicle. Journal of Terramechanics, Volume 98, 2021, Pages 42-49, ISSN 0022-4898.

Yuta, S., 2017. Development of Remotely Controlled Semi-Underwater Heavy Carrier Robot for Unmanned Construction Works, Journal of Disaster Research Vol.12 No.3, pp. 432-445.

7603 | Evaluation of off-road uncrewed ground vehicle mobility using discrete element method and scalability investigation

Ayush Nuwal [a,*], Ajay Kumar [b,*], John Economou [b]

[a] Former Cranfield University PhD Student; currently at Bahrain Defence Force, General Headquarters, Bahrain

[b] Centre for Defence Engineering, Cranfield Defence and Security, Cranfield University, UK

* Corresponding author: ajay.kumar@cranfield.ac.uk and Ayush.Nuwal@bdf.bh

ABSTRACT

Modern armed forces are exploring the teaming of military vehicles with smaller uncrewed ground vehicles (UGVs), to improve the success of operations in complex and demanding off-road terrains. Smaller scale UGVs can be used to perform initial mobility testing on soft soils, to predict the go/no-go performance of larger crewed / uncrewed vehicles. Because of the variation in the sizes of the UGV and military vehicle, it is imperative whether the scalability of tyre-soil interaction exists or not. The scalability assumes that similar systems behave in a similar manner at different dimensional scales. Dimensional analysis is carried out to determine similarity between the systems and identify design parameters affecting scalability.

In this study, the lightweight vehicles (FED Alpha) are considered as the full-scale systems (as upper boundary) and UGVs (Husky or Warthog) as scaled system. The 335/65R22.5 tyre with operational range of loading for full scale vehicle is considered. The smaller UGV tyres (0.7, 0.5 and 0.25 scale) represent scaled system. The 2NS and fine-grain sands were modelled using the DEM (EpAM contact model). The direct shear and pressure-sinkage tests were simulated to calibrate the soil model (cone index from 14.79-149 kPa).

Validated simulations of tyre-soil interaction, show that 'drawbar-pull vs slip' and 'tractive-efficiency vs slip' are scalable, within given size and loading conditions. However, the prediction is dependent on soil parameters and size of the scaled systems (0.7 and 0.5 scale demonstrated the scalability clearly). The prediction was better in 2NS sand due to higher cone index. Up to 0.5 scale-system can predict the full- scale system's mobility performance on sandy soils. This result can be used to develop lighter UGVs to support full-scale vehicles in the off-road terrains.

Keywords
Scalability
Uncrewed Ground Vehicles
Mobility
Discrete Element Method

1. Introduction

Speed and mobility offer key tactical advantages to the armed forces on the battlefields. The rapid movement of mechanized land forces concentrated at the enemy's weak point often provides dominance on the modern battlefield. This is one of the popular military tactics in the history of warfare. Mechanized land warfare significantly depends on the use of infantry fighting vehicles, self-propelled artillery, and other combat vehicles, as well as other support units. However, climatic conditions and soil texture significantly challenge the vehicle mobility. Once the vehicle loses its ability to move and becomes immobile due to soil conditions or any other factors, then it becomes a vulnerable asset. Therefore, vehicle mobility in off-road conditions with suitable critical analysis can help to increase the vehicle mobility and therefore survivability.

With the advancements in modern military warfare, traditional mechanized land vehicles are now being integrated with autonomous systems. The NATO technical report highlights that there is a need for significant development to quantitatively model the mobility performance metrics of autonomous systems, particularly in the context of Uncrewed Ground Vehicles (UGVs) (NATO TR SAS – 097, 2018). UGVs are military robots that primarily operate off-road without human presence onboard, as depicted in Fig. 1. The increasing use of UGVs in present-day military operations, along with their expected greater deployment in the coming decades, raises questions about how to correlate the mobility of UGVs with that of full-scale military vehicles across various terrains and different climatic conditions encountered in NATO operations. As Science and Technology (S&T) continue to advance, future operating environments could become even more complex and uncertain, featuring high slippage areas, dense vegetation, or radioactive environments. Future UGVs capability can have the potential to significantly redefine the way modern warfare is conducted.

Fig. 1. REX UGV (Egozi, 2018)

Lower downside risk and higher confidence in the success of missions are two strong motivators for the continuous expansion of UGVs across a broad spectrum of warfighting and peacetime missions. During such missions, UGVs can team up with full-scale military vehicles to ensure three critical tasks as mentioned below.

- UGVs can be utilized for initial mobility testing on soft soil during missions to determine the go/no-go conditions for full-scale military vehicles. This capability would assist defense forces in preventing full-scale military vehicles from becoming immobilized during

https://doi.org/10.56884/RT9571XM

tactical operations on soft soils or off-road terrains, as depicted in Fig. 2. Furthermore, it would contribute to terrain mobility analysis, thereby improving its fuel economy and extending the operational vehicle range.

- Full-scale military vehicles frequently operate with heavy weapon systems and other heavy payloads on soft soils. The payload is a critical factor that significantly impacts mobility performance on soft soils. UGVs can be employed for initial mobility testing, allowing for the assessment of scaled payloads that simulate the conditions of full-scale vehicle payloads. Consequently, this will aid in determining the maximum payload carrying capacity of full-scale military vehicles on soft soils.
- Pre-planning the correct UGVs to team up with full-scale military vehicles for tactical operations is of utmost importance. The predictive capability of a UGV is essential for effectively monitoring and guiding the full-scale military vehicle, ensuring its safety while mitigating mission constraints such as no-go areas and communication limitations, and maximizing performance metrics such as speed and mobility. Therefore, the current study aims to facilitate the decision-making process for determining the suitable mobility specifications of supporting UGVs (e.g., tyre specifications and payload), scaled in accordance with full-scale military vehicles.

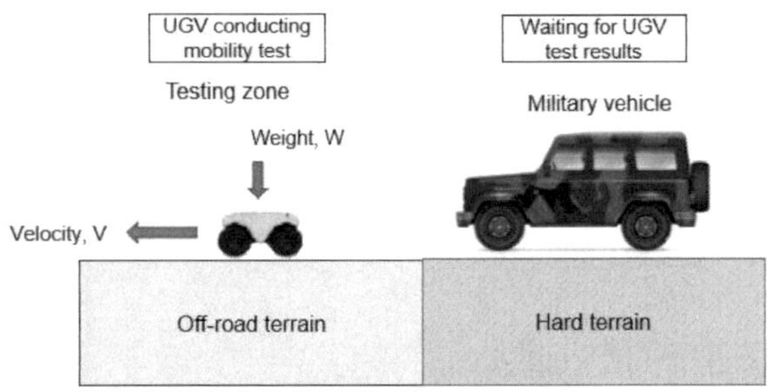

Fig. 2. UGV conducting initial mobility testing for the full-scale military vehicle to decide go/no-go condition on off-road terrain (or soft soils)

Historically, the assessment of wheeled vehicle mobility on off-road terrain has relied heavily on full-scale physical testing conducted in natural or prepared soil conditions. Over the past five decades, a significant portion of terramechanics research has focused primarily on light utility military vehicles, resulting in the development of various methodologies for predicting mobility. These methodologies have employed empirical approaches that involve resource-intensive and time-consuming experimental testing and have gained wide adoption within the defense research community (Jayakumar et al., 2013).

However, due to their empirical nature, it remains an open question whether the mobility performance models obtained from these methods can be used to accurately represent the mobility of small, lightweight Uncrewed Ground Vehicles (UGVs). Therefore, the concept of scalability in tyre-soil interaction is being proposed. Scalability is defined based on the assumption that it is possible to establish similar systems that exhibit similar behavior at different model scale (Freitag, 1966). Scale-model testing can be utilized to establish scalability in tyre-soil interaction system.

Further, to reduce the dependency on full-scale experimental testing, a novel simulation contact model using Discrete Element Method (DEM) has been implemented in EDEM software, to predict vehicle mobility based on vehicle and soil characteristics (DEM, 2011).

In the present study, the mobility analysis focuses on lightweight Armoured Personnel Carrier (APC) military vehicle tyres, such as the Land Rover (7.5R16) or FED Alpha tyre (335/65R22.5), as representative examples of full-scale systems. In contrast, the mobility analysis of small, lightweight UGVs considers tyres such as the Warthog (24" Argo) as representative examples of scaled systems.

This paper is organized as follows. Section 2 discusses the scale model testing and establishes scalability in the context of the tyre-soil interaction system. Section 3 discusses the mobility performance parameters of the system. Section 4 uses the dimensional analysis technique to non-dimensionalise the performance parameters of the system. Section 5 discusses the discrete element method (DEM) simulation technique with the Edinburgh elastic-plastic adhesion (EpAM) contact model. Sections 6, 7 and 8 discuss the numerical simulation test procedure, results, and conclusion.

2. Scale model testing

Scale-model testing is an indispensable tool in the aerospace and shipbuilding industry. Small-scale modelling of full-scale aircraft and ships made it easy to address aerodynamics and hydrodynamics challenges over time within a controlled environment. It has proven useful in checking analytical work and providing empirical and semi-empirical solutions to challenges, which withstand the analytical treatment. The high costs of constructing and testing full-scale aircraft and ship models also made scale-model testing an economic necessity in both industries. Scale-model testing also provides the opportunity to perform controlled condition experiments, which are hazardous to operators while performing full-scale model testing. Aside from the economic benefits of small-scale model testing, it can solve immediate and specific technical problems. It is undoubtedly true in predicting changes in the full-scale model's mobility performance concerning design and environmental conditions using scale model testing. It can also yield design solutions to more complex challenges rapidly and economically.

Full-scale model testing has been the backbone of vehicle mechanics studies over the years. It is primarily, because of their relatively small size and ease of testing compared to aircraft and ships. Most of the past research focused on studying vehicle tyre interaction with uniformly prepared homogeneous soil test sections that incurred a substantial cost in order to generate an extensive database of the mobility performance parameters (e.g., drawbar pull vs slip, tractive efficiency) (Jayakumar et al., 2013). Therefore, scalability of tyre-soil interaction system can utilize the existing database of full-scale model testing to design and develop, cost-effective lighter UGVs to support full-scale military vehicles on the battlefield.

Swanson (1973) conducted scale model testing at the U.S. Army, Engineer Waterways Experiment Station (WES) to assess the tractive performance of different scale tyres on dual-layered terrain. The 2.50-4 tyre, representing a scale model that was approximately 1/5 times smaller than the full-scale standard military tyre, 11.00-20 of $2\frac{1}{2}$ ton truck, was chosen as the scale tyre for the tests. The experiments were performed on clay soil, considering two different cone index values: 160 kPa and 55 kPa. The scale tyre loading was varied between 111 N to 667 N with tyre deflection ranging from 15% and 35%. The experiments were performed at constant 20% slip for all tyre conditions. Fifteen different combinations of wheel load, tyre deflection and soil strength were tested with three replications. It was found that sixty-eight percent of the mobility performance data of scale tyre was in the scatter band of 1-standard deviation of the full-scale tyre system. Therefore, the mobility performance of scale models follows normal distribution with respect to mobility performance of full-scale model.

Freitag (1966) performed the scale model testing on an approximately one-half scale model of full-scale tyre. It quotes "*The application of dimensional analysis to the experimental study of the vehicle-soil system is based on the fundamental assumption that similar systems that behave in a similar manner can be established. Thus, if the generalisations obtained by a systematic variation of the independent parameters are to be used with confidence, it must be shown that the observed tyre – soil relations do not vary with the size of the system.*"

The 4.00-7 tyre, representing a scale model that was approximately 1/2 times smaller than the full-scale tyre, 9.00-14, was chosen as the scale tyre for the tests. Both tyres were tested at several different loads at several different loading conditions. The same forward speed was used for both tyres in all tests. Two types of similarities were established among both systems: Geometric similarity and dynamic similarity, using dimensional analysis. The geometric similarity ensures that both tyres have similar geometry while interacting with soil surface. It was achieved by keeping the same deflection number *(δ/h)* between the full-scale and scale systems. The dynamic similarity assures that the nature of interaction (e.g., transfer of forces and friction) is the same between the full-scale and scale systems. It was achieved by keeping the same clay loading number $\left(\frac{CI.d^2}{W}\right)$ between the full-scale and scale systems. It was found that the mobility performance parameters vs clay number datasets of the scale system were in close relation to the full-scale system curves.

Bekker (1956) proposed the use of dimensional analysis to develop an analytical model for the tyre-soil interaction system. This approach involves utilizing non-dimensional parameters to create similar systems. He introduced the concepts of geometric and dynamic similarity in tyre-soil interaction. Geometric similarity is achieved by equating the ratio of tyre physical parameters, such as tyre diameter to tyre width *(b/d)*, between the full-scale system and scale systems. Dynamic similarity is established when the full-scale system and prototype systems experience similar net forces.

However, in addition to performance parameters like drawbar pull force, Bekker identified separable force components that can interact with the tyre-soil interaction system. These components need to be considered in the analysis of the system. The stated force components are:

- Forces generated due to soil friction, cohesion (c) arising from the acceleration of the soil particles
- Forces from elastic deformation and viscous shearing of the soil
- Forces resulted from applied force and input torque
- Forces generated by soil to wheel friction

Bekker derived all the above separable force components into the non-dimensional group at the criterion of safe load as shown below. $\emptyset$ is internal angle of friction, F is force term, l is length dimension and γ_s is specific weight.

$$(\emptyset, \frac{cl^2}{F}, \frac{\gamma_s l^3}{F})$$

Therefore, if the full-scale tyre and scale tyre are tested on the same soil model under identical soil conditions, it becomes possible to balance these non-dimensional parameters in both systems.

3. Tyre-soil interaction system

3.1. Mobility performance parameters

3.1.1. Longitudinal Slip Ratio (LSR)

Slip can be defined as the reduction in distance travelled or speed because of flexing in the tractive device (tyre or wheel) or shearing within the soil (R. He et al., 2020). It generally occurs between the overlapping surfaces of the tractive device and the terrain. The wheel slip ratio in the longitudinal direction of travel of the tractive device is considered as longitudinal slip ratio (LSR). The angle formed between the direction of travel and the line of intersection of the longitudinal central plane of tractive device is known as a slip angle (Wong, 2008). In this study, wheel slip ratio is defined as longitudinal slip ratio (LSR) at zero slip angle.

Mathematically, it can be defined as:

$$S = 1 - \frac{V_a}{\omega r} \tag{1}$$

where, S = Longitudinal slip ratio (LSR)
ω = angular velocity of the wheel, rad/s
r = rolling radius of the wheel on a hard surface, m
V_a = actual velocity of the wheel, m/s

3.1.2. Motion Resistance

In broad terms, motion resistance (MR) can be defined as the resistive force offered to the motion of a vehicle due to internal and external stimulus (R. He et al., 2020). The internal stimulus (R_{int}) is primarily due to the hysteresis developed into the tyre material because of carcass deflection and its impact inside the wheel or track while rolling. The external stimulus is an aggregate of compact resistance (R_c), bulldozing resistance (R_b), aerodynamic motion resistance (R_a), inertial resistance (R_i), slope resistance (R_s) and obstacle resistance (R_{ob}). The motion resistance is therefore given by Eq. 2.

$$MR = R_{int} + R_c + R_b + R_a + R_i + R_s + R_{ob} \tag{2}$$

3.1.3. Gross traction

Gross traction (GT) tends to be a thrust force generated due to powered wheel interaction with terrain (R. He et al., 2020). The powered wheel signifies a tyre with an input torque from the engine or motor. During this interaction, the powered running wheel applies force on the soil, trying to break it. Therefore, as per Newton's third law of motion, soil generates an equal and opposite force to wheel force. This opposite soil force generated is the thrust force or gross traction (GT). It helps the vehicle to overcome motion resistance and produce useful work.

3.1.4. Drawbar pull

Drawbar pull (D) is the excessive force produced by the vehicle at the drawbar to do the external work in the direction parallel to vehicle motion, as shown in Fig. 3 (R. He et al., 2020). Theoretically, drawbar pull defined as a difference of gross traction and motion resistance, as shown in Eq. 3. Hence, a positive drawbar pull indicates mobility, a zero pull indicates self-operation, and a negative pull indicates immobility.

$$D = GT - MR \tag{3}$$

3.1.5. Tractive efficiency

The term 'tractive efficiency' refers to the vehicle's capability in transforming the input axle power from the engine to the usable power available at the drawbar (R. He et al., 2020). Theoretically, it is defined as the drawbar power (P_d) available to the input power (P_{in}) by vehicle engine with sprocket torque, T and angular velocity of the drive wheel, ω as shown in Eq. 4. θ_r and θ_f are the rear and forward contact angles of the tyre-soil interface, as shown in Fig. 3. W is the normal load and N is the normal reaction force on tyre from surface.

$$\eta = \frac{P_d}{P_{in}} = \frac{D.V}{T.\omega} \quad (4)$$

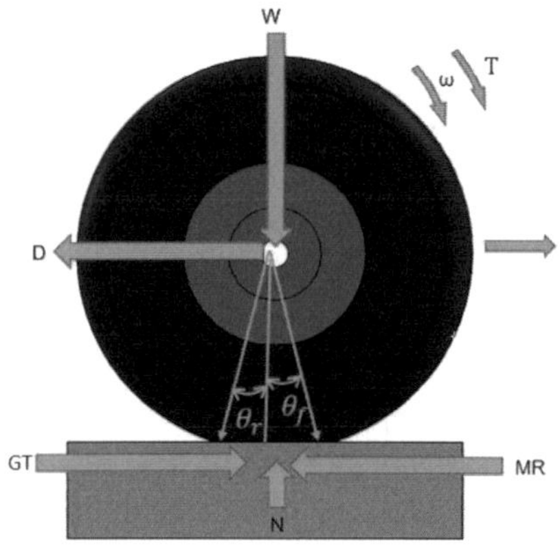

Fig. 3. Free body diagram of tyre-soil interaction system

3.2. Soil system

3.2.1. Cone Index

The cone index (CI), within the empirical methodology framework, is the most widely used mechanical property to determine soil strength. Experimentally, it is measured using a standard instrument known as cone penetrometer. A Cone penetrometer (Fig. 4) is a device, which has a conical surface or probe at its head with a tip downward and a cylindrical shaft as its body. A load cell (or dial indicator) is present at the top of the shaft within a proving ring, which indicates the force applied axially to the penetrometer. There are circumferential bands around the shaft that indicate the depth of penetration (R. He et al., 2020). The cone penetrometer's penetration velocity is kept slow and steady to maintain dynamic equilibrium to measure the cone index. This soil penetration resistance is the cone index. Theoretically, the cone index represents the force per unit base area that is necessary to push a cone probe into the soil at a steady rate.

Many of the existing empirical models use the cone index as a soil parameter to study soil's influence on vehicle mobility. This view is supported by the fact that the cone index profile is a composite property, reflecting both the soil's cohesive and frictional nature. However, it is difficult to separate the contribution from cohesion and friction between soil particles to soil's strength. Therefore, the cone index is used to analyze soil strength effect on the vehicle mobility in present study.

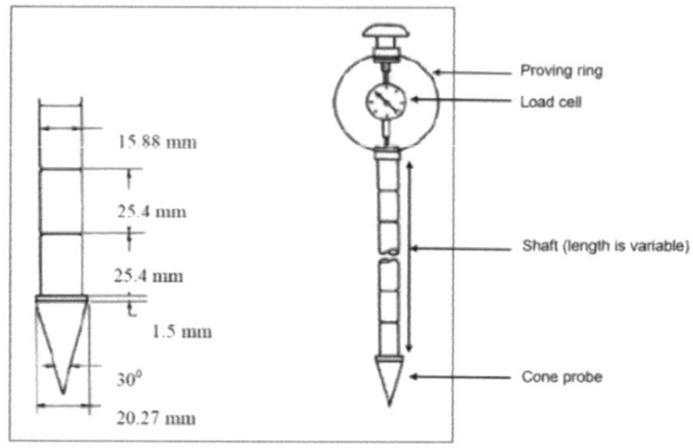

Fig. 4. Standard cone penetrometer (ASABE standard, 2006)

4. Dimensional analysis

Dimensional analysis is a powerful tool to reduce complex physical problems to the simplest form. In this approach, a non-dimensional relationship is developed by analyzing the fundamental dimensions of the physical parameters involved in the process. In this section, dimensional analysis is conducted on the mobility performance parameters of the tyre-soil interaction.

Tyre – soil interaction can be expressed as a function of nine independent variables as shown in Table 1 and three dependent variables as shown in Table 2 (Bekker, 1958; Freitag, 1966 and Swanson, 1973). However, the effect of the tyre's velocity is neglected by keeping it the same for both the full-scale and scale models testing (Freitag, 1966). Therefore, wheel torque, T can be expressed as a function of pertinent parameters shown in Table 1 and 2.

$$T = f(d, b, r, \delta, h, W, S, MR, D, CI) \quad (5)$$

Table 1
Independent variables

Parameter	Symbol	Dimension
Tyre diameter	d	L
Tyre width	b	L
Section height	h	L
Rolling radius	r	L
Tyre deflection	δ	L
Cone index	CI	$ML^{-1}T^{-2}$
Normal load	W	MLT^{-2}
Slip	S	-
Velocity	V	LT^{-1}

Table 2
Dependent variables

Parameter	Symbol	Dimension
Torque	T	ML^2T^{-2}
Motion resistance	MR	MLT^{-2}
Drawbar pull	D	MLT^{-2}

Using the Buckingham Pi theorem, the wheel torque, T can be written in nine functional Pi terms as shown in Eq. 6.

$$\frac{T}{rW} = \left\{ \left(\frac{d}{r}\right), \left(\frac{b}{r}\right), \left(\frac{\delta}{r}\right), \left(\frac{h}{r}\right), (S), \left(\frac{MR}{W}\right), \left(\frac{D}{W}\right), \left(\frac{CI\, r^2}{W}\right) \right\} \quad (6)$$

The last Pi term $\left(\frac{CI.r^2}{W}\right)$ can be multiplied with $\left(\frac{d}{r}\right)$and $\left(\frac{b}{r}\right)$ Pi terms which are constant for any specific tyre at constant load.

Therefore, a more appropriate Pi term can be derived i.e., $\left(\frac{CI.b.d}{W}\right)$ which helps to understand the relationship between tyre-soil interaction. Therefore, equation 6 can be re-written as Eq. 7.

$$\frac{T}{rW} = \left\{\left(\frac{b}{d}\right), \left(\frac{r}{d}\right), \left(\frac{\delta}{h}\right), \left(\frac{h}{d}\right), (S), \left(\frac{MR}{W}\right), \left(\frac{D}{W}\right), \left(\frac{CI\ bd}{W}\right)\right\} \quad (7)$$

However, the relationship between dependent variables is as shown in Eq. 8.

$$\frac{MR}{W} = \frac{T}{rW} - \frac{D}{W} \quad (8)$$

The $\left(\frac{r}{d}\right)$ ratio is constant for most of the tyres at optimal inflation pressure, and thus, this term can be neglected in the tyre - soil interaction analysis. However, $\left(\frac{h}{d}\right)$ ratio between tyre parameters can be re-written as Eq. 9.

$$\frac{h}{d} = \frac{1 - \frac{2r}{d}}{\frac{\delta}{h}} \quad (9)$$

Therefore, the dependent parameters can be re-written, as shown below.

$$\frac{T}{rW} = f\left(\frac{b}{d}, \frac{\delta}{h}, S, \frac{CI\ bd}{W}\right) \quad (10)$$

$$\frac{D}{W} = g\left(\frac{b}{d}, \frac{\delta}{h}, S, \frac{CI\ bd}{W}\right) \quad (11)$$

where f and g are two different functions.

Therefore, all four Pi terms should be kept the same for similar systems. It governs the similar tyre-soil interaction for both full-scale and scale tyres. The Pi terms *(b/d)* should be the same during selection of scale tyre and full-scale tyre. The Pi term *(δ/h)* is the indicator of the stiffness of a tyre. It is also a function of tyre inflation pressure and tyre loading. As per the Engineering Design Handbook, automotive series, United States Army Material Command (1967) and Wong (2010) proposes that if the tyre inflation pressure is greater than the soil strength, then the tyre-soil contact area is round, and the tyre behaves like a rigid wheel. Moreover, the terrain deformation will be much greater than tyre deformation. Therefore, in this case, pneumatic tyre can be regarded as a rigid wheel (Wong, 2010).

The third Pi term is wheel slip. The mobility performance parameters are expressed as a function of wheel slip. Therefore, it is not considered a variable. The fourth Pi term $\left(\frac{CI\ bd}{W}\right)$ contain the significant soil parameter, CI and the load, W. Both these parameters are independent variables and involves the concept of force. Therefore, for a constant tyre geometry, only changes in the cone index, CI and load, W can cause changes in the dependent variables. In this paper, the cone index was kept the same as both full-scale and scale tyres were tested on the same soil conditions.

Therefore, considering the tyre-soil interaction system based on tyre behaviour as rigid wheel and similar soil penetration consistency (or same soil conditions), the geometric and dynamic similarity criterion can be reduced to the term $\frac{W}{bd}$ which is the pressure term. However, the only pressure term acting in the tyre-soil interaction is the ground pressure. It is important to note that the average ground pressure is generated due to the transfer of forces from the tyre to the soil system. On equating the above parameter for full-scale and scale systems in the above conditions, the average ground pressure is the same for both systems. Therefore, similarity can be established between full-scale and scale tyre-soil systems by maintaining the same ground pressure term.

5. Discrete Element Method

5.1. Algorithm

Discrete Element Method (DEM) simulates the mechanical behaviour of a collection of arbitrarily shaped and arranged particles compromising in a system. A typical DEM material model schematic overview is shown in Fig. 5 (EDEM, 2011). It contains choice of particle shape, particle size distribution, particle parameters (e.g., particle density) and contact model. The particles act as independent entities and interact only at the contacts and interfaces between them. The algorithm involves the application of Newton's 2nd law of motion to particles and force-displacement law contact models at the interfaces. Newton's 2nd law of motion is used to determine the particle's motion arising from the contact and body forces on it. The force-displacement model is used to update the contact forces acting from each contact's relative motion.

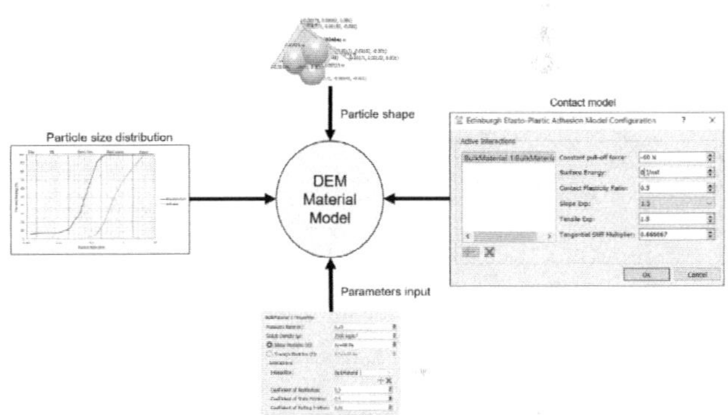

Fig. 5. A schematic overview of the DEM Material model (DEM, 2011)

The calculation cycle requires the repeated application of both the laws at each particle and relative contact, respectively, as shown in Fig. 6. Initially, the particles settle under gravity. The boundary condition is auto updated with the boundary of the box where the simulation takes place in the EDEM software as shown in Fig. 7. At the start of each cycle, the contacts are determined from the known particle positions. The force-displacement contact model is further applied to each contact to update the contact forces based on the relative motion, damping coefficient, and overlapping between the two entities at each time step, typically less than 1% of particle diameter. Later, the law of motion is applied to each particle to update its position and velocity based on the resultant forces and moments from the constitutive contact model and body forces acting on the particle. Further, the simulations were performed in the Altair EDEM software purchased under the academic license.

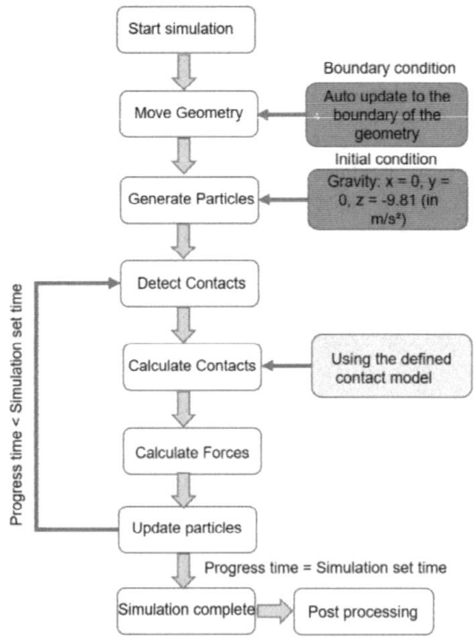

Fig. 6. Calculation cycle for DEM

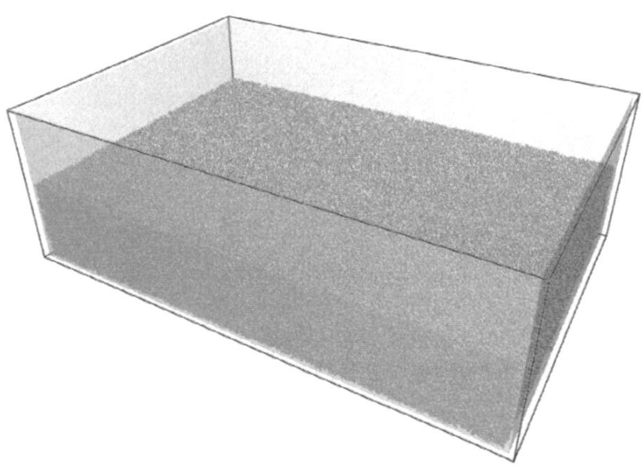

Fig. 7 Boundary condition of the box 6 m x 4 mx 2 m with red marking gets auto updated in Altair EDEM software

5.2. Edinburgh elastic-plastic adhesion contact model

The Edinburgh elastic-plastic adhesion (EpAM) contact model is based on the physics phenomenon observed in adhesive contacts between different nature and sizes of particles and entities. When two particles or entities are pressed together, they experience elastic and plastic deformation. It is assumed that the adhesive (or pull off) strength increases with an increase in the plastic contact area. Therefore, a non-linear contact model that accounts for both elastic and plastic deformation and contact area dependent adhesion is proposed. A schematic diagram of particle contact and normal force overlap (f_n-δ) curve for this model are in Fig. 8 (a) and (b).

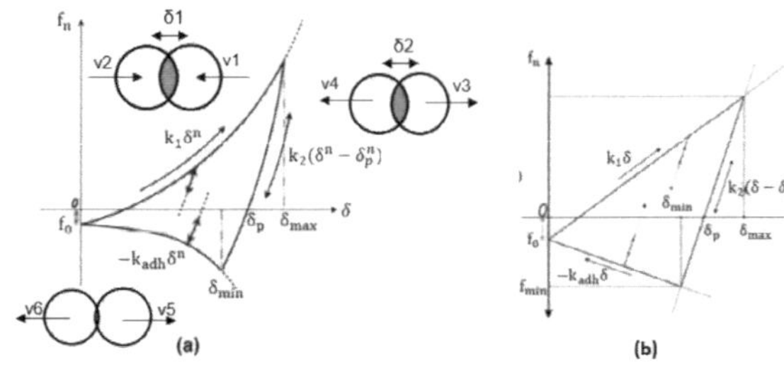

Fig. 8. EpAM normal force – displacement relationship (a) non-linear (b) linear (DEM, 2011)

The loading, unloading/reloading, and adhesive branches are characterised by seven parameters used in the model: the virgin loading stiffness parameter k_1, the unloading and re-loading stiffness parameter k_2, the contact plasticity ratio, λ_p, the constant adhesion force f_0, the adhesion energy parameter, $\Delta\gamma$, the adhesion exponent χ and the stiffness exponent n. The linearity and non-linearity of the loading and unloading branches are controlled by parameter n, and all becomes linear when n is equal to 1. The adhesion exponent χ controls the rate at which adhesion forces drop during contact separation; the higher the value, the sharper is the drop when plastically flattened contacts separate.

During the initial loading, the contact is determined using virgin loading path, k_1 and later, upon unloading of the contact; the contact switches to unloading/reloading path k_2, reaching zero force at a specific overlap which is also described as the plastic overlap, δ_p. If reloading occurs, then initially, the contact follows the unloading path k_2 but later switches to the virgin loading path k_1 when the previous maximum loading force is reached. Unloading along the k_2 path below the plastic overlap δ_p results in the maximum adhesive force development, f_{min} at $-f_0$-$k_{adh}\delta_{min}$. Further, unloading past this point reduces both the normal overlap and the attractive forces until the separation occurs at $\delta = 0$. If reloading of the contact occurs while on the adhesion branch, the contact follows the k_2 path (there are an infinite number of k_2 paths depending on the point of initial unloading) parallel to the initial unloading/reloading path until reaching the k_1 path. Further, loading takes place on the virgin loading path, k_1. When the particles are separated, the contact information is lost.

The default EDEM rolling friction model is based on a contact independent directional constant torque model. The total applied torque τ_i is given by Eq. 12.

$$\tau_i = -\mu_r f_{hys} R_i \omega_i \tag{12}$$

where, μ_r is the coefficient of rolling friction between two particles, and a particle and a wall. μ_r is the ratio of the rolling resistance force *(F)* and the normal reaction force *(N)*. R_i is the distance between the contact point and particle centre of mass and ω_i is the angular velocity at the contact point.

6. Numerical simulations

6.1. Soil data

Soil data represents the soil texture and strength parameters, which differentiate between types of soils. In the present study, two soil samples selected from the NATO NRMM CDT (2018). They are 2NS sand and fine-grained sand (FGS). Both sands are selected because of the availability of the soil test data (e.g., grain distribution, moisture content and bulk density).

2NS sand is a poorly graded medium sand but contains a fraction closer to the coarse sand (NATO CDT-308, 2020). It also has more angular sand particles. There are gravel particles in the sand, which makes it a little stiffer (Table 3). The fine-grained sand has cohesiveness due to the presence of silt and clay. It is sandy silt soil of low plasticity. Figure 9 shows the grain size distribution for both soils. The moisture content for 2NS sand is 1.13%, and FGS is 18.55%.

Two different soil models of 2NS sand and fine-grained soil of 6m x 4m x 2m were simulated using the EpAM contact model. The dimensions of the box were selected such that there is no boundary wall effect during the simulations. The models were validated using the experimental data of average cone index values of both sands. Table 3 shows the input parameters of the EpAM contact model. Further, the average cone index of 2NS sand is 149 kPa, and FGS is 22 kPa. The average bulk density of 2NS sand is 2687.66 kg/m^3, and FGS is 1466.64 kg/m^3.

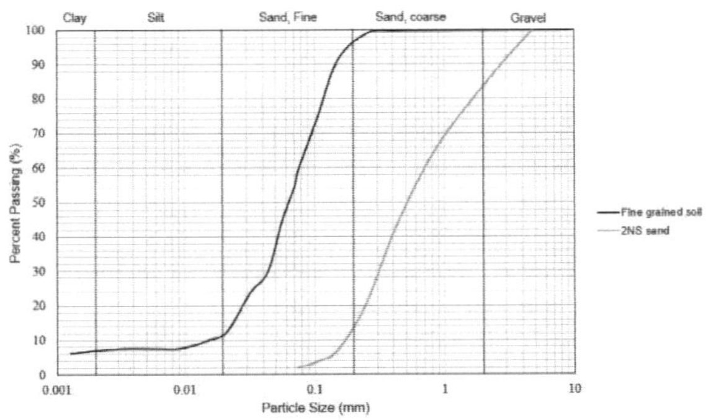

Fig. 9 Grain size analysis of 2NS sand and FGS (NATO CDT-308, 2020)

Table 3
Input simulation parameters of soil models

Input parameter	2NS sand	FGS
Particle density (kg/m3)	2500	2500
Particle size (mm)	10	10
Shear modulus, G (Pa)	5E+6	5E+8
Poisson's ratio, ϑ	0.4	0.25
Restitution, e	0.6	0.6
Static friction	0.5	0.5
Rolling friction	0.01	0.01
Constant pull off force, f_0 (N)	-60	-2
Surface energy $\Delta\gamma$ (J/m2)	0	7
Contact plasticity ratio λ_p	0.5	0.9
Slope exponent, n	1.5	1.5
Tensile exponent	1.5	1.5
Tangential stiffness multiplier, χ	0.67	0.67

6.2. Test tyre

The military vehicle, FED Alpha tyre (335/65R22.5), is considered a full-scale tyre (NATO CDT-308, 2020). The full-scale tyre is designated as Scale = 1. Figure 10 depicts the 3D CAD model of the exterior of the full-scale tyre, which was developed in SOLIDWORKS, v2018. The engineering datasheet from the Goodyear Technical Centre was utilized during the development process.

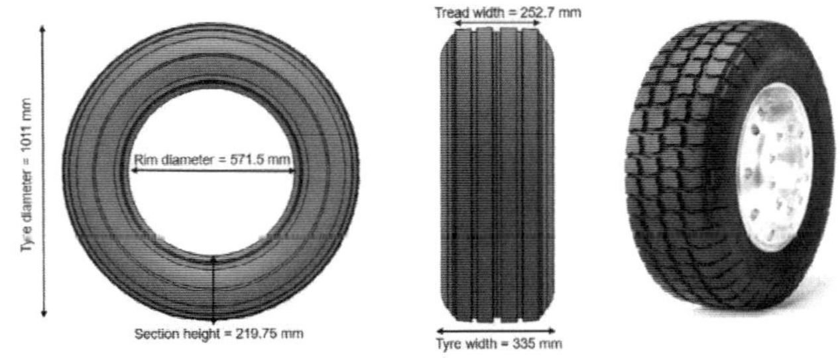

Fig. 10. Tyre model, 335/65R22.5

Additionally, three scaled prototypes of the full-scale tyre were constructed in SOLIDWORKS, with scaling factors of 0.7, 0.5, and 0.25. The scaling was done on geometrical dimensions (e.g., tyre diameter and width), while keeping the same non-dimensional parameter, tyre width/diameter *(b/d)* (Eq. 11). The tyre specifications of full-scale and scale tyres are shown in the Table 4.

Table 4
Tyre models specifications

Scale	d [mm]	b [mm]	r [mm]	Rim size	[b/d]
1	1011	335	505	22.5	0.33
0.7	708	234	354	16	0.33
0.5	505	167	253	11	0.33
0.25	253	84	126	5.5	0.33

Baranawoski et al. (2012) derived the tyre material properties experimentally to be used in the DEM modelling. The same material properties are used in the current study for all four tyres to ensure the same material stiffness. The tyre-soil interaction is defined using the EpAM contact model. Table 5 shows the input parameters for tyre model in DEM modelling.

Table 5
Input simulation parameters of tyre models

Input parameter	Value
Density (kg/m3)	1173
Shear modulus, G (Pa)	4.828E+06
Young's modulus, E (Pa)	14E+06
Poisson's ratio, ϑ	0.45
Restitution, e	0.6
Static friction	0.5
Rolling friction	0.01
Constant pull off force, f_0 (N)	0
Surface energy $\Delta\gamma$ (J/m2)	0.0045
Contact plasticity ratio λ_p	0.5
Slope exponent, n	1.5
Tensile exponent	1.5
Tangential stiffness multiplier, χ	0.67

6.3. Test procedure

Initially, the soil bin, a rectangular box of dimension 6 m long, 4 m wide and 4 m deep was generated for 2NS sand and FGS respectively. The soil particles (uniform angular shapes and sizes) were generated in this box and were allowed to settle under the gravity force till the kinetic energy of system becomes zero (Fig. 11). The particle size is upscaled to improve the simulation time and the discrete particle size does not significantly affect the simulation results. The mean particle size is kept constant, and scale of size variation was also kept constant to develop a homogeneous soil system. After settling, the thickness of soil bed was 2 m, and 2 m high void region was provided above the soil bed surface so that the soil could flow without constraints (Fig. 7).

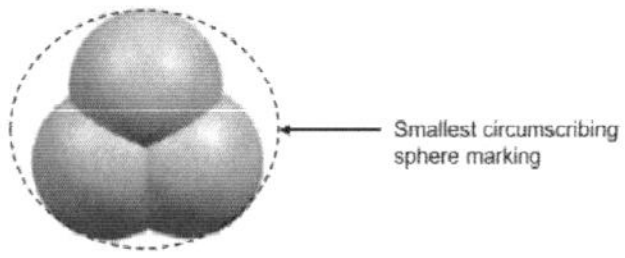

Fig. 11. 3D – 3 sphere particle model

A defined tyre model was imported into the EDEM software and was allowed to settle under gravity on the soil surface. A normal load was applied at the tyre center. Initially, the wheel was driven in a self-propelled condition, characterized by a longitudinal slip ratio of zero, where the circumferential velocity equals the wheel's translational velocity. This self-propelled condition was sustained for an initial duration of 0.5 seconds, during which no additional force was necessary to maintain the vehicle's current speed. In this state, a translational speed of 2.76 m/s and a rotational speed of 52.26 RPM (or 5.47 rad/s) were attained for the full-scale tyre model. The normal load of 13438 N was applied on the full-scale tyre model (NATO CDT-308, 2020).

Following the self-propelled condition, the wheel's rotational speed remained constant at 52.26 RPM (or 5.47 rad/s) for the full-scale tyre model. However, the translational speed was gradually decreased from its initial value of 2.76 m/s to zero over a period of 2 seconds. As a result, the slip varied from 0% to 100% over a duration of 2 seconds. Figures 12 and 13 show the full-scale tyre-soil interaction models.

However, the simulation procedure is similar to NATO CDT-308 (2020). However, in addition to that, this approach offers other advantages, such as maintaining a constant rotational speed and gradually reducing the translational speed. These benefits lead to a reduction in the distance traveled in the forward direction and also contribute to a decrease in computational time by reducing the number of test runs. Moreover, they have reduced the processing time from 10 days on a 6 GB RAM PC to 36 hours on 64 GB RAM PC.

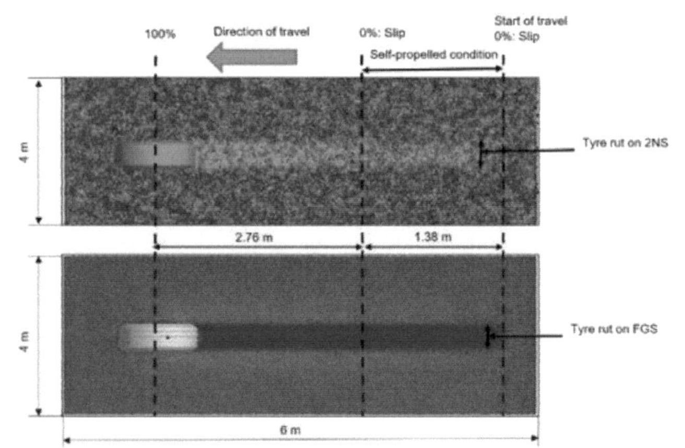

Fig. 12 Tyre testing procedure

Fig. 13 3D model of Tyre-soil interaction system

A similar approach was adopted for the other three scaled tyres with scaling factors of 0.7, 0.5, and 0.25. The rotational speed of 52.26 RPM (or 5.47 rad/s) was maintained consistently for all tyres. Therefore, tyres have different initial translational speed to achieve respective self-propelled conditions. The normal loads on scale tyres were calculated by equating the pressure term $\frac{W}{bd}$ between the full-scale tyre system and scale tyre system as shown in Eq. 13. Table 6 shows the test matrix of full-scale and scale tyre models.

$$\left(\frac{W}{bd}\right)_{full-scale} = \left(\frac{W}{bd}\right)_{scaled} \quad (13)$$

Table 6
Test matrix

Scale	d [mm]	b [mm]	W [N]	[bd/W]
1	1011	335	13438	25.20
0.7	708	234	6585	25.20
0.5	505	167	3359	25.20
0.25	253	84	840	25.20

7. Results

7.1. Model validation

The full-scale tyre soil interaction model was validated with the experimental datasets of drawbar pull vs slip and NATO Reference Mobility Model (NRMM) (NATO CDT-308, 2020). Figures 14 and 15 compare the drawbar pull vs slip relation obtained from simulation model with the experimental datasets. The trend observed in the drawbar pull results aligns with previous studies conducted by Johnson et al. (2017) and Johnson et al. (2015), specifically in the case of a solid wheel operating on soft soil. A quantitative comparison between the simulation and experimental results revealed a mean error of 12% and 9% in the drawbar pull relationship for the 2NS sand and fine-grained sand models, respectively. The simulation results also fall within the prediction band of the NRMM model with an error of less than 10%. Therefore, the proposed simulation techniques, which employs the discrete element method with the Edinburgh elastic-plastic adhesion contact model, to accurately represent experimental conditions in software.

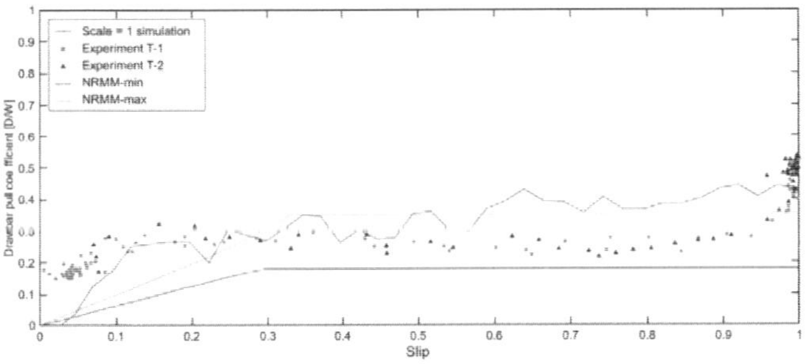

Fig. 14. Drawbar pull vs slip relation of full-scale tyre on 2NS sand

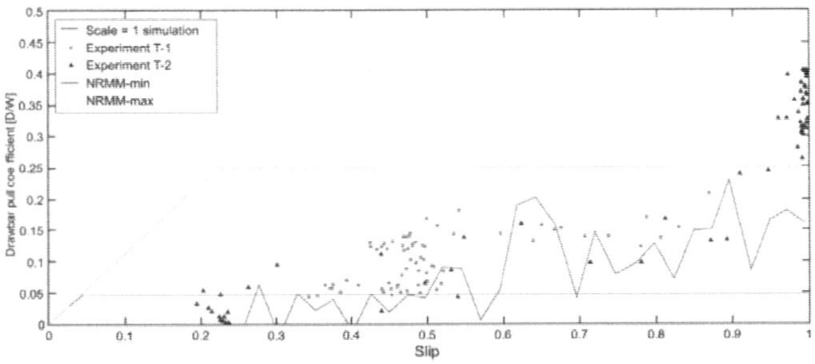

Fig. 15. Drawbar pull vs slip relation of full-scale tyre on FGS

7.2. Scalability analysis

Three independent parameters were incorporated to evaluate the mobility performance of powered-driven wheels. They are drawbar pull coefficient [D/W], gross traction coefficient [T/rW] and tractive efficiency [η]. Figures 16 to 18 shows the comparison of independent parameters of scale tyres (scale = 0.7, 0.5 and 0.25) with the full-scale tyre (scale = 1) on 2NS sand model. Figures 19 to 21 shows the comparison of independent parameters of scale tyres (scale = 0.7, 0.5 and 0.25) with the full-scale tyre (scale = 1) on FGS model. A one-standard-deviation scatter band, depicted using red dashed and star lines, was created around the datasets of the full-scale tyre. This scatter band serves to indicate the accepted error limits and facilitate the comparison of the datasets of the scaled tyres. Tables 7 and 8 present the percentage of data points from the scaled tyre systems (scale = 0.7, 0.5, and 0.25) that fall within the scatter band representing one standard deviation of the performance data points from the full-scale tyre (scale = 1).

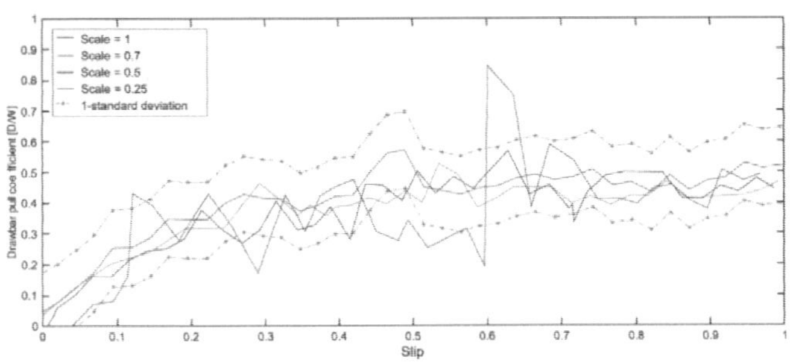

Fig. 16. Scalability of drawbar pull coefficient in 2NS sand

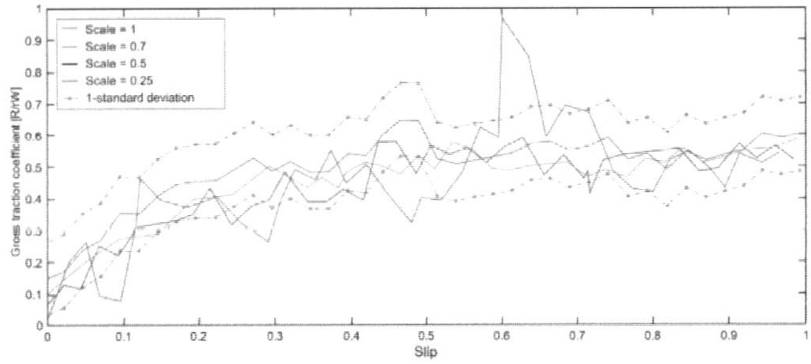

Fig. 17. Scalability of gross traction coefficient in 2NS sand

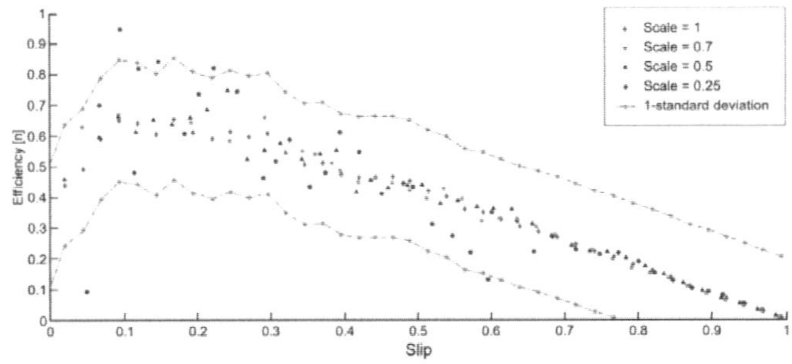

Fig. 18. Scalability of tractive efficiency in 2NS sand

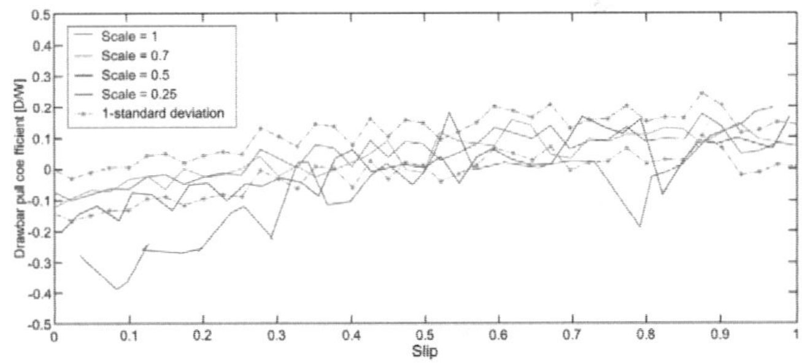

Fig. 19. Scalability of drawbar pull coefficient in FGS

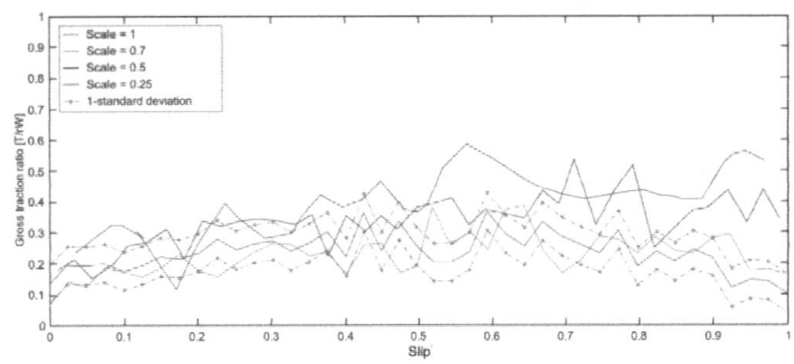

Fig. 20. Scalability of gross traction coefficient in FGS

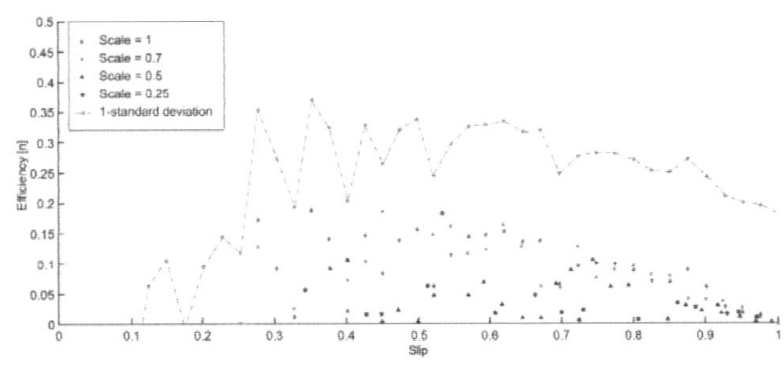

Fig. 21. Scalability of tractive efficiency in FGS

Table 7
Scalability analysis on 2NS sand

Scale	Parameter	Percentage (%)
0.7	D/W	100
	T/rW	97.1
	η	100
0.5	D/W	93
	T/rW	86.1
	η	97.7
0.25	D/W	75.6
	T/rW	80.5
	η	87.8

Table 8
Scalability analysis on FGS

Scale	Parameter	Percentage (%)
0.7	D/W	85.7
	T/rW	78.0
	η	63.4
0.5	D/W	70.7
	T/rW	41.4
	η	56.1
0.25	D/W	17.1
	T/rW	9.8
	η	41.5

The accuracy of scalability is better in 2NS sand than FGS. It might be due to the dependency of scalability on soil characterization and cone index. It was found that scale = 0.7 predicted better performance parameters of full-scale tyre system followed by scale = 0.5 and then by scale = 0.25 in both sands. It might be because of the increase in the rate of change of tyre curvature from 0.7 to 0.5 to 0.25 scaled tyre systems. The tyre curvature plays a significant role in predicting tyre-soil ground contact pressure/forces for tyre diameter less than 50 cm (Griffith and Spenko, 2011, 2013, 2014).

It can be seen that tractive efficiency is significantly affected by a decrease in the scale of tyres in FGS compared to 2NS sand. FGS is a low-bearing capacity soil with a lower cone index than 2NS sand. Therefore, it might be due to the increase in tailing pile build-ups in height (due to the digging effect) which increases the bulldozing resistance, as the scale reduces in FGS, thus reducing the drawbar pull (Fig. 19). There is a non-linear increase in tyre-soil contact area due to digging, which also increases the gross traction ratio (Fig. 20). Therefore, tractive efficiency is reduced in FGS with the scale.

8. Conclusion

The key findings of this study demonstrates that the scalability of tyre-soil system exists well in 2NS sand and Fine Grain Sand (FGS). The similarity between the different tyre-soil systems can be established by equating the pressure term, *(W/bd)*, for the same soil type. However, the accuracy of scalability decreases with the decrease in the scales of the tyre – soil systems. Up to 0.5 scale-system can predict the full- scale system's mobility performance on sandy soils. Furthermore, the accuracy depends on various factors such as the soil characterization, soil type and cone index.

According to the presented simulation technique, the discrete element method utilizing the Edinburgh elastic-plastic adhesion contact model (EpAM), provides a better prediction of the experimental drawbar pull versus slip relationship for the full-scale tyre on both soil types than NRMM. Therefore, the EpAM model can be utilized to design and optimize tyre performance on various soil types and improve our understanding of the tyre-soil interaction system.

Future research will investigate the applicability of scalability in heterogeneous soil models of 2NS sand and FGS. Further, the effect of parameters such as cone index, normal load and tyre deflection will be analysed.

9. Data Availability Statement

Experimental data supporting this study is openly available at https://www.mtu.edu/cdt/

10. References

ASABE Standards, 2006. ASAE S313.3 FEB04 Soil cone penetrometer. St. Joseph, Michigan, ASABE.

Baranowski, P., Bogusz, P., Gotowicki, P., Małachows, J., 2012. Assessment of mechanical properties of offroad vehicle tyre: coupons testing and - FE model development. Acta Mechanica et Automatica, 6(2), pp.17-22.

Bekker, M. G., 1956. Theory of land locomotion. The University of Michigan Press.

Bekker, M.G., 1969. Introduction to terrain-vehicle systems, vol. I. The University of Michigan Press.

DEM Solutions Ltd., 2011. EDEM 2.4 Theory Reference Guide. Edinburgh, Scotland, UK.

Egozi, A., 2018. US Army pursues Israeli robots. Breaking Defense, November 12. Available at: https://breakingdefense.com/2018/11/us-army-pursues-israeli-robots/ (Accessed: 2024.04.09).

Engineering Design Handbook, 1967. Automotive series, United States Army Material Command.

Freitag, D.R., 1965. A dimensional analysis of the performance of pneumatic tyres on soft soils. Technical report number: 3-688, USAE Waterways Experiment Station.

Griffith, M. G. and Spenko, M., 2011. A modified pressure–sinkage model for small, rigid wheels on deformable terrains. Journal of Terramechanics, 48 (2), pp. 149-155.

Griffith, M. G. and Spenko, M., 2013. A pressure-sinkage model for small-diameter wheels on compactive, deformable terrain. Journal of Terramechanics, 50(1), pp. 37-44.

Griffith, M. G. and Spenko, M., 2014. Development and experimental validation of an improved pressure-sinkage model for small-wheeled vehicles on dilative, deformable terrain. Journal of Terramechanics, 51, pp.19-29.

Jayakumar, P, Melanz, D, MacLennan, J, Gorsich, D, Senatore, C and Iagnemma, K, 2014. Scalability of classical terramechanics models for lightweight vehicle applications incorporating stochastic modeling and uncertainty propagation. Journal of Terramechanics, 54, pp. 37-57.

Johnson, B., Duvoy, X., Kulchitsky, V., Creager, C. and Moore, J., 2017. Analysis of Mars Exploration Rover wheel mobility processes and the limitations of classical terramechanics models using discrete element method simulations. Journal of Terramechanics, 73, pp. 61-71.

Johnson, B., Kulchitsky, V., Duvoy, P, Iagnemma, K, Senatore, C, Raymond, E. A. and Jeffery, M., 2015. Discrete element method simulations of Mars Exploration Rover wheel performance. Journal of Terramechanics, 62, pp. 31-40.

NATO CDT – 308, 2020. North Atlantic Treaty Organization STO Technical report Cooperative Demonstration of Technology (CDT- 308): Next-Generation NATO Reference Mobility Model (NG-NRMM), Research and Technology Organization.

NATO NRMM CDT, 2018. North Atlantic Treaty Organization STO Technical report AVT-248: Next-Generation NATO Reference Mobility Model (NRMM) Development, Research and Technology Organization.

NATO TR SAS – 097, 2018. North Atlantic Treaty Organization STO Technical report: AC/323 (SAS-097) TP/776, Research and Technology Organization.

Rui He, Corina Sandu, Hoda Mousavi, Mohit N. Shenvi, Kirsten Braun, Ray Kruger, P. Schalk Els, 2020. Updated Standards of the International Society for Terrain-Vehicle Systems, Journal of Terramechanics, Volume 91, 2020, Pages 185-231.

Swanson, G D, 1973. Scale model Tyre Tests in clay. Journal of Terramechanics, 10 (3), pp. 21-27.

Wong J., 2008. Theory of ground vehicles, fourth ed., Wiley.

Wong, J., 2010. Terramechanics and Off-Road Vehicle Engineering, second ed., Wiley.

7829 | Development of Foot in Balloon Biped Robot using Buoyancy Force for Traveling Soft Ground

Noriaki Mizukami [a,*], Masanori Aragane [a], Mami Nishida [a]

[a] *International Professional University of Technology in Tokyo*

* *Corresponding author: mizukami.noriaki@t.iput.ac.jp*

ABSTRACT

In planetary exploration mission, an exploration robot has to travel on unknown and uneven, soft ground. Thus, a locomotion mechanism is an important, and the locomotion systems are discussed three types such as wheel, crawler and legs. The leg type has advantages of adaptability and an effective locomotion for terrains. Because a walking robot is able to avoid rocks and depressed areas. However, the biped robot is structurally complex and fall down easily. In this research, we developed the biped robot that is made of a soft balloon body fulfilled by helium. Since the body of the robot is always floating by helium balloon, the robot cannot fall down while walking. The leg structure for walking is made of an artificial muscle actuator. When the control command applies to the actuators, the legs swing up lightly. When the control command stops applying, the legs swing down slowly.

The robot is able to move forward when the legs swing up and down repeat alternately. We have achieved the walking movement by a manual ON-OFF control of actuators on firm grounds (Mizukami et al. 2023). The purpose of this research is to clarify the interaction between the leg mechanism and soft soil.

In this paper, we performed walking experiments on soft soil to estimate states of walking movements on flat soft soil using original flat shape of foots. It was confirmed that the foots did not sink and slip and the robot moved forward. Furthermore, we propose a new type of foot parts that is able to travel on uneven, soft or slope grounds, and perform experiments for verifying effectiveness.

Keywords
Soft biped walking robot,
Planetary surface exploration,
Soft robotics

1. Introduction

An exploration robot travels on unknown and uneven, soft ground and able to return soil samples from wide and different areas in planetary exploration mission. Thus, it is important to design a locomotion mechanism. The type of locomotion mechanism is mainly discussed three types such as wheels, crawlers and legs. However, since planetary surfaces are covered with soft soils, wheels on a planetary rover easily slip and get stuck. On the other hand, the biped robot also has an issue in risk of falling down, because of complex structure and weight on the robot. The walking robot with legs is able to avoid rocks and depressed areas. Thus, in these locomotion systems, the leg types locomotion has advantages of adaptability and effective for uneven terrains such as soft soil terrains and uneven terrains, where there are a lot of rocks and depression areas. In these terrains, the wheel or crawler locomotion is not able to travel. The biped robot, which is same size of human and generally mounts motors or servomechanisms, has been developing. For example, Tesla has been developing Tesla Bot (Tesla, Inc.), and Boston Dynamics developed Atlas using hydraulic actuators (Boston Dynamics). These robots are very heavy for loading transportation such as rockets.

We have been developing a biped robot with a soft balloon body filled with helium. The robot cannot fall down while walking, because the body of the robot is always floating by helium balloon. The leg structure for walking consists of an artificial muscle actuator: the shape memory metal and soft thin expanded polystyrene board. We have achieved the walking movement by a manual ON-OFF voltage command of actuators on firm surfaces. When the voltage command applies to the actuators, the legs swing up lightly. When the voltage command stops applying, the legs swing down slowly. The robot is able to move forward when the legs swing up and down repeat alternately. The balloon robot is developed at light in weight and low cost, thus, it is possible to mount a lot of the robots on a rocket. It is expected that a surface exploration and cooperative work by plurality of robots can achieve, and the robots are able to climb soft slopes and avoid rocks.

In this research, we aim to achieve a walk stabilization control by a simple control method that takes advantage of special feature in the proposed robot. And the robot is able to walk on uneven and soft soil surface and cannot fall down because of buoyancy. However, the interaction between a leg and a foot mechanisms and soft soil has not been clarified yet. The purpose of this research is to clarify the interaction between the leg and foot mechanisms and soft soil. In this paper, we performed walking experiments on soft

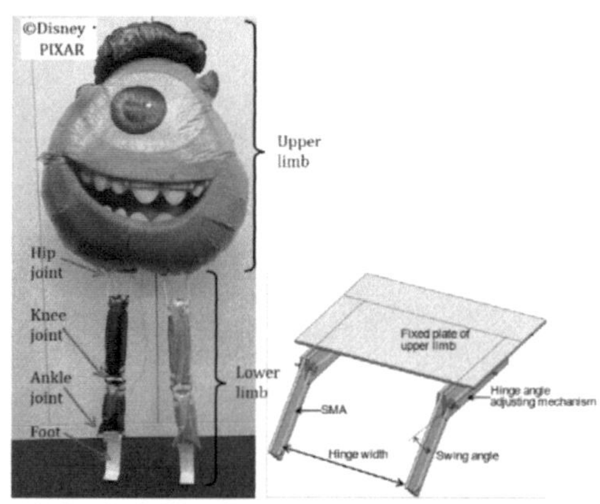

(a) Balloon biped walking robot (b) Joint of lower limb

Fig. 1. Balloon biped walking robot and Joint of lower limb

https://doi.org/10.56884/ZGWJUUA3

soil to estimate states of walking movements on flat soft soil using original flat shape of foots. It was confirmed that the foots did not sink and slip and the robot moved forward. Furthermore, we propose a foot design that is able to travel on uneven, soft or slope grounds, and perform experiments for verifying effectiveness.

2. Description of balloon biped robot

2.1. Robot structure

Figure 1(a) shows an aspect of a balloon biped robot that has been developing in this research (Mizukami et al. 2023). The robot has both upper limb and lower limb. A total height of the robot is about 160[cm] and a mass is about 50[g]. Because of a helium balloon as an upper side body in the robot, a buoyancy force works and adjusts a contacting weight to become about 7[g]. Thus, because the robot is floating, the robot achieves walking without fall down when the robot faces an external force. The structure of the robot is soft lightweight materials for keeping safe because of avoiding the robot damages that is caused by falling down and collision. The upper limb is made of a soft aluminum balloon and connects to the lower limb by a hinge joint structure as shown in Fig.1(b). The lower limb also consists of a soft lightweight material; the aluminum balloon. The hinge joint structure and foot parts are made of a thin expanded polystyrene board, and a shape memory metal(SMA) is used as an actuator.

2.2. Hinge Joint in Lower Limb and Walking Mechanism

When the robot walks, the leg swing motion is achieved by the hinge joint as shown in Fig.1(b). The hinge joints are mounted both right leg and left leg. The shape memory metal (SMA) shrinks with warm when a current flows, and return a natural length with cold when the current does not flows. A swing angle of leg is controlled by features of expansion and contraction in SMA. The walking motion mechanism describes as follows. The SMA of the hinge joint shrinks by current flows when the application of the voltage command is ON. Because the hinge joint swings up, the foot moves from the floor and the motion of leg becomes a flexion in a swing phase. When the application of the voltage command is OFF and no current flows, the SMA of the hinge joint returns an initial length, and the motion of leg becomes an extension in a support phase. When the application of the voltage command changes from OFF to ON, the foot moves from the floor and becomes the swing phase from the support phase. A center of gravity in the robot moves forward, because the balloon body receives a force of swing leg and a reaction force from the support leg when the SMA the leg returns the initial length.

2.3. Foot Structure

In a movement function of a foot part, a force that is the gravity force acting on the balloon body becomes a rotational motion by a fulcrum in the sole of the foot while walking. This movement causes the principle of swinging lever. The gravity force changes a propulsion force by the rotational motion (Götz-Neumann et al. 2005). This is called a rocker function, and it describes a forward movement of body using a rocking chair. The rocker function has three phases as shown in Fig.2. A heel rocker is a rotational movement by a fulcrum of the heel, and a leg moves forward. An ankle rocker is a rotational movement by a fulcrum of the ankle, and the center of gravity moves forward. A forefoot rocker is a rotational movement by a fulcrum of the toes, and the center of

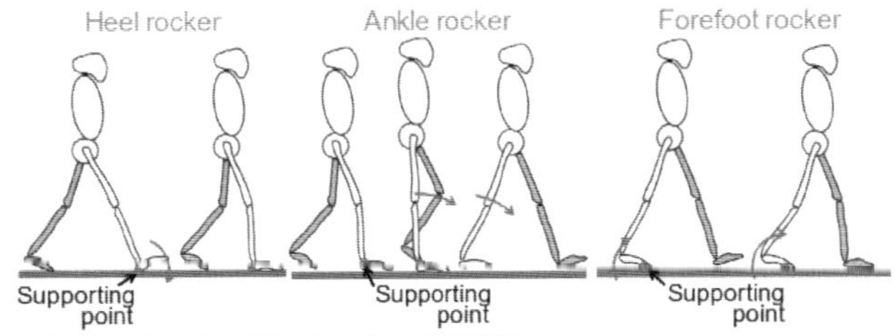

Fig. 2. Rocker function (Mizukami et al. 2024)

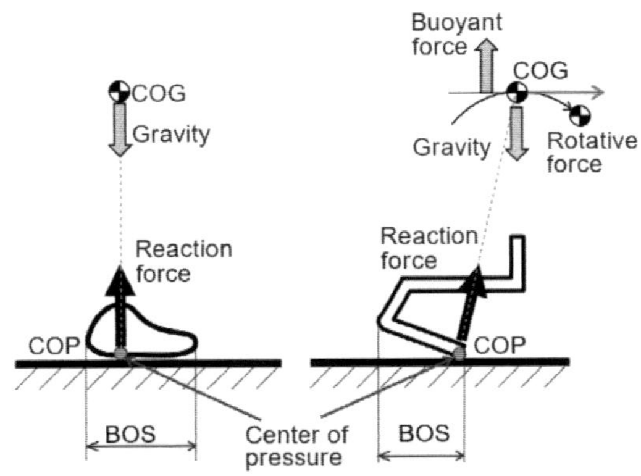

(a) General biped robot (b) Balloon biped robot

Fig. 3. Rocker function (Mizukami et al. 2024)

gravity moves further forward.

It is important for biped walking robots to move the center of gravity forward. In the conventional biped walking robots, a center of pressure (COP) is positioned only in a base of support (BOS) as shown in Fig.3(a). When the center of gravity (COG) moves out of BOS, a rotative force becomes uncontrollable and the robot falls down. On the other hand, a vertical motion by the gravity is not considered in the proposed balloon biped walking robots, because of the buoyancy provided. A walking pattern is generated by focusing on the horizontal of upper limb and the rotational motion, because a shape of foot as shown in Fig.3(b) makes the rocker function work. The foot is a blade mechanism that is inspired by the characteristic in prosthetic leg for competition. It bends strongly and is able to kick back stronger.

3. Walking experiments on soft soil surface

The robot walks from a left side to a right side as shown in Fig.4(a)~(f). The right leg is the supporting leg and left leg is the swing leg. The left leg moves forward from a state of double stance as shown in Fig.4(a) to a state of double stance again as shown in Fig.4(f). The gravity of the robot moves up and down while walking.

3.1. Description of experiments

It was confirmed that the balloon biped robot walked on the firm surface such as a desk and a carpet. The issues of walking on the soft surface have not been clarified yet. Thus, to clarify effects of a propulsion force of walking, a slip and a sink of legs when a contact force of leg on the soft surface is small, the experiments are performed. Figure 5 shows an experiment environment. The size of a sand container is 790mm in the long side and 480mm in the short side. The soft soil is a silica sand No.5 and the sand depth is about 10mm in the container. In the experiments, we change the

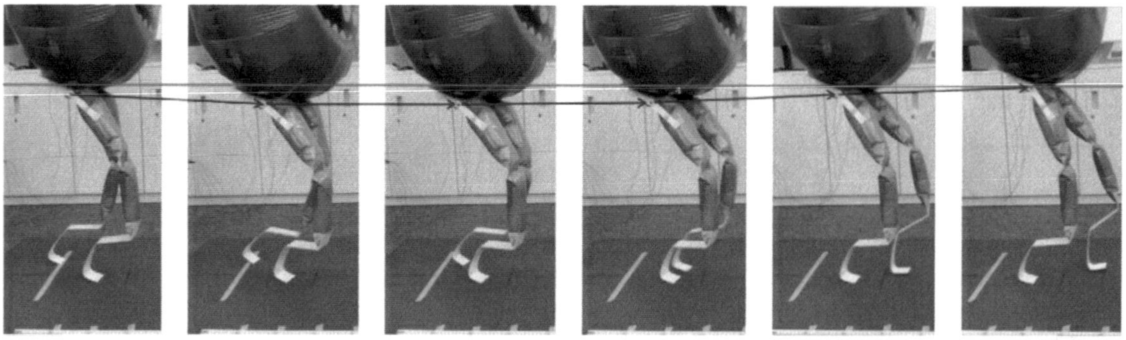

Fig. 4. Walking movement on firm surfaces (Mizukami et al. 2024)

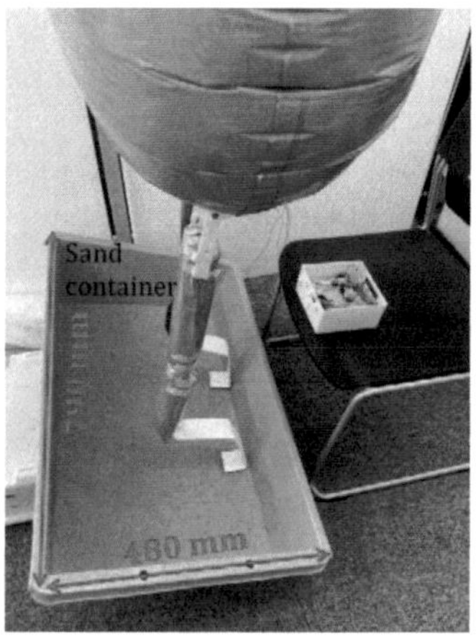

Fig. 5. Experiment environment

slope angle of the sand container and use two types of the foot parts. The robot walked by a manual ON-OFF voltage command.

3.2. Experimental result

The robot without lugs walked a long distance; 790mm on a flat surface from the right side to the left side as shown in Fig.6(e)-(a). The robot without lugs also climbed the long distance; 790mm on a slope surface (The slope angle is 12.5 degree) from the right side to the left side as shown in Fig.7(e)-(a). However, the robot without lugs could not climb on a slope surface (The slope angle is 30 degree). When the slope angle is about 20 degree, the robot without lugs climbed the middle on a short distance; 480mm as shown in Fig.8(e)-(a). To improve climbing ability, we considered a foot part that has a 10mm lug on a bottom of foot as shown in Fig.9. The robot with lugs also climbed the middle on the short distance; 480mm as shown in Fig.9(e)-(a). In experiments, we did not confirm the sink of foot on both the flat and the slope sand surface. However, we confirmed that the slip of foot on the slope surface.

4. Discussion

It was confirmed that the robot without lugs is able to walk and climb on the soft soil surface, however, the bigger an inclination of the slope is, it becomes more difficult to climb. Since there is a friction force between the bottom of the foot and soil surface and the legs support the robot on the soil surface, the robot walks forward and climb up. The bigger the inclination of slope is, the friction force is not large enough and the legs slip. Thus, the supporting leg could not keep the foot grip on the soil surface. In the results of Fig.8 and 9, the robot with 10mm lugs was expected to climb longer distance than the robot without lug, however, the distance of climbing was same. Because a penetration of lugs and the contact between the foot surface and the soft soil surface is not enough, the friction force is not sufficient to climb the bigger slope.

The result shows that the friction force and sufficient penetration of lug are important factors to climb on soft soil surface. Since the contact force of foots are very small, the balloon robot does not get enough large friction force and reaction force from lugs. Thus, it is required that the foot presses soft soil surface by the ankle joint control for getting enough contact force.

5. Conclusion

In this paper, we explained about an experiment verification by the balloon biped robot. The walking and climbing experiments were performed on the flat and slope soft soil surface. The robot is able to walk on the flat soft soil surface and climb on slope. However, the bigger inclination of slope is, it became more difficult to climb. It was assumed that the penetration of lugs and the contact between the foot surface and the soft soil surface is not enough, the friction force is not sufficient and then the foots slip. The friction force and sufficient penetration of lug are important factors to climb on soft soil surface

In the future work, we will develop the ankle joint device that the foot is able to press on the soft soil surface. And then we will perform experiment to evaluate abilities of walking and climbing.

References

Boston Dynamics, https://bostondynamics.com/atlas/ (accessed 2024-05-27)

Götz-Neumann. K., (translator) Tsukishiro. K. et al., 2005, Gehen verstehen: Ganganalyse in der Physiotherapie, IGAKU-SHOIN Ltd., pp. 9-16.

Mizukami. N., Aragane. M. and Nishida. M., 2023, Validity Evaluation and Modeling of Walking Motion in Balloon Biped Robot, Announcement of 28th Robotics Symposia, 1D2.

Mizukami. N., Aragane M. and Nishida. M., 2024, Verification of Walk Stabilization in Balloon Biped Robot by Soft Foot Landing Motion, Announcement of 29th Robotics Symposia, 6C3.

Tesla, Inc., https://www.tesla.com/AI (accessed 2024-05-27)

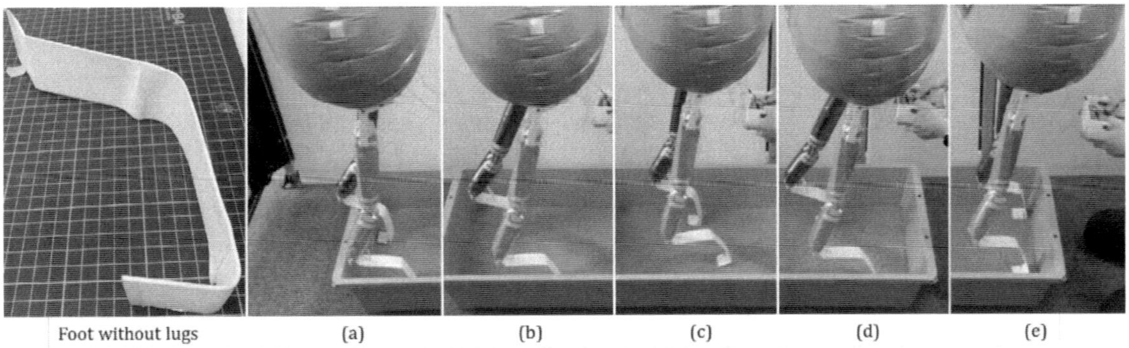

Fig. 6. Walking movement from the right side to the left side on soft soil flat surface: no Lug legs

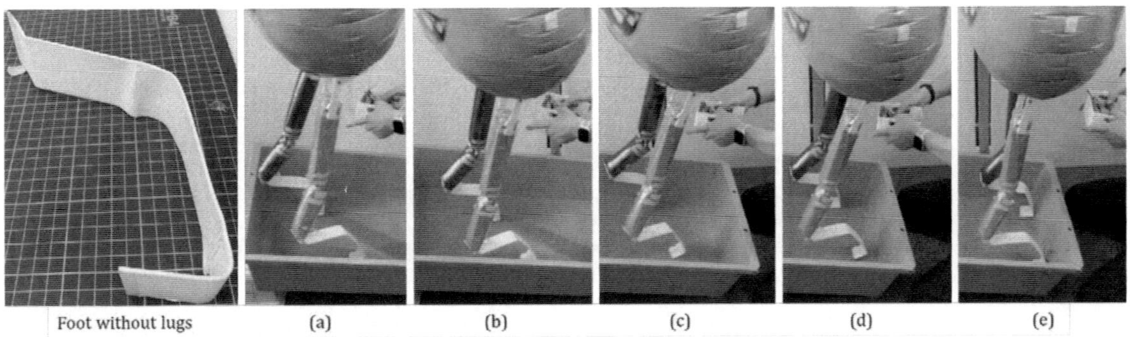

Fig. 7. Walking movement from the right side to the left side on soft soil slope surface (angle 12.5 degree): no Lug legs

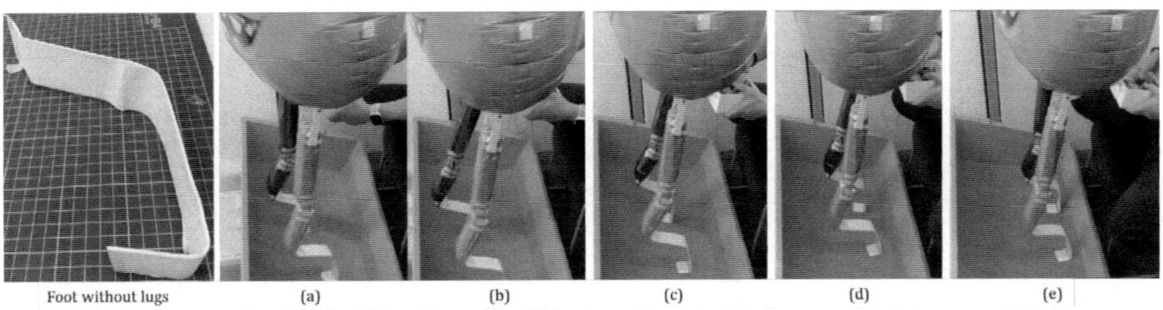

Fig. 8. Walking movement from the right side to the left side on soft soil slope surface (angle 20 degree): no Lug legs

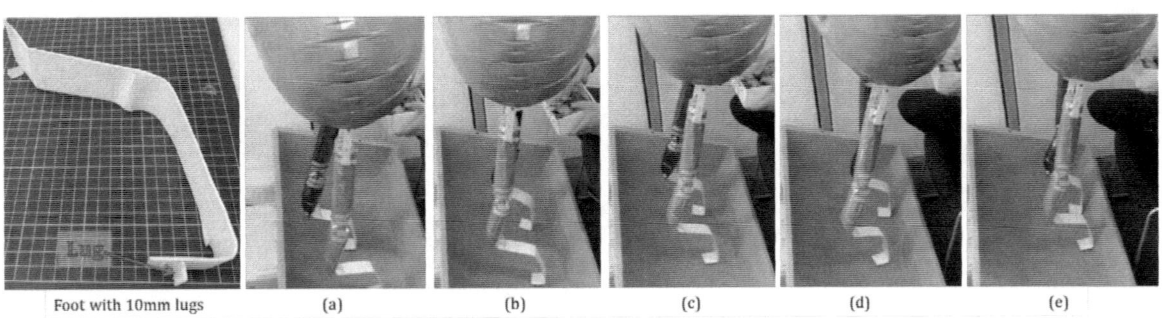

Fig. 9. Walking movement from the right side to the left side on soft soil slope surface (angle 20 degree): 10mm Lug on legs

8084 | Log detection for autonomous forwarding using auto-annotated data from a real-time virtual environment

Mattias Lehto [a,*], Håkan Lideskog [a],
Magnus Karlberg [a]

[a] *Luleå University of Technology, 97187 Luleå, Sweden*

* *Corresponding author: mattias.lehto@ltu.se*

ABSTRACT

An integral part of autonomous forestry is the ability of the vehicles, e.g., forwarders and harvesters, to perceive their environment. At Luleå University of Technology, object detectors have previously been developed, allowing forestry vehicles to detect and position important objects in forestry, such as tree stumps, stones, and logs. These detectors have been developed by training on physical manually annotated data, which is both time-consuming and costly. Training on virtual data allows for significant time- and cost reductions. Since the ground truth in the virtual model is known, the training data can be auto-annotated, allowing for the creation of larger training datasets, at a lower cost. In this work, a virtual environment in Unity is used in co-simulation with a real-time digital twin of a physical forestry vehicle, to generate auto-annotated training data, as captured by an onboard stereo camera. A detailed emulation of the stereo camera is used to achieve realistic results. First, a log detector trained on physical manually annotated data, is evaluated on virtually created data. It is shown that the log detector trained on physical data can detect logs in the virtual environment. Second, new detectors are trained, using different shares of physical and virtual data. It is shown that a detector trained using only virtual data, can learn to detect logs in the physical world. Moreover, virtual pre-training is shown to improve the performance of physically trained and tested detectors, both at low availability of physical training data, and in terms of domain generalization. Furthermore, the real-time capable virtual models also enable future machine learning tasks utilizing different levels of Hardware-in-the-Loop.

Keywords
Transfer learning
Domain generalization
Virtual training
Auto-annotation
Real-time
Co-simulation
Logging
Tree harvesting
Forwarder
Cut-to-length
CTL

1. Introduction

The Nordic cut-to-length (CTL) tree harvesting system is currently the world's most technologically advanced. Recent developments seek to increase the level of automation (LOA) of the forestry vehicles used within CTL, such as forwarders and harvesters (Lindroos et al., 2017). To increase the level of automation, machine learning (ML) can be applied to enable the vehicles to autonomously perform different tasks or subtasks. In the related automotive industry, ML has been used for many years to successfully enable vehicle automation (Pandharipande et al., 2023), although with fewer Degrees-of-Freedom to control compared to forestry machines (cf. lateral/longitudinal in Kuutti et al. (2021)) and in substantially different environments. ML in forestry is studied to a lesser extent, nevertheless, autonomous navigation of forestry vehicles has been studied (Jelavic et al., 2021), and ML used for forestry crane control was studied by Andersson et al. (2021). An integral part of autonomous forestry operations, is the ability of the vehicles to perceive their environment, this has been studied to some extent (Condat et al., 2024; Li and Lideskog, 2023, 2021). The previous work on machine learning within forestry has almost exclusively been performed either in a fully virtual setting (Andersson et al., 2021) or in a fully physical setting (Condat et al., 2024; Li and Lideskog, 2023, 2021). When working in a fully virtual setting, the question of whether the learned abilities can be applied in the physical world remains unanswered (Choi et al., 2021). Some degree of transferability from virtual training to the physical world, within a forestry context, was claimed by (Grondin et al., 2022) for their instance segmentation of trees, but no numbers on the degree of transfer were provided. Using a fully physical setting for machine learning has other clear disadvantages, for example with regards to time consumption, costs, and risks. Training of object detectors in a physical environment usually involves going out to the environment in question, taking a large number of pictures in different conditions, and thereafter manually annotating the pictures before starting the training. This practice can be highly time consuming and costly, especially for generating large datasets. Virtual models, on the other hand, have a known ground truth. This allows for virtual training data to be auto-annotated, which enables the creation of larger training datasets, at a lower cost. In addition, virtual data generation can be done without transporting personnel and equipment to potentially remote areas.

If an accurate real-time capable digital model could be used for machine learning tasks within forestry, many of the disadvantages of using a fully physical environment could be mitigated. The real time capability would also allow for incorporating different parts of the physical vehicle into the simulation loop, in a Hardware-in-the-Loop (HiL) or Vehicle-in-the-Loop (ViL) fashion (Fathy et al., 2006). This could be useful for increasing the model accuracy by incorporating parts of the physical hardware in the simulation, saving modeling time, or for incorporating physical black box devices into the simulation, such as a proprietary traction control system, as an example. One of the two main questions, with regards to such an approach, is whether a sufficiently accurate digital model for this purpose will still be real-time capable. The second main question is to what degree the abilities learned using the real-time digital model can be applied to the physical world. The goal of this paper is to answer these two questions. In order to achieve this, a real-time digital twin of a forestry vehicle, previously described in (Lehto et al., 2023), will be used to generate auto-annotated RGB camera training data for object

https://doi.org/10.56884/XD21D6FR

detection of logs. Thereafter the transferability of the virtually trained detector to the physical world will be evaluated. HiL implementation will not be extensively explored in this paper, but the potential for more extensive future training which incorporates HiL is an exciting future extension.

2. Method and materials

In this work, auto-annotated RGB camera training data for object detection was generated through a co-simulation between the real-time multi-physics simulation software Mevea (Mevea) and the game engine Unity (Unity). This section describes the details of the simulation model, data acquisition, and subsequent object detector training.

2.1. Vehicle model and co-simulation master

The real-time multi-physics simulation software Mevea was used as the master in the co-simulation with Unity, and the connection was made using the Mevea Unity API. All the physics calculations were performed in the Mevea simulation model, thereafter the positions of each object, at each time step, were sent to Unity for visualization. The Mevea simulation model contained a detailed real-time digital twin of a full-scale physical forestry vehicle, as well an environment model, and the models of five different types of birch logs. The modeled forestry vehicle weighs 10 tons and was developed and built at Luleå University of Technology (Lideskog et al., 2015). The vehicle serves as a modular research platform for hardware and software testing and validation. It also performed the world's first fully autonomous forwarding of full-scale logs in 2021 (La Hera et al., 2024). The real-time digital twin of the forestry vehicle features multibody simulation, hydraulics, contacts and collisions, flexible tires, emulated sensors, and various co-simulation capabilities. The simulated hydraulics include a driveline with hydraulic wheel motors, hydraulic pendulum arms, and a hydraulic articulated joint. The hydraulics simulation in Mevea is based on lumped fluid theory (Watton, 1989). For more information about the forestry vehicle real-time digital twin, see Lehto et al. (2023).

2.2. Environment and log models

The environment and log visualization models were supplied by the company Repli5 (Repli5), and created using their proprietary modeling software. The environment model was made to have high similarity to a physical test site in Hörnefors, Sweden, shown in Fig. 1. The physical test site in Hörnefors was previously used for generating training and validation data for another log detector, which was trained on physical manually annotated data (Li and Lideskog, 2023). For the virtually trained log detector described in this study, the previous log detector was to serve as a reference for comparison. The same physical validation dataset from Hörnefors was chosen to be used as the main validation set also for the new virtually trained log detector. In order for the virtually trained log detector to perform well on the chosen main validation set, it was assumed that it could be advantageous if the virtual and physical environments were relatively similar.

The environment model and log models had separate collision and visualization graphics. The visualization graphics were used for the emulated camera in Unity and had a high level of detail to increase realism. The ground model was separated into 27 separate flat surfaces, consisting of 54 triangles in total. Each of the 27 ground surfaces had their own texture, where each texture was 4096x4096 pixels. In addition to the ground graphics there were also visualization graphics for 102 different objects such as trees, stumps, stones, and shrubs in the environment. These consisted of 3.1 million triangles in total and each object had a texture of either 2048x2048 or 4096x4096 pixels.

Five different birch log models were used, ranging between 49 000 and 171 000 triangles, and they had one texture each. The log textures ranged from 1024x1024 to 4096x4096 pixels in size. In addition to the textures, all objects also had a normal map of the same pixel size as their texture.

The collision graphics were made as simple as possible to reduce collision calculation time. Since the test site was flat and rectangular it could have theoretically been simplified to consist of only 2 triangles. But because of the large size of the environment, 150x33 m, this made the points too sparse for the collision calculations in Mevea to work properly. Therefore, a single flat surface consisting of 24 triangles was iteratively chosen instead. This worked well for the purpose of placing the logs on the surface.

The log collision graphics were made to enable rotating the logs to many different positions around their main axis, and thereby give the detector many different visual inputs from the same log. As a compromise between simplicity and number of possible rotational positions around the main axis, the log collision graphics were made as a sweep between two octagons of different size, and consisted of 28 triangles per log. No collision graphics were prepared for the other objects, such as stumps, stones, trees, and shrubs since it was considered outside the scope of this work.

The virtual environment in Unity is shown in Fig. 2. The five different types of birch logs scattered in the virtual environment is shown in Fig. 3.

Fig. 1. The physical test site in Hörnefors, Sweden

Fig. 2. The virtual environment in Unity

Fig. 3. The virtual environment in Unity with five different types of birch logs scattered

2.3. Data acquisition

The physical forestry vehicle has a Stereolabs ZED 2 stereo camera mounted at the front. For object detection only one of the two cameras in the stereo camera is used. In order to emulate this camera well in Unity, data about the physical camera was gathered from datasheets and input as parameters for the Unity camera component. By selecting the "physical camera" option the focal length and sensor dimensions could be used to achieve the correct field of view (FOV) in both the horizontal and vertical directions. Thereafter the correct output image size of 1280x720 pixels was set in the game mode display options. The Unity Perception package was used to auto-annotate the RGB camera images.

In its current configuration the physical stereo camera's angle with respect to the ground plane is manually adjustable, but fixed during operation. However, the previous physical training and validation data was gathered by using a handheld ZED 2 stereo camera. This was done both for overall simplicity and to facilitate the capturing of images at multiple different angles, which was expected to make the resulting detector more robust. In order to achieve the same thing for the virtual training data, the camera component in Mevea was equipped with an object mover component, making the camera in Unity pan up and down with a period of 5 s.

In order to make the resulting detector more robust, multiple configurations of the virtual environment were used. The positions and angles of the logs were adjusted multiple times, and many different combinations of the five different birch log models were used. Same as for the physical dataset, 4-5 logs were spread out in the environment at the same time. The environment was also adjusted multiple times, sometimes all the additional objects, such as stones, trees, shrubs, and stumps were included in the simulation, other times they were not. For the previous detector, trained on physical data, it was seen that a common false positive was caused by log-like shadows cast on the ground by trees or standing deadwood. These kinds of shadows can be seen in Fig. 1, Fig. 2 , and Fig. 3. To make the virtually trained detector learn that these shadows are not logs, as well as to learn that standing birch stems are not logs, six standing birch stems were added to the virtual environment, two of these can be seen in Fig. 2. Some, or all of them, were occasionally moved or removed to increase the detector robustness.

The virtual RGB camera data was collected at a rate of 2 Hz, while manually steering the vehicle around the virtual environment at a speed of 0.9 m/s. The rate of 2 Hz was chosen as it was perceived to generate non-redundant images at a suitable rate in this scenario. As a reference, the previous physical image data was collected at a rate of 1 Hz for the same reasons, although with a slower stride during collection.

The main virtual dataset, before augmentation, consisted of 10 112 annotated images with 16 079 annotated logs. These images came from 14 different configurations, as described above. The number of images from each configuration ranged from 340 to 1130, although most of the configurations generated around the average of 766 images. As a reference, the previous physical dataset consisted of 1739 images. Out of the 10 112 images 90% were used for training, while 5% were reserved for validation and 5% for testing, leaving 9099 images for training. Although the detectors were mainly validated against a physical dataset, virtual validation still served other purposes, such as evaluating the quality of the virtual training dataset. The virtual test set was not yet utilized, but kept for any future scenario where it could be needed.

2.4. Training

The previous log detector, trained on physical data, which was chosen as a reference for this work (Li and Lideskog, 2023), used YOLO v3-Tiny (Adarsh et al., 2020) and the open-source neural network framework Darknet (Darknet). To enable making a fair comparison to the previous detector, while using virtual training data instead of physical, the YOLO v3-Tiny model and Darknet were used for this work as well. The same training settings were also used, except the number of batches, which was always adjusted to ensure that the training did not stop until the validation results had stopped improving. The validation results were assessed by measuring the AP50 (average Precision at an intersection over union threshold of 0.5) (Salton and McGill, 1983), since only one class is used this is equal to the mAP50 (mean average Precision at an intersection over union threshold of 0.5). During training, the weights corresponding to the best validation score were continuously saved, so that running the training a bit too long would not result in a worse detector. Additional methodology and the number of batches for each different training case are specified in *3. Results.* The YOLO v3-Tiny training settings: angle, saturation and exposure were all set to 0, the remaining training settings are shown in Table 1.

Table 1
Training settings

Width	Height	Momentum	Decay	Hue	Learning rate	Burn in
608	608	0.9	0.0005	0.1	0.001	1000

Other than the AP50, the Precision and Recall scores of the detectors were also evaluated. Precision and Recall are calculated according to Eq. 1 and Eq. 2, where TP is the number of true positives, FP is the number of false positives and FN is the number of false negatives.

$$Precision = \frac{TP}{TP + FP} \qquad (1)$$

$$Recall = \frac{\mathrm{TP}}{TP + FN} \tag{2}$$

Before training, the input data needed to be prepared. Same as for the previous detector, Roboflow (Roboflow) was used to resize the pictures from 1280x720 pixels to 608x608 pixels. Black edges above and below the images were used to change the aspect ratio without distorting the images. Roboflow was also used to add augmentation to the datasets. Different levels of augmentation were evaluated, these are further specified in *3. Results.* In addition, the Unity Perception package outputs the auto-annotated data in a proprietary format, which comprises of a folder structure with .png, and .json files. Since Roboflow supports this proprietary format, the Unity Perception package datasets were imported into Roboflow, where they were resized, augmented, and thereafter exported in the format supported by YOLO v3-Tiny, which is .jpg files combined with .txt files describing the bounding boxes.

2.5. A new physically trained log-detector

The previous physically trained log detector has an AP50 of 80.5% on the physical validation dataset (Li and Lideskog, 2023). This was also confirmed by retesting it. However, during the course of this work, it was seen that training a new detector, using same physical training and validation data as for the previous detector, as well as the same training settings, was resulting in better detectors with AP50 above 90%. The only known differences to the previous training was that a newer Darknet version (Darknet v2.0-157-g37a95745) and repository (Hank-ai/Darknet) were used, but since the model is still YOLO v3-Tiny, no significant differences would be expected because of that. It was not possible to conclude why such improvements were possible, within the scope of this study. However, in order to make a fair comparison between training on physical and virtual data, it was concluded that the best (in terms of AP50 score) corresponding physically trained detector needed to be used as a reference for comparison. Therefore, all uses of a physically trained log detector in this work were done using the new best physically trained log detector, except for the first real-time test in Section 3.1, which was done prior to discovering that improvements were possible. However, since no comparisons to virtually trained detectors were made in Section 3.1, the conclusions of Section 3.1 are not affected by the possibility of even better physically trained detectors. The new physically trained log detector was trained using the settings shown in Table 1. For more data on the new physically trained log detector, see Section 3.3.

3. Results

3.1. HiL-capable virtual environment for machine learning

The multi-physics digital twin of the forestry vehicle in Mevea was possible to run in co-simulation with the RGB camera emulation environment in Unity, in real-time on a single gaming laptop. This was possible also while performing auto-annotation of virtual RGB camera data and saving it. The time-steps that were achievable on the gaming laptop for this, were 1.6 ms for the main simulation and 64 μs for the hydraulics simulation. The utilized laptop had an Intel i7-10750H processor (6 cores, 12 threads), a Nvidia RTX 2070 MaxQ graphics card, and 16 GB of RAM. On a more capable desktop PC the achievable timesteps are expected to be significantly lower. Auto-annotation of logs during co-simulation between Mevea and Unity is shown in Fig. 4.

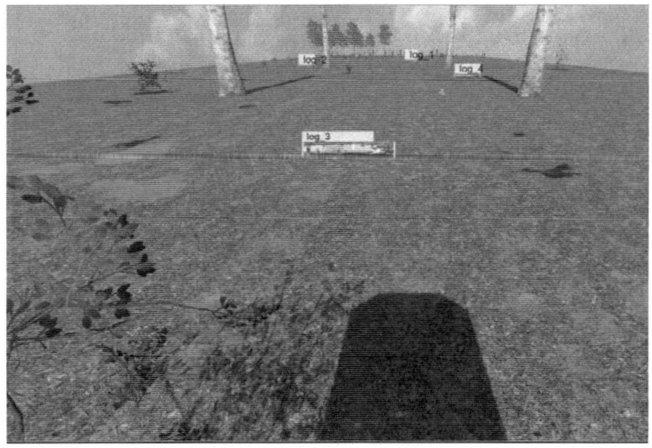

Fig. 4. Auto-annotation of logs during co-simulation

Furthermore, instead of performing auto-annotation, another setup was tested, where a 10 Hz camera feed (since this was the frequency most recently used on the physical vehicle) from Unity was shared as a ROS-topic (Robot Operating System) (ROS), which was subscribed to by a second computer in the network. This second computer was a laptop with an Intel i7-6820HQ processor (4 cores, 8 threads), a Nvidia Quadro M2000M graphics card, and 16 GB of RAM. The second computer was running the physically trained log detector, and was able to detect logs in the virtual environment, in real-time, just as if it was running on the vehicle while driving in the real world. By observing the bounding boxes and the corresponding confidence scores on the live camera feed in the ROS visualizer Rviz, it was found that the detector, trained on only physical data, could detect logs in the virtual environment. This was done with no modifications to the codebase, detector model, or ROS-environment, which were previously used for performing the same task on the physical vehicle (La Hera et al., 2024). This is considered the first proof of concept of HiL-implementation of this virtual environment for machine learning.

3.2. Physically trained detector in a virtual environment

The first test of whether there was some transferability between the virtual and the physical datasets, was to test the detector, trained on physical data, in the virtual environment. The real-time detector test in Section 3.1, using ROS communication, showed that there was some transferability, since the detector, trained on only physical data, was observed to detect logs in the virtual environment. However, the test in Section 3.1 could not quantify the performance of the detector in the virtual environment, since no statistics or data were stored. Therefore, two different virtual test datasets of different difficulty, were used to evaluate the best physically trained detector (see Section 2.5) on virtual data. The simpler test dataset, named "Simple", consisted of images of a single log, initially a few meters ahead of the vehicle, like the closest log in Fig. 4, and subsequently being passed by the vehicle. The more difficult test dataset, named "Difficult", consisted of 5% of the main virtual dataset, described in Section 2.3, before augmentation. The test results for the two different datasets are shown in Table 2.

Table 2
Physically trained detector evaluated on two virtual datasets

Test set	Number of images	Precision [%]	Recall [%]	AP50 [%]
Simple	160	89.4	87.4	90.5
Difficult	503	44.5	28.0	25.3

The main observation from the results in Table 2 is that there is transfer between the virtual and physical datasets, especially for the simpler case. This shows promise for virtual training to be effective.

Another observation is that the transfer to the simple virtual dataset is good, while the transfer to the difficult virtual dataset is significantly worse. This seems to indicate that the physical training dataset is somehow inadequate for what is encountered in the more difficult virtual dataset. The big discrepancy could also indicate that the transfer between virtual and physical datasets could be good in some cases, while being far worse in other cases.

3.3. Virtually trained detector in a physical environment

The dataset of 9099 training images from the virtual environment, described in Section 2.3, was used to train three different detectors, trained exclusively on virtual data. This was done to evaluate three different levels of image augmentation, to assess whether more augmentation would, in general, improve the transferability of a virtually trained detector to the physical world. This could be expected based on the work by (Tobin et al., 2017). The three different levels of augmentation are described in Table 3. The columns in Table 3 are, from left to right: augmentation level, rotation, shear (horizontal and vertical), grayscale (% of images), hue, saturation, brightness, exposure, blur, noise, and outputs (number of outputs per initial image). The lowest augmentation level was chosen to equal what was used for the physically trained log detectors. The noise augmentation parameter ("salt and pepper noise", which replaces random pixels with black or white pixels) was not increased, since it was already considered to be relatively high, based on both visual inspection and the augmentation parameter recommendations in Roboflow.

Table 3
The three different levels of image augmentation

Augm. Level	Rot. [°]	Shear [°]	Graysc. [%]	Hue [°]	Sat. [%]	Brightn. [%]	Exp. [%]	Blur [px]	Noise [%]	Outp.
Low	±3	0	0	0	±5	0	±5	0.5	2	4x
Medium	±3	0	0	±20	±5	±20	±5	0.5	2	4x
High	±15	±5	25	±25	±40	±35	±20	1.0	2	6x

The test results of the three virtually trained detectors, and the best physically trained detector as a reference, are shown in Table 4. All four detectors were trained for 48 000 batches, which was more than enough for validation AP50 to stop improving (see Section 2.4), all the detectors had their best score before 24 000 batches.

A comparison to the physically trained detector is not fair though, since all detectors are evaluated against the validation part of the dataset that the physically trained detector was trained on. So, although the physically trained detector was not trained on those specific images, they are still images of the same logs, from the same environment, using the same camera, at the same occasion.

Table 4
Three virtually trained detectors, and the best physically trained detector, all validated against the validation part of the physical dataset

Detector (incl. augm. level)	Number of training images (incl. augm.)	Precision [%]	Recall [%]	AP50 [%]
Physical-Low	6272	81.3	94.0	91.3
Virtual-Low	37421	57.1	12.0	18.2
Virtual-Medium	37421	54.1	23.7	24.9
Virtual-High	54618	85.5	28.2	35.9

The main observation from Table 4 is that detectors trained only in a virtual environment were able to learn to detect logs in the physical world. Another important observation from Table 4 is that higher levels of augmentation seem to, in general, lead to better transferability of virtually trained detectors to the physical world. A third observation is that the virtually trained detectors, when evaluated on physical data, have better Precision scores than Recall scores. The same observation was made by (Grondin et al., 2022), although they did not provide any quantification of it. The Precision score of the best virtually trained detector is relatively good. However, the Recall and AP50 scores of all the virtually trained detectors are significantly lower than the physically trained detector, but seem to improve with increased augmentation. A physical log detected by the best virtually trained log detector, along with the confidence score, is shown in Fig. 5. The image is part of the physical validation dataset.

Fig. 5. A physical log, in an image from the physical validation set, detected by the best virtually trained log detector, including confidence score

3.4. Virtual pre-training at low data levels

One potential use case for virtual training could be as pre-training, especially in the case of a low number of available annotated physical images. In the current scenario, it seemed that the 1568 physical training images (6272 after augmentation) were more than enough for AP50 validation on the corresponding physical validation set. This was concluded since using only 25% of the 1568 images for training, only reduced the AP50 score from 91.3% to 88.6%. Therefore, using the best virtually trained model "Virtual-High" (see Table 4) as pre-training, before training on the full physical training set, unsurprisingly did not yield any significant improvements on the physical validation set. Although the AP50 score was improved from 91.27% to 91.45% by employing the pre-training, the improvement was not significant and could be a result of run-to-run variance.

However, using the best virtually trained model "Virtual-High" as pre-training, at lower numbers of physical training images, did produce better results. The results from using only 5% of the 1568 physical training images, still using the "Low" augmentation level from Table 3, with either: no pre-training, general MSCOCO pre-training, or the "Virtual-High" detector weights as pre-training, are shown in Table 5. The MSCOCO dataset of 80 classes is a commonly used pre-training for YOLO v3-Tiny, the pre-trained weights are available where the Darknet and YOLO v3-Tiny repositories are provided (github.com/hank-ai/darknet). Based on the logic detailed in Section 2.4, the number of batches were 24 000 for each of the trained models in Table 5.

The results in Table 5 show that virtual pre-training can improve AP50-scores on a physical dataset, at low amounts of physical training data. The virtual pre-training was also more effective than general MSCOCO pre-training for this purpose.

Table 5
The effects of pre-training when training on only 5% of the physical training images

Pre-training	Number of training images (incl. augm.)	Precision [%]	Recall [%]	AP50 [%]
None	406	84.6	70.9	71.9
MSCOCO	406	78.4	75.4	74.3
Virtual-High	406	82.7	78.7	78.7

3.5. Virtual pre-training for domain generalization

Another potential use case for virtual pre-training that was considered, was for improving the domain generalization of a trained detector, to have it work better on a completely new dataset. In order to assess this potential, models with virtual pre-training and no pre-training were compared on a new physical dataset that was collected in two new environments, using a different camera (MYNTEYE D-1000-120-FOV). The new physical dataset consisted of 318 annotated images of logs, where 23 images were taken at a sand extraction site, and the remaining 295 images were taken on the university campus in Luleå, Sweden.

Since there seemed to be some random variance in how well the detectors performed on the new dataset, 5 different detectors of each type were trained and evaluated to reduce the impact of this variance. Different numbers of batches between 24 000 and 48 000 were also tried, the number of batches inside this span did not seem to matter, but for a conservative comparison the detectors without pre-training were always trained for as many, or more, batches as the pre-trained ones. The other training settings and data were always the same for all detectors: "Low" augmentation level (see Table 3), the training settings from Section 2.4, and training on the full original physical training set of 1568 images (6272 after augmentation).

The results of this evaluation, presented in Table 6, show that virtual pre-training has a positive impact when testing on a new dataset. The virtually pre-trained detectors had 3.9 pp (percentage points) higher mean Precision, 5.3 pp higher mean Recall, 6.2 pp higher minimum AP50, 2.6 pp higher maximum AP50, and 3.9 pp higher mean AP50, when compared to the detectors that were not pre-trained. This indicates that virtually pre-trained detectors seem to have an advantage when it comes to domain generalization.

Table 6
The effects of virtual pre-training on physically trained detectors evaluated on a new physical dataset

Pre-training	Number of detectors	Mean Precision [%]	Mean Recall [%]	Min AP50 [%]	Max AP50 [%]	Mean AP50 [%]
None	5	77.0	43.2	41.2	49.7	46.5
Virtual-High	5	80.9	48.5	47.4	52.3	50.4

4. Conclusions and future work

A HiL-capable virtual environment for machine learning was realized. While the virtual environment was computationally efficient enough to be HiL-capable, it was still accurate enough for the virtual learning to be applicable in the physical world. The main conclusion regarding application of the virtual training in the physical world, was that an object detector trained only on auto-annotated data from the virtual environment, could learn to detect logs in the physical world. Another conclusion regarding application in the physical world, was that virtual pre-training could improve the performance of physically trained and tested detectors, both at low availability of physical training data, and in terms of domain generalization.

Although the virtual training already proved useful in various ways, if the virtually trained detector could reach higher Recall and AP50 values, the usefulness in all the above-mentioned use cases would likely increase further. Since this work is more of a pilot study, there should be a lot of room for improvement of the Recall and AP50 values. Especially since it was also concluded that higher amounts of image augmentation, in general, appeared to increase both the Recall and AP50 scores.

This leads to the topic of future work. Further testing with even higher amounts of image augmentation would be highly interesting, since this appeared to be an intervention that was improving the results. Also, optimizing each individual image augmentation parameter (see Table 3) should be further investigated. There is likely a point for each augmentation parameter, where it will start detracting from the results instead of improving them. Since the augmentation was applied in bulk, some of the parameters might already be past that point. Also, it's not certain that all the applied augmentation parameters make the results better, some of them could even prove to be detrimental.

Other types of augmentation that should be explored are adjustments of the Unity simulation where, for example, lighting conditions and weather could be automatically varied to create a more diverse training dataset. In addition, the effects of larger, or in other ways adjusted, training datasets should be explored. A much larger virtual training dataset could easily be created, since auto-annotation removes the bottleneck of annotating the images. The detection results on individual images could also be analyzed, to discern if there are certain types of situations where the virtually trained detector fails, and then investigate supplying more training images of a similar type.

Exploration of virtual training for use-cases other than log detection, such as obstacle avoidance, among others, is of interest. Other use-cases might also require object segmentation, for which manual annotation of training data is even more time-consuming than for bounding boxes, since it is done at the pixel level. Since the Unity Perception package supports generation of object segmentation data, this is another area with potential for improvements in terms of time- and cost-efficiency.

Finally, since the simulation setup has a high level of detail, both with regards to the vehicle dynamics model and the visual perception, while still being computationally efficient enough to run in a HiL or ViL setup, this combination would be interesting to leverage. For example, machine learning could be applied in the development of offroad mobility features, such as an active pendulum arm control that leverages visual input. Conducting such training in a HiL setup instead of a physical environment could facilitate the generation of large training datasets, while still allowing the inclusion of parts of the physical driveline and/or physical black-box devices, such as a proprietary traction control system, into the training scenario.

5. Acknowledgements

This research was conducted within the project Sustainable Autonomous Material Handling (SAMHand), funded by the EU and Region Norrbotten through Interreg Aurora, as well as by The Kempe Foundation through the Jubilee Fund at Luleå University of Technology, and Skogstekniska Klustret (The Cluster of Forest Technology).

6. Declaration of competing interest

The authors declare that they have no known competing financial interests or personal relationships that could have appeared to influence the work reported in this paper.

7. References

Adarsh, P., Rathi, P., Kumar, M., 2020. YOLO v3-Tiny: Object Detection and Recognition using one stage improved model. Presented at the 2020 6th International Conference on Advanced Computing and Communication Systems (ICACCS), pp. 687–694. https://doi.org/10.1109/ICACCS48705.2020.9074315

Andersson, J., Bodin, K., Lindmark, D., Servin, M., Wallin, E., 2021. Reinforcement Learning Control of a Forestry Crane Manipulator, in: IEEE International Conference on Intelligent Robots and Systems. pp. 2121–2126. https://doi.org/10.1109/IROS51168.2021.9636219

Choi, H.S., Crump, C., Duriez, C., Elmquist, A., Hager, G., Han, D., Hearl, F., Hodgins, J., Jain, A., Leve, F., Li, C., Meier, F., Negrut, D., Righetti, L., Rodriguez, A., Tan, J., Trinkle, J., 2021. On the use of simulation in robotics: Opportunities, challenges, and suggestions for moving forward. Proceedings of the National Academy of Sciences of the United States of America 118. https://doi.org/10.1073/pnas.1907856118

Condat, R., Vasseur, P., Allibert, G., 2024. Focusing on Object Extremities for Tree Instance Segmentation in Forest Environments. IEEE Robotics and Automation Letters 9, 5480–5487. https://doi.org/10.1109/LRA.2024.3393212

Darknet. Available at: https://pjreddie.com/darknet (accessed 3 June 2024).

Fathy, H.K., Filipi, Z.S., Hagena, J., Stein, J.L., 2006. Review of hardware-in-the-loop simulation and its prospects in the automotive area, in: Proceedings of SPIE - The International Society for Optical Engineering. Presented at the Proceedings of SPIE - The International Society for Optical Engineering. https://doi.org/10.1117/12.667794

Grondin, V., Pomerleau, F., Giguère, P., 2022. Training Deep Learning Algorithms on Synthetic Forest Images for Tree Detection. ArXiv.

Hank-ai/Darknet. Available at: https:// github.com/hank-ai/darknet (accessed 3 June 2024).

Jelavic, E., Jud, D., Egli, P., Hutter, M., 2021. Towards Autonomous Robotic Precision Harvesting. ArXiv.

Kuutti, S., Bowden, R., Jin, Y., Barber, P., Fallah, S., 2021. A Survey of Deep Learning Applications to Autonomous Vehicle Control. IEEE Transactions on Intelligent Transportation Systems 22, 712–733. https://doi.org/10.1109/TITS.2019.2962338

La Hera, P., Mendoza-Trejo, O., Lindroos, O., Lideskog, H., Lindbäck, T., Latif, S., Li, S., Karlberg, M., 2024. Exploring the feasibility of autonomous forestry operations: Results from the first experimental unmanned machine. Journal of Field Robotics 41, 942–965. https://doi.org/10.1002/rob.22300

Lehto, M., Tikanmäki, A., Lindbäck, T., Lideskog, H., Karlberg, M., 2023. Modeling and verification of a full-scale forestry vehicle real-time multi-physics digital twin. Presented at the 16th European-African Regional Conference of the ISTVS. https://doi.org/https://doi.org/10.56884/TSOG1043

Li, S., Lideskog, H., 2023. Realization of Autonomous Detection, Positioning and Angle Estimation of Harvested Logs. Croatian Journal of Forest Engineering 44, 369–383. https://doi.org/10.5552/crojfe.2023.2056

Li, S., Lideskog, H., 2021. Implementation of a system for real-time detection and localization of terrain objects on harvested forest land. Forests 12. https://doi.org/10.3390/f12091142

Lideskog, H., Karlberg, M., Bergsten, U., 2015. Development of a Research Vehicle Platform to Improve Productivity and Value-extraction in Forestry, in: Procedia CIRP. Presented at the Procedia CIRP, pp. 68–73. https://doi.org/10.1016/j.procir.2015.07.014

Lindroos, O., La Hera, P., Häggström, C., 2017. Drivers of advances in mechanized timber harvesting - a selective review of technological innovation. Croatian Journal of Forest Engineering 38, 243–258.

Mevea. Available at: https://mevea.com (accessed 3 June 2024).

Pandharipande, A., Cheng, C.-H., Dauwels, J., Gurbuz, S.Z., Ibanez-Guzman, J., Li, G., Piazzoni, A., Wang, P., Santra, A., 2023. Sensing and Machine Learning for Automotive Perception: A Review. IEEE Sensors Journal 23, 11097–11115. https://doi.org/10.1109/JSEN.2023.3262134

Repli5. Available at: https://repli5.com (accessed 3 June 2024).

Roboflow. Available at: https://roboflow.com (accessed 3 June 2024).

ROS. Available at: https://ros.org/ (accessed 3 June 2024).

Salton, G., McGill, M.J., 1983. Introduction to Modern Information Retrieval. McGraw-Hill.

Tobin, J., Fong, R., Ray, A., Schneider, J., Zaremba, W., Abbeel, P., 2017. Domain randomization for transferring deep neural networks from simulation to the real world, in: IEEE International Conference on Intelligent Robots and Systems. pp. 23–30. https://doi.org/10.1109/IROS.2017.8202133

Unity. Available at: https://unity.com (accessed 3 June 2024).

Watton, J., 1989. Fluid power systems: Modeling, simulation, analog, and microcomputer control. Prentice Hall International, Hemel Hempstead.

8590 | Modelling of Terrain Deformation by a Grouser Wheel for Lunar Rover Simulation

Junnosuke Kamohara [a,*,†], Vinicius E. Ares [b,†], James Hurrell [a], Keisuke Takehana [a], Antoine Richard [c], Shreya Santra [a], Kentaro Uno [a], Eric Rohmer [b], Kazuya Yoshida [a]

[a] *The Space Robotics Lab. (SRL) in Department of Aerospace Engineering, Tohoku University, Sendai, Miyagi, Japan*

[b] *The Advanced Robotics Lab. (AdRoLab) in School of Electrical and Computer Engineering, State University of Campinas, Brazil*

[c] *The Space Robotics Research Group at the Interdisciplinary Research Center for Security reliability and Trust (SnT), University of Luxembourg, Luxembourg*

** Corresponding author: kamohara.junnosuke.t6@dc.tohoku.ac.jp*

† These authors contributed equally.

ABSTRACT

Simulation of vehicle motion in planetary environments is challenging. This is due to the modeling of complex terrain, optical conditions, and terrain-aware vehicle dynamics. One of the critical issues of typical simulators is that they assume terrain is a rigid body, which limits their ability to render wheel traces and compute the wheel-terrain interactions. This prevents, for example, the use of wheel traces as landmarks for localization, as well as the accurate simulation of motion. In the context of lunar regolith, the surface is not rigid but granular. As such, there are differences in the rover's motion, such as sinkage and slippage, and a clear wheel trace left behind the rover, compared to that on a rigid terrain. This study presents a novel approach to integrating a terramechanics-aware terrain deformation engine to simulate a realistic wheel trace in a digital lunar environment. By leveraging the Discrete Element Method simulation results alongside experimental single-wheel test data, we construct a regression model to derive deformation height as a function of contact normal force. The region of interest in a height map is retrieved from the wheel poses. The elevation values of corresponding pixels are subsequently modified using contact normal forces and the regression model. Finally, we apply the determined elevation change to each mesh vertex to render wheel traces during runtime. The deformation engine is integrated into our ongoing development of a lunar simulator based on NVIDIA's Omniverse IsaacSim. We hypothesize that our work will be crucial to testing perception and downstream navigation systems under conditions similar to outdoor or terrestrial fields. A demonstration video is available here: https://www.youtube.com/watch?v=TpzD0h-5hv4

Keywords
Robotics
Lunar Simulator
Wheel trace

Fig. 1. Simulation of a rover with generated wheel traces.

1. Introduction

1.1. Background

The Lunar and Martian surfaces have seen a resurgence of interest as a focus of recent space explorations. Recently, India, Japan, and the United States have accomplished landings on the lunar surface. In the upcoming years, space agencies and private companies are looking toward long-term resource utilization on the lunar surface.

Robotic systems play a vital role in these future missions as they can reach hazardous areas that humans cannot or yet explore cost-effectively. These systems are expected to autonomously complete various tasks such as exploration, sample return, and construction. For robotic systems to operate effectively on a planetary surface, their hardware and software must be thoroughly tested in advance. A typical software test consists of two stages: the first is conducted using simulators, and the second in analog environments.

The simulation ranges from dedicated simulation of mechanical parts to end-to-end mission simulation. Particularly in robotics, a simulator can play a role in multiple areas: perception, navigation, mobility, radiation simulation, and mission planning. Today, there are several robotic simulators, such as Gazebo, CoppeliaSim (Rohmer et al. 2013), Unity, Unreal Engine, and IsaacSim. These simulators have the fundamental capability to simulate multiple sensors and vehicle dynamics with built-in rigid-body physics engines. However, one of the critical limitations of typical simulators is that they assume terrain as a rigid body. Therefore, they cannot represent wheel traces and compute wheel-terrain interactions (i.e., terramechanics). In reality, lunar regolith is not rigid but granular, so the granular material properties must be simulated.

In this work, we extend our previous work on the lunar simulator, OmniLRS (Richard et al., 2024), and focus on providing a terramechanics-aware visual representation of wheel trace in lunar environments. The simulation of wheel traces is essential in validating vision-based systems, such as Visual Inertial SLAM, with similar visual observation as the real-world sensory data. By incorporating a terramechanics-aware visual representation of wheel traces into the simulator, we provide a more accurate and valuable tool for the development and testing of lunar exploration rovers.

1.2. Significance of simulating wheel traces

On uneven terrains such as the lunar and Martian surfaces, wheel slippage during rover traversal can occur, potentially resulting in the rover becoming stuck. To minimize slippage and

https://doi.org/10.56884/FRYX2UHE

design an effective vehicle control and body, it is imperative to study terramechanics, which considers soil deformation.

Terramechanics is the field of study that addresses the interaction between soil and machines, focusing primarily on the characteristics traversing over unpaved, deformable soil. Research in this field has been progressing since the 1950s when Bekker developed fundamental models of wheel mechanics (Bekker, 1956). Subsequently, traction models were developed and refined by Wong et al. (Wong et al., 1967, 2008).

In recent years, research in the field of terramechanics has focused on the trace formed by wheels passing over soil. These traces can reflect the state of the wheel when they are created. In particular, the slip condition, which is crucial for rover control, can be estimated from the spacing of these traces. Saku et al. use machine learning to estimate slip conditions from the trace formed during single-wheel tests (Saku et al., 2019). Similarly, Reina et al. have proposed a vision-based method for estimating sideslip angles by observing the wheel traces, useful for the rover's localization and control systems (Reina et al., 2008). Therefore, the development of simulators that consider soil deformation plays a critical role in estimating the driving state.

2. Related Works

In the field of planetary robotics, there are usually two types of simulators: one focusing more on sensor simulation (Allan et al., 2019; Crues et al., 2022; Bingham et al., 2023; Richard and Kamohara et al., 2024; Müller et al., 2021; Pieczyński et al., 2023), and the other on accurate physics computation (Tasora et al., 2016; Krenn and Hirzinger, 2009; Zhou et al., 2023; Wang et al., 2022; Serban et al., 2023). The former is usually built upon a game engine, featuring realistic illumination and shadow rendering thanks to GPU-accelerated rendering. The latter typically supports a dedicated physics engine for wheel-terrain interactions or terramechanics.

2.1. Lunar ground simulators

As of today, there are various lunar ground simulators developed using different rendering engines. Allan et al. describe a Gazebo-based simulator aimed at simulating NASA Ames' VIPER missions. Although it incorporates many crucial features for lunar exploration, such as large-scale terrain management, realistic surface illumination, realistic shadows, and the opposition effect, the rendering of wheel traces is still limited to a primitive method. In their work, the wheel traces are simulated by manipulating the normal map, one of the texture images. The same approach is used by Pieczyński et al. to realize wheel traces in UnrealEngin4 (Pieczyński et al., 2023). However, these approaches have limitations in simulating other types of sensors such as LiDAR and Radar, which require mesh geometry.

Digital Lunar Exploration Sites (DLES) by Crues et al., is another simulator developed by NASA Johnson Space Center to simulate landing and rover exploration in the lunar south pole. It thoroughly explains the procedure of pre-processing digital elevation maps and managing large-scale terrain. Furthermore, it presents a database to efficiently store and query the coordinates and sizes of craters and boulders. High-resolution terrain models, along with surface details, rocks, and craters from DLES, are incorporated into the DLES Unreal Simulation Tool (DUST) by Bingham et al. to simulate the lunar south pole. However, these works still lack terramechanics consideration, particularly realistic wheel traces for vision-based systems.

OmniLRS is a robotic simulator for lunar analog environments, based on Nvidia's Omniverse Isaac-Sim. It provides two types of virtual environments: the Lunalab, a lunar analogue at the University of Luxembourg, and the Lunaryard, a procedurally generated lunar terrain. With its synthetic data generation pipeline, it demonstrated sim-to-real transfer from the virtual environment to the real-world lab environment in rock segmentation tasks. However, it still lacks features as a comprehensive robotics simulator, such as visual-inertial odometry, SLAM, and navigation. Additionally, it lacks wheel traces simulation, which limits its ability to replicate real-world scenarios and test perception and navigation systems in such settings.

2.2. Terramechanics in simulators

Another essential simulation tool is the simulator dedicated to terramechanics. This type of simulator focuses on accurate terrain-wheel modeling and the computation of the forces and torques applied to the wheels.

The Discrete Element Method (DEM), originally formalized by Cundall and Strack (1979), is a numerical approach used to study the interaction between particulate materials and solid objects, such as rover wheels. DEM provides detailed insights into the complex dynamics of wheel-soil interaction. Its utility in terramechanics is well-documented in the literature, such as in O'Sullivan (2011). Additionally, Lichtenheldt et al. (2016) highlight its relevance in planetary exploration vehicle design and its ability to simulate terrain deformation to form wheel traces, also known as rutting. Wasfy et al. (2014) explored the coupling of multibody dynamics with DEM for vehicle mobility on granular terrains, highlighting the detailed physical interactions. However, it does not address real-time simulation involving control and navigation in the loop because of the high computational cost.

Because of DEM's heavy computational demand, less accurate, but more feasible empirical terramechanics models are on the rise. Chrono (Tasora et al.) is an open-source C++ framework designed for physics-based modeling and simulation. The library simulates various multibody systems, such as wheeled and tracked vehicles on deformable terrains, robots, mechatronic systems, compliant mechanisms, and fluid-solid interactions. Particularly, Chrono supports deformable terrain, modeled by a soil contact model (SCM) (Krenn and Hirzinger), and allows computing force between terrain and the arbitrary contact surface, and deforming mesh in real-time. Serban et al. improved the computational efficiency of Chrono SCM by minimizing the number of ray-casting via the moving patch approach and parallelizing ray-casting. However, its usage is limited to a few-meter ordered field, and there is no study on perception and navigation using the deformable terrain.

There is also an effort to integrate terramechanics into common robotics simulators. MarsSim (Zhou et al.) is a simulator built upon Gazebo that supports rendering of the Martian landscape and includes a custom-built terramechanics plugin. The plugin is based on a semi-empirical model, both from the theory and experimental data, capable of obtaining drawbar pull, wheel slip, and wheel sinkage. In one of their experiments, they tested the transition from soft sand to hard bedrock. Unfortunately, their work does not consider the rendering of wheel traces, thus its application in vision systems is still limited.

In this work, we focus on integrating soil deformation in the previously developed photorealistic lunar simulator based on Nvidia IsaacSim. Our contribution is twofold: first, we model deformation depth and geometry in a data-driven manner using

DEM simulation. Second, we integrate this deformation engine into our previous lunar simulator, OmniLRS, allowing for rover operation with ROS1 or ROS2 while rendering wheel traces, as shown in Fig. 1. Computationally intensive DEM simulations provide accurate deformation data that is used to build a linear regression model, that can then run in real-time. We evaluate the quality of the simulated wheel traces by comparing them with images from hardware experiments.

3. Method

In this work, we propose a method to simulate wheel traces in a virtual lunar environment. Our system is composed of the following modules: the DEM simulation, a wheel trace model, real-time mesh deformation in IsaacSim. Figure 2 provides an overview of our system, with each component described in detail in the following sections.

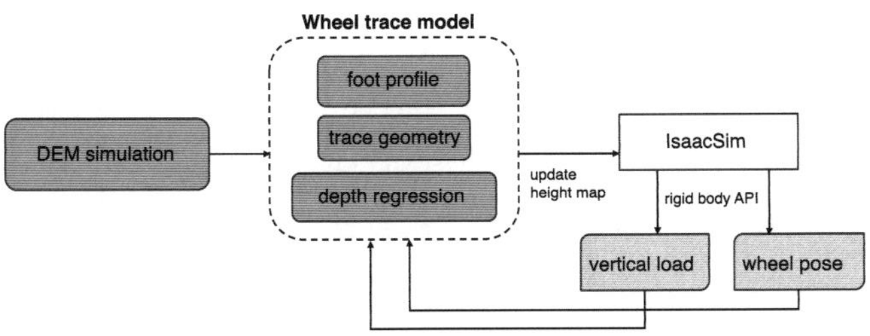

Fig. 2. The figure shows the entire architecture of the proposed method. It takes vertical load and wheel pose to update the height map during simulation.

3.1. Discrete Element Method

A data-driven approach is employed to model soil deformation using the Discrete Element Method (DEM) simulation, validated against experimental results from single-wheel testing. We utilize DEM to retrieve trace information, including trace geometry and resulting depth values for vertical loads acting on the wheel.

We use the EDEM (Altair, 2024), DEM simulation software, to compute wheel trace dimensions from input vertical load. The DEM simulations replicate a single-wheel test (SWT) experiment, where a wheel is driven with controlled velocities in a sandbox under a given load and output the trace geometry. Inside the EDEM, SWT is performed for a set of vertical loads, and the resulting wheel traces are measured to build the wheel trace model that can be used in IsaacSim.

We use the material called Toyoura sand (Ozaki et al., 2023) which is known for its low cohesion value and a narrow size distribution with a similar average particle size to lunar regolith. To reduce simulation time, particles size is scaled by a factor of ten and modelled as single spheres with the scaled average particle size. As for the contact model between particles and objects, we employ a standard Hertz-Mindilin model (Coetzee, 2017) with a Type-C rolling contact model (Ai et al., 2011). The simulation parameters are calibrated following our previous study (Hurrell et al., 2023) that uses the Angle of Repose (AoR) experiment to calibrate the parameters.

3.2. Wheel Trace Geometry

To simulate wheel traces, we implement (1) a foot profile generator, (2) *xz* and *yz* cross-section generators, and (3) a trace depth regression model. As for the coordinate frame, we use three types: the local coordinate of the robot's link, the global coordinate, and the pixel coordinate of the reference height map. The local coordinate follows the FLU convention, in which the x-axis faces robot's forward motion direction and z-axis faces upward.

3.2.1. Foot profile

First, the foot profile is created based on the geometry and dimensions of the links in contact with the surface (these links are wheels in our case). From now on, we call these links in contact with the surface *target links*. The foot profile is a set of two-dimensional points, defined in each target link's local coordinate, occupied by their projections. In the case of a wheeled robot, this profile is a set of points inside rectangles which are the projections of the wheels. We assume the width and height of the rectangular profile are in the same dimensions as the width (w) and diameter of the wheels ($2r$). These sets of points inside the foot profile, expressed in local coordinates, are indicated in Eq. 1. In the end, the derived points are transformed from their local coordinates to global coordinates and then to pixel coordinates. These transformed points indicate the pixel positions where deformation is applied.

$$FP = \left\{(x,y) \mid -r \le x \le r, -\frac{w}{2} \le y \le \frac{w}{2}\right\} \quad (1)$$

3.2.2. xz, yz cross-section

Once the foot profile is computed, we define the distribution of the deformation profile. This profile tells us the scattering distribution of the wheel trace $\{z = f(x,y) \mid (x,y) \in \mathrm{FP}\}$. In our work, we assume the trace distribution is a combination of two distributions. Based on the results from DEM simulations, we propose using a trigonometric function and a trapezoidal curve for $h(x)$ and $g(y)$, as they approximate the DEM results well. For the xz and yz cross-sections, we use the functions in Eq. 2 and Eq. 3.

$$h(x) = \cos(n\pi x); n = \frac{N}{2\pi r(1-s)} \quad (2)$$

$$g(y) = \begin{cases} \tan(\theta)\frac{2y}{w} - \tan(\theta) & \left(1 - \frac{1}{\tan(\theta)} < \frac{2y}{w} \le 1\right) \\ -1 & \left(-1 + \frac{1}{\tan(\theta)} \le \frac{2y}{w} \le 1 - \frac{1}{\tan(\theta)}\right) \\ -\tan(\theta)\frac{2y}{w} - \tan(\theta) & \left(-1 \le \frac{2y}{w} \le -1 + \frac{1}{\tan(\theta)}\right) \end{cases} \quad (3)$$

Here, n and θ are the frequency of the wave computed from the number of grousers (N) and the slip ratio (s), and the internal angle of repose of the soil material. In the end, the deformation profile is computed as shown in Eq. 4 by combining Eq. 2 and Eq. 3. $A(F_z)$ and $\mu(F_z)$ are computed from the vertical load - depth regression model discussed in the following section.

$$z = g(y) \cdot \left(A(F_z) \cdot h(x) + \mu(F_z)\right) \quad (4)$$

3.2.3. Depth regression model

The last component in the trace geometry generator is the depth regression model. This regression model, as a function of vertical load, determines the depth value of the amplitude $A(F_z)$ and the mean depth $\mu(F_z)$ of the wheel trace. We use pairs of vertical load and deformation profiles retrieved from DEM simulations to find the slope and y intercept of the linear regression model.

3.3. Integration in IsaacSim

The final step is to integrate our deformation engine in IsaacSim to simulate the robot locomotion with wheel traces. We manipulate the reference digital elevation map and update the terrain mesh during runtime to create wheel traces in Isaac.

3.3.1. Mesh deformation

As discussed in previous sections, the deformation engine requires the vertical load applied on the terrain and the pixel positions of the target links. We use Isaac's rigid body API (*RigidPrimView*) to compute the vertical load applied on the terrain and the pose of the wheels. Taking these values as input, our deformation engine computes a set of pixel indices to apply deformation and the elevation change. Then, the digital elevation map is updated, and the z position of the mesh vertices is changed accordingly.

In practice, we only update the position of the mesh vertices, but not the collision. This is because IsaacSim needs to encode the mesh into a specific data structure so that its physics engine can understand it to compute collisions. However, this encoding process takes a lot of time and pauses the simulation. Therefore, accurate physics computation is not within the scope of our current work.

3.3.2. The vehicle dynamics

For the simulation, we use a multi-body robot model of the four-wheeled custom rover (Rodríguez-Martínez et al., 2024). The model is exported from a CAD model to the Universal Robot Description Format (URDF) and the Universal Scene Description (USD), which is compatible with IsaacSim. The resulting model includes a set of rigid bodies, joints (fixed, revolute, and prismatic), and a collider that represents mesh collision.

When simulating the robot, the robot state, such as pose, velocity, and acceleration, is updated by the PhysX5 rigid body engine, taking actuator commands as inputs. These actuator commands consist of the angular velocity of the four driving motors and the steering angle of the two front steering motors. These low-level actuator commands are computed using the Ackermann steering model (Simionescu and Beale, 2002), with inputs of forward velocity and steering angle. The control inputs are then applied to the robot using a PD controller implemented in Isaac.

4. Implementation

Primary applications of wheel trace simulation include vision-based tasks such as Visual-Inertial-SLAM and vision-based navigation. Wheel traces are crucial for simulating outdoor terrains in high fidelity because they are common landmarks seen in the field. We demonstrate that our simulator can simulate a robot's locomotion with visually realistic traces. As a qualitative evaluation, we compare the images from our simulator with those from outdoor experiments.

4.1. DEM simulation settings

Inside the DEM simulation, we used the wheel of the Rashid rover (Almaeeni, Els and Almarzooqi, 2024), designed for the Emirates lunar mission to explore the Atlas crater. The wheel has a diameter of 200 mm, a width of 80 mm, and 14 grousers of 20 mm in length. The virtual SWT sandbox measures 650 mm in length, 140 mm in width, and 120 mm in depth. Particles are modeled as spheres with a homogeneous size distribution. Due to the sub-millimeter size of Toyoura sand particles and the high computational cost, a scaling factor of ten was used, resulting in particles with a 1 mm radius. For terrain parameters, we used those from Hurrell et al. Table 1 shows the list of terrain parameters used in DEM simulation.

Table 1
List of terramechanics parameters used in DEM simulation.

Wheel and particle geometry	Value
Wheel diameter	200 mm
Wheel width	80 mm
Grouser length	20 mm
Grouser count	14
Sandbox length	650 mm
Sandbox width	140 mm
Sandbox depth	120 mm
Particle geometry	sphere
Particle diameter	2 mm
Particle volume	4.189 mm^3
Geometry discretization and particle counts	
Wheel facets	15.6×10^3
Wheel nodes	7,802
Sandbed box facets	10
Sandbed box nodes	8
Particle count	1.38×10^6
Contacts count total	2.948×10^6
Particle-particle contact count	2.901×10^6
Wheel-particle contact count	1,582
Sandbox-particle contact count	45,200
Material properties and contact model	
Particle material	Toyoura sand
Poisson ratio	0.25
Shear modulus	2.0×10^7Pa
Density	2,650 kg/m^3
Wheel material	Aluminum
Poisson ratio	0.33
Shear modulus	2.71×10^{10}Pa
Density	2,810 kg/m^3
Particle-particle interaction	
Coefficient of restitution	0.40
Coefficient of static friction	0.65
Coefficient of rolling friction	0.15
Wheel-particle interaction	
Coefficient of restitution	0.40
Coefficient of static friction	0.60
Coefficient of rolling friction	0.40
Contact Model	Hertz-Mindlin
Coefficient of rolling stiffness	3
Rolling viscous damping ratio	0.3

The SWT DEM simulation begins by filling the sandbox with particles to form the sand-bed and allowing the particles to settle. Next, the depth, bulk density, and level surface are checked, removing excess particles if necessary to ensure they match the desired conditions. The wheel is then inserted with the required load and allowed to settle for one second. The wheel is driven with the set horizontal and rotational velocities, and the simulation is allowed to process until the total time is reached and the travel is finished. After the simulation, a cross-section is created in the plane y=0 to reveal the wave pattern of the wheel trace. Finally, the heights of the trough and peak of the wave shape are measured relative to the original height of the sand-bed.

4.2. Simulation in IsaacSim

For the simulation, we used a procedurally generated lunaryard as explained in Richard et al. We chose the synthetic lunar terrain because it provides more control over the size and resolution of the terrain. The terrain was generated with a size of 20 m × 20 m and a resolution of 1.5 cm/texel. As for the texture, we used sand-like materials as well as finer gravel materials from Polyhaven and Nvidia's Base Material collection.

To simulate the dynamics of rovers, we must have a realistic model of robot. As mentioned in the previous section, we used the EX1 rover (Rodríguez-Martínez et al., 2024), a four-wheeled rover with steering developed by our group. The wheels of the EX1 rover are scaled to match the dimensions of those on the Rashid rover. To observe the wheel traces from the rover's point of view, two cameras were attached to the front and back of the rover. Since the actual EX1 hardware uses the Intel D435 camera (Keselman, 2017), we used the same camera intrinsic parameters.

5. Results and Discussion

To integrate wheel traces in IsaacSim, we create the wheel trace model from DEM simulation. Then, we leverage this model to compute how much terrain mesh deforms and render traces during runtime. First, we present the results from the DEM simulation and introduce a regression model that computes deformation depth. Finally, we compare simulated and actual images from the outdoor field test.

5.1. DEM simulation

Fig.3 presents *xz* and *yz* cross-sections of wheel traces from DEM simulation. To measure the dimensions of wheel traces, we used a ruler tool available inside EDEM. When looking at *yz* cross-section, it resembles the shape of a trapezoid, making Eq. 3 a reasonable assumption. From several profiles in the *xz* plane, the heights of the trough and peak of wheel traces were measured for a set of loads, as presented in Table 2.

Table 2
Measured values of trough, crest, mean, and amplitude of the wheel trace in the SWT simulation for each applied vertical load.

Vertical Load [N]	Trough height [mm]	Crest height [mm]	Mean height [mm]	Amplitude [mm]
3.0	-6.5	1.5	-2.50	8.0
5.0	-8.0	0.5	-3.75	8.5
8.6	-9.5	-1.0	-5.25	8.5
12.9	-12.0	-3.0	-7.50	9.0
17.9	-13.5	-4.5	-9.00	9.0

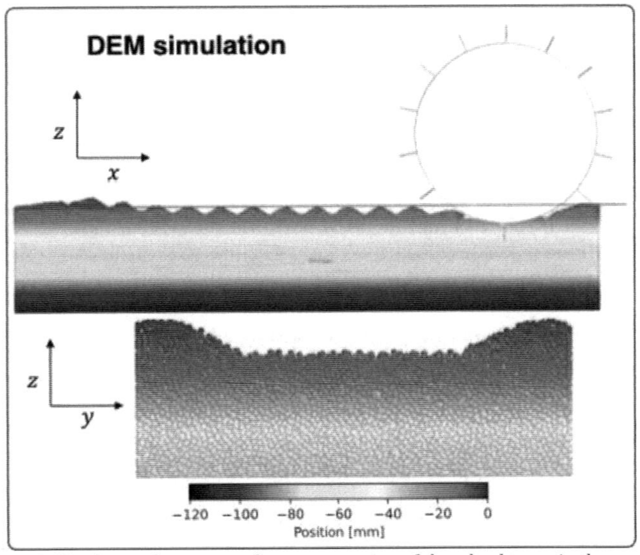

Fig. 3. The figure shows xz and yz cross-section of the wheel trace in the EDEM simulation.

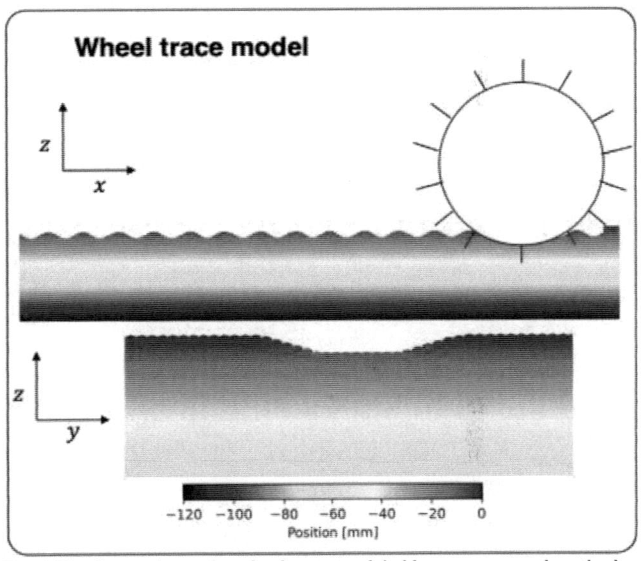

Fig. 4. The figure shows the wheel trace modeled by our proposed method. Same as Fig.3, the top image indicates xz cross-section and the bottom image represents yz cross-section.

The values of average height and amplitude are used to build linear regression models to predict average height and amplitude as functions of the vertical load. Eq. 5 and Eq. 6 show the computed linear regression models for average height and amplitude. The fitted model for average height yielded a mean absolute error of 0.16 mm and a mean percentage error of 3.5%. In the case of amplitude, it had a mean absolute error of 0.14 mm and a mean percentage error of 1.6%. The regression models can then be used in Eq. 4 to render wheel traces in the next section.

$$\mu(F_z) = -1.32 - 0.458Fz \qquad (5)$$

$$A(F_z) = 7.98 + 0.066Fz \qquad (6)$$

5.2. Qualitative evaluation of wheel trace rendering

Inside IsaacSim, we drove the EX1 rover on flat terrain and recorded the rendering with wheel traces. The rover was operated inside a 20 m × 20 m procedurally generated lunar terrain under an illumination source with 3,000 intensity. Fig. 5 shows the rendering of wheel traces and images from the outdoor field test at a sand dune in Tottori, Japan. The image comparison suggests that the grouser-shaped wheel traces are well simulated in our simulator. However, we still have difficulty rendering the trace with more surface roughness and sharpness. This is because our current texture is in 4,000-pixel resolution, which is not high enough to represent higher surface roughness. The texture problem will be addressed in future work.

Another advantage of our method is its scalability to other types of wheeled robots, as shown in Fig. 6. This is because our method has control over the dimensions of the foot profile, which is adjustable through configuration files. In the future, we will support other types of robots, such as quadrupedal robots, by considering arbitrary contact foot geometry.

Fig. 5. Comparison between the rendering of wheel traces in IsaacSim and the images from the outdoor field test. Top two images show rendering in simulation and the bottom two show images from the outdoor field test.

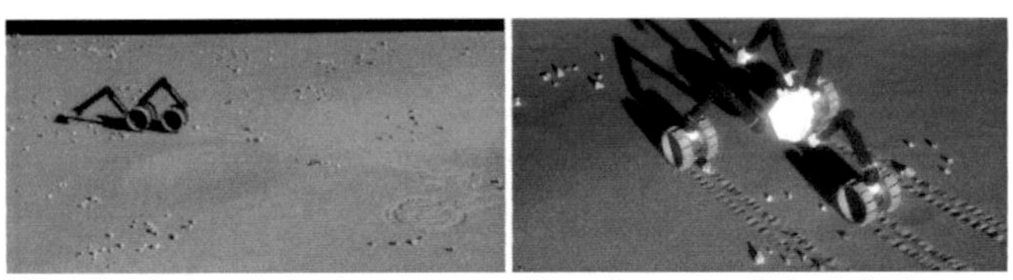

Fig. 6. Rendering of wheel traces created by other types of wheeled robots.

5.3. Computation time and memory complexity of simulation

In the end, we analyzed the mesh update time and VRAM consumption for different terrain sizes and resolutions. The terrain sizes chosen were 10m × 10m, 20m × 20m, and 40m × 40m, with resolutions of 0.1m/texel, 0.2m/texel, 0.25m/texel, and 0.5m/texel. To compare mesh update time and GPU memory consumption against the mesh dimensions, we used the product of mesh size and the inverse of squared resolution (i.e., texels) as mesh dimension metrics. Figure 7 shows the plot of computation time and GPU usage on the vertical axis and texels on the horizontal axis. The figure suggests that both mesh update time and GPU memory consumption linearly increase. According to the figure, we can assume that a reasonable mesh size is up to one million texels (20m × 20m at 0.02m/texel), which results in 12.9ms. Furthermore, the GPU usage decreases as the number of texels increases. This is happening as the update of the mesh is happening at the CPU level, as the number of texels increases, the time it takes to update the mesh gets increasingly longer reducing the overall GPU usage.

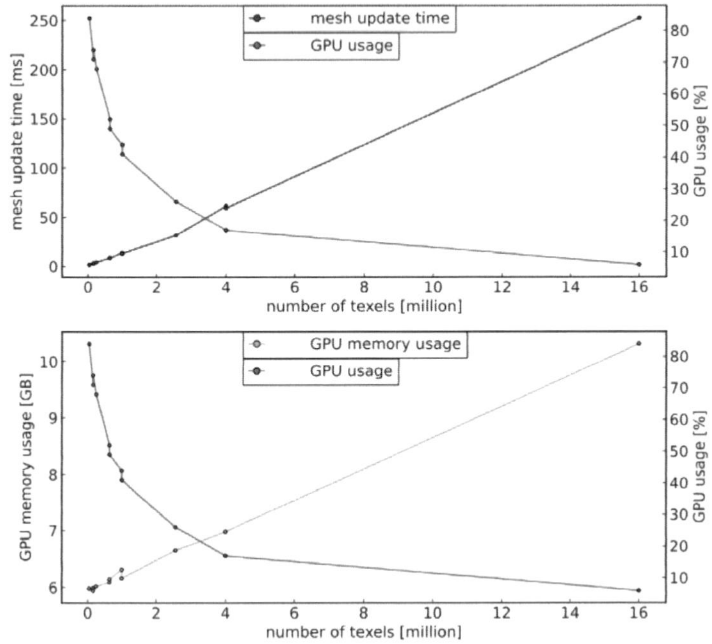

Fig. 7. The diagram shows the plot of mesh update time, GPU memory consumption, and GPU usage against the number of texels.

5.4. Future work

As for future work, we will include the wheel trace in large-scale, kilometer-order terrain and simulate long-range traversal. Then, we will study if visual SLAM can leverage wheel traces as additional landmarks to increase the accuracy of localization.

In addition, the mesh deformation time is upper bounded by the streaming of mesh vertex points done on the CPU. A faster method to stream vertices will reduce the computation time, making our method scalable to finer and larger terrains.

6. Conclusion

This study proposed a method to simulate wheel traces for lunar rovers using a developed terramechanics-aware deformation engine. By utilizing DEM simulations, we developed a regression model to compute deformation height based on contact normal forces and integrated it into IsaacSim. Our method overcomes the limitations of rigid terrain assumptions in existing simulators and realizes real-time wheel traces in the simulation loop.

Our approach is scalable to different types of wheeled robots, rendering wheel traces with specified contact surface geometry and dimensions. Future work will extend our simulator to be able to handle large-scale terrain and simulate long-range traverses with wheel traces. This feature allows us to test visual inertial SLAM with and without wheel traces, and study if SLAM can leverage wheel traces to increase localization accuracy. Furthermore, visual analysis of the wheel traces left by the rover can provide valuable insights into the vehicle's state and performance during its traversal, which will eventually aid in mission planning.

7. Acknowledgements

The authors would like to acknowledge Kenta Sawa, Masahiro Uda, and Kohta Naoki for kindly providing the images recorded from the EX1 rover at the Tottori sand dunes. This work was supported by JST Moonshot R&D Program, Grant Number JPMJMS223B. This study was financed in part by the Coordenação de Aperfeiçoamento de Pessoal de Nível Superior (CAPES) - PrInt Program- Brazil (Finance Code 001 - Process no. 1041304P), of which author V.E. Ares is a beneficiary.

8. Declaration of competing interest

The authors declare that they have no known competing financial interests or personal relationships that could have appeared to influence the work reported in this paper.

9. References

Ai, J., Chen et al., 2011. Assessment of rolling resistance models in discrete element simulations, 206(3), pp. 69-282. doi: https://doi.org/10.1016/j.powtec.2010.09.030.

Allan, M. et al., 2019. Planetary rover simulation for lunar exploration missions, in Proc. 2019 IEEE Aerospace Conf., 2019, pp. 1-19.

Almaeeni, S. et al., 2021. To a dusty Moon: Rashid's mission to observe lunar surface processes close-up, in Proc. 52nd Lunar and Planetary Science Conf., LPI Contrib. No. 2548, 2021. Available: https://www.hou.usra.edu/meetings/lpsc2021/pdf/1906.pdf.

Altair Engineering Inc., 2024. EDEM. Accessed on 2024-05-19. Available: https://altair.com/edem

Bekker, M. G., 1956. Theory of Land Locomotion, University of Michigan Press.

Bingham, L. et al., 2023. Digital lunar exploration sites unreal simulation tool (DUST), in Proc. 2023 IEEE Aerospace Conf., 2023, pp. 1-12.

Coetzee, C.J., 2017. Review: Calibration of the discrete element method, Power Technology, 310, pp. 104-142. doi: https://doi.org/10.1016/j.powtec.2017.01.015

Crues, E. Z. et al., 2022. Digital lunar exploration sites (DLES), in Proc. 2022 IEEE Aerospace Conf., 2022, pp. 1-13.

Hurrell, J. et al., 2023. Lunar Rover Discrete Element Method Study and Calibration, in Proc. 16th European-African Regional Conference of the ISTVS (ISTVS '23), Lublin, Poland.

Keselman, L. et al., 2017. Intel(R) RealSense(TM) Stereoscopic Depth Cameras, in Proc. 2017 IEEE Conference on Computer Vision and Pattern Recognition Workshops (CVPRW), pp. 1267-1276. doi:10.1109/CVPRW.2017.167

Krenn, R., Hirzinger, G., 2009. SCM -- A SOIL CONTACT MODEL FOR MULTI-BODY SYSTEM SIMULATIONS, in Proc. 11th European Regional Conf. of the Int. Society for Terrain-Vehicle Systems - ISTVS 2009, Oct. 2009. https://elib.dlr.de/60535/

Lichtenheldt, R. et al., 2016. Wheel-Ground Modeling in Planetary Exploration: From Unified Simulation Frameworks Towards Heterogeneous, Multi-tier Wheel Ground Contact Simulation, in Multibody Dynamics: Computational Methods and Applications, Font-Llagunes, J. M., Ed., Springer International Publishing, Cham, pp. 165-192. doi:10.1007/978-3-319-30614-8_8. Available: https://doi.org/10.1007/978-3-319-30614-8_8

Müller, M. G., et al., 2021. A photorealistic terrain simulation pipeline for unstructured outdoor environments, in Proc. 2021 IEEE/RSJ Int. Conf. on Intelligent Robots and Systems (IROS), 2021, pp. 9765-9772.

O'Sullivan, C., 2011. Particulate Discrete Element Modelling: A Geomechanics Perspective, CRC Press. doi:10.1201/9781482266498. Available: https://doi.org/10.1201/9781482266498

Ozaki, S. et al., 2023. Granular flow experiment using artificial gravity generator at International Space Station, npj Microgravity, 9(1). doi:10.1038/s41526-023-00308-w. Available: https://www.scopus.com/inward/record.uri?partnerID=HzOxMe3b&scp=85167503625&origin=inward

Pieczyński, D. et al., 2023. LunarSim: Lunar Rover Simulator Focused on High Visual Fidelity and ROS 2 Integration for Advanced Computer Vision Algorithm Development, Appl. Sci., 13(22), 12401. doi:10.3390/app132212401

Reina, G. et al., 2008. Vision-based estimation of slip angle for mobile robots and planetary rovers, IEEE International Conference on Robotics and Automation, Pasadena, CA, USA, 2008, pp. 486-491, doi: 10.1109/ROBOT.2008.4543254.

Richard, A. et al., 2024. OmniLRS: A Photorealistic Simulator for Lunar Robotics, International Conference on Robotics and Automation, arXiv preprint arXiv:2309.08997.

Rodríguez-Martínez et al., 2024. Enabling Faster Locomotion of Planetary Rovers With a Mechanically-Hybrid Suspension, IEEE Robotics and Automation Letters, 9(1), pp. 619-626. doi:10.1109/LRA.2023.3335769

Rohmer, E. et al., "CoppeliaSim (formerly V-REP): a Versatile and Scalable Robot Simulation Framework", in Proc. of The International Conference on Intelligent Robots and Systems (IROS)", 2013

Saku, Y., Ishigami, G., Slip Estimation Using Machine Learning of Wheel Trace Patterns for Mobile Robot on Loose Soil, 15th Europian-African Regional Conference of the International Society of Terrain Vehicle Systems (ISTVS), Prague, Czech Republic, September 2019

Serban, R., Taves, J., Zhou, Z., 2023. Real-time Simulation of Ground Vehicles on Deformable Terrain, Journal of Computational and Nonlinear Dynamics, 18(8), pp. 081007. American Society of Mechanical Engineers.

Simionescu, P.A., Beale, D., 2002. Optimum synthesis of the four-bar function generator in its symmetric embodiment: the Ackermann steering linkage, Mechanism and Machine Theory, 37(12), pp. 1487-1504. https://doi.org/10.1016/S0094-114X(02)00071-X.

Tasora, A. et al., 2016. Chrono: An Open Source Multi-physics Dynamics Engine, in Proc. 2016, pp. 19-49. doi:10.1007/978-3-319-40361-8_2

Wang, S. et al., 2022. Collision-Free Trajectory Planning for a 6-DoF Free-Floating Space Robot via Hierarchical Decoupling Optimization, IEEE Robotics and Automation Letters, 7(2), pp. 4953-4960.

Wasfy, T.M., Wasfy, H.M., Peters, J.M., 2014. Coupled Multibody Dynamics and Discrete Element Modeling of Vehicle Mobility on Cohesive Granular Terrains, in Proc. International Design Engineering Technical Conferences and Computers and Information in Engineering Conference (IDETC/CIE 2014), pp. V006T10A050. American Society of Mechanical Engineers.

Wong, J.-Y., Reece, A. R., 1967. Prediction of rigid wheel performance based on the analysis of soil-wheel stresses part I. Performance of driven rigid wheels, Journal of Terramechanics, 4(1), pp. 81-98. doi:https://doi.org/10.1016/0022-4898(67)90105-X. Available: https://www.sciencedirect.com/science/article/pii/002248986790105X

Wong, J. Y., 2008. Theory of Ground Vehicles, John Wiley & Sons, Ltd.

Zhou, R. et al., 2023. MarsSim: A High-Fidelity Physical and Visual Simulation for Mars Rovers, IEEE Trans. on Aerospace and Electronic Systems, 59(2), pp. 1879-1892. doi:10.1109/TAES.2022.3207705.

8774 | Proposal of hybrid locomotion lunar rover with crawling mechanism

Paweł Tomiło [a,*]

[a] *Lublin University of Technology, Lublin, Poland*

* *Corresponding author: p.tomilo@pollub.pl*

ABSTRACT

The Moon's surface is covered with a layer of loose, weathered rock (fine regolith) several meters thick. This type of surface can create problems for the movement of small lunar rovers, especially in sloping areas. Legged robots are able to perform complex operations, but their design requires appropriate sensors to maintain balance, on the other hand, in the case of wheeled robots there is a risk of slipping. In order to combine the advantages of both, it was decided to create a hybrid robot.

The main purpose of the article is to present the design and testing of a prototype lunar rover. The design of a hybrid robot, which combines the characteristics of wheeled and legged robots, is shown in the article. The center of gravity may be dynamically adjusted thanks to the use of two robot arms with wheels attached. Furthermore, the robot can "crawl" on sloping areas where a wheeled vehicle could slip owing to loose ground. Crawling is possible through the use of a mechanism for swinging the blades, which are mounted on the arms. The article describes the design of the rover, describes the results of research regarding the performance of crawling mechanism, and presents the limitations of the current design of the rover and presents possible improvements and further development work.

Keywords
Crawling mechanism
Lunar rover
Hybrid-locomotion

1. Introduction

Researchers are always searching for creative ways to overcome the difficulties presented by varied and frequently treacherous terrain in an attempt to improve the capabilities of rovers for planetary exploration. Seeing the necessity for a flexible and effective exploration platform, scientists looked at a number of rover prototypes with different locomotion mechanisms. These consist of walking and wheeled rovers, spherical and transforming rovers, hybrids that combine wheeled and walking locomotion, and advanced wheeled rovers with unique weight distribution systems and enhanced wheel geometry.

Wheeled rovers have long been a mainstay of planetary exploration missions, renowned for their speed and simplicity of movement mechanics. They easily traverse level and smooth surfaces, providing a stable platform for scientific instruments and data collection. However, their effectiveness can be reduced in difficult or unpredictable terrain, where the limitations of wheel geometry are apparent. To overcome limitations a researchers designed a different types of suspension's systems. For example NASA uses rocker-bogie design for their rovers. A rocker is an intermediate element between the vehicle's frame and wheels. It is a kind of pendulum that allows balancing uneven terrain. This allows the structure to adapt to changing surfaces, maintaining stability and providing traction. Bogie is a suspension system that allows the wheels to move freely in the vertical plane. It consists of interconnected spring elements and bushings that absorb shocks and allow the vehicle to traverse uneven terrain for a smooth ride (Harrington and Voorhees, 2004). A diagram of the rocker-bogie system and its implementation in the perseverance rover is shown in Figure 1.

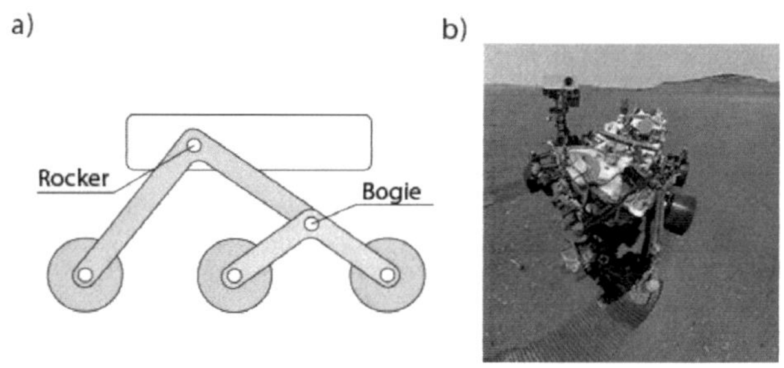

Fig. 1. a) Rocker-boogie suspension, b) Perseverance rover with rocker-boogie suspension ("Images from the Mars Perseverance Rover - NASA Mars," n.d.)

Researchers (Cordes et al., 2014) have designed a planetary rover with an active suspension system. The rover has 4 legs, where each leg independently can be lifted up (lift) and rotated relative to the vertical axis (pan). Each of the rover's legs ends with a wheel and a mechanism that allows to steer it (rotate with respect to the vertical axis). The design uses linear actuators, which are located in such a way that tensile forces are transferred to them during contact of the wheel with the ground, which ensures the rigidity of the system. A diagram of the leg of the rover in question is shown in Figure 2.

https://doi.org/10.56884/JJVQDZLM

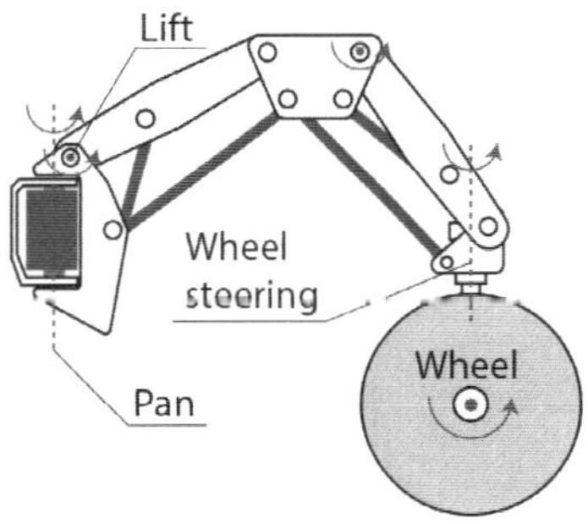

Fig. 2. Design of the arm of Sherpa rover (Cordes et al., 2014)

On the other hand, walking rovers are an attractive alternative, using a more dynamic movement mechanism to navigate challenging landscapes. Their ability to adapt to different types of terrain offers a distinct advantage, demonstrating greater stability on unstable ground compared to their wheeled counterparts. However, walking rovers also have their disadvantages, which include a complex structure, slower movement speed, increased energy consumption and the need to use additional mechanisms and data processing units (Fan et al., 2024; Mikolajczyk et al., 2022).

Researchers (Yin et al., 2022) have proposed the hexapod robot for Lunar exploration. The locomotion system of the discussed robot is based on 6 legs. A single leg uses a four-bar linkage mechanism and an extension mechanism to increase stability and maintain smooth motion. The developed robot design features reduced degrees of freedom relative to the standard hexapod design, and thus reduces the complexity of the algorithms controlling the robot's movement. By installing an extension mechanism on the robot foot, the instability caused by the foot sinking during walking is efficiently addressed by creating a larger support area. The task of the extension mechanism is, at the moment of contact between the activator and the ground, to unfold the umbrella shaped foot tip. A schematic of the discussed hexapod leg design of the planetary rover is shown in Figure 3.

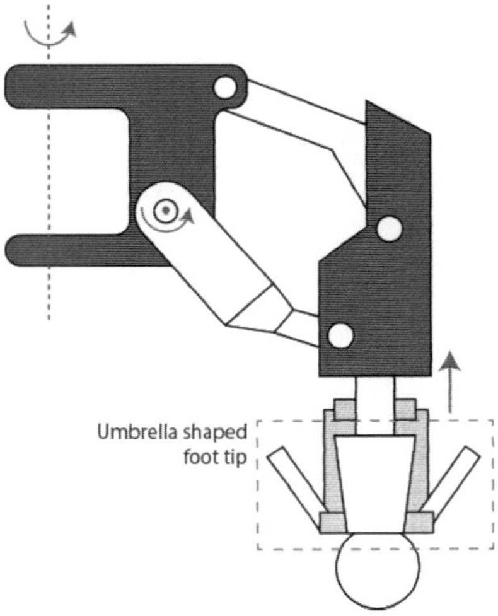

Fig. 3. Hexapod rover leg design (Yin et al., 2022)

One of the solutions used in the design of planetary rovers is the push-pull concept. The concept's primary goal is to produce thrust force by maintaining one part of the vehicle immobile in relation to the ground while actively articulating another segment to a new place. The entire vehicle is translated as a result of the positioning of each vehicle segment. In this cycle, the vehicle is effectively being pushed or pulled while grasping the ground by the fixed elements in contact with the terrain.

One variation of the pull-push concept is inch-worming. An example of this is the Scarab rover, its design consists of 4 wheels, each installed at the end of an arm. The arms allow to vary the wheel base. In the discussed example, the concept of inch-worming is to fix the position of the rear wheels in place relative to the ground and spread the arms (change the distance between the front and rear axles). The front wheels are then held in place and the rear wheels are driven while reducing the wheel base at the same time. Push-pull concepts are also used in the rovers, which are equipped with arms, similar in design to hexapods. However, wheels are mounted on the ends of the rover's legs. (Creager et al., 2011).

Apart from planetary rovers in hexapod systems, researchers (Arm et al., 2019) have developed a robot based on a quadruped system, which has been named SpaceBok. The rover was created to study the dynamics of legged rovers under low-gravity conditions. The robot's design was optimized for high jumps. The robot's leg design uses two actuated degrees of freedom per leg, both actuators in single leg are contributing equally in the acceleration phase of a jump. Tests conducted by the researchers have shown that pronking gait (robot jumps with all four legs at the same time) is a good solution for a low-gravity environment as it can jump continuously on flat terrain.

Quite an interesting direction in the development of space rovers are spherical rovers and transforming rovers. Spherical rovers in their design are characterized by the placement of all the rover's components inside wheels whose shape is similar to a half-sphere. In this type of rover, each side of the sphere is able to rotate independently, allowing the rover to perform a turning maneuver. Quite often, researchers add various kinds of solutions to spherical designs. One example is SphereX rover with the use of an additional hopping mechanism. This mechanism allows the rover to jump high from places where it could get stuck. Studies have shown that this mechanism allows a hop of 8 - 10 cm in simulated Martian gravity, and a jump of 16 - 20 cm in lunar gravity (Raura et al., 2017). The design of the robot developed by the Japan Aerospace Exploration Agency (JAXA) Sora-Q also follows the spherical rover concept. Its design also incorporates a transformation capability that allows the sides of a sphere to be folded inward, which translates into a reduction in the robot's size and provides greater structural rigidity. In its unfolded state, the robot performs an eccentric rotation of its wheels, and has a tail that provides stability. ("JAXA | Palm-Sized Lunar Excursion Vehicle 2(LEV-2)," n.d.). A schematic representation of both designs of the rovers in question is shown in Figure 4.

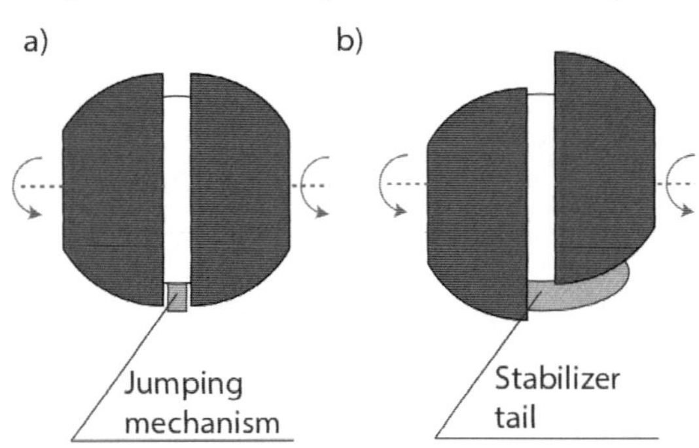

Fig. 4. Conceptual scheme: a) Sphere X, b) Sora-Q

The advantages and disadvantages of walking and wheeled rovers outlined previously support the idea of developing a robot with a hybrid locomotion system that takes advantage of both solutions. An additional improvement in the rover's mode of locomotion will be a crawling mechanism. In this paper, I examine the idea of hybrid locomotion in rover design, examining the benefits, difficulties, and possible uses of combining walking and wheeled vehicle capabilities. Through investigating rover prototype and its mobility systems.

2. Rover design

The robot prototype was constructed with an emphasis on the performance of the crawling mechanism, for this reason the front nor rear wheels are not driven. Rover has two front limbs, which consist of an arm, forearm and crawling mechanism. A wheel is installed at the end of the front limbs, one for each limb. The robot's arms are moved by servo mechanisms. In order to maintain the robot's stability, a "tail" ended by a wheel was used. A visualization of the prototype robot is shown in Figure 5.

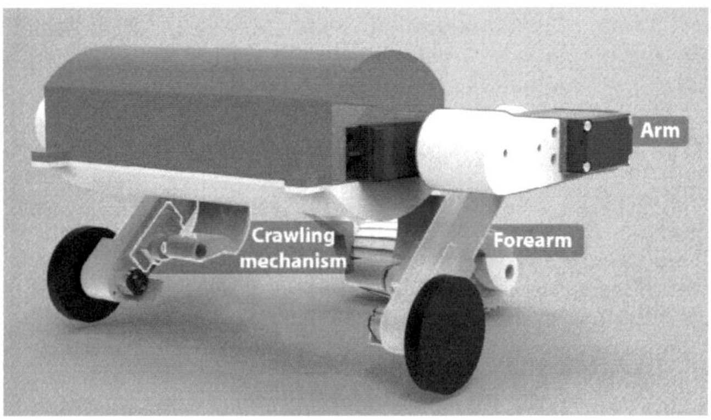

Fig. 5. Visualization of rover design

A total of 3 servos are used in the design of the rover's foreleg, the first is connected to the robot's body and is responsible for moving the arm, another servo is attached to the arm to move the forearm. The last servo is attached to the forearm and its task is to deploy and retract the blades. At the end of the forearm is installed DC motor responsible for driving wheels. In addition, a sharp elbow was used at the end of the arm to increase the ability to crawl up steep slopes. The limb is designed so that in default mode the wheel is under the point, which is the center of rotation of the arm. A diagram of the rover's front limb is shown in Figure 6.

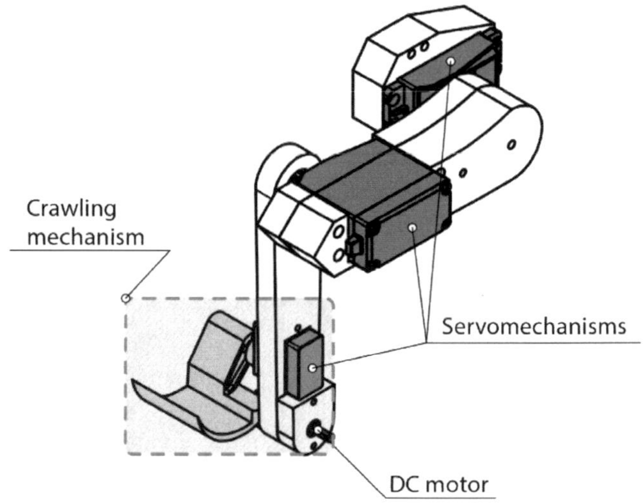

Fig. 6. Diagram of the rover's front limb

The crawling mechanism has been designed so that in its inactive state it does not affect the rover's driving capabilities. The blade of the crawling mechanism in the active state, extends and locks against the support plate. The purpose of the plate is to assist in maintaining the position of the blade. Na Figure 7 shows a visualization of the crawling mechanism.

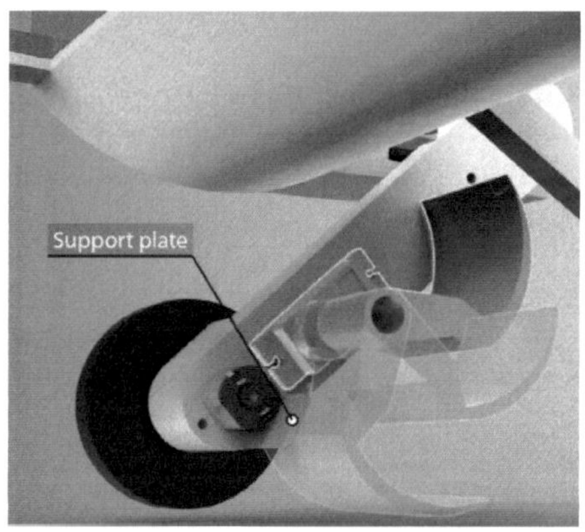

Fig. 7. Visualization of the crawling mechanism

The rover was designed in such a way that it could be entirely printed using 3D printing technology. 3D printing allowed the creation of parts with complex shapes and structures, which are necessary for the rover to function properly. The components of the rover in question, are shown collectively in Figure 8.

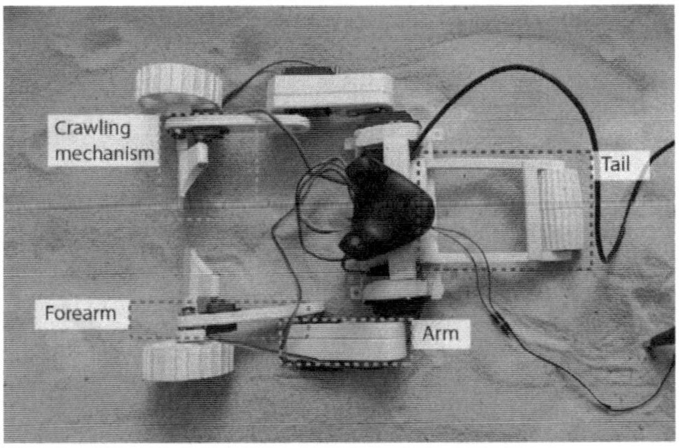

Fig. 8. Photo of the rover with the main components highlighted

At the moment, the crawling motion is defined based on manually determined frames. The rover's default state is driving mode (frame 1). The rover transitions from driving mode to crawling mode by extending the front limbs forward and leaning on the sharp elbow (frame 2). In the next phase, the blades are extended (frame 3). In frame 4, the forearms are lowered down, and then one arm is straightened (frame 5). In the last frame (6), the robot pulls up with both forearms until the elbows are over the point, which is the center of the arm's rotation. The last frame in the sequence (7) is the same as frame 1. A diagram of the crawling motion in operation is shown in Figure 9 - keyframe by keyframe.

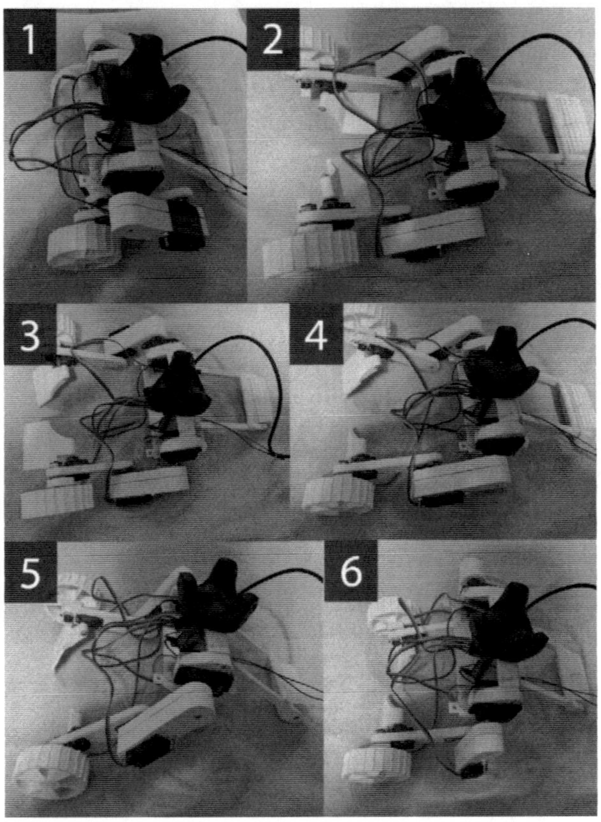

Fig. 9. Schematic of crawling motion in operation shown keyframe by keyframe

The exact same motion pattern, previously shown in Figure 9, was used for all measurements taken in this study.

3. Test stand

In order to test the rover's ability to crawl up a slope, a test stand consisting of a container filled with silica sand was made. The sand container is lifted upward using a hydraulic actuator. A lighthouse system was used to determine the rover's location. Lighthouse system makes use of drum-driven infrared (IR) light planes from Lighthouse base stations (Taffanel et al., 2021). The signal from the base station goes to the receiver, which is mounted on the rover. Research carried out by (Bauer et al., 2021) has demonstrated that the lighthouse system exhibits a repeatability of several millimeters. Two lighthouse transmitters were mounted near the test stand. A schematic of the stand and a photo are shown in Figure 10.

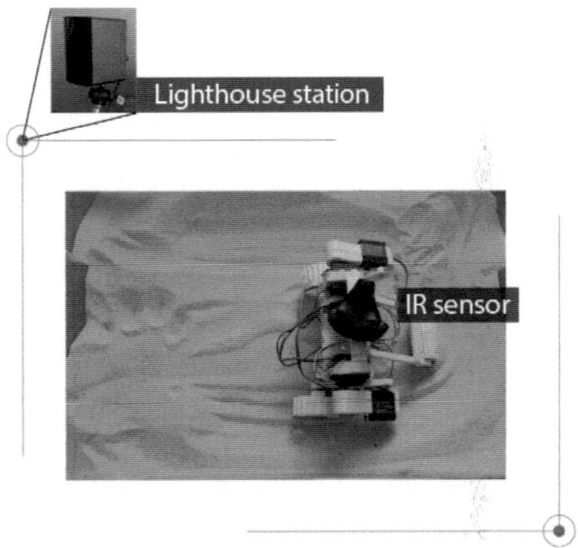

Fig. 10. Test stand

The test bed was designed so that data from the IR sensor mounted on the rover goes through the use of radio frequency (RF) communication to the main unit. The main unit is also responsible for managing the rover's movement. Two-way communication is used between the microcontroller, which manages the movement of the servos, and the main unit. The data exchange is about sending information about the execution of a given frame and when the frame has ended. The scheme of communication between the various systems is shown in Figure 11.

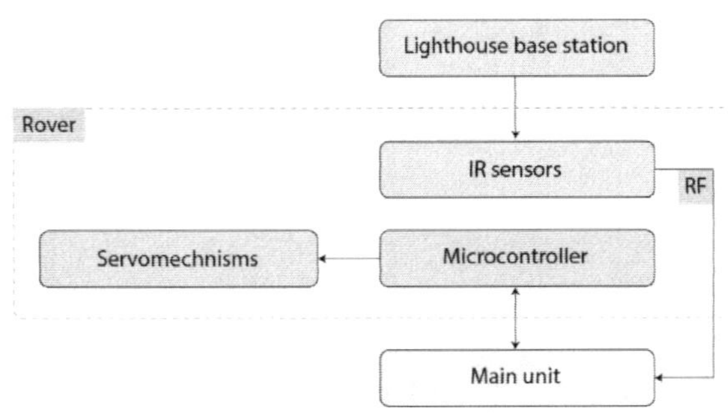

Fig. 11. Block diagram of the communication process

In order to determine the distance traveled by the rover, a metric was used in the form of the Euclidean distance between the average starting point and the average ending point. In both cases, the average is calculated from 30 samples, which translates into a set of samples from a time period of 0.5s. During this time, the rover made no movement.

4. Slope climbing test

The tests were conducted in such a way that the rover was in a crawling position and performed a single crawling motion. The tests were carried out at a slope angle of 0 to 20 degrees with a 5-degree increase in slope. For each slope angle, 10 measurements were made, from which average values were calculated. The largest average shift of the rover was achieved at an angle of 0 degrees and was 12.05 cm, while the smallest value (3.09 cm) was for 15-degree. This value is close to the average shift for 20-degree, which was 3.11 cm, and had a smaller standard deviation (0.75 cm) than for 15 degrees (1.02 cm). A full analysis of the data including the mean value, standard deviation, minimum and maximum values, as well as the first second and third Quartile (Q) is presented in Table 1.

Table 1. Statistical analysis of crawling movement

Slope [deg]	0	5	10	15	20
Mean [cm]	12,05	7,85	5,59	3,09	3,11
Std [cm]	0,76	2,38	2,32	1,02	0,75
Min [cm]	10,87	2,34	2,68	1,46	1,86
Q1 [cm]	11,83	7,19	3,73	2,69	2,66
Q2 [cm]	12,07	8,00	5,69	3,17	3,40
Q3 [cm]	12,19	9,34	6,83	4,01	3,71
Max [cm]	13,56	10,62	9,23	4,20	3,77

Due to the potential occurrence of outliers, it was decided to create a linear regression model using the RANSAC (RANdom SAmple Consensus) algorithm - RANSAC regressor. This algorithm is an iterative method that allows for the estimation of model parameters from data that contain outliers, and therefore it allows for their detection (Kaspi et al., 2017). The detected outliers and the regression line are shown in Figure 12.

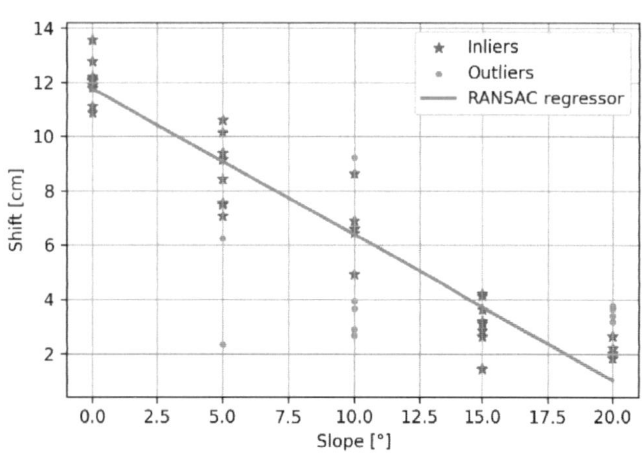

Fig. 12. Results of RANSAC regressor

The RANSAC regressor model shows that for the current rover design, for each degree increase in slope, the distance covered with a single climbing motion decreases by approximately 0.45cm. The linear regression equation shown in Equation 1.

$$r_s = -0.45s_a + 11.43 \quad (1)$$

where:

s_d - slope angle [°];

r_s – rover position shift [cm].

5. Improvements

Based on the tests, the following limitations of the current rover design have been identified:

- The current design makes it impossible to go from the driving state to the crawling state without the robot pushing back. This translates into a slide down a slope.
- The rear wheel is not driven which definitely reduces the rover's ability to climb steep slopes. A possible additional improvement would be to use an additional servo to adjust the tilt of the rover's "tail."
- Servos with a range of motion of 0 - 180 degrees are insufficient for optimal crawling motion of the rover.
- With a slope in the range of more than 15 degrees, there is a significant drop in the rover's crawling capability.

To overcome the limitations the design of the rover's front limbs must be altered. The limb cannot touch the ground when performing the crawling motion again, for this reason, additional servomechanism must be used to provide lateral movement. An overview diagram of the design of the new rover's front limb is shown in Figure 13.

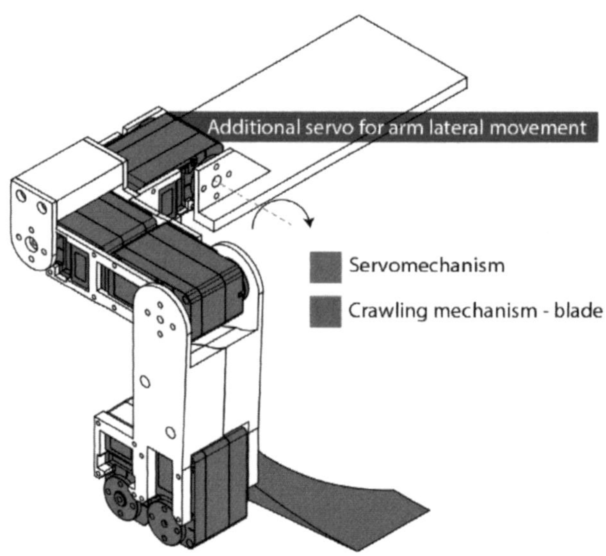

Fig. 13. Schematic of the new rover's front limb

The rover's body, or more precisely, its lower part, needs to be redesigned so that when the crawling motion is performed again, it contacts the ground in order to reduce the rover's tendency to slip down a slope. In addition, the tail of the robot in the current prototype is fix-mounted. Therefore, at higher slopes, the front part of the rover's lower body digs into the ground making crawling motion difficult. In the next robot prototype, the tail will be redesigned so that it will be moved by a servo mechanism, allowing the tail to adaptively adjust relative to the slope. In addition, the tail will be equipped with a mechanism to counteract sliding down if the use of rear driven wheels is insufficient.

6. Conclusions and future work

The current design of the rover and the defined crawling motion allows it to crawl a slope up to 20 degrees, where for flat ground the robot's center moves an average of 10.97cm, and for a slope of 20 degrees it moves an average of 3.11cm in a single movement. For the time being, the developed rover's constitution prevents optimal re-execution of the crawling motion. In further work, the layout of the rover's arms will be redesigned with servos that allow rotation of more than 180 degrees, and an additional servo will be used to allow lateral movement of the front limb. This solution will allow the crawling process to be carried out in a cycle. At present, the robot's design, when switching back to crawling mode, causes the robot to push off, resulting in a slide down a slope. Another improvement will be to change the design of the robot's tail and add a driven wheel or 2 wheels. The use of a rear drive can translate into crawling performance. In addition, the robot's tail has been equipped with 2 servos and an additional locking mechanism, whose task is to ensure optimal crawling conditions.

7. Declaration of competing interest

The authors declare that they have no known competing financial interests or personal relationships that could have appeared to influence the work reported in this paper.

8. References

Arm, P., Zenkl, R., Barton, P., Beglinger, L., Dietsche, A., Ferrazzini, L., Hampp, E., Hinder, J., Huber, C., Schaufelberger, D., Schmitt, F., Sun, B., Stolz, B., Kolvenbach, H., Hutter, M., 2019. SpaceBok: A dynamic legged robot for space exploration. Proc IEEE Int Conf Robot Autom 2019-May, 6288–6294. https://doi.org/10.1109/ICRA.2019.8794136

Bauer, P., Lienhart, W., Jost, S., 2021. Accuracy Investigation of the Pose Determination of a VR System. Sensors 2021, Vol. 21, Page 1622 21, 1622. https://doi.org/10.3390/S21051622

Cordes, F., Oekermann, C., Babu, A., Kuehn, D., Stark, T., Kirchner, F., 2014. An Active Suspension System for a Planetary Rover, in: I-SAIRAS.

Creager, C.M., Moreland, S.J., Skonieczny, K., Johnson, K., Asnani, V., Gilligan, R., 2011. Benefit of "Push-pull" Locomotion for Planetary Rover Mobility.

Fan, Y., Pei, Z., Wang, C., Li, M., Tang, Z., Liu, Q., 2024. A Review of Quadruped Robots: Structure, Control, and Autonomous Motion. Advanced Intelligent Systems 2300783. https://doi.org/10.1002/AISY.202300783

Harrington, B.D., Voorhees, C., 2004. The Challenges of Designing the Rocker-Bogie Suspension for the Mars Exploration Rover. 37th Aerospace Mechanisms Symposium.

Images from the Mars Perseverance Rover - NASA Mars [WWW Document], n.d. URL https://mars.nasa.gov/mars2020/multimedia/raw-images/ (accessed 4.18.24).

JAXA | Palm-Sized Lunar Excursion Vehicle 2(LEV-2) [WWW Document], n.d. URL https://global.jaxa.jp/activity/pr/jaxas/no088/03.html (accessed 4.20.24).

Kaspi, O., Yosipof, A., Senderowitz, H., 2017. RANdom SAmple Consensus (RANSAC) algorithm for material-informatics: application to photovoltaic solar cells. J Cheminform 9. https://doi.org/10.1186/S13321-017-0224-0

Mikolajczyk, T., Mikołajewska, E., Al-Shuka, H.F.N., Malinowski, T., Kłodowski, A., Pimenov, D.Y., Paczkowski, T., Hu, F., Giasin, K., Mikołajewski, D., Macko, M., 2022. Recent Advances in Bipedal Walking Robots: Review of Gait, Drive, Sensors and Control Systems. Sensors 2022, Vol. 22, Page 4440 22, 4440. https://doi.org/10.3390/S22124440

Raura, L., Warren, A., Thangavelautham, J., 2017. Spherical planetary robot for rugged terrain traversal. IEEE Aerospace Conference Proceedings. https://doi.org/10.1109/AERO.2017.7943758

Taffanel, A., Rousselot, B., Danielsson, J., McGuire, K., Richardsson, K., Eliasson, M., Antonsson, T., Hönig, W., 2021. Lighthouse Positioning System: Dataset, Accuracy, and Precision for UAV Research.

Yin, K., Qi, C., Gao, Y., Sun, Q., Gao, F., 2022. Landing control method of a lightweight four-legged landing and walking robot. Frontiers of Mechanical Engineering 17, 1–20. https://doi.org/10.1007/S11465-022-0707-1/METRICS

8812 | Energy method to compare performance of new types of sugar cane transport equipment

Jorge Lopera [a,*], Carlos Daniel Muñoz [b], Fernando Casanova [b], Adolfo Leon Gomez [b]

[a] *PACIFIC+, Cali, Colombia*

[b] *School of Mechanical Engineering, Universidad del Valle , Cali , Colombia*

* *Corresponding author: jlopera@pacificplus.co*

ABSTRACT

Mobility indexes (MI) have been used to understand and compare the performance for ground vehicles, which must complete tasks or follows paths for several types of environments. The MI research has been studied by several authors (*Wong 2008, Larin 2007, Chudakov, Vantsevich 2022*). There are multiple methods that supports the evaluation of mobility, from experimental, numerical, and simulated (VCI, MMP, NTVPM). To estimate the MI is a key feature in an initial stage of concept and design of new types of ground vehicles. At the agriculture industry, the challenges for a MI estimation could be like the ones at military industry, but with different points of interest (soil conservation, reduce load transfer to soil, crop protection). For the sugar cane industry at the Valle del Cauca region (Colombia-South America), these considerations have become important since the sugar cane in that region, is a whole year crop growing, that means harvesting activities are developed no matter if there is dry or wet season. The conservation concerns for the industry at wet seasons appears, because the sugar cane supply equipment (harvester, cranes, trailers) with wheels and even with tracks, reduce the reliability, with the sinkage and compaction phenomena with heavy vehicles. This work proposes the application of mobility index concept for a new type of vehicle to transport the sugar cane on the field. For the sugar cane sector, the adoption for these methodologies would help reducing the risk to deploy unreliable technologies and to compare design alternatives at conceptual phases.

Keywords
Off-road mobility
Soil Impact
Sugar cane industry
Mobility index
Tracked Systems

1. Introduction

1.1 Food and energy requirements.

The global population is steadily increasing, leading to heightened demand for food and energy production within the agricultural sector, these demands could be three times bigger than under a hypothetical scenario of all countries growing at the same rate. *(Emiko Fukase a, 2020*). This drives a focus on enhancing productivity in the food supply chain, often with the same or even fewer resources such as soil and water. Certain crops, like sugar cane, have undergone evolution from being solely dedicated to food production, primarily raw sugar, to becoming integral to a diversified agro-industry, yielding multiple products, the same plants can be harvested for several years, although yields decline over time. In addition to sugar and ethanol, sugarcane can also be used to produce derivatives such as electricity (through bagasse surplus) and bioplastics *(OECD and Food and Agriculture Organization of the United Nations, 2020)*).

1.2 Impact of agricultural equipment to the environment

To tackle this challenge, agro-industries like this one have embraced mechanization and specialized equipment to enhance supply chain performance. This includes harvesters, tipping wagons, and trains pulled by trucks with multiple wagons (*Fig. 1*).

Fig. 1 Sugar cane Harvester with auto-rolling trailer

These machines require energy to conduct field activities, typically powered by internal combustion engines, which then transfer mechanical energy via hydraulic systems or geared drivelines directly to the wheels or tracks. This energy transfer generates forces and velocities that interact with the soil *Fig. 2.*

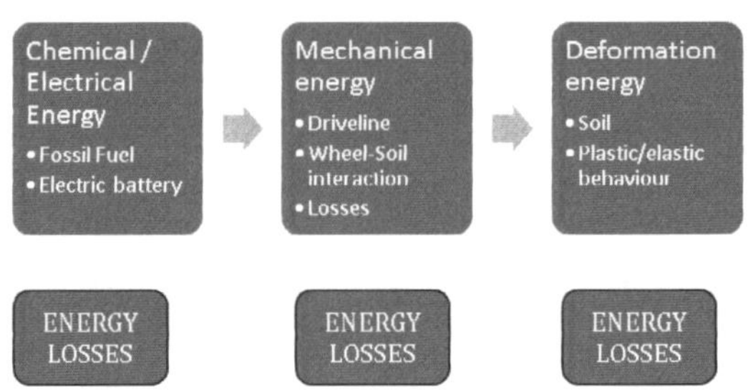

Fig. 2 Energy Flow for motion of an agricultural equipment

https://doi.org/10.56884/TU3WEI8X

With this energy flow, the agro-equipment R&D challenges aims to move more crop and biomass products with greater speed, leading to increased power and consequently exerting more weight on the soil. The load transfer to ground poses sustainability concerns for industries relying on the same soil for each crop cycle *(R.L. Raper, 2005)*. Additionally, factors such as track dimensions and steering capabilities of vehicles play crucial roles. Depending on these dimensions, the turning radius and wheel distance may cause damage to plant growth. Research on route planning for agricultural machines has been carried out, in order to increase their field efficiency *(D.D. Bochtis, 2008)*

Taking into account these considerations, the sugar industry must continue to pursue the development of new types of equipment that can boost crop productivity while preserving soil conditions.

1.3 Mobility Indexes.

To face the R&D challenges about vehicle development the mobility index (MI) has been developed and partially adopted. These indexes have been used to evaluate and compare the performance of a vehicle in a specific task.

Depending on the approach, the MI requires vehicle data, and soil conditions to estimate and compare operational performance. As an empirical estimation, the MI could have inherent limitations and uncertainties *(J.Y. Wong, 2020)*.

Vehicle cone index (VCI) and Mean Maximum Pressure (MMP) are two examples of these methods. VCI is used as a metric to evaluate the cross-country mobility of military ground vehicles on fine-grained soils and on muskeg (organic soils) *(Ahlvin, 1992)*. MMP is defined as the mean value of the maxima pressure occurring under all the wheel stations of a vehicle. These indexes, normally should not be extrapolated beyond the conditions upon which they were originally derived *(J.Y. Wong, 2020), (Wong, 2009)*

,

The VCI is the minimum cone index of a soil in the critical layer that permits a given vehicle to make a specific number of passes without immobilization, when the traffic of vehicles is increased, the soil become harder to penetrate, consequently, the index will increase. It was observed that the relative proportions of the shear, compressive and tensile strengths reflected by the cone index vary with moisture content. As soil moisture content increases, the cone index becomes increasingly insensitive to changes in shear or compressive strength *(Wong, 2009)*.

With the advancement of programming and electronics, other models have been developed to evaluate the mobility for different conditions applied to military vehicles. The NATO Reference Mobility Model (NG-NRMM) model helps NATO armed forces to compare vehicles with different soil conditions, looking a standardization between countries to compare and to share military knowledge and equipment. *(Ahlvin, 1992; Michael McCullough, 2017)*.

The published findings of several studies identify more accuracy when these methods are accompanied by simulations to validate their performances. Methods like NVTPM (for tracks) and NRMM (For wheels) finds more relation with physical measure related with the loads transferred to soil *(J.Y. Wong, 2018)*

For a new development of ground vehicle with agricultural applications, the R&D team could use indexes for comparison as a short method to assess the potential impact and to anticipate, if this vehicle is a sugar cane transporter or other crop transporter could face limitations. Other studies *(Salavat Mudarisov, 2020)* searches for the condition where the load transfer to soil could be controlled changing the suspension parameters for the setup of rollers in a tracked vehicle. This type of studies it helps when the vehicle it has been developed, but not when it is at conceptual stage of development.

To measure the impact of vehicle traffic on field activities, commercial technologies were developed to measure and create maps of the soil conditions and helps the farmers to take decisions *(VERIST TECHNOLOGIES, 2024)*. Those technologies help to take decisions on the field, before or after the field work.

Despite of the uncertainties expressed between methods, it becomes important to create and deploy methods for comparison of new alternatives of vehicles design at early stages, but not only to identify if a vehicle could deal with a type of soil, for agricultural and sustainability requirements, but also is important to reduce the impact of traffic of vehicles.

It could be helpful for decision makers roles, to use low-cost methods, to evaluate the potential impact of a tailor-made vehicles design for field activities. This method could consider:

1. Geometric and Weight considerations of vehicle.
2. Soil conditions (humidity, wet or dry season,)
3. Harvesting cycle (number of passes)
4. Energy approach

2. Methodology

After review the mobility indexes (MI), a comparison was performed for typical MI's, from a comprehensive point of view, not only looking for a performance for the vehicle, but looking how a type of motion would impact the soil health and crop condition, this evaluation has been expressed at the ***Errore. L'origine riferimento non è stata trovata.***.

Table 1
Mobility Index Qualitative Comparison

MOBILITY INDEX	Main field of application	Take account soil health to evaluate	Energy Analysis
VCI	Military	No	No
MMP	Military	No	No

In this work a methodology is used to propose an index to evaluate the agricultural equipment motion on a field. This MI doesn't focus only on motion to overcome an obstacle, for the agriculture it takes into account how the vehicle could harm or damage the environment, measured principally by the soil health.

This index could have interest for the sugar cane industry, to increase the sustainable relation of the soil and harvester operations on the field. The evaluation of different energy states for the relation should be taken into account.

This analysis begins with the conservative energy principle, widely applied by engineering, for example, thermodynamics models *(Çengel, 2008)* and fluid dynamics *(Pritchard, 2015)*, uses the conservation energy principles to be applied in a control volume. For its purpose the control volume requires theoretical and abstract conditions to be applied for a particular physical phenomenon. But the control volume could be used to understand and evaluate the relation between a vehicle and its surroundings.

The *Fig. 3* shows a layout of a control volume where the analysis could integrate several aspects, not only the vehicle motion with its kinetic energy. In the case of an internal combustion engine (ICE), the fuel chemical energy at fuel, has a transformation into mechanical energy and losses, the mechanical energy is converted into the energy required to gain motion ($\boldsymbol{E_k}$) with the interaction of soil and wheel/track ($\boldsymbol{E_w}$). The strain energy ($\boldsymbol{E_s}$), which produces the soil compaction for each time the vehicle passes through the same point.

Regarding, to the vehicle and the soil, the energy approach allows to identify each element separately inside a control volume, how could be the impact in terms of energy, which could be transmitted in deformation. At (1) , $\boldsymbol{E_k}$is the kinetic energy to move a vehicle, $\boldsymbol{E_l}$ are the energy losses that appears at the engine and drivelines, despite of the type (ICE or electric drive), $\boldsymbol{E_w}$ is the energy that interacts with soil and wheels, principally are the shear forces that are developed at the interaction between the soil and the wheels/tracks. and $\boldsymbol{E_s}$, are the energy that remains at the soil, where with each pass, compact the soil is compacted.

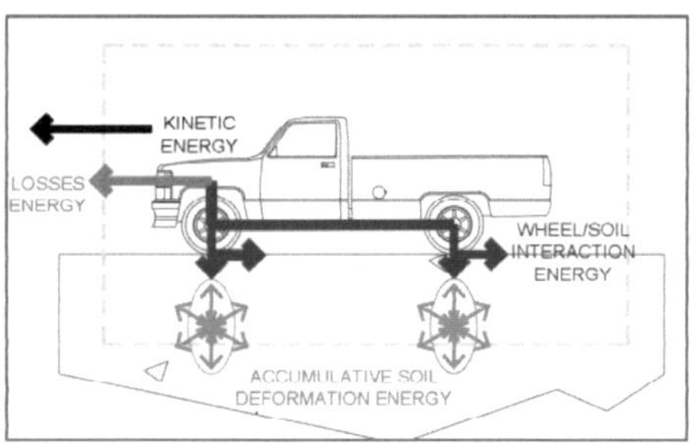

Fig. 3 Energy Flows for a Ground Vehicle in a Soil for crop

$$E_k - E_l \cong E_w + E_s + E_{ASDE} \tag{1}$$

$$E_k = \frac{mV^2}{2} \tag{2}$$

$$E_l = \eta_e \eta_d E_l = \eta_e \eta_d \frac{mV^2}{2} \tag{3}$$

$$E_w = (cA_c + W\ tan\phi)\Delta_x \tag{4}$$

$$E_s = \frac{EA}{L}\Delta_z \tag{5}$$

Where.

Ek: Kinetical energy

El: Kinetical Energy losses due to internal combustion engine and drivelines.

Ew: Shear resistance energy of soil applied to a portion of longitudinal distance (Δ_x). The shear resistance is estimated with the cohesion (c), Area of contact (A_c), weight (W) and friction angle (ϕ).

$\boldsymbol{E_{ASDE}}$: *Accumulative Soil Deformation Energy* is a measure of the amount of energy that remains at soil after a motion of a vehicle.

Manipulating the (1), it can be found the amount of energy that is transmitted to the soil due to the vehicle motion.

$$E_{ASDE} \cong E_k - E_l - E_w - E_s \tag{6}$$

The proposal method of this work is to find a quick method to estimate under a consideration of operation the accumulative energy that the soil receives from the energy flow. If this amount of energy remains at low level, the combination of vehicle, operation and soil could be balanced to repeat the operation, but if the soil deformation energy begins to increase, the probability for compaction appears.

With this estimation, it could be more useful to compare the soil deformation energy with the kinetic energy. This analysis could be deeper, if it separates the transient stage when the vehicle accelerates or decelerates in a motion task. This condition increases the energy loaded into the soil, considering the load transfer between axles due to the location of center of gravity of vehicle.

3. Application to a Conceptual Tracked Vehicle For Sugar Cane Transportation On The Field

Despite of the previous, it could be useful to apply the proposed methodology and identify considerations for future application to R&D of agricultural equipment.

At ***Errore. L'origine riferimento non è stata trovata.*** it could be seen a concept of proposal of tracked vehicle to transport sugarcane at field, this idea is no new, authors like *(Wu, 2021)*, explored this configuration of vehicle to transport the sticks of sugar cane from the field .

A previous work *(Delgado, 2023)*, proposes a tracked system to replace the wheel system. This is a good sample to evaluate the impact of this concept as it was considered previously.

Fig. 4 Sugar Cane Transporter with Tracked system (Delgado, 2023)

At the ***Errore. L'origine riferimento non è stata trovata.*** it shows the initial data used for the application of the estimation of $\boldsymbol{E_{ASDE}}$

Table 2 Properties to evaluate Sugar cane transporter

PROPERTY (letter)	VALUE	UNIT
Mass (m)	14000	Kg
Weight (N)	1.37E+05	N
Cohesion (C) (Tagar, 2017) Dry 70% 80% 70%	127.5 30.4 25.5 21.6	kPa
Friction angle (ϕ) (Tagar, 2017)	15.8 18.1 17.2 14.3	degrees
Young's Module (E)	100 40 25 15	MPa
Diesel ICE Efficiency (η_e)	0.45	
Driveline efficiency (η_d)	0.8	
Sinkages values (Δ_z)	0.05 0.1 0.15	m
Contact Area (A_c)	2.376	m2
Length of Soil between surface and zone of interest (L)	1	m

At ***Errore. L'origine riferimento non è stata trovata.***, it can be seen the ASDE vs. longitudinal velocity for a configuration of vehicle (**Errore. L'origine riferimento non è stata trovata.**), and ***Errore. L'origine riferimento non è stata trovata.***. The *Fig 5* shows a single chart showing with several operation conditions for different type of soils.

4. **Analysis**

This study looks to increase the R&D teams performance, that works on new vehicle development projects. To accomplish this challenge, the tools develop must visualize the impact and compare what happen when a specific vehicle road in a range of soil types

At *Fig. 5* it shows how could be the construction of a family of curves, using the parameters from ***Errore. L'origine riferimento non è stata trovata.***, this chart could be crossed with the soil condition when the soil health is limited by the trafficability. About the MI previously analyzed, it is important to recognize the effort for military operations this index, but the presented energy method, has a view integrating view for the operation of vehicles at field. At the literature reviewed it is common to analyze the compaction with the relation of other properties, like moisture content *(Haim Kalman, 2021)*. Or multi-pass effect for military convoys *(Jeremy P. Gray, 2015)*.

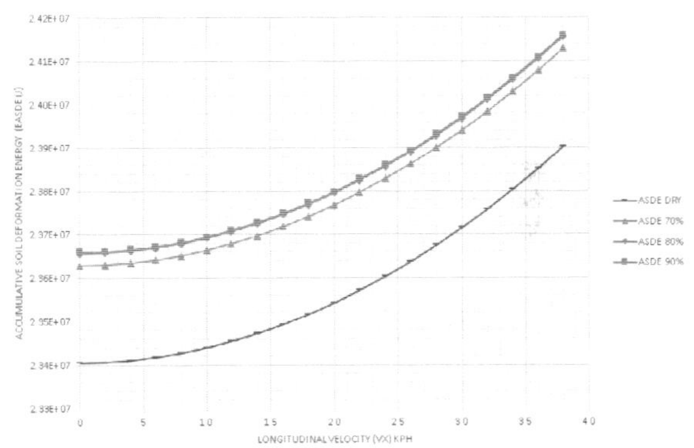

Fig. 5 Energy vs Velocity at different types of soils, using parameters of proposal tracked sugar cane transporter.

At *Table 3*, it shows a sample of variation of parameters to use for three types of conceptual vehicles, changing weight and contact area for tracks.

VEHICLE	MASS (kg)	CONTACT AREA (Ac) m2
V1	12.000	2.5
V2	12.000	3
V3	12000	3.5

Table 3 Design variation to test the method

Applying the exposed method, it could be possible to see the design variation that remains more energy at field (V3), with this sample, it would be used other parameters, like sinkages, driveline efficiency, cohesion, shear angle.

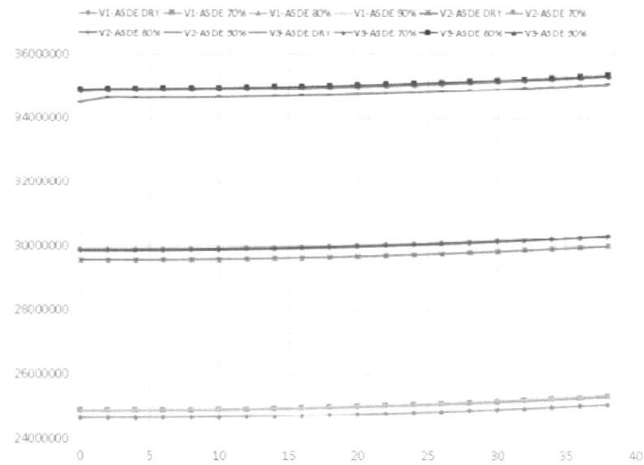

Fig. 6 Comparison for three version of conceptual vehicle $(\boldsymbol{E_{ASDE}}\ (J)\, vs\ \boldsymbol{V_x}(\boldsymbol{kph}))$

5. Conclusions

The mobility index creates a path to evaluate the performance and risk for military operation. For sugar cane industry and agricultural equipment development, it's not recommended to use it.

The main difficult to use, is that mobility index only take account the soil state for an operation condition, but not to an agro-industrial task a sugar cane field.

The energy approach presented in this work, settled the challenge to consider the energy flows at vehicle performance for agricultural equipment development.

It's important to notice, that this work open a discussion about the use mobility index but to take account the wide discussed and documented impact from the agricultural traffic to the environment.

6. Future Work

For future work, it is expected to evaluate this methodology through an experimental vehicle. Each energy model recreated could be validated, even changing the soil energy estimation or the kinetic energy to move the vehicle.

It is necessary to propose and to build prototypes of tracked vehicles, this could be leaded it with new methods for prototyping reducing the risk and to gain confidence for agro-industrial investors, stakeholders for this type of projects.

To take advantage with new technologies, it could be possible to use Virtual Reality systems (VR) or Augmented reality (AR), with this technology it could be possible to recreate the whole vehicle and soil and simulates the energy flows for a prototype or an existing commercial vehicle. It is possible to correlate a motion with the loads transferred to the soil and to represents the potential accumulated damage that appears at soil. This could help for sugar cane industry and others agro industries increase the reliability, sustainability, and results communication for research community.

7. Declaration of competing interest

The authors declare that they have no known competing financial interests or personal relationships that could have appeared to influence the work reported in this paper.

8. Acknowledgements

The authors thankfully acknowledge to CENICAÑA under the project "CATE II" for the support and information for the project.

9. References

Ahlvin, R. H. P., 1992. NATO Reference Mobility Model Edition II, NRMM II, Vicksburg MI: Waterways Experimental Station.

Çengel, Y. A., 2008. Thermodynamics : an engineering approach.. Boston: McGraw-Hill Higher Education.

D.D. Bochtis, S. V., 2008. Minimising the non-working distance travelled by machines operating in a headland field pattern. Biosystems Engineering, pp. Volume 101, Issue 1.

Delgado, C. D. M., 2023. DISEÑO DE UN SISTEMA DE TRACCIÓN TIPO ORUGA ADAPTADO A UN VAGÓN AUTOPROPULSADO PARA EL TRANSBORDO DE CAÑA DE AZÚCAR. CALI: UNIVERSIDAD DEL VALLE.

Emiko Fukase a, W. M., 2020. Economic growth, convergence, and world food demand and supply. World Development, p. Volume 132.

Haim Kalman, D. P., 2021. Analyzing bulk density and void fraction: B. Effect of moisture content and compression pressure. Powder Technology, Volumen Volume 381, pp. Pages 285-297.

J.Y. Wong, P. J. E. T. J. P.-T., 2018. Journal of Terramechanics Comparison of simulation models NRMM and NTVPM for assessing military tracked vehicle cross-country performance, pp. Pages 31-48.

J.Y. Wong, P. J. E. T. J. P.-T., 2020. A review of mobility metrics for next generation vehicle mobility models. Journal of Terramechanics, pp. Pages 11-20.

Jeremy P. Gray, V. V. V., 2015. Multi-vehicle convoy mobility in severe terrain conditions: Factor impact analysis, estimation and control strategy. Journal of Terramechanics, Volumen Volume 61, pp. Pages 43-61.

Michael McCullough, P. J. J. D. D. G., 2017. The Next Generation NATO Reference mobility model development. Journal of Terramechanics, pp. Pages 49-60.

OECD and Food and Agriculture Organization of the United Nations, 2020. OECD-FAO Agricultural Outlook 2020-2029. [En línea] Available at: https://doi.org/10.1787/1112c23b-en

Pritchard, P. J. &. M. J. W., 2015. Fox and McDonald's introduction to fluid mechanics (9th ed.).. 9th ed. s.l.:John Wiley & Sons..

R.L. Raper, 2005. Agricultural traffic impacts on soil. Journal of Terramechanics, pp. Volume 42, Issues 3–4 Pages 259-280.

Rowland, D., 1972. Tracked vehicle ground pressure and its effect on soft ground. Stockholm-Kiruna, Sweden,, s.n., p. pp. 353–384..

Salavat Mudarisov, I. G. I. G. E. H. I. F., 2020. Soil compaction management: Reduce soil compaction using a chain-track tractor. Journal of Terramechanics, Volumen 89, pp. 1-12.

Tagar, A. G. M. A. J. L. N. &. S. A., 2017. Assessment of implement efficiency and soil structure under different conventional tillage implements and soil moisture contents in a silty loam soil. Agricultural and Food Sciences, Environmental Science, pp. 413-420.

TECHNOLOGIES, V., 2024. The Perfect Platform for Every Operation. [En línea] Available at: https://www.veristech.com/sensing

Wong, J., 2009. Terramechanics and Off-Road Vehicle Engineering. s.l.:Butterworth-Heinemann.

Wu, T. W. H. L. Q. e. a., 2021. Design and Development of Tracked Sugarcane Transporter. Sugar Tech 23, p. 1137–1146.

9028 | Uncertainty quantification for wheeled locomotion machine learning predictions on soft soil

Vladyslav Fediukov [a,b,#,*], Jana Huhne[b,#], Felix Dietrich [b], Fabian Buse [a]

[a] *Institute of System Dynamics and Control, German Aerospace Center (DLR), Weßling, Germany*

[b] *School of Computation, Information and Technology, Technische Universität München, Garching b. München, Germany*

 # These authors contributed equally to this work
** Corresponding author: vladyslav.fediukov@dlr.de*

ABSTRACT

Surrogate modeling with machine learning (ML) techniques is becoming increasingly popular in engineering and physical fields. Models based on statistical inference often lack uncertainty measures, which are crucial for comprehensive predictions. Uncertainty quantification (UQ) addresses these challenges, especially in tasks lacking analytical solutions or extensive experimental data, such as modeling wheel locomotion on soft soils. High-fidelity data from real experiments or precise particle-level simulations are scarce, adding inherent uncertainty to statistical models. In our paper we analyzed the UQ aspect of the terramechanical surrogate modeling. Our surrogate model leverages the probabilistic nature of Gaussian processes to facilitate uncertainty calculation and make further analysis easier. We extend UQ analysis into a new multi-fidelity model for wheel locomotion. Our work aims to improve the model's interpretability and optimization through uncertainty propagation, sensitivity analysis and uncertainty decomposition.

Keywords
Rover locomotion
Surrogate modeling
Machine learning
Multi-fidelity
Uncertainty quantification

1. Introduction

Surrogate modeling is gaining popularity in engineering and physical fields, where more and more models are constructed using machine learning (ML) techniques. Models based on statistical inference are often incomplete without the uncertainty measure that accompanies the prediction. Uncertainty quantification (UQ) is a field which is dedicated to overcome such challenges. UQ is especially important in tasks where a wide variety of analytical solutions or a sufficiently large set of experimental data are not available. One example of such a challenge is modeling the wheel locomotion on soft soils, which does not have an analytical solution and therefore requires a large number of numerical simulations and approximations, i.e. with ML and statistical inferences. In addition to that, high-fidelity data from real experiments or precise particle-level simulations are scarce and therefore any statistical model will inherently have a comparatively high level of uncertainty.

A detailed analysis and classification of different aspects of UQ was given by Hüllermeier & Waegeman (2021), where they split uncertainty into two broader categories: epistemic and aleatoric. The former represents the part of uncertainty produced by the lack of knowledge about the behavior of a target function or underlying physical processes, but which could be compensated with more observation, while the latter corresponds to the part of uncertainty introduced by fundamental randomness of the process or by our inability to understand it. Precise decoupling of uncertainties into these two categories is ambiguous, because different factors often interfere with each other and their disentanglement is impossible.

UQ of surrogate modeling is an important part of model analysis and performance assessment. As shown by Yudin et al. (2023), Bergmann et al. (2023) and Hechinger et al. (2024), UQ significantly increases the interpretability of a model and opens a possibility for further optimization, potentially increasing the overall performance. In the field of terramechanics the UQ always plays an important role in the modeling and analysis pipeline. Works of Gallina et al. (2016) and Jayakumar et al. (2014) are good examples of the importance of the UQ for terramechanical tasks. One advantage of our surrogate model is the direct probabilistic nature of Gaussian processes, which makes it easier to calculate uncertainties and conduct further analysis.

The first step in uncertainty analysis is uncertainty propagation. This includes a proper calculation of resulting statistical moments, here, mean and variance. In the case of multi-fidelity modeling, where we have hierarchical surrogate models, it is important to assess how uncertainty is propagated through levels. The second step is to determine how input variables and their variance contribute to the posterior total variance. The third step is a decomposition of total uncertainty into uncertainty introduced by numerical approximation and inherent mathematical uncertainty of the model.

In this paper we will analyze the uncertainties of the terramechanical surrogate model and continue the work of Huhne (2023) by extending this analysis to a new multi-fidelity model for wheel locomotion presented in Ravi et. al (2024). We aim to provide the uncertainty analysis for our terramechanical surrogate model, including uncertainty propagation, sensitivity analysis and uncertainty decomposition. This is the first step for increasing the interpretability and further optimization of the model.

https://doi.org/10.56884/6VTE9FAQ
ISBN 978-1-942112-55-6

2. Background

2.1. Surrogate modeling with Gaussian process

Surrogate modeling is a common approach when it comes to physical simulations. A Gaussian process (GP) is an efficient and flexible framework and has proven itself well when it comes to UQ. GP is defined by the mean and covariance functions K. Marginal and conditional distributions of Gaussian are also Gaussian distributions, which simplifies the statistical inference. Gaussian process regression (GPR) is a statistical inference over unobserved points X* given a set of observed (training) points X, where $f(X^*)$ and $f(X)$ are evaluations of a target function on unobserved and observed points respectively. As shown by Williams and Rasmussen (2006), we construct the joint distribution

$$K\begin{bmatrix} f(X) \\ f(X^*) \end{bmatrix} \sim N\left(0, \begin{matrix} K(X,X) & K(X,X^*) \\ K(X^*,X) & K(X^*,X^*) \end{matrix}\right). \tag{1}$$

The predictive distribution over an unobserved point is defined with the posterior mean $E(f(X^*))$ and covariance $cov(f(X^*))$, which are formulated as

$$\begin{aligned} E(f(X^*)) &= K(X,X^*)^T K(X,X)^{-1} f^{-1}(X), \\ cov(f(X^*)) &= K(X^*,X^*) \\ &\quad - K(X,X^*)^T K(X,X)^{-1} K(X^*,X)^T. \end{aligned} \tag{2}$$

And $cov(f(X^*))$ defines the uncertainties produced by the GPR surrogate models.

Table 1
Variables description that are collected on different levels of fidelity.

Variable name	Specification	Description
$\overrightarrow{p_w}$	Input	Wheel position w.r.t. the surface
$\overrightarrow{v_w}$	Input	Wheel velocity
$\overrightarrow{w_w}$	Input	Angular velocity of the wheel
$\overrightarrow{n_g}$	Input	Gravity direction
$\overrightarrow{F_w}$	Output	Forces acting on the wheel
$\overrightarrow{T_w}$	Output	Torques acting on the wheel

2.2. Terramechanical data description

In order to build a surrogate terramechanical model, we utilized existing terramechanical simulations, developed at DLR's Institute of System Dynamics and Control. These simulations have different fidelities. For this work, we designated SCM simulation introduced by Buse (2022) as a lower-fidelity data source. The TROLL testbed described by Buse et al. (2018), was used as a high-fidelity data source. This is a robotic arm with a mounted wheel on top and sensors scanning the surface and recording all the related signals. One must keep in mind that this categorization into low and high fidelity is relative and can vary depending on available data sources. In both cases, we gather the information about the wheel's trajectory, velocity, acting forces and torques. Figure 1 depicts the scheme of a wheel and data which is collected during its movement. The features are described in Table 1. Each of the input and output variables is 3-dimensional. The high-fidelity dataset was designed in order to capture as many potential wheel movements as possible. Experimental design and motivation behind them was described in the previous publication Fediukov et al. (2022). In this study we created 50 runs in the testbed, with various changes to wheel's tilt, steering, slippage and surface conditions.

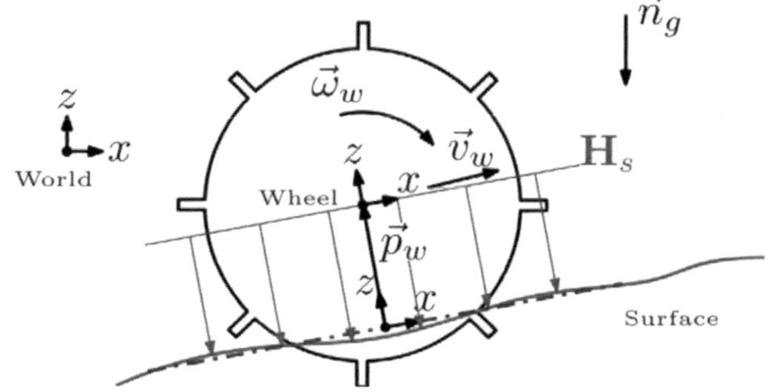

Figure 1. This scheme illustrates the wheel and measurements used in the ML model, with signals designed for easy collection during a real mission. Consequently, the trained ML surrogate model will be suitable for real-world applications. Features description is provided in Table 1. Visualization is adapted from Fediukov et al. (2022).

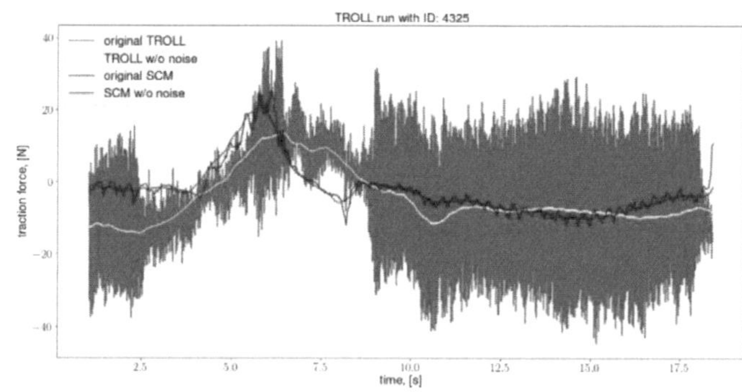

Figure 2: An example of a traction force recorded from one TROLL testbed run, which involved changing the steering angle between 0.7 and -0.7 radians, rapid deceleration, backward movement and with constant -20% slippage. The original signal (green) is visibly noisy, therefore we extract the noise generated by the robotic arm, filter out all the frequencies higher than 5 Hz and apply moving average. Resulting signal (yellow) is used as a target variable for the model's training. The same procedure is conducted for the SCM signal (noisy blue and filtered purple signals).

After conducting the high-fidelity experiments, we run the SCM simulation on trajectories reconstructed from the TROLL data, to have a direct replication of high-fidelity data in lower-fidelity simulations. We also use that the lower-fidelity model is relatively cheap regarding memory and computational time. Signal preprocessing involves frequency filtering and smoothing. An example or raw and preprocessed data is shown on Fig. 2. With a simulation of 20 seconds taking approximately one minute to run on a personal computer and several megabytes to store. This mean we can run a large number of simulation runs with different scenarios and thus can capture even greater variety of scenarios, compared to higher-fidelity runs. In this study we are constructing surrogate models for traction force, normal force and pitch. All outputs are important for wheel-soil interaction modeling.

2.3. Multi-fidelity

The Multi-Fidelity (MF) model leverages multiple sources of information, balancing the limitations of lower-fidelity data with the precision of higher-fidelity data. It compensates for the scarcity of expensive high-fidelity data by utilizing the abundance of lower-fidelity data. The concept behind MF is to incorporate the outputs of lower-fidelity functions as additional dimensions to the input space. This aims to create a mapping function between fidelities that is expected to be smoother than the mapping between the original inputs and the high-fidelity model, and thus easier to learn with fewer training points. Lower-fidelity functions can either be the same as the higher-fidelity function but at a lower resolution, or entirely different simulations, i.e. simplified formulas that approximate the high-fidelity function with reduced accuracy. In our previous study Ravi et al. (2024), we evaluated the performance of various multi-fidelity frameworks. The Non-linear Auto-regressive Deep Gaussian process (NARDGP) model, first introduced by Cutajar et al. (2019), demonstrated the best performance, particularly in predicting the traction force from TROLL simulations. Consequently, this model will be used as a surrogate in our sensitivity analysis, in Section 3.1.

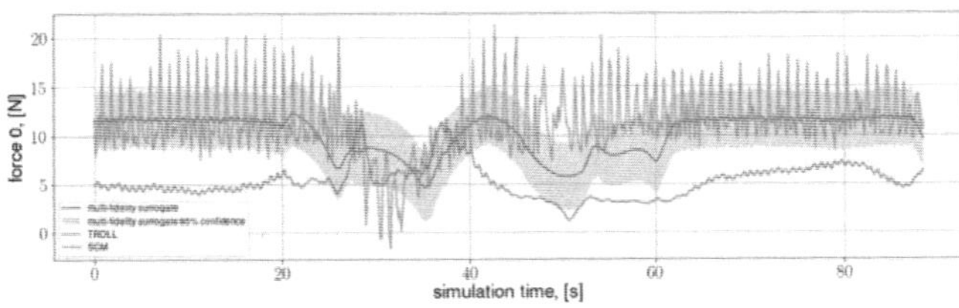

Figure 3: Example of the multi-fidelity surrogate prediction for traction force. TROLL data in orange is a high-fidelity test run, which we want to recreate. The multi-fidelity surrogate was trained on different high-fidelity runs and has not observed this particular run. We can see that the approximation capabilities are better than with the semi-empirical SCM method.

NARDGP is a subclass of Deep Gaussian Processes (DGP) proposed by Damianou & Lawrence (2013), inspired by multi-layer neural network architectures, compose multiple GPs with each other in a way that an output of one GP serves as an input for the next one. In contrast to NARGP, in DGP we replace the function evaluations from the previous fidelity level directly with the GP from previous level

$$f_l(X) = g_l(X, g_{l-1}(\dots)). \tag{3}$$

However, since certain parts of uncertainty analysis is challenging with DGP models, we will employ the Non-linear auto-regressive Gaussian process (NARGP) model, introduced by Perdikaris et al. (2017), for uncertainty decomposition analysis. This model also showed reasonable performance in certain scenarios and could potentially improve provided more high-fidelity points are included in the dataset.

In NARGP we consider learning non-linear function g: $R^{n+1} \to R$, which maps the original n-dimensional input X with lower-fidelity output $f_l(X)$ to the higher-fidelity

$$f_h(X) = g(X, f_l(X)). \tag{4}$$

An example of a multi-fidelity surrogate prediction is depicted in Fig. 3. In this case we used NARGP prediction to approximate a traction force from the test run recorded by the TROLL testbed. The blue shaded area is the uncertainty measure of the prediction, which is provided by the posterior covariance matrix, as shown by Eq. 2.

2.4. Uncertainty propagation in multi-fidelity Gaussian processes

In UQ we are trying to understand both the model's performance as well as the confidence of the model in its own predictions. In GP's, probabilistic predictions are simple to obtain. Inference with GP includes posterior mean predictions as well as a variance, as illustrated by Eq. 2. This posterior variance is used as a measure for the model's uncertainty. Posterior variance highly depends on the GP's kernel function.

A more comprehensive uncertainty quantification would include measurement noise and perform uncertainty propagation from input uncertainty to the output. In practice, researchers often consider noise-free inputs and omit the step of input noise propagation. The train process of GP with noise input is described by McHutchon & Rasmussen (2011) and by Dallaire et al. (2009). To incorporate input noise into the training procedure, we need to either assume the noise distribution as Gaussian or evaluate the empirical distribution of the noise, which can be challenging.

When we build multi-fidelity models, we must choose if we include predicted variances from the lower-fidelity surrogate in the multi-fidelity approach. In NARGP, we build a cascade of Gaussians. If we omit the variance from lower fidelity levels, we just use a mean prediction as one of the inputs and there is no uncertainty propagation through the fidelity levels.

To incorporate lower-fidelity variance, the posterior distribution, which is used for prediction on unobserved test data point looses its Gaussian form, because (see Eq. 4) the posterior prediction will be done for the test point $(x_*, f_{t-1}(x_*))$ in a new space with an additional dimension, where the Gaussianity is not guaranteed. Therefore, an analytical inference as in Eq. 2 is not available. Subsequently, as shown by Perdikaris et al. (2017), we estimate the resulting distribution by integration

$$\begin{aligned} p(f_{*_t}(x_*)) &= p\big(f_t(x_*, f_{t-1}(x_*)) | f_{*_{t-1}}, x_*, x_t, y_t\big) \\ &= \int p(f_t(x_*, f_{t-1}(x_*)) | \, x_*, x_t, y_t) p(f_{*_{t-1}}(x_*)) dx_* \end{aligned} \tag{5}$$

where f_t , f_{t-1} are surrogates from current and previous fidelity level, respectively and x_*, x_t, y_t are unobserved test points, observed training points and observed training target variables respectively.

In NARDGP a posterior distribution from $f_l(X)$ is also analytically intractable. Therefore to build a cascade of GPs, we assume that the prior for the fidelity level l is a GP and we train DGP by minimizing Kullback-Leibler (KL) divergence $KL(q||p)$ between variational distribution q and the real posterior distribution p on each fidelity level l. An exact evaluation of KL is infeasible and we solve a maximization problem for the Evidence Lower Bound (ELBO). The process is described in detail by Salimbeni & Deisenroth (2017). After obtaining the posterior distribution, we can have a posterior variance, which serves as an uncertainty measure of the surrogate.

Another important part of uncertainty propagation is sensitivity analysis, which is a method used to determine how different input variables affect the output uncertainty of the model, which we cover in the next section.

2.5. Sensitivity analysis

Uncertainty propagation processes requires more than actual quantification of resulting uncertainties. It is essential to also analyze the influence of input variables on the resulting uncertainty of a surrogate model. This is important for model's validation and interpretation. We can evaluate what variables contribute the most to the surrogate and the proportions between those contributions. We can both validate if the model's behavior has a physical explanation and can simplify it by removing unimportant variables.

With these techniques we want to explore what part of the resulting output uncertainty is explained by uncertainties of the individual inputs or its combinations. There is a variety of approaches for sensitivity analysis, both local sensitivity analysis (LSA) and global sensitivity analysis (GSA). As discussed by Homma & Saltelli (1996), LSA methods are focusing on the local changes in the output function after small changes, i.e. taking a derivative, in one input variable, while fixing all other variables. In contrast, GSA focuses on analyzing the output uncertainty over the whole range of input variables. Given that we want to analyze physical processes with non-linear dependency between all their inputs and output, estimating a derivatives only for one arbitrary variable will drastically narrow down parameter space. This may not give us information about the behavior of the function in the entire space. Therefore we will only consider GSA in this paper.

The most popular method for GSA is Sobol' indices, introduced by Sobol' (1990). This method estimates sensitivity in terms of changes in the variance of the model output due to changes in input variables.

Sobol' indices of the first order identify the fraction of the overall output variance due to not knowing the value of input i with respect to the entire input space. They are calculated as

$$S_i = \frac{V[E[Y|X_i]]}{V[Y]}. \tag{6}$$

Higher order Sobol' indices similarly represent the output variance due to ineraction effects between two or more variables, that go beyond the sum of individual effects. The second order Sobol' index is defined as

$$S_{ij} = \frac{V[E[Y|X_i, X_j]] - V[E[Y|X_i]] - V[E[Y|X_j]]}{V[Y]}. \tag{7}$$

Often, the overall influence of an input variable is of interest, for example when deciding if features can be excluded from the model. For this purpose, the total Sobol' index S_{T_i} can be calculated as

$$S_{T_i} = 1 - \frac{V[E[Y|X_{-i}]]}{V[Y]}. \tag{8}$$

It represents the variance left unexplained in the posterior surrogate variance $V[Y]$ if all features are known except X_i, combining the first order index and all higher order contributions through interactions. Therefore, an input variable can be considered noninfluential if its total Sobol' index is close to zero.

Correlations among variables complicate the interpretation of Sobol' indices, as correlated variables and their interactions mutually influence each other's results. For instance, the first-order Sobol' index of input X_i might also depend on X_j because of their correlation, which is never the case for independent variables. To address this, Jacques et al. (2006) suggest grouping correlated variables together and computing Sobol' indices for each group. This can be done by treating the combined variables in a group as one multidimensional variable in the equations. As a result, the influence of variables across groups can be compared without interference from correlation effects.

With Gaussian processes as surrogate models we can utilize not only the mean prediction, but also consider the probabilistic structure of the posterior GP

$$Z(X) \sim GP(\mu, K), \tag{9}$$

and as a result, Sobol' indices will become random variables

$$\widetilde{S_i} = \frac{V[E[Z(X)|X_i]]}{V[Z(X)]}, \tag{10}$$

where $E[Z(X)|X_i]$ is also a Gaussian process, because it is a linear transformation of a Gaussian process. On the contrary, $V[Z(X)]$ is not a linear transformation, therefore the analytical solution is unavailable and we should solve an integral via sampling.

As shown by Le Gratiet et al. (2017), by capturing the variance of the full Gaussian process, we can account for both epistemic and aleatoric uncertainties of the predictive distribution. Taking into account the entire Gaussian process instead of its mean prediction requires extensive computational resources and we will defer this to future work.

2.6. Computational Uncertainty

To fully understand the uncertainty in surrogate model predictions, it is necessary to consider sources beyond the training data. One such source is computational uncertainty due to approximations during training and evaluation of the model. This is especially relevant for Gaussian Processes, since the computational effort scales cubically with the number of training points, making approximate methods necessary to handle large amounts of data. However, these methods introduce additional uncertainty that is not reflected in the predictive uncertainty of the approximate model. To address this problem, Wenger et al. (2022) developed a method that quantifies computational uncertainty and directly incorporates it into the predictions. Their framework can be applied to several of the most common GP approximations.

The key idea is to model the computationally expensive part of the GP posterior, the *representer* weights

$$v^* = (K(X,X) + \sigma^2 I)^{-1} \left(y - m(X)\right) \tag{11}$$

as a random variable, which represents the uncertainty about their exact value. Marginalization yields the resulting uncertainty in the model outputs:

$$p\left(f(X^*)\right) = \int p\left(f(X^*)|v^*\right)p(v^*)dv^*. \tag{12}$$

The information about v^* gained by performing computations is then represented by Bayesian updates. To do this, we first define a Gaussian prior over v^*, so that the marginal distribution in Eq. 12 yields the GP prior. This represents the state of no information about v^*. Next, we write the approximation algorithm under consideration as a series of update steps, each using a one-dimensional projection of the current residual

$$r_{i-1} = (K(X,X) - \sigma^2 I)(v^* - v_{i-1}). \tag{13}$$

These projections $\alpha_i = s_i^\top r_{i-1}$ are defined by actions s_i that are chosen according to the specific approximation algorithm used. With this information, we can update the belief about v^* with the information gained in step i using the distribution from the previous update step as prior:

$$p(v^*|\{\alpha_j\}_{j=1}^{i}, \{s_j\}_{j=1}^{i}) = \frac{p(\alpha_i|s_i, v^*)p(v^*|\{\alpha_j\}_{j=1}^{i-1}, \{s_j\}_{j=1}^{i-1})}{\int p(\alpha_i|s_i, v^*)p(v^*|\{\alpha_j\}_{j=1}^{i-1}, \{s_j\}_{j=1}^{i-1})\, dv^*}. \quad (14)$$

Due to the Gaussian prior, the resulting distribution is again Gaussian. After marginalization, the predictive distribution at step i is

$$p(f(X^*)) = N(m(X^*) + K(X^*, X)v_i, \\ K(X^*, X^*) - K(X^*, X)C_i K(X, X^*)), \quad (15)$$

where

$$C_i := S_i\left(S_i^\top(K(X,X) + \sigma^2 I)S_i\right)^{-1} S_i^\top \quad (16)$$

is a rank-i matrix.

The predictive variance entails both mathematical and computational uncertainty. In theory, these two can be decomposed as

$$K(X^*, X^*) - K(X^*, X)(K(X,X) + \sigma^2 I)^{-1}K(X, X^*) \\ + K(X^*, X)\Sigma_i K(X, X^*), \quad (17)$$

where the term in the first line is the mathematical uncertainty and the term in the second line is the computational uncertainty. Σ_i is the covariance of the *representer* weights after i iterations. However, doing so requires computing the exact posterior, so in practice only the combined uncertainty is known.

We illustrate the application of the algorithm with two GP approximation methods, Conjugate Gradients (CG) and Sparse Variational Gaussian Processes (SVGP).

As a widely used method for solving systems of linear equations, CG is a common choice for GP approximation, as discussed in Bartels & Hennig (2020). Since it iteratively updates the solution with projections of the residual, it can be easily integrated into the algorithm by Wenger et al. (2022). Choosing conjugate gradients as actions s_i recovers the CG algorithm for an initial vector $v_0 = 0$ exactly. This implies that the approximate posterior mean remains the same as in regular CG and only the variance as measure of uncertainty will differ. Moreover, in both cases, the asymptotic runtime is $O(in^2)$, where i denotes the number of iterations and n the number of training points. This means that considering computational uncertainty does not introduce additional computational cost in this scenario.

In contrast, SVGP introduced by Hensman et al. (2013) is more closely tied to GPs and not as readily integrable into the framework as CG. Instead of using the full set of training data, the method employs a smaller set of pseudo-inputs, also called inducing points, that do not have to coincide with actual data points. The positions of the inducing points and their prior distribution are optimized using a variational approach. SVGP is especially well-suited for large amounts of data, since its computational cost scales linearly with the size of the training data. Using actions $s_i = K(X, z_i)$, with z_i being the inducing points, leads to an approximate posterior that is conceptually close to that of SVGP. While the computational cost remains quadratic, the algorithm of Wenger et al. (2022) allows for a more accurate representation of both the posterior mean and variance.

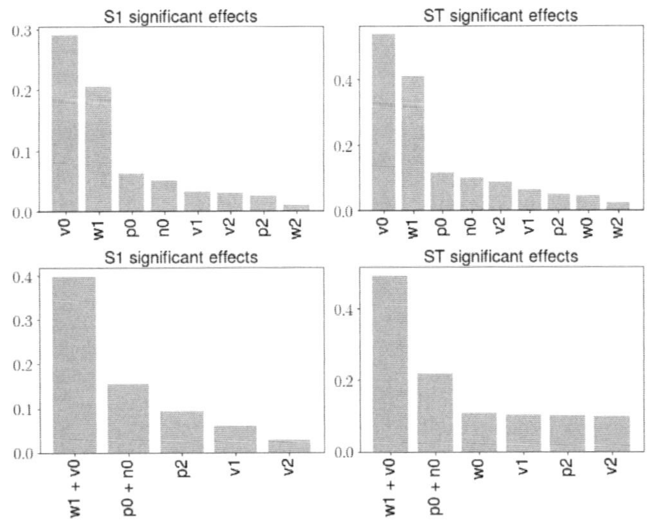

Figure 4: (Upper row) Sobol' indices for the multi-fidelity surrogate model for traction force prediction. The first-level Sobol' indices (S1) are showing what part of total resulting variance the variance of marginalized mean can explain. While first-order indices fail to capture the potential influence of the respective feature in different combinations, the total Sobol' indices (ST) show what proportion of variance is lost when the respective feature is not included. In this situation, we can also capture the influence of the respective feature in general, including all indirect influences. (Bottom row) The first (S1) and total (ST) Sobol' indices are calculated for the grouped variables. As our variables correlate, we can see their combined contribution to the total variance.

3. Numerical experiments

3.1. Sensitivity analysis of a surrogate model

Sensitivity analysis is a first step to understand what input variables are influencing the model's uncertainty in its predictions. In practice, SA is deployed to increase interpretability of a model and to reduce the input space, based on the importance of the input variables. In this study, we analyzed the sensitivity of the multi-fidelity NARDGP surrogate model.

In order to calculate the sensitivity of the model, we want to calculate first-level and total Sobol' indices (SI), which characterize the model's sensitivity from different perspectives. Since some of the input variables are correlated, we also partition them into independent groups before computing the SI. Input variables are treated as independent if their correlation coefficient in the training data is less than 0.4. Individual Sobol' indices are also considered, but should be viewed with caution as they assume independent input variables

Equation 6 shows first-level SI and in order to calculate it, we take the variance of target function $V[Y]$, where Y in this example is a traction force. By the nature Gaussian processes, variance is implicitly provided by the surrogate model. Next, we calculate $V[E[Y|X_i]]$. We use Monte Carlo in order to sample the conditional distribution of mean $E[Y|X_i]$, which is a mean surrogate model's predictions Y w.r.t. to a certain feature X_i, i.e. translational velocity *v0* or sinkage *p2*, etc. and then take the variance of the distribution of mean predictions $V[E[Y|X_i]]$, which are changing due to the changes in X_i input variable. By calculating the ration between $V[Y]$ and $V[E[Y|X_i]]$ as in Eq. 6, we can find how strongly or weakly the changes in a certain input variable X_i (for example translational velocity) can affect the changes in the surrogate's prediction of the target variable Y (for example traction force).

Grouped and total SI can be calculated in the similar manner and are shown in Eq. 7 and 8 respectively.

The sensitivity analysis was done using open-source python library SALib developed by Herman & Usher (2017) and Iwanaga et al. (2022), which can be accessed on Github: https://github.com/SALib/SALib.

3.1.1. *Sensitivity analysis of traction force*

Figure 4 depicts first-level and total SI, showing the influence of each feature on the resulting prediction of the model, which predicts the traction force of a wheel. We can see there that both indicators show the importance of *v0*, i.e. translational velocity in x-direction, and *w1* *i.e.* angular velocity in y-direction (pitch). The same situation can be observed for grouped variables, where closely-correlated variables, like *w1* and *v0*, contribute the biggest part when we are trying to predict traction force acting on the wheel. In the second place there is a combination of *n0*, gravity vector in x-direction, and the *p0*, wheel's center shift compared to the ground. Changes in these variables indicate changes in the surface, for example going over a hill. This result shows a good connection with the physical intuition behind the wheel-soft soil interaction and dependency of the models predictions on the variables that are physically connected with the traction force.

3.1.2. *Sensitivity analysis of normal force*

Next we analyze the sensitivity of the normal force surrogate. For the surrogate predicting the normal force, we created a separate lower-fidelity dataset covering specific wheel movements. Our initial dataset, which was designed to learn the traction force, contains wheel movements that do not cover a big variety in the normal force. This is because during steering, acceleration and uphill or downhill movement, we do not have enough variation in normal force. To learn the correct physical behavior of the normal force acting on the wheel, we need to represent the canonical pressure-sinkage relationship. As shown in Fig. 5, the original dataset has not captured this relationship, therefore we had to collect a separate dataset, where this relationship was more distinct. However we were not able to use this new dataset for training the traction force, because the wheel's movements were artificial and did not contain any meaningful variety in either traction force, nor in the driving torque.

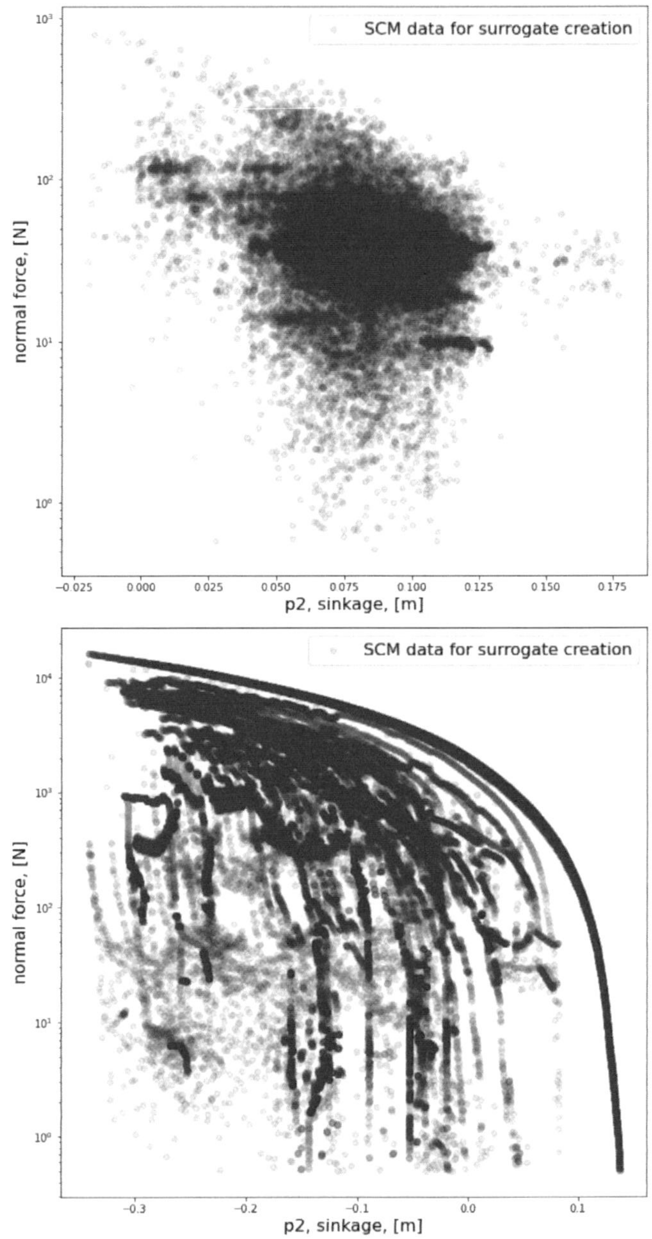

Figure 5: As was shown by Bekker (1960), the sinkage-normal force curve should follow an exponential form in the general case, with differences introduced by the soil characteristics. The dataset which was collected using standard scenarios only, which might be seen as standard wheel movements, like moving forward, backward, steering, etc. does not form at least slightly visible exponential dependency between sinkage and normal force. Therefore we conducted a set of artificial simulations, like burying the wheel deep inside the ground and varying the wheel's mass, so that it replicates the functionality of a Bevameter and can capture the pressure-sinkage dependency more properly. The normal force axis is logarithmic for clarity.

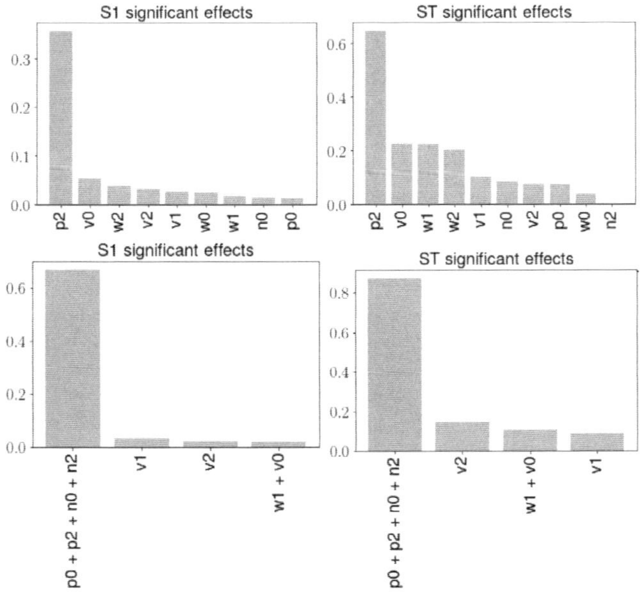

Figure 6: (Upper row) Sensitivity analysis for the surrogate model of the normal force shows very big dependency of the resulting variance on the p2 feature, representing the sinkage of a wheel. (Bottom row) Sobol' indices of grouped variables for the normal force surrogate.

After training our multi-fidelity surrogate on this new dataset, we analyze its sensitivity. Figure 6 shows Sobol' indices for the normal force surrogate. We can see that among first-level Sobol' indices the most important variable by far is sinkage (*p2*). Variables with far smaller impact are translational velocity and angular velocities. Still, total Sobol' indices assign them a slightly higher importance. Which means that those features have additional impact on the posterior model's variance, but rather indirect, through combinations with other features. If we analyze the grouped variables, which are highly correlated with one another, we see that the group of variables representing sinkage (*p2*), vertical (*n2*) and horizontal (*n0*) gravity components and wheel's center shift (*p0*), contribute the most to the variance of the normal force prediction. This can serve as a verification of the trained surrogate, as it has mainly learned that the normal force is mostly described by the sinkage variable, however considering some other, less relevant factors.

3.1.3. Sensitivity analysis of drive torque

The last experiment of the surrogate's sensitivity was conducted for the pitch torque surrogate. In this case model's most important variables should resemble variables from the traction force surrogate, due to high dependency between traction force and drive torque. Figure 7 shows that translational velocity and the angular velocity are the most important variables describing the pitch torque variability. These results also validate the learned behavior of the surrogate and show that surrogate's behavior is physically intuitive and therefore can be regarded not as a "black box", but is rather "gray box", where individual contributions of each physical input to the surrogates' predictions are explainable.

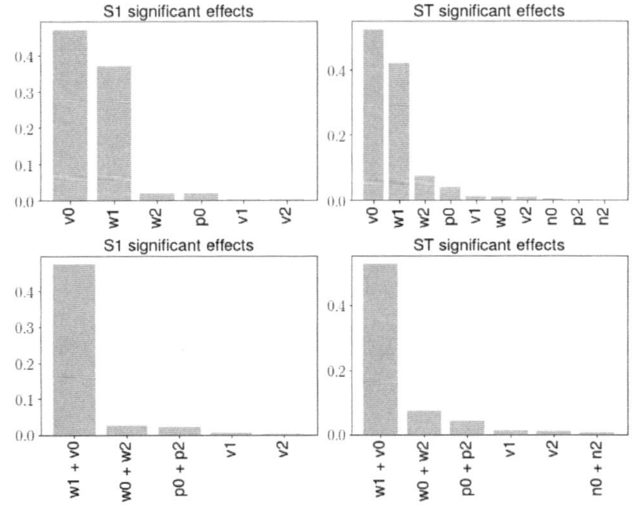

Figure 7: (Upper row) First-level and total Sobol' indices for the surrogate model predicting drive torque. (Bottom row) Sobol' indices of grouped variables for the drive torque surrogate.

3.2. Examples of computational uncertainty

To investigate the effect of limited computation on the uncertainty, we apply the algorithm described in section 2.6, called IterGP, to our model and compare the results to the corresponding regular versions of CG and SVGP. Because IterGP is not directly applicable to the sparse variational method used in deep GPs, we use a NARGP model predicting the traction force for these experiments.

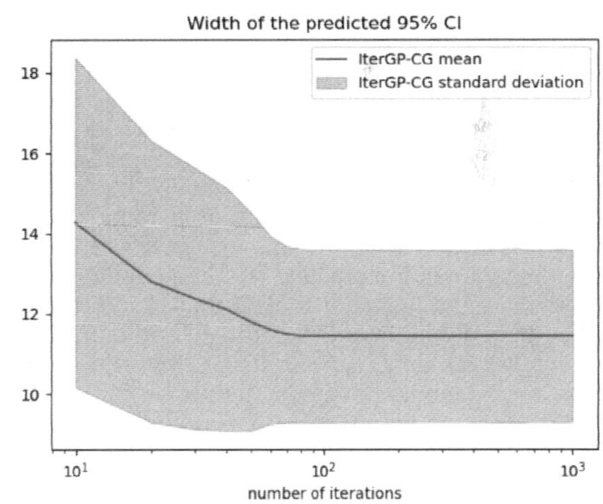

Figure 8: Mean and standard deviation of the width of the 95% CI over all test data, for the CG version of IterGP for an increasing number of iterations.

Figure 8 shows the mean width of the 95% confidence interval (CI) as a measure of the overall uncertainty for the CG version of IterGP across all test data. The uncertainty decreases with an increasing number of iterations, particularly up to around 100 iterations. Afterwards, it remains stable, indicating that the combined uncertainty has converged to the mathematical uncertainty and the computational part of the uncertainty is close

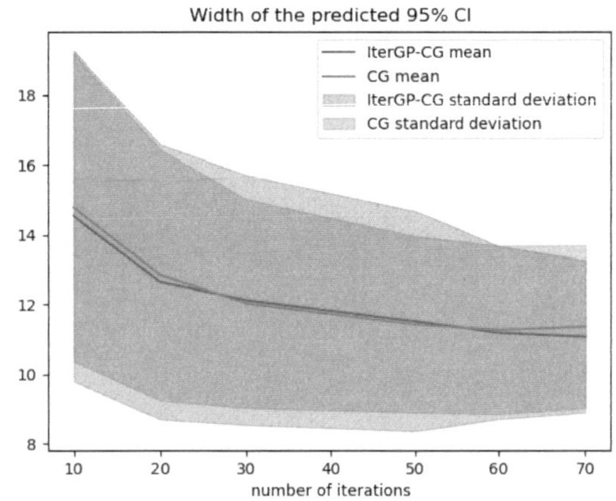

Figure 9: Mean and standard deviation of the width of the 95% CI over all test data, for the CG version of IterGP compared to the regular CG method for an increasing number of iterations.

to zero. Additional iterations beyond this threshold contribute minimally to further reducing the uncertainty.

Figure 9 compares the CG version of IterGP to an equivalent implementation of CG (kindly provided by Jonathan Wenger) that does not consider computational uncertainty. Interestingly, the difference between the two approaches is minimal for all tested numbers of iterations. Thus, the regular CG method also provides a good measure for the combined mathematical and computational uncertainty of our model. However, it is important to note that the regular CG method lacks the interpretability of the result as such due to not explicitly considering both sources of uncertainty. IterGP therefore provides valuable insights about the uncertainty even if the results themselves are very similar. The comparison between SVGP and the corresponding IterGP version in Figure 10 shows larger differences between the predicted uncertainties. IterGP maintains a narrower average CI width than regular SGVP by roughly a half. Thus, SVGP consistently overestimates the posterior uncertainty even compared to the combined mathematical and computational uncertainty of IterGP. Therefore, the uncertainty estimate of IterGP is significantly more accurate compared to the mathematical posterior than the one from SVGP. In terms of accuracy, the two methods are more similar, as can be seen in Fig. 11. While IterGP tends to have a lower RMSE, the benefit is not as large as in the cases tested by Wenger et al. (2022). However, we could not investigate the trend of the accuracy for larger numbers of inducing points, since the IterGP algorithm experienced numerical instability for more than 80 inducing points.

4. Conclusion

In this paper, we address uncertainty quantification and its analysis for a terramechanical multi-fidelity surrogate model. We discuss uncertainty propagation in different multi-fidelity models and in different scenarios. Then, we performed sensitivity analysis of the resulting variance, to determine how much each input variable contributes to the final prediction variance. Our results add interpretation to the model, making its predictions more understandable for its user. SA results can also serve as a validation tool, because we can check if variable importance in the model corresponds to physical intuition. We showed how final

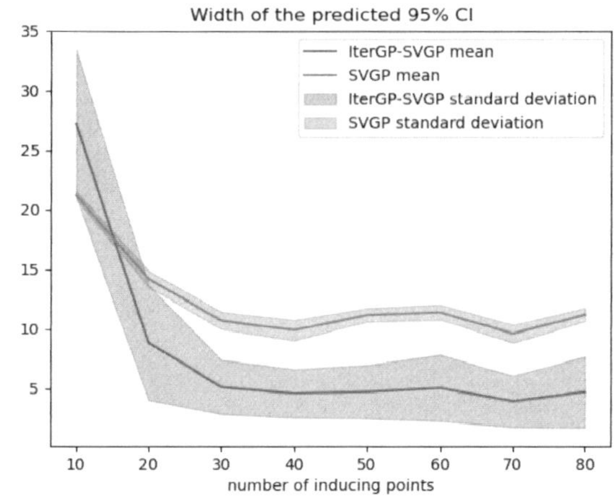

Figure 10: Mean and standard deviation of the width of the 95% CI over all test data, for the SVGP version of IterGP compared to the regular SVGP method for an increasing number of iterations.

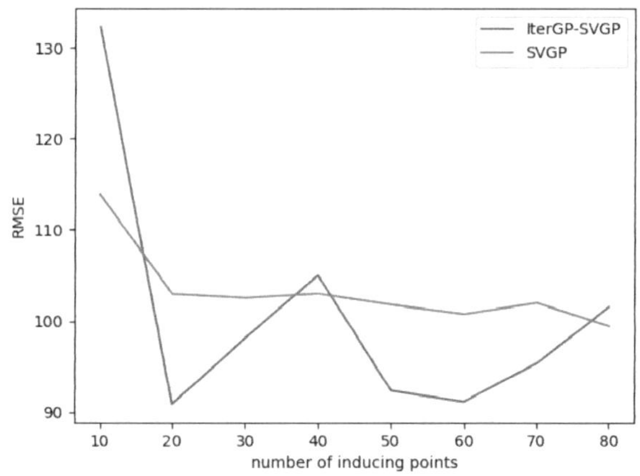

Figure 11: RMSE of SVGP and the corresponding IterGP version for different numbers of inducing points.

uncertainty can be decomposed into computational and mathematical parts, which opens a new perspective on the model's optimization process.

As future work, we see more expensive experiments as a promising path for deeper analysis, for example doing SA directly with the samples from the predicted distribution, not only from the mean prediction. Another important improvement would be the incorporation of input (measurement) noise as an uncertainty of the input variables directly into the training process. This should provide a more comprehensive UQ and also increase the model's performance.

Acknowledgements

Research by Vladyslav Fediukov is funded by the Helmholtz Association under the "Munich School for Data Science - MuDS". Felix Dietrich acknowledges funding by the DFG project no. 468830823, and also association to DFG-SPP-229. We would like to thank Jonathan Wenger for kindly sharing the code implementation from his article.

References

Bartels, S., & Hennig, P. (2020). Conjugate gradients for kernel machines. Journal of Machine Learning Research, 21(55), 1-42.

Bekker, M. G. (1960). Off-the-road locomotion. Research and development in terramechanics.

Bergmann, M., Ravi, K., Fischer, R., Angioni, C., Höfler, K., Molina Cabrera, P., et al. (2023). Introduction and Uncertainty Quantification of Kinetic Models in the Integrated Data Analysis Framework. Talk presented at 86. Jahrestagung der DPG und DPG-Frühjahrstagung der Sektion Materie und Kosmos (SMuK). Dresden. 2023-03-20 - 2023-03-24

Buse, F., Bellmann, T., Lichtenheldt, R., & Krenn, R. (2018). The DLR Terramechanics Robotics Locomotion Lab.

Buse, F. (2022). Development and Validation of a Deformable Soft Soil Contact Model for Dynamic Rover Simulations (Doctoral dissertation, Tohoku University).

Cutajar, K., Pullin, M., Damianou, A., Lawrence, N., & González, J. (2019). Deep gaussian processes for multi-fidelity modeling. arXiv preprint arXiv:1903.07320.

Dallaire, P., Besse, C., & Chaib-Draa, B. (2009). Learning Gaussian process models from uncertain data. In Neural Information Processing: 16th International Conference, ICONIP 2009, Bangkok, Thailand, December 1-5, 2009, Proceedings, Part I 16 (pp. 433-440). Springer Berlin Heidelberg.

Damianou, A., & Lawrence, N. D. (2013, April). Deep gaussian processes. In Artificial intelligence and statistics (pp. 207-215). PMLR.

Fediukov, V., Dietrich, F., & Buse, F. (2022, September). Multi-fidelity machine learning modeling for wheel locomotion. In 11th Asia-Pacific Regional Conference of the International society for terrain-vehicle systems, ISTVS 2022. ISTVS.

Gallina, A., Krenn, R., & Schäfer, B. (2016). On the treatment of soft soil parameter uncertainties in planetary rover mobility simulations. Journal of Terramechanics, 63, 33-47.

Gratiet, L. L., Marelli, S., & Sudret, B. (2016). Metamodel-based sensitivity analysis: polynomial chaos expansions and Gaussian processes. In: Handbook of Uncertainty Quantification. Ed. by R. Ghanem, D. Higdon, and H. Owhadi. Cham: Springer International Publishing, 2017, pp. 1289–1325

Hechinger, K., Zhu, X. X., & Kauermann, G. (2024). Categorising the world into local climate zones: towards quantifying labelling uncertainty for machine learning models. Journal of the Royal Statistical Society Series C: Applied Statistics, 73(1), 143-161.

Hensman, J., Fusi, N., & Lawrence, N. D. (2013). Gaussian Processes for Big Data. In Uncertainty in Artificial Intelligence (p. 282).

Herman, J. and Usher, W. (2017) SALib: An open-source Python library for sensitivity analysis. Journal of Open Source Software, 2(9). doi:10.21105/joss.00097

Homma, T., & Saltelli, A. (1996). Importance measures in global sensitivity analysis of nonlinear models. Reliability Engineering & System Safety, 52(1), 1-17.

Huhne, J. (2023). Uncertainty Quantification for Gaussian Processes.

Hüllermeier, E., & Waegeman, W. (2021). Aleatoric and epistemic uncertainty in machine learning: An introduction to concepts and methods. Machine learning, 110(3), 457-506.

Iwanaga, T., Usher, W., & Herman, J. (2022). Toward SALib 2.0: Advancing the accessibility and interpretability of global sensitivity analyses. Socio-Environmental Systems Modelling, 4, 18155. doi:10.18174/sesmo.18155

Jacques, J., Lavergne, C., & Devictor, N. (2006). Sensitivity analysis in presence of model uncertainty and correlated inputs. Reliability Engineering & System Safety, 91(10), 1126-1134. The Fourth International Conference on Sensitivity Analysis of Model Output (SAMO 2004). issn: 0951-8320. doi: https://doi.org/10.1016/j.ress.2005.11.047.

Jayakumar, P., Melanz, D., MacLennan, J., Gorsich, D., Senatore, C., & Iagnemma, K. (2014). Scalability of classical terramechanics models for lightweight vehicle applications incorporating stochastic modeling and uncertainty propagation. Journal of Terramechanics, 54, 37-57.

McHutchon, A., & Rasmussen, C. (2011). Gaussian process training with input noise. Advances in neural information processing systems, 24.

Perdikaris, P., Raissi, M., Damianou, A., Lawrence, N. D., & Karniadakis, G. E. (2017). Nonlinear information fusion algorithms for data-efficient multi-fidelity modelling. Proceedings of the Royal Society A: Mathematical, Physical and Engineering Sciences, 473(2198), 20160751.

Ravi, K., Fediukov, V., Dietrich, F., Neckel, T., Buse, F., Bergmann, M., & Bungartz, H. J. (2024). Multi-fidelity Gaussian process surrogate modeling for regression problems in physics. arXiv preprint arXiv:2404.11965.

Salimbeni, H., & Deisenroth, M. (2017). Doubly stochastic variational inference for deep Gaussian processes. Advances in neural information processing systems, 30.

Sobol, I. (1990): On sensitivity estimation for nonlinear mathematical models. Mathematical Modeling 2(1), 112–118

Wenger, J., Pleiss, G., Pförtner, M., Hennig, P., & Cunningham, J. P. (2022). Posterior and computational uncertainty in Gaussian processes. Advances in Neural Information Processing Systems, 35, 10876-10890.

Williams, Christopher KI and Rasmussen, Carl Edward (2006). Gaussian processes for machine learning, volume 2. MIT press Cambridge, MA.

Yudin, Y., Coster, D., von Toussaint, U., & Jenko, F. (2023, June). Epistemic and aleatoric uncertainty quantification and surrogate modelling in high-performance multiscale plasma physics simulations. In International Conference on Computational Science (pp. 572-586). Cham: Springer Nature Switzerland.

9295 | Granular Scaling Laws for Accurate Prediction of Wheel Mobility on Slopes in Low-gravity Environments

Takuya Omura [a,*], Genya Ishigami [a]

[a] *Department of Mechanical Engineering, Keio University, Yokohama, Kanagawa, Japan*

* *Corresponding author: ta_hu823tyann@keio.jp*

ABSTRACT

Analyzing the mobility of wheeled rovers on loose sand in low-gravity environments remains a significant challenge. Among several experimental techniques, such as parabolic flight and reduced-weight tests, granular scaling laws (GSL) have recently been proposed to predict wheel mobility under low-gravity conditions via earth-gravity tests. Although the GSL accurately predicts wheel mobility on flat terrain in low-gravity environments, its capability to predict wheel mobility on slopes in such environments still needs to be verified. This study developed a GSL and investigated its accuracy in predicting wheel mobility on slopes in low-gravity environments. The discrete element method (DEM) was utilized to test wheel mobility at various slope angles under Earth's gravity. Subsequently, by applying a multiple scaling function, the GSL converted the results from the Earth-gravity tests to predict wheel mobility under lunar gravity. The GSL-based predictions were compared with DEM simulations conducted under lunar gravity conditions. The results indicated that the wheel mobility under lunar gravity predicted by the GSL closely corresponded to that calculated via the DEM. These findings indicate that the GSL can accurately predict wheeled-rover mobility on slopes in low-gravity environments.

Keywords
Granular scaling laws
Wheel mobility prediction
Low gravity
Lunar regolith simulant

1. Introduction

The release of the Global Exploration Roadmap (International Space Exploration Coordinate Group, 2018) has increased interest in space exploration. This roadmap calls for extending beyond traditional scientific expeditions to include missions demonstrating technologies for sustained activities on celestial bodies such as the Moon and Mars. The updated space exploration scenario builds on the success of previous missions and illustrates the importance of robotics. In particular, wheeled rovers have proven to have high mobility efficiency on flat terrains, despite their simple mechanism (Sanguino, 2017). They will continue to be among the most widely used tools in future space exploration. Moreover, missions such as NASA's VIPER (Colaprete, 2021) will explore the Moon's craters for water resources. Therefore, the mobility of the rover must be improved to allow it to travel on loose sands and slopes.

Researchers have analyzed wheel–soil interactions on slopes. Sutoh et al. (2012) and Inotsume et al. (2019) studied wheel mobility on slopes with various wheel designs and grousers. Du et al. (2018) and Watanabe et al. (2023) improved terramechanics models by considering soil deformation. However, the effects of gravity were not considered. Therefore, modeling wheel–soil interactions on slopes in low-gravity environments remains an open problem.

The similarity law is valuable for rover design because it can predict wheel mobility under low-gravity conditions through Earth-gravity tests (Kuroda et al., 2005; Li et al., 2012). Slonaker et al. (2017) recently proposed granular scaling laws (GSL) that account for time scaling and employ multiple dimensionless numbers. Thoesen et al. (2020a, 2020b) and Daca et al. (2022, 2023) verified that the GSL accurately predicts wheel mobility on flat terrain in low-gravity environments. Furthermore, Zhang et al. (2020) proposed an expanded GSL for predicting the slope wheel mobility and verified its effectiveness for environments with stronger gravity than Earth.

In the present study, we simulated wheel mobility under various driving conditions in a low-gravity environment and evaluated the prediction accuracy of a developed GSL that combines the concepts proposed by Zhang et al. (2020) and Daca et al. (2022, 2023). Wheel mobility was tested at various slope angles under Earth's gravity using the discrete element method (DEM). The proposed GSL then converted the Earth-gravity test results to predict wheel mobility under lunar gravity. Subsequently, the GSL-based predictions were compared with the results of DEM simulations under lunar gravity conditions. The wheel mobilities under lunar gravity predicted by the GSL closely agreed with those calculated via the DEM. This study contributes to the field of planetary rover mobility in several ways.

- The proposed GSL was comprehensively verified under various driving conditions. The results confirmed the applicability of the GSL for predicting wheel mobility on slopes in a low-gravity environment.
- This study confirmed that the GSL can accurately predict wheel mobility under various slip conditions. In particular, this study is the first to verify the high accuracy of GSL prediction under negative slip ratios, demonstrating the applicability of the GSL even under previously untested and challenging conditions.
- The findings also confirmed the capability of the GSL for predicting time-series data, including transient states. Further, the GSL can accurately predict the power consumption for the rover driving, which is essential for planning and managing their operational time.

https://doi.org/10.56884/DM46X2U9

The remainder of this paper is organized as follows. Section 2 describes the GSL theory and related works. Section 3 presents the numerical experimental methods using the DEM and the simulation setup. In Section 4, the simulation results are presented, the prediction accuracy is evaluated, and the effectiveness of the GSL is demonstrated. Finally, Section 5 concludes the paper and suggests directions for future research.

2. Granular Scaling Laws

Slonaker et al. (2017) derived the original GSL inspired by dynamical similarity in fluid mechanics. The foundational concept of GSL is to provide a physical basis for directly relating different granular locomotion problems for the same soil without additional experiments or simulations, as shown in Fig. 1. Importantly, the GSL is a similarity law that explicitly addresses gravitational acceleration. Therefore, it is a powerful tool for space exploration because it can predict the dynamic behavior of rovers in different gravitational environments. The original GSL is a similarity law composed of multiple dimensionless numbers, as expressed by Eq. 1.

$$\left[\frac{V}{\sqrt{Dg}},\frac{P}{Mg\sqrt{Dg}}\right]=\Psi_1\left(\sqrt{\frac{g}{D}}t,f,\frac{g}{D\omega^2},\frac{\rho BD^2}{M}\right) \tag{1}$$

Here, Ψ_1 is a function formulated in dimensionless variables with four inputs and two outputs. Among the output variables, V represents the wheel's horizontal velocity; P represents the power consumed to drive the wheel; and g represents the gravitational acceleration. The wheel parameters M, D, and B represent the mass, diameter, and width, respectively; f is a set of points defining the wheel shape; ω represents the wheel's angular velocity; ρ represents the soil density; and t represents time.

Assuming the same wheel shape f and the same soil, and the test input condition is given as (g, M, D, B, ω), then the other test should be conducted under the condition $(g', M', D', B', \omega') = (\alpha g, \beta M, \gamma D, \beta\gamma^{-2}B, \alpha^{1/2}\gamma^{-1/2}\omega)$ for any positive scalars α, β, and γ. Correspondingly, the outputs V and P scale such that $(V', P') = (\alpha^{1/2}\gamma^{1/2}V, \alpha^{3/2}\beta\gamma^{1/2}P)$. In addition, the time at prediction is scaled by $\alpha^{1/2}\gamma^{-1/2}$ because the dimensionless time term is $\sqrt{g/D}\,t$. Thus, Eq. 1 is a similarity law that considers the gravity timescale. Slonaker et al. (2017) verified that two outputs could be accurately predicted under low-speed driving conditions.

Thoesen et al. (2020a, 2020b) and Daca et al. (2022, 2023) enhanced the original GSL. Thoesen et al. (2020a, 2020b) confirmed that the GSL could predict the dynamics of an entire vehicle through coupled multibody dynamics and DEM simulations. Daca et al. (2022, 2023) proposed a GSL with three outputs, which is expressed by Eq. 2.

$$\left[\frac{F_{DP}}{Mg},\frac{h}{D},\frac{P}{Mg\sqrt{Dg}}\right]=\Psi_2\left(\sqrt{\frac{g}{D}}t,f,\frac{g}{D\omega^2},\frac{\rho BD^2}{M},\frac{V}{\sqrt{Dg}}\right) \tag{2}$$

Here, Ψ_2 is a function with five inputs and three outputs. F_{DP} represents the drawbar pull, and h represents the wheel sinkage. The horizontal velocity of the wheel was controlled to achieve steady slip ratios; thus, the dimensionless number for the velocity V could be moved to the input side. Their contribution was to increase the number of physical quantities that could be predicted and confirm that dimensionless numbers could be moved to either the input or output side.

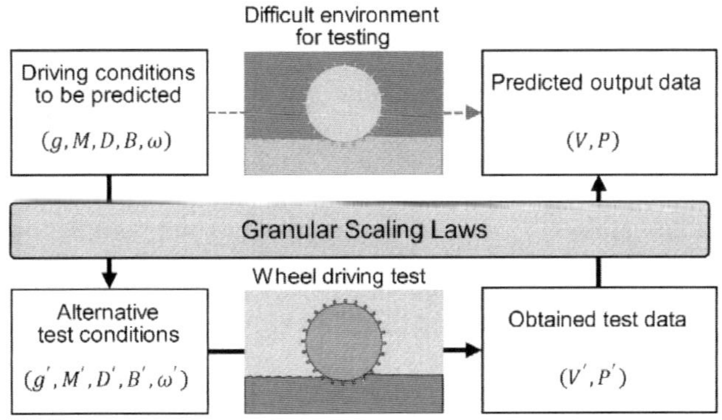

Fig. 1. Conceptual diagram of the original Granular Scaling Laws

Furthermore, Zhang et al. (2020) added the slope angle of the terrain to the input side of Eq. 1, as given by Eq. 3.

$$\left[\frac{V}{\sqrt{Dg}},\frac{P}{Mg\sqrt{Dg}}\right]=\Psi_3\left(\sqrt{\frac{g}{D}}t,f,\frac{g}{D\omega^2},\frac{\rho BD^2}{M},\theta\right) \tag{3}$$

Ψ_3 is a function with five inputs and two outputs, and θ represents the slope angle. Quasi-three-dimensional DEM simulations were used to verify that Eq. 3 is valid for driving conditions in which the slip ratio is >0.5. Note that the gravitational acceleration exceeds that of the Earth.

In this paper, an enhanced GSL is proposed, as expressed by Eq. 4, that combines the concepts in Eqs. 2 and 3 to predict various wheel mobilities during driving on slopes in low-gravity environments.

$$\left[\frac{V}{\sqrt{Dg}},s,\frac{h}{D},\frac{P}{Mg\sqrt{Dg}}\right]=\Psi_4\left(\sqrt{\frac{g}{D}}t,f,\frac{g}{D\omega^2},\frac{\rho BD^2}{M},\theta\right) \tag{4}$$

Here, Ψ_4 is a function with five inputs and four outputs, and s represents the slip ratio. The slip ratio is strongly correlated with the wheel's horizontal velocity V. However, this ratio is included as one of the output terms because it is a crucial parameter for evaluating wheel mobility in terramechanics. Assuming the same wheel geometry f and the same soil, if the wheel driving data are predicted in the condition $(g, M, D, B, \omega, \theta)$, the experiment should be conducted in the condition $(g', M', D', B', \omega', \theta') = (\alpha g, \beta M, \gamma D, \beta\gamma^{-2}B, \alpha^{1/2}\gamma^{-1/2}\omega, \theta)$. Then, the output data of interest, such as (V, s, h, P), can be predicted as $(\alpha^{-1/2}\gamma^{-1/2}V', s', \gamma^{-1}h', \alpha^{-3/2}\beta^{-1}\gamma^{-1/2}P')$ using the obtained experimental data (V', s', h', P') and the positive scalars α, β, and γ.

3. Numerical method

3.1. Analysis method

The DEM is a numerical simulation method that models granular materials as a collection of discrete particles. By solving for the motion of individual particles at each instant, it allows the analysis of the overall behavior of granular materials and the particle–particle and particle–wheel interactions under various gravity conditions. In this study, the commercial software ANSYS Rocky (ANSYS Inc., 2023) was used to acquire wheel-driving data in low-gravity environments and evaluate the prediction accuracy of the GSL for wheel mobility.

For the force models governing the interactions between particles and between particles and a wheel, the Hertzian spring–dashpot and the Mindlin–Deresiewicz models were adopted for the normal and tangential directions, respectively. These models accurately simulate the dynamic responses of granular materials because they are nonlinear spring–dashpot systems. In addition to these fundamental force models, a rolling resistance model was introduced to simulate the actual shape of a granular material and its resistance to rotation. This made the simulation more realistic by accounting for the intrinsic characteristics of real-world particles. Particles are subject to attractive forces such as van der Waals and electrostatic forces. The DEM can collectively model these forces as adhesive forces. This study adopted a constant-adhesive force model.

Furthermore, the ANSYS DEM provides coarse-grained modeling (CGM), as proposed by Bierwisch et al. (2009), to increase computational efficiency. The CGM reduces the total number of particles in the simulation by representing groups of small particles as larger particles, which are known as parcels. It involves scaling up each parcel using a specific factor referred to as f_{CGM}, which helps maintain the overall system dynamics. The scaling process adjusts the contact and adhesive models to ensure that the dynamics of the scaled-up system are consistent with those of the original particle system. These modifications are made using only the scaling factor, f_{CGM}, to eliminate the need to recalibrate the models. This approach allows efficient simulations while preserving the essential dynamics and interactions observed in the original system. Appendix A details the modeling aspect.

3.2. DEM parameters

The sand with characteristics of a lunar regolith simulant called FJS-1 was considered. Ozaki et al. (2023) measured the characteristics of FJS-1 and found moderate cohesion (Table 1). The DEM parameters were tuned such that the minimum bulk density and direct shear tests performed using the DEM reproduced the characteristics of FJS-1 (Table 1).

The DEM parameters used to reproduce the characteristics of FJS-1 are listed in Table 2. The particle density was considered as a fixed DEM parameter of 2.890 g/cm^3, which was the measured value of FJS-1, and the particle diameter was set to the average grain size of FJS-1. A scaling factor, $f_{CGM} = 6$, was used during the calibration to reproduce the characteristics of FJS-1. For computational efficiency, a particle size of $f_{CGM} = 15$ was used in the wheel mobility analysis, which maintained the same driving data as $f_{CGM} = 6$.

3.3. Wheel design and soil bed

A rigid grouser wheel was used to examine the effectiveness of the GSL. Figure 2 shows the properties of the wheels. The wheel size was the same as that used by Kobayashi et al. (2010) in their parabolic flight experiments, with a wheel diameter *D* of 150 mm and width *B* of 80 mm. In addition, the wheel model was designed with 24 grousers of 8.5 mm height to increase the traction force, based on Inotsume et al. (2019). The thickness of the grousers was 2.0 mm. Furthermore, the sandbox was 2400 mm long, 200 mm wide, and 300 mm deep to provide sufficient travel length and neglect the effect of the sandbox walls. The slope angle θ was set by inclining the entire sandbox, as shown in Fig. 2.

Table 1
FJS-1 soil characteristics and soil test results of the discrete element method

Parameter	Measurement results (Ozaki et al., 2023)	DEM soil test results
Minimum bulk density [g/cm^3]	1.549–1.560	1.554
Peak cohesion [kPa]	2.5 (Dr 70%)	2.1
Peak internal friction angle [°]	44.1 (Dr 70%)	43.0
Residual cohesion [kPa]	0.2–1.9	1.0
Residual internal friction angle [°]	34.1–37.2	36.0

Table 2
Discrete element method parameters for FJS-1

Density [g/cm^3]		2.890
Particle diameter [μm]		236 (sphere)
Scaling factor f_{CGM} [-]		6 (soil test calibration)
		15 (wheel driving simulation)
Young's modulus [GPa]		0.150
Poisson's ratio [-]		0.30
Rolling resistance coefficient [-]		0.40
	Particle–particle	Particle–wheel
Static friction coefficient [-]	0.40	0.50
Dynamic friction coefficient [-]	0.40	0.50
Restitution coefficient [-]	0.60	0.30
Adhesion coefficient [-]	1.0	0.0
Distance threshold [μm]	5.0	0.0

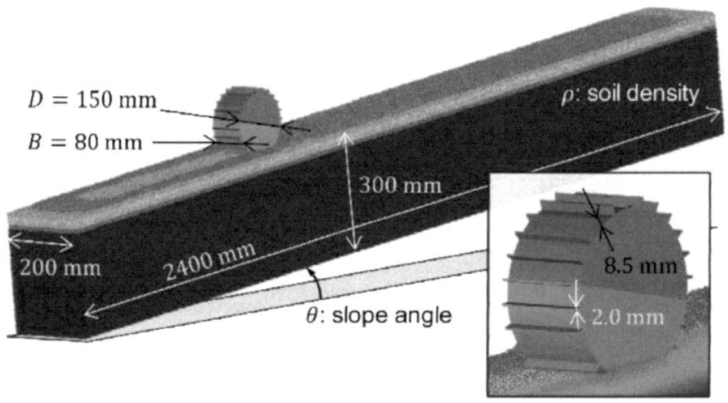

Fig. 2. Wheel properties and slope terrain simulation overview

4. GSL-based prediction results for wheel mobility

10 driving conditions comprising two different angular velocities ω and five different slope angles θ were tested. The predicted targets were the wheel mobilities in lunar gravity (i.e., 1/6 of the Earth's gravity). The test and experimental conditions for Earth's gravity obtained from Eq. 4 are listed in Table 3.

High wheel angular velocities were considered to verify that the GSL can predict wheel mobility even under dynamic conditions, that is, when sand behavior is complex. In addition, rovers used in future planetary explorations will be faster from the perspective of exploratory efficiency. Agarwal et al. (2021) observed that while a linear relationship between the wheel's horizontal velocity and wheel angular velocity is maintained in the quasi-static domain for $\omega < \pi$ rad/s, this proportionality does not hold at higher angular velocities. Therefore, this study defined $\omega \geq \pi$ rad/s as the dynamic domain and selected π and 2π rad/s from this domain.

Table 3
Simulation conditions scaled with scalars $\alpha = 6$, $\beta = 1$, and $\gamma = 1$ based on granular scaling laws

	Gravity g [m/s^2]	Mass M [kg]	Wheel diameter D [mm]	Wheel width B [mm]	Wheel angular velocity ω [rad/s]	Slope angle θ [°]
Predicted targets	1.635	10.2309	150	80	$\{\pi, 2\pi\}$	$\{0, \pm5, \pm10\}$
Experimental conditions under Earth's gravity	9.810	10.2309	150	80	$\{\sqrt{6}\pi, 2\sqrt{6}\pi\}$	$\{0, \pm5, \pm10\}$

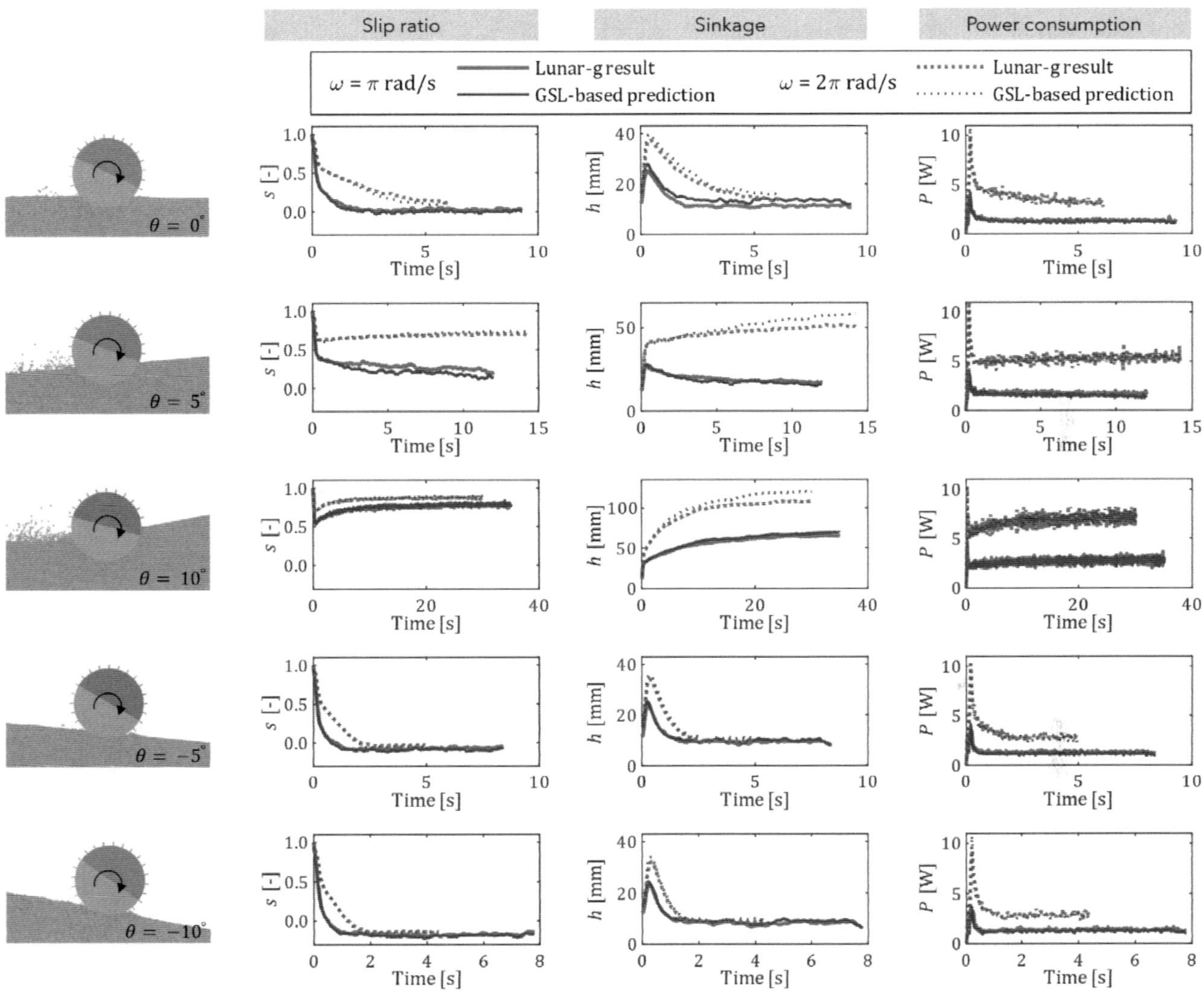

Fig. 3. Simulation results for the wheel driving on various slopes

Figure 3 shows the simulation results at slope angles ranging from -10° to 10°. The red lines represent the wheel mobilities when the wheel is driven under lunar gravity, and the black lines represent the predicted wheel mobilities under lunar gravity based on the GSL from the results obtained under Earth's gravity. Therefore, the black lines represent the converted results of the data (V', s', h', P') from the Earth-gravity tests based on Eq. 4. The results for the wheel's horizontal velocity V are presented in Appendix B owing to space limitations.

The results in Fig. 3 indicate that the GSL can accurately predict the wheel mobilities on slopes in a low-gravity environment, including the slip ratio s, wheel sinkage h, and power consumption P, regardless of the slope angle and wheel angular velocity. The wheel experienced a slip ratio of approximately 0.9 when climbing a 10° slope at a wheel angular velocity of 2π rad/s. Moreover, the wheel experienced a slip ratio of approximately -0.2 when descending a 10° slope at a wheel angular velocity of π rad/s. This suggests that the GSL can accurately predict wheel mobility even under high or negative slip ratio conditions, which can complicate wheel-soil interactions (Johnson et al., 2017). It is particularly noteworthy that the GSL accurately predicts the transient states that begin at the start of wheel driving and settle down to a steady state. This is the first study in which the GSL predictability was determined in transient states, as the DEM provides a stable gravity environment for verification.

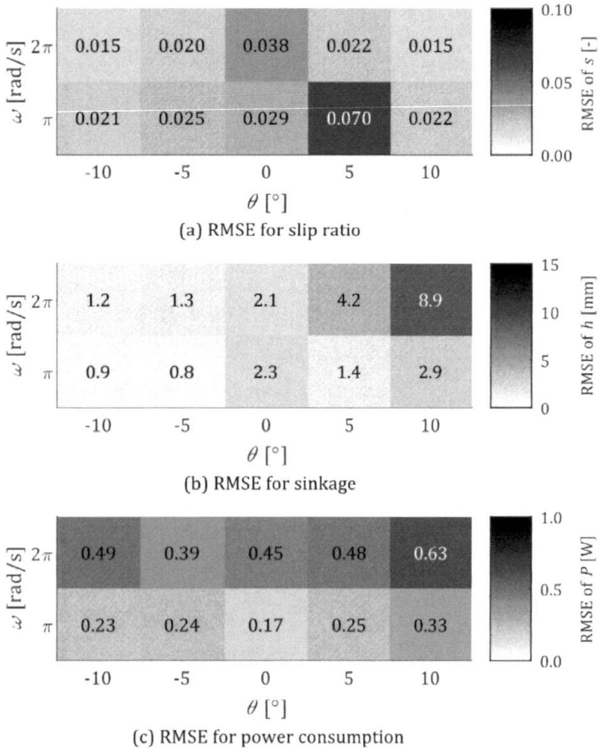

Fig. 4. Accuracy evaluation of granular scaling laws based on root-mean-square error (RSME)

Figure 4 shows a comparison of the root-mean-square error (RMSE) values for the lunar-g results and the predictions based on the GSL. Fig. 4(a) confirms that the GSL has a high prediction accuracy for the slip ratio. The findings show that the RMSE is as small as 0.015 even under the conditions of $\theta = 10^{\circ}$ and $\omega = 2\pi$ rad/s, when the slip ratio is at its highest ($s \approx 0.9$). In addition, the results quantitatively demonstrated that the GSL is effective for predicting wheel mobility on slopes in low-gravity environments because the RMSE values for climbing/descending a slope are lower than those for driving on flat terrain, except for climbing a 5°slope at an angular velocity of π rad/s.

Figure 4(b) shows the RMSE values for wheel sinkage. The error was larger when the wheel climbed slopes at $\omega = 2\pi$ rad/s than when it drove on flat terrain at the same speed. However, the scaled-up grain size was approximately 3.5 mm; thus, the magnitude of the error was a few sand grains. Furthermore, the prediction results of wheel sinkage based on the GSL exceed the lunar-gravity results, as shown in Fig. 3. This indicates that the GSL tends to predict worse performance than actual observations, thus providing a margin of safety. Such conservative estimates are beneficial in designing wheel systems and planning rover missions. This conservatism of the GSL corresponds with the results verified by Daca et al. (2022, 2023) in their parabolic flight experiments.

Finally, the RMSE values for the power consumed to drive the wheel are presented in Fig. 4(c). The RMSE is larger with a higher angular velocity of the wheel. This is attributed to more power being consumed. The results indicated highly accurate predictions of power consumption at different slopes and wheel speeds.

5. Conclusions and future work

This study simulated the wheel mobilities under various slope angles and wheel angular velocities and evaluated the prediction accuracy to demonstrate the effectiveness of an enhanced GSL that combines the concepts proposed by Zhang et al. (2020) and Daca et al. (2022, 2023). The DEM simulation reproduced sand with soil characteristics similar to those of a lunar regolith simulant to create an environment that closely resembled the actual lunar surface. Subsequently, wheel mobility was tested at slope angles ranging from -10° to 10°under Earth's gravity. In the analysis, the enhanced GSL was applied to convert the Earth-gravity test results to predict the wheel mobility under lunar gravity. These GSL-based predictions were compared with the wheel mobility results obtained from DEM simulations under lunar gravity conditions. The comparison results indicated that the GSL can accurately predict the wheel mobility for the slip ratio, sinkage, and power consumption needed to drive the wheel. In addition, the accuracy of wheel-mobility prediction for sloped driving was comparable to that for flat terrain. These results verify that the GSL effectively predicts the wheel slope-driving performance in low-gravity environments. Additionally, the simulation results demonstrated for the first time that the GSL can be applied over a wide range of slip conditions, from a negative slip ratio of approximately −0.2 to a high slip ratio of approximately 0.9. Furthermore, the simulation results suggest that the GSL can be applied to time-series prediction, including transient states.

Overall, the GSL is an effective method for predicting complex wheel mobility in low-gravity environments, on a wide range of slopes, and at high speeds. Therefore, the GSL is essential for planning and executing space exploration missions. In future work, the prediction accuracy should be investigated when using scaled-down wheels to increase the cost efficiency of the rover design process.

6. Nomenclature

B	Wheel width	[mm]
D	Wheel diameter	[mm]
f	Wheel shape function	[-]
f_{CGM}	Scaling factor for coarse-grain model	[-]
F_{DP}	Drawbar pull	[N]
g	Gravitational acceleration	[m/s^2]
h	Wheel sinkage	[mm]
M	Wheel mass	[kg]
P	Power consumption to drive a wheel	[W]
s	Slip ratio	[-]
t	Time	[s]
V	Wheel horizontal velocity	[mm/s]
α	Scaling factor for gravitational acceleration	[-]
β	Scaling factor for wheel mass	[-]
γ	Scaling factor for wheel diameter	[-]
θ	Slope angle	[°]
ρ	Soil density	[g/cm^3]
Ψ_1	GSL function of Slonaker et al. (2017)	[-]
Ψ_2	GSL function of Daca et al. (2022, 2023)	[-]
Ψ_3	GSL function of Zhang et al. (2020)	[-]
Ψ_4	GSL function of this study	[-]
ω	Wheel angular velocity	[rad/s]

7. Acknowledgements

This study was partially supported by JSPS KAKENHI (grant number: JP23KJ1901).

8. Declaration of competing interest

The authors declare that they have no competing financial interests or personal relationships that could have influenced the work reported herein.

9. References

Agarwal, S., Karsai, A., Goldman, D. I., Kamrin, K., 2021. Surprising simplicity in the modeling of dynamic granular intrusion. Sci. Adv. 7(17), eabe0631.

Ansys Inc., 2023. DEM technical manual: release 2023 R2.

Bierwisch, C., Kraft, T., Riedel, H., Moseler, M., 2009. Three-dimensional discrete element models for the granular statics and dynamics of powders in cavity filling. J. Mech. Phys. Solids, 57(1), 10-31.

Colaprete, A., 2021. Volatiles investigating polar exploration rover (viper).

Daca, A., Skonieczny, K., 2022. Evaluating 1-G testing methods for Predicting Planetary Rover Mobility in reduced gravity. In 16th Symposium on Advanced Space Technologies in Robotics and Automation (ASTRA) 2022, Noordwijk, The Netherlands.

Daca, A., Tremblay, D., Skonieczny, K., 2023. Expansion and experimental evaluation of scaling relations for the prediction of wheel performance in reduced gravity. Microgravity Sci. Technol. 35, 59.

Du, Y., Gao, J., Jiang, G., Zhang, Y., 2018. Development and numerical validation of an improved prediction model for wheel-soil interaction under multiple operating conditions. J. Terramech. 79, 1-21.

Inotsume, H., Moreland, S., Skonieczny, K., Wettergreen, D., 2019. Parametric study and design guidelines for rigid wheels for planetary rovers. J. Terramech. 85, 39-57.

International Space Exploration Coordination Group, 2018. ISECG Global Exploration Roadmap (3rd edition). https://www.globalspaceexploration.org/wordpress/wp-content/isecg/GER_2018_small_mobile.pdf (accessed 26 March 2024).

Johnson, J. B., Duvoy, P. X., Kulchitsky, A. V., Creager, C., Moore, J., 2017. Analysis of Mars exploration rover wheel mobility processes and the limitations of classical terramechanics models using discrete element method simulations. J. Terramech. 73, 61-71.

Kobayashi, T., Fujiwara, Y., Yamakawa, J., Yasufuku, N., Omine, K., 2010. Mobility performance of a rigid wheel in low gravity environments. J. Terramech. 47, 261-274.

Kuroda, Y., Teshima, T., Sato, Y., Kubota, T., 2005. Mobility performance evaluation of planetary rovers in consideration of different gravitational acceleration. In 2005 IEEE/RSJ International Conference on Intelligent Robots and Systems, pp. 2991-2996.

Li, M., Gao, F., Sun, P., 2012. Prediction method of lunar rover's tractive performance based on similitude methodology. In 2012 Fifth International Conference on Intelligent Computation Technology and Automation, pp. 686-689.

Ozaki, S., Ishigami, G., Otsuki, M., Miyamoto, H., Wada, K., Watanabe, Y., Nishino, T., Kojima, H., Soda, K., Nakao, Y., Sutoh, M., Maeda, T., Kobayashi, T., 2023. Granular flow experiment using artificial gravity generator at International Space Station. NPJ Microgravity 9, 61.

Sanguino, T. D. J. M., 2017. 50 years of rovers for planetary exploration: A retrospective review for future directions. Robot. Auton. Syst. 94, 172-185.

Sutoh, M., Yusa, J., Nagatani, K., Yoshida, K., 2021. Traveling performance evaluation of planetary rovers on loose soil. J. Field Robot. 29(4), 648-662.

Slonaker, J., Motley, D. C., Zhang, Q., Townsend, S., Senatore, C., Iagnemma, K., Kamrin, K., 2017. General scaling relations for locomotion in granular media. Phys. Rev. E, 95(5), 052901.

Thoesen, A., McBryan, T., Mick, D., Green, M., Martia, J., Marvi, H., 2020a. Comparative performance of granular scaling laws for lightweight grouser wheels in sand and lunar simulant. Powder Technol. 373, 336–346.

Thoesen, A., McBryan, T., Mick, D., Green, M., Martia, J., Marvi, H., 2020b. Granular scaling laws for helically driven dynamics. Phys. Rev. E 102, 032902.

Watanabe, Y., Nakano, S., Suzuki, H., Ozaki, S., 2023. Numerical analysis of wheel locomotion on soft soil using the extended terramechanics model based on cellular automata. J. Terramech. 109, 9-20.

Zhang, Q., Townsend, S., Kamrin, K., 2020. Expanded scaling relations for locomotion in sloped or cohesive granular beds. Phys. Rev. Fluids. 5(11), 114301.

Appendix A. DEM models

In this study, nonlinear spring-dashpot models were used as the contact-force models between particles and between particles and a wheel. The Hertzian spring-dashpot model was considered for the normal contact force F_n, as expressed by Eq. (A.1).

$$F_n = k_n s_n^{\frac{3}{2}} + c_n s_n^{\frac{1}{4}} \dot{s_n} \tag{A.1}$$

Here, k_n and c_n represent the normal contact stiffness and damping coefficient, respectively. s_n represents the contact normal overlap, and $\dot{s_n}$ represents the time derivative of the contact normal overlap.

The Mindlin–Deresiewicz model calculates the tangential force F_τ given by Eq. (A.2).

$$F_\tau = -k_\tau F_n \left(1 - \varsigma^{\frac{3}{2}}\right) + c_\tau F_n^{\frac{1}{2}} \varsigma^{\frac{1}{4}} \dot{s_\tau} \tag{A.2}$$

$$\varsigma = 1 - \frac{\min\left(s_\tau,\ s_{\tau,\max}\right)}{s_{\tau,\max}} \tag{A.3}$$

Here, k_τ and c_τ represent the tangential contact stiffness and the normal damping coefficient, respectively. s_τ represents the tangential relative displacement at the contact, and $\dot{s_\tau}$ represents the tangential component of the relative velocity at the contact. $s_{\tau,\max}$ represents the maximum relative tangential displacement at which particles begin to slide, and ς represents the proportion of acceptable tangential displacement before the particles begin to slide.

An attractive normal force to the repulsive contact force described above was included in the model to represent the cohesive force of sand. A constant-adhesive force model was used. This model is the simplest adhesion model, where a constant force $F_{n,\mathrm{adh}}$ is applied between particles closer than a certain distance δ_{adh}. This model is expressed by Eq. (A.4).

$$F_{n,\mathrm{adh}} = \begin{cases} 0 & \text{if} - s_n \geq f_{\mathrm{CGM}} \delta_{\mathrm{adh}} \\ \dfrac{f_{\mathrm{adh}} m g}{f_{\mathrm{CGM}}} & \text{if} - s_n < f_{\mathrm{CGM}} \delta_{\mathrm{adh}} \end{cases} \tag{A.4}$$

Here, f_{adh} is the adhesion coefficient, and δ_{adh} is the distance threshold determining whether adhesion occurs between particles. m represents the particle mass.

Using a rolling resistance model is practical for reproducing the nonspherical shape of particles and the resistance to rotation caused by this shape. The rolling resistance moment M_τ^t for the linear spring rolling limit used is expressed by Eq. (A.5).

$$M_\tau^t = \min\left(M_r^t, \mu_r \frac{r F_n}{2}\right) \tag{A.5}$$

Here, μ_r is the rolling resistance coefficient, and r represents the rolling radius of the particles. M_r^t is a purely elastic rolling resistance parameter given by Eq. (A.6).

$$M_r^t = M_r^{t-\Delta t} - k_r \omega_{\mathrm{rel}} \Delta t \tag{A.6}$$

Here, $M_r^{t-\Delta t}$ represents the rolling resistance moment at the previous time; k_r represents the rolling stiffness; ω_{rel} represents the relative angular velocity, which is defined as the difference between the angular velocities of two particles in contact; and Δt represents the simulation timestep.

Appendix B. Additional simulation results

Figure B.1 shows a comparison of the wheel's horizontal velocities in lunar gravity and the predicted ones based on the GSL in the DEM simulation. The results indicate that the GSL can accurately predict the horizontal velocity V, as shown in Fig. B.1.

The prediction accuracy for the horizontal velocity is related to that for the slip ratio. The slip ratio s is expressed in terms of the horizontal velocity V and the input angular velocity ω, as given by Eq. (B.1). Therefore, if the GSL accurately predicts the slip ratio s for an appropriately scaled input angular velocity ω', it can also predict the horizontal velocity V correctly. This can be explained by a variant of the equation, as given by Eq. (B.2).

$$s = 1 - \frac{V}{\frac{D}{2}\omega} \tag{B.1}$$

$$\begin{aligned}\frac{V'}{\sqrt{D'g'}} &= \frac{\frac{D'}{2}\omega'(1-s')}{\sqrt{D'g'}} = \frac{\frac{\gamma D}{2}\sqrt{\frac{\alpha}{\gamma}}\omega(1-s)}{\sqrt{\gamma D \alpha g}} \\ &= \frac{\frac{D}{2}\omega(1-s)}{\sqrt{Dg}} = \frac{V}{\sqrt{Dg}}\end{aligned} \tag{B.2}$$

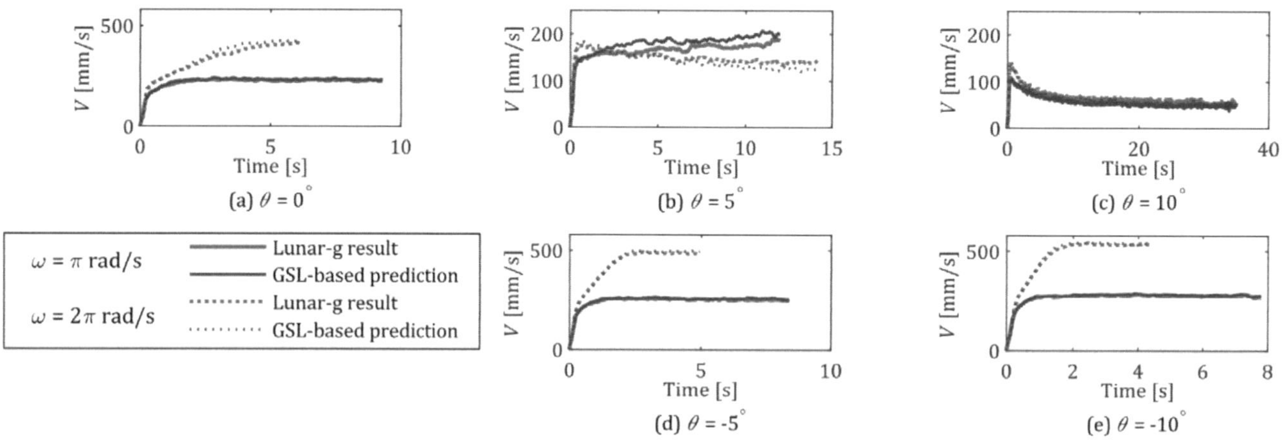

Fig. B.1. Simulation results for the wheel's horizontal velocity

9663 | Year-round measurements of the soil cone index on grass airfields for ground performance of airplane

Jarosław Pytka [a,*], Anna Zalewska-Tytłak [a], Paweł Tomiło [a], Michał Kuszneruk [a], Dawid Tatarynow [a]

[a] *Lublin University of Technology, Lublin, Poland*

* *Corresponding author: j.pytka@pollub.pl*

ABSTRACT

The paper concerns the performance of aircraft operating from grass airfields, in particular the take-off and landing distances. These performances are strictly dependent on the conditions on the runway, and in the case of a grass runway, they depend on the influence of weather factors. Year-long measurements of the soil cone index were carried out on the runways of 5 grass airfields, which are characterized by different ground soils. The measurements were carried out in the period from July 2019 to June 2020.

The paper presents a statistical analysis of the measurement results. Generally, it was found that weather factors have a significant impact on the values of the cone index and changes in the CI value reach up to 400% (the highest relative to the lowest). Moreover, it was observed that the type of soil underlying the grass on the runway had a significant impact on the CI index values. The most sensitive ground was marl, with a high chalk content, which constitutes the foundation of one of the tested grass airfields.

The results will be used to parameterize and verify the model describing the ground performance of the aircraft.

Keywords
Cone index
Grass airfields
Airplane performance
Weather impact

1. Introduction

The Cone Index (CI) is a versatile indicator that determines the mechanical properties of soil that are important for predicting wheel performance (Kumar et al 2012, Shoop et al 2008). In particular, it concerns the compactness of the soil, which can be indirectly used to determine tractive forces acting on a wheel. One of the methods is to use the CI index to determine the so-called wheel numeric (Wismer and Luth, 1973). Another approach allows rolling resistance to be determined based on the CI readings.

The ground performance of airplane operating from a grass airfield is the result of many factors, including the mechanical parameters of the soil, its humidity and the presence of green (aboveground) vegetation (van Es 1998). Research in this area focuses on estimating the take-off and landing distance (Pytka et al 2022), depending on the rolling resistance coefficient of the aircraft landing gear wheel. To properly determine the rolling resistance of a wheel on the ground, it is possible to measure it directly using a wheel force transducer or load cells (Shoop et al 1994, Pytka et al 2019). These methods are accurate, but their disadvantage is the need to use complicated measurement equipment. When examining the influence of weather factors, repeated measurements of rolling resistance are troublesome and expensive (Burhard and Calame 1998, Kuchler and Schrupp 2002, Loh and Nohl 1992)

In the context under consideration, it is much simpler to determine rolling resistance based on CI values, which are measured using a portable, easy-to-use measuring device. The subject literature contains publications reporting measurements of CI and models that convert the cone index into the rolling resistance coefficient. However, it is problematic to determine the characteristics of the cone index depending on soil moisture (Anderson, 1983).

The current study is an element of the GARFIELD project, initiated at the Lublin University of Technology and continued in cooperation with scientific and research institutions from European countries (Pytka et al 2019a). Project GARFIELD aims to develop an online information system about the condition of the surface of grass airfields within the European Union. Available to any interested user, the GARFIELD system is intended to be a navigation aid that allows determining the ground performance of a given aircraft with a declared load (crew, baggage, fuel) taking into account temporary meteorological conditions. Based on soil maps, the system will return information for any grass airport in the EU (Pytka et al 2019a).

As part of the GARFIELD system, a NOTAM-style (Notice to Airmen) navigation message was developed regarding grass airfields and based on the mechanical parameters of the runway surface, compactness and adhesion (Pytka et al 2019b).

Following the above mentioned, the authors of this paper decided to conduct a year-long study of the soil condition on the runways of 5 grass airfields in Poland.

2. Methods

The year-round experiment method involved measuring selected parameters describing the physic-mechanical properties of the soil during one year, at 4-week intervals and were performed from July 2019 to June 2020. The choice of individual measurement days was an important factor, as it may happen that the day selected for measurements based on the calendar algorithm (e.g. every Monday, every four weeks) will differ

https://doi.org/10.56884/H1VZQ8F8

significantly from the long-term average in terms of weather conditions. Therefore, the calendar algorithm was adjusted if significant weather anomalies occurred.

Below is some information about the measuring instruments as well as about the grass and soils on which the experiment was planned.

2.1. Measuring equipment

An electronic cone penetrometer has been used in the study. For all measurements we've used a standard cone and the soil on the runway was penetrated down to 25 or 30cm. Readings of the penetration pressure were collected in the electronic module of the penetrometer. Figure 1 shows the penetrometer maintained by a technician.

Simultaneously we've also performed measurements of soil moisture and temperature. For this purpose a handheld TDR (Time Domain Reflectometry) meter has been used (Pytka et al 2019). The use of the TDR meter is simple: two metal electrodes (steel rods) were driven into soil down to 10-15cm and the result of the measurement, soil moisture and temperature were displayed immediately on a screen.

During each measurement, 5 points on the grass runway were chosen, located several meters apart. The read values were then saved immediately. The TDR soil moisture meter allows to read soil moisture expressed as a percentage and soil temperature expressed in degrees Celsius. Figure 2 shows this measuring device.

Fig. 1. The cone penetrometer used in the study

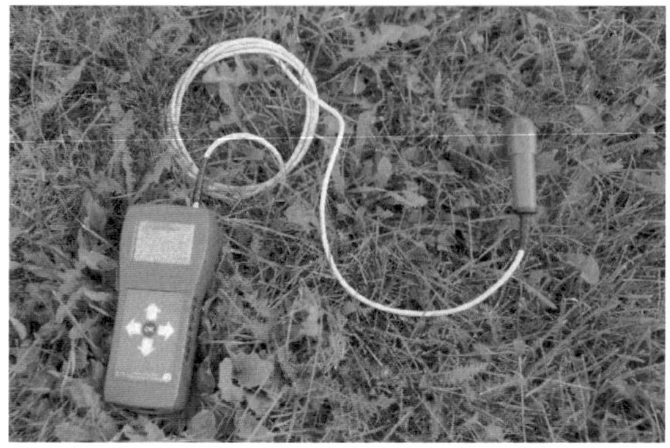

Fig. 2. A device for measuring soil humidity and temperature based on TDR technology

2.2. Grass airfields

Measurements were carried out at 5 different grass airfields located in eastern Poland. For this study we've selected airfields built on different soils. The chart in Fig. 3 shows the location of each of the five measurement points. Below is a brief description of the soils at individual airfields (Zalewska-Tytłak 2024).

The Radawiec airfield has a subsoil based on tawny soil. Luvisors are soils consisting of loess, aqueous dust, various types of clays, primarily boulder clays, and sands lying on the clay. This type of soil is characterized by a well-developed, diversified profile.

The Depułtycze Królewskie airfield is located near the city of Chełm. At this airfield, there are Rendzina-type soils, i.e. soils formed from various carbonate rocks such as limestone, marl, dolomites, marble and gypsum. A characteristic feature of these soils is the presence of rock fragments, the number of which increases with depth. In the upper layers of Rendzina there is a humus level below which the proportion of skeleton gradually increases until it transforms into the parent rock.

The airfield in Żłobek Mały is located in a place where podsolic and rusty soils occur. Podsolic soils are formed from various rocks with a predominance of sands, rocks without carbonates, granites, gneisses, quartzites and sandstones, while rusty soils consist of sand, gravel, granite and sandstone deposits.

The Biała Podlaska airfield is located in the area of swampy soils, which are soils characterized by high humidity resulting from the high level of groundwater. A characteristic feature of these soils is the rich content of organic substances in the form of peat and silt.

The Stara Wieś airfield is located in the area of brown soil. The main components of brown soils are boulder clays, sands, sandstones, flysch shales and loess.

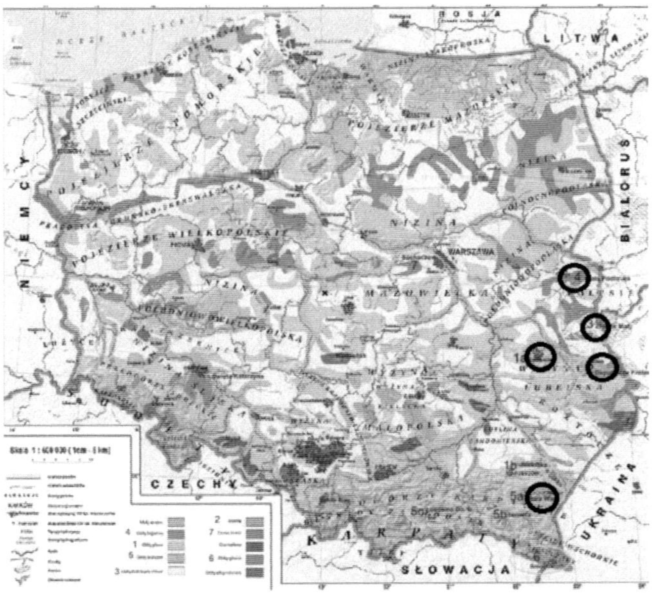

Fig. 3. Location of grass airfields subjected to the experiment on Poland's chart. ! - Radawiec, 2 - Depułtycze Królewskie, 3 - Żłobek Mały, 4 - Biała Podlaska, 5 - Stara Wieś.

2.3. Data reduction

For each variant of measurements, a minimum of 10 repetitions of the penetrometer measurement and a minimum of 5 soil moisture measurements were performed. In order to facilitate the analysis of the results for each measurement site, the average value resulting from each measurement series was calculated for each measurement day (Zalewska-Tytłak, 2024).

3. Results

The synthetic results of the conducted research are presented in Figures 4 - 8, which correspond to individual airfields. Digits 1., 2., 3. thru 12. on the graphs denote year's months: January, February, March thru December. For a concise presentation, the CI measurement results for a depth of 5 cm are shown. Obviously have the authors also measurement results for other depths (10, 15, 20 and 25 cm).

Analyzing the results globally, it can be seen that the airfield with the highest conciseness values throughout the year was the airfield in Radawiec.

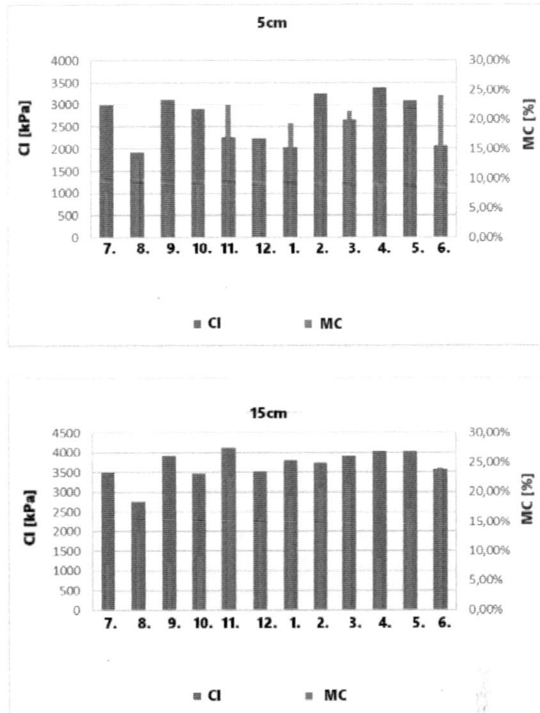

Fig. 4. Results of measurements of the CI index and soil moisture in individual months for Radawiec airfield

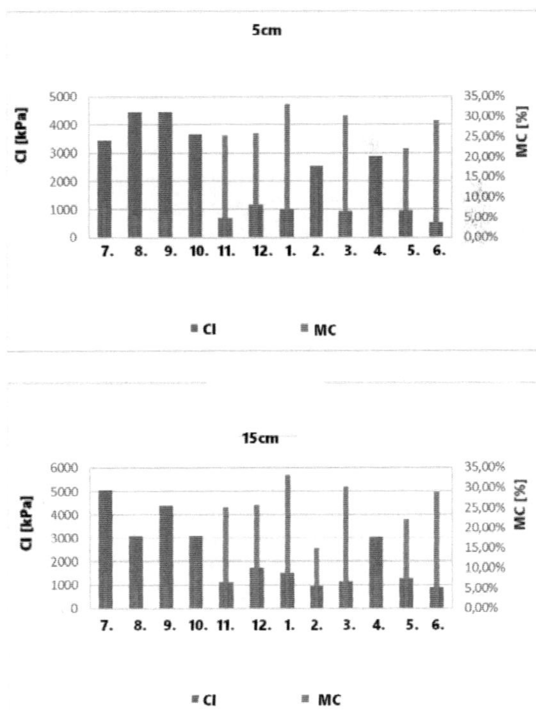

Fig. 5. Results of measurements of the CI index and soil moisture in individual months for Depułtycze Królewskie airfield

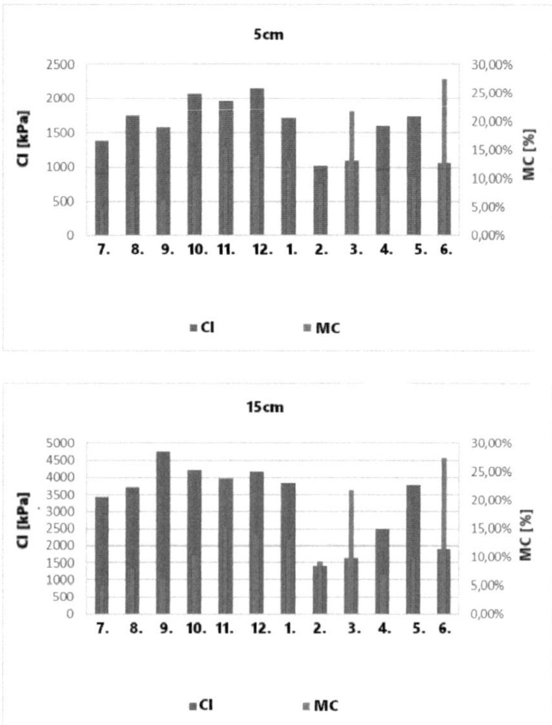

Fig. 6. Results of measurements of the CI index and soil moisture in individual months for Złobek Mały airfield

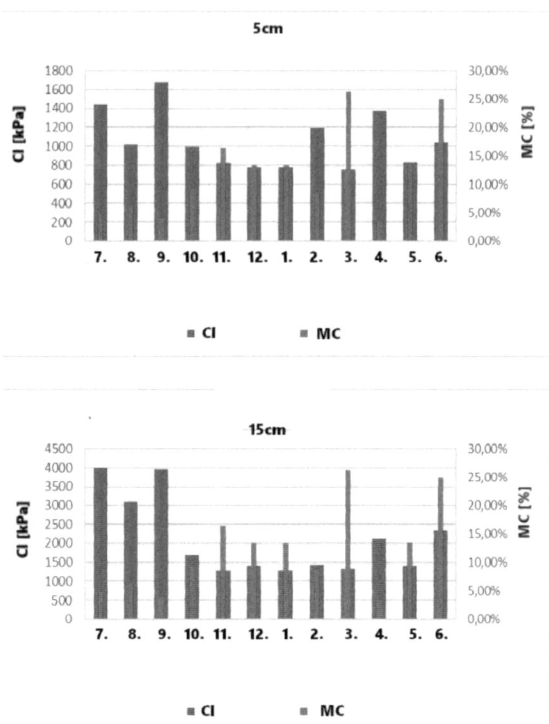

Fig. 7. Results of measurements of the CI index and soil moisture in individual months for Biala Podlaska airfield

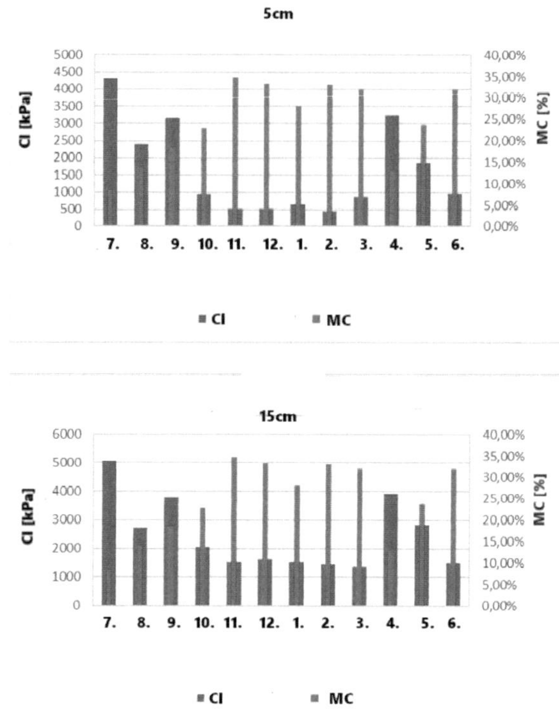

Fig. 8. Results of measurements of the CI index and soil moisture in individual months for Stara Wieś airfield

In the case of the Radawiec airfield, this type of regularity was observed both for the measurement depth of 5 cm and for the depth of 15 cm. The airfield in Radawiec was also characterized by a relatively low amplitude of the soil compactness value for a depth of 5 cm, and the lowest amplitude in relation to the measurement depth of 15 cm (Fig. 4). In this way, we can conclude that due to the highest values of soil compactness and its low annual amplitudes, this airfield provides relatively the best and most predictable operating conditions.

Another of the analyzed airfields was the Depułtycze Królewskie, the measurement results are presented in Figure 5. It is worth noting that in the case of all airfields examined and based on the measurements obtained at them, it follows that the soil compactness is higher for measurements obtained at a depth of 15 cm than at a depth of 5 cm. cm. The exception are measurements made at the airfield in Depułtycze Królewskie.

On a year-round basis, this airfield was characterized by a relatively high average value of soil compactness at a depth of 5 cm, while at the same time the lowest value for a depth of 15 cm among all airfields. The indicator that most distinguishes the airfield in Depułtycze Królewskie is the highest amplitude of the soil compactness value. This is visible in Fig. 5 in the soil compaction and moisture diagram for the airfield in Depułtycze Królewskie, the soil compactness at this airfield decreased dramatically when there was a high level of humidity. In connection with the above, it can be concluded that the specificity of the described airfield requires great caution in relation to the possible assessment of the conditions of its use.

After a heavy rainfall, the airfield in Depułtycze Królewskie is closed due to too high soil moisture. In such cases, the turf covering the airfield poses a great danger to landing and taking off planes by getting the wheel stuck in the ground and possible nose-down position.

Analyzing the results obtained at the Żłobek Mały airfield in comparison to the others, it was observed that in the case of this airfield, a high average value of soil compactness was recorded at a depth of 15 cm, while the soil compactness at a depth of 5 cm was around half of the scale adopted for all airfields (CI=1595kPa). In relation to the amplitude of the compactness measurements at 5 cm, its value was low, while in the case of a measurement depth of 15 cm, the amplitude reached a relatively high value. In the case of the airfield in Żłobek Mały, an interesting relationship was observed, which can be seen in Fig. 6. Such a relationship can also be observed in the case of the airfield in Radawiec. It can be noticed that increases in soil moisture values did not cause significant decreases in the values showing its compactness, but the compactness increased with increases in soil moisture. This is a very characteristic phenomenon occurring only in these two airfields. This relationship was already observed during the measurements and its existence was confirmed by this analysis of the results. This phenomenon is somewhat opposite to those observed at other airfields.

Another airfield included in the research is the airstrip in Biała Podlaska. The analysis of the conducted research (Fig. 7) shows that the landing site in Biała Podlaska was characterized by the lowest values of average soil compactness, both measured at a depth of 5 and 15 cm. Compared to the airfield in Radawiec, these values are even more than twice lower. Additionally, in the case of this landing site, the lowest value of the amplitude of the compactness value was recorded at a depth of 5 cm, which means that throughout the year, despite the prevailing weather conditions, changes in the compactness value at a depth of 5 cm change to the smallest extent among all the mentioned airfields. In the case of a measurement depth of 15 cm, the amplitude value is comparably higher, but only reaches a value higher than that observed at the airfield in Radawiec. This means that in the case of the landing site in Biała Podlaska, the aspect related to the amplitude of changes in the soil compactness value can be described as favorable. However, it should be remembered that in this place there are comparably lowest values of soil compactness regardless of the measurement depth, which in turn is a phenomenon that is definitely unfavorable from the utility point of view.

In the case of the Stara Wieś airfield, what is most noticeable is the fact that there is a very large amplitude of the soil compactness value at this landing site, comparable to that observed at the airfield in Depułtycze Królewskie. Particularly drastic decreases in the average soil compactness values were observed in the autumn and winter months, when the highest soil moisture values occurred. This type of phenomenon can be described as definitely unfavorable from a practical point of view. It should also be noted that in the case of both airfields the highest values of both average soil moisture and the amplitude of changes in its value were recorded. In the summer months, during the drought, low soil moisture determined the increase in soil compactness (Fig. 8), but in the autumn and winter months, the increasing humidity resulted in drastic decreases in the compactness value. Despite the high humidity in Stara Wieś, there was no case during the research in which the turf conditions did not allow light aircraft to take off and land.

4. Conclusions and future work

The presented measurements of soil moisture and Cone Index were performed year-round, cyclically at airfields located in the area of various soils, allow to see the relationships determining the variability of individual operational parameters. Among all the tested airfields, the highest CI-values and very low amplitudes of these values were obtained at the airfield in Radawiec, located in the area of luvisory soils. Therefore, this airfield has relatively the best and most predictable conditions for aircraft operation. The airfield in Depułtycze Królewskie, located on Rendzina-type soils, is characterized by very large amplitudes of soil CI values. The occurrence of high humidity caused a drastic decrease in the CI at this airfield, which requires special caution during flight operation (takeoff, landing and taxiing). Additionally, the airfield in Depułtycze Królewskie is the only one among those examined to have higher soil CI values at a depth of 5 cm than at a depth of 15 cm, which further worsens the predictability of the surface behavior in the event of changing weather conditions.

The airfield in Żłobek Mały, located in the zone of podsolic and rusty soils, has favorable characteristics in the field of measurement depth of 15 cm, while the measurement depth of 5 cm is characterized by values around half of the scale adopted for all airfields. In the case of the airfield in Żłobek Mały and the airfield in Radawiec, a tendency for the CI value to increase with increasing humidity was observed, which was only observed in the case of these two airfields. The airfield in Biała Podlaska, located in the area of swampy soils, is characterized by the comparatively lowest values of soil CI throughout the year, which can be assumed as an unfavorable relationship from the point of view of operation. Despite this, it is also characterized by the smallest amplitude of changes in the soil compactness value, which is a beneficial phenomenon because it affects the high predictability of the conditions of use. The airfield in Stara Wieś, located on brown soil, is characterized by the highest amplitudes of changes in the value of soil compactness and very large decreases in this value due to increases in humidity. The low moisture content of this soil in the dry months resulted in large increases in its compactness values. Although the described relationships are not favorable due to the operational parameters, during the research there was no day when the conditions did not allow the use of the airfield.

Future research in this area will consist in statistical analysis of the obtained results in order to demonstrate the relationship between CI, soil moisture and the time (month) of measurement as well as the effect of soil substrate of the tested airfields (Zalewska-Tytłak, 2024).

Moreover, an important aspect of the presented topic is to determine the impact of green vegetation (grass) on the aircraft's performance. As Shoop et al 2015 pointed out, green parts of plants have a significant impact on road wheel performance and should not be neglected when modeling aircraft or ground vehicle performance. The first analyzes and measurements have been made and their results will be included in the doctoral dissertation of the co-author of this work (Zalewska-Tytłak, 2024).

5. Acknowledgements

This work was prepared within the project PM/SP/0065/2021/1entitled "Method for airplane takeoff and landing distance measurement", financed by the Ministry of Education and Science (Poland) as a part of the Polish Metrology Programme.

6. Declaration of competing interest

The authors declare that they have no known competing financial interests or personal relationships that could have appeared to influence the work reported in this paper.

7. References

Anderson M. On the applicability of soil water finite diference models to operational trafficability models, Journal of Terramechanics, Vol. 20 No. ¾, p. 139-152, 1983.

Burkard H., Calame C. Rotating Wheel Dynamometer with High Frequency Response. Tire Technology International, 1998, p.154

van Es, G.H.W., *Rolling resistance of aircraft tires in dry snow*, NLR Report TR-98165, National Aerospace Laboratory, 1998.

Kuchler M. and Schrupp R. Mehrkomponenten-Motorradmessnabe / Multiaxial Motorcycle Wheel Load Transducer. VDI-FVT Jahrbuch 2002, VDI Verlag Dusseldorf, pp. 91-119 (In German with English summary)

Kumar A., Chen Y, Sadek A., Rahman S. Soil cone index in relation to soil texture, moisture content, and bulk density for no-tillage and conventional tillage, Agric Eng Int: CIGR Journal, March, 2012

Loh R., Nohl F.W. Mehrkomponenten-Radmessnabe. Einsatzmoeglichkeiten und Ergebnisse. (in German with English summary). ATZ (Automobiltechnische Zeitschrift) 94 (1992), 1, pp. 44-53

Pytka J., Budzyński P., Józwik J., Łyszczyk T., Gnapowski E., Laskowski J. GARFIELD information system – old problems and new perspectives. Proceedings of the 6th International Workshop on Metrology for Aerospace (MetroAeroSpace), Torino, Italy, 19–21 June, 2019.

Pytka J., Budzyński P., Józwik J., Łyszczyk T., Gnapowski E., Laskowski J. GRASSTAM – an idea of a notice on grassy runway condition. Proceedings of the 6th International Workshop on Metrology for Aerospace (MetroAeroSpace), Torino, Italy, 19–21 June 2019.

Pytka, J, Budzyński, P., Kamiński, M., Łyszczyk, T., Józwik, J. Application of the TDR Moisture Sensor for Terramechanical Research. Sensors, Vol. 19, No. 9, p. 2116, 2019

Pytka J., Józwik J., Budzyński P., Łyszczyk T., Tofil A., Gnapowski E. and Laskowski J. Wheel dynamometer system for aircraft landing gear testing, Measurement (148) 2019, 106918

Pytka J., Budzyński P., Tomiło P., Michałowska J., Błążejczak D., Gnapowski E., Pytka J.D., Gierczak K., 2022. Measurement of aircraft ground roll distance during takeoff and landing on a grass runway, Measurement 195, 111130.

Shoop, S., Young, B., Alger, R., Davis, J. Effect of test method on winter traction measurements, J Terramech 31(3), 153-161, 1994.

Shoop, S.A., Diemand. D., Wieder, W.L., Mason, G. and Seman P.M. Predicting California Bearing Ratio From Trafficability Cone Index Values. Opportune Landing Sites Program. Technical Report, ERDC/CRREL TR-08-17, 2008.

Shoop, S.A., Coutermarsh B., Cary T., Howard H., 2015. Quantifying vegetation biomass impacts on vehicle mobility. Journal of Terramechanics, 61, 63-76, 2015.

Wismer R.D, Luth H.J. Off-road traction prediction for wheeled vehicles. Journal of Terramechanics, Nr. 10, 49-61, 1973

Zalewska-Tytłak A. Identyfikacja mechanicznych właściwości podłoża nieutwardzonego w aspekcie osiągów trakcyjnych koła jezdnego (Identification of the mechanical properties of unpaved soil surface in terms of the traction performance of the road wheel). Doctoral Thesis, Lublin University of Technology, 2024.

9747 | Application of a rockie bogie suspension for a new type of sugar cane transport system

Jorge Lopera [a,*]

[a] PACIFIC+, Cali, ,Colombia

* Corresponding author: jlopera@pacificplus.co

ABSTRACT

The rockie-bogie suspension arm is a topic fully studied for experimental rovers. These types of suspensions allow vehicles to increase the stability and move across with multiple obstacles on the ground. It has been evaluated this suspension system for some academic studies for the agricultural research topics, but it does not have yet an industrial deployed application for agriculture sector. The agricultural sector is looking for solutions to increase the productivity without harming the environment and to improve the health of soils. This global challenge must be faced with improved technological developments. For instance, the sugar cane industry at Colombia, South America has been challenged to adopt new technological system to complain about. At the Cauca River Valley (24MTon/year sugar cane production), 75% of the crop fields to harvester and transport the sugar cane stalks is used with commercial harvesters and a system of trailers to move the sugar cane from the field to the mill factories. The main surface it has low slopes, otherwise. for the remain 25%, the industry faces with high slopes terrains, additionally with the lower capacity of the soil at the rainy and wet seasons. This condition increases the risk with commercial agricultural machinery (harvester, trailers, and tractors). This work is inspired with the rockie-arm suspension, to propose a simplified method to use this type of suspension into a conceptual transport sugar cane vehicle. A CAD Model has been developed to understand the kinematical and kinetics for the suspension and compared with other types of suspensions considered for this industry. This work aims to explore new possibilities to adopt technologies developed from other industries to solve the current agriculture challenges.

Keywords
Agricultural Equipment
Sugar cane transport
Rockie bogie suspension

1. Introduction

The rockie-arm suspension has been adopted for space exploration projects. The remarkable features to avoid multiple terrain conditions drives the space agencies to adopt this suspension type for exploration of planets with gravity acceleration. Space programs have been developed different rovers and suspensions to understand the behavior of this type of rovers suspensions (RCL-C, RCL-D, RCL-E, CRAB, 3-bogie, V-bogie and Rocker bogie), previous researches *(Nildeep Patel, 2010)* have documented the evolution of these vehicles development. The challenges premises are different for each one.

The R&D agricultural roles have multiple development challenges to face for new developments. Heavy loads, soil health, increase productivity, energy efficiency, are related topics to be faced for teams, the ability to validate theoretical and experimentally models for new developments. Even the robotics adoption for new developments, must exhibits adaptability and speed. With new developments, it could be adopted interdisciplinary collaboration (biology, robotics, control engineering, AI) to face the harvester and post-harvester new developments. (*Balvinder Kaur, 2023*).

The R&D for new agricultural equipment must balance the engineering method to develop, back-and-forward and iteration models using CAx technology (CAD, CAE, CAM) to manage the product lifecycle. This balance must enhance the interaction between disciplines and systems engineering.

Designers and engineers are deploying CAD software to meet mechatronic engineering challenges and to further enhance efficiency. Such factors are resulting in an increasing demand for CAD software at the global level, the CAD market is estimated to grow annually >6% (*Yi-Hsin Lin, 2023*) . If this market grows with that force, struggles the R&D roles, to deliver more results with competitive methods to achieve their design goals.

1.1 The challenges for sugar cane industry

The sugar cane industry at Colombia, has developed an agro-industrial model for harvesting and transport sugar cane, this operation it works 24 hours by 7 days a week. Natural increasing productivity are asked for the equipment to engage the supply chain, and with new added features with the demands for precision agriculture (sensors, AI, robotics, IOT) *(Shivani Garg, 2023*) & *(Georg Steinberger, 2009*). To harvest the sugar cane can be used two methods, a manual method using a hand tool with a wide blade sharpened (*pacora o machete*) and mechanized harvester, where it uses a harvester equipment, a self-roll trailer system and a tractor to pull the self-roll trailer is required (*Fig. 1*).

With this configuration, when the tractor needs to pull the tipping trailer, could damage the growing sugar cane (steering radius > 0). With more tipping wagons, the damage could increase.

https://doi.org/10.56884/47B4US9J

Fig. 1 Harvester with auxiliary equipment for harvester and transport sugar cane at field (CASE, 2024)

The natural response to increase productivity at the agricultural industry is to introduce more and bigger equipment. For instance, it has been collected the historical uses of tractors at Northern Italy (Bacenetti, 2022), this study shows how the average power of tractors has increased through time (60 kW at 1997 to 160 kW at 2017) almost 2.5 times in 25 years and the working time at field (250 h/y at 1997 to 1000 h/y at 20217), both condition implies more powerful and heavier equipment, and more impact to the soil.

With an entire world analysis, the World Bank publish the growth for agricultural tractors (23.8 to 25 million of tractors from 1992 to 2000 year) (THE WORLD BANK, 2024). This confirms the growth crop products demands and in consequences the increase for more equipment working at field.

Focus on sugar cane culture, the debate is not closed, there are researches that supports that full mechanized systems for sugar cane crops are not harmful for the soil in some cases (Jian Xiao, 2023), otherwise, there are studies that supports the opposite, the soil reduces the capacity when heavy equipment works at sugar cane fields (Boase Bangita, 2012), (Wanderlei Bieluczyk, 2023).

Despite of discussion, as a goal, to reduce the impact on environment must be keep it, this means to lead research and development for new type of vehicle; to reduce the load upon the soil and to improve the life span for the plants at field.

1.2 CAD/CAE Modelling.

With the development of software for CAD and CAE environments, there is a natural race for competition about the features for each brand of software. There are multiple brands that offers the combination for CAD+CAE simulation for dynamic mechanisms analysis, this study does not evaluate CAD/CAE technologies, it uses a commercial software, focusing in reducing the number of tools and steps to achieve a feasible model at conceptual stages of development.

2. Methodology

The proposed method has two main goals: to use the findings at exploratory rover's development applied for a sugar cane transporter and to reduce the effort for conceptual stage using simple CAD modelling. With this work, it hopes to add value for the team management leaders at R&D areas, and to find a balance between the CAD tools capabilities and effort dedicated at conceptual stages for new agricultural equipment. Using a CAD tool for a large proportion of the working day is not always the most effective environment for idea generation and to add value at the new product development (B.F. Robertson, 2009).

The proposed method has been made it with the creation of three type of models: a schematical 2D model, (2D-SM), a schematical 3D model (3D-SM) and a functional 3D CAD Model (3D-FM). Each model has a level of detail that offers to the R&D team quick insights to understand the kinematical and kinetics behavior. To simplify the interaction with CAD environment, this work uses the CAD tool Autodesk Inventor ®. *Table 1* describes the description for each model.

Table 1
Types of Conceptual Models

MODEL	DESCRIPTION
2D-SM	Two 2D sketches; one is a plane sketch using ZX for longitudinal motion and the other one is XY oriented for steer mechanism motion. (It is adopted SAE vehicle axis references) The length for the linkages is driven by CAD parameters. It uses a part file(.ipt) to analysis the main results for kinematical and kinetics analysis. To view the behavior for the velocity, acceleration it uses a vectorial approach in a 2D sketch. Each position could be view on charts, without replicating the geometry.
3D-SM	3D model uses the 2D-SM for the length of linkages and layout of components. Each linkage represented by a prismatic simple part. This configuration allows to create range of motion for the suspension. It could be visualized and charted to identify the full range of motion
3D-FM	3D model taking length of linkages from the 2D-SM. This model has a more detailed geometry for the linkages. It shows the relationship between the suspension and the steer mechanism proposed for the transporter. It could be a gate to a first physical rapid prototype.

The geometrical and weight restrictions for the sugar cane transporter is listed at table *(Table 2)*. For this, it has been identified three random commercial suppliers (AZUTEC, n.d.) (CHENHAN TECHNOLOGY, n.d.) (SITEFAB, n.d.) , each one, offers currently similar tipping trailers, for the study, it has been named as S1, S2 and S3, to avoid uses the commercial names for the study.

Table 2
Geometrical and Weight Requirements

NAME	S1-S2-S3	REQUIREMENT USED	UNITS
PAYLOAD	10-11-8	10.000	t
CAGE LENGTH(x-axis)	4.06	4	m
CAGE WIDTH (y-axis)	3.4-2.78 - 3.66	3.5	m
LOADING HEIGHT (z-axis)	3.6-3.45-3.5	3.45	m
VOLUME CAPACITY	24.3-20-24	24	m^3
WEIGHT	7.2-5-5.97	6	t

3. 2D-Schematical Model

For this model it has represented the most simply geometry that defines the vehicle and the suspension assembled. It is represented only with lines, and with the restrictions established at *Table 2*. The *Fig. 2*, it shows a lateral view for the concept model, the names of each length and point of interest. The Front Arm is named because it is assumed the main motion for vehicle from left to right, and this arm face first the obstacles at the road, but the vehicle could move any longitudinal direction. The front arm geometry is defined by dimensions L1, L2 and L3. The Front arm is connected to the body through the connection lever at point D. This point D has a degree of freedom at vertical direction, this condition allows the entire system to be adapted at any terrain geometrical condition and helps to keep the chassis body in a horizontal position when appears obstacles. The rear arm is defined by the dimensions L4, L5 and WB1 and WB2 represents the wheelbases. Each center of wheel is named with letters A, B, C in order from right to left.

This simple model allows to recreate different motions conditions. *Fig. 3* shows the basic longitudinal motion with a defined slope angle θ, the *Fig. 4*, shows the condition where the vehicle face an obstacle, this obstacle could be a geometry convex, flat, or concave. With this it could be possible to identify the range of motion for the suspension system, and *Fig. 5*, is the model where it could be possible to see the behavior of the vehicle crossing theoretical sinusoidal terrain model.

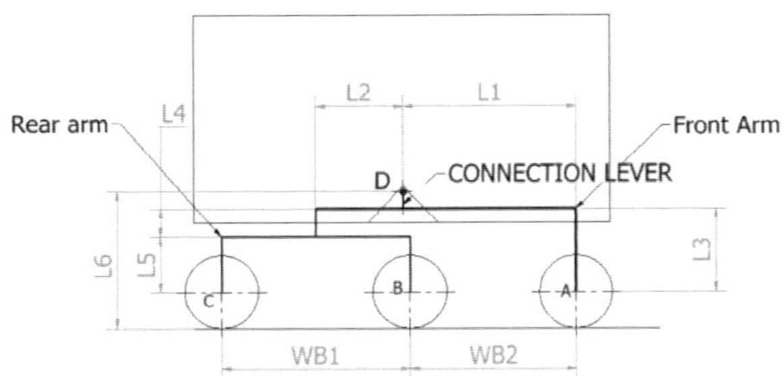

Fig. 2 Definition of parameters of linkages for suspension

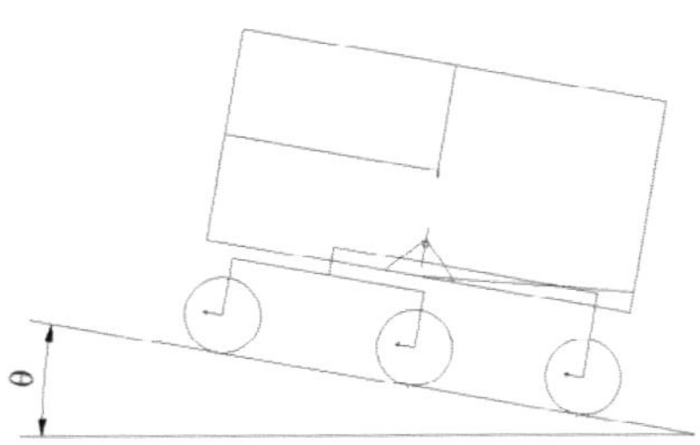

Fig. 3 2D SM. Longitudinal motion

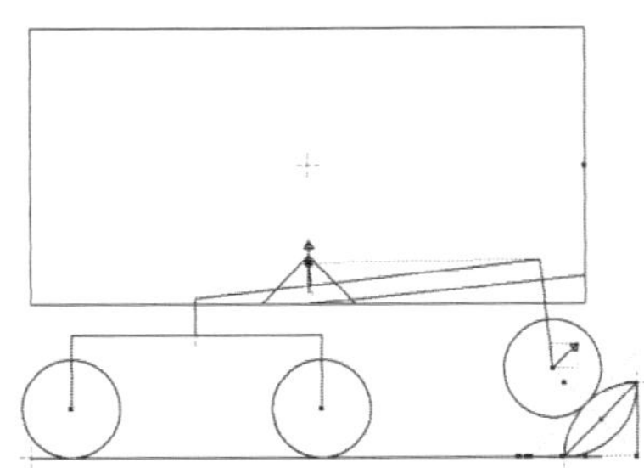

Fig. 4 2D SM. Single Convexe - Concave Obstacle

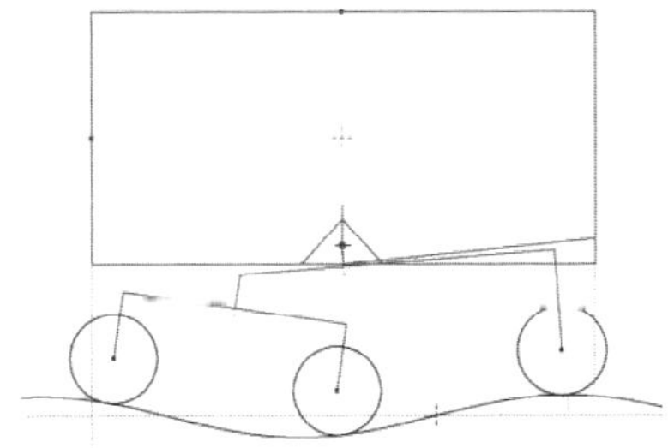

Fig. 5 2D-SM. Sinusoidal Terrain

At the *Fig. 3, Fig. 4, Fig. 5*, it is possible to see the body chassis keeps its horizontal level position, this desired feature allows the vehicle to face obstacle without increasing the oscillation for the rolling angles or balance angles that appears when the body of vehicle reacts to the interaction of soil, with this condition approximately controlled, the suspension must face the obstacles without increasing the acceleration at the chassis and therefore to reduce the risk of lateral rolling.

To control this degree of freedom at point D, is necessary includes a linear or rotational actuator to allows the motion, but with controlled conditions.

At the same 2D-SM is created another sketch feature for the model, it is the steering mechanism for each wheel of the suspension system. These four bars linkages it is based on the principle of self-propelled module transport (SPMT) developed by Cometto® (Belousov, 2014). Where a single linear actuator acts on a linkages mechanism and controls the rotation for a wheel. When the linear actuator acts, the points G and J does not move, so the bar FGH rotates with center on G, and the HI bar pulls the point I where make rotate the IJ that it is moving at same time with the wheel.

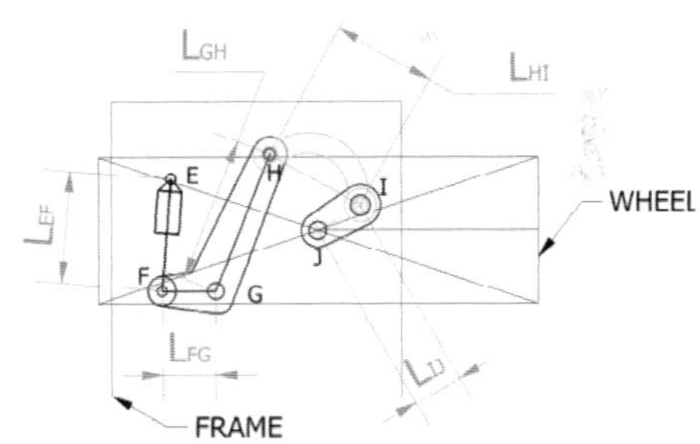

Fig. 6 Steering Proposed Mechanism 2D-SM Top View

4. Schematical 3D Model (3D-SM)

This model is a natural evolution from the 2D-SM. With the help of features from the parametric model tool selected, it could be possible to create a 3D model linked to the 2D-SM.
This connection allows to the R&D to go back and forward and with fewer steps, it could take a deep virtual validation understanding for the motion, using 3D simple models for each lever and linkages.

This 3D-SM model does not have a special or specific condition for the joints, or manufacturing assembly specification (assembly tolerance).

At *Fig. 7* it can be seen the 3D-SM model, with the geometrical representation. At this model, only is visualized the longitudinal motion, the steer layout and 3d components are displaced forward for the 3D-FM model.

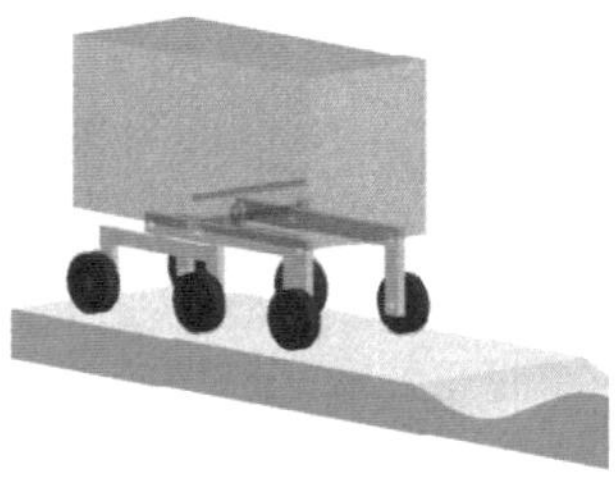

Fig. 7 Schematical 3D Model

With the 3D-SM model it could be possible to estimate the Power requirement and to visualize the behavior for the concept suspension in a range of soils. The tool used to model the 3D-SM was Autodesk Inventor®, this software allows to recreate the tentative geometries for each linkage. With the dynamic simulation module, it can be possible to recognizes each constraint like a dynamic joint, this means that can be translate the kinematical and kinetic properties for each component. This module does not take account the flexibility of materials, only consider the dynamic of components as a rigid body. The inertial properties for the bars and wheels are assumed with similar mass weight of components from trailers.

To simulate a model first, its necessary set up a terrain modeled, for the purpose of this work it has modeled five proposed types for geometric soils type. It is represented at *Table 3.*

Table 3
Models of Geometry of Soils used for 3D-SM.

NAME	IMAGE	CHARACTERISTICS
S0	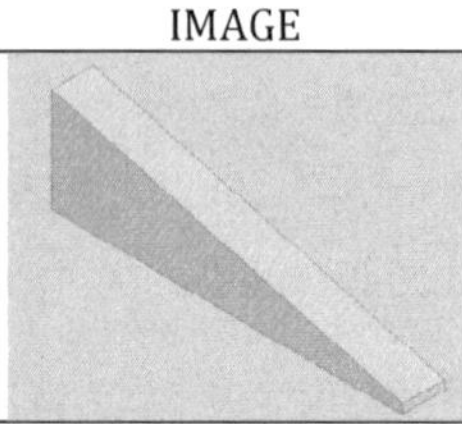	Flat surface with a slope of 15° (26.8%) to evaluate gradeability
S1		Sinusoidal surface without slope. $\sin\left(100\frac{\pi x}{1000}\right)$ Two waves with length of 13000mm
S2	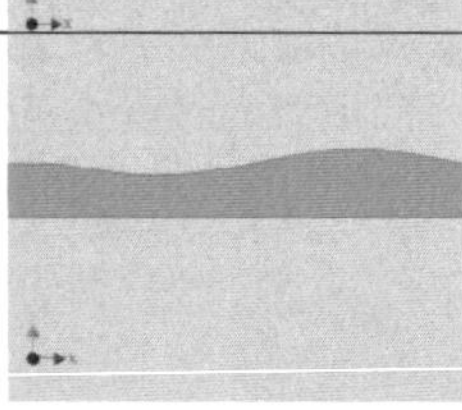	Sinusoidal surface without slope. $\sin\left(200\frac{\pi x}{1000}\right)$ Two waves with length of 13000mm
S3	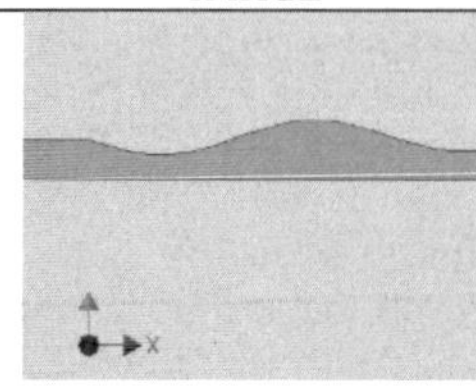	Sinusoidal surface without slope. $\sin\left(350\frac{\pi x}{1000}\right)$ Two waves with length of 13000mm

To observe different conditions, for each type of soil it has introduced dynamic parameters for the condition of contact between the soil and each wheel. These parameters are represented at the *Table 4*, the goal is to simulate the motion with different simulated soils, only with the stiffness of soil. It is known that there are more parameters to characterize a soil and to recreate the interaction with wheels, for this work it is proposed this approach to reduce the number or restrictions. In a future work, it must be necessary to validate this premises with an experimental setup and adjust the soil model with validated data from the sugar cane fields or other fields (geometrical and approximated stiffness and damping)

Table 4
Contact parameters to approximate to soil conditions.

SOIL NAME	Stiffness (N/mm)	Damping (N s/mm)	Friction
K1	5	1	1
K2	10	1	1
K3	30	1	1
K4	50	1	1

To move the model, it has been introduced an angular velocity condition for the joint that represents the wheel with each linkage, this velocity condition, it is simulated with a ramp curve to accelerate from static condition. With two seconds as a fix parameter to all models, to achieve a cruise velocity for the simulation, depending on that cruise velocity it is defined the simulation lapsed time. This velocity simulates a movement, and time could be stopped when the simulation does not have any representative change. With this model it could be extracted a required torque for the motion at each wheel. At *Fig. 8* it can be seen the ramp of angular velocity used for one of the simulated conditions.

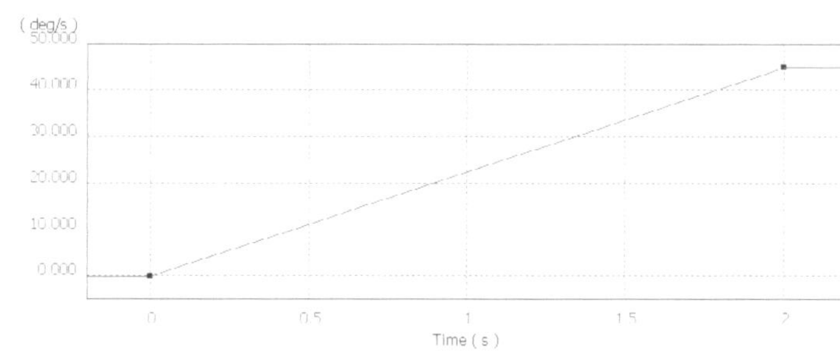

Fig. 8 Angular velocity ramp w vs time ($\omega \frac{deg}{s}$ vs t sec)

With this model, it could be possible to simulate different velocities and different approximated soils conditions, in other terms to find the energy required for each type of vehicle with a specific type of rockie-arm suspension. To express the velocity, it is expressed using the linear motion velocity with the theoretical radius of wheel (1).

$$\omega = \frac{V_x}{r} \tag{1}$$

For the simulated model, it has been used three values for velocity: 3, 15, and 25 kilometers by hour (kph) for the S0 soil and 3, 5, 8 kph for the S1, S2 and S3 model.

To complement the model, it has been introduced a spring damper model at joint "D" and connected with the rear arm as it shows at Fig. 9. For the spring damper model, it has been used a stiffness of 66 N/mm and a damping coefficient of 0.010 N s/mm. The spring it is required to stabilize the rear arm at the wheel B, and force that wheel to keeps the contact with ground.

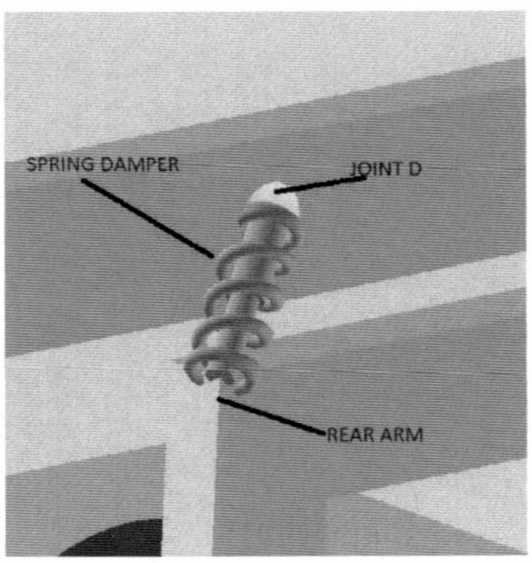

Fig. 9 Spring Damper Model at S3M

With the previous considered, it has been elaborated different models for different conditions expressed at *Table 5.*

Table 5
Model codification with parameters variation

MODEL CODE	SOIL CAD MODEL	SOIL STIFFNESS	Linear velocity (kph)
M1	S0	K1	3
M2	S0	K1	15
M3	S0	K1	25
M4	S0	K2	3
M5	S0	K2	15
M6	S0	K2	25
M7	S0	K3	3
M8	S0	K3	15
M9	S0	K3	25
M10	S0	K4	3
M11	S0	K4	15
M12	S0	K4	25
M13	S1	K1	3
M14	S1	K1	5
M15	S1	K1	8
M16	S2	K1	3
M17	S2	K1	5
M18	S2	K1	8
M19	S3	K1	3
M20	S3	K1	5
M21	S3	K1	8

For each model, it could be possible to extract tabulated data., with a manipulation in a spreadsheet. For this work, it has been used Microsoft Excel ® to analyze the results. At *Table 6*, it shows the type of data extracted for each model.

Table 6
Data extracted from Simulated models.

VARIABLE NAME	USES FOR	UNITS
Maximum Penetration (pmax)	It measures how a wheel penetrates the surface model	Mm
Angular velocity (w)	Velocity extracted from each wheel joint.	Deg/s
Moment at wheel (My)	Torque measure at each wheel joint	N-mm
Position at z for the point D	It shows how the system could impact the vertical motion for the sugar cane cage	Mm

To reduce the simulation time and taking the advantage of symmetry for the model, only the model for simulation the 3D-SM is the half of the model, even, the body chassis is not represented, to reduce the simulation time. The weight of chassis is replaced by a vector force applied at the center of body model, which represents 1000 kg, this force could be oriented with the gravity vector consider the gradeability condition. It could be estimated the total power required for a vehicle in terms of weight in a basis of tons (equation 2). This analysis does not consider the pull thrust force and energy developed to compare tractors, because this proposal it is to be self-propelled vehicle.

$$P_v = \frac{3M_y \cdot \omega}{\eta_t} \ (W/ton \tag{2}$$

Where P_v, represents the totally power required in Watts for a vehicle by ton. η_t, represents the combined driveline efficiency and electric conversion to the shaft ($\eta_t = 0.7 - 0.8$)

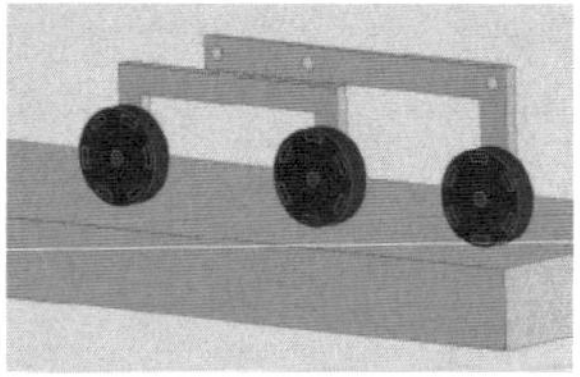

Fig. 10 Half model to run the simulation model.

5. Functional 3D model (3D-FM)

After analysis the modelling with the S2M and S3M, it is proposed a functional 3D Model (3D-FM), this model increase the accuracy with additional features.

The R&D teams must at this stage to communicate the design intention, so This 3D-FM is a linked bridge between the conceptual engineering models and an improved communication model. It serves like a point of decision for product development process, to receive feedback before a physical prototype phase, so in this case, it could be useful to have more features detailed. At this stage it is not defined yet the material and manufacturing process to scale the idea, so in this stage only is stablished a geometry that summarizes the findings from previous stages.

As a functional geometry, the model is built in the same software (Autodesk Inventor) to simplify CAD translations for the team. The *Fig. 11* it shows the representation for this 3D-FM. This is represented in the conceptual lanes where the vehicle must road inside the crop field. These lines are pyramidal to simulate the ground and to avoid the wheels to touch this soil portions.
The chassis (blue) is the component that supports the suspensions subsystem (rear arm-green and front arm yellow) and supports the sugar cane cage. Each suspension arm, uses a DSS (Drive, suspension, and steering system), is a module that increases the performance of the vehicle and the maneuverability. With the linkage mechanism inside conceptualized at *Fig. 6*, it could be simulated the range of motion, for longitudinal and steering system, The DSS is proposed to be electric powered module, so receives energy from a potentially battery system. This DSS has three actuators (gearbox electric motor, electric linear actuator for suspension and electrical linear actuator for steering mechanism). This study does not cover the selection for these components.

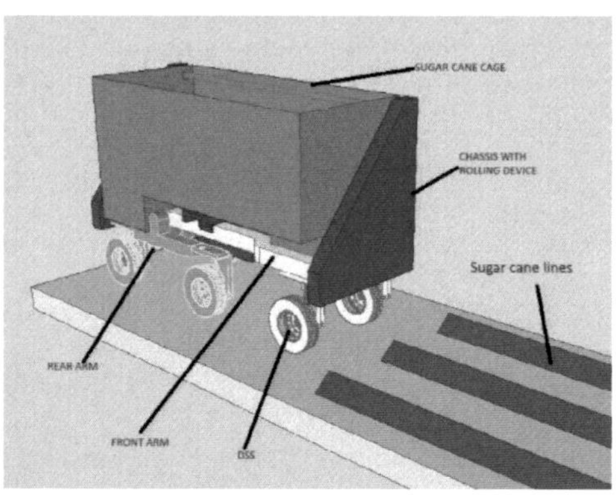

Fig. 11 3D-FUNCTIONAL MODEL

To evolute the engineering method, the 3D-FM allows the R&D team to dig deeper than the other previous models, one of the challenges at this model is to refine the inertial properties of the model and to integrate the full functionality for the steering system.

6. Results

The results of this work it could divided against the three models achieved (2D-SM, 3D-SM, and 3D-FM).

For the 2D-SM, one of the results is to iterate graphically, a first approximation for the length for each lever of suspension linkage. At *Fig. 12,* it can be seen the dimensions selected for the analysis, these dimensions are based on requisites expressed at Table 2. At *Fig. 13*, it can be seen a sample evaluation for the front arm and impact on the motion when face an obstacle. At *Fig. 14*, it can be seen when the 2D-SM moves through a sine terrain how the connection rod of point D (*Fig. 2*) must displace to compensate the oscillation for the front arm and rear arm.

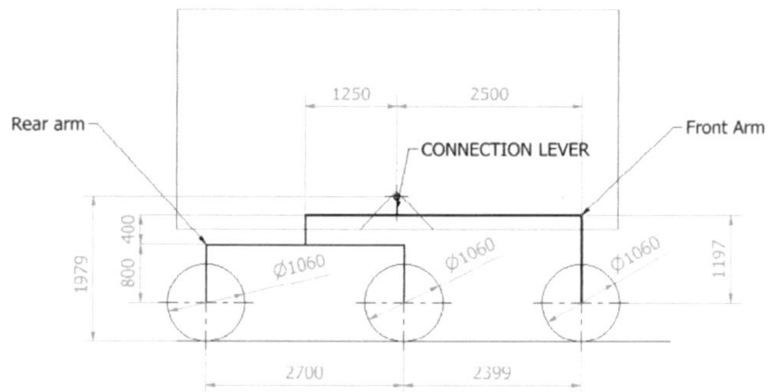

Fig. 12 2D-SM with dimensions evaluated.

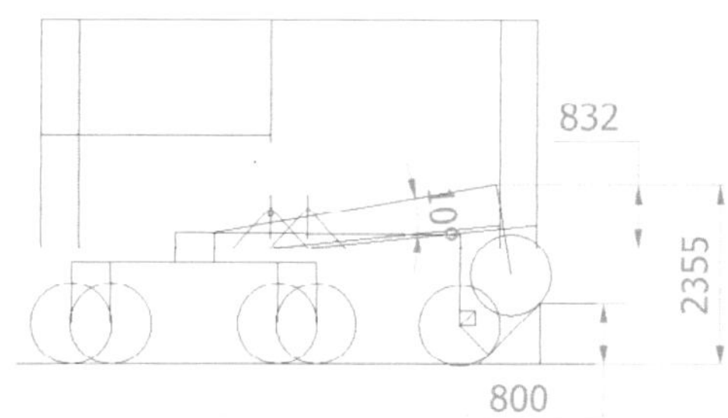

Fig. 13 2D-SM facing an obstacle- suspension elevation estimation.

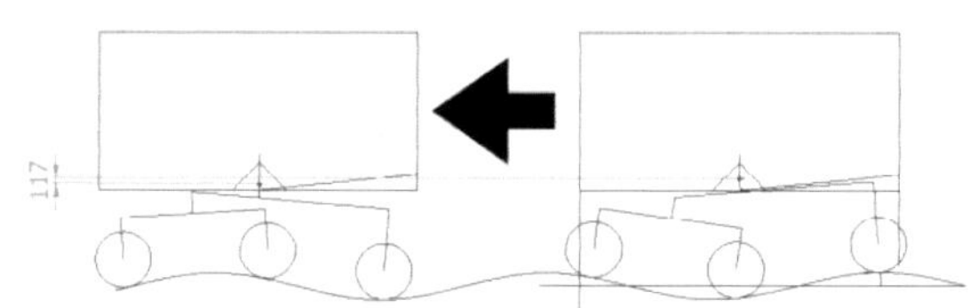

Fig. 14 Height variation for the joint connecting the rockie-arm suspension with chassis.

For the 3D-SM, the first insight is to compare the vertical oscillation for the connection rod at point D, this analysis represented at *Fig. 15* it helps to determinate the maximum velocity when the vehicle must pass terrain with sinusoidal geometrical similar surface, it represents too the spring-damping requirements. Because the point D represents the body chassis.

At *Fig. 16*, it shows the power requirement by tons for diverse types of soil and velocity in a ramp (S0). With this diagram, it could be possible to limit the operational velocity and find a gradeability for high slopes. At *Fig. 17*, it can be used to estimate how a vehicle could sink in a simulated soil condition, this sinkage is related with the condition that could face when slip the wheels.

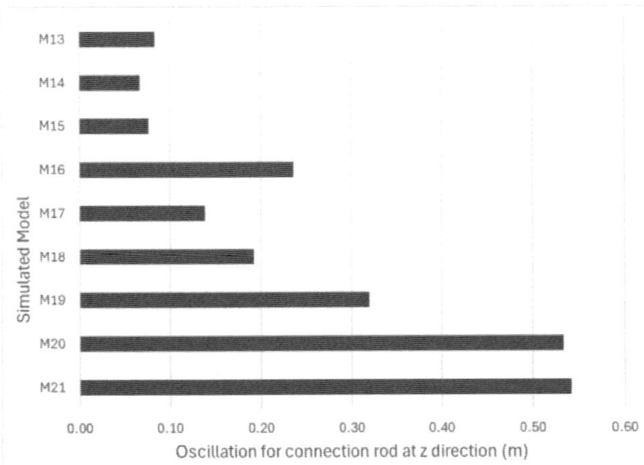

Fig. 15 Height Oscillation for connection rod at different simulation models.

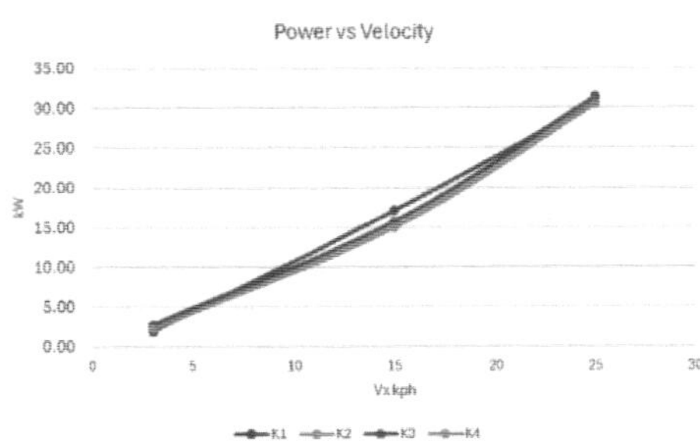

Fig. 16 Power/ton required vs linear velocity using four diverse types of soils stiffness.

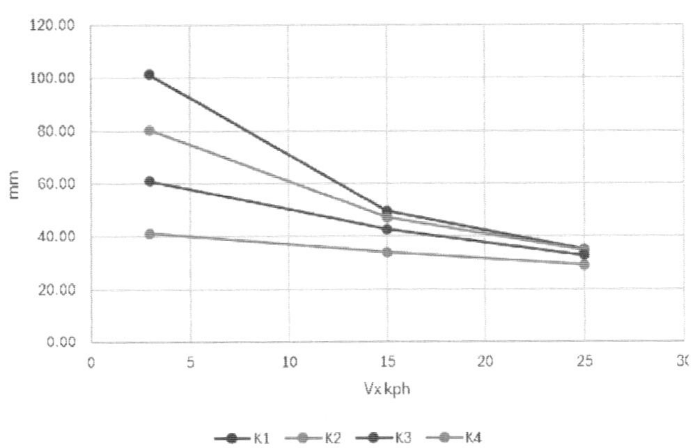

Fig. 17 Vertical Sinkage for different simulated soils and velocities

7. Conclusion and analysis

This work presents a method to be used at R&D teams, this method is a different approach for the estimation of new concept vehicles dynamic performance.

With the three models approach allows to answer question with more reliable results depending on the moment.

It is possible to create a model that represents more restrictions, but an experimental setup must be developed to validate the variables introduced. This method could be a bridge between the research findings and the applied engineering teams.

8. Future work

For the purposes of this work, it has been limited other parameters that could be included, for example a graphical curve for the transmitted loads at the soils. For a future work, the load transmitted to soil could be visualized as a parametric function.

To represents accurate the geometrical soil conditions, it could be represented the geometrical variation for the grounds, connecting a point cloud with a surface CAD model, this could help to increase the accuracy for this method.

To gain advantage of the CAD Modelling it could be important and necessary to build physical prototypes, to validate the premises for the rockie-arm suspension applied for a sugar cane transport system. This opens other prototyping methodologies that must be faced.

9. Declaration of competing interest

The authors declare that they have no known competing financial interests or personal relationships that could have appeared to influence the work reported in this paper.

10. References

Ahlvin, R. H. P., 1992. NATO Reference Mobility Model Edition II, NRMM II, Vicksburg MI: Waterways Experimental Station.

AZUTEC, s.f. VAGON AUTO VOLTEO. [En línea] Available at: https://irp.cdn-website.com/43f59c57/files/uploaded/15-05-VAGON%20AUTOVOLTEO%20DE%2025%20m3%20UNIFICADO%20COMPRIMIDO.pdf

B.F. Robertson, D. R., 2009. Impact of CAD tools on creative problem solving in engineering design. Computer-Aided Design, Volume 41(Issue 3), pp. Pages 136-146.

Bacenetti, J., 2022. How does annual utilisation can affect the environmental impact of tractors? A life-cycle assessment comparing hypothetical scenarios for farmers and agricultural contractors in Northern Italy. Biosystems Engineering, Volumen Volume 213, pp. Pages 63-75.

Balvinder Kaur, M. S. D. J. S. S. M. N. C. T. K. B. S. S. P. S. S. A. D. U. Y. A. S. A. M. A. R. B. N. V. K. A. K., 2023. Insights into the harvesting tools and equipment's for horticultural crops: From then to now. Journal of Agriculture and Food Research, Volumen Volume 14.

Belousov, B. &. P. S., 2014. Heavy-Duty Wheeled Vehicles: Design, Theory, Calculations. Warrendale, PA, USA: SAE.

Boase Bangita, B. R. R., 2012. Impacts of compaction relief treatments on soil physical properties and performance of sugarcane (Saccharum spp.) under zonal tillage system. Geoderma, Volumen Volumes 189–190, pp. Pages 351-356.

CASE, 2024. Austoft 4000 Sugar Cane Harvester. [En línea] Available at: https://www.caseih.com/en/africamiddleeast/products/harvesting/austoft-4000-sugarcane-harvesters

Çengel, Y. A., 2008. Thermodynamics : an engineering approach.. Boston: McGraw-Hill Higher Education.

CHENHAN TECHNOLOGY, s.f. 7YGS 6T/8T Sugarcane Self Tipping Trailer. [En línea]

Available at: https://www.chenhantech.com/product/7ygs-20-sugarcane-transportation-trailer/

D.D. Bochtis, S. V., 2008. Minimising the non-working distance travelled by machines operating in a headland field pattern. Biosystems Engineering, pp. Volume 101, Issue 1.

Delgado, C. D. M., 2023. DISEÑO DE UN SISTEMA DE TRACCIÓN TIPO ORUGA ADAPTADO A UN VAGÓN AUTOPROPULSADO PARA EL TRANSBORDO DE CAÑA DE AZÚCAR. CALI: UNIVERSIDAD DEL VALLE.

Emiko Fukase a, W. M., 2020. Economic growth, convergence, and world food demand and supply. World Development, p. Volume 132.

Georg Steinberger, M. R. H. A., 2009. Mobile farm equipment as a data source in an agricultural service architecture. Computers and Electronics in Agriculture, pp. Pages 238-246.

J.Y. Wong, P. J. E. T. J. P.-T., 2018. Journal of Terramechanics Comparison of simulation models NRMM and NTVPM for assessing military tracked vehicle cross-country performance, pp. Pages 31-48.

J.Y. Wong, P. J. E. T. J. P.-T., 2020. A review of mobility metrics for next generation vehicle mobility models. Journal of Terramechanics, pp. Pages 11-20.

Jian Xiao, T. L. S. Y. H. T., 2023. Do full mechanized management strategies destroy soil health and fertility in sugarcane fields?. CATENA, Volumen Volume 224.

Michael McCullough, P. J. J. D. D. G., 2017. The Next Generation NATO Reference mobility model development. Journal of Terramechanics, pp. Pages 49-60.

Nildeep Patel, R. S. J. C., 2010. The ExoMars rover locomotion subsystem. Journal of Terramechanics, Volume 47(Issue 4), pp. Pages 227-242.

OECD and Food and Agriculture Organization of the United Nations, 2020. OECD-FAO Agricultural Outlook 2020-2029. [En línea] Available at: https://doi.org/10.1787/1112c23b-en

Pritchard, P. J. &. M. J. W., 2015. Fox and McDonald's introduction to fluid mechanics (9th ed.).. 9th ed. s.l.:John Wiley & Sons..

R.L. Raper, 2005. Agricultural traffic impacts on soil. Journal of Terramechanics, pp. Volume 42, Issues 3–4 Pages 259-280.

Rowland, D., 1972. Tracked vehicle ground pressure and its effect on soft ground. Stockholm-Kiruna, Sweden,, s.n., p. pp. 353–384..

Salavat Mudarisov, I. G. I. G. E. H. I. F., 2020. Soil compaction management: Reduce soil compaction using a chain-track tractor. Journal of Terramechanics, Volumen 89, pp. 1-12.

Shivani Garg, ,. N. P. R. S. R., 2023. Smart agriculture and nanotechnology: Technology, challenges, and new perspective. Advanced Agrochem.

SITEFAB, s.f. [En línea] Available at: https://sitefab.co.za/trailers/

Tagar, A. G. M. A. J. L. N. &. S. A., 2017. Assessment of implement efficiency and soil structure under different conventional tillage implements and soil moisture contents in a silty loam soil. Agricultural and Food Sciences, Environmental Science, pp. 413-420.

TECHNOLOGIES, V., 2024. The Perfect Platform for Every Operation. [En línea] Available at: https://www.veristech.com/sensing

THE WORLD BANK, 2024. Agricultural machinery, tractors. [En línea] Available at: https://data.worldbank.org/indicator/AG.AGR.TRAC.NO?end=2009&start=1961&view=chart&year=2002

Wanderlei Bieluczyk, L. F. M. M. R. C. L. W. M. J. A. B. R. R. R. P. B. d. C. W. H. v. d. P. S. M. T., 2023. Forest restoration rehabilitates soil multifunctionality in riparian zones of sugarcane production landscapes. Science of The Total Environment, Volumen Volume 888.

Weihua Li, L. D. H. G. Z. D. N. L., 2013. ROSTDyn: Rover simulation based on terramechanics and dynamics. Journal of Terramechanics, 50(3), pp. Pages 199-210.

Wong, J., 2009. Terramechanics and Off-Road Vehicle Engineering. s.l.:Butterworth-Heinemann.

Wu, T. W. H. L. Q. e. a., 2021. Design and Development of Tracked Sugarcane Transporter. Sugar Tech 23, p. 1137–1146.

Yi-Hsin Lin, F.-J. L., 2023. An assessment framework for the purchase of 3D CAD software in manufacturing industries. Technological Forecasting and Social Change, Volumen 192.

ISTVS

Abstract-only submissions

0178 | Model to estimate the force exerted on the sphere rolling on the sandy road

Takeshi Fukumoto [a,*], Ken Yamamoto [a], Makoto Katsura [a], Hiroaki Katsuragi [a]

[a] *Osaka University, Japan*

* *Corresponding author: t.fukumoto@ess.sci.osaka-u.ac.jp*

The importance of the rover for exploring the planetary surface is recently increasing. It is important to move off-road (e.g. regolith) ground safely without getting stuck. To do that, it is necessary to predict the force exerted on the tire while moving on an off-road environment. The traditional terramechanics model has been widely used to predict the force since a half-century ago by many researchers. However, there are some drawbacks to that model. There are many parameters (~10) to predict the force and it takes a lot of time and effort to determine the reasonable parameter values. We would like to establish a way of predicting the force easily. This research aims to develop a semi-empirical model to predict the force with few parameters based on the traditional terramachanics model. In this work, we carried out a simplified experiment to obtain the relation between soil and tire. Here, we experimentally investigated the dynamics of the spheres (without driving force and their radius is 6.4 mm) rolling up a sandy slope (consisting of 0.8 mm glass beads). We systematically varied the density of the sphere (0.93, 1.4, 2.6, 3.9, and 7.8 g/cm3), the slope angle (from 0 to 20 deg), and the initial velocity (0.2-0.7 m/s). As a result, the dynamics of the sphere show a constant deceleration in both translational and rolling motions. We assume there are 2 forces (normal force and tangential force) exerted on the sphere at one point with these constant acceleration. The ratio between normal force and tangential force is constant, and the point angle is about 0.4 times the angle of the sinking depth. This relationship suggests that we can compute the forces exerted on the object (sphere, tire) with the relations we obtain from this research.

0376 | Improvement of formulations of Clegg impact hammer and Rammsonde penetrometer for use in compacted snow

Mohit Shenvi [a,*], Corina Sandu [a,*], Costin Untaroiu [a]

[a] *Virginia Tech, United States*

* *Corresponding author: mshenvi@vt.edu, csandu@vt.edu*

From the terramechanics viewpoint, assessing sinkage and penetration resistance in a deformable terrain is crucial for on-site terrain characterization. Existing devices often encounter limitations when characterizing compacted snow, as documented in the literature. The Clegg impact hammer and the Russian snow penetrometer (utilizing the Rammsonde principle) are extensively employed tools for gauging the penetration resistance of compacted snow. Despite the widespread use of the Clegg impact hammer, there are relatively few studies, primarily conducted by CRREL, focusing on its application in snowy conditions. This study aims to propose enhancements to classical literature formulations of two devices that have been used in compacted snow conditions to evaluate snow characteristics. One such device viz. the Clegg impact hammer has outputs correlated to the evaluation of sinkage and Young's modulus. Modifications to the classical methodologies have been proposed in this work. The in-house device yielded consistent measurements even when the CTI gauge exhibited variations on a commercial test track used for winter tire evaluations. In this case, formulations have been developed that contribute to a more analytical assessment of the resistance pressure. Additionally, these formulations may aid in evaluating the compacted zone forming in front of the cone as it penetrates deeper into the terrain. The comparison results offer insights into the potential variation in calculating these parameters, with a subsequent discussion of the findings from a physics-based perspective. Future work in this field could entail evaluating the resistance pressure across various terrain types to validate the proposed hypotheses or offer correction factors based on field testing results.

0677 | Adaptive particle refinement in terramechanical DEM simulation

Markus Pogulis [a,*], Martin Servin [a]

[a] *Umeå University, Sweden*

* *Corresponding author: markus.pogulis@umu.se*

DEM is computationally intensive for granular dynamics simulation, leading to a need for efficient strategies. This study explores using local particle refinement, scaling particle size based on expected spatial resolution needs, inspired by adaptive mesh refinement in FEM. Finer particles are used where intense interaction occurs, and coarser particles further away. We hypothesize this method can maintain good accuracy while reducing particle count and computational effort. Fine particles are used on the soil bed's top, with coarser particles at greater depth, creating a particle size gradient. By adjusting the gradient we introduce a "scaling aggressiveness", allowing control over the trade-off between efficiency and accuracy. We use triaxial tests to verify that the method is scale invariant. Pressure-sinkage and shear-displacement tests are then used to evaluate the method's effectiveness and accuracy in terramechanics applications. All beds were compared to a reference bed with homogenous particle size, where the mean static sinkage was 1.25 mm for a 50 kPa load. The dynamic sinkage was 73 mm for the full simulation time. For quasi-2d simulations, mild scaling aggressiveness reduced the particle count by 2-4 times with relative error up to 4% for dynamic sinkage (11% for static sinkage). For medium aggressiveness, 4-6 times reduction with relative error of 4% (19% static). For highest aggressiveness, 6-8 times reduction with relative error of 7% (29% static). The internal friction proved to be very resistant to gradient changes, with errors within 1%. When extending the model to full 3D, we estimate up to a reduction in particle count of up to a factor 25.

0815 | Using simulation to characterize RASSOR excavation in lunar environments

Luning Bakke [a,*], Zhenhao Zhou [a], Alexandra Kissel [a], Ruochun Zhang [a], Dan Negrut [a,*]

[a] University of Wisconsin-Madison, United States

* Corresponding author: lfang9@wisc.edu, negrut@engr.wisc.edu

We describe modeling and simulation approaches used to investigate the operation of NASA's Regolith Advanced Surface Systems Operations Robot (RASSOR) excavator while operating in lunar gravitational environments. With a total mass of 66 kg, RASSOR is equipped with four wheels and two arms, each of the latter connected with a rotating bucket drum for soil excavation. The design compensates for the low mass and low gravitational pull by digging using two counter-rotating bucket drums. The rover can raise its drums when transporting the soil, and it unloads the material by reversing the drum's rotation at destination. We simulate digging operations at two levels of fidelity using Project Chrono, an open-source multi-physics simulation platform that provides terramechanics modeling, robot modeling, sensor models, and a ROS2 autonomy stack for synthesizing RASSOR's autonomy in simulation. We discuss two terrain modeling techniques – the Continuum Representation Method (CRM) and the Discrete Element Method (DEM). CRM employs the μ(I)-rheology model for describing the elasto-plastic behavior of the granular material, with the partial differential equations for the mass, momentum, and Cauchy rate of change spatially-discretized via the Smoothed Particle Hydrodynamics (SPH) method. The interaction between the implement and terrain is captured using Boundary Conditions Enforcement (BCE) markers. Unlike CRM, DEM represents the terrain discretely, by describing frictional contact forces at particle level. In this study, we compare the CRM and DEM results in terms of accuracy and time to solution.

1499 | Development of evaluation technology for tire traveling performance on soft soil of lunar surface

Kei Tsuchiya [a,*], Kon Seiji [a], Masahiro Katayama [a], Tomoya Arai [b], Shingo Ozaki [b]

[a] Bridgestone Corporation, Japan

[b] Faculty of Engineering, Yokohama National University, Japan

* Corresponding author: kei.tsuchiya@bridgestone.com

Bridgestone Corporation is taking on the challenge of developing tires for lunar vehicles. In this development, evaluating the tire traveling performance on the soft soil of the lunar surface is important because it is nearly impossible to conduct pre-testing using actual tires under actual conditions. Therefore, we are developing both experimental and simulation technologies for performance evaluation. In this presentation, we will explain two testing devices developed by Bridgestone Corporation for evaluating the performance, including examples of measurement results. The first device is an indoor installation type that uses scaled-down model tires. The device incorporates special features to simulate various traveling conditions such as constant slip ration mode, cornering, and climbing. The second device is an outdoor testing type that allows for testing with tires equivalent in size to those used on lunar vehicles. This device is portable and can be taken to various testing sites. Additionally, while conducting outdoor tests, it can reproduce steady-state traveling conditions which are important for accurate evaluations of tire traveling performance. Furthermore, in this presentation, we will briefly introduce the evaluation technology that combines experiments and simulation using the extended terramechanics model proposed by the Ozaki Laboratory at Division of System Research, Faculty of Engineering, Yokohama National University, with whom we are conducting joint research.

2520 | Efficient tire-terrain interaction modelling: effect of flexibility on traction

Mahdi Maleki [a,*], Jozsef Kovecses [a,*]

[a] McGill University, Montreal, Canada

* Corresponding author: mahdi.maleki@mail.mcgill.ca, jozsef.kovecses@mcgill.ca

The analysis of tire dynamics is essential in the simulation of vehicle behaviours. The forces exerted on a tire depend on the tire and terrain interaction, and the tire structure directly influences this interaction. Complex models with a high number of degrees of freedom, such as lumped parameter models or finite element models, are typically required to represent tire flexibility appropriately. These models, however, can lead to high computational costs that can be a significant challenge for real-time simulation. In this work, we developed a reduced model for the flexible tire that can represent the effects of tire flexibility with low computational costs in the simulation. By calculating the effective stiffness of a flexible tire model (base model) and augmenting it with a model representing the rigid body motion of the wheel, we could represent the flexibility of the tire more efficiently. The tire deformation affects the tire contact patch size, influencing the traction forces acting on the tire. By having the effective stiffness at hand, we are able to calculate the contact patch size at each instant of time. The traction forces acting on the wheel depend on the size of the contact patch since a larger contact patch would be capable of carrying more tangential load. In order to observe the effect of the contact patch size on the traction forces, we scaled the friction coefficient based on the size of the contact patch. This way, we could take the effect of tire deformation on traction forces into account. Our simulations show efficient real-time performance while maintaining accuracy. By integrating effective stiffness and adjusting friction, we effectively capture tire dynamics, making them a practical solution for diverse vehicle simulation applications.

2800 | Granular flow in reduced gravity: analysis and insights from centrifuge experiments at the international space station

Genya Ishigami [a,*], Shingo Ozaki [b], Masatsugu Otsuki [c], Hideaki Miyamoto [d], Koji Wada [e], Masataku Sutoh [c], Takao Maeda [f], Taizo Kobayashi [g]

[a] *Keio University, Japan*

[b] *Faculty of Engineering, Yokohama National University, Japan*

[c] *Japan Aerospace Exploration Agency, Japan*

[d] *The University of Tokyo, Japan*

[e] *Chiba Institute of Technology, Japan*

[f] *Tokyo University of Agriculture and Technology, Japan*

[g] *Ritsumeikan University, Japan*

* *Corresponding author: ishigami@mech.keio.ac.jp*

Gravity-dependent characteristics of regolith, fine-grained granular media covering extraterrestrial surfaces, are essential for reliable design and feasible operation of space probes. Parabolic flight or drop tower facilities for simulating reduced gravity experiments on Earth can only perform short test durations and a limited number of tests with less quality artificial gravity. A numerical simulation requires an accurate interaction model of space probe and regolith. Further, the model parameters must be carefully identified for verification and validation. Therefore, the experimental dataset of granular media under stable reduced gravity is essential for solving the abovementioned issues. We performed a granular flow experiment under varied artificial gravity generated by a centrifuge on the International Space Station. An hourglass-shaped apparatus containing granular media was used to observe the granular flow in high-quality, long-term, and stable gravity conditions. We also performed a numerical simulation that calculates the granular flow in both artificial and natural gravity environments. The simulation verifies that the granular flow at the hourglass's orifice in artificial gravity is equivalent to natural gravity. The mass flow rate of the granular media measured from the experiment follows a well-known physics-based law, while some deviations are found in low- and micro-gravity conditions. The deviation implies that the bulk density of the granular media decreases as the gravity decreases. This finding provides a useful insight that improved simulation of space probes in reduced gravity can be realized by reducing the bulk density of the granular media, resulting in reliable design and analysis of the space probes.

3071 | Study on applicability of extended terramechanics model to various traveling modes of wheels

Tomoya Arai [a,*], Kei Tsuchiya [b], Seiji Kon [b], Masahiro Katayama [b], Shingo Ozaki [a,*]

[a] *Yokohama National University, Japan*

[b] *Bridgestone Corporation, Japan*

* *Corresponding author: arai-tomoya-sk@ynu.jp, ozaki-shingo-xd@ynu.ac.jp*

To promote lunar and planetary exploration missions, the development and operation of exploration rovers is essential. However, it is difficult to evaluate prototypes of wheels and vehicles under environments corresponding to extraterrestrial gravity and atmosphere conditions. Therefore, systematic evaluation of vehicle traveling characteristics using a numerical simulation is required. Under these circumstances, an extended terramechanics model based on cellular automaton that considers terrain surface deformation was proposed by Yokohama National University group, and its validity was confirmed through comparison with the results of traveling experiments using a rigid wheel. In this study, we propose a model that can handle various traveling modes to improve the versatility of numerical analysis simulations that implement extended terramechanics theory. First, we conduct single-wheel experiments using a rigid wheels and obtain data for model verification of both straight traveling and braking. Then, we extend the extended terramechanics model by implementing braking logic, etc., and establish a simulation model for a single rigid wheel using the commercial software package Simscape. In addition, systematic simulations are performed under forced-slip condition in which the angular velocity is fixed, while the translational velocity is varied. Here, the wheel specification and traveling conditions are the same as those of the experiment equipment. Finally, we demonstrate the effectiveness of the proposed model by comparing it with the results of experiment under various conditions.

3304 | Development of an unmanned exploration robot for lunar surface geotechnical investigation

Taizo Kobayashi [a,*], Atsushi Kakogawa [a], Shinichi Ito [a], Masafumi Nakagawa [b], Takeshi Tsuji [c], Shingo Ozaki [d], Satoshi Matsumura [e], Akihiko Kondo [e], Masanori Takigawa [f], Keitaro Kitamura [f], Takahito Hiramatsu [f], Hisatoshi Sano [f], Takeshi Yoneoka [g], Junichiro Odaka [g], Kenji Hosobori [g], Taichi Ikenaga [h], Ryu Taniguchi [h]

[a] *Ritsumeikan University, Japan*

[b] *Shibaura Institute of Technology, Japan*

[c] *University of Tokyo, Japan*

[d] *Faculty of Engineering, Yokohama National University, Japan*

[e] *Port and Airport Research Institute, Japan*

[f] *Asia Air Survey Co., Ltd., Japan*

[g] *Kiso-Jiban Consultants Co., Ltd., Japan*

[h] *Soil and Rock Engineering, Co., Ltd., Japan*

* *Corresponding author: kobat@fc.ritsumei.ac.jp*

The lunar surface is covered by a thick layer of soil known as regolith, which remains largely unexplored from a soil mechanics perspective. Additionally, significant uncertainties exist in geology and topography, posing risks to the safety and efficiency of lunar surface operations. Effective geotechnical risk assessment and management are essential for the safe planning and execution of lunar activities. To address these challenges, we are developing an unmanned exploration robot, named the Robotic Geotechnical Investigation System (RGIS), aimed at gathering critical data on the lunar surface. The RGIS is equipped with four key components: (1) a positioning and surveying system for detailed micro-topography measurements, (2) an active seismic survey system to investigate subsurface stratigraphy and bulk density distribution, (3) a radio isotope density meter for precise soil density measurement, and (4) a plate loading and shear testing system to evaluate the deformation and strength characteristics of lunar regolith. Data collected by the RGIS will enable the construction of a three-dimensional geological and geotechnical map of the lunar surface. This map will aid in predicting the behavior of exploration vehicles, construction robots, and designing earthworks such as excavation, filling, leveling, and module/structure installation. This presentation offers a comprehensive overview and outlines the current development status of RGIS.

4046 | Study on mobility and strategy of Mars rover with faulty-driven wheel

Zhicheng Jia [a,*], Jingfu Jin [a], Xinju Dong [a], Yingchun Qi [a], Meng Zou [a], Lianbin He [a,*]

[a] *Jilin University, China*

* *Corresponding author: jiazc22@mails.jlu.edu.cn, helb@jlu.edu.cn*

Due to the particularity of planetary exploration mission, the Mars rover cannot be recovery and repaired in time when it fails like a conventional field robot. The failure of its core components such as wheels may directly lead to the interruption of detection tasks and even the loss of its own mobility. In this study, the motion state and model of the Mars rover with faulty-driven wheel are analyzed, and an emergency movement mode based on active suspension wheel lifting driving and faulty wheel dragging driving is established. In order to restore the hidden danger scene on the ground and accumulate experience in advance, the mobility test was carried out using the Zhurong Mars rover prototype based on the simulated Martian terrain. The test results show that the mobility of the Mars rover under the two emergency mobility modes is reduced. The average driving power of active suspension wheel lifting driving increases by about 30.1 %, the average current increases by about 24.8 %, the sinkage is larger and the terrain adaptability becomes weaker. The average driving power of faulty wheel dragging driving is increased by about 167.6 %, the average current is increased by about 112.6 %, the bulldozing resistance is larger and the heading deflection occurs. To maintain the subsequent basic detection capability of the Mars rover, this study concludes by summarizing and proposing an emergency control system that integrates fault movement mode switching strategy and intermittent heading correction. These test results on ground can provide valuable reference for Mars rover operators and related researchers, and help them take timely and effective measures when facing similar accidents.

4375 | Development of a multi-legged robot capable of ground stiffness detection

Tomoya Nishiyama [a,*], Tatsuki Honjo [a], Yugo Hosizawa [b], Ryota Hayashi [b], Koji Yoshida [b], Tetsuya Kinugasa [b,*]

[a] *Graduate School of Science and Engineering, Okayama University of Science, Japan*

[b] *Department of Mechanical Systems Engineering, Faculty of Engineering, Okayama University of Science, Japan*

* *Corresponding author: r23smv5rw@ous.jp, kinugasa@ous.ac.jp*

Centipedes move by generating retrograde waves in their legs in an antiphase pattern left and right, and they increase their stride by undulating their trunk as their speed increases. On branches, they move with the legs on both sides in phase, stretching and contracting their trunk. Furthermore, they can swim by relaxing their legs and undulating their trunk on the water, demonstrating high mobility in various environments.

The changes in gait adapted to the environment shown by such animals are interesting and provide essential clues for realizing vehicles that can move in multiple environments. Terramechanics focuses on propulsion mechanisms on terrain, especially on soft terrain. Understanding the movement strategies animals adopt in various environments and how they achieve their gait patterns are significant in ground vehicles. Hence, the objective of this study is to analyze the gait of centipedes on sand and to develop a prototype multi-legged robot, which is designed to emulate the observed gait. First, it was confirmed that centipedes change to a gait with legs moving in phase on both sides on the sand. It was also confirmed that, whereas the swing phase is longer than the propulsive phase on rigid terrain, the opposite is true on sand, with a longer propulsive phase. Next, to realize a robot that can recognize the ground stiffness and alter its gait accordingly, a system was introduced that has a bending sensor in the flexible legs and can actively rotate each leg. Finally, it was clarified that the flexibility and rotational angle of the legs recognize the terrain stiffness.

4695 | An image analysis method for obtaining wheel performance of rover using wheel tread traces remained on lunar simulant soil

Yujin Lim [a,*], Viet D. Le [b]

[a] *Paichai University, South Korea*

[b] *Kumoh National Institute of Technology, South Korea*

* *Corresponding author: yujin@pcu.ac.kr*

An image analysis procedure was proposed and adapted in this study to obtain rover wheel performance by analyzing wheel tread images remained on the surface of lunar simulant soil. Wheel-soil interaction is usually simulated by using Bevameter test results and adapting sinkage model such as Bekker's and Wong and Reece methods. However, the pressure-sinkage models obtained from the Bevameter test do not clearly provide in-detail information about sinkage and slip of the wheel when a rover runs on loose and fine dry soil. The wheel tread images captured by a camera mounted on the single wheel tracking device were analyzed successfully by using a specific pattern recognition procedure that is composed of Gabor wavelet filter, Principal Component Analysis (PCA), and Support Vector Machine (SVM). Next, We do successfully develop a required scheme of machined learning that was used to obtain an improved calculated torque of the wheel that is required for the next movement of the rover running on lunar terrain.

4998 | Dynamical modeling of the power hop phenomenon in an agricultural tractor with front axle suspension

Masahisa Watanabe [a,*], Keisuke Kazama [b], Kenshi Sakai [a]

[a] *Tokyo University of Agriculture and Technology, Japan*

[b] *Nihon University, Japan*

* *Corresponding author: corresponding@email.domain*

Power hop refers to the coupled oscillation of vertical, longitudinal, and pitch motions observed when four-wheel-drive tractors tow moderate to high draft loads on dry soils or operate on slippery roads or slopes. The severe vibrations resulting from power hop reduce operational precision, ride comfort, and tractor stability, while also increasing soil compaction and damage to operators and tractor body. Unlike forced oscillations, power hop is a self-excited oscillation caused by the nonlinear dynamics of agricultural tractors, including stick-slip and impact dynamics. Our previous paper investigated how these nonlinear elements contribute to the power hop phenomenon in farm operations. In this study, we explore the efficacy of front axle suspension in mitigating the occurrence of power hop and its associated vibrations. Although front axle suspension is generally employed to improve ride comfort during tractor operation at higher speeds, its efficacy on power hop have not been investigated thus far. We newly developed a power hop model for an agricultural tractor equipped with front axle suspension and conducted numerical simulations varying model parameters of the developed model. Our results demonstrate that front axle suspension can effectively suppress power hop even under load and soil conditions that would induce power hop in a tractor without such suspension. However, we also observed occurrences of power hop in front axle suspension tractors under more severe conditions including higher draft loads and drier soils.

6325 | Proposal of a new manual for telescopic penetrometer

Klara Cibulova [a,*], Martin Priesner [b]

[a] *University of Defence Kounicova 65, 662 10 Brno, Czech Republic, Czechia*

[b] *Engineer Regiment, Czech Republic, Czechia*

* *Corresponding author: klara.cibulova@unob.cz*

Mobility is solved in almost all areas of our lives. The importance of transportation of people and material is increasing every day. Mobility has many subareas, but the authors of this article focus on the mobility of wheeled vehicles in the terrain. Off-road mobility is necessary not only in crisis situations such as road damage, floods, but also in everyday life in agriculture, forestry and it is an essential part of the military sphere as well. In order to avoid getting stuck, it is important to be able to evaluate whether the given route is trafficable or not. Penetrometers are used for such evaluation; they measure the resistance of penetration to the ground - the bearing capacity of the soil. In the Czech army, a telescopic penetrometer is used for this evaluation. Based on many years of field measurements, this instrument measures reliably but it has unreliable evaluation system. Therefore, the authors decided to propose a new evaluation manual. To develop it, a comparison of existing evaluation methods was made. Next, the authors analysed the parameters influencing trafficability of wheeled vehicles and decided which ones to include in the procedure. Based on their previous research, they also decided to include a new driver parameter. New forms and evaluation tables were created. The whole manual was verified by the field measurements. The contribution of the work is the creation of a reliable manual for the evaluation of the trafficability for wheeled vehicles by the telescopic penetrometer, which is already established in the Army of Czech Republic.

6865 | Extended terramechanics model considering ground surface deformation and its application to wheel traveling analysis

Shingo Ozaki [a,*], Tomoya Arai [a], Mai Shimizu [a]

[a] *Yokohama National University, Japan*

* *Corresponding author: ozaki-shingo-xd@ynu.ac.jp*

Based on the pioneering work by Bekker, Wong, and Reece, terramechanics models have been used to evaluate the traveling performance of off-road vehicles of a wide range of scales, from small robots to mining dump trucks. Recently, it has also been used to study the performance of lunar and planetary exploration rovers. Meanwhile, multi-body dynamics analysis implementing the terramechanics model is a typical method to study the performance of off-road vehicles on soft ground. This approach can be a powerful tool not only for evaluating the overall vehicle behavior, but also for evaluating safe work plans. The quality of a multi-body dynamics analysis of an off-road vehicle depends on the terramechanics model that describes the interaction between the driving parts and the ground. In this study, the effectiveness of an extended terramechanics model (xTerramechanics model) is demonstrated, which considers soil deformation actions based on cellular automata, for the evaluation of traveling performance of a rigid wheel. First, the results of single-wheel traveling analysis are compared with experimental results under the forced-slip condition, and it is shown that drawbar-pull and sinkage are represented with good accuracy. We then apply the xTerramechanics model under the self-propelled traveling condition at a constant towing load and slope climbing. The model successfully reproduced the well-known "difference in traveling performance depending on traveling conditions."

7116 | DEM-SPH analysis for interaction mechanics of tracked vehicle on wet sand

Hiroki Yanagawa [a,*], Genya Ishigami [a]

[a] *Keio University, Japan*

* *Corresponding author: yanahiro0214@keio.jp*

Tracked vehicles are extensively employed in unstructured environments, traversing on uneven and deformable terrains. The mobility facilitated by the track unit which is composed of interconnected metal plates, is pivotal for enhancing vehicle mobility and safety. In the context of construction machinery, there is a paramount need for stable traversability across sand imbued with moisture. Despite this, the existing research on evaluating tracked vehicle performance in wet sand conditions is significantly limited in contrast to the extensive research conducted on dry sand. This study aims to develop a DEM-SPH-based simulation methodology for assessing tracked vehicle traversability on water-laden soil. The DEM, or Distinct Element Method, is a proven technique for soil-machinery interaction analysis, but it inaccurately represents water-infused sand dynamics. Incorporating the Smoothed Particle Hydrodynamics (SPH) addresses these difficulties, allowing it to improve the simulation accuracy of moist soil behavior. Our approach involves calibrating and validating DEM-SPH parameters through cone penetration tests across various soil moisture levels and resistance force variations. Subsequent cross-validation of the calibration was performed by examining the deposition angles of differently moistened sand in a cylindrical container. By calibrating DEM-SPH parameters, our simulations of track unit vehicles on moist sand considering various slip ratios of the track provide insights into the complex dynamics of vehicle-soil interactions. This research highlights the potential of DEM-SPH in delivering precise analyses of tracked vehicle performance on moist sand.

7555 | Practical applications of hybrid terramechanics model using machine learning

Eric Karpman [a,*], Jozsef Kovecses [a], Marek Teichmann [a]

[a] *McGill University, Montreal, Canada*

[b] *CM Labs Simulation Inc., Montreal, Canada*

* *Corresponding author: eric.karpman@mail.mcgill.ca*

Real-time wheel-soil models in terramechanics primarily rely on traditional semi-empirical terramechanics models as a foundation. Because of the steady-state assumption that these models are formulated with, their accuracy can suffer in dynamic simulations. Methods such as the Finite Element Method (FEM) and the Discrete Element Method (DEM) can capture transient effects that traditional semi-empirical models cannot, but their computational cost is prohibitive for real-time applications. Using a machine-learning (ML) approach in combination with the semi-empirical models, the authors have previously shown that it is possible to create a hybrid model that can run in real time while capturing transient wheel-soil behaviour. This was achieved by generating training data in the form of DEM simulations and using the resulting forces to compute the difference between the force prediction of the semi-empirical and DEM models. A neural network was trained to predict the difference between these forces. This work builds off the authors' previous work to expand the scope of this modelling approach so that it can be used in a wider range of practical applications. This is achieved by studying the ideal network input and output parameters and creating DEM simulation scenarios that result in high quality training data. Notably, the idea of having the neural network predict a force per unit width of the wheel is explored as strategy for saving computational resources and creating neural networks that work for a wider range of problems. Detailed description of the creation of training data and network training procedure as well as various examples of the trained network implemented as part of a hybrid wheel-soil model in real-time dynamic simulations will be presented.

8330 | Semi-empirical terramechanics model for variable terrain height in 3D

Eric Karpman [a,*], Wing Hang Ho [a], Jozsef Kovecses [a], Marek Teichmann [b]

[a] *McGill University, Montreal, Canada*

[b] *CM Labs Simulation Inc., Montreal, Canada*

* *Corresponding author: eric.karpman@mail.mcgill.ca*

When employing semi-empirical terramechanics models, dynamic simulations typically rely on the assumption that all terrain, even rough terrain, can be approximated as a plane in any given simulation time step. This assumption is made necessary by the fact that traditional semi-empirical terramechanics models are formulated to compute wheel-soil interaction forces for a wheel travelling over a flat plane, and adapting these models to compute the reaction forces for a wheel travelling on a sloped plane is straightforward. In many cases, approximating the contact between a wheel and a complex terrain mesh as contact between a wheel and a sloped plane - whose normal direction is determined by the terrain nodes that intersect with the collision geometry of the wheel - can give a reasonable approximation for the contact forces. However, there are conceivable scenarios where important terrain features, such as gaps in the wheel-terrain contact patch, can be overlooked when using this approach. The authors have previously proposed an alternative method for adapting the traditional semi-empirical models for rough terrain without the need to simplify the contact problem to a wheel on a sloped plane. This is accomplished by treating the terrain as a height field and integrating the contact stresses along the wheel's rim by computing the sinkage at each point along the rim based on the un-deformed height field height at that position in space rather than based on the wheel's position relative to an approximate contact plane. In this work, the previous 2-D implementation of the proposed approach is extended to three dimensions to illustrate how it can be used in full-scale dynamic simulations.

8506 | Introduction of Hourglass mission to investigate characteristics of granular materials in low gravity environment

Masatsugu Otsuki [a,*], Shingo Ozaki [b], Genya Ishigami [c], Takao Maeda [d], Masataku Sutoh [a], Taizo Kobayashi [e]

[a] *Japan Aerospace Exploration Agency, Japan*

[b] *Faculty of Engineering, Yokohama National University, Japan*

[c] *Keio University, Japan*

[d] *Tokyo University of Agriculture and Technology, Japan*

[e] *Ritsumeikan University, Japan*

* *Corresponding author: otsuki.masatsugu@jaxa.jp*

The Hourglass mission has been conducted to investigate the gravitational dependence of basic parameters for reproduction of behavior of granular materials such as regolith and ground sand, and to obtain information that contributes to future spacecraft design. In the Hourglass mission, the behaviors of regolith and ground sand in an arbitrary gravity environment are observed with an artificial gravity generator included in the Cell Biology Experiment Facility (CBEF) in the Kibo module of the International Space Station (ISS). The purpose of this mission is to investigate the effect of low gravity on the properties of granular materials. An hourglass-type container and a measuring-cylinder-type container including particles such as simulated regolith of planets and ground sand are packed into a sealed metal box mounted on the artificial gravity generator. The behavior of particles is observed with an optical camera while the containers are periodically flipped under arbitrary low gravity. Eight kinds of specimens are employed for the target samples, and dynamic behavior and sedimentation state (bulk density, angle of repose, etc.) of these granular materials are evaluated. Hourglass mission would have the contribution of understanding of the celestial growth process, provision of basic data for the construction of terramechanics on celestial bodies, optimization of design for future landers, exploration rovers, automatic construction machines on the lunar surface and manned pressurized rover for lunar exploration, and the appeal of the value and ability of Kibo artificial gravity environment. This presentation outlines the Hourglass mission, from start-up to development, and provides application examples of the results obtained.

8861 | An efficient and high-fidelity track model for dynamic simulation of off-road tracked vehicles

Oz Ben Yosef [a,*], Dror Rubinstein [a,*]

[a] *Ariel University, Department of Mechanical Engineering and Mechatronics, Israel*

* *Corresponding author: ozby@ariel.ac.il, drorru@ariel.ac.il*

A high-fidelity simulation model of tracked vehicles is required for proper prediction of the mobility of tracked vehicles traveling over soft soils. The vehicle components can be modelled using standard tools of multi-body programs. A track model was developed and successfully worked together with Altair's MotionSolve multi-body program. The model based on classic soil mechanics equations. The grousers, which are a significant part of many types of track-links, are taking into the account. The plasticity and viscosity properties of the soil are considered in the model. Verification tests were conducted in an agricultural field in the Jezreel Valley. The tests were run over soils with varying mechanical prop-erties, achieved through irrigation and tillage. The chosen test vehicle was an M113 armoured carrier. Several drawbar pull loads were applied on each soil condition. Reasonable correlation between the tests and simulation results were achieved. However, this model is not efficient and require consumption of large amount of CPU time. On the other hand, the more efficient models are based on simplifying assumptions with lack of accuracy. This work proposes a method for creating an efficient high-fidelity model. This can be implemented by representing the interaction between the track and the ground according to the previous work. The solution for the track-links will be done independently to the solu-tion of the entire vehicle. The solution of the track link will be obtained through a solver that will be developed for this purpose. The solution of the entire vehicle will be done by the solver of the multibody program (Altair MotionSolve). Good correlation between the efficient model and the previous model were achieved.

9688 | Evaluation of the dynamic sinkage and its effect on the compaction force prediction of off-road vehicles

Yang Jiao [a,*], Jozsef Kovecses [a,*], Marek Teichmann [b,*]

[a] *McGill University, Montreal, Canada*

[b] *CM Labs Simulation Inc., Montreal, Canada*

* *Corresponding author: yangjiao0706@gmail.com, jozsef.kovecses@mcgill.ca, marek@cm-labs.com*

In off-road wheel/tracked vehicle operations, shearing and compaction actions of- ten occur simultaneously, with both contributing to the total sinkage of the vehicle. The sinkage induced by the vehicle's compaction is typically referred to as static sinkage. This type of sinkage can be accurately quantified using pressure-sinkage relations such as the Bekker equation or the Wong and Reece equation. The sinkage resulting from the shearing behavior of the wheel and track is referred to as slip-sinkage (or dynamic sinkage in some literature). Compared with the first part, the slip sinkage is usually neglected in simulation. However, neglecting slip-sinkage can lead to inaccuracies in predicting motion resistance and traction, both of which are closely tied to vehicle performance. In past literature, slip sinkage has typically been calculated based on the slip ratio and static sinkage. In this research, the Discrete Element Method (DEM) virtual experiments were conducted to investigate the physical causes of wheel slip sinkage and identify the main contributing factors. The experimental results demonstrated that, in steady-state conditions, slip-sinkage increased with higher slip ratios, while also being influenced by the shear velocity of the wheel. In transient states, the primary contributor appears to be shear displacement rather than the slip ratio. The DEM virtual experimental results were utilized to develop an empirical equation for calculating slip sinkage, incorporating shear displacement, shear velocity, and slip ratio through symbolic regression. The normal force calculation based on the pressure sinkage equation was also modified to better accommodate slip-sinkage, thus improving the accuracy and stability of the simulation.

ISTVS

Author index

Name	Affiliation	Submissions
Ahmed, Ahmed Elawad Eltayeb	Mechanical Engineering Doctoral School - Hungarian University of Agriculture and Life Sciences (MATE)., Hungary	4476 | Effect of initial soil color on sandy soil moisture content determination
Aragane, Masanori	International Professional University of Technology in Tokyo, Japan	7829 | Development of foot in balloon biped robot using buoyancy force for traveling soft ground
Arai, Tomoya	Yokohama National University, Japan	1499 | Development of evaluation technology for tire traveling performance on soft soil of lunar surface *(abstract only)* 3071 | Study on applicability of extended terramechanics model to various traveling modes of wheels *(abstract only)* 6865 | Extended terramechanics model considering ground surface deformation and its application to wheel traveling analysis *(abstract only)*
Ares, Vinicius	State University of Campinas, Brazil	8590 | Modeling of terrain deformation by a grouser wheel for lunar rover simulator
Bakke, Luning	University of Wisconsin - Madison, United States	0815 | Using simulation to characterize RASSOR excavation in lunar environments *(abstract only)*
Becker, Carl	University of Pretoria, South Africa	7322 | Lateral tyre characterization: rolling tyre vs static tyre testing
Bell, Haley	U.S. Army Engineer Research and Development Center, United States	2651 | Rapid assessment tools for estimating trafficability on low-volume roads
Ben-Yosef, Oz	Ariel University, Department of Mechanical Engineering and Mechatronics, Israel	8861 | An efficient and high-fidelity track model for dynamic simulation of off-road tracked vehicles *(abstract only)*
Bheora, Jasleen Kaur	Department of Civil and Environmental Engineering, Virginia Tech, Blacksburg, VA, United States	4093 | The role of tire-soil interface characteristics on performance parameters through experimental and numerical investigation
Borrero, Micah M.	Department of Aeronautics and Astronautics, MIT; US Army ERDC-Cold Regions Research and Engineering Laboratory (CRREL), United States	3649 | Modeling the resistive forces on vehicles in deep snow
Buse, Fabian	German Aerospace Center (DLR), Germany	9028 | Uncertainty quantification for wheeled locomotion machine learning predictions on soft soil
Candalot, Arthur	Tohoku University, Department of Aerospace Engineering, Space Robotics Laboratory, Japan	6422 | Sinkage study in granular material for space exploration legged robot gripper
Capek, Jaromir	University of Defence, Kounicova 65, 66210 Brno, Czechia	7453 | The impact of changes in the river regime on the mobility of off-road vehicles
Casanova, Fernando	Universidad del Valle, Colombia	8812 | Energy method to compare performance of new types of sugar cane transport equipment
Casir Ricano, Jorge Ruben	Kyushu Institute of Technology, Japan	1649 | Dynamic mode decomposition for PIV-based sand flow field beneath traveling wheel

Name	Affiliation	Submissions
Castellanos, Bernardo	Department of Civil and Environmental Engineering, Virginia Tech, Blacksburg, VA, United States	4093 | The role of tire-soil interface characteristics on performance parameters through experimental and numerical investigation
Chen, Guangming	Nanjing University of Aeronautics and Astronautics, China	6412 | Traversing abilities simulation of a biomimetic robot on granular soil terrain
Chiarabelli, Damiano	CNR-STEMS, National Research Council, Inst. of Sci. and Tech. for Sustainable Energy and Mobility, Ferrara, Italy	5334 | Enhanced open-loop control of automatic gear shifting in hydromechanical CVT for agricultural tractors
Chiba, Teiichirou	Komatsu Ltd., Japan	1241 | Study on classification of excavated soil using internal sensor data of hydraulic excavator
Cibulova, Klara	University of Defence Kounicova 65, 662 10 Brno, Czech Republic, Czechia	6325 | Proposal of a new manual for telescopic penetrometer *(abstract only)*
Dietrich, Felix	Technische Universität München (TUM), Germany	9028 | Uncertainty quantification for wheeled locomotion machine learning predictions on soft soil
Dohnal, Filip	University of Defence, Kounicova 65, 66210 Brno, Czechia	7453 | The impact of changes in the river regime on the mobility of off-road vehicles
Dong, Xinju	Jilin University, China	4046 | Study on mobility and strategy of Mars rover with faulty-driven wheel *(abstract only)*
Economou, John	Cranfield University, United Kingdom	7603 | Evaluation of off-road uninhabited ground vehicle mobility using discrete element method and scalability investigation
Edwards, Lulu	U.S. Army Engineer Research and Development Center, United States	2651 | Rapid assessment tools for estimating trafficability on low-volume roads
Els, P. Schalk	University of Pretoria, South Africa	1710 | A novel measurement method to aid development of soft soil tyre-terrain interaction models
		4715 | UAV-based three-dimensional rough terrain modelling
		4966 | Real-time implementation of non-linear controllers and predictors for off-road vehicle dynamics on embedded systems
		7322 | Lateral tyre characterization: rolling tyre vs static tyre testing
Endo, Daisuke	National Research and Development Agency Public Works Research Institute, Japan	1241 | Study on classification of excavated soil using internal sensor data of hydraulic excavator
Eto, Ryosuke	National Defense Academy of Japan, Japan	5620 | Step-climbing motion acquisition of tracked robot with flippers without using environment information by reinforcement learning
Fediukov, Vladyslav	German Aerospace Center (DLR), Technische Universität München (TUM), Germany	9028 | Uncertainty quantification for wheeled locomotion machine learning predictions on soft soil
Frankenstein, Susan	ERDC-CRREL, United States	3769 | Mobility during the transition seasons in the arctic

Name	Affiliation	Submissions
Fujii, Takatomo	Sakai Heavy Industries, Ltd., Japan	2369 | Experimental verification of particle behavior during crushing and mixing of deteriorated asphalt pavement layers by stabilizer
Fujiwara, Daisuke	Suwa University of Science, Japan	2411 | Proposing turning motion of the small and lightweight push-rolling rover with mimimal configuration
		5720 | Suppress slip while crossing loose slopes using reverse rotation behavior of rovers with function of independent contraction/expansion mechanism
Fukami, Koichiro	NARO Kyushu Okinawa Agricultural Research Center, Japan	4142 | Root system analysis after vibratory roller compaction in dry direct seeding of rice field
Fuke, Takahiro	Department of Mechanical Engineering, Keio university, Yokohama, Japan	2919 | Accurate rover mobility analysis using HILS-DRFT with real-time parameter tuning approach
Fukumoto, Takeshi	Osaka University, Japan	0178 | Model to estimate the force exerted on the sphere rolling on the sandy road *(abstract only)*
Gessi, Silvia	CNR-STEMS, National Research Council, Inst. of Sci. and Tech. for Sustainable Energy and Mobility, Ferrara, Italy	5334 | Enhanced open-loop control of automatic gear shifting in hydromechanical CVT for agricultural tractors
Gomez Perlaza, Adolfo Leon	Universidad del Valle, Colombia	8812 | Energy method to compare performance of new types of sugar cane transport equipment
Gorsich, David	U.S. Army DEVCOM Ground Vehicle Systems Center, Warren, MI 48397, United States	3181 | Assessing sensitivities of off-road pneumatic tire on clay: A finite element investigation on tire operational and design parameters
		4093 | The role of tire-soil interface characteristics on performance parameters through experimental and numerical investigation
Guan, Dehai	Nanjing University of Aeronautics and Astronautics, China	6412 | Traversing abilities simulation of a biomimetic robot on granular soil terrain
Guthrie, A. Glenn	University of Pretoria, South Africa	4715 | UAV-based three-dimensional rough terrain modelling
		4966 | Real-time implementation of non-linear controllers and predictors for off-road vehicle dynamics on embedded systems
Hamersma, Herman	University of Pretoria, South Africa	4715 | UAV-based three-dimensional rough terrain modelling
Hansen, Brad	U.S. Army Engineer Research and Development Center, United States	6476 | Improved trafficability over soft soils using ground matting
Hashim, Malik-Manel	Technische Universität Berlin, Germany	6422 | Sinkage study in granular material for space exploration legged robot gripper
Hashimoto, Takeshi	National Research and Development Agency Public Works Research Institute, Japan	1241 | Study on classification of excavated soil using internal sensor data of hydraulic excavator

Name	Affiliation	Submissions
Hayashi, Ryota	Professor, Department of Mechanical Systems Engineering, Faculty of Engineering, Okayama University of Science, Japan	4375 | Development of a multi-legged robot capable of ground stiffness detection *(abstract only)*
He, Lianbin	Jilin University, China	4046 | Study on mobility and strategy of Mars rover with faulty-driven wheel *(abstract only)*
Hickey, Brigid	University at Buffalo, NY, United States	6422 | Sinkage study in granular material for space exploration legged robot gripper
Hidekawa, Mutsumi	Kobashi Industries Co. Ltd., Japan	1204 | Measurement and visualization of soil cutting and throwing behavior by a rotary tillage blade
Hirai, Yasumaru	Kyushu University, Japan	1204 | Measurement and visualization of soil cutting and throwing behavior by a rotary tillage blade
Hiramatsu, Takahito	Asia Air Survey Co., Ltd., Japan	3304 | Development of an unmanned exploration robot for lunar surface geotechnical investigation *(abstract only)*
Hirayama, Yuta	Kyushu University, Japan	1204 | Measurement and visualization of soil cutting and throwing behavior by a rotary tillage blade 2369 | Experimental verification of particle behavior during crushing and mixing of deteriorated asphalt pavement layers by stabilizer
Ho, Wing Hang	McGill University, Montreal, Canada	8330 | Semi-empirical terramechanics model for variable terrain height in 3D *(abstract only)*
Honjo, Tatsuki	System Science Major, Graduate School of Science and Engineering, Okayama University of Science, Japan	4375 | Development of a multi-legged robot capable of ground stiffness detection *(abstract only)*
Hosizawa, Yugo	Department of Mechanical Systems Engineering, Faculty of Engineering, Okayama University of Science, Japan	4375 | Development of a multi-legged robot capable of ground stiffness detection *(abstract only)*
Hosobori, Kenji	Kiso-Jiban Consultants Co., Ltd., Japan	3304 | Development of an unmanned exploration robot for lunar surface geotechnical investigation *(abstract only)*
Hubacek, Martin	University of Defence, Kounicova 65, 66210 Brno, Czechia	7453 | The impact of changes in the river regime on the mobility of off-road vehicles
Huhne, Jana	Technische Universität München, Germany	9028 | Uncertainty quantification for wheeled locomotion machine learning predictions on soft soil
Hurrell, James	Tohoku University, Japan	3773 | Evaluation of the multi-pass effect of an exploration rover by single wheel testing assuming lunar gravity and soil 6422 | Sinkage study in granular material for space exploration legged robot gripper 8590 | Modeling of terrain deformation by a grouser wheel for lunar rover simulator

Name	Affiliation	Submissions
Iikawa, Naoki	Osaka University, Komatsu Ltd., Japan	4927 | Tip angle dependence for resistive force into dry granular materials at shallow cone penetration
Iizuka, Kojiro	Shibaura Institute of Technology, Japan	0906 | Experimental investigation of increased bearing capacity when imparting vibration to loose ground in low atmospheric pressure 2411 | Proposing turning motion of the small and lightweight push-rolling rover with mimimal configuration 5720 | Suppress slip while crossing loose slopes using reverse rotation behavior of rovers with function of independent contraction/expansion mechanism
Ikenaga, Taichi	Soil and Rock Engineering, Co., Ltd., Japan	3304 | Development of an unmanned exploration robot for lunar surface geotechnical investigation *(abstract only)*
Imanishi, Masaya	Komatsu.Ltd., Japan	1241 | Study on classification of excavated soil using internal sensor data of hydraulic excavator
Ishigami, Genya	Keio University, Japan	2800 | Granular flow in reduced gravity: analysis and insights from centrifuge experiments at the international space station *(abstract only)* 2919 | Accurate rover mobility analysis using HILS-DRFT with real-time parameter tuning approach 4213 | DEM-based analysis and optimization of an excavation bucket drum for in-situ resource utilization 7116 | DEM-SPH analysis for interaction mechanics of tracked vehicle on wet sand *(abstract only)* 8506 | Introduction of Hourglass mission to investigate characteristics of granular materials in low gravity environment *(abstract only)* 9295 | Granular scaling laws for accurate prediction of wheel mobility on slopes in low-gravity environments
Ishikawa, Sora	Department of Mechanical Engineering, Keio university, Yokohama, Japan, Japan	2919 | Accurate rover mobility analysis using HILS-DRFT with real-time parameter tuning approach
Ito, Shinichi	Ritsumeikan University, Japan	3304 | Development of an unmanned exploration robot for lunar surface geotechnical investigation *(abstract only)*
Jaksa, Mark	Lunar Construction Group, Andy Thomas Centre for Space Resources, University of Adelaide, SA 5005, Australia	0356 | Hammering energy requirements for surveying lunar surface with dynamic cone penetrometers and seismic methods
Jia, Zhicheng	Jilin University, China	4046 | Study on mobility and strategy of Mars rover with faulty-driven wheel *(abstract only)*

Name	Affiliation	Submissions
Jiao, Yang	McGill University, Montreal, Canada	9688 | Evaluation of the dynamic sinkage and its effect on the compaction force prediction of off-road vehicles *(abstract only)*
Jin, Jingfu	Jilin University, China	4046 | Study on mobility and strategy of Mars rover with faulty-driven wheel *(abstract only)*
Kabutz, Thomas Bernd	University of Pretoria, South Africa	1710 | A novel measurement method to aid development of soft soil tyre-terrain interaction models
Kaida, Kenji	Kobashi Industries Co. Ltd., Japan	1204 | Measurement and visualization of soil cutting and throwing behavior by a rotary tillage blade
Kakogawa, Atsushi	Ritsumeikan University, Japan	3304 | Development of an unmanned exploration robot for lunar surface geotechnical investigation *(abstract only)*
Kamohara, Junnosuke	Georgia Institute of Technology, United States	8590 | Modeling of terrain deformation by a grouser wheel for lunar rover simulator
Karlberg, Magnus	Luleå University of Technology, Sweden	8084 | Log detection for autonomous forwarding using auto-annotated data from a real-time virtual environment
Karpman, Eric	McGill University, Montreal, Canada	7555 | Practical applications of hybrid terramechanics model using machine learning *(abstract only)*
		8330 | Semi-empirical terramechanics model for variable terrain height in 3D *(abstract only)*
Katayama, Masahiro	Bridgestone Corporation, Japan	1499 | Development of evaluation technology for tire traveling performance on soft soil of lunar surface *(abstract only)*
		3071 | Study on applicability of extended terramechanics model to various traveling modes of wheels *(abstract only)*
Katou, Toyohiro	Kyushu University, Japan	1204 | Measurement and visualization of soil cutting and throwing behavior by a rotary tillage blade
		2369 | Experimental verification of particle behavior during crushing and mixing of deteriorated asphalt pavement layers by stabilizer
Katsura, Makoto	Osaka University, Japan	0178 | Model to estimate the force exerted on the sphere rolling on the sandy road *(abstract only)*
Katsuragi, Hiroaki	Osaka University, Japan	0178 | Model to estimate the force exerted on the sphere rolling on the sandy road *(abstract only)*
		4927 | Tip angle dependence for resistive force into dry granular materials at shallow cone penetration

Name	Affiliation	Submissions
Kazama, Keisuke	Nihon University, Japan	4998 | Dynamical modeling of the power hop phenomenon in an agricultural tractor with front axle suspension *(abstract only)*
Kinugasa, Tetsuya	Professor, Department of Mechanical Systems Engineering, Faculty of Engineering, Okayama University of Science, Japan	4375 | Development of a multi-legged robot capable of ground stiffness detection *(abstract only)*
Kiss, Peter	Hungarian University of Agriculture and Life Sciences, Hungary	1525 | Photogrammetry based mobility mapping for small scale model vehicle
		1902 | The combination of exhaust gas recirculation and water injection in a modern diesel engine
		4476 | Effect of initial soil color on sandy soil moisture content determination
Kissel, Alexandra	University of Wisconsin-Madison, United States	0815 | Using simulation to characterize RASSOR excavation in lunar environments *(abstract only)*
Kitamura, Keitaro	Asia Air Survey Co., Ltd., Japan	3304 | Development of an unmanned exploration robot for lunar surface geotechnical investigation *(abstract only)*
Kobayashi, Taizo	Ritsumeikan University, Japan	2800 | Granular flow in reduced gravity: analysis and insights from centrifuge experiments at the international space station *(abstract only)*
		3304 | Development of an unmanned exploration robot for lunar surface geotechnical investigation *(abstract only)*
		8506 | Introduction of Hourglass mission to investigate characteristics of granular materials in low gravity environment *(abstract only)*
Kon, Seiji	Bridgestone Co., Ltd, Japan	3071 | Study on applicability of extended terramechanics model to various traveling modes of wheels *(abstract only)*
Kondo, Akihiko	Port and Airport Research Institute, Japan	3304 | Development of an unmanned exploration robot for lunar surface geotechnical investigation *(abstract only)*
Körmöczi, Dávid	Hungarian University of Agriculture and Life Sciences, Hungary	1525 | Photogrammetry based mobility mapping for small scale model vehicle
Kovarik, Vladimir	University of Defence, Kounicova 65, 66210 Brno, Czechia	7453 | The impact of changes in the river regime on the mobility of off-road vehicles

Name	Affiliation	Submissions
Kovecses, Jozsef	McGill University, Montreal, Canada	2520 | Efficient tire-terrain interaction modelling: effect of flexibility on traction *(abstract only)* 7555 | Practical applications of hybrid terramechanics model using machine learning *(abstract only)* 8330 | Semi-empirical terramechanics model for variable terrain height in 3D *(abstract only)* 9688 | Evaluation of the dynamic sinkage and its effect on the compaction force prediction of off-road vehicles *(abstract only)*
Kumar, Ajay	Cranfield University, United Kingdom	7603 | Evaluation of off-road uninhabited ground vehicle mobility using discrete element method and scalability investigation
Kuramoto, Ohta	Kyushu Institute of Technology, Japan	1649 | Dynamic mode decomposition for PIV-based sand flow field beneath traveling wheel
Kurosu, Takashi	Sakai Heavy Industries, Ltd., Japan	2369 | Experimental verification of particle behavior during crushing and mixing of deteriorated asphalt pavement layers by stabilizer
Kusnets, Tamur	Estonian Military Academy, Estonia	4859 | Trafficability conditions for military wheeled trucks on cultivated fields
Kuszneruk, Michal	Lublin University of Technology, Poland	9663 | Year-round measurements of the soil cone index on grass airfields for ground performance of airplane
Laine, Mickael	Tohoku University, Department of Aerospace Engineering, Space Robotics Laboratory, Japan	6422 | Sinkage study in granular material for space exploration legged robot gripper
Le, Viet D.	Kumoh National Institute of Technology, South Korea	4695 | An image analysis method for obtaining wheel performance of rover using wheel tread traces remained on lunar simulant soil *(abstract only)*
Lehto, Mattias	Luleå University of Technology, Sweden	8084 | Log detection for autonomous forwarding using auto-annotated data from a real-time virtual environment
Lei, Xiang	Nanjing University of Aeronautics and Astronautics, China	6412 | Traversing abilities simulation of a biomimetic robot on granular soil terrain
Lideskog, Håkan	Luleå University of Technology, Sweden	8084 | Log detection for autonomous forwarding using auto-annotated data from a real-time virtual environment
Lim, Yujin	Paichai University, South Korea	4695 | An image analysis method for obtaining wheel performance of rover using wheel tread traces remained on lunar simulant soil *(abstract only)*
Lodewijks, Gabriel	University of Newcastle, Australia	6412 | Traversing abilities simulation of a biomimetic robot on granular soil terrain

Name	Affiliation	Submissions
Lopera, Jorge	PACIFIC+, Cali, Colombia	8812 | Energy method to compare performance of new types of sugar cane transport equipment 9747 | Application of a rockie bogie suspension for a new type of sugar cane transport system
Maeda, Takao	Tokyo University of Agriculture and Technology, Japan	2800 | Granular flow in reduced gravity: analysis and insights from centrifuge experiments at the international space station *(abstract only)* 8506 | Introduction of Hourglass mission to investigate characteristics of granular materials in low gravity environment *(abstract only)*
Maleki, Mahdi	McGill University, Montreal, Canada	2520 | Efficient tire-terrain interaction modelling: effect of flexibility on traction *(abstract only)*
Marani, Pietro	CNR-STEMS, National Research Council, Inst. of Sci. and Tech. for Sustainable Energy and Mobility, Ferrara, Italy	5334 | Enhanced open-loop control of automatic gear shifting in hydromechanical CVT for agricultural tractors
Martelli, Massimo	CNR-STEMS, National Research Council, Inst. of Sci. and Tech. for Sustainable Energy and Mobility, Ferrara, Italy	5334 | Enhanced open-loop control of automatic gear shifting in hydromechanical CVT for agricultural tractors
Mason, Destiny	Department of Mechanical Engineering, Virginia Tech, Blacksburg, VA, United States, United States	3181 | Assessing sensitivities of off-road pneumatic tire on clay: A finite element investigation on tire operational and design parameters 4093 | The role of tire-soil interface characteristics on performance parameters through experimental and numerical investigation
Matsumura, Satoshi	Port and Airport Research Institute, Japan	3304 | Development of an unmanned exploration robot for lunar surface geotechnical investigation *(abstract only)*
Matsuo, Naoki	NARO Kyushu Okinawa Agricultural Research Center, Japan	4142 | Root system analysis after vibratory roller compaction in dry direct seeding of rice field
Mitsuoka, Muneshi	Kyushu University, Japan	1204 | Measurement and visualization of soil cutting and throwing behavior by a rotary tillage blade
Miyamoto, Hideaki	The University of Tokyo, Japan	2800 | Granular flow in reduced gravity: analysis and insights from centrifuge experiments at the international space station *(abstract only)*
Mizukami, Noriaki	International Professional University of Technology in Tokyo, Japan	7829 | Development of foot in balloon biped robot using buoyancy force for traveling soft ground
Mochizuki, Tsukasa	Shibaura Institute of Technology, Japan	5720 | Suppress slip while crossing loose slopes using reverse rotation behavior of rovers with function of independent contraction/expansion mechanism

Name	Affiliation	Submissions
Morisawa, Naoki	Komatsu.Ltd., Japan	1241 | Study on classification of excavated soil using internal sensor data of hydraulic excavator
Muñoz Delgado, Carlos Daniel	Universidad del Valle, Colombia	8812 | Energy method to compare performance of new types of sugar cane transport equipment
Nagaoka, Kenji	Kyushu Institute of Technology, Japan	1649 | Dynamic mode decomposition for PIV-based sand flow field beneath traveling wheel 1882 | Spatio-temporal analysis of sand-density distribution beneath traveling wheel based on particle image velocimetry 1928 | A novel soil stress estimation method of wheel-soil interaction using photoelasticity
Nakagawa, Masafumi	Shibaura Institute of Technology, Japan	3304 | Development of an unmanned exploration robot for lunar surface geotechnical investigation *(abstract only)*
Nakano, Keiko	NARO Kyushu Okinawa Agricultural Research Center, Japan	4142 | Root system analysis after vibratory roller compaction in dry direct seeding of rice field
Nakano, Tomoyasu	Keio University, Japan	4213 | DEM-based analysis and optimization of an excavation bucket drum for in-situ resource utilization
Nakatani, Koki	Kobashi Industries Co. Ltd., Japan	1204 | Measurement and visualization of soil cutting and throwing behavior by a rotary tillage blade
Negrut, Dan	University of Wisconsin-Madison, United States	0815 | Using simulation to characterize RASSOR excavation in lunar environments *(abstract only)*
Nishida, Mami	International Professional University of Technology in Tokyo, Japan	7829 | Development of foot in balloon biped robot using buoyancy force for traveling soft ground
Nishiyama, Tomoya	System Science Major, Graduate School of Science and Engineering, Okayama University of Science, Japan	4375 | Development of a multi-legged robot capable of ground stiffness detection *(abstract only)*
Noren, Charles	Carnegie Mellon University, United States	2812 | An interaction-aware two-level robotic planning and control system for vegetation override
Nuwal, Ayush	Cranfield University, Bahrain	7603 | Evaluation of off-road uninhabited ground vehicle mobility using discrete element method and scalability investigation
Odaka, Junichiro	Kiso-Jiban Consultants Co., Ltd., Japan	3304 | Development of an unmanned exploration robot for lunar surface geotechnical investigation *(abstract only)*
Oikawa, Osamu	Sakai Heavy Industries, Ltd., Japan	2369 | Experimental verification of particle behavior during crushing and mixing of deteriorated asphalt pavement layers by stabilizer

Name	Affiliation	Submissions
Okayasu, Takashi	Kyushu University, Japan	1204 | Measurement and visualization of soil cutting and throwing behavior by a rotary tillage blade 2369 | Experimental verification of particle behavior during crushing and mixing of deteriorated asphalt pavement layers by stabilizer
Omura, Takuya	Keio University, Japan	4213 | DEM-based analysis and optimization of an excavation bucket drum for in-situ resource utilization 9295 | Granular scaling laws for accurate prediction of wheel mobility on slopes in low-gravity environments
Ordaz, Margarita	U.S. Army Engineer Research and Development Center, United States	2651 | Rapid assessment tools for estimating trafficability on low-volume roads
Otsuki, Masatsugu	Japan Aerospace Exploration Agency, Japan	2800 | Granular flow in reduced gravity: analysis and insights from centrifuge experiments at the international space station *(abstract only)* 8506 | Introduction of Hourglass mission to investigate characteristics of granular materials in low gravity environment *(abstract only)*
Ozaki, Shingo	Faculty of Engineering, Yokohama National University, Japan	1499 | Development of evaluation technology for tire traveling performance on soft soil of lunar surface *(abstract only)* 2800 | Granular flow in reduced gravity: analysis and insights from centrifuge experiments at the international space station *(abstract only)* 3071 | Study on applicability of extended terramechanics model to various traveling modes of wheels *(abstract only)* 3304 | Development of an unmanned exploration robot for lunar surface geotechnical investigation *(abstract only)* 6086 | Traveling analysis of wheel for lunar exploration rover based on extended terramechanics model: Examination of similarity law of gravity 6865 | Extended terramechanics model considering ground surface deformation and its application to wheel traveling analysis *(abstract only)* 8506 | Introduction of Hourglass mission to investigate characteristics of granular materials in low gravity environment *(abstract only)*

Name	Affiliation	Submissions
Parker, Michael	U.S. Army Engineer Research and Development Center, United States	3769 | Mobility during the transition seasons in the arctic 6476 | Improved trafficability over soft soils using ground matting
Peenze, Andries J.	University of Pretoria, South Africa	4966 | Real-time implementation of non-linear controllers and predictors for off-road vehicle dynamics on embedded systems
Pillinger, György	Hungarian University of Agriculture and Life Sciences, Hungary	4476 | Effect of initial soil color on sandy soil moisture content determination
Põdersalu, Priit	Estonian Military Academy, Estonia	4859 | Trafficability conditions for military wheeled trucks on cultivated fields
Pogulis, Markus	Umeå University, Sweden	0677 | Adaptive particle refinement in terramechanical DEM simulation *(abstract only)*
Priesner, Martin	Engineer Regiment, Czech Republic, Czechia	6325 | Proposal of a new manual for telescopic penetrometer *(abstract only)*
Pytka, Jarosław	Lublin University of Technolog, Poland	9663 | Year-round measurements of the soil cone index on grass airfields for ground performance of airplane
Qi, Yingchun	Jilin University, China	4046 | Study on mobility and strategy of Mars rover with faulty-driven wheel *(abstract only)*
Ravichandran, Nikhil	Department of Mechanical Engineering, Terramechanics, Multibody, and Vehicle Systems (TMVS) Laboratory, Virginia Tech, United States	3303 | Soil instrumentation for measuring normal stress distribution under off-road tire
Richard, Antoine	University of Luxembourg, Luxembourg	8590 | Modeling of terrain deformation by a grouser wheel for lunar rover simulator
Rohmer, Eric	State University of Campinas, Brazil	8590 | Modeling of terrain deformation by a grouser wheel for lunar rover simulator
Rubinstein, Dror	Ariel University, Department of Mechanical Engineering and Mechatronics, Israel	8861 | An efficient and high-fidelity track model for dynamic simulation of off-road tracked vehicles *(abstract only)*
Rushing, John	U.S. Army Engineer Research and Development Center, United States	2651 | Rapid assessment tools for estimating trafficability on low-volume roads 6476 | Improved trafficability over soft soils using ground matting
Rybansky, Marian	University of Defence, Kounicova 65, 66210 Brno, Czechia	7453 | The impact of changes in the river regime on the mobility of off-road vehicles
Sakai, Kenshi	Tokyo University of Agriculture and Technology, Japan	4998 | Dynamical modeling of the power hop phenomenon in an agricultural tractor with front axle suspension *(abstract only)*
Sakai, Seigo	Kyushu Institute of Technology, Japan	1882 | Spatio-temporal analysis of sand-density distribution beneath traveling wheel based on particle image velocimetry

Name	Affiliation	Submissions
Sakata, Takaaki	Shibaura Institute of Technology, Japan	5720 | Suppress slip while crossing loose slopes using reverse rotation behavior of rovers with function of independent contraction/expansion mechanism
Sandu, Corina	Virginia Tech, United States	0376 | Improvement of formulations of Clegg impact hammer and Rammsonde penetrometer for use in compacted snow *(abstract only)*
		3181 | Assessing sensitivities of off-road pneumatic tire on clay: A finite element investigation on tire operational and design parameters
		3303 | Soil instrumentation for measuring normal stress distribution under off-road tire
		4093 | The role of tire-soil interface characteristics on performance parameters through experimental and numerical investigation
		5295 | Model-based online optimal control for vehicles in reduced gravity.
Sano, Hisatoshi	Asia Air Survey Co., Ltd., Japan	3304 | Development of an unmanned exploration robot for lunar surface geotechnical investigation *(abstract only)*
Santra, Shreya	Tohoku University, Japan	0412 | Evaluation of grouser wheel traction performance at high-speed by single-wheel test
		8590 | Modeling of terrain deformation by a grouser wheel for lunar rover simulator
Sato, Hayatake	National Defense Academy of Japan, Japan	5620 | Step-climbing motion acquisition of tracked robot with flippers without using environment information by reinforcement learning
Sawa, Kenta	Tohoku University, Japan	0412 | Evaluation of grouser wheel traction performance at high-speed by single-wheel test
Scherer, Sebastian	Carnegie Mellon University, United States	2812 | An interaction-aware two-level robotic planning and control system for vegetation override
Scott, Brendan	Lunar Construction Group, Andy Thomas Centre for Space Resources, University of Adelaide, SA 5005, Australia, Australia	0356 | Hammering energy requirements for surveying lunar surface with dynamic cone penetrometers and seismic methods
Sebeck, Katherine	U.S. Army DEVCOM Ground Vehicle Systems Center, Warren, MI 48397, United States	3181 | Assessing sensitivities of off-road pneumatic tire on clay: A finite element investigation on tire operational and design parameters
		4093 | The role of tire-soil interface characteristics on performance parameters through experimental and numerical investigation
Seiji, Kon	Bridgestone Corporation, Japan	1499 | Development of evaluation technology for tire traveling performance on soft soil of lunar surface *(abstract only)*

Name	Affiliation	Submissions
Servin, Martin	Umeå University, Sweden	0677 | Adaptive particle refinement in terramechanical DEM simulation *(abstract only)*
Shenvi, Mohit	Virginia Tech, United States	0376 | Improvement of formulations of Clegg impact hammer and Rammsonde penetrometer for use in compacted snow *(abstract only)*
Shimizu, Mai	Yokohama National University, Japan	6086 | Traveling analysis of wheel for lunar exploration rover based on extended terramechanics model: Examination of similarity law of gravity 6865 | Extended terramechanics model considering ground surface deformation and its application to wheel traveling analysis *(abstract only)*
Shirose, Burhanuddin	Carnegie Mellon University, United States	2812 | An interaction-aware two-level robotic planning and control system for vegetation override
Slumba, Karlis	Lunar Construction Group, Andy Thomas Centre for Space Resources, University of Adelaide, SA 5005, Australia	0356 | Hammering energy requirements for surveying lunar surface with dynamic cone penetrometers and seismic methods
Sonalkar, Chaitanya	Department of Mechanical Engineering, Terramechanics, Multibody, and Vehicle Systems (TMVS) Laboratory, Virginia Tech, United States	3303 | Soil instrumentation for measuring normal stress distribution under off-road tire 5295 | Model-based online optimal control for vehicles in reduced gravity.
Suga, Tatsuhiko	Shibaura Institute of Technology, Japan	5720 | Suppress slip while crossing loose slopes using reverse rotation behavior of rovers with function of independent contraction/expansion mechanism
Sutoh, Masataku	JAXA, Japan	2800 | Granular flow in reduced gravity: analysis and insights from centrifuge experiments at the international space station *(abstract only)* 8506 | Introduction of Hourglass mission to investigate characteristics of granular materials in low gravity environment *(abstract only)*
Swamy, Varsha S	Department of Mechanical Engineering, Virginia Tech, Blacksburg, VA, United States, United States	3181 | Assessing sensitivities of off-road pneumatic tire on clay: A finite element investigation on tire operational and design parameters 4093 | The role of tire-soil interface characteristics on performance parameters through experimental and numerical investigation
Szőllősi, Dániel	Hungarian University of Agriculture and Life Sciences, IBIDEN Hungary Kft., Hungary	1902 | The combination of exhaust gas recirculation and water injection in a modern diesel engine
Takahashi, Kimiyasu	NARO Kyushu Okinawa Agricultural Research Center, Japan	4142 | Root system analysis after vibratory roller compaction in dry direct seeding of rice field

Name	Affiliation	Submissions
Takehana, Keisuke	Tohoku University, Japan	0412 | Evaluation of grouser wheel traction performance at high-speed by single-wheel test 3773 | Evaluation of the multi-pass effect of an exploration rover by single wheel testing assuming lunar gravity and soil 8590 | Modeling of terrain deformation by a grouser wheel for lunar rover simulator
Takigawa, Masanori	Asia Air Survey Co., Ltd., Japan	3304 | Development of an unmanned exploration robot for lunar surface geotechnical investigation *(abstract only)*
Tanaka, Tomomi	Tohoku University, Japan	3773 | Evaluation of the multi-pass effect of an exploration rover by single wheel testing assuming lunar gravity and soil
Taniguchi, Ryu	Soil and Rock Engineering, Co., Ltd., Japan	3304 | Development of an unmanned exploration robot for lunar surface geotechnical investigation *(abstract only)*
Tatarynow, Dawid	Lublin University of Technology, Poland	9663 | Year-round measurements of the soil cone index on grass airfields for ground performance of airplane
Teichmann, Marek	CM Labs Simulation Inc., Montreal, Canada	7555 | Practical applications of hybrid terramechanics model using machine learning *(abstract only)* 8330 | Semi-empirical terramechanics model for variable terrain height in 3D *(abstract only)* 9688 | Evaluation of the dynamic sinkage and its effect on the compaction force prediction of off-road vehicles *(abstract only)*
Tomiło, Paweł	Lublin University of Technology, Poland	8774 | Proposal of hybrid locomotion lunar rover with crawling mechanism 9663 | Year-round measurements of the soil cone index on grass airfields for ground performance of airplane
Tõnutare, Tõnu	Estonian University of Life Sciences, Estonia	4859 | Trafficability conditions for military wheeled trucks on cultivated fields
Travers, Matthew	Carnegie Mellon University, United States	2812 | An interaction-aware two-level robotic planning and control system for vegetation override
Tsuchiya, Kei	Bridgestone Corporation, Japan	1499 | Development of evaluation technology for tire traveling performance on soft soil of lunar surface *(abstract only)* 3071 | Study on applicability of extended terramechanics model to various traveling modes of wheels *(abstract only)*

Name	Affiliation	Submissions
Tsuji, Takeshi	University of Tokyo, Japan	3304 | Development of an unmanned exploration robot for lunar surface geotechnical investigation *(abstract only)*
Tsukimoto, Yukinori	Sakai Heavy Industries, Ltd., Japan	2369 | Experimental verification of particle behavior during crushing and mixing of deteriorated asphalt pavement layers by stabilizer
Uno, Kentaro	Tohoku University, Japan	0412 | Evaluation of grouser wheel traction performance at high-speed by single-wheel test 3773 | Evaluation of the multi-pass effect of an exploration rover by single wheel testing assuming lunar gravity and soil 8590 | Modeling of terrain deformation by a grouser wheel for lunar rover simulator
Untaroiu, Costin	Virginia Tech, United States	0376 | Improvement of formulations of Clegg impact hammer and Rammsonde penetrometer for use in compacted snow *(abstract only)*
van Aswegen, Christian	University of Pretoria, South Africa	4715 | UAV-based three-dimensional rough terrain modelling
Vennik, Kersti	Estonian Military Academy, Estonia	4859 | Trafficability conditions for military wheeled trucks on cultivated fields
Verulkar, Adwait	Virginia Polytechnic Institute and State University, United States	5295 | Model-based online optimal control for vehicles in reduced gravity.
Vilsan, Alexandru	Department of Mechanical Engineering, Terramechanics, Multibody, and Vehicle Systems (TMVS) Laboratory, Virginia Tech, United States	3303 | Soil instrumentation for measuring normal stress distribution under off-road tire
Vundurthy, Bhaskar	Carnegie Mellon University, United States	2812 | An interaction-aware two-level robotic planning and control system for vegetation override
Wada, Koji	Chiba Institute of Technology, Japan	2800 | Granular flow in reduced gravity: analysis and insights from centrifuge experiments at thc international space station *(abstract only)*
Watanabe, Masahisa	Tokyo University of Agriculture and Technology, Japan	4998 | Dynamical modeling of the power hop phenomenon in an agricultural tractor with front axle suspension *(abstract only)*
Watanabe, Tomohiro	Niigata University, Japan	0906 | Experimental investigation of increased bearing capacity when imparting vibration to loose ground in low atmospheric pressure
Welling, Orian	US Army ERDC-Cold Regions Research and Engineering Laboratory (CRREL), United States	3649 | Modeling the resistive forces on vehicles in deep snow
Wheeler, Allan	ERDC-CRREL, United States	3769 | Mobility during the transition seasons in the arctic
Witte, Clifford	ERDC-CRREL, United States	3769 | Mobility during the transition seasons in the arctic

Name	Affiliation	Submissions
Ximenes, Samuel	Exploration Architecture Corp., San Antonio, TX 78205, United States	0356 | Hammering energy requirements for surveying lunar surface with dynamic cone penetrometers and seismic methods
Yamakawa, Junya	National Defense Academy of Japan, Japan	5620 | Step-climbing motion acquisition of tracked robot with flippers without using environment information by reinforcement learning
Yamamoto, Hiroshi	Komatsu.Ltd., Japan	1241 | Study on classification of excavated soil using internal sensor data of hydraulic excavator
Yamamoto, Ken	Osaka University, Japan	0178 | Model to estimate the force exerted on the sphere rolling on the sandy road *(abstract only)*
Yanagawa, Hiroki	Keio University, Japan	7116 | DEM-SPH analysis for interaction mechanics of tracked vehicle on wet sand *(abstract only)*
Yanagishita, Masaki	Komatsu.Ltd., Japan	1241 | Study on classification of excavated soil using internal sensor data of hydraulic excavator
Yerro, Alba	Department of Civil and Environmental Engineering, Virginia Tech, Blacksburg, VA, United States	3181 | Assessing sensitivities of off-road pneumatic tire on clay: A finite element investigation on tire operational and design parameters
		4093 | The role of tire-soil interface characteristics on performance parameters through experimental and numerical investigation
Yoneoka, Takeshi	Kiso-Jiban Consultants Co., Ltd., Japan	3304 | Development of an unmanned exploration robot for lunar surface geotechnical investigation *(abstract only)*
Yoshida, Kazuya	Tohoku University, Japan	0412 | Evaluation of grouser wheel traction performance at high-speed by single-wheel test
		3773 | Evaluation of the multi-pass effect of an exploration rover by single wheel testing assuming lunar gravity and soil
		6422 | Sinkage study in granular material for space exploration legged robot gripper
		8590 | Modeling of terrain deformation by a grouser wheel for lunar rover simulator
Yoshida, Koji	Professor, Department of Mechanical Systems Engineering, Faculty of Engineering, Okayama University of Science, Japan	4375 | Development of a multi-legged robot capable of ground stiffness detection *(abstract only)*
Yoshida, Yuto	Kyushu Institute of Technology, Japan	1928 | A novel soil stress estimation method of wheel-soil interaction using photoelasticity
Zalewska-Tytłak, Anna	Lublin University of Technology, Poland	9663 | Year-round measurements of the soil cone index on grass airfields for ground performance of airplane
Zhang, Ruochun	University of Wisconsin-Madison, United States	0815 | Using simulation to characterize RASSOR excavation in lunar environments *(abstract only)*

Name	Affiliation	Submissions
Zhang, Yan	Nanjing University of Aeronautics and Astronautics, China	6412 | Traversing abilities simulation of a biomimetic robot on granular soil terrain
Zhou, Zhenhao	University of Wisconsin-Madison, United States	0815 | Using simulation to characterize RASSOR excavation in lunar environments *(abstract only)*
Zhou, Zhenwen	Nanjing University of Aeronautics and Astronautics, China	6412 | Traversing abilities simulation of a biomimetic robot on granular soil terrain
Zou, Meng	Jilin University, China	4046 | Study on mobility and strategy of Mars rover with faulty-driven wheel *(abstract only)*

International Society for Terrain-Vehicle Systems
www.istvs.org

ISBN 979-8-3313-1394-4

CURRAN ASSOCIATES INC.
proceedings.com